INTEGRABILITY: FROM STATISTICAL SYSTEMS TO GAUGE THEORY

Lecture Notes of the Les Houches Summer School:
Volume 106, 6 June – 1 July 2016

Integrability: From Statistical Systems to Gauge Theory

Edited by

Patrick Dorey, Gregory Korchemsky, Nikita Nekrasov,
Volker Schomerus, Didina Serban, Leticia Cugliandolo

Great Clarendon Street, Oxford, OX2 6DP,
United Kingdom

Oxford University Press is a department of the University of Oxford.
It furthers the University's objective of excellence in research, scholarship,
and education by publishing worldwide. Oxford is a registered trade mark of
Oxford University Press in the UK and in certain other countries

© Oxford University Press 2019

The moral rights of the authors have been asserted

First Edition published in 2019

All rights reserved. No part of this publication may be reproduced, stored in
a retrieval system, or transmitted, in any form or by any means, without the
prior permission in writing of Oxford University Press, or as expressly permitted
by law, by licence or under terms agreed with the appropriate reprographics
rights organization. Enquiries concerning reproduction outside the scope of the
above should be sent to the Rights Department, Oxford University Press, at the
address above

You must not circulate this work in any other form
and you must impose this same condition on any acquirer

Published in the United States of America by Oxford University Press
198 Madison Avenue, New York, NY 10016, United States of America

British Library Cataloguing in Publication Data

Data available

Library of Congress Control Number: 2018967557

ISBN 978–0–19–882815–0

DOI: 10.1093/oso/9780198828150.001.0001

Links to third party websites are provided by Oxford in good faith and
for information only. Oxford disclaims any responsibility for the materials
contained in any third party website referenced in this work.

Previous sessions

I	1951	Quantum mechanics. Quantum field theory
II	1952	Quantum mechanics. Statistical mechanics. Nuclear physics
III	1953	Quantum mechanics. Solid state physics. Statistical mechanics. Elementary particle physics
IV	1954	Quantum mechanics. Collision theory. Nucleon-nucleon interaction. Quantum electrodynamics
V	1955	Quantum mechanics. Non equilibrium phenomena. Nuclear reactions. Interaction of a nucleus with atomic and molecular fields
VI	1956	Quantum perturbation theory. Low temperature physics. Quantum theory of solids. Ferromagnetism
VII	1957	Scattering theory. Recent developments in field theory. Nuclear and strong interactions. Experiments in high energy physics
VIII	1958	The many body problem
IX	1959	The theory of neutral and ionized gases
X	1960	Elementary particles and dispersion relations
XI	1961	Low temperature physics
XII	1962	Geophysics; the earths environment
XIII	1963	Relativity groups and topology
XIV	1964	Quantum optics and electronics
XV	1965	High energy physics
XVI	1966	High energy astrophysics
XVII	1967	Many body physics
XVIII	1968	Nuclear physics
XIX	1969	Physical problems in biological systems
XX	1970	Statistical mechanics and quantum field theory
XXI	1971	Particle physics
XXII	1972	Plasma physics
XXIII	1972	Black holes
XXIV	1973	Fluids dynamics
XXV	1973	Molecular fluids
XXVI	1974	Atomic and molecular physics and the interstellar matter
XXVII	1975	Frontiers in laser spectroscopy
XXVIII	1975	Methods in field theory
XXIX	1976	Weak and electromagnetic interactions at high energy

XXX	1977	Nuclear physics with heavy ions and mesons
XXXI	1978	Ill condensed matter
XXXII	1979	Membranes and intercellular communication
XXXIII	1979	Physical cosmology
XXXIV	1980	Laser plasma interaction
XXXV	1980	Physics of defects
XXXVI	1981	Chaotic behavior of deterministic systems
XXXVII	1981	Gauge theories in high energy physics
XXXVIII	1982	New trends in atomic physics
XXXIX	1982	Recent advances in field theory and statistical mechanics
XL	1983	Relativity, groups and topology
XLI	1983	Birth and infancy of stars
XLII	1984	Cellular and molecular aspects of developmental biology
XLIII	1984	Critical phenomena, random systems, gauge theories
XLIV	1985	Architecture of fundamental interactions at short distances
XLV	1985	Signal processing
XLVI	1986	Chance and matter
XLVII	1986	Astrophysical fluid dynamics
XLVIII	1988	Liquids at interfaces
XLIX	1988	Fields, strings and critical phenomena
L	1988	Oceanographic and geophysical tomography
LI	1989	Liquids, freezing and glass transition
LII	1989	Chaos and quantum physics
LIII	1990	Fundamental systems in quantum optics
LIV	1990	Supernovae
LV	1991	Particles in the nineties
LVI	1991	Strongly interacting fermions and high T_c superconductivity
LVII	1992	Gravitation and quantizations
LVIII	1992	Progress in picture processing
LIX	1993	Computational fluid dynamics
LX	1993	Cosmology and large scale structure
LXI	1994	Mesoscopic quantum physics
LXII	1994	Fluctuating geometries in statistical mechanics and quantum field theory
LXIII	1995	Quantum fluctuations
LXIV	1995	Quantum symmetries
LXV	1996	From cell to brain
LXVI	1996	Trends in nuclear physics, 100 years later
LXVII	1997	Modeling the earths climate and its variability
LXVIII	1997	Probing the Standard Model of particle interactions
LXIX	1998	Topological aspects of low dimensional systems
LXX	1998	Infrared space astronomy, today and tomorrow
LXXI	1999	The primordial universe
LXXII	1999	Coherent atomic matter waves

LXXIII	2000	Atomic clusters and nanoparticles
LXXIV	2000	New trends in turbulence
LXXV	2001	Physics of bio-molecules and cells
LXXVI	2001	Unity from duality: Gravity, gauge theory and strings
LXXVII	2002	Slow relaxations and nonequilibrium dynamics in condensed matter
LXXVIII	2002	Accretion discs, jets and high energy phenomena in astrophysics
LXXIX	2003	Quantum entanglement and information processing
LXXX	2003	Methods and models in neurophysics
LXXXI	2004	Nanophysics: Coherence and transport
LXXXII	2004	Multiple aspects of DNA and RNA
LXXXIII	2005	Mathematical statistical physics
LXXXIV	2005	Particle physics beyond the Standard Model
LXXXV	2006	Complex systems
LXXXVI	2006	Particle physics and cosmology: the fabric of spacetime
LXXXVII	2007	String theory and the real world: from particle physics to astrophysics
LXXXVIII	2007	Dynamos
LXXXIX	2008	Exact methods in low-dimensional statistical physics and quantum computing
XC	2008	Long-range interacting systems
XCI	2009	Ultracold gases and quantum information
XCII	2009	New trends in the physics and mechanics of biological systems
XCIII	2009	Modern perspectives in lattice QCD: quantum field theory and high performance computing
XCIV	2010	Many-body physics with ultra-cold gases
XCV	2010	Quantum theory from small to large scales
XCVI	2011	Quantum machines: measurement control of engineered quantum systems
XCVII	2011	Theoretical physics to face the challenge of LHC
Special Issue	2012	Advanced data assimilation for geosciences
XCVIII	2012	Soft interfaces
XCIX	2012	Strongly interacting quantum systems out of equilibrium
C	2013	Post-Planck cosmology
CI	2013	Quantum optics and nanophotonics
Special Issue	2013	Statistical physics, optimization, inference and message-passing algorithms
CII	2014	From molecules to living organisms: An interplay between biology and physics
CIII	2014	Topological aspects of condensed matter physics
CIV	2015	Stochastic processes and random matrices
CV	2015	Quantum optomechanics and nanomechanics
CVI	2016	Integrability: from statistical systems to gauge theory
CVII	2016	Current trends in atomic physics

Publishers

- Session VIII: Dunod, Wiley, Methuen
- Sessions IX and X: Herman, Wiley
- Session XI: Gordon and Breach, Presses Universitaires
- Sessions XII–XXV: Gordon and Breach
- Sessions XXVI–LXVIII: North Holland
- Session LXIX–LXXVIII: EDP Sciences, Springer
- Session LXXIX–LXXXVIII: Elsevier
- Session LXXXIX– : Oxford University Press

Contents

List of Participants		xv
1	Integrability in statistical physics and quantum spin chains *Jesper Lykke Jacobsen*	1
2	A guide to two-dimensional conformal field theory *Jörg Teschner*	60
3	Lectures on the holographic duality of gauge fields and strings *Gordon W. Semenoff*	121
4	Introduction to scattering amplitudes *David Kosower*	183
5	Integrability in sigma-models *K. Zarembo*	205
6	Integrability in 2D fields theory/sigma-models *Sergei L. Lukyanov and Alexander B. Zamolodchikov*	248
7	Applications of integrable models in condensed matter and cold atom physics *Fabian H.L. Essler*	319
8	Introduction to integrability and one-point functions in $\mathcal{N}=4$ supersymmetric Yang–Mills theory and its defect cousin *Marius de Leeuw, Asger C. Ipsen, Charlotte Kristjansen, and Matthias Wilhelm*	352
9	Spectrum of $\mathcal{N}=4$ supersymmetric Yang–Mills theory and the quantum spectral curve *Nikolay Gromov*	400
10	Three-point functions in $\mathcal{N}=4$ supersymmetric Yang–Mills theory *Shota Komatsu*	449
11	Localization and $\mathcal{N}=2$ supersymmetric field theory *Vasily Pestun*	501
Appendix A: Lectures given at the school but no written contribution		541
	Slava Rychkov Conformal field theory in higher dimensions	541

Leonardo Rastelli
Introduction to 4d N = 2 SUSY theories   542

Benjamin Basso and Pedro Vieira
Integrability and amplitudes in N = 4 SYM   543

Paul Fendley
Integrability from topology   545

Andrei Okounkov
Random partitions in gauge and string theory   546

Appendix B: Seminars given during the school   547

Vladimir Bazhanov
Quantum geometry   547

Vladimir Kazakov
New integrable 3D and 4D QFTs from strongly twisted N = 4 SYM   548

Ivan Kostov
Correlation functions of heavy states in N = 4 SYM   549

Paul Zinn-Justin
Symmetric functions and quantum integrability   550

List of Participants

Organizers

Leticia CUGLIANDOLO

Patrick DOREY
Department of Mathematical Sciences, Durham University, Lower Mountjoy, Stockton Road, Durham DH1 3LE, United Kingdom

Gregory KORCHEMSKY
Institut de Physique Théorique, Saclay, UMR3681 CNRS, bat 774 Saclay, F-91191 Gif-sur-Yvette, France

Nikita NEKRASOV
Simons Center for Geometry and Physics, Stony Brook University, Stony Brook NY 11794-3636 USA

Volker SCHOMERUS
DESY Theory Group, DESY Hamburg, Notkestraße 85, D-22607 Hamburg, Germany

Didina SERBAN
Institut de Physique Théorique, Saclay, UMR3681 CNRS, bat 774 Saclay, F-91191 Gif-sur-Yvette, France

Lecturers

Benjamin BASSO
Laboratoire de Physique Théorique, École Normale Supérieure, 24 Rue Lhomond, 75005 Paris, France

Vladimir BAZHANOV
Department of Theoretical Physics, Research School of Physics and Engineering, Australian National University, Canberra, ACT 2601, Australia

Marius DE LEEUW
Niels Bohr Institute, Copenhagen University Blegdamsvej 17, DK-2100 Copenhagen

Fabian H.L. ESSLER
The Rudolf Peierls Centre for Theoretical Physics, Oxford University, Oxford, OX1 3NP, United Kingdom

Paul FENDLEY
The Rudolf Peierls Centre for Theoretical Physics, Oxford University, Oxford, OX1 3NP, United Kingdom

Nikolay GROMOV
Mathematics Department, King's College London, The Strand, London WC2R 2LS, United Kingdom

Asger C. IPSEN
Niels Bohr Institute, Copenhagen University Blegdamsvej 17, DK-2100 Copenhagen

Jesper Lykke JACOBSEN
Laboratoire de Physique Théorique, École Normale Supérieure, 24 Rue Lhomond, 75005 Paris, France

Vladimir KAZAKOV
Laboratoire de Physique Théorique, École Normale Supérieure, 24 Rue Lhomond, 75005 Paris, France

Shota KOMATSU
School of Natural Sciences, Institute for Advanced Study, Princeton, New Jersey USA

Ivan KOSTOV
Institut de Physique Théorique, Saclay, UMR3681 CNRS, bat 774 Saclay, F-91191 Gif-sur-Yvette, France

David KOSOWER
Institut de Physique Théorique, Saclay, UMR3681 CNRS, bat 774 Saclay, F-91191 Gif-sur-Yvette, France

Charlotte KRISTJANSEN
Niels Bohr Institute, Copenhagen University, Blegdamsvej 17, 2100 Copenhagen, Denmark

Sergei L. LUKYANOV
NHETC, Department of Physics and Astronomy, Rutgers University, Piscataway, NJ 08855-0849, USA

Andrei OKOUNKOV
Columbia University, New York, USA

Vasily PESTUN
Institut des Hautes Études Scientifiques, Bures-sur-Yvette, France

Leonardo RASTELLI
C. N. Yang Institute for Theoretical Physics, Stony Brook University, Stony Brook, 11794, NY, USA

Slava RYCHKOV
Laboratoire de Physique Théorique, École Normale Supérieure, 24 Rue Lhomond, 75005 Paris, France

Gordon W. SEMENOFF
Department of Physics and Astronomy, University of British Columbia, Vancouver, BC V6T 1Z1 Canada

Jörg TESCHNER
Department of Mathematics, University of Hamburg, Bundesstrasse 55, 20146 Hamburg, Germany

Pedro VIEIRA
Perimeter Institute for Theoretical Physics, Waterloo, Ontario N2L 2Y5, Canada

Matthias WILHELM
Niels Bohr Institute, Copenhagen University Blegdamsvej 17, DK-2100 Copenhagen

Alexander B. ZAMOLODCHIKOV
NHETC, Department of Physics and Astronomy, Rutgers University, Piscataway, NJ 08855-0849, USA

Konstantin ZAREMBO
Nordita, KTH Royal Institute of Technology and Stockholm University, Roslagstullsbacken 23, SE-106 91 Stockholm, Sweden

Paul ZINN-JUSTIN
School of Mathematics and Statistics, The University of Melbourne, Parkville, Victoria 3010, Australia

Students and Auditors

Jeremias AGUILERA DAMIA
Instituto de Fsica de La Plata (IFLP), CONICET, Universidad Nacional de La Plata, Argentina

Mikhail ALFIMOV
École Normale Supérieure, Laboratoire de Physique Théorique, 24 rue Lhomond, 75231 Paris, France and Institut de Physique Théorique CEA Saclay, Orme des Merisiers bâtiment 774, Point courrier 136, F-91191 Gif-sur-Yvette Cedex, France

Rodrigo ALVES PIMENTA
Centro de Ciências Extas e de Tecnologia-Universidade Federal de São Carlos Rodovia Washington Luis, km 235, S ao Carlos-SP-Brasil

Raphaël BELLIARD
CEA Saclay, Institut de Physique Théorique, Orme des Merisiers bâtiment 774, Point courrier 136, F-91191 Gif-sur-Yvette Cedex, France

Lorenzo BIANCHI
II. Institut für Theoretische Physik, Universität Hamburg, Luruper Chaussee 149, 22761 Hamburg, Germany

Isak BUHL-MORTENSEN
The Niels Bohr Institute, Blegdamsvej 17, 2100 København, Denmark

Dmitry BYKOV
Max-Plank-Institut fuer Gravitationsphysik, Am Muehlenberg 1, D-1476, Postdam-Golm, Germany

Alessandra CAGNAZZO
NORDITA, Roslagstullsbacken 23, 106 91 Stockholm, Sweden and DESY, Theory Group Notkestrasse 8, Bldg.2a D-22607 Hamburg, Germany

Robert CARCASSEÉS QUEVEDO
Universidade do Porto, Faculdade de Ciências, Rua do Campo Alegre 1021/1055, 4169-007 Porto, Portugal

Andrea CAVAGLIA
Physics Department, University of Torino, Dipartimento di Fisica, via Pietro Giuria 1, 10125, Torino, Italy

Vsevolod CHESTNOV
DESY, Notkestraße 85, D-22607 Hamburg, Germany

Martina CORNAGLIOTTO
DESY Notkestraße 85 22607 Hamburg, bad. 2A room 306, Germany

Frank CORONADO
Perimeter Institute for Theoretical Physics, 31 Caroline Street North, Waterloo, ON N2L 2Y5, Canada

Alejandro DE LA ROSA GOMEZ
University of York, Heslington, York YO10 5DD, United Kingdom

Alejandro DE LA ROSA GOMEZ
University of York, Heslington, York YO10 5DD, United Kingdom

Olga DIMITROVA
Landau Institute for Theoretical Physics, Kosygina Street 2, 119334 Moscow, Russia

Stefan DRUC
University of Southampton, School of Physics and Astronomy, 46/4119 University Rd, Southampton SO17 1BJ, United Kingdom

Gyorgy FEHER
LPTHE-CNRS-UPMC, Boîte 126, T13-14 4ème étage, 4 place Jussieu 75252 Paris CEDEX 05, France

Rouven FRASSEK
Department of Mathematical Sciences, Durham University, Lower Mountjoy, Stockton Road, Durham DH1 3LE, United Kingdom

Oleksandr GAMAYUN
Niels Bohrweg 2, Leiden, NL-2333 CA, The Netherlands

Alexandr GARBALI
ACEMS, School of Mathematics and Statistics, The University of Melbourne, Victoria 3010, Australia

Pavlo GAVRYLENKO
7, Vavilova str., Moscow, Russia, 117312

Victor GIRALDO RIVERA
ICTP Strada Costiera 11, 34151, Trieste, Italy

Tamás GOMBOR
Wigner Research Centre for Physics of H.A.S, 29-33 Konkoly Thege Miklós út, Budapest, XII., H-1121 and Institute of Theoretical Physics, Eötvös Loránd University, 1518 Budapest, Pf.32.Hungary

Lucia GOMEZ CORDOVA
Perimeter Institute for Theoretical Physics, 31 Caroline Street North, Waterloo, ON N2L 2Y5, Canada

Guzman HERNANDEZ
Center for Cosmology and Particle Physics, Physics Department, New York University, 4 Washington Place, Room 424, New York, NY 10003, USA

Lorenz HILFIKER
Universität Hamburg, Department of Mathematics, Bundesstrasse 55, 20146 Hamburg, Germany

Asger IPSEN
The Niels Bohr Institute, University of Copenhagen, Blegdamsvej 17, 2100 Copenhagen, Denmark

Mikhail ISACHENKOV
Department of Particle Physics and Astrophysics, Weizmann Institute of Science, Rehovot 7610001, Israel

Saebyeok JEONG
Department of Physics and Astronomy, Story Brook University, Stony Brook, NY 11790, USA

Yunfeng JIANG
Institute for Theoretical Physics, ETH Zurich, Wolfgang-Pauli Strasse 27, 8093 Zurich, Switzerland

Rob KLABBERS
II. Institute for Theoretical Physics, Hamburg University, Germany

Gleb KOTOUSOV
Rutgers University, NHETC, Rutgers, The State University of New Jersey, 126 Frelinghuysen Rd, Piscataway, NJ 08854-8019, USA

Sylvain LACROIX
Laboratoire de Physique (UMR CNRS 5672), École Normale Supérieure de Lyon, 46, allée d'Italie, F-69364 F-69364 Lyon, France

Jules LAMERS
Institute for Theoretical Physics, Utrecht University, Leuvenlaan 4, 3584 CE Utrecht, The Netherlands

Fedor LEVKOVICH-MASLYUK
King's College London, Mathematics Department, The Strand, WC2R 2LS London, United Kingdom

Andrii LIASHYK
National Research University Higher School of Economics, Faculty of Mathematics, Mathematical Physics Dept., 7 Vavilova Str., Moscow, Russia

Christian MARBOE
Trinity College Dublin, College Green 2, Dublin 2, Ireland

Marko MEDENJAK
Faculty of Mathematics and Physics, University of Ljubljana, Jadranska 19, SI-1000 Ljubljana, Slovenia

Daniel Ricardo MEDINA RINCON
Nordita-Nordic Institute for Theoretical Physics, Roslagstullsbacken 23, 106 91 Stockholm, Sweden

David MEIDINGER
Institut für Mathematik und Institut für Physik, Humboldt-Universität zu Berlin, IRIS Gebäude, Zum Großen Windkanal 6, 12489 Berlin, Germany

Vladimir MITEV
Institut für Physik, WA THEP, Johannes Gutenberg University Mainz, Staudingerweg 7, 55128 Mainz, Germany

Dennis MÜLLER
Humboldt-Universität zu Berlin, Institut für Physik, AG Quantenfeld-und Stringtheorie, Zum Großen Windkanal 6, D-12489 Berlin, Germany

Hagen MÜNKLER
Humboldt-University Berlin, Germany

Juan MIGUEL NIETO GARCÍA
Institut de Physique Théorique, Orme des Merisiers batiment 774, Point courrier 136, F-91191 Gif-sur-Yvette Cedex, France

Baptiste PEZELIER
Laboratoire de Physique de l'ENS de Lyon, 46 allée d'Italie, F-69364 Lyon, France

Elli POMONI
DESY, Theory Group, Notkestrasse 85, Bldg.2a, D-22607 Hamburg, Germany

Daria RUDNEVA
Math Department at Higher School of Economics, Vavilova street 7, Moscow, Russia

Naveen SUBRAMANYA PRABHAKAR
C.N.Yang Institute for Theoretical Physics, Department of Physics and Astronomy, Stony Brook University, Stony Brook, NY 11794-3800, USA

István M. SZÉCSÉNYI
Department of Mathematical Sciences, Durham University, Lower Mountjoy, Stockton Road, Durham DH1 3LE, United Kingdom

Vuong-Viet TRAN
Department of Mathematical Sciences, Durham University, Lower Mountjoy, Stockton Road, Durham DH1 3LE, United Kingdom

Adam VARGA
City University London, Northampton Square, London EC1V 0HB, United Kingdom

Erik WIDÉN
Nordita, Roslagstullsbacken 23, 106 91 Stockholm, Sweden

Ruidong ZHU
University of Tokyo, 7-3-1 Hongo, Bunkyo, Tokyo, Japan

Leonard ZIPPELIUS
Institut für Physik, Humboldt Universität zu Berlin, Unter den Linden 6, 10099 Berlin, Germany

1
Integrability in statistical physics and quantum spin chains

Jesper Lykke JACOBSEN

Laboratoire de Physique Theorique, Ecole Normale Superieure,
24 Rue Lhomond, 75005 Paris, France

Chapter Contents

1 Integrability in statistical physics and quantum spin chains   1
Jesper Lykke JACOBSEN

	Preface	3
1.1	Potts model	3
1.2	Quantum integrability	14
1.3	Algebraic Bethe ansatz	24
1.4	Thermodynamic limit	36
1.5	Nineteen-vertex models	50
	References	58

Preface

The goal of these lectures is to illustrate some basic concepts of quantum integrable systems on two important models of statistical physics: the Q-state Potts model and the $O(n)$ model. Both models have a geometric formulation as lattice models of fluctuating loops, hence making contact with the Temperley–Lieb algebra and its dilute generalization, respectively. They possess a conformally invariant continuum limit, that can be easily visualized in the Coulomb gas formalism by viewing the loops as level lines of a deformed Gaussian field theory. For particular values of Q and n, the continuum limit is more subtle and gives rise to logarithmic conformal field theories [1]. Important challenges remain to be solved in this case; this is however beyond the scope of these introductory lectures.

We set out by transforming the Potts model into a loop model, and further into a six-vertex model. Both formulations unveil the underlying Temperley–Lieb algebra, and hence permit us to identify the quantum integrable R-matrix. This leads to the solution of the model by the algebraic Bethe ansatz technique. Extracting information about the continuum limit from the resulting Bethe ansatz equations calls instead for techniques of analysis. We focus on the critical case $-1 < \Delta < 1$, and derive the ground state free energy in detail. Then we discuss elementary excitations and establish the relationship with the Coulomb gas.

A final section provides a survey of some more recent material. We discuss the spin-one Izergin–Korepin model and establish its relation with an $O(n)$ loop model. We focus on the so-called regime III, which gives rise to an unusual, non-compact continuum limit, in which the spectrum of critical exponents contains both discrete and continuous components. The $n \to 0$ limit in this regime describes a collapse transition of polymers due to the critical self-attraction between monomers, an example of a so-called Θ-point.

These lectures are based on sundry material, including standard textbooks and reviews on integrability [2–5] and conformal field theory [6], my lecture notes for the AIMES course [7] given at the ICFP graduate school at the Ecole Normal Supérieure in Paris since 2011, my review of CFT applied to loop models [8], a few theses [9, 10], and a lot of original research articles. The main text cites some of the latter, with no attempt at exhaustiveness, the principal aim being to provide the reader with a few entry points to a large and ever growing body of literature.

I take the opportunity here of expressing my deep gratitude towards my students, colleagues, and collaborators over the years for all they have taught me. I also thank the students of the Summer School for their suggestions to improve these lecture notes.

1.1 Potts model

In this section we define the Potts model and transform it into a loop model. By orientating the loops we then exhibit the equivalence with a six-vertex model. Both formulations provide a relation to the Temperley–Lieb algebra, which is the basis for solving the model by techniques of quantum integrability.

Even though we are mainly interested in the model defined on a square lattice, the main steps are valid on more general graphs [11]. Since it is hardly more complicated—and a lot more instructive—to work in the 'correct' generality, we shall choose to do so and specialize only when needed.

1.1.1 Spin representation

Let $G = (V, E)$ be an arbitrary connected graph with vertex set V and edge set E. The Q-state Potts model [12] is initially defined by assigning a spin variable σ_i to each vertex $i \in V$. Each spin can take Q different values, by convention chosen as $\sigma_i = 1, 2, \ldots, Q$. We denote by σ the collection of all spin variables on the graph. Two spins i and j are called nearest neighbours if they are incident on a common edge $e = (ij) \in E$. In any given configuration σ, a pair of nearest neighbour spins is assigned an energy $-\mathcal{J}$ if they take identical values, $\sigma_i = \sigma_j$. The Hamiltonian (dimensionless energy functional) of the Potts model is thus

$$\mathcal{H} = -K \sum_{(i,j) \in E} \delta(\sigma_i, \sigma_j), \tag{1.1}$$

where the Kronecker delta function is defined as

$$\delta(\sigma_i, \sigma_j) = \begin{cases} 1 & \text{if } \sigma_i = \sigma_j \\ 0 & \text{otherwise} \end{cases} \tag{1.2}$$

and $K = \mathcal{J}/k_B T$ is a dimensionless coupling constant (interaction energy).

The case $Q = 2$ corresponds to the Ising model. Indeed, if $S_i = \pm 1$ we have

$$2\delta(S_i, S_j) = S_i S_j + 1. \tag{1.3}$$

The second term amounts to an unimportant shift of the interaction energy, and so the models are equivalent if we set $K_{\text{Potts}} = 2K_{\text{Ising}}$.

The thermodynamic information about the Potts model is encoded in the partition function

$$Z = \sum_\sigma e^{-\mathcal{H}} = \sum_\sigma \prod_{(ij) \in E} e^{K\delta(\sigma_i, \sigma_j)} \tag{1.4}$$

and in various correlation functions. By a correlation function we understand the probability that a given set of vertices are assigned fixed values of the spins.

In the ferromagnetic case $K > 0$ the spins tend to align at low temperatures ($K \gg 1$), defining a phase of ferromagnetic order. Conversely, at high temperatures ($K \ll 1$) the spins are almost independent, leading to a paramagnetic phase where entropic effects prevail. On physical grounds, one expects the two phases to be separated by a critical point K_c where the effective interactions between spins becomes long ranged.

For certain regular planar lattices K_c can be determined exactly by duality considerations [12, 14]. Moreover, K_c will turn out to be the locus of a second order phase transition if $0 \leq Q \leq 4$ [13]. In that case the Potts model enjoys conformal invariance in the limit of an infinite lattice, allowing its critical properties to be determined exactly by a variety of techniques. These properties turn out to be *universal*, i.e. independent of the lattice used for defining the model microscopically.

1.1.2 Fortuin–Kasteleyn cluster representation

The initial definition (1.1) of the Potts model requires the number of spins Q to be a positive integer. It is possible to rewrite the partition function and correlation functions so that Q appears only as a parameter [15]. This makes its possible to assign to Q arbitrary real (or even complex) values.

Notice first that by (1.2) we have the identity

$$e^{K\delta(\sigma_i,\sigma_j)} = 1 + v\delta(\sigma_i,\sigma_j), \tag{1.5}$$

where we have defined $v = e^K - 1$. Now, it is obvious that for any edge-dependent factors h_e one has

$$\prod_{e \in E}(1 + h_e) = \sum_{E' \subseteq E} \prod_{e \in E'} h_e, \tag{1.6}$$

where the subset E' is defined as the set of edges for which we have taken the term h_e in the development of the product $\prod_{e \in E}$. In particular, taking $h_e = v\delta(\sigma_i, \sigma_j)$ we obtain for the partition function (1.4)

$$Z = \sum_{E' \subseteq E} v^{|E'|} \sum_{\sigma} \prod_{(ij) \in E'} \delta(\sigma_i, \sigma_j) = \sum_{E' \subseteq E} v^{|E'|} Q^{k(E')}, \tag{1.7}$$

where $k(E')$ is the number of connected components in the graph $G' = (V, E')$, i.e. the graph obtained from G by removing the edges in $E \setminus E'$. Those connected components are called *clusters*, and (1.7) is the Fortuin–Kasteleyn *cluster representation* of the Potts model partition function. The sum over spins σ in (1.4) has now been replaced by a sum over edge subsets, and Q appears as a parameter in (1.7) and no longer as a summation limit.

In (1.7)—and in all that follows—it presents no added complication to consider arbitrary edge-dependent couplings v_e. In that sense, $v^{|E'|}$ is just a short-hand notation for $\prod_{e \in E'} v_e$. We shall need this inhomogeneous generalization in section 1.1.7.

1.1.3 Duality of the partition function

Consider now the case where $G = (V, E)$ is a connected *planar* graph. Any planar graph possesses a *dual graph* $G^* = (V^*, E^*)$ which is constructed by placing a dual vertex $i^* \in V^*$ in each face of G, and connecting a pair of dual vertices by a dual edge $e^* \in E^*$ if and only if the corresponding faces are adjacent in G. In other words, there is a bijection between edges and dual edges, since each edge $e \in E$ intersects precisely one dual edge $e^* \in E^*$. Note that by the Euler relation,

$$|V| + |V^*| = |E| + 2. \tag{1.8}$$

By construction, the dual graph is also connected and planar. Note also that duality is an involution, i.e. $(G^*)^* = G$.

The Euler relation can easily be proved by induction. If $E = \emptyset$, since G was supposed connected we must have $|V| = |V^*| = 1$, so (1.8) indeed holds. Each time a further edge is added to E, there are two possibilities. Either it connects an existing vertex to a new vertex, in which case $|V|$ increases by one and $|V^*|$ is unchanged. Or it connects two existing vertices, meaning that a cycle is closed in G. In this case $|V|$ is unchanged and V^* increases by one. In both cases (1.8) remains valid.

Recalling the cluster representation (1.7)

$$Z_G(Q, v) = \sum_{E_1 \subseteq E} v^{|E_1|} Q^{k(E_1)}$$
$$Z_{G^*}(Q, v^*) = \sum_{E_2 \subseteq E^*} (v^*)^{|E_2|} Q^{k(E_2)} \tag{1.9}$$

we now claim that it is possible to choose v^* so that

$$Z_G(Q, v) = k Z_{G^*}(Q, v^*), \tag{1.10}$$

where k is an unimportant multiplicative constant.

To prove this claim, we show that the proportionality (1.10) holds term by term in the summations (1.9). To this end, we first define a bijection between the terms by $E_2 = (E \setminus E_1)^*$, i.e. an edge is present in E_1 if its dual edge is absent from E_2, and vice versa. This implies

$$|E_1| + |E_2| = |E|. \tag{1.11}$$

We have moreover the topological identity for the induced (not necessarily connected) graphs $G_1 = (V, E_1)$ and $G_2 = (V^*, E_2)$,

$$k(E_1) = |V| - |E_1| + c(E_1) = |V| - |E_1| + k(E_2) - 1, \tag{1.12}$$

where $k(E_1)$ and $c(E_1)$ are respectively the number of connected components and the number of independent cycles[1] in the graph G_1.

The proof of (1.12) is again by induction. If $E_1 = \emptyset$, we have $k(E_1) = |V|$, $c(E_1) = 0$, and $k(E_2) = 1$. Each time an edge is added to E_1 there are two possibilities. Either $c(E_1)$ stays constant, in which case $k(E_1)$ is reduced by one and $k(E_2)$ is unchanged. Or $c(E_1)$ increases by one, in which case $k(E_1)$ is unchanged and $k(E_2)$ increases by one. In both cases (1.12) remains valid.

Combining (1.11)–(1.12) gives

$$v^{|E_1|} Q^{k(E_1)} = k (v^*)^{|E_2|} Q^{k(E_2)}, \tag{1.13}$$

where we have defined

$$k = Q^{1-|V^*|} v^{|E|} = Q^{|V|-|E|-1} v^{|E|} \tag{1.14}$$

and $v^* = Q/v$. Comparing (1.13) with (1.9) completes the demonstration of (1.10) and furnishes the desired duality relation

$$vv^* = Q. \tag{1.15}$$

The duality relation (1.15) is particularly useful when the graph is selfdual, $G^* = G$. This is the case of the regular square lattice. Assuming the uniqueness of the phase transition, the critical point is given by the selfdual coupling

$$v_c = \pm\sqrt{Q} \quad \text{(square lattice)} \tag{1.16}$$

---

[1] The number of independent cycles—also known as the circuit rank, or the cyclomatic number—is the smallest number of edges to be removed from a graph in order that no graph cycle remains.

1.1.4 Special cases

One of the strengths of the Q-state Potts model is that it contains a large number of interesting special cases. Many of those make manifest the geometrical content of the partition function (1.7). The equivalence between $Q=2$ and the Ising model has already been discussed. We shall concentrate here on a couple of other subtle equivalences, that explicitly exploit the fact that Q can now be used as a continuous variable.

1.1.4.1 Bond percolation

For $Q=1$ the Potts model is seemingly trivial, with partition function $Z=(1+v)^{|E|}$. Instead of setting $Q=1$ brutally, one can however consider taking the *limit* $Q \to 1$. This leads to the important special case of *bond percolation*.

Let $p \in [0,1]$ and set $v = p/(1-p)$. We then consider the rescaled partition function

$$\tilde{Z}(Q) \equiv (1-p)^{|E|} Z = \sum_{E' \subseteq E} p^{|E'|} (1-p)^{|E|-|E'|} Q^{k(E')}. \tag{1.17}$$

We have of course $\tilde{Z}(1) = 1$. But formally, what is written here is that each edge is present in E' (i.e. percolating) with probability p and absent (i.e. non-percolating) with probability $1-p$. Appropriate correlation functions and derivatives of $\tilde{Z}(Q)$ in the limit $Q \to 1$ furnish valuable information about the geometry of the percolation clusters. For instance

$$\lim_{Q \to 1} Q \frac{d\tilde{Z}(Q)}{dQ} = \langle k(E') \rangle \tag{1.18}$$

gives the average number of clusters.

1.1.4.2 Trees and forests

Using (1.12), and defining $w = \frac{Q}{v}$, one can rewrite (1.7) as

$$Z = \sum_{E_1 \subseteq E} \left(\frac{Q}{w}\right)^{|V|+c(E_1)-k(E_1)} Q^{|V|-|E_1|+c(E_1)}$$
$$= v^{|V|} \sum_{E_1 \subseteq E} w^{k(E_1)-c(E_1)} Q^{c(E_1)}. \tag{1.19}$$

Take now the limit $Q \to 0$ and $v \to 0$ in such a way that the ratio $w = Q/v$ is fixed and finite, and consider the rescaled partition function $\tilde{Z} = Zv^{-|V|}$. The limit $Q \to 0$ will suppress any term with $c(E_1) > 0$, and we are left with

$$\tilde{Z} = \sum_{E_1 \subseteq E}{}' w^{k(E_1)}, \tag{1.20}$$

where the prime indicates that the summation is over edge sets such that the graphs $G_1 = (V, E_1)$ have no cycles, $c(E_1) = 0$. Such graphs are known as forests, or more precisely (since the vertex set V is that of G), *spanning forests* of G. Each connected component carries a weight w.

For $w \to 0$, the surviving terms are *spanning trees*, i.e. forests with a single connected component. Note that the critical curve on the square lattice (1.16) goes through the point $(Q, v) = (0, 0)$ with a vertical tangent (i.e. $w \to 0$) and thus describes spanning trees. General values of w have also been studied in detail [16–18].

1.1.5 Loop representation

We now transform the Potts model defined on a planar graph G into a model of self-avoiding loops [19] on a related graph $\mathcal{M}(G)$, known in graph theory as the *medial graph*. Each term E' in the cluster representation (1.7) is in bijection with a term in the loop representation. The correspondence is, roughly speaking, that the loops turn around the connected components in $G' = (V, E')$ as well as their elementary internal cycles. More precisely, the loops separate the clusters from their duals.

To make this transformation precise, first notice that we can draw a quadrangle around each pair of intersecting edges, $e \in E$ and $e^* \in E^*$. This set of quadrangles defines the quadrangulation $\hat{G}$ (see Figure 1.1b). The medial lattice is just the dual of this quadrangulation: $\mathcal{M}(G) = \hat{G}^*$ (see Figure 1.1c). Each vertex of $\mathcal{M}(G)$ thus stands at the intersection between e and e^*.

In the case where G is the square lattice, the medial $\mathcal{M}(G)$ is just another (tilted) square lattice.

Figure 1.1 *(a) A planar graph G (black circles and solid lines) and its dual graph G^* (white circles and dashed lines). (b) The plane quadrangulation $\hat{G} = \mathcal{M}(G)^*$. (c) The medial graph $\mathcal{M}(G) = \mathcal{M}(G^*)$.*

We recall that the partition function $Z_G(Q,v)$ and its dual $Z_{G^*}(Q,v^*)$ were given in (1.9) as sums over mutually dual subsets of edges E_1 and E_2. The last step of the transformation is to split the vertices of $\mathcal{M}(G)$ in the following way:

$$\text{If } e \in E_1, e \notin E_2: \qquad \text{If } e \notin E_1, e \in E_2: \qquad (1.21)$$

In concrete terms, this definition means that the loops bounce off all edges in E_1 and E_2, or, equivalently, they separate the FK clusters from their duals.

To complete the transformation, note that the number of loops $l(E_1)$ is the sum of the number of connected components $k(E_1)$ and the number of independent cycles $c(E_1)$,

$$l(E_1) = k(E_1) + c(E_1). \tag{1.22}$$

Inserting this and the topological identity (1.12) into (1.7) we arrive at

$$Z = Q^{|V|/2} \sum_{E_1 \subseteq E} x^{|E_1|} Q^{l(E_1)/2}, \tag{1.23}$$

where we have defined $x = vQ^{-1/2}$.

This is the *loop representation* of the Potts partition function. It importance stems from the fact that the loops, their connectivity properties, and the non-local quantity $l(E_1)$ all admit an algebraic interpretation within the Temperley–Lieb algebra [20].

In terms of the x variables the duality relation (1.15) reads simply

$$xx^* = 1. \tag{1.24}$$

In the case of the square lattice, the self-dual points are $x_c = \pm 1$, and the usual critical point is $x_{c+} = 1$. The loop model (1.23) then becomes extremely simple: there is just a weight $n = \sqrt{Q}$ for each loop.

1.1.6 Vertex model representation

In the definition of the Q-state Potts model, Q was originally positive integers. However, in the corresponding loop model (1.23) it appears as formal parameters and may thus take arbitrary complex values. The price to pay for this generalization is the appearance of a non-locally defined quantity, the number of loops l. The locality of the model may be recovered at the expense of introducing complex Boltzmann weights, as we now show.

The following argument supposes that $G = (V, E)$ is a (connected) planar graph. Most applications however suppose a regular lattice, a situation to which we shall return shortly.

Consider any model of self-avoiding loops defined on G (or some related graph, such as the medial graph $\mathcal{M}(G)$ for the Potts model). The Boltzmann weights are supposed to consist of a local piece—depending on if and how the loops pass through a given vertex—and a non-local piece of the form n^l, where n is the loop weight and l is the number of loops. In the case of the Potts model we have $n = \sqrt{Q}$.

In a first step, each loop is independently decorated by a global orientation, which by planarity and self-avoidance can be described as either anticlockwise or clockwise. Let us give a weight $\exp\left(i\gamma \frac{\alpha}{2\pi}\right)$ whenever an orientated loop turns an angle α in the positive (trigonometric) direction. Summing over orientations this gives

$$n = e^{i\gamma} + e^{-i\gamma} = 2\cos\gamma. \tag{1.25}$$

Note that in the expected critical regime, $n \in [-2, 2]$, we have $\gamma \in [0, \pi]$.

The loop model is now transformed into a *local vertex model* by assigning to each edge traversed by a loop the orientation of that loop. The total vertex weight equals the above local loop weights summed over the possible splittings of oriented loops which are compatible with the given edge orientations. In addition, one must multiply this by any loop-independent local weights, such as x in (1.23).

1.1.7 Six-vertex model

To see how this is done, we finally specialize to the Potts model defined on the square lattice G. The loop model is defined on the corresponding medial lattice $\mathcal{M}(G)$ which is another (tilted) square lattice. Each edge of the lattice is visited by a loop, and two loop segments (possibly parts of the same loop) meet at each vertex. In the orientated loop representation, each vertex is therefore incident on two outgoing and two ingoing edges.

It is convenient for the subsequent discussion to make the couplings of the Potts model anisotropic. In its original spin formulation (1.4) we therefore let K_1 (resp. K_2) denote respectively the dimensionless coupling in the horizontal (resp. vertical) direction of the square lattice, and we let

$$x_1 = \frac{e^{K_1} - 1}{\sqrt{Q}}, \qquad x_2 = \frac{e^{K_2} - 1}{\sqrt{Q}} \tag{1.26}$$

be the corresponding parameters appearing in the loop representation (1.23).

The six possible configurations of arrows around a vertex of the medial lattice $\mathcal{M}(G)$ are shown in Figure 1.2. The corresponding vertex weights are denoted ω_p (resp. ω_p') on the even (resp. odd) sublattice of $\mathcal{M}(G)$. By definition, a vertex of the even (resp. odd) sublattice of $\mathcal{M}(G)$ is the mid point of an edge with coupling K_1 (resp. K_2) of the original spin lattice G. With respect to Figure 1.2 we define the even sublattice to be such that

12 *Integrability in statistical physics and quantum spin chains*

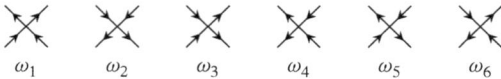

Figure 1.2 *The allowed arrow arrangements (top) around a vertex that define the six-vertex model, with the corresponding particle trajectories (bottom).*

an edge $e \in E$ runs horizontally, and the corresponding dual edge $e^* \in E^*$ is vertical. For the odd sublattice, exchange e and e^*.

Using (1.23) we then have

$$Z = Q^{|V|/2} \sum_{\text{arrows}} \prod_{p=1}^{6} (\omega_p)^{N_p} (\omega'_p)^{N'_p}, \quad (1.27)$$

where the sum is over arrow configurations satisfying the constraint 'two in, two out' at each vertex, and N_p (resp. N'_p) is the number of vertices on the even (resp. odd) sublattice with arrow configuration p. Thus, the square-lattice Potts model has been represented as a *staggered six-vertex model*.[2] The weights read explicitly

$$\omega_1, \ldots, \omega_6 = 1, 1, x_1, x_1, e^{i\gamma/2} + x_1 e^{-i\gamma/2}, e^{-i\gamma/2} + x_1 e^{i\gamma/2} \quad (1.28)$$

$$\omega'_1, \ldots, \omega'_6 = x_2, x_2, 1, 1, e^{-i\gamma/2} + x_2 e^{i\gamma/2}, e^{i\gamma/2} + x_2 e^{-i\gamma/2} \quad (1.29)$$

To see this, note that configurations $i = 1, 2, 3, 4$ are compatible with just one linking of the oriented loops:

$$\underset{\omega_1}{\times} = \underset{1}{)(} , \quad (1.30)$$

whereas $i = 5, 6$ are compatible with two different linkings (and the weight is obtained by summing over these two):

$$\underset{\omega_5}{\times} = \underset{e^{i\gamma/2}}{)(} + \underset{x_1 e^{-i\gamma/2}}{\asymp} . \quad (1.31)$$

Note that the even and odd sublattices are related by a $\pi/2$ rotation of the vertices in Figure 1.2. This rotation interchanges configurations $(\omega_1, \omega_2) \leftrightarrow (\omega'_3, \omega'_4)$ and $\omega_5 \leftrightarrow \omega'_6$. On the level of the weights it corresponds to $x_1 \leftrightarrow x_2$.

---

[2] The term *staggered* means that the weights alternate between sublattices.

The staggered six-vertex model is not exactly solvable in general. However, if we impose that the couplings be mutually dual,

$$x_2 = (x_1)^{-1}, \tag{1.32}$$

we have $\omega'_i = (x_1)^{-1}\omega_i$ for any $i = 1,2,\ldots,6$. The factors $(x_1)^{-1}$ from each ω'_i can be taken outside the summation in (1.27) and we have effectively $\omega'_i = \omega_i$. The six-vertex model then becomes homogeneous.

The homogeneous six-vertex model turns out to be solvable when the weights ω_i are invariant under a global arrow reversal. The resulting weights are traditionally denoted

$$a = \omega_1 = \omega_2, \qquad b = \omega_3 = \omega_4, \qquad c = \omega_5 = \omega_6. \tag{1.33}$$

The constraint $\omega_5 = \omega_6$ is actually not necessary. Indeed, the corresponding vertices act as sources and sinks of arrows in either lattice direction, so with appropriate (periodic) boundary conditions there is an equal number of vertices of either type. Therefore Z depends on ω_5, ω_6 only via their product, and we might as well set $c = (\omega_5\omega_6)^{1/2}$.

In the study of the six-vertex model, a special role is played by the so-called anisotropy parameter

$$\Delta = \frac{a^2 + b^2 - c^2}{2ab}. \tag{1.34}$$

In our case we have simply

$$\Delta = -\cos\gamma. \tag{1.35}$$

We shall see that the Bethe ansatz equations depend *only* on Δ.

The other independent ratio among a, b, c is essentially x_1. We shall use it to parameterize the so-called spectral parameter u. It is seen to control the spatial anisotropy of the 2D statistical model (the anisotropy corresponding to Δ refers instead to the equivalent 1D quantum spin chain).

We stress once more that the square-lattice Potts model is solvable at its selfdual point, but not at arbitrary temperatures.[3] This is in contrast with the Ising model, which is solvable at any temperature [24]. In that sense the Ising model is a rather untypical integrable model.

---

[3] It is however possible to solve it also at the antiferromagnetic transition [21–23].

1.2 Quantum integrability

Vertex models are statistical models in which integer-valued states are defined on each edge of some lattice, and the interaction takes place at the vertices. For a lattice of coordination number z with s possible states on each edge, each vertex can see $k_0 = z^s$ possible arrangements of its incident edges. The Boltzmann weight at a vertex is taken to depend on this arrangement. Usually only $k \leq k_0$ arrangements correspond to a non-zero weight. We refer to the corresponding statistical model as a *k-vertex model*.

1.2.1 *R*-matrix

It is natural to think of the propagation through a vertex in a transfer matrix formalism. When z is even, we can define a transfer direction so that $\frac{z}{2}$ consecutive edges define the in-state, and the remaining $\frac{z}{2}$ edges the out-state. The Boltzmann weights can then be regrouped in a $(k_0)^{1/2}$ dimensional matrix, called the *R*-matrix, with precisely k non-zero entries.

Figure 1.2 defines the 6-vertex model on a square lattice ($z = 4$ and $s = 2$). We take the transfer direction to be upwards, so the two edges on the bottom (resp. top) define the in-state (resp. out-state). The state of an edge supporting a down-arrow (resp. an up-arrow) is denoted $|1\rangle$ (resp. $|0\rangle$). The state $|1\rangle$ can be interpreted as the presence of a particle. The particles are conserved by the time evolution, because the allowed vertices have two ingoing and two outgoing arrows.

The basis of in-states can now be written

$$(\mathbb{C}^2)^2 = \{|00\rangle, |01\rangle, |10\rangle, |11\rangle\}, \tag{1.36}$$

where the left in-state refers to the leftmost edge on the bottom. The out-states can be labelled similarly, but two different conventions are possible. In the first convention, the left out-state refers to the *rightmost* edge on the top, so that the labelling of spaces follows the lines (which intersect at the vertex). This defines the *R*-matrix:

$$R = \begin{bmatrix} \omega_1 & 0 & 0 & 0 \\ 0 & \omega_3 & \omega_6 & 0 \\ 0 & \omega_5 & \omega_4 & 0 \\ 0 & 0 & 0 & \omega_2 \end{bmatrix}. \tag{1.37}$$

In the second convention, the left out-state refers to the leftmost edge on the top, so that in-states and out-states have the same left/right convention. This defines the $\check{R}$-matrix:

$$\check{R} = \begin{bmatrix} \omega_1 & 0 & 0 & 0 \\ 0 & \omega_5 & \omega_4 & 0 \\ 0 & \omega_3 & \omega_6 & 0 \\ 0 & 0 & 0 & \omega_2 \end{bmatrix}. \tag{1.38}$$

Obviously $\check{R} = PR$, where P is the operator that permutes the two spaces.

1.2.2 Spectral parameter

Let us parameterize the weights of the six-vertex model as follows:

$$\begin{aligned}
\omega_1 &= \omega_2 = \sin(\gamma - u), \\
\omega_3 &= \omega_4 = \sin u, \\
\omega_5 &= e^{-i(u-\eta)} \sin \gamma, \\
\omega_6 &= e^{i(u-\eta)} \sin \gamma.
\end{aligned} \qquad (1.39)$$

We have then $\Delta = -\cos\gamma$. The gauge parameter η can be chosen at will, since vertices of type 5 and 6 appear in pairs, and only the value of $\sqrt{\omega_5 \omega_6}$ enters the computation of the partition function. The correspondence (1.28) with the selfdual Potts model then requires

$$x_1 = \frac{1}{x_2} = \frac{\omega_3}{\omega_1} = \frac{\sin u}{\sin(\gamma - u)}. \qquad (1.40)$$

We note that u parameterizes the spatial anisotropy of the coupling constants. The isotropic point at which $x_1 = x_2 = 1$ corresponds to $u = \frac{\gamma}{2}$.

More generally, we wish to formalize some useful properties of the R-matrix for vertex models defined on a so-called *Baxter lattice* [25]. By this we mean any lattice that can be drawn in the plane as a collection of lines (one can think of them as straight lines, but this is not necessary) that undergo only pairwise intersections. We attribute a fixed orientation to each line in the lattice, and associate a so-called *spectral parameter* $u \in \mathbb{C}$ with each oriented line.

The Boltzmann weights for a vertex where two lines with spectral parameters u and v intersect are supposed to have the *difference property*: they depend only on the difference $u - v$. To be precise, when viewing the vertex along the forward direction of the two oriented lines, u (resp. v) is the spectral parameter of the line to the left (resp. right) of the observer. In particular, inverting the direction of the line corresponds to $u \to -u$.

It is useful to turn the vertices through an angle $-\frac{\pi}{4}$, so that time flows in the north-easterly direction, rather than upwards. This is shown for the six-vertex model in Figure 1.3 where we also give the corresponding trajectories of particles (the state $|1\rangle$). The elements of the R-matrix can then be written,

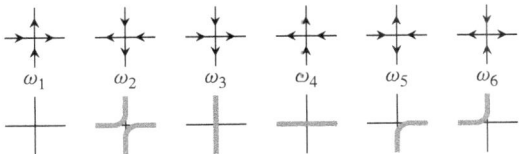

Figure 1.3 *The allowed arrow arrangements (top) around a vertex that define the six-vertex model, with the corresponding particle trajectories (bottom).*

$$\mu_i \xrightarrow{u} \begin{array}{c}\beta_i \\ \uparrow \\ \\ \alpha_i\end{array} \mu_{i+1} \;=\; R^{\mu_{i+1}\beta_i}_{\mu_i \alpha_i}(u-v) \;=\; {}_a\langle \mu_{i+1}| \otimes {}_i\langle \beta_i | R_{ai}(u-v) |\mu_i\rangle_a \otimes |\alpha_i\rangle_i. \qquad (1.41)$$

We have oriented the horizontal line towards the right and called the corresponding space a (for *auxiliary* space). The vertical line is oriented upwards, and its space is labelled i (for i'th *quantum* space). The indices μ_i, α_i and μ_{i+1}, β_i label states living on the lattice edges. The last notation $R_{ai}(u-v)$ makes explicit both the labelling of spaces and the difference property of the spectral parameters.

More formally, the R-matrix is a linear operator (endomorphism)

$$R_{ai} \,:\, V_a \otimes V_i \mapsto V_a \otimes V_i, \qquad (1.42)$$

where the vector spaces V_a (auxiliary) and V_i (quantum) carry the edge degrees of freedom. For instance, in the six-vertex model they are both equal to the spin-$\tfrac{1}{2}$ representation space $\mathbb{C}^2$, since each arrow can be in two possible states; the R-matrix is then a 4×4 matrix.

1.2.3 Transfer matrix

We now wish to define the row-to-row transfer matrix for a system of width L with periodic boundary directions in the horizontal direction.

The transfer matrix t is an endomorphism on the tensor product of all quantum spaces,

$$t : V_1 \otimes V_2 \otimes \cdots \otimes V_L \mapsto V_1 \otimes V_2 \otimes \cdots \otimes V_L. \qquad (1.43)$$

It can be written as

$$t = \mathrm{Tr}_a \, (R_{aL} R_{aL-1} \cdots R_{a2} R_{a1}), \qquad (1.44)$$

where Tr_a denotes the trace over the auxiliary space V_a. For simplicity we have not written the dependence on the spectral parameters. Indeed, one has the possibility of taking *different* spectral parameters for each quantum space V_i, and also for V_a, which will correspond to a completely inhomogeneous lattice model. The matrix elements of t can be written very explicitly as

$$\langle \beta | t | \alpha \rangle = \sum_{\mu_1, \dots, \mu_L} R^{\mu_1 \beta_L}_{\mu_L \alpha_L} R^{\mu_L \beta_{L-1}}_{\mu_{L-1} \alpha_{L-1}} \cdots R^{\mu_3 \beta_2}_{\mu_2 \alpha_2} R^{\mu_2 \beta_1}_{\mu_1 \alpha_1}. \qquad (1.45)$$

We remark that μ_1 appears both in the rightmost and the leftmost factor, so we indeed perform the operator Tr_a.

Notice that although the individual R-matrices evolve the system in the north-easterly direction, the result of the trace is that t evolves the system upwards.

1.2.4 Commuting transfer matrices

The R-matrix formalism presented here makes sense on any lattice of orientated lines that make only pairwise intersections. Such a lattice is called a *Baxter lattice*. It is possible to impose periodic boundary conditions on some of the lines. Notice that the lines are not required to be straight, nor to be disposed in any regular fashion.

A statistical model defined on a Baxter lattice is said to be integrable provided its R-matrix satisfies the Yang–Baxter equation and the inversion relation. The Yang–Baxter relation reads pictorially

$$\tag{1.46}$$

The algebraic transcription is

$$R_{12}(u)R_{13}(u+v)R_{23}(v) = R_{23}(v)R_{13}(u+v)R_{12}(u), \tag{1.47}$$

where we have set $u = u_1 - u_2$ and $v = u_2 - u_3$, so that $u_1 - u_3 = u + v$. We stress again that the spectral parameters u_i always follow the lines of the Baxter lattice. The same is true for the labels of the representation spaces, that appear as subscripts for the R-matrix. It is sometimes convenient for these labels to stay well ordered in space (i.e. with 1 on the left, 2 in the middle, and 3 on the right) at all times (in the diagram time flows upwards). In that case one uses instead the $\check{R}$-matrix, for which the Yang–Baxter equation reads

$$\check{R}_{23}(u)\check{R}_{12}(u+v)\check{R}_{23}(v) = \check{R}_{12}(v)\check{R}_{23}(u+v)\check{R}_{12}(u). \tag{1.48}$$

The inversion relation can be represented pictorially as

$$\tag{1.49}$$

and reads algebraically

$$R_{21}(u)R_{12}(-u) \propto I. \tag{1.50}$$

The constant of proportionality could of course be set to unity by a suitable rescaling of R. Note also how the sign convention for spectral parameters comes into use when writing (1.50).

The relations (1.47) and (1.50) imply the commutation of two transfer matrices corresponding to different choices of spectral parameters on the auxiliary lines. This is best demonstrated graphically:

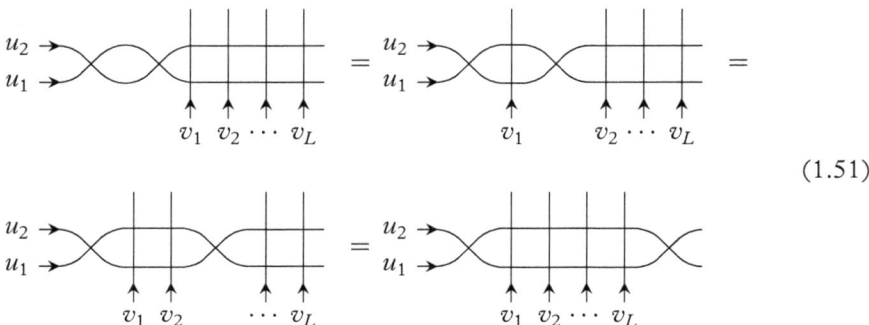

$$\tag{1.51}$$

The first picture represents the product $t(u_2)t(u_1)$, since the two crossings to the left amount to the identity by (1.50).[4] In the second picture we have used (1.47) to push the v_1 line to the left. This is repeated in the third picture for the next v_2 line. Repeating this operation L times, we finally arrive at the last picture, which represents the product $t(u_1)t(u_2)$, apart from the crossings on the left and right. But the right crossing can be taken around the periodic boundary condition (more formally, we are using the cyclicity of the trace), and using once more (1.50) the two crossings annihilate. Summarizing, we have shown that

$$t(u_2)t(u_1) = t(u_1)t(u_2). \tag{1.52}$$

The existence of an infinite family of commuting transfer matrices has important consequences. Indeed the Bethe ansatz technique permits us to diagonalize all these transfer matrices simultaneously.

---

[4] The transfer matrices depend also on the spectral parameters $v_1, v_2, \ldots, v_L$ of the quantum spaces, but we omit this dependence for notational convenience.

Moreover, we can take derivatives of (1.52) with respect to u_2. All these derivatives commute with $t(u_1)$, hence are conserved by the time evolution process. In other words, an integrable system has an infinite number of conserved quantities. The first few derivatives can be identified with the Hamiltonian, the momentum operator, and so on. We shall present explicit examples later.

Note also that the various vector spaces in which the R-matrices act need not be isomorphic. In particular, one can have different representations on the quantum and auxiliary spaces. From a basic integrable model—such as the six-vertex model in the spin-$\frac{1}{2}$ representation—one can construct higher-spin solutions by appropriate fusions of representation spaces. One speaks in that case of descendent models. An example will be given in Chapter 6.

1.2.5 Six-vertex model

It is an instructive exercise to verify that the R-matrix of the six-vertex model indeed satisfies (1.47) and (1.50). Choosing the gauge $\eta = 0$ in the parameterization (1.39), the $\check{R}$-matrix (1.38) reads

$$\check{R}(u) = \begin{bmatrix} \sin(\gamma - u) & 0 & 0 & 0 \\ 0 & e^{-iu}\sin\gamma & \sin u & 0 \\ 0 & \sin u & e^{iu}\sin\gamma & 0 \\ 0 & 0 & 0 & \sin(\gamma - u) \end{bmatrix}. \quad (1.53)$$

We identify the uniformizing parameter u with the difference of spectral parameters.

In tensor notation (1.48) reads in the space $(\mathbb{C}^2)^3$

$$(I \otimes \check{R}(u))(\check{R}(u+v) \otimes I)(I \otimes \check{R}(v)) = (\check{R}(v) \otimes I)(I \otimes \check{R}(u+v))(\check{R}(u) \otimes I). \quad (1.54)$$

This identity between 8×8 matrices is greatly simplified by the symmetries of the problem. Firstly, the number of particles is conserved. Secondly, the weights are invariant under a global negation ($0 \leftrightarrow 1$) of the occupation numbers.

The only equations to be verified thus concern a 1×1 matrix (in the 0-particle space $|000\rangle$) and a 3×3 matrix (in the 1-particle space $|100\rangle, |010\rangle, |001\rangle$). Only the latter gives rise to non-trivial equations. They turn out to be verified, upon application of trigonometric identities.

The inversion relation (1.50) can be verified similarly. In particular, we obtain the proportionality factor,

$$\check{R}(u)\check{R}(-u) = \sin(\gamma - u)\sin(\gamma + u)I. \quad (1.55)$$

There is an alternative route to deriving the integrable R-matrix of the six-vertex model, which has the advantage of revealing a rich underlying algebraic structure. To achieve this we first need a few definitions.

1.2.6 Temperley–Lieb algebra

The Temperley–Lieb algebra $TL_N(n)$ is a unital associative algebra over $\mathbb{C}$. Its $N-1$ generators are denoted E_m for $m = 1, 2, \ldots, N-1$. They satisfy the relations [20]

$$(E_m)^2 = nE_m,$$
$$E_m E_{m\pm 1} E_m = E_m,$$
$$E_m E_{m'} = E_{m'} E_m \text{ for } |m - m'| > 1. \qquad (1.56)$$

As for any abstract algebra, $TL_N(n)$ can be represented in different ways. We shall be particularly interested in its *loop-model representation*, since this permits us to make contact with Section 1.1.5 where the Potts model was formulated as a loop model. In this representation, $TL_N(n)$ is viewed as an algebra of diagrams acting on N numbered vertical strands (for convenience depicted inside a dashed box) as

$$E_m = \quad \underset{1 \quad 2 \quad\quad m \ \ m+1 \quad\ N-1 \ N}{\text{[diagram]}} \qquad (1.57)$$

Multiplication in $TL_N(n)$ is defined by stacking diagrams vertically. More precisely, the product of two generators $g_2 g_1$ is defined by placing the diagram for g_2 above the diagram for g_1, identifying the bottom points of g_2 with the top points of g_1. The resulting diagram is considered up to smooth isotopies that keep fixed the surrounding box, and any closed loop is replaced by the factor n.

In this way we have for instance (omitting strands on which the action is trivial)

$$(E_m)^2 = \quad \text{[diagram]} \quad = n \ \text{[diagram]} \quad = n E_m$$

and

$$E_m E_{m+1} E_m = \quad \text{[diagram]} \quad = \ \text{[diagram]} \quad = E_m \ .$$

It is thus readily seen that all the defining relations (1.56) are satisfied. Moreover, for generic values of n no further relations hold; the loop-model representation is faithful.

The connection with the Potts model is provided by (1.31): the left-hand diagram is nothing but the generator E_m of (1.57), while the right-hand diagram is the identity operator. We recall that according to (1.25) the loop weight is $n = 2\cos\gamma$.

1.2.6.1 Integrable Ř-matrix in the loop representation

Starting from first principles, we now construct an integrable model based on the TL algebra. Let us suppose that the Ř-matrix has the form

$$\check{R}_{m,m+1}(u) = f(u)I + g(u)E_m, \tag{1.58}$$

where $f(u)$ and $g(u)$ are some functions of the spectral parameter u to be determined. Inserting this into the Yang–Baxter equation (1.48) yields

$$(f(u)I + g(u)E_2)(f(u+v)I + g(u+v)E_1)(f(v)I + g(v)E_2) =$$
$$(f(v)I + g(v)E_1)(f(u+v)I + g(u+v)E_2)(f(u)I + g(u)E_1). \tag{1.59}$$

Using the algebraic relations (1.56) we can expand both sides of (1.59). The left-hand side produces

$$f(u)f(u+v)f(v)I + f(u)g(u+v)f(v)E_1 +$$
$$g(u)g(u+v)f(v)E_2E_1 + f(u)g(u+v)g(v)E_1E_2 +$$
$$[g(u)g(v)(g(u+v) + nf(u+v)) + f(u+v)(f(u)g(v) + f(v)g(u))]E_2,$$

and the right-hand side becomes

$$f(v)f(u+v)f(u)I + f(v)g(u+v)f(u)E_2 +$$
$$f(v)g(u+v)g(u)E_2E_1 + g(v)g(u+v)f(u)E_1E_2 +$$
$$[g(u)g(v)(g(u+v) + nf(u+v)) + f(u+v)(f(u)g(v) + f(v)g(u))]E_1.$$

These expressions must be identical in $TL_3(n)$, and so we can identify the coefficients for each of the five possible words in the algebra. The relations resulting from the words I, E_1E_2, and E_2E_1 are trivial. The relations coming from E_1 and E_2 are identical—related via an exchange of the left- and right-hand sides—and read

$$g(u)g(v)(g(u+v) + nf(u+v)) + f(u+v)(f(u)g(v) + f(v)g(u)) =$$
$$f(u)f(v)g(u+v). \tag{1.60}$$

The functional relation (1.60) is a typical outcome of this way of solving the Yang–Baxter equations. It is in general not easy to solve this type of relation, and even if one finds solutions it is often difficult to make sure that one has found *all* the solutions. Worse, in more complicated cases than the one considered here the ansatz for the Ř-matrix will involve more terms and the functions $f(u), g(u),\ldots$ must satisfy several coupled functional equations.

It is useful to rewrite (1.60) in terms of the parameters $z = e^{iu}$, $w = e^{iv}$ and $q = e^{i\gamma}$. That is, instead of the *additive* spectral parameters u, v we have now *multiplicative* spectral parameters z, w. Thus,

$$g(z)g(w)\left(g(zw)+(q+q^{-1})f(zw)\right)+f(zw)\left(f(z)g(w)+f(w)g(z)\right)=f(z)f(w)g(zw). \tag{1.61}$$

It is tempting to set $f(z)=1$, since the overall normalization of the $\check{R}$-matrix is unimportant, but in general this is *not* a good idea. A time proven strategy is to suppose that $f(z)$ and $g(z)$ are polynomials of some small degree in the variables z, z^{-1}, q and q^{-1}. (In some cases one needs to try fractional powers of q as well). In this case we are lucky, there is a solution of degree one,

$$f(z) = \frac{q}{z} - \frac{z}{q}, \tag{1.62}$$

$$g(z) = z - z^{-1}. \tag{1.63}$$

Going back to additive spectral parameters, we thus have a trigonometric solution of (1.59),

$$f(u) = \sin(\gamma - u), \qquad g(u) = \sin(u). \tag{1.64}$$

In general, solutions to the Yang–Baxter equation turn out to be polynomial, trigonometric, or elliptic (in order of increasing difficulty).

Summarizing, we have found the integrable $\check{R}$-matrix

$$\check{R}(u) = \sin(\gamma - u)I + \sin(u)E. \tag{1.65}$$

1.2.6.2 Back to the vertex model

Combining (1.65) with (1.53) we obtain the TL generator in the vertex-model representation,

$$E = \begin{bmatrix} 0 & 0 & 0 & 0 \\ 0 & e^{-i\gamma} & 1 & 0 \\ 0 & 1 & e^{i\gamma} & 0 \\ 0 & 0 & 0 & 0 \end{bmatrix}. \tag{1.66}$$

By forming tensor products, one can of course verify that the defining relations (1.56) are indeed satisfied.

It is convenient to make manifest the spin-$\frac{1}{2}$ nature of the six-vertex model by re-expressing things in terms of the Pauli matrices

$$\sigma^x = \begin{bmatrix} 0 & 1 \\ 1 & 0 \end{bmatrix}, \qquad \sigma^y = \begin{bmatrix} 0 & -i \\ i & 0 \end{bmatrix}, \qquad \sigma^z = \begin{bmatrix} 1 & 0 \\ 0 & -1 \end{bmatrix}. \tag{1.67}$$

The arrow conservation then means that the transfer matrix $t(u)$ commutes with the total magnetization

$$S^z = \frac{1}{2}\sum_{m=1}^{L} \sigma_m^z. \tag{1.68}$$

The Temperley–Lieb generator can be written as

$$E_m = \frac{1}{2}\left[\sigma_m^x \sigma_{m+1}^x + \sigma_m^y \sigma_{m+1}^y - \cos\gamma\, (\sigma_m^z \sigma_{m+1}^z - I) - i\sin\gamma\, (\sigma_m^z - \sigma_{m+1}^z)\right]. \tag{1.69}$$

1.2.7 Spectral parameter and anisotropy

The physical meaning of the spectral parameter u is that it controls the spatial anisotropy of the system. To see this qualitatively, note that in the $u \to 0$ limit, the $\check{R}$-matrix is proportional to the identity by (1.65). The transfer matrix $t(u)$ thus acts on a state just by shifting all spins one unit to the right (with periodic boundary conditions); note that this follows from the fact that time propagates in the north-easterly direction.

In a 1+1 dimensional quantum mechanical analogy, the $u \to 0$ limit thus means that interactions between spins happen very slowly. Equivalently, the time direction has been stretched with respect to the spatial direction. A homogeneous system can be retrieved by rescaling time by a certain anisotropy factor $\zeta(u)$. Determining $\zeta(u)$ requires some more work: the result is [26]

$$\zeta(u) = \sin\left(\frac{\pi u}{\gamma}\right). \tag{1.70}$$

This predicts that the isotropic point $\zeta(u) = 1$ occurs for $u = \frac{\gamma}{2}$, which is in accord with (1.65).

1.2.8 Spin chain Hamiltonian

Using (1.65) we thus see that in the completely anisotropic limit $u \to 0$ the transfer matrix becomes

$$t(0) = \sin^L(\gamma)\, e^{-iP}, \tag{1.71}$$

where e^{-iP} is the shift operator that translates the lattice sites one unit to the right. Equivalently, P can be interpreted as the momentum operator.

We know from the path-integral formalism that the transfer matrix (the time evolution operator) is the exponential of the quantum Hamiltonian. To make things completely precise, note that to first order in u, one may omit one of the factors $\sin(\gamma - u)I$ in (1.65) and take $\sin(u)E$ instead. The correct development in the limit $u \to 0$ therefore reads

$$t(u) \simeq t(0) \exp\left[-\frac{u}{\sin\gamma} H\right], \tag{1.72}$$

where H is the Hamiltonian of the spin chain. Equivalently

$$H = -\sin\gamma \left.\frac{\partial}{\partial u} \log t(u)\right|_{u \to 0} = -\sin\gamma\, t(0)^{-1} t'(0). \tag{1.73}$$

Here the inverse $t(0)^{-1} = (\sin\gamma)^{-L} e^{iP}$ is just the shift in the opposite (left) direction. The derivative $t'(0)$ gives L terms, one for each of the factors in the product (1.45). Using (1.65) we have $\check{R}'(0) = -\cos\gamma\, I + E$. Therefore

$$H = L \cos\gamma\, I - \sum_{m=1}^{L} E_m. \tag{1.74}$$

Inserting the expression (1.69) for the TL generators in terms of Pauli matrices, the piece in $i\sin\gamma(\sigma_m^z - \sigma_{m+1}^z)$ simplifies by telescopy. With open boundary condition it would become a surface magnetic field acting on the first and last spins. We consider instead periodic boundary conditions, so this term vanishes altogether. One is left with

$$H = -\frac{1}{2}\sum_{m=1}^{L}\left[\sigma_m^x \sigma_{m+1}^x + \sigma_m^y \sigma_{m+1}^y + \Delta(\sigma_m^z \sigma_{m+1}^z + I)\right], \tag{1.75}$$

where we recall that $\Delta = -\cos\gamma$.

We thus arrive at the Hamiltonian of a Heisenberg-type spin chain, where however the interaction is anisotropic along the z-direction. For that reason, this is called the *XXZ spin chain* with anisotropy parameter Δ.[5]

Let us emphasize that due to the commutativity of transfer matrices, the eigenvectors of the six-vertex model transfer matrix and of the XXZ spin chain Hamiltonian are *identical*. It is thus equivalent to diagonalize one or the other, and in that sense the two models are equivalent.

1.3 Algebraic Bethe ansatz

We have seen that the existence of an integrable $\check{R}$-matrix entails an infinite number of conserved quantities. This makes us suspect that the corresponding statistical model—our main example being the selfdual Potts model on a square lattice, or the equivalent XXZ quantum spin chain—may be exactly solvable in some sense.

The algebraic Bethe ansatz [27] provides the fulfilment of these expectations. It provides a formalism in which the partition function and correlation functions can be exactly computed, at least in the thermodynamic limit. Recent years have also seen much

---

[5] We are here referring to an anisotropy between the different components of the interaction in the space direction. This should not be confused with the space–time anisotropy linked with the spectral parameter u.

progress on the computation of certain correlation functions in finite size and/or at finite separation of the points, but this topic is beyond the scope of these lectures.

1.3.1 Monodromy matrix

We define the monodromy matrix $T(u)$ as the same product over R-matrices that was used in defining the transfer matrix (1.44), but without the trace over the auxiliary space,

$$T(u) = R_{aL} R_{aL-1} \cdots R_{a2} R_{a1}. \tag{1.76}$$

Thus $T(u)$ is an endomorphism on the auxiliary space V_a and we have

$$t(u) = \mathrm{Tr}_a T(u). \tag{1.77}$$

When several auxiliary spaces are involved we shall sometimes use the notation $T_a(u)$ to make clear what space is involved. We shall also denote matrix elements of $T(u)$ using the same convention as for the R-matrix, and sometimes represent them graphically as

$$T_i^j(u) = \quad i \mathrel{-\!\!\parallel\!\!-} j\,. \tag{1.78}$$

These matrix elements are operators acting in the quantum spaces, here shown symbolically as a double line.

We can repeat the reasoning of (1.51),

$$\tag{1.79}$$

to establish that

$$R_{ab}(u-v)\,(T_a(u)\,T_b(v)) = (T_b(v)\,T_a(u))\,R_{ab}(u-v). \tag{1.80}$$

This identity is known popularly as the RTT relation. Using the double line convention of (1.78) it can also be written pictorially

$$
R_{13} \underset{\substack{u_1 \\ 1}}{\diagdown} \underset{\substack{u_2 \\ 2}}{\overset{T_{12}}{\Big\|}} \underset{\substack{u_3 \\ 3}}{\overset{T_{23}}{\Big\|}} = \underset{\substack{u_1 \\ 1}}{\overset{T_{23}}{\Big\|}} \underset{\substack{u_2 \\ 2}}{\overset{T_{12}}{\Big\|}} R_{13} \underset{\substack{u_3 \\ 3}}{\diagdown} \tag{1.81}
$$

1.3.2 Co-product and Yang–Baxter algebra

A Yang–Baxter algebra $\mathcal{A}$ is a couple (R, T) satisfying the RTT relation (1.80). Its generators are the matrix elements $T_i^j(u)$. It is equipped with a product, obtained graphically by stacking two monodromy matrices along a common quantum space (represented as a double line).

In addition to this product, $\mathcal{A}$ is also equipped with a co-product Δ,[6] obtained graphically by gluing together two monodromy matrices along a common auxiliary space (represented as a single line). We have

$$\Delta(i \;\substack{\| \\ —} \; j) = \sum_k \; i \;\substack{\| \\ —} \; k \;\substack{\| \\ —} \; j. \tag{1.82}$$

The co-product thus serves to map the algebra $\mathcal{A}$ into the tensor product $\mathcal{A} \otimes \mathcal{A}$:

$$\begin{aligned} \Delta &: \mathcal{A} \to \mathcal{A} \otimes \mathcal{A} \\ T_i^j(u) &\mapsto \sum_k T_i^k(u) \otimes T_k^j(u) \end{aligned} \tag{1.83}$$

while preserving the algebraic relations of $\mathcal{A}$.

In particular, the co-product ΔT_i^j must again satisfy the RTT relation (1.80). It is a nice exercise to understand what this means and to prove it.

An algebra equipped with a product and a co-product is called a *bi-algebra*. To be precise, we need a little more structure (co-associativity, existence of a co-unit, ...). If in addition we have an antipode (and if various diagrams commute) one arrives at a Hopf algebra.

---

[6] This Δ should not be confused with the anisotropy parameter of the six-vertex model (XXZ spin chain).

1.3.3 Six-vertex model

When the auxiliary space is $\mathbb{C}^2$, the matrix elements of the monodromy matrix are usually denoted as follows:

$$T^0_0(u) = A(u), \quad T^0_1(u) = B(u), \quad T^1_0(u) = C(u), \quad T^1_1(u) = D(u). \tag{1.84}$$

Recall that the structure constants of a Lie algebra provide a representation, known as the adjoint. In the same way, the R-matrix provides a representation of dimension 2 of the Yang–Baxter algebra. Indeed, in the special case where the double line is just a single line, the monodromy matrix reduces to the R-matrix,

$$\left(T^j_i(u)\right)^k_l = R^{jk}_{il}(u). \tag{1.85}$$

The RTT relation is then nothing but the Yang–Baxter relation for the R-matrix.

The notation (1.84) just amounts to reading the R-matrix as a 2×2 matrix of blocks of size 2×2. According to (1.37) we have

$$R = \begin{bmatrix} \omega_1 & 0 & 0 & 0 \\ 0 & \omega_4 & \omega_5 & 0 \\ 0 & \omega_6 & \omega_3 & 0 \\ 0 & 0 & 0 & \omega_2 \end{bmatrix} = \begin{bmatrix} A(u) & B(u) \\ C(u) & D(u) \end{bmatrix}. \tag{1.86}$$

We recall the usual weights a, b, c (which now depend on the spectral parameter u), and we take the gauge $\eta = u$ in (1.53):

$$a(u) = \sin(\gamma - u), \quad b(u) = \sin u, \quad c(u) = \sin \gamma. \tag{1.87}$$

We have then in explicit notation, and in terms of Pauli matrices,

$$A(u) = \begin{bmatrix} a(u) & 0 \\ 0 & b(u) \end{bmatrix} = \frac{a(u) + b(u)}{2} I + \frac{a(u) - b(u)}{2} \sigma^z,$$

$$B(u) = \begin{bmatrix} 0 & 0 \\ c(u) & 0 \end{bmatrix} = \frac{c(u)}{2}(\sigma^x - i\sigma^y) = c(u)\sigma^-,$$

$$C(u) = \begin{bmatrix} 0 & c(u) \\ 0 & 0 \end{bmatrix} = \frac{c(u)}{2}(\sigma^x + i\sigma^y) = c(u)\sigma^+,$$

$$D(u) = \begin{bmatrix} b(u) & 0 \\ 0 & a(u) \end{bmatrix} = \frac{a(u) + b(u)}{2} I - \frac{a(u) - b(u)}{2} \sigma^z. \tag{1.88}$$

Note that $B(u)$ (resp. $C(u)$) acts as a creation (resp. annihilation) operator on the quantum space, with respect to the pseudo-vacuum in which all spins are up. We shall see later that this interpretation remains valid when taking co-products: $B(u)$ transforms n particle states into $n+1$ particle states (and vice versa for $C(u)$).

1.3.3.1 Co-product

Establishing how the co-product acts on the operators $A(u)$, $B(u)$, $C(u)$, and $D(u)$ will turn out to be an important ingredient in the sequel. In more formal terms, we wish to obtain a representation of the six-vertex Yang–Baxter algebra $\mathcal{A}$ on the space $V^{\otimes L}$.

Let us begin by examining the case $L=2$ in details. Consider for instance the construction of $\Delta B(u)$. We have

$$\Delta B(u)|00\rangle = \underbrace{1 \overset{1}{\underset{0}{-0-}} 0 \overset{0}{\underset{0}{-}}}_{c(u)a(u)|10\rangle} + \underbrace{1 \overset{0}{\underset{0}{-1-}} 0 \overset{1}{\underset{0}{-}}}_{b(u)c(u)|01\rangle} \tag{1.89}$$

Here the left and right indices define $B(u) = T_1^0(u)$, and the co-multiplication implies a sum over the middle index. The bottom (resp. top) indices define the in-state (resp. out-state) of the quantum spaces, here denoted as kets.

Proceeding in the same way for the three other in-states, we find that $\Delta B(u)$ can be written in the basis $\{|00\rangle, |01\rangle, |10\rangle, |11\rangle\}$ as

$$\Delta B(u) = \begin{bmatrix} 0 & 0 & 0 & 0 \\ b(u)c(u) & 0 & 0 & 0 \\ c(u)a(u) & 0 & 0 & 0 \\ 0 & c(u)b(u) & a(u)c(u) & 0 \end{bmatrix}$$

$$= \begin{bmatrix} 0 & 0 & 0 & 0 \\ 0 & 0 & 0 & 0 \\ c(u)a(u) & 0 & 0 & 0 \\ 0 & c(u)b(u) & 0 & 0 \end{bmatrix} + \begin{bmatrix} 0 & 0 & 0 & 0 \\ b(u)c(u) & 0 & 0 & 0 \\ 0 & 0 & 0 & 0 \\ 0 & 0 & a(u)c(u) & 0 \end{bmatrix}$$

$$= B(u) \otimes A(u) + D(u) \otimes B(u). \tag{1.90}$$

It is actually simpler to avoid specifying the states of the quantum spaces altogether. Applying (1.82) directly one then obtains

$$\Delta^{L-1}B(u) = \underbrace{1 \,\|\, 0 \,\|\, 0}_{\Delta^{L_1-1}B(u) \otimes \Delta^{L_2-1}A(u)} + \underbrace{1 \,\|\, 1 \,\|\, 0}_{\Delta^{L_1-1}D(u) \otimes \Delta^{L_2-1}B(u)} \tag{1.91}$$

Note that this derivation applies for any bipartition $L_1 + L_2 = L$, and not only for $L_1 = L_2 = 1$.

Repeating the working for the three other operators, the complete co-multi-plication table reads

$$\Delta A(u) = A(u) \otimes A(u) + C(u) \otimes B(u),$$
$$\Delta B(u) = B(u) \otimes A(u) + D(u) \otimes B(u),$$
$$\Delta C(u) = C(u) \otimes D(u) + A(u) \otimes C(u),$$
$$\Delta D(u) = D(u) \otimes D(u) + B(u) \otimes C(u). \tag{1.92}$$

To generalize this construction from $L=2$ to arbitrary L it suffices to use the associativity of the co-multiplication. Indeed for $L \geq 2$ we have

$$\Delta^{L-1} : \mathcal{A} \to \mathcal{A}^{\otimes L}$$
$$\Delta^{L-1} \mapsto (I^{\otimes L-2} \otimes \Delta)\Delta^{L-2}. \tag{1.93}$$

When making this definition, we have chosen to insert new tensorands from the right. Inserting them from the left would make no difference to the result, since in any case it can also be computed directly along the lines of (1.89). In the latter case, one has to sum over all $L-1$ intermediate indices, of the type k in (1.82). It is a useful exercise to compute $\Delta^2 B(u)$ for $L=3$ in all three ways and check that one obtains identical results.

In the following we shall simplify the notation and write, for example, $B(u)$ instead of $\Delta^{L-1} B(u)$. Thus $B(u)$ is an operator that acts on all L spaces in the tensor product $V^{\otimes L}$. Using (1.92)–(1.93) repeatedly it can be expanded in fully tensorized form, as an expression with 2^{L-1} terms. This expanded form is (1.92) for $L=2$, and the expression for $L=3$ is contained in the above exercise. The factors entering each term in the expanded form act on a single space V.

1.3.3.2 Commutation relations

The operators $A(u)$, $B(u)$, $C(u)$, and $D(u)$ satisfy a set of commutation relations which follow as a direct consequence of the RTT relation (1.80).

To see in details how this works, we first write out the RTT relation in component form,

$$\sum_{j_1, j_2} R_{j_1 j_2}^{k_1 k_2}(u-v) T_{i_1}^{j_1}(u) T_{i_2}^{j_2}(v) = \sum_{j_1, j_2} T_{j_2}^{k_2}(v) T_{j_1}^{k_1}(u) R_{i_1 i_2}^{j_1 j_2}(u-v). \tag{1.94}$$

This gives a relation for each choice of (k_1, k_2, i_1, i_2). Consider for instance the choice $(0,0,1,0)$:

$$R_{00}^{00}(u-v) T_1^0(u) T_0^0(v) = T_1^0(v) T_0^0(u) R_{10}^{01}(u-v) + T_0^0(v) T_1^0(u) R_{10}^{10}(u-v). \tag{1.95}$$

Insert now the R-matrix elements from (1.86)–(1.87) and the monodromy matrix elements from (1.84), recalling that the former are just scalars, whereas the latter are (non-commuting) operators. This gives

$$a(u-v)B(u)A(v) = c(u-v)B(v)A(u) + b(u-v)A(v)B(u). \tag{1.96}$$

Among all the possible commutation relations we shall actually only need a few. Firstly, for two operators of the same type we have simply

$$A(u)A(v) = A(v)A(u), \quad B(u)B(v) = B(v)B(u),$$
$$C(u)C(v) = C(v)C(u), \quad D(u)D(v) = D(v)D(u). \tag{1.97}$$

Secondly, to push an A or a D past a B we have

$$A(u)B(v) = \frac{a(v-u)}{b(v-u)}B(v)A(u) - \frac{c(v-u)}{b(v-u)}B(u)A(v),$$
$$D(u)B(v) = \frac{a(u-v)}{b(u-v)}B(v)D(u) - \frac{c(u-v)}{b(u-v)}B(u)D(v). \tag{1.98}$$

The first of these relations follows from (1.96) after a relabelling $u \leftrightarrow v$ and some rearrangement. The second relation is obtained from a similar computation.

1.3.3.3 Algebraic Bethe ansatz

We now have all necessary ingredients to treat the six-vertex model using the algebraic Bethe ansatz.

As in the coordinate Bethe ansatz, one starts from the pseudo-vacuum, or reference state, in which all spins point up and no particle world-lines are present. We denote this state as

$$|\Uparrow\rangle = |\uparrow\uparrow \cdots \uparrow\rangle = |00 \cdots 0\rangle. \tag{1.99}$$

Recall that $B(u)$ creates a particle (or equivalently, flips down one spin), whereas $C(u)$ annihilates a particle. Thus, an n-particle state (i.e. with n down spins) can be constructed as follows:

$$|\Psi_n\rangle = \prod_{i=1}^{n} B(u_i)|\Uparrow\rangle. \tag{1.100}$$

The states (1.100) are called algebraic Bethe ansatz states.

Our goal is to diagonalize the transfer matrix

$$t(u) = \mathrm{Tr}_a T(u) = A(u) + D(u). \tag{1.101}$$

This means solving the eigenvalue equation

$$t(u)|\Psi_n\rangle = [A(u) + D(u)] \prod_{i=1}^{n} B(u_i)|\Uparrow\rangle = \Lambda_n(u;\{u_i\}) \prod_{i=1}^{n} B(u_i)|\Uparrow\rangle. \quad (1.102)$$

This can obviously only be done if the parameters $\{u_i\}$ satisfy certain conditions, called the *Bethe ansatz equations*, that we shall derived shortly.

To compute $[A(u) + D(u)] \prod_{i=1}^{n} B(u_i)|\Uparrow\rangle$ we use the commutation relations (1.98) to push $A(u)$ and $D(u)$ to the right, past the string of B's. When they have been pushed completely to the right, one applies the relations

$$A(u)|\Uparrow\rangle = a(u)^L|\Uparrow\rangle, \qquad D(u)|\Uparrow\rangle = b(u)^L|\Uparrow\rangle. \quad (1.103)$$

Note that (1.103) follows from the first and last lines of (1.92), generalized for $L=2$ to arbitrary L. Consider for instance $\Delta A(u)$. It is easy to see that the right-hand side will contain a single term $A(u)^{\otimes L}$, and all remaining terms will contain at least one factor $C(u)$ in the tensor product. But this $C(u)$ will annihilate $|\Uparrow\rangle$, so the only contribution is $a(u)^L|\Uparrow\rangle$ indeed.

Each time we push $A(u)$ one position towards the right, we obtain two contributions from the right-hand side of (1.98). The unique term obtained by always choosing the first contribution is a *wanted A-term*. In this term, the arguments of the $B(u_i)$ remain unchanged, and $A(u)$ simply 'goes through'. The remaining $2^n - 1$ terms are *unwanted A-terms*. In those terms, at least one of the arguments u_i of the B's has been changed into u, and so the state is not of the form (1.100). Similarly, there is one wanted and $2^n - 1$ unwanted D-terms.

The wanted A-term and the wanted D-term produce the expression for the eigenvalue of the transfer matrix,

$$\Lambda_n(u;\{u_i\}) = a(u)^L \prod_{i=1}^{n} \frac{a(u_i - u)}{b(u_i - u)} + b(u)^L \prod_{i=1}^{n} \frac{a(u - u_i)}{b(u - u_i)}. \quad (1.104)$$

The condition that the unwanted A-terms cancel the unwanted D-terms leads to the Bethe ansatz equations (BAE),

$$\left(\frac{a(u_i)}{b(u_i)}\right)^L = \prod_{\substack{j=1 \\ j \neq i}}^{n} \frac{a(u_i - u_j)b(u_j - u_i)}{a(u_j - u_i)b(u_i - u_j)}. \quad (1.105)$$

Proof of (1.105). Let us abbreviate $A_0 \equiv A(u)$ and $A_i \equiv A(u_i)$ for $i = 1, 2, \ldots, n$, and similarly for the other types of operators. We also set

$$\alpha_{ij} \equiv \frac{a(u_j - u_i)}{b(u_j - u_i)}, \qquad \beta_{ij} \equiv -\frac{c(u_j - u_i)}{b(u_j - u_i)}, \quad (1.106)$$

so that the commutation relations (1.98) can be rewritten

$$A_i B_j = \alpha_{ij} B_j A_i + \beta_{ij} B_i A_j,$$
$$D_i B_j = \alpha_{ji} B_j D_i + \beta_{ji} B_i D_j. \tag{1.107}$$

The unwanted A-terms (resp. D-terms) are those where A_0 (resp. D_0) exchanges its spectral parameter with one or more of the B's and hence becomes some A_i (resp. D_i) with $i \geq 1$ as it is pushed to the right of $\prod_{i=1}^n B_i$. The sum of these unwanted A-terms is

$$\sum_{i=1}^n \bar{a}_i \left(\prod_{\substack{j=0 \\ j \neq i}}^n B_j \right) A_i | \Uparrow \rangle = \sum_{i=1}^n \bar{a}_i \, a(u_i)^L \left(\prod_{\substack{j=0 \\ j \neq i}}^n B_j \right) | \Uparrow \rangle. \tag{1.108}$$

At first sight, it may appear complicated to compute the coefficients $\bar{a}_i$, since the A-operator might exchange its rapidity with the B's several times, as it is moved through the product. However, we can simplify the computation of $\bar{a}_i$ dramatically by using (1.97) to rewrite (1.100) as

$$|\Psi_n\rangle = B_i \prod_{\substack{j=1 \\ j \neq i}}^n B_j | \Uparrow \rangle. \tag{1.109}$$

The action of A_0 on this can then only produce an A_i on the right if the exchange of spectral parameter happens when A_0 is commuted through the very first factor B_i in (1.109). Therefore,

$$\bar{a}_i = \beta_{0i} \prod_{\substack{k=1 \\ k \neq i}}^n \alpha_{ik}. \tag{1.110}$$

By this simple trick, the total number of unwanted A-terms has been reduced from $2^n - 1$ to just n.

Similarly, the unwanted D-terms read

$$\sum_{i=1}^n \bar{d}_i \left(\prod_{\substack{j=0 \\ j \neq i}}^n B_j \right) D_i | \Uparrow \rangle = \sum_{i=1}^n \bar{d}_i b(u_i)^L \left(\prod_{\substack{j=0 \\ j \neq i}}^n B_j \right) | \Uparrow \rangle \tag{1.111}$$

with

$$\bar{d}_i = \beta_{i0} \prod_{\substack{k=1 \\ k \neq i}}^n \alpha_{ki}. \tag{1.112}$$

Because of (1.87) we have $\beta_{0i} = -\beta_{i0}$. Therefore the sum of (1.108) and (1.111) vanishes provided that

$$\left(\frac{a(u_i)}{b(u_i)}\right)^L = \prod_{\substack{k=1 \\ k \neq i}}^{n} \frac{\alpha_{ki}}{\alpha_{ik}}. \tag{1.113}$$

Plugging back (1.106) we arrive at (1.105). $\square$

Alternatively (1.105) follows also from the form (1.104) of the eigenvalue, as we now argue. We set $z = e^{2iu}$ and $q = e^{i\gamma}$, and we define the shifted eigenvalue $\tilde{\Lambda} = (2ie^{iu})^L \Lambda_n(u, \{u_i\})$. Elementary computations then bring (1.105) into the form

$$\tilde{\Lambda} = \left(q - q^{-1}z\right)^L \prod_{i=1}^{n} \frac{q^{-1}z_i - qz}{z - z_i} + (z-1)^L \prod_{i=1}^{n} \frac{qz_i - q^{-1}z}{z - z_i}. \tag{1.114}$$

Defining the polynomials

$$Q(z) = \prod_{i=1}^{n}(z - z_i), \qquad \phi_L(z) = (z-1)^L, \tag{1.115}$$

we obtain

$$\tilde{\Lambda} Q(z) = (-q)^{L-n} \phi_L(q^{-2}z) Q(q^2 z) + (-q)^N \phi_L(z) Q(q^{-2}z). \tag{1.116}$$

Whenever the spectral parameter equals one of the Bethe roots, we have $Q(z_j) = 0$ on the left-hand side of (1.116). Therefore the right-hand side must also vanish. Working backwards through the change of variables then produces the BAE (1.105).

1.3.4 Coordinate Bethe ansatz

A more direct approach consists in considering the action of transfer matrix t on n-particle states giving the positions of the world-lines of $|1\rangle$ states within one row of the lattice. Such states can be written $|x_1, x_2, \ldots, x_n\rangle$, where we have assumed $x_1 < x_2 < \cdots < x_n$, and the world-lines are depicted in the bottom part of Figure 1.3.

This *coordinate Bethe ansatz* (CBA) approach [28] is often the first strategy one would try out on a new problem, but it lacks some of the power and elegance of the algebraic Bethe ansatz (ABA). On the other hand, in some cases, problems which are solvable by CBA do not admit the ABA approach.[7] This might be due, e.g., to the failure to make

---

[7] For example, the ABA version of the biquadratic model [29] is presently unknown.

sense of the spectral parameter u, or because a proper pseudo vacuum $|\Uparrow\rangle$ cannot be identified.

In the CBA, one wishes to construct n-particle states

$$|\Psi_n\rangle = \sum_{1 \leq x_1 < \cdots < x_n \leq L} g(x_1,\ldots,x_n)|x_1,\ldots,x_n\rangle, \qquad (1.117)$$

which are eigenvectors of t,

$$t|\Psi_n\rangle = \Lambda_n|\Psi_n\rangle. \qquad (1.118)$$

To this end one tries an ansatz of the form

$$g(x_1,\ldots,x_n) = \sum_{p \in \mathfrak{S}_n} A_p z_{p(1)}^{x_1} z_{p(2)}^{x_2} \cdots z_{p(n)}^{x_n}, \qquad (1.119)$$

where the sum runs over all permutations $p \in \mathfrak{S}_n$ of the particle labels $\{1,2,\ldots,n\}$. The complex numbers z_j are related to the so-called quasi-momenta k_j through the relation $z_j = \exp(ik_j)$. Loosely speaking, the ansatz (1.119) can be interpreted as coupled plane waves.

The link between the two approaches is provided by relating the Bethe roots $\{u_j\}$ with the quasi-momenta $\{z_j\}$. This can be done by comparing the one-particle states, which read in the CBA

$$|\Psi_1\rangle = \sum_{x=1}^{L} z^x |x\rangle, \qquad (1.120)$$

while in the ABA we have

$$|\Psi_1\rangle = \Delta^{L-1} B(u)|\Uparrow\rangle. \qquad (1.121)$$

Expanding out $\Delta^{L-1} B(u)$, using (1.92)–(1.93), and using that any $C(u)$ tensorand annihilates $|\Uparrow\rangle$, we see that the ABA expression for $|\Psi_1\rangle$ also has precisely L non-zero terms (rather than 2^L), each one characterized by one flipped spin. Identifying the position of the flipped spin with $|x\rangle$ and matching coefficients, we arrive at

$$z_j = e^{ik_j} = \frac{a(u_j)}{b(u_j)}. \qquad (1.122)$$

The Bethe ansatz equations (1.105) can the be written in the suggestive form

$$z_j^L = \prod_{\substack{l=1 \\ l \neq j}}^n \hat{S}_{lj}(z_l, z_j) \quad \text{for } j = 1, 2, \ldots, n,\tag{1.123}$$

where we have introduced the scattering phases

$$\hat{S}_{ji} = \frac{a(u_j - u_i)b(u_i - u_j)}{a(u_i - u_j)b(u_j - u_i)} = -\frac{1 - 2\Delta z_i + z_i z_j}{1 - 2\Delta z_j + z_i z_j}.\tag{1.124}$$

We stress that the BAE depend *only* on the six-vertex weights a, b, c via the combination Δ. One consequence of this is that the universality class (critical exponents) will depend on Δ, but not on the anisotropy given by (1.40).

The form (1.123) of the BAE can be interpreted physically as follows. When the particle j is taken around the periodic direction and back to its original position, it picks up a scattering phase $\hat{S}_{lj}$ each time it crosses another particle l.

1.3.5 Energy and momentum

We can compute the energy E of the Bethe ansatz state (1.100). To this end we just need to recall the link (1.73) between the transfer matrix $t(u)$ and the Hamiltonian H. Taking expectation values with respect to the state (1.100) the operator H gets replaced by its expectation value E, and $t(u)$ gets replaced by the eigenvalue $\Lambda(u; \{u_i\})$. Therefore,

$$E_n(\{u_i\}) = -\sin \gamma \left. \frac{\partial}{\partial u} \log \Lambda_n(u; \{u_i\}) \right|_{u \to 0}.\tag{1.125}$$

In (1.104) only the first term contributes in the $u \to 0$ limit:

$$\Lambda_n \simeq \sin^L(\gamma) \prod_{i=1}^n \frac{a(u_i - u)}{b(u_i - u)}.\tag{1.126}$$

Taking the derivative we arrive at

$$E_n(\{u_i\}) = L \cos \gamma + \sum_{i=1}^n \epsilon(u_i),\tag{1.127}$$

where the energy of a single particle with quasi-momentum (1.122) is

$$\epsilon(u_i) = -\frac{\sin^2(\gamma)}{\sin(u_i)\sin(\gamma - u_i)}.\tag{1.128}$$

In the same way we can express the momentum of the Bethe ansatz state:

$$-iP = \log\left(\frac{t(0)}{\sin^L(\gamma)}\right) = \sum_{i=1}^{n} \log\left(\frac{a(u_i)}{b(u_i)}\right), \qquad (1.129)$$

meaning that the quasi-momenta (1.122) just add up.

> Note that the constant (i.e. independent of u_i) term in (1.127) is consistent with that of (1.74). If we normalize the Hamiltonian as $H = -\sum_{m=1}^{L} E_m$, the n-particle energy reads simply $E_n(\{u_i\}) = \sum_{i=1}^{n} \epsilon(u_i)$.

1.4 Thermodynamic limit

The first word in the term *algebraic* Bethe ansatz should of course not conceal the fact that once the Bethe ansatz equations for a given model have been found, there is usually a fair amount of *analytical* work to be done in order to derive the free energy in the thermodynamic limit and the critical exponents.

In this section we sketch the phase diagram of the six-vertex model, and for the critical regime $|\Delta| < 1$ we derive the free energy in the thermodynamic limit starting from the BAE; the exposition largely follows [2]. The key technical point is to solve the Lieb equation (1.146) by Fourier transformation.

Then we briefly discuss the treatment of elementary excitations, first with zero and then with non-zero magnetization. We outline the Coulomb gas (CG) formalism for the continuum limit, and use the exact result for the magnetic excitations to fix the CG coupling constant in terms of microscopic parameters.

1.4.1 Phase diagram

The six-vertex model has a non-trivial phase diagram, and it is hardly surprising that its thermodynamic limit depends on the parameter Δ. In fact the Bethe ansatz equations (1.123) show that the limit depends *only* on Δ. Although the phase diagram can be derived in detail, let us first discuss a few qualitative arguments.

First we define a *transition cycle* as a closed path on the lattice along which all arrows are consistently orientated. It is easy to see that the least possible change to any configuration consists in reversing all arrows along a transition cycle.

If either a or b is large compared to the other weights, the predominant configurations are those in which all vertical arrows and all horizontal arrows point in the same direction; see Figure 1.4a. For an $L \times L$ lattice, all transition cycles for such a configuration have length L. The energy needed to reverse a transition cycle therefore grows linearly with L, so in the $L \to \infty$ limit the system is frozen into the given state. It is a non-trivial fact that this freezing occurs whenever $\Delta > 1$. The largest eigenvalue is that of the $n = 0$ particle sector, whence trivially $\Lambda_{\max} = a^L + b^L$.

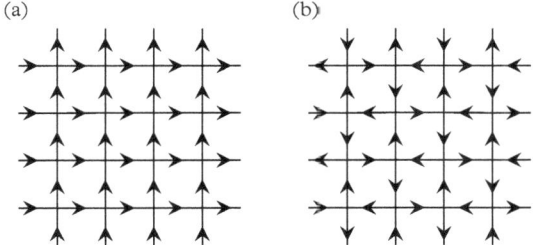

Figure 1.4 *Dominant configurations of the six-vertex model in (a) the limit $\Delta \gg 1$ and (b) the limit $\Delta \ll -1$.*

If, on the contrary, c is very large compared to the other weights, the predominant configurations are those where all vertices on the even (resp. odd) sublattice are of the type ω_5 (resp. ω_6); see Figure 1.4b. However, there are a large number of transition cycles of finite length (e.g. L^2 transition cycles of length four). Therefore the system is not frozen for finite values of c, meaning that it exhibits fluctuations around the predominant configuration. One would expect non-critical behaviour as long as c remains reasonably large. This is indeed the case: when $\Delta < -1$ the system is in a non-critical phase.

The most interesting phase occurs for $-1 < \Delta < 1$. In that range the six-vertex model is critical, and it turns out that the critical exponents depend continuously on Δ. This is an interesting counter-example to naive ideas of universality.

We shall concentrate most of the subsequent discussion on the critical case $-1 < \Delta < 1$, where the free energy can be expressed in terms of Fourier integrals. (The non-critical case $\Delta < -1$ can also be worked out in detail and calls instead for the use of Fourier series.)

1.4.2 Thermodynamic limit for $\Delta < 1$

We return to the geometry of a semi-infinite cylinder of circumference N, and we consider the Bethe ansatz equations (1.123) in the sector with n particles. It is not known how to solve the BAE for finite n and N. This situation is quite common in the study of integrable systems.

Nevertheless, it turns out that the six-vertex model is exactly solvable in the thermodynamic limit. By this we mean precisely that the free energy $f = -\frac{1}{\beta N} \log \Lambda_{\max}$, or equivalently the ground state energy in the spin chain, can be determined analytically for $N \to \infty$. The same is true for the low-lying excitations, but for the moment we concentrate on the ground state.

1.4.2.1 Location of the quasi-momenta

The BAE (1.123) possess many solutions for the quasi-momenta z_j. It is not a priori clear which one corresponds to the ground state. In what follows we shall admit the following fact:

- The solution of the BAE (1.123) that maximizes the eigenvalue Λ is such that $z_1, z_2, \ldots, z_n$ are distinct, lie on the unit circle, are distributed symmetrically about unity, and are packed as closely as possible.

This can actually be proved quite rigorously, using some lengthy analysis [30]. An easier method, that usually works quite well for more general integrable models, is to study numerically the solutions for low values of N—confronting the results with exact diagonalizations of the transfer matrix—until the pattern has become clear [26].

1.4.2.2 Transformation to a set of real equations

We introduce the momenta $k_j \in \mathbb{R}$ and the function $\Theta(p,q)$ so that the quasi-momenta and the scattering phases read

$$z_j = \exp(ik_j), \tag{1.130}$$

$$-\hat{S}_{ij}(z_i, z_j) = \exp(-i\Theta(k_j, k_i)). \tag{1.131}$$

By (1.124) we have then

$$e^{-i\Theta(p,q)} = \frac{1 - 2\Delta e^{ip} + e^{i(p+q)}}{1 - 2\Delta e^{iq} + e^{i(p+q)}}. \tag{1.132}$$

To see that $\Theta(p,q)$ is a real function, it suffices to notice that

$$\tan\left(\frac{1}{2}\Theta(p,q)\right) = -i\frac{1 - e^{-i\Theta(p,q)}}{1 + e^{-i\Theta(p,q)}} = \frac{\Delta \sin\left(\frac{p-q}{2}\right)}{\cos\left(\frac{p+q}{2}\right) - \Delta \cos\left(\frac{p-q}{2}\right)},$$

where the right-hand side is manifestly real.

Including the term $l = j$ obviously leaves the right-hand side of (1.123) unchanged, so we can rewrite it as

$$\exp(iNk_j) = (-1)^{n-1} \prod_{l=1}^{n} \exp(-i\Theta(k_j, k_l)),$$

where now both sides of the equation are unimodular. Taking logarithms we have

$$Nk_j = 2\pi I_j - \sum_{l=1}^{n} \Theta(k_j, k_l), \tag{1.133}$$

where I_j ranges between $\pm\left(\frac{n-1}{2}\right)$, hence is an integer (resp. half an odd integer) if n is odd (resp. even). Note that both sides of this equation are real.

The hypothesis that $k_1, k_2, \ldots, k_n$ be distinct, symetrically distributed about the origin, and packed as closely as possible implies that the ground state is obtained by choosing

$$I_j = j - \frac{1}{2}(n+1), \quad \text{for } j = 1, 2, \ldots, n. \tag{1.134}$$

1.4.2.3 Continuum limit

The thermodynamic limit is obtained by sending $n, N \to \infty$, while keeping the ratio n/N fixed and finite. This ratio describes the (fixed) ratio of up-pointing arrows in each row of the lattice. The distribution function $\rho(k)$ of Bethe roots is defined so that $N\rho(k)dk$ is the number of k_j lying between k and $k + dk$. By assumption $\rho(k)$ has support on a symmetric interval $[-Q, Q]$, where Q will be determined later. Thus,

$$\int_{-Q}^{Q} \rho(k) dk = \frac{n}{N}. \tag{1.135}$$

For a given value k_j of k, the quantity $I_j + \frac{1}{2}(n+1) = N \int_{-Q}^{k} \rho(k') dk'$ is the number of momenta k_l with $l < j$. Passing from sums to integrals in (1.133)—and denoting k_j simply as k—then produces

$$Nk = -\pi(n+1) + 2\pi N \int_{-Q}^{k} \rho(k') dk' - N \int_{-Q}^{Q} \Theta(k, k') \rho(k') dk'.$$

Taking derivatives with respect to k, and dividing by N, then leads to a linear integral equation for $\rho(k)$,

$$2\pi \rho(k) = 1 + \int_{-Q}^{Q} \frac{\partial \Theta(k, k')}{\partial k} \rho(k') dk'. \tag{1.136}$$

Let us define

$$\begin{aligned} L_i = L(z_i) &= \frac{a(u_i - u)}{b(u_i - u)} = \frac{\sin(\gamma - (u_i - u))}{\sin(u_i - u)}, \\ M_i = M(z_i) &= \frac{a(u - u_i)}{b(u - u_i)} = \frac{\sin(\gamma - (u - u_i))}{\sin(u - u_i)}, \end{aligned} \tag{1.137}$$

so that the transfer matrix eigenvalue (1.104) can be written as

$$\Lambda = a^N L_1 L_2 \cdots L_n + b^N M_1 M_2 \cdots M_n. \tag{1.138}$$

The free energy is then given by (1.138) as

$$f = -\frac{1}{\beta}\max\left\{\log a + \frac{1}{N}\sum_{j=1}^{n}\log L(z_j), \log b + \frac{1}{N}\sum_{j=1}^{n}\log M(z_j)\right\}.$$

In the thermodynamic limit this becomes

$$f = -\frac{1}{\beta}\max\left\{\log a + \int_{-Q}^{Q}[\log L(e^{ik})]\rho(k)dk,\right.$$
$$\left.\log b + \int_{-Q}^{Q}[\log M(e^{ik})]\rho(k)dk\right\}. \qquad (1.139)$$

1.4.3 Free energy for $-1 < \Delta < 1$

A natural strategy for solving the linear integral equation (1.136) would be to use Fourier transformation. This is however only possible if we can find a transformation to a difference kernel. Fortunately this is possible.

1.4.3.1 Difference kernel transformation

For $-1 < \Delta < 1$ we parameterize [cf. (1.35)]

$$\Delta = -\cos\mu, \quad \text{with } 0 < \mu < \pi \qquad (1.140)$$

and we trade k for a new variable α defined by

$$e^{ik} = \frac{e^{i\mu} - e^{\alpha}}{e^{i\mu+\alpha} - 1}. \qquad (1.141)$$

Note that $\alpha \in \mathbb{R}$. Indeed, supposing this, it is easily seen that $|e^{ik}|^2 = 1$ from the right-hand side of (1.141), as is consistent with the hypothesis that $k \in \mathbb{R}$.

Differentiating logarithmically—i.e. using $\frac{d}{d\alpha}e^{ik} = ie^{ik}\frac{dk}{d\alpha}$ to isolate $\frac{dk}{d\alpha}$—we find

$$\frac{dk}{d\alpha} = \frac{\sin\mu}{\cosh\alpha - \cos\mu}. \qquad (1.142)$$

This proves in particular that $k(\alpha) \in \mathbb{R}$ is a monotonically increasing function (since $0 < \mu < \pi$), and by (1.141) it maps the interval $(-\infty,\infty)$ onto $(\mu-\pi, \pi-\mu)$. It follows directly from (1.141) that $k(-\alpha) = -k(\alpha)$, i.e. the function is odd.

The change of variables (1.141) is not as miraculous as it may first appear. Indeed consider the map $z : \mathbb{C} \mapsto \mathbb{C}$ given by

$$z(w) = \frac{a_{11}w + a_{12}}{a_{21}w + a_{22}}. \qquad (1.143)$$

This transformation, variously known as a projective map or a Möbius transformation. It is a global conformal transformation, and as such it has the property that it preserves the set of circles and straight lines (the latter being considered circles of infinite radius). After a normalization, usually taken as det $a_{ij} = 1$, it depends on three complex parameters.

To make contact with (1.141) we set $z = e^{ik}$ and $w = e^{\alpha}$. We then have $|z| = 1$ (a circle), whereas $w \in \mathbb{R}_+$ (a straight line). We also abbreviate the free parameter in (1.141) as $\rho = e^{i\mu}$. The desired symmetry property $k(-\alpha) = -k(\alpha)$ can be rewritten $z(\frac{1}{w}) = \frac{1}{z(w)}$. Using (1.142), this is solved by $a_{11} = a_{22}$ and $a_{12} = a_{21}$. If we now fix the global scale by setting $z(0) = \frac{a_{12}}{a_{11}} = \frac{1}{z(\infty)} = -\rho$, we obtain precisely (1.141).

Let us define $p = k(\alpha)$ and $q = k(\beta)$, so that

$$e^{ip} = \frac{e^{i\mu} - e^{\alpha}}{e^{i\mu+\alpha} - 1}, \qquad e^{iq} = \frac{e^{i\mu} - e^{\beta}}{e^{i\mu+\beta} - 1}.$$

Inserting this into (1.132) the scattering phase becomes

$$e^{-i\Theta(p,q)} = \frac{e^{\alpha-\beta} - e^{2i\mu}}{e^{\beta-\alpha} - e^{2i\mu}}. \qquad (1.144)$$

Crucially, this depends only on the difference $\alpha - \beta$ (and on the constant μ).

We shall need the root density function $R(\alpha)$ transformed to the α variable (and renormalized by $\frac{1}{2\pi}$ for later convenience),

$$R(\alpha)d\alpha = 2\pi\rho(k)dk. \qquad (1.145)$$

Plugging this into (1.136) leads to

$$R(\alpha) = \frac{dk}{d\alpha} + \frac{1}{2\pi} \int_{-Q_1}^{Q_1} \frac{\partial \Theta(\alpha,\beta)}{\partial \beta} R(\beta)\,d\beta,$$

where we note that there is a new integration range $(-Q_1, Q_1)$ corresponding to the α variable. Computing the derivatives from (1.143) and (1.144) finally leads to

$$R(\alpha) = \frac{\sin\mu}{\cosh\alpha - \cos\mu} - \frac{1}{2\pi}\int_{-Q_1}^{Q_1}\frac{\sin(2\mu)}{\cosh(\alpha-\beta)-\cos(2\mu)}R(\beta)\,\mathrm{d}\beta, \quad (1.146)$$

and the normalization condition (1.135) for the root density function now reads

$$\frac{1}{2\pi}\int_{-Q_1}^{Q_1} R(\alpha)\,\mathrm{d}\alpha = \frac{n}{N}. \quad (1.147)$$

1.4.3.2 Parameterization

We have already parameterized $\Delta = -\cos\mu$ in (1.140). The Bethe ansatz equations (1.123)—and hence the universality class of the six-vertex model—depend only on the initial vertex weights a, b, c through $\Delta = \frac{a^2+b^2-c^2}{2ab}$. We must therefore choose a parameterization of the two independent ratios $a:b:c$ that respects this latter constraint (we 'uniformize the spectral curve'). This can be done in this case using trigonometric functions:

$$a:b:c = \sin\left(\frac{\mu-w}{2}\right) : \sin\left(\frac{\mu+w}{2}\right) : \sin\mu, \quad -\mu < w < \mu \quad (1.148)$$

defining another parameter w.

The eigenvalues of the transfer matrix are determined by (1.138) through the functions $L(z)$ and $M(z)$ given by (1.137). Recalling that $z = \mathrm{e}^{ik}$ is parameterized by (1.141), the parametric form of these functions now becomes

$$L(\mathrm{e}^{ik}) = \frac{\mathrm{e}^{i(w+\mu)} - \mathrm{e}^{\alpha-i\mu}}{\mathrm{e}^\alpha - \mathrm{e}^{iw}},$$

$$M(\mathrm{e}^{ik}) = \frac{\mathrm{e}^{i(w-\mu)} - \mathrm{e}^{\alpha+i\mu}}{\mathrm{e}^\alpha - \mathrm{e}^{iw}}. \quad (1.149)$$

1.4.3.3 Solution by Fourier integrals

The integral equation (1.146) that determines the root density function now has a difference kernel. One can therefore solve it by Fourier transformation, provided that $Q_1 = \infty$. Let us suppose that this is so, and justify the following assumption. The Fourier transformed root density function then reads

$$\tilde{R}(x) = \frac{1}{2\pi}\int_{-\infty}^{\infty} R(\alpha)\mathrm{e}^{ix\alpha}\,\mathrm{d}\alpha. \quad (1.150)$$

Let us define the function

$$\phi_\mu(\alpha) = \frac{\sin\mu}{\cosh\alpha - \cos\mu}, \quad (1.151)$$

which is often referred to as the *source term* of the Bethe ansatz equations. The difference kernel equation (1.146) then decouples upon Fourier transformation, since the Fourier transform of a convolution is the product of Fourier transforms. Explicitly, multiplying both sides of (1.146) by $\frac{1}{2\pi}e^{ix\alpha}$ and integrating over α leads to

$$\tilde{R}(x) = \tilde{\phi}_\mu(x) - \tilde{\phi}_{2\mu}(x) \cdot \tilde{R}(x). \tag{1.152}$$

Exercise: Show that the Fourier transform of the source term is

$$\tilde{\phi}_\mu(x) = \frac{\sinh((\pi - \mu)x)}{\sinh(\pi x)}. \tag{1.153}$$

Inserting this yields

$$\tilde{R}(x) = \frac{\sinh((\pi - \mu)x)}{\sinh(\pi x)} - \frac{\sinh((\pi - 2\mu)x)}{\sinh(\pi x)}\tilde{R}(x) \tag{1.154}$$

and we can then finally isolate

$$\tilde{R}(x) = \frac{1}{2\cosh(\mu x)}. \tag{1.155}$$

The normalization condition (1.147) is such that $\tilde{R}(0) = \frac{n}{N}$, and evaluating (1.155) we arrive at

$$\frac{n}{N} = \frac{1}{2}. \tag{1.156}$$

This simple result justifies the assumption $Q_1 = \infty$ a posteriori. Indeed, the largest sector of the transfer matrix precisely corresponds to the case where there are as many up-pointing as down-pointing arrows. By a simple entropic reasoning, this is also the ground state sector.[8]

From (1.149) one obtains

$$|L(e^{ik})|^2 = \frac{\cos(w + 2\mu) - \cosh\alpha}{\cos w - \cosh\alpha},$$
$$|M(e^{ik})|^2 = \frac{\cos(w - 2\mu) - \cosh\alpha}{\cos w - \cosh\alpha}.$$

This implies that $|L| > |M|$ for $w < 0$, and $|L| < |M|$ for $w > 0$.

---

[8] A variant argument is obtained by examining (1.138).

Suppose in the sequel that $w < 0$; a similar calculation for $w > 0$ can be shown to lead to exactly the same end result. The free energy is then given by the first term in (1.139),

$$f = -\frac{1}{\beta}\left(\log a + \frac{1}{2\pi}\int_{-\infty}^{\infty}[\log|L(e^{ik})|]R(\alpha)d\alpha.\right) \tag{1.157}$$

Using (1.155) and the fact that parity is conserved by Fourier transformation, we see that $R(\alpha)$ is an even function. Under the integral we can therefore replace the other factor $\log|L(e^{ik})|$ by its even part, which is also its real part. From (1.149) we get

$$\text{Re } L(e^{ik}) = -\cos\mu + \frac{\sin\mu \sin w}{\cos w - \cosh\alpha}, \tag{1.158}$$

and the Fourier transform of $\log|L(e^{ik})|$ becomes

$$\frac{1}{2\pi}\int_{-\infty}^{\infty} e^{ix\alpha}\log|L(e^{ik})|d\alpha = \frac{\sinh((\mu+w)x)\sinh((\pi-\mu)x)}{x\sinh(\pi x)}. \tag{1.159}$$

Exercise: Detail this computation.

To compute (1.157) we can use that the Fourier transform of a product is the convolution of Fourier transforms. The end result follows by combining (1.155) and (1.159):

$$f = -\frac{1}{\beta}\left(\log a + \int_{-\infty}^{\infty}\frac{\sinh((\mu+w)x)\sinh((\pi-\mu)x)}{2x\cosh(\mu x)\sinh(\pi x)}dx\right). \tag{1.160}$$

As already stated, exactly the same result is found for $w > 0$. We have therefore found, for any $w \in (-\mu, \mu)$, the free energy of the six-vertex model in the critical region $\Delta \in (-1, 1)$.

1.4.4 Elementary excitations

It is convenient to introduce the shifted Bethe roots

$$\lambda_j = i\left(\frac{\gamma}{2} - u_j\right) \tag{1.161}$$

and to rewrite the logarithmic form (1.133) of the BAE as

$$Nk(\lambda_j) = 2\pi I_j - \sum_{l=1}^{n}\Theta(\lambda_j - \lambda_l), \tag{1.162}$$

where the quasi-momentum and scattering phase are written as

$$k(\lambda) = f\left(\frac{\gamma}{2}, \lambda\right) \qquad (1.163)$$

$$\Theta(\lambda) = -f(\gamma, \lambda) \qquad (1.164)$$

in terms of the function

$$f(\gamma, \lambda) = -i \log\left(\frac{\sinh(i\gamma - \lambda)}{\sinh(i\gamma + \lambda)}\right) = 2\arctan(\tanh\lambda \cotan\gamma). \qquad (1.165)$$

By taking the limit $\lambda_j \to \infty$, we see that the Bethe integers I_j must satisfy

$$-I_{\max} < I_1 < I_2 < \cdots < I_{n-1} < I_n < I_{\max} \qquad (1.166)$$

with

$$I_{\max} = \frac{N-n}{2}. \qquad (1.167)$$

The ground state configuration (1.134) in the n-particle sector provides but one choice of the Bethe integers compatible with (1.166). Other choices correspond to excited states, still with total magnetization $S^z = \frac{N}{2} - n$.

It is also possible to find solutions of the BAE in which the roots λ_j are not real. In particular, some solutions contain one or more pairs of conjugate roots of the form $\lambda_* \pm \frac{i\gamma}{2}$. Such pairs are called two-strings. It is possible to manipulate the BAE in order to obtain a single equation for λ_*. We shall however not discuss such complex solutions any further here.

The so-called Lieb equation (1.146) is instrumental also when discussing excited states. It can be rewritten as

$$k'(\lambda) = 2\pi\rho(\lambda) - \int_{-Q(n)}^{Q(n)} d\mu\, \rho(\mu) K(\lambda - \mu), \qquad (1.168)$$

where the kernel is $K(\lambda) = \Theta'(\lambda)$. In the sector with $S^z = 0$ (hence $n = \frac{N}{2}$) we have seen that $Q(\frac{N}{2}) = \infty$, and we can rewrite the root density, whose Fourier transform is given by (1.155), as

$$\rho^{(0)}(\lambda) = s(\lambda) \equiv \frac{1}{2\gamma \cosh\left(\frac{\pi\lambda}{\gamma}\right)}. \qquad (1.169)$$

1.4.4.1 Zero magnetization

We now consider perturbing the ground state configuration (1.134) by taking out a particle at the Bethe integer $I_h + 1$ and moving it just above the last occupied position. In other words,

$$I_1 = -\frac{n-1}{2},$$
$$I_{j+1} - I_j = 1 + \delta_{j+1,h}, \quad \text{for } j = 1, 2, \ldots, n-1. \tag{1.170}$$

We say that we have excited a particle just above the *Fermi level*, leaving behind a *hole* at $I_h + 1$.

Note that the Bethe integers corresponding to holes also satisfy the BAE. In particular, (1.162) can be applied with $I_j = I_h + 1$, and we call $\tilde{\lambda}$ the corresponding solution for λ:

$$Nk(\tilde{\lambda}) = 2\pi I_j - \sum_{l=1}^{n} \Theta(\tilde{\lambda} - \lambda_l). \tag{1.171}$$

We are interested in observable $A[\rho]$, such as the momentum or energy per site, that depend on the continuum distribution $\rho(\lambda)$ of Bethe roots λ via a local density $a(\lambda)$, through

$$A[\rho] = \int_{-\infty}^{\infty} d\lambda \, \rho(\lambda) a(\lambda). \tag{1.172}$$

For example, the continuum limit of the energy density (1.127) reads, up to an unimportant additive constant,

$$\lim_{N \to \infty} \frac{E_n}{N} = \int_{-Q(n)}^{Q(n)} d\lambda \, \rho(\lambda) \epsilon(\lambda), \tag{1.173}$$

where the one-particle energy density is given by (1.128).

We mention without proof (see [31]) that the variation of $A[\rho]$ with respect to its ground state value reads, for $N \to \infty$,

$$A[\rho] - A\left[\rho^{(0)}\right] = -\frac{1}{N}\tilde{a}(\tilde{\lambda}), \tag{1.174}$$

where $\tilde{a}(\lambda)$ is called the *dressed function* corresponding to the observable A. It can be found from the integral equation

$$\tilde{a}(\lambda) = a(\lambda) + \frac{1}{2\pi} \int_{-\infty}^{\infty} d\mu \, K(\lambda - \mu) \tilde{a}(\mu), \tag{1.175}$$

which takes into account the combined effect that a single-particle contribution is missing at the position of the hole, and that the Bethe roots of the remaining particles have been shifted due to the perturbation (a sort of screening phenomenon).

In particular, the dressed momentum and energy can be found by this method as

$$\tilde{k}(\lambda) = 2\pi \int_0^\lambda d\mu \, s(\mu), \qquad (1.176)$$

$$\tilde{\epsilon}(\lambda) = -2\pi \sin(\gamma) \, s(\lambda), \qquad (1.177)$$

where $s(\lambda)$ is defined in (1.169). If $\tilde{\lambda}$ is close to the Fermi level, here $Q(n) = \infty$, we obtain a linear dispersion relation,

$$\tilde{\epsilon}(\lambda) \sim v_F \, \tilde{k}(\lambda), \qquad \text{as } \lambda \to \infty, \qquad (1.178)$$

with the proportionality constant

$$v_F = \frac{\pi \sin \gamma}{\gamma} \qquad (1.179)$$

being known as the *Fermi velocity*. Conformal invariance includes rotational invariance as a special case, and in particular a conformal field theory (CFT) must have space and time scale in the same manner. This invariance has been disturbed by taking the completely anisotropic (Hamiltonian) limit $u \to \infty$; see Section 1.2.8. But isotropy is easily reinstated, in the continuum limit, by dividing time by an appropriate factor, which is precisely v_F. This effect is reminiscent of the anisotropy factor $\zeta(u)$ that we discussed in Section 1.2.7.

1.4.4.2 Non-zero magnetization

If the magnetization $S^z = \frac{N}{2} - n$ is non-zero, the integration limits in the Lieb equation (1.168) are finite: $0 < Q(n) < \infty$. This means that we cannot solve it directly by Fourier transform. We now briefly outline an alternative method [30] that can deal with this situation.

One first introduces the function $\tilde{J}(\mu)$ via the Fourier transformed scattering kernel $\tilde{K}(\omega)$,

$$\left[1 + \tilde{J}(\omega)\right]\left[1 - \frac{\tilde{K}(\omega)}{2\pi}\right] = 1, \qquad (1.180)$$

and the shifted root density function $g(\lambda) = \rho(Q + \lambda)$. The Lieb equation,

$$s(Q + \lambda) = g(\lambda) + \int_0^\infty d\mu \, g(\mu) J(\lambda - \mu) + \int_0^\infty d\mu \, g(\mu) J(2Q + \lambda + \mu), \qquad (1.181)$$

can then be analysed to first order in the small quantity $\zeta = \exp(-\frac{\pi\lambda}{\gamma})$. The last term in (1.181) is then $\sim \zeta/Q^2$ and can be neglected, so that (1.181) becomes an equation of the Wiener–Hopf type. After some analysis one finds that

$$\frac{E_n - E_n^{(0)}}{N} = \frac{2\pi v_F X}{N^2} + o(N^{-2}), \qquad (1.182)$$

where

$$X = \frac{1}{4}\left[1 - \frac{\tilde{K}(0)}{2\pi}\right](S^z)^2. \qquad (1.183)$$

Show that

$$\frac{\tilde{K}(\omega)}{2\pi} = \frac{\sinh\left((\gamma - \frac{\pi}{2})\omega\right)}{\sinh\left(\frac{\pi\omega}{2}\right)}. \qquad (1.184)$$

Because of a fundamental CFT result, X in (1.182) can be interpreted as the critical exponent (scaling dimension) associated with the excitation described by the non-zero value of S^z.

1.4.5 Relation with the Coulomb gas formalism

In Section 1.5.2 we have shown the equivalence between the Q-state Potts model and a particular six-vertex model. The latter is dual to a *height model*, in which a scalar variable $h(\mathbf{x})$ is attached to each lattice face. When traversing a left-pointing (resp. right-pointing) arrow, this height goes up (resp. down) by an amount $a = \pi$; this is well defined because of the six-vertex constraint of two ingoing and two outgoing arrows. In other words, the level lines of $h(\mathbf{x})$ are precisely the oriented loops of the Potts model.

In the thermodynamic limit, and for the critical theory with $|\Delta| \leq 1$ (hence $0 \leq Q \leq 4$), the height can be argued to converge to a compactified scalar field,

$$h(\mathbf{x}) \to \phi(\mathbf{x}) \in \mathbb{R}/(2\pi\mathbb{Z}). \qquad (1.185)$$

The continuum-limit partition function then takes the form of a functional integral,

$$Z = \int \mathcal{D}\phi(\mathbf{x}) \exp\left(-S[\phi(\mathbf{x})]\right), \qquad (1.186)$$

where the Euclidean action is essentially that of a free boson,

$$S_E = \frac{g}{4\pi} \int d^2\mathbf{x} (\nabla \phi)^2, \quad (1.187)$$

with coupling constant $g > 0$. This is—in embryonic form, and leaving out a number of fine details—the Coulomb gas approach to CFT [8, 32].

Consider placing this theory on a semi-infinite cylinder of circumference N. We fix $S^z = \frac{m}{2} > 0$, so that a defect consisting of m open, oriented loop strands runs from the bottom to the top of the cylinder. This means that the zero-divergence constraint on the height is violated at the cylinder extremities. Let us write $\phi(\mathbf{x}) = \phi(x,t)$ with $x \in [0, N)$, and $t \in [-\frac{T}{2}, \frac{T}{2}]$, with $T \to \infty$. The boundary condition is then $\phi(x+N, t) = \pi m + \phi(x, t)$. We can gauge away the defect by setting

$$\phi(x,t) = \tilde{\phi}(x,t) - \frac{\pi m x}{N}, \quad (1.188)$$

where now $\tilde{\phi}$ is a true Gaussian field with periodic boundary conditions. The corresponding change in the (free) energy per site is readily evaluated from (1.187) as

$$\frac{S_E(S^z) - S_E(0)}{NT} = \frac{g}{4\pi} \left(\frac{\pi m}{N}\right)^2 = \frac{2\pi X}{N^2}, \quad (1.189)$$

and comparison with (1.183) fixes the coupling constant,

$$g = \frac{1}{2}\left[1 - \frac{\tilde{K}(0)}{2\pi}\right] = 1 - \frac{\gamma}{\pi}, \quad (1.190)$$

so that it gets related to the loop weight,

$$n_{\text{loop}} = -2\cos(\pi g). \quad (1.191)$$

It is also possible to fix g by a purely field-theoretical argument [33, 34], using that the operator conjugate to the loop weight $\sqrt{Q}$ must be *exactly marginal* in the renormalization group sense, because the same weight is attributed to a loop independently of its size. This leads directly to (1.190) without applying the exact solvability of the model.

Apart from the 'magnetic type' defects just discussed, the Coulomb gas theory also contains 'electric type' vertex operators $V_e = \exp(ie\phi)$. Inserted as a pair of oppositely charged vertex operators at the cylinder extremities amounts to modifying the weighting of loops that wind around the cylinder—or imposing twisted periodic boundary conditions on the spin chain. Such excitations can be realized—and their critical exponents can be computed—within the Bethe ansatz formalism by backscattering e particles from the left to the right Fermi level. This means choosing the Bethe integers as follows:

$$I_1,\ldots,I_n = -\frac{n-1}{2}+e,\ldots,\frac{n-1}{2}+e, \qquad (1.192)$$

i.e. packed as densely as possible, but centred around e instead of around the origin.

1.5 Nineteen-vertex models

The algebraic Bethe ansatz solution of the six-vertex model can essentially be generalized in two different ways, namely, by increasing the rank, or by increasing the spin.

The six-vertex model and the Potts model are closely related to the algebra $su(2)$. To be more precise, we have seen that the Potts model can be formulated in terms of loops with weight $n = q + q^{-1}$, where we have set $q = e^{i\gamma}$. For a system of size N strands and non-periodic (reflecting) boundary conditions, these loops provide a representation of the Temperley–Lieb algebra $TL_N(n)$. This representation is faithful whenever q is not a root of unity (i.e. $\gamma \notin \pi\mathbb{Q}$) [1]. It can then be shown [35] that $TL_N(n)$, and in particular the spin chain Hamiltonian $H = -\sum_{m=1}^{N-1} E_m$ obtained in the anisotropic limit $u \to 0$, commutes with the quantum algebra $U_q(su(2))$. This algebra is a deformation of the usual $su(2)$, with modified commutation relations that depend on q. The classical $su(2)$ is recovered for $q \to 1$.

In the first generalization, one replaces $su(2)$ by a Lie algebra (or super Lie algebra) of higher rank r, such as $su(k)$ with $k \geq 3$. In many cases, the transfer matrices of the corresponding models may be diagonalized by Bethe ansatz. Generally, a recursive construction is possible, so that after an initial application of the Bethe ansatz one is left with a system of rank one lower, which can in turn be solved by the Bethe ansatz. This is known as the nested Bethe ansatz approach (see [2] for a review). One is left with r distinct Bethe equations for r types of particles, each coming with their own quasi-momenta.

In the second generalization, one keeps the algebra unchanged, but considers representations with higher spin. In the $su(2)$, one first step in this direction is to increase the spin from $1/2$ to 1, such that each edge can be in three different states: $S^z = +1, 0, -1$, rather than $S^z = +\frac{1}{2}, -\frac{1}{2}$. We can represent $S^z = \pm 1$ as the usual up- and down-pointing arrows, whereas $S^z = 0$ is a vacancy (no arrow). Imposing the conservation of S^z, as in the six-vertex case, then leads to nineteen possible arrow configurations around each vertex. In this final section we shall survey a few aspects of this nineteen-vertex model, connect it to a statistical model of dilute loops, and discuss some recent results with applications to polymer physics.

1.5.1 *R*-matrices

In the basis

$$|++\rangle, |+0\rangle, |+-\rangle, |0+\rangle, |0,0\rangle, |0-\rangle, |-+\rangle, |-0\rangle, |--\rangle, \qquad (1.193)$$

the R-matrix with spectral parameter λ can be written

$$R(\lambda) = \begin{pmatrix} x_1 & 0 & 0 & 0 & 0 & 0 & 0 & 0 & 0 \\ 0 & y_5 & 0 & x_2 & 0 & 0 & 0 & 0 & 0 \\ 0 & 0 & y_7 & 0 & y_6 & 0 & x_3 & 0 & 0 \\ 0 & x_2 & 0 & x_5 & 0 & 0 & 0 & 0 & 0 \\ 0 & 0 & \epsilon y_6 & 0 & \epsilon x_4 & 0 & \epsilon x_6 & 0 & 0 \\ 0 & 0 & 0 & 0 & 0 & y_5 & 0 & x_2 & 0 \\ 0 & 0 & x_3 & 0 & x_6 & 0 & x_7 & 0 & 0 \\ 0 & 0 & 0 & 0 & 0 & x_2 & 0 & x_5 & 0 \\ 0 & 0 & 0 & 0 & 0 & 0 & 0 & 0 & x_1 \end{pmatrix}. \quad (1.194)$$

The signs $\epsilon = \pm 1$ refer to a Z_2-grading $p(\alpha)$ (with $\alpha = 1, 2, 3$) of each vector space $V = (\mathbb{C})^3$, such that $p(1) = p(3) = 0$ and $(-1)^{p(2)} = \epsilon$. The multiplication rules in the graded tensor product space $V \otimes V$ read in component form,

$$(A \otimes B)^{\gamma\delta}_{\alpha\beta} = (-1)^{p(\beta)(p(\alpha)+p(\gamma))} A_{\alpha\gamma} B_{\beta\delta}. \quad (1.195)$$

There are three known integrable R-matrices—solutions of the Yang–Baxter equation—of the form (1.194) with trigonometric Boltzmann weights x_i and y_j (for brevity we omit the known expressions for the weights). The first two of these are ungraded ($\epsilon = 1$).

The first solution is known as the Fateev–Zamolodchikov (FZ) or $a_1^{(1)}$ model [37]. It can be constructed simply from the six-vertex model by a procedure called *fusion*: regroup the spin-$\frac{1}{2}$ quantum spaces two by two, and project pairs of spaces on the triplet (spin-1) representation.

The second solution, the Izergin–Korepin (IK) or $a_2^{(2)}$ model [38], cannot be constructed by fusion in this sense. It can however be related to a spin-1 loop model, that can be obtained after an ingenious series of transformations by regrouping pairs of spin-$\frac{1}{2}$ oriented loops on an appropriate lattice [39]. More directly, one can obtain the spin-1 loop model equivalent to (1.194) by summing over orientations, just like we did when relating the six-vertex and Potts loop models. This leads to the following nine allowed loop configurations around a vertex on the square lattice [40]:

$$\underset{\rho_1}{\times} \underset{\rho_2}{\times} \underset{\rho_3}{\times} \underset{\rho_4}{\times} \underset{\rho_5}{\times} \underset{\rho_6}{\times} \underset{\rho_7}{\times} \underset{\rho_8}{\times} \underset{\rho_9}{\times} \quad (1.196)$$

with trigonometric weights ρ_i (with $i = 1, 2, \ldots, 9$) for each diagram, and an overall weight n_{loop} for each loop. A typical configuration on a large lattice is shown in Figure 1.5.

Finally, the third model is graded ($\epsilon = -1$) and corresponds to the $osp(1|2)$ super Lie algebra in the fundamental representation. It was found by Bazhanov and Shadrikov [41].

Figure 1.5 *Real-life application of the IK loop model, here at the dilute polymer point $n_{\text{loop}} \to 0$ in regime I ($\gamma = \frac{3\pi}{4}$). Photo from the shopping mall EKZ Wien Mitte (Vienna, Austria).*

All three models can be solved in a unified way using coordinate Bethe ansatz [42]. The algebraic Bethe ansatz is also known [42, 43]. We henceforth focus on the second, Izergin–Korepin model.

1.5.2 Algebraic Bethe ansatz for the IK model

The monodromy matrix (1.78) for the IK model is now a 3×3 matrix,

$$T(u) = \begin{bmatrix} A_1(u) & B_1(u) & B_2(u) \\ C_1(u) & A_2(u) & B_3(u) \\ C_2(u) & C_3(u) & A_3(u) \end{bmatrix}, \quad (1.197)$$

where each entry is an operator on the quantum spaces $(\mathbb{C}^3)^{\otimes N}$ for a chain of length N. Starting from the pseudo-vacuum $|\Uparrow\rangle$, we can produce algebraic Bethe ansatz states as in (1.100), but acting now with the *three* types of creation operators $B_j(u_i)$. We note that $B_1(u_i)$ and $B_3(u_i)$ each create *one* particle ($|\uparrow\rangle \to |0\rangle$ or $|0\rangle \to |\downarrow\rangle$), while $B_2(u_i)$ creates *two* particles ($|\uparrow\rangle \to |\downarrow\rangle$). Similarly, the $C_j(u_i)$ are annihilation operators of one or two particles.

The aim is now to construct n-particle states which are eigenstates of the transfer matrix

$$t(u) = A_1(u) + A_2(u) + A_3(u). \tag{1.198}$$

To this end, one uses the RTT relation (1.80) to produce the commutation relations among the entries in (1.197). These are quite a bit more complicated than in the six-vertex case. Omitting all details, one finds in the end that the the eigenvalue of $t(u)$ reads

$$\Lambda_n(u;\{u_i\}) = x_1(u)^N \prod_{i=1}^n z(u_i - u)$$
$$+ x_2(u)^N \prod_{i=1}^n \frac{z(u-u_i)}{\omega(u-u_i)} + x_3(u)^N \prod_{i=1}^n \frac{x_2(u-u_i)}{x_3(u-u_i)}, \tag{1.199}$$

provided that the u_i satisfy the Bethe ansatz equations

$$\left(\frac{x_1(u_i)}{x_2(u_i)}\right)^N = \prod_{\substack{j=1 \\ j \neq i}}^n \frac{z(u_i - u_j)}{z(u_j - u_i)} \omega(u_j - u_i). \tag{1.200}$$

We have here used the short-hand notations

$$z(u) = \frac{x_1(u)}{x_2(u)}, \qquad \omega(u) = \frac{x_1(u)x_3(u)}{x_3(u)x_4(u) - x_6(u)y_6(u)}. \tag{1.201}$$

To discuss the $O(n_{\mathrm{loop}})$ loop model with configurations (1.196), we parameterize the loop weight as

$$n_{\mathrm{loop}} = -2\cos(2\gamma), \qquad \text{with } \gamma \in [0,\pi]. \tag{1.202}$$

The BAE (1.200) can then be rewritten as

$$\left(\frac{\sinh\left(\lambda_j - i\frac{\gamma}{2}\right)}{\sinh\left(\lambda_j + i\frac{\gamma}{2}\right)}\right)^N = \prod_{\substack{i=1 \\ i \neq j}}^n \frac{\sinh(\lambda_j - \lambda_i - i\gamma)}{\sinh(\lambda_j - \lambda_i + i\gamma)} \frac{\cosh\left(\lambda_j - \lambda_i + i\frac{\gamma}{2}\right)}{\cosh\left(\lambda_j - \lambda_i - i\frac{\gamma}{2}\right)}. \tag{1.203}$$

A spin chain Hamiltonian H can be defined via the completely anisotropic limit $u \to 0$, as in (1.73). We shall multiply it by a factor $\mathcal{N} = \pm 1$ and study both signs. The eigenenergies of H then take the form

$$E = -\mathcal{N} \sum_{i=1}^n \frac{\sin\gamma}{\cosh 2\lambda_i - \cos\gamma}. \tag{1.204}$$

1.5.2.1 Regimes

The IK spin chain admits three different *regimes* [44], a feature that we have not seen in the six-vertex model. In each regime, the Bethe roots corresponding to the ground state have different structural properties, meaning that they are situated on different curves in the complex plane. Similarly, the different types of elementary excitations call for various modifications of the Bethe root configurations, which are proper to each regime. This implies that—even though the BAE themselves are unchanged—thermodynamic quantities such as the free energy per site will be described by different analytical expressions in each regime. More generally, the continuum limit of each regime is described by a distinct CFT.

The easiest case is called regime I. It corresponds to $\mathcal{N} = -1$, and the ground state has the roots λ_j lying on a line with imaginary part $\Im \lambda_j = \frac{\pi}{2}$. The continuum limit is again the Coulomb gas CFT, discussed in Section 1.5.5, but with

$$n_{\text{loop}} = -2\cos(\pi g), \quad \text{and } 0 < g \le 2. \tag{1.205}$$

The range $0 < g \le 1$ gives the same CFT has for the Potts model, and accordingly the loops will fill up the space in a dense way. The analytical continuation to $1 \le g \le 2$ gives a physically different phase, in which loops are dilute in space. In particular, the case $g = \frac{3}{2}$ describes a polymer problem ($n_{\text{loop}} \to 0$) that can be identified with self-avoiding random walks.

Regime II corresponds to $\mathcal{N} = 1$ and $\gamma \in [\frac{\pi}{3}, \pi]$. Its continuum limit is essentially the same as regime I, but with an extra fermionic (Ising-like) degree of freedom.

Finally, regime III corresponds to $\mathcal{N} = 1$ and $\gamma \in [0, \frac{\pi}{3}]$. It is arguably the most interesting regime, and its continuum limit has only been (partially) understood very recently [45]. The ground state is made of complexes λ_j (called 2-strings) with imaginary part $\Im \lambda_j$ approximately equal to $\pm \frac{1}{4}(\pi - \gamma)$:

$$\lambda_j = \Re \lambda_j \pm \frac{i}{4}(\pi - \gamma). \tag{1.206}$$

By 'approximately' is meant that $\Im \lambda_j$ approaches this value in the $N \to \infty$ limit, but in finite size the 'lines' of Bethe roots have some curvature. This feature alone makes the problem more difficult to understand, both numerically and analytically.

1.5.3 Continuum limit of regime III

We now briefly outline the proposed continuum limit for regime III of the IK model [45]. Recall that in the Coulomb gas CFT (corresponding here to regime I) the boson was compactified on a circle. Quite differently—and highly unusually for the spin chains previously studied in statistical physics—the continuum limit of regime III is *non-compact*. To be more precise, the continuum limit CFT is defined in terms of one compact and one non-compact boson. This implies that the 'spectrum' of critical exponents—which is usually a discrete set of numbers, as in (1.183) for instance— contains both a discrete and a continuous part.

The situation is vaguely reminiscent of the quantum mechanics of a free particle confined to a box. The energy spectrum is discrete, but when the size of the box (the 'target space') goes to infinity, one recovers plane wave solutions and a continuous spectrum.

1.5.3.1 Black hole sigma model

Using a combination of analytical and numerical arguments—among which the numerical resolution of the BAE (1.203) has played a prominent role—it has been shown that the continuum limit of the IK spin chain in regime III coincides with the so-called $SL(2,\mathbb{R})_k/U(1)$ Euclidean black hole σ-model. This CFT was first introduced in string theory [46] as a toy model for the scattering of a string on a two-dimensional black hole.

The black hole σ-model is defined by the action

$$S = \frac{k}{4\pi} \int d^2x \sqrt{h} h^{ij} \left(\partial_i r \partial_j r + \tanh^2 r \partial_i \theta \partial_j \theta \right) - \frac{1}{8\pi} \int d^2x \sqrt{h} \Phi(r,\theta) \mathcal{R}. \qquad (1.207)$$

Here h^{ij} denotes the (fixed) world-sheet metric, while $\theta \in [0, 2\pi)$ and $r \in [0, \infty)$ are fluctuating bosonic fields of compact and non-compact nature, respectively. The target space metric is

$$ds^2 = \frac{k}{2} d\sigma^2, \qquad \text{with } d\sigma^2 = (dr)^2 + \tanh^2 r (d\theta)^2. \qquad (1.208)$$

This describes a cigar-shaped surface (whence also the nickname *cigar model* for this CFT) in three dimensions, with radius $\tanh r$, where $r \geq 0$ denotes the geodesic distance from the origin.

The first term in (1.207) is the classical action, with the parameter k called the level. The second term couples the dilaton field $\Phi(r,\theta) = 2\ln \cosh r + \Phi_0$ to the Gaussian curvature $\mathcal{R}$ of the world-sheet. It is this term which makes the theory gapless at the quantum level, and hence conformally invariant.

The central charge at the quantum level is $c = 2 + \frac{6}{k-2}$, naively different from the result one would expect for the untwisted spin chain (namely $c = 2$).

The spectrum of the quantum theory has the following conformal weights (critical exponents) [47]:

$$h(\bar{h}) = -\frac{\mathcal{J}(\mathcal{J}+1)}{k-2} + \frac{(n \pm kw)^2}{4k} \qquad (1.209)$$

of the primary fields $\Phi^{\mathcal{J}}_{nw}(z,\bar{z})$. These fields are characterized by several quantum numbers. In the compact θ-direction, $n \in \mathbb{Z}$ is the angular momentum of rotations around the cigar axis, while w are winding modes that can be identified with the twist of spin chain. In the non-compact r-direction, one writes $\mathcal{J} = -\frac{1}{2} + is$, where s is the momentum along the cigar. It is required that $s \in \mathbb{R}$ for the eigenfunctions to be normalizable.

We note in particular that the identity field ($\mathcal{J} = 0$) is *not* normalizable, and hence does not belong to the spectrum. Rather, the lowest conformal weight is found for $\mathcal{J} = -\frac{1}{2}$ and $n = w = 0$. By (1.209) it reads $x = h + \bar{h} = \frac{1}{2(k-2)}$, so that the effective central charge is $c_{\text{eff}} = c - 12x = 2$ indeed.

The black hole σ-model also allows for *discrete states*, corresponding to eigenfunctions localized near the cigar tip. For these states to be normalizable, one must require [48, 49]

$$\mathcal{J} \in \left[\frac{1-k}{2}, -\frac{1}{2}\right] \cap \left(\mathbb{N} - \frac{1}{2}|kw| + \frac{1}{2}|n|\right). \tag{1.210}$$

To identify [45] this CFT with the IK loop model in regime III, we must first give the weight $n_{\text{loop}} = 2\cos\gamma$ with $\gamma = \frac{2\pi}{k}$ to contractible loops (those not winding around the periodic direction on the cylinder). We further identify the winding w with the electric charge e that we have already encountered in the Coulomb gas setup. Therefore non-contractible loops have the weight $\tilde{n}_{\text{loop}} = 2\cos\varphi$ with $w = \frac{\varphi}{2\pi}$. This corresponds to twisting the corresponding spin chain.

One crucial point in the identification comes from the discrete states. In the early treatment of the IK loop model, two distinct expressions were found for the central charge [44]:

$$c^* \equiv 2 - 6k\left(\frac{\varphi}{2\pi}\right)^2, \qquad \text{for } \varphi \leq \frac{2\pi}{k},$$
$$c_1 \equiv -1 + \frac{3k}{k-2}\left(\frac{\varphi - \pi}{\pi}\right)^2, \qquad \text{for } \frac{2\pi}{k} \leq \varphi.$$

We now see that while c^* is the genuine central charge for the twisted chain, the second expression c_1 matches the effective central charge corresponding to the first discrete state. The fact that c_1 is observed as the ground state only for a twist $\varphi \geq \frac{2\pi}{k}$ is a nice confirmation of (1.210); indeed, for smaller twists the corresponding discrete state is not normalizable.

In numerical studies it was also observed [44, 45] that the convergence of the numerical data to c_1 is fast, in terms of the spin chain size N, as for usual models with a compact continuum limit. On the other hand, the convergence to c^* is very slow. The correspondence with the black hole σ-model also explains these observations. Indeed, c^* is but the bottom of a continuum of non-compact excitations, implying that the finite-size correction contains a term of the form $A/(B + \log N)^2$. This precise form is indeed verified by numerically solving the BAE. More generally, all of the discrete states predicted by (1.210) are observed numerically, for the predicted ranges of the twist φ [45].

1.5.3.2 Theta point of collapsing polymers

Regime III of the IK loop model contains a special point, $\gamma = \frac{\pi}{4}$ corresponding to $n_{\text{loop}} \to 0$, that has important implications for the physics of polymers that collapse to a dense phase as a consequence of the mutual attraction between monomers [50]. The connection with a CFT devised to model the scattering of a string on a black hole is rather remarkable in this context.

By a suitable choice of the spectral parameter u, the weights in (1.196) can be made isotropic. The weights can then be interpreted as a monomer fugacity K, a bending rigidity p, and a monomer self-attraction τ: see Figure 1.6a. This is known as the vertex-interacting self-avoiding walk (VISAW) model. Its schematic phase diagram is shown in Figure 1.6b. It possesses a multicritical line, along which two exactly solvable points, Θ_{DS} (with $p=0$) and Θ_{BN}, describe different variants of the collapse transition. Regime III is precisely Θ_{BN}, and, crucially, the corresponding Boltzmann weights are all positive:

$$p_{\text{BN}} = \sqrt{2}\sin\left(\frac{\pi}{16}\right) \approx 0.275899,$$

$$K_{\text{BN}} = \left[2\cos\left(\frac{\pi}{16}\right)\left(1 + \frac{1}{\sqrt{2}}\tan\left(\frac{\pi}{16}\right)\right)\right]^{-1} \approx 0.446933, \qquad (1.211)$$

$$\tau_{\text{BN}} = \frac{1}{2}\left(2 + \sqrt{2} + \sqrt{2+\sqrt{2}}\right) \approx 2.630986.$$

The two-point function in the polymer problem can be written

$$G(r_1, r_2) = \sum_{\text{VISAW}: r_1 \to r_2} K^{\#\text{ monomers}} \, \tau^{\#\text{ self-contacts}} \, p^{\#\text{ straight}}. \qquad (1.212)$$

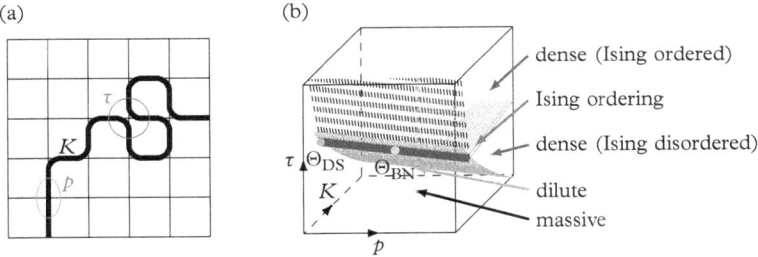

Figure 1.6 *VISAW model: (a) Typical configuration and Boltzmann weights, and (b) schematic phase diagram.*

In the case of Θ_{DS}, the continuum limit is a usual, compact CFT with a discrete spectrum. Therefore,

$$G(r_1, r_2) \propto \left(\frac{\epsilon}{|r_1 - r_2|}\right)^{2x_1} + C\left(\frac{\epsilon}{|r_1 - r_2|}\right)^{2x'_1} + \ldots, \quad (1.213)$$

takes the form of a power law with a correction-to-scaling term ($x'_1 > x_1$); here ϵ denotes a cut-off corresponding to the lattice spacing. However, for Θ_{BN} we have instead a non-compact continuum limit with a continuous spectrum. We known the dependence of the critical exponent on the non-compact quantum number s from (1.209), so we would expect (1.213) to be replaced by

$$G(r_1, r_2) \approx \int_0^\infty ds \frac{f_1(s)}{|r_1 - r_2|^{2x_1 + 2a_1 s^2}}, \quad (1.214)$$

where $f_1(s)$ is the density of excitations with a given value of s. This function is unfortunately unknown.

If one makes the assumption

$$f_1(s) = s^z [b_0 + s b_1 + \ldots], \quad \text{for } s \to 0, \quad (1.215)$$

one obtains

$$\log G = -2x_1 \log|r_1 - r_2| + A - B \log \log|r_1 - r_2| \quad (1.216)$$

for certains constants A and B. The scaling form turns out to be in excellent agreement with a Monte Carlo simulation of the VISAW model [50].

References

[1] A.M. Gainutdinov, J.L. Jacobsen, N. Read, H. Saleur, and R. Vasseur, J. Phys. A: Math. Theor. **46**, 494012 (2013).
[2] R.J. Baxter, *Exactly Solved Models in Statistical Mechanics* (Academic Press, London, 1982).
[3] V.E. Korepin, N.M. Bogoliubov, and A.G. Izergin, *Quantum Inverse Scattering Method and Correlation Functions* (Cambridge University Press, 1993).
[4] C. Gómez, M. Ruiz-Altaba, and G. Sierra, *Quantum Groups in Two-Dimensional Physics* (Cambridge University Press, 1996).
[5] L.D. Faddeev, How the algebraic Bethe Ansatz works for integrable models. In A. Connes, K. Gawędzki, and J. Zinn-Justin (eds.), *Symétries quantiques: Les Houches, session LXIV, 1 août – 8 septembre 1995*, 149–219 (North-Holland, Amsterdam, 1998); hep-th/9605187.
[6] P. Di Francesco, P. Mathieu, and D. Sénéchal, *Conformal Field Theory* (Springer-Verlag, New York, 1997).
[7] J.L. Jacobsen, *Algèbres, intégrabilité et modèles exactement solubles*, lecture notes (unpublished, 2011).

[8] J.L. Jacobsen, Conformal field theory applied to loop models. In A.J. Guttmann (ed.), *Polygons, Polyominoes and Polycubes*, Lecture Notes in Physics **775**, 347–424 (2009).
[9] Y. Ikhlef, *Résultats exacts sur les modèles de boucles en deux dimensions*, Ph.D. thesis (Université Paris Sud, Orsay, 2007).
[10] A. Garbali, *The Izergin-Korepin model*, Ph.D. thesis (Université Pierre et Marie Curie, Paris, 2015).
[11] F.Y. Wu, J. Stat. Phys. **52**, 99 (1988).
[12] R.B. Potts, Math. Proc. Cambridge Phil. Soc. **48**, 106 (1952).
[13] R.J. Baxter, J. Phys. C: Solid State Phys. **6**, L445 (1973).
[14] J.L. Jacobsen, J. Phys. A: Math. Theor. **47**, 135001 (2014).
[15] C.M. Fortuin and P.W. Kasteleyn, Physica **57**, 536 (1972).
[16] J.L. Jacobsen, J. Salas, and A.D. Sokal, J. Stat. Phys. **119**, 1153 (2005).
[17] S. Caracciolo, J.L. Jacobsen, H. Saleur, A.D. Sokal, and A. Sportiello, Phys. Rev. Lett. **93**, 080601 (2004).
[18] J.L. Jacobsen and H. Saleur, Nucl. Phys. B **716**, 439 (2005).
[19] R.J. Baxter, S.B. Kelland, and F.Y. Wu, J. Phys. A Math. Gen. **9**, 397 (1976).
[20] H.N.V. Temperley and E.H. Lieb, Proc. Roy. Soc. London A **322**, 251 (1971).
[21] R.J. Baxter, Proc. Roy. Soc. London A **383**, 43 (1982).
[22] J.L. Jacobsen and H. Saleur, Nucl. Phys. B **743**, 207 (2006).
[23] Y. Ikhlef, J.L. Jacobsen, and H. Saleur, Nucl. Phys. B **789**, 483 (2008).
[24] L. Onsager, Phys. Rev. **65**, 117 (1944).
[25] R.J. Baxter, Stud. Appl. Math. **50**, 51 (1971).
[26] F.C. Alcaraz, M.N. Barber, and M.T. Batchelor, Ann. Phys. **182**, 280 (1988).
[27] L.D. Faddeev, E.K. Sklyanin, and L.A. Takhtajan, Theor. Math. Phys. **40**, 86 (1979).
[28] H. Bethe, Zeitschrift für Physik **71**, 205 (1931).
[29] J.B. Parkinson, J. Phys. C: Solid State Phys. **21**, 3793 (1988).
[30] C.N. Yang and C.P. Yang, Phys. Rev. **150**, 321 (1966); *ibid.* **150**, 327 (1966).
[31] H.J. de Vega and F. Woynarovitch, Nucl. Phys. B **251**, 439 (1985).
[32] B. Nienhuis, J. Stat. Phys. **34**, 731 (1984).
[33] J. Kondev, Phys. Rev. Lett. **78**, 4320 (1997).
[34] J.L. Jacobsen and J. Kondev, Nucl. Phys. B **532**, 635 (1998).
[35] V. Pasquier and H. Saleur, Nucl. Phys. B **330**, 523 (1990).
[36] O. Babelon, *A short introduction to classical and quantum integrable systems*, lecture notes (Cours de Physique Théorique du SPhT, 2007).
[37] A.B. Zamolodchikov and V.A. Fateev, Sov. J. Nucl. Phys. **32**, 298 (1980).
[38] A.G. Izergin and V.E. Korepin, Commun. Math. Phys. **79**, 303 (1981).
[39] B. Nienhuis, Phys. Rev. Lett. **49**, 1062 (1982).
[40] H.W.J. Blöte and B. Nienhuis, J. Phys. A: Math. Gen. **22**, 1415 (1989).
[41] V.V. Bazhanov and A.G. Shadrikov, Theor. Math. Phys. **73**, 1302 (1987).
[42] A. Lima-Santos, J. Phys. A: Math. Gen. **32**, 1819 (1999).
[43] V.O. Tarasov, Theor. Math. Phys. **76**, 793 (1988).
[44] S.O. Warnaar, M.T. Batchelor and B. Nienhuis, J. Phys. A: Math. Gen. **25**, 3077 (1992).
[45] E. Vernier, J.L. Jacobsen and H. Saleur, J. Phys. A: Math. Theor. **47**, 285202 (2014).
[46] E. Witten, Phys. Rev. D **44**, 314 (1991).
[47] R. Dijkgraaf, H. Verlinde, and E. Verlinde, Nucl. Phys. B **371**, 269 (1992).
[48] A. Hanany, N. Prezas, and J. Troost, JHEP **04** 014 (2002).
[49] S. Ribault and V. Schomerus, JHEP **02** 019 (2004).
[50] E. Vernier, J.L. Jacobsen, and H. Saleur, J. Stat. Mech.: Theor. Exp. P09001 (2015).

2
A guide to two-dimensional conformal field theory

Jörg Teschner

Department of Mathematics, University of Hamburg,
Bundesstrasse 55, 20146 Hamburg, Germany and
DESY theory, Notkestrasse 85, 20607 Hamburg, Germany

Chapter Contents

2 **A guide to two-dimensional conformal field theory**   60
    Jörg TESCHNER

    Preface   62
2.1   Conformal symmetry   63
2.2   Bootstrap   80
2.3   Relations to integrable models[+]   101
    Acknowledgements   117
    References   117

Preface

Starting from the pioneering works on conformal field theory (CFT) from the 1970s and early 1980s in the 20th century, the subject has grown rather large. It has found many different applications to various parts of theoretical physics on the one hand, and it has inspired several new directions of mathematical research on the other. However, as often happens, the original unity of the subject has got lost over time, it has branched out in various directions to such an extent that communication may be hard between different branches of research on the subject of conformal field theory.

In such a situation it may be helpful to create a 'basic operating system' for CFT by identifying an efficient formalism that reaches as far as possible which can easily be extended by 'downloading suitable apps' from the literature for more specialized or more advanced topics. This is what we are trying to do in this chapter.

More specific goals are to:

(i) Revisit 2d CFT from a more general perspective than is common. Non-generic features are often assumed from the outset, here, we'll emphasize whatever seems to hold more generically.

(ii) Create bridges between the physicist's point of view and (some of) the mathematical approaches to CFT, wherever this seems beneficial for one or both disciplines.

(iii) Discuss some connections between CFT and the theory of integrable models, a subject of growing importance which is rarely discussed in the existing literature.

Although the author is making efforts to achieve a balanced presentation, a certain bias due to expertise and the particular interests of the author is probably unavoidable. There are at least two important areas about which almost nothing will be found in this chapter. One is the subject of boundary CFT, for which [1] offers a starting point, the other is the operator-algebraic approach to CFT, reviewed e.g. in [2]. Limitations of space and time have forced me to give a fairly concise presentation, in which many important aspects could not be covered, and many details have had to be omitted. As *a guide to the reader*, as previously indicated, the idea behind this chapter is to present a 'skeleton' of conformal field theory, consisting of some aspects that are most essential for many applications, and most basic of mathematical theory. Starting from this skeleton one can hopefully access more easily whatever more specialized topic one is interested in. The author is looking for a reasonable compromise between being mathematically precise, easily accessible for readers with a physics background, and easy to adapt to more general situations.

This chapter may serve two tasks. The main part may be used as a largely self-contained introductory course if the reader is studying the text with some care, reading concentratedly, filling in the gaps, and solving the exercises. One may, on the other hand, use it for getting an overview, a first idea about more advanced topics, and pointers towards further literature on the subject. This is reflected in the structure of the chapter. Sections having a title marked with an plus sign contain supplementary or more advanced material that could be omitted in a very first reading. Sections marked with a double plus are offering a guide to the literature on some further directions.

The chapter consists of three parts. The first part discusses the notion of conformal symmetry and how the corresponding constraints on correlation functions get encoded into the representation-theoretic definition of the conformal blocks as solutions to the conformal Ward identities. Extensions of conformal symmetry are briefly discussed at the end of this part.

The second part discusses the conformal bootstrap. Starting from the gluing construction of conformal blocks we identify consistency conditions for the construction of physical correlation functions from conformal blocks. Minimal models and Liouville theory are discussed as examples where this programme has been fully realized. Some approaches to the construction of more general CFTs are indicated.

The third part outlines some of the known relations between CFT and integrable models, using the connections to the isomonodromic deformation problem as main example. It is shown that CFT can be understood as a quantization of the isomonodromic deformation problem. A few relations to quantum integrable models are very briefly pointed out at the end.

On some occasions we offer shortcuts to some key results which seem to be hard to extract from the existing literature. This concerns especially the material in Sections 2.1.8 and 2.3.3.

2.1 Conformal symmetry

A two-dimensional (2d) QFT will be called a (2d) conformal field theory if it has a symmetry group with Lie algebra containing the Virasoro algebra.

2.1.1 States

The Virasoro algebra Vir_c is the infinite-dimensional Lie algebra with generators L_n, $n \in \mathbb{Z}$ and relations

$$[L_n, L_m] = (n-m)L_{n+m} + \frac{c}{12}n(n^2 - 1)\delta_{n+m,0}. \tag{2.1}$$

A 2d QFT has Virasoro symmetry if

(i) its Hilbert space $\mathcal{H}$ is a unitary representation of the product $\mathsf{Vir}_c \times \overline{\mathsf{Vir}}_c$ with generators[1] denoted by $L_n, \bar{L}_n, n \in \mathbb{Z}$, and
(ii) the Hamiltonian H is given in terms of the operators L_0 and $\bar{L}_0$ as

$$\mathsf{H} = L_0 + \bar{L}_0 + (\text{const.}). \tag{2.2}$$

---
[1] We use the same notation for generators of the Lie algebra Vir_c and for the operators representing them.

It follows from (i), (ii) that $\mathcal{H}$ has the form

$$\mathcal{H} = \bigoplus_{R',R''} M_{R',R''} \otimes R' \otimes R'', \qquad (2.3)$$

where R' and R'' are unitary irreducible representations of the Virasoro algebras generated by L_n and $\bar{L}_n$, respectively, and $M_{R'R''}$ is a multiplicity space transforming trivially under $\mathrm{Vir}_c \times \overline{\mathrm{Vir}}_c$. In order to have stability ($\mathsf{H} > 0$) we need to have representations R' containing a vector of lowest energy, usually called highest weight representations, containing a vector $e_0 \in R'$ such that

$$L_n e_0 = \Delta_{R'} \delta_{n,0} e_0, \qquad n \geq 0. \qquad (2.4)$$

Vectors of the form $L_{-n_1} \cdots L_{-n_k} e_0$ with $k \geq 0$ and $n_l > 0$ for $l = 1, \ldots, k$ span R', but can be linearly dependent for some values of c and Δ_0. The situation is analogous for R''. The representation theory of the Virasoro algebra is discussed in many references, including [3].

We will furthermore mostly assume[2] that the Hilbert space $\mathcal{H}$ contains a distinguished vector $|0\rangle$ of lowest energy called vacuum. The vacuum vector $|0\rangle$ satisfies $L_k|0\rangle = 0$, $k = 0, \pm 1$.

2.1.2 Fields and correlation functions

Conformal field theories form a special class of quantum field theories (QFTs). We will mostly work in the framework of Euclidean QFT on a two-dimensional cylinder with coordinates $w = \tau + i\sigma$, $w^* = \tau - i\sigma$, $\sigma \sim \sigma + 2\pi$. A Euclidean QFT on the cylinder is characterized by its set of fields $\{\Phi_v^{\mathrm{cyl}}(w, \bar{w}); v \in \mathcal{F}\}$, with $v \in \mathcal{F}$ being a label for elements of a basis for the set of fields, together with the collection of its Schwinger functions $\mathcal{Z}_{\mathbf{v}}^{\mathrm{cyl}}(\mathbf{w}, \mathbf{w}^*) = \left\langle \prod_{r=1}^n \Phi_{v_r}^{\mathrm{cyl}}(w_r, w_r^*) \right\rangle$, using tuple notations $\mathbf{w} = (w_1, \ldots, w_n)$, $\mathbf{v} = (v_1, \ldots, v_n)$ etc. We will furthermore mostly restrict attention to QFTs having only bosonic fields. Key properties of the Schwinger functions are their single-valuedness and permutation symmetry, and the invariance under exchange of all variables with indices r and s, for $r, s = 1, \ldots, n$.

Under certain conditions on the Schwinger functions, the most important being called reflection positivity, one may reconstruct[3] from the collection of all Schwinger functions $\mathcal{Z}_{\mathbf{v}}^{\mathrm{cyl}}(\mathbf{w}, \mathbf{w}^*)$ a Hilbert space $\mathcal{H}$ with vacuum vector $|0\rangle$ acted upon by a family of operators $\Phi_v^{\mathrm{cyl}}(w, \bar{w})$, such that the Schwinger functions $\mathcal{Z}_{\mathbf{v}}^{\mathrm{cyl}}(\mathbf{w}, \mathbf{w}^*)$ with time-ordered arguments $\tau_n > \tau_{n-1} > \cdots > \tau_1$ can be represented as vacuum expectation

---

[2] This assumption may be weakened by allowing vacuum vectors $|0\rangle$ in the distributional sense, as the example of Liouville theory shows [4].

[3] This follows from a variant of the Osterwalder–Schrader reconstruction theorem. See [5] for an account tailored to two-dimensional CFT.

values $\langle 0|\Phi^{\text{cyl}}_{v_n}(w_n, w_n^*)\ldots\Phi^{\text{cyl}}_{v_1}(w_1, w_1^*)|0\rangle$. We are using the notation $\langle v_1|v_2\rangle$ for the scalar product of two vectors $v_1, v_2 \in \mathcal{H}$.

Conformal symmetry will allow us to relate fields $\Phi^{\text{cyl}}_v(w, w^*)$ on the Euclidean cylinder to fields $\Phi_v(z, \bar{z})$ on the complex plane $\mathbb{C}$ having coordinates $z = e^w$. In any 'reasonable' CFT there should exist an analytic continuation of $\mathcal{Z}_{\mathbf{v}}(\mathbf{z}, \mathbf{z}^*) = \langle\prod_{r=1}^n \Phi_{v_r}(z_r, z_r^*)\rangle$ to a multivalued analytic function $\mathcal{Z}_{\mathbf{v}}(\mathbf{z}, \bar{\mathbf{z}})$ of independent variables $z_r, \bar{z}_r, r = 1, \ldots, n$, having an expansion in z_r/z_{r+1} and $\bar{z}_r/\bar{z}_{r+1}$ which is convergent for $|z_r/z_{r+1}| < 1$ and $|\bar{z}_r/\bar{z}_{r+1}| < 1$, $r = 1, \ldots n$. The functions $\mathcal{Z}_{\mathbf{v}}(\mathbf{z}, \bar{\mathbf{z}})$ defined in this way should have an analytic continuation to multivalued analytic functions on $\mathcal{C}_n \times \mathcal{C}_n$, $\mathcal{C}_n = \mathbb{C}^n \setminus \{z_r = z_s : r \neq s; r, s = 1, \ldots, n\}$ which are single-valued when restricted to the so-called Euclidean domain $\bar{z}_r = z_r^*, r = 1, \ldots, n$, with z_r^* being the complex conjugate of z_r.

A further basic feature of conformal field theories is the state-operator correspondence, an isomorphism: $v \mapsto \Phi_v$ between $\mathcal{H}$ and the space of fields, which may be formulated in terms of the fields $\Phi_v(z, \bar{z})$ on the complex plane conveniently as

$$\lim_{z, \bar{z} \to 0} \Phi_v(z, \bar{z}) |0\rangle = v \in \mathcal{H}. \tag{2.5}$$

The special fields

$$T(z) \equiv \Phi_{L_{-2}|0\rangle}(z, \bar{z}) = \sum_{k \in \mathbb{Z}} L_n z^{-k-2}, \quad \bar{T}(z) \equiv \Phi_{\bar{L}_{-2}|0\rangle}(z, \bar{z}) = \sum_{k \in \mathbb{Z}} \bar{L}_n z^{-k-2}, \tag{2.6}$$

represent the only non-vanishing components of the energy–momentum tensor. The translations L_{-1} of the complex plane are realized on all fields $\Phi_v(z, \bar{z})$ in a particularly simple way,

$$[L_{-1}, \Phi_v(z, \bar{z})] = \partial_z \Phi_v(z, \bar{z}). \tag{2.7}$$

The conditions on the Schwinger functions thus formulated are necessary for getting a physically reasonable CFT. The bootstrap programme, to be discussed, is designed to construct examples satisfying these requirements. Our next step will be to formulate the conformal invariance conditions in terms of the Schwinger functions.

2.1.3 Conformal Ward identities

On the level of correlation functions one may express the conformal symmetry of a CFT using the conformal Ward identities [6]

$$0 = \left\langle \int_{\mathcal{C}_t} dy\, \eta(y) T(y) \prod_{r=1}^n \Phi_{v_r}(z_r, \bar{z}_r) \right\rangle \tag{2.8}$$

$$= \sum_{r=1}^n \left\langle \Phi_{v_n}(z_n, \bar{z}_n) \cdots \left(\int_{\mathcal{C}_{z_r}} dy\, \eta(y) T(y) \Phi_{v_r}(z_r, \bar{z}_r) \right) \cdots \Phi_{v_1}(z_1, \bar{z}_1) \right\rangle,$$

where $\mathcal{C}_t$ is a small circle on $\mathbb{P}^1$ not encircling any element of $\{z_1,\ldots,z_n\}$ which can be deformed into the sum of circles $\mathcal{C}_r$ encircling only z_r for $r=1,\ldots,n$. The identity (2.8) is required to hold for all meromorphic functions $\eta(y)$ on $\mathbb{C}$ that are allowed to have poles of arbitrary order only at z_r, $r=1,\ldots,n$, and behave for $y \to \infty$ as $\eta(y) = \mathcal{O}(y^2)$. Such functions $\eta(y)$ are in one-to-one correspondence with vector fields $\eta(y)\partial_y$ on $\mathbb{P}^1 \setminus \{z_1,\ldots,z_n\}$.

Exercise 1 Derive (2.8) using the analytic properties of the Schwinger functions mentioned above.

In order to discuss the consequences of the conformal Ward identities (2.8), let us note the following simple operator product expansion:

$$T(y)\Phi_v(z,\bar{z}) = \sum_{k\in\mathbb{Z}}(y-z)^{k-2}\Phi_{L_{-k}v}(z,\bar{z}), \qquad (2.9)$$

following from the state-operator correspondence. By considering functions $\eta(y)$ in (2.8) having poles only at one point z_r, it is easy to see that (2.8) allows us to rewrite the action of L_{-k} on v_r in terms of the action of L_m, $m \geq -1$ on v_s, $s \neq r$.

Equations (2.8) are a linear system of equations relating the different correlation functions of a CFT. This suggests looking for the most general solution to these equations, and constructing the correlation functions as an expansion over a basis for the space of solutions to (2.8). Let us note that (2.8) only involves the copy of the Virasoro algebra having generators L_n. There is of course a similar identity for the other copy with generators $\bar{L}_n$. This suggests that $\mathcal{Z}_\mathbf{v}(\mathbf{z},\bar{\mathbf{z}})$ can be expanded as

$$\mathcal{Z}_\mathbf{v}(\mathbf{z},\bar{\mathbf{z}}) = \sum_{\beta',\beta''} C_{\beta'\beta''\mathbf{m}} \mathcal{F}_{\beta',\mathbf{v}'}(\mathbf{z}) \mathcal{F}_{\beta'',\mathbf{v}''}(\bar{\mathbf{z}}), \qquad (2.10)$$

assuming that $\mathbf{v} = (v_1,\ldots,v_n)$ with $v_r = m_r \otimes v'_r \otimes v''_r \in \mathcal{H}$, and that $\{\mathcal{F}_{\beta,\mathbf{v}}(z); \beta \in \mathcal{I}\}$ is a basis for the space of solutions to the conformal Ward identities (2.8). In order to realize these ideas more precisely let us clarify the mathematical meaning of the conformal Ward identities.

2.1.4 A mathematical reformulation

Let us consider the Riemann surface $C \equiv C_{0,n} = \mathbb{P}^1 \setminus \{z_1,\ldots,z_n\}$, and let $\mathcal{R} = \bigotimes_{r=1}^n R_r$ be the tensor product of representations R_r associated to the points z_r, respectively. We may define an action of the Lie algebra $\mathsf{Vect}(C)$ of vector fields on C by setting

$$T_\xi := \sum_{r=1}^n \sum_{k\in\mathbb{Z}} \xi_k^{(r)} L_k^{(r)}, \qquad (2.11)$$

where $\xi_k^{(r)}$ are the Laurent expansion coefficients of the vector field $\xi \in \text{Vect}(C)$, $\xi = \xi(y)\partial_y$ around z_r, defined by

$$\xi(y) = \sum_{k \in \mathbb{Z}} \xi_k^{(r)} (y - z_r)^{k+1}, \qquad (2.12)$$

and $L_k^{(r)}$ acts only on the r-th tensor factor of $\mathcal{R}$,

$$L_k^{(r)} = \text{id} \otimes \cdots \otimes \text{id} \otimes \underset{\text{r-th}}{L_k} \otimes \text{id} \otimes \cdots \otimes \text{id}. \qquad (2.13)$$

Using these notations we will define conformal blocks as linear maps $f_C : \mathcal{R} \to \mathbb{C}$ satisfying

$$f_C(T_\xi v) = 0, \qquad \forall\, \xi \in \text{Vect}(C),\ \forall\, v \in \mathcal{R}. \qquad (2.14)$$

The set of linear equations (2.14) defines a subspace $\text{CB}(C, \mathcal{R})$ in the dual $\mathcal{R}'$ of the vector space $\mathcal{R}$. The vector space $\text{CB}(C, \mathcal{R})$ is called the space of conformal blocks.

It is not hard to verify that the Ward identities (2.8) will hold provided that the functions $\mathcal{F}_{\beta,\mathbf{v}}(z)$ appearing in (2.10) are related to conformal blocks f_C^β satisfying (2.14) as

$$f_C^\beta(v) := \mathcal{F}_{\beta,\mathbf{v}}(z), \quad \text{where } v = v_1 \otimes \cdots \otimes v_n \in \mathcal{R} \text{ if } \mathbf{v} = (v_1, \ldots, v_n). \qquad (2.15)$$

The definition of conformal blocks via (2.14) is the formulation of the conformal Ward identities that has become customary in the mathematical literature, see e.g. [7] and references therein.

Exercise 2

(a) Verify in detail that (2.10), (2.14), and (2.15) imply (2.8).

(b) Given a conformal block $\hat{f}_{C_{0,n+1}} \in \text{CB}(C_{0,n+1}, V_0 \otimes \mathcal{R})$ show that the definition

$$f_{C_{0,n}}(v_n \otimes \cdots \otimes v_1) := \hat{f}_{C_{0,n+1}}(e_0 \otimes v_n \otimes \cdots \otimes v_1) \qquad (2.16)$$

yields a conformal block $f_{C_{0,n}} \in \text{CB}(C_{0,n}, \otimes \mathcal{R})$. Use this to conclude that there exists an isomorphism $\text{CB}(C_{0,n+1}, V_0 \otimes \mathcal{R}) \simeq \text{CB}(C_{0,n}, \mathcal{R})$. This isomorphism is often referred to as *propagation of vacua*.

Let us try to get a somewhat more concrete idea how the spaces of conformal blocks look. First note that by using the vector fields $\xi = (y - z_r)^{1-k} \partial_y$, $k = 2, 3, \ldots$, one may express the values of f_C on arbitrary vectors v in terms of its values on vectors w of the form $w = \bigotimes_{r=1}^n (L_{-1})^{l_r} e_r$, where the vectors e_r satisfy the highest weight property $L_n e_r = \delta_{n,0} \Delta_r e_r$, $n \geq 0$. One may next use the vector fields $\xi = y^k \partial_y$, $k = 0, 1, 2$, to express the values of f_C on arbitrary vectors v in terms of its values on vectors $w_\mathbf{n}$ of the form $w_\mathbf{n} = e_1 \otimes e_2 \otimes e_3 \otimes \bigotimes_{r=4}^n (L_{-1})^{\nu_r} e_r, \mathbf{n} = (\nu_4, \ldots, \nu_n) \in \mathbb{Z}_{\geq 0}^{n-3}$. This means that a conformal block f_C is completely characterized by the infinite collection of complex numbers $f_C(w_\mathbf{n})$, $\mathbf{n} \in \mathbb{Z}_{\geq 0}^{n-3}$, which may be reformulated as the statement that $\mathrm{CB}(C, \mathcal{R}) \simeq (\mathcal{R}_{-1})'$, where $(\mathcal{R}_{-1})'$ is the dual of $\mathcal{R}_{-1} = \bigotimes_{r=4}^n \mathcal{R}_{-1,r}$, with $\mathcal{R}_{-1,r} = \mathrm{Span}\{L_{-1}^\nu e_r; \nu \in \mathbb{Z}_{\geq 0}\}$. Note, in particular, that for $n = 3$ it follows that the space of conformal blocks is at most one-dimensional. For $n \geq 4$ one finds an infinite-dimensional space of conformal blocks, in general.

It should be noted however, that the dimension of the space of conformal blocks may depend sensitively on the choices of representations R_r in $\mathcal{R} = \bigotimes_{r=1}^N R_r$. For special choices of the representations R_r one may get a finite-dimensional space $\mathrm{CB}(C, \mathcal{R})$ of conformal blocks.

Example 1 *As an example let us consider the case $n = 4$, $R_r = V_{\alpha_r}$, where V_α is the irreducible highest weight representation of Vir_c generated from a highest weight vector e_α satisfying $L_n e_\alpha = \delta_{n,0} \alpha(Q - \alpha) e_\alpha, n \geq 0$ if $c = 1 + 6Q^2$. If Q is parameterized as $Q = b + b^{-1}$, and α_2 is chosen to be equal to $-b/2$, one finds that there is the relation $(b^2 L_{-2} + L_{-1}^2) e_{\alpha_2} = 0$ in V_{α_2}.*

Exercise 3 Demonstrate that this relation, combined with the conformal Ward identities (2.14) implies that the space of conformal blocks is two-dimensional in this case.

Conformal field theories where all spaces of conformal blocks are finite-dimensional are called rational conformal field theories. Such CFTs are technically easier to study, but represent only a small subclass of the CFTs of interest for theoretical physics and mathematics.

2.1.5 Variations of insertion points

The definition of the conformal blocks is not yet quite complete.

2.1.5.1 Completing the definition of conformal blocks

We shall complete the definition of the conformal blocks by adding the requirement that

$$f_C(L_{-1}^{(r)} v) = \partial_{z_r} f_C(v). \tag{2.17}$$

This is necessary to get

$$\partial_z \Phi_v(z, \bar{z}) = [L_{-1}, \Phi_v(z, \bar{z})] = \Phi_{L_{-1} v}(z, \bar{z}). \tag{2.18}$$

The second equality in (2.18) is a consequence of the state–operator correspondence together with $L_{-1}|0\rangle = 0$. Consistency with (2.10) requires that we adopt (2.17).

Note that the operation mapping a conformal block f_C to the conformal block taking values $f_C(L_{-1}^{(r)}v)$ on vectors $v \in \mathcal{R}$ is a linear operator on $\mathrm{CB}(C,\mathcal{R})$. One may therefore read (2.17) as the definition of a flat connection on the bundle of conformal blocks over $\mathcal{M}_{0,n}$.

2.1.5.2 Consequences

Let us introduce the *chiral partition function* $\mathcal{Z}_f(C)$ as the value $f_C(e)$ of f_C on $e = \bigotimes_{r=1}^n e_r$: the product of highest weight vectors. Keeping in mind that the value of the conformal blocks f_C can be expressed in terms of the values f_C has on vectors w of the form $w = \bigotimes_{r=1}^n (L_{-1})^{l_r} e_r$, one sees that the conformal Ward identities supplemented by the definition (2.17) allow us to express the values $f_C(v)$ on arbitrary $v \in \mathcal{R}$ as multiple derivatives of the chiral partition functions $\mathcal{Z}_f(C)$.

Exercise 4 Show that for $C_{0,n} = \mathbb{P}^1 \setminus \{z_n, \ldots, z_1\}$, $C_{0,n+1} = C_{0,n} \setminus \{y\}$, we have

$$f_{C_{0,n+1}}(L_{-2}e_0 \otimes e_n \otimes \cdots \otimes e_1) = \sum_{r=1}^n \left(\frac{\Delta_r}{(y-z_r)^2} + \frac{1}{y-z_r}\partial_r \right) \mathcal{Z}_f(C). \quad (2.19)$$

The resulting Ward identities for non-chiral correlation functions can be written in the form

$$\left\langle T(y) \prod_{r=1}^n \Phi_{v_r}(z_r, \bar{z}_r) \right\rangle = \sum_{r=1}^n \left(\frac{\Delta_r}{(y-z_r)^2} + \frac{1}{y-z_r}\partial_r \right) \left\langle \prod_{r=1}^n \Phi_{v_r}(z_r, \bar{z}_r) \right\rangle, \quad (2.20)$$

that one often finds in the literature.

Whenever the space of conformal blocks is finite-dimensional due to special properties of the representations involved, one gets systems of differential equations, as the following example illustrates.

Example 1 (continued) *As an example let us return to the case* $n=4$, $R_r = V_{\alpha_r}$, *where* V_α *is the irreducible highest weight representation of* Vir_c *generated from a highest weight vector* e_α *satisfying* $L_n e_\alpha = \delta_{n,0}\alpha(Q-\alpha)e_\alpha$, $n \geq 0$ *if* $c = 1 + 6Q^2$. *For* $Q = b + b^{-1}$, *and* $\alpha_2 = -b/2$ *we had previously noted the relation* $(b^2 L_{-2} + L_{-1}^2)e_{\alpha_2} = 0$ *in* V_{α_2}. *We may, without loss of generality, assume that* $(z_1, z_2, z_3, z_4) = (0, z, 1, \infty)$ *and denote* $C_z = \mathbb{P}^1 \setminus \{0, z, 1, \infty\}$. *Use the conformal Ward identities to show that the chiral partition functions* $\mathcal{Z}_f(z) \equiv \mathcal{Z}_f(C_z)$ *satisfy the differential equation* $\mathcal{D}_{\mathrm{BPZ}} \mathcal{Z}_f(z) = 0$, *where*

$$\mathcal{D}_{\mathrm{BPZ}} = \frac{1}{b^2}\frac{d^2}{dz^2} + \frac{2z-1}{z(1-z)}\frac{d}{dz} + \frac{\Delta_1}{z^2} + \frac{\Delta_3}{(1-z)^2} + \frac{\kappa}{z(1-z)}, \quad (2.21)$$

using the notations $\kappa = \Delta_1 + \Delta_2 + \Delta_3 - \Delta_4$ *and* $\Delta_r = \alpha_r(Q - \alpha_r)$ *for* $r = 1, \ldots, 4$. *Show that the equation* $\mathcal{D}_{\mathrm{BPZ}} \mathcal{Z}_f(z) = 0$ *can be reduced to the hypergeometric differential* equation $z(1-z)F'' + [c-(a+b+1)z]F' - abF = 0$, and that the two-dimensional space of solutions can be identified with the space of conformal blocks in this case.

We shall later use this example further.

2.1.5.3 Geometric meaning

To reformulate this observation in a more geometric way, let us introduce the space

$$\mathcal{M}_{0,n} = \{(z_1,\ldots,z_n) \in (\mathbb{P}^1)^n; z_r \neq z_s, \ r,s = 1,\ldots,n\}/\mathrm{PSL}(2,\mathbb{R}), \qquad (2.22)$$

where elements of the group $\mathrm{PSL}(2,\mathbb{R})$ act via Möbius transformations $z_r \to \frac{az_r+b}{cz_r+d}$. The space $\mathcal{M}_{0,n}$ can be identified with the moduli space of Riemann surfaces $C_{0,n}$ of genus 0 and n marked points. The universal cover $\widetilde{\mathcal{M}}_{0,n}$ of the space $\mathcal{M}_{0,n}$ can be identified with the Teichmüller space $\mathcal{T}_{0,n}$, the space of deformations of the complex structures on $C_{0,n}$.

One may observe that the values $f_C(\mathbf{w_n})$ characterizing a conformal block f_C can be identified with the Taylor expansion coefficients of $\mathcal{Z}_f(z)$ around the values $z = (z_1,\ldots,z_n)$ defining the given Riemann surface C. Considering the case $n=4$ as an example, let us recall that the conformal blocks f are in this case fully characterized by the infinite collection of complex numbers $\mathcal{Z}_k = f(e_4 \otimes e_3 \otimes L^k_{-1} e_2 \otimes e_1)$. We may represent $C_{0,4}$ as $C_z \simeq \mathbb{P}^1 \setminus \{\infty, 1, z, 0\}$, allowing us to identify the parameter z in this description as a coordinate for $\mathcal{M}_{0,4}$. The definition (2.17) relates the numbers $\mathcal{Z}_k$ to the derivatives $\partial_z^k \mathcal{Z}_f(z)$. Any locally defined function $\mathcal{Z}(z)$ on $\mathcal{T}_{0,4}$ defines a conformal block in this way. This observation is easily generalized to $n > 4$.

Remark 1. The converse to this statement is not true, in general. Given the collection of complex numbers f_k characterizing a conformal block on $C_{0,4}$, one gets a function $\mathcal{Z}_f$ defined in a neighbourhood of the point with coordinate z in $M_{0,4}$ only if the Taylor series $\sum_{k=0}^{\infty} t^k \mathcal{Z}_k/k!$ converges, which will not be the case for arbitrary solutions of the conformal Ward identities. However, for physical applications one will *not* be interested in the most general solution of the conformal Ward identities in the purely algebraic sense, but rather in solutions for which the corresponding chiral partition functions can be analytically continued over $\mathcal{T}_{0,n}$. The space of all such 'well-behaved' conformal blocks generates a subspace of the space of all algebraic solutions to the conformal Ward identities (2.14).

2.1.6 Conformal blocks versus vertex operators

Considering the special case $(z_3, z_2, z_1) = (\infty, z, 0)$, $\mathcal{R} = R_3 \otimes R_2 \otimes R_1$, one may use the conformal blocks $f_{C_{0,3}}$ to define families of operators $V_\rho(v_2, z) : R_1 \to R_3$ labelled by vectors $v_2 \in R_2$ and triples $\rho = \begin{bmatrix} {}_{R_3}{}^{R_2}{}_{R_1} \end{bmatrix}$ of representations, such that

$$f_{C_{0,3}}(v_3 \otimes v_2 \otimes v_1) = \langle v_3, V_\rho(v_2, z) v_1 \rangle_{R_3}. \qquad (2.23)$$

The operators $V_\rho(v_2, z)$ defined in this way are called *chiral vertex operators*. When this becomes relevant we will make the dependence on the triple of representations ρ involved more explicit, writing $V\begin{bmatrix}R_3 & R_2 & R_1\end{bmatrix}(v_2, z)$ instead of $V_\rho(v_2, z)$.

Exercise 5

(a) Demonstrate that the definition (2.23) together with the conformal Ward identities imply the following commutation relations:

$$L_n V_\rho(v_2|z) - V_\rho(v_2|z) L_n = \sum_{k=-1}^{\infty} \binom{n+1}{k+1} z^{n-k} V_\rho(L_k v_2|z). \quad (2.24)$$

(b) Demonstrate that the vertex operators $V_\rho(v|z)$ are uniquely determined by the relations (2.24) up to multiplication by a constant that may depend on ρ and $v \in R_2$.

(c) Use the conformal Ward identities to demonstrate that the vertex operators $V_\rho(v|z)$ furthermore satisfy the relations

$$V_\rho(L_{-2}v|z) = T_<(z) V_\rho(v|z) + V_\rho(v|z) T_>(z), \quad (2.25a)$$
$$V_\rho(L_{-1}v|z) = \partial_z V_\rho(v|z), \quad (2.25b)$$

where $T_<(z)$ and $T_>(z)$ are defined as

$$T_<(z) = \sum_{n \leq -2} L_n z^{-n-2}, \qquad T_>(z) = \sum_{n > -2} L_n z^{-n-2}. \quad (2.26)$$

Relations (2.25) allow us to express $V_\rho(v|z), v \in R_2$ in terms of $V_\rho(z) \equiv V_\rho(e_2|z)$, where e_2 is the highest weight vector of the representation R_2. It follows that the chiral vertex operators $V_\rho(v|z)$ are uniquely defined by (2.24) and (2.25) up to a constant that may depend on ρ.

(d) Verify that any vertex operator satisfying (2.24) and (2.25) defines a conformal block via (2.23).

In the special case where $v = e_2$, the highest weight vector of the representation R_2, one calls the vertex operator $V_\rho(z) \equiv V_\rho(e_2|z)$ a chiral *primary field*. The relations (2.24) simplify to

$$L_n V_\rho(z) - V_\rho(z) L_n = z^n (z \partial_z + \Delta_{R_2}(n+1)) V_\rho(z), \quad (2.27)$$

using (2.17) and $L_0 e_2 = \Delta_{R_2} e_2$.

The three-point functions of the physical vertex operators $\Phi_v(z, \bar{z})$ satisfy the conformal Ward identities. It follows that $\Phi_v(z, \bar{z})$ can be decomposed as a sum over products of chiral vertex operators for the representations of the Virasoro algebras generated by L_n and $\bar{L}_n$, respectively.

Exercise 6 Assume that the Hilbert space $\mathcal{H}$ has the form

$$\mathcal{H} = \bigoplus_{R \in \mathcal{O}} \mathcal{H}_R, \qquad \mathcal{H}_R = M_R \otimes R, \tag{2.28}$$

where $\mathcal{O}$ is a set of unitary highest weight representations of the Virasoro algebra containing each representation R only once, and M_R is a multiplicity space on which the Virasoro generators L_n, $n \in \mathbb{Z}$ act trivially (the $\bar{L}_n$ may act non-trivially on M_R). Demonstrate the 'conformal Wigner–Eckart theorem': Let $\rho = \begin{bmatrix} R_2 \\ R_3 & R_1 \end{bmatrix}$ and $v = \mu_2 \otimes v_2 \in \mathcal{H}_{R_2}$. The operator $\Phi_v(z,\bar{z})$ then admits a decomposition into CVOs of the following form:

$$\Pi_{R_3} \cdot \Phi_v(z,\bar{z}) \cdot \Pi_{R_1} = \Xi^{\mu_2}_{R_3,R_1} \otimes V_\rho(v_2,z), \tag{2.29}$$

where Π_R is the orthogonal projection onto $\mathcal{H}_R$ and the operator $\Xi^{\mu_2}_{R_3,R_1} : M_{R_1} \to M_{R_3}$ is z-independent (but will depend on $\bar{z}$, in general).

Another point of view is sometimes helpful. The commutation relations (2.24) suggest defining an action of Virasoro algebra on a tensor product $R_2 \otimes R_1$ of representations by setting

$$L_n(v_2 \otimes v_1) = v_2 \otimes (L_n v_1) + \sum_{k=-1}^{\infty} \binom{n+1}{k+1} z^{n-k}(L_k v_2) \otimes v_1. \tag{2.30}$$

We will denote the Virasoro module defined using (2.30) as $R_2 \boxtimes R_1$. The chiral vertex operators $V_\rho(v_2|z)$ are thereby identified as close relatives of the Clebsch–Gordan maps $C_z[R_3|R_2,R_1]$ intertwining the Virasoro module $R_2 \boxtimes R_1$ defined using (2.30) with the standard action on R_3. Note that we have chosen not to include the dependence on z, manifest in (2.30), in the notation $R_2 \boxtimes R_1$, as this dependence will not be of interest whenever we use this point of view, and since it could be restored quite easily if needed.

This point of view puts conformal field theory into useful analogy with group representation theory, as first emphasized in [8, 9]. Similar constructions can be introduced for extensions of the conformal symmetry like current algebras. They have been used to exhibit profound relations between conformal field theory and quantum group representation theory [10].

2.1.7 Localizing conformal blocks[+]

It is interesting and sometimes useful to use an alternative representation in which conformal blocks in $\mathrm{CB}(C,\mathcal{R})$ are represented by elements of the dual V'_0 to the vacuum representation V_0 characterized by a modified invariance condition. This means that all information on the conformal blocks can be encoded in a vector in V'_0.

To see this, let us first recall (from Exercise 2b) that the conformal Ward identities allow us to represent conformal blocks with insertions of the vacuum representation in terms of the conformal blocks without such insertions. Let us consider

$\mathrm{CB}(C_{0,n+1}, \mathcal{R} \otimes V_0)$, with $C_{0,n+1} = C_{0,n} \setminus \{z_0\}$ and vacuum representation V_0 associated to z_0. We had previously seen that the values of $f_{C_{0,n+1}}$ on arbitrary vectors can be expressed in terms of the values $f_{C_{0,n+1}}(L_{-1}^{k_n} e_n \otimes \cdots \otimes L_{-1}^{k_1} e_1 \otimes v)$, with $v \in V_0$ and $k_r \in \mathbb{Z}_{\geq 0}$, $r = 1, \ldots, n$. Using the vector fields

$$\xi_r(y) \frac{\partial}{\partial y} = \left(\frac{y - z_0}{z_r - z_0}\right)^{3-r} \prod_{\substack{s=1 \\ s \neq r}}^{n} \frac{y - z_s}{z_r - z_s} \frac{\partial}{\partial y}, \tag{2.31}$$

one may compute $f_{C_{0,n+1}}(L_{-1}^{k_n} e_n \otimes \cdots \otimes L_{-1}^{k_1} e_1 \otimes v)$ in terms of $f_{C_{0,n+1}}(e_n \otimes \cdots \otimes e_1 \otimes v')$ for some $v' \in V_0$. The functional $g : V_0 \to \mathbb{C}$ defined by

$$g(v) := f_{C_{0,n+1}}(e_n \otimes \cdots \otimes e_1 \otimes v) \qquad \forall v \in V_0, \tag{2.32}$$

satisfies a variant of the conformal Ward identities of the form

$$g(T_\xi v) + \sum_{r=1}^{n} \xi_0^{(r)} \Delta_r g(v) = 0, \tag{2.33}$$

for all vector fields $\xi \in \mathsf{Vect}(\mathbb{P}^1 \setminus \{z_0\})$ which vanish at z_r for all $r = 1, \ldots, n$. Using the above observations it is not hard to show that equations (2.33) define a subspace of the dual V_0' of V_0 isomorphic to $\mathrm{CB}(C_{0,n+1}, \mathcal{R} \otimes V_0)$, which is furthermore isomorphic to $\mathrm{CB}(C_{0,n}, \mathcal{R})$ by the propagation of vacua.

One may furthermore notice that the conformal Ward identities (2.14) specialized to the vector fields ξ_r defined in (2.31) combined with (2.17) imply that g satisfies identities of the form

$$\partial_{z_r} g(v) = g(\mathsf{L}_{-1}^{(r)} v), \tag{2.34}$$

where $\mathsf{L}_{-1}^{(r)}$ are operators on V_0 satisfying $[\mathsf{L}_{-1}^{(r)}, \mathsf{L}_{-1}^{(s)}] = 0$ for all $r, s = 1, \ldots, n$.

Exercise 7

(a) Let $t_g(x) \equiv g(T(x - z_0)e_0)$ be the expectation value of the energy–momentum tensor. Given that g satisfies (2.33) and (2.34), demonstrate that $t_g(x)$ extends to a function that is meromorphic on $C_{0,n}$ satisfying

$$t_g(x) = \sum_{r=1}^{n} \left(\frac{\Delta_r}{(x - z_r)^2} + \frac{1}{x - z_r} \partial_r\right) g(e_0). \tag{2.35}$$

(b) Let $t_f(x) = f(e_n \otimes \cdots \otimes e_2 \otimes T(x - z_1)e_1)$. Show that $t_f(x) = t_g(x)$ if g is defined from a conformal block $f_{C_{0,n}}$ as above.

We conclude that there is a one-to-one correspondence between linear functionals $g : V_0 \to \mathbb{C}$ satisfying (2.33) and conformal blocks $f_{C_{0,n}} \in \mathrm{CB}(C_{0,n}, \mathcal{R})$. All information on the conformal block $f_{C_{0,n}}$ on the n-punctured sphere $C_{0,n}$ can be 'localized' within the element $g \in V_0'$ assigned to an arbitrary point z_0 on $C_{0,n}$. This type of representation for the conformal blocks will be referred to as one-point localization. The operators $\mathsf{L}_{-1}^{(r)}$ appearing in (2.34) realize an infinitesimal motion of the puncture at z_r within the one-point localization.

2.1.8 Higher genus conformal blocks[+]

We will now briefly discuss how the conformal Ward identities can be generalized when $C_{0,n}$ is replaced by a Riemann surface $C \equiv C_{g,n}$ of higher genus g. For simplicity we will here restrict attention to the case where $n = 1$ with vacuum representation V_0 inserted at a point $P \in C$.

2.1.8.1 Conformal Ward identities

Let (C, P, x) be a Riemann surface C with a marked point P, and a coordinate x on a disc D around P such that $x(P) = 0$. We may in this case define the conformal blocks as elements of the dual V_0' of V_0 satisfying

$$f(T_\xi v) = 0, \quad \forall \xi \in \mathsf{Vect}(C \setminus P), \quad \forall v \in V_0, \tag{2.36}$$

where T_ξ is defined as

$$T_\xi = \sum_{n \in \mathbb{Z}} \xi_n L_n, \tag{2.37}$$

if $\xi = \xi(x)\frac{\partial}{\partial x} \in \mathsf{Vect}(C \setminus P)$, and $\xi(x) = \sum_{n \in \mathbb{Z}} \xi_n x^{n+1}$ is the Laurent expansion around $x = 0$. The space of conformal blocks, defined as the solutions to (2.36), may be finite-dimensional in some cases (minimal models, see below), but will be infinite-dimensional in general.

In order to understand the consequences of (2.36) more concretely, let us use the following consequence of the Riemann–Roch theorem, in this form proven in [11, Appendix B]. For generic[4] points P on C there exist bases for $H^0(C \setminus P, K^2)$ and $H^0(C \setminus P, K^{-1}) \equiv \mathsf{Vect}(C \setminus P)$, respectively, generated by elements having Laurent expansions around $x = 0$ of the form

$$q_n(x)(dx)^2 = \left(x^{n-2} + \sum_{m \geq h+2} Q_{nm} x^{m-2} \right) (dx)^2, \quad n \leq h+1 \tag{2.38}$$

---

[4] If P is not a Weierstrass point, see [11].

$$\xi_n(x)\frac{\partial}{\partial x} = \left(x^{n+1} + \sum_{m \geq -h-1} V_{nm} x^{m+1}\right)\frac{\partial}{\partial x}, \qquad n \leq -h-2, \qquad (2.39)$$

where $h = 3g - 3$. The elements satisfy

$$\int_C q_n \xi_m = 0, \qquad \forall\, m \leq -h-2, \quad \forall\, n \leq h+1, \qquad (2.40)$$

where C is a small circle around P.

It then follows from (2.36) that we can express the values of $f(v)$ on arbitrary $v \in V_0$ in terms of the special values $f_{\mathbf{n}} := f(L_{-h-1}^{n_h} L_{-h}^{n_{h-1}} \cdots L_{-2}^{n_1} e_0)$, where $\mathbf{n} = (n_1, \ldots, n_h)$.

2.1.8.2 Variations of the conformal blocks

We are now going to show that variations of the defining data (C, P, x) can be represented using the Virasoro action on V_0. Let us consider general infinitesimal variations of a conformal block f defined as

$$(\delta_\zeta f)(v) = f(T_\zeta v), \qquad \zeta \in \mathsf{Vect}(A), \qquad (2.41)$$

where A is the annulus $D \setminus P$, and T_ζ is defined by replacing ξ by ζ in (2.37). Our goal will be to identify conditions allowing us to interpret $g := (1 + \delta_\zeta) f$ as a conformal block associated to an infinitesimal variation of the data (C, P, x) defining f. To this aim let us note that

$$[T_\zeta, T_\xi] = -T_{[\zeta,\xi]} + \frac{c}{12}(\zeta,\xi), \qquad (\zeta,\xi) := \frac{1}{2\pi i}\int_C dx\, \zeta'''(x)\xi(x), \qquad (2.42)$$

leading to

$$g(T_\xi v) = f(T_\zeta T_\xi v) = -f(T_{[\zeta,\xi]}v) + \frac{c}{12}(\zeta,\xi) f(v). \qquad (2.43)$$

If we choose ζ such that the bilinear form (ξ, ζ) vanishes for all $\xi \in \mathsf{Vect}(C \setminus P)$, we find that g satisfies

$$g(T_{\xi + [\zeta,\xi]} v) = 0 + \mathcal{O}(\zeta^2), \qquad (2.44)$$

the conformal Ward identity for an infinitesimal variation of the vector fields ξ generated by the adjoint action of a vector field ζ.

We may next notice that variations of the data (C, P, x) will induce variations of the $\xi \in H^0(C \setminus P, K^{-1})$ appearing in (2.36) which can be represented by the adjoint action of suitable vector fields $\zeta \in \mathsf{Vect}(A)$. Infinitesimal variations of (P, x) can be represented by a vector field ζ that is holomorphic on D. In order to see that all variations of the complex structure of C can be represented in this way one may use the Virasoro uniformization theorem describing the Teichmüller spaces $\mathcal{T}(C)$ in terms of vector fields on an annulus. This theorem states that the Teichmüller space $\mathcal{T}(C)$ can be represented as double quotient,

$$\mathcal{T}(C) = \mathsf{Vect}(C\setminus P) \setminus \mathsf{Vect}(A) / \mathsf{Vect}(D). \tag{2.45}$$

A proof can be found e.g. in [7, Section 17] or [12, Section 2.4].

It remains to investigate the conditions for having $(\zeta,\xi) = 0$ for all $\xi \in \mathsf{Vect}(C\setminus P)$ in (2.43). According to (2.40) this will be the case if $\zeta'''(x)(dx)^2 \in H^0(C\setminus P, K^2)$. This condition is satisfied if ζ is an element of the subspace $\mathsf{V}_\mathcal{T}$ of $\mathsf{Vect}(A)$ spanned by the vector fields

$$\zeta_n(x)\frac{\partial}{\partial x} = \left(\frac{x^{n+1}}{v_n} + \sum_{m\geq h+2} Q_{nm}\frac{x^{m+1}}{v_m}\right)\frac{\partial}{\partial x}, \quad 2 \leq |n| \leq h+1, \tag{2.46}$$

where $v_n = n(n^2 - 1)$ and the coefficients Q_{nm} have been introduced in (2.38). The subspace $\mathsf{V}_\mathcal{T}$ is $6g - 6$-dimensional and carries a non-degenerate symplectic form given by the restriction of the form $(.,.)$. The Virasoro uniformization theorem identifies $\mathcal{T}(C)$ as a quotient of $\mathsf{V}_\mathcal{T}$ by the subspace generated by the ζ_n with $2 \leq n \leq h+1$. This subspace is isomorphic to the space $H^0(C, K^2)$ of quadratic differentials on C, which is canonically isomorphic to the cotangent fibre $T^*\mathcal{T}(C)$ of $\mathcal{T}(C)$. We thereby identify $\mathsf{V}_\mathcal{T}$ with the total space of the cotangent bundle $T^*\mathcal{T}(C)$.

There is an interesting reformulation of the definitions above in terms of the chiral partition function $\mathcal{Z}_f := f(e_0)$, and the expectation value of the energy–momentum tensor $\langle T(x)\rangle_f = f(T(x)e_0)/f(e_0)$. It can be shown [7, Section 9.2] that the defining Ward identity (2.36) is *equivalent* to the condition that $t_f(x) \equiv \langle T(x)\rangle_f$ defines a holomorphic projective c-connection on C, which means that the transformation of $t_f(x)$ from one patch on C with coordinate x to another patch with coordinate y is represented as

$$t_f(x) = (y'(x))^2 \tilde{t}_f(y(x)) + \frac{c}{12}\{y,x\}, \quad \{y,x\} = \frac{y'''}{y'} - \frac{3}{2}\left(\frac{y''}{y'}\right)^2. \tag{2.47}$$

One may furthermore note that the definition of the canonical connection (2.41) relates multiple derivatives of the chiral partition function $\mathcal{Z}_f$ with respect to the complex structure moduli of C to the defining data of the conformal blocks, the values of f on a sufficiently large collection of vectors in V_0. First-order derivatives of $\mathcal{Z}_f$, in particular, are given by the expectation values $\langle T_\zeta\rangle_f = \frac{1}{2\pi i}\int_C dx\, t_f(x)\zeta(x)$.

2.1.8.3 Projective flatness

We have seen that we may describe variations of the complex structure of C in the form (2.41). This defines a connection on the bundle of conformal blocks over $\mathcal{M}(C)$, the moduli space of complex structures on C. The connection is not flat, but only projectively flat, as the form (ζ_1, ζ_2) will be non-vanishing for general elements ζ_1, ζ_2 of $V_\mathcal{T}$. We are now going to discuss the implications of the projective flatness in a little more detail.

The variations δ_ζ may[5] be integrable if one restricts the choice of the vector fields ζ to Lagrangian subspaces $\mathsf{L}_\mathcal{T}$ of $\mathsf{V}_\mathcal{T}$ spanned by elements satisfying $(\zeta_1, \zeta_2) = 0$ for all $\zeta_1, \zeta_2 \in \mathsf{L}_\mathcal{T}$. It follows from the above that such Lagrangian subspaces are isomorphic to $\mathcal{T}(C)$, but not canonically so. The definition of chiral partition functions $\mathcal{Z}_f$ by means of integration of the canonical connection therefore depends on the choice of a Lagrangian subspace in $\mathsf{V}_\mathcal{T}$. One should note that the resulting ambiguity affects the chiral partition functions of all conformal blocks in the same way. Modifying the choice of a Lagrangian subspace will modify the chiral partition functions of all conformal blocks by multiplication with the same locally defined function.

Having chosen families of Lagrangians in $\mathsf{V}_\mathcal{T}$, varying holomorphically over open subsets $M \subset \mathcal{M}(C)$ may allow us to define chiral partition functions $\mathcal{Z}_{f,M}$ on M by integrating the parallel transport defined using (2.41). The Lagrangians used to define the parallel transport in two neighbourhoods M and N within $\mathcal{M}(C)$ may differ on the overlap $M \cap N$. Keeping in mind that we have $\delta_\zeta \log \mathcal{Z}_f = \langle T_\zeta \rangle_f$, and noting that a change of the choice of Lagrangians in $\mathsf{V}_\mathcal{T}$ changes the expectation values $t_f(x)$ for all conformal blocks f by addition of the same quadratic differential, it is easy to see that the partition functions $\mathcal{Z}_{f,M}$ and $\mathcal{Z}_{f,N}$ defined in this way in two neighbourhoods M and N differ by an overall factor, $\mathcal{Z}_{f,M} = \chi_{MN} \mathcal{Z}_{f,N}$, with χ_{MN} being a function on $M \cap N \subset \mathcal{M}(C)$ independent of the choice of f. The function χ_{MN} is defined by the choices used to define $\mathcal{Z}_{f,M}$ and $\mathcal{Z}_{f,N}$ up to an overall multiplicative constant. On triple overlaps we can therefore only require a weakened form of the usual consistency condition, $\chi_{MN} \chi_{NO} \chi_{OM} = \eta_{MNO}$, with η_{MNO} being a constant which is easily seen to be representable in the form $\eta_{MNO} = e^{\pi i c v_{MNO}}$. The collection of functions χ_{MN} associated to a cover of $\mathcal{M}(C)$ defines what is called a projective line bundle $\mathcal{E}_c$ over $\mathcal{M}(C)$ in [13].

By integrating the canonical connection, one can locally define bases for the bundle of conformal blocks spanned by horizontal sections. One may thereby define a holomorphic vector bundle $\mathcal{V}_c$ over $\mathcal{M}(C)$. More canonically defined is the projective vector bundle $\mathcal{W}_c$ over $\mathcal{M}(C)$ obtained from $\mathcal{V}_c$ by taking the tensor product with the projective line bundle $\mathcal{E}_c^{-1}$. This removes the ambiguities from the choices in the local integration of the canonical connection; the price to pay is that the consistency conditions for transition functions of $\mathcal{W}_c$ on triple overlaps are satisfied only up to a constant multiple of the identity.

The above discussion, together with its continuation in Section 2.2.6.4, reproduces the key ingredients of the perspective on conformal field theory proposed in [13]. The ambiguity observed in the definition of the horizontal sections can be physically interpreted as a generically unavoidable dependence on the choice of a renormalization scheme in the definition of the energy–momentum tensor on higher genus surfaces [13].

---

[5] Another condition for being integrable is well-behavedness of the conformal blocks in the sense explained in Remark 1 at the end of Section 2.1.5.3, which will be assumed in the following.

2.1.9 Extensions of conformal symmetry++

We will see that conformal symmetry alone provides too little information to solve conformal field theories completely, in general. There are, however, several cases where the conformal symmetry is extended to a larger algebraic structure, called a vertex operator algebra (VOA), allowing us to obtain further information or even a complete solution of more complicated CFTs. We shall not attempt to give a complete treatment, but mention a few examples, and try to indicate the main idea behind the definition of VOAs.

Extensions of the conformal symmetry can be generated by the Laurent expansion coefficients of a set of fields $W^{(i)}(z)$,

$$W^{(i)}(z) = \sum_{m \in \mathbb{Z}} W_m^{(i)} z^{-m-\Delta^{(i)}}, \tag{2.48}$$

with i being an index labelling the different fields, and the parameter $\Delta^{(i)}$ is called the spin of $W^{(i)}(z)$. The set of fields contains the energy–momentum tensor $T(z)$, and it is assumed that

$$[L_n, W_m^{(i)}] = (n(\Delta^{(i)} - 1) - m) W_{n+m}^{(i)}, \tag{2.49}$$

for the modes of all fields $W^{(i)}(z)$ except $T(z)$.

2.1.9.1 Examples

(i) *Free boson algebra* Generated by modes a_n, $n \in \mathbb{Z}$ satisfying $[a_n, a_m] = \frac{n}{2}\delta_{n+m,0}$. Out of a representation of the free boson algebra one may construct a one-parameter family of representations of the Virasoro algebra using

$$\begin{aligned} L_n &= \mathrm{i}(n+1)Q a_n + \sum_{k \in \mathbb{Z}} a_k a_{n-k}, \quad n \neq 0, \\ L_0 &= a_0^2 + \mathrm{i} Q a_0 + 2 \sum_{k > 0} a_{-k} a_k. \end{aligned} \tag{2.50}$$

The representations of the Virasoro algebra defined in this way will have central charge $c = 1 + 6Q^2$. If $\mathcal{F}_\alpha$ is a representation in which the central element a_0 is represented as multiplication by $-\mathrm{i}\alpha$ times the identity operator one gets a representation of the Virasoro algebra with highest weight $\Delta_\alpha = \alpha(Q - \alpha)$.

(ii) *Affine Lie algebra* Generated by fields $\mathcal{J}^a(z)$ with spin 1, having Laurent modes with relations

$$[\mathcal{J}_n^a, \mathcal{J}_m^b] = \mathrm{i} f^{abc} \mathcal{J}_{n+m}^c + \mathsf{k} \eta^{ab} n \delta_{n,-m}, \tag{2.51}$$

where f^{abc} are the structure constants of the semi-simple Lie algebra $\mathfrak{g}$ with generators T^a, relations $[T^a, T^b] = \mathrm{i} f^{abc} T^c$, and invariant bilinear form $(T^a, T^b) = \eta^{ab}$. $\mathbf{k}$ is the central element. Fixing $\mathbf{k}$ to a value k defines the affine Lie algebra $\hat{\mathfrak{g}}_k$. The Virasoro algebra gets embedded into the universal enveloping algebra $\mathcal{U}(\hat{\mathfrak{g}}_k)$ of the affine Lie algebra by means of the Sugawara construction.

(iii) *W algebras* The algebras $\mathcal{W}_N$, $N \geq 3$ are generated by the Laurent modes of fields $W^{(i)}(z)$, $i = 2, \ldots, N$ with $W^{(2)} = T(z)$, having complicated commutation relations. In the case $N = 3$ we have, for example [14],

$$[W_n, W_m] = \frac{c}{3 \cdot 5!}(n^2 - 1)(n^2 - 4)\delta_{n,-m} + \frac{16}{22 + 5c}\Lambda_{n+m} \qquad (2.52)$$
$$(n - m)\left(\frac{1}{15}(n+m+2)(n+m+3) - \frac{1}{6}(n+2)(m+3)\right)L_{n+m},$$

where $\Lambda_n = \frac{1}{5}x_n L_n + \sum_{k \in \mathbb{Z}} : L_{n-k} L_k :$, $x_{2l} = 1 - l^2$, $x_{2l+1} = (2+l)(1-l)$. We see from the example (2.52) that the modes of the $\mathcal{W}_3$-algebra do not generate a Lie algebra. The algebras $\mathcal{W}_N$ for $N > 3$ are even more complicated, and therefore best defined using the quantum Drinfeld–Sokolov reduction, see [7] and references therein.

2.1.9.2 Vertex operator algebras[++]

A flexible framework for the description of extended chiral symmetries in CFT is provided by the notion of a vertex operator algebra (VOA). The concept of a VOA may be motivated by the state–operator correspondence. Given an extended symmetry algebra, it is natural to consider the vector space $\mathcal{A}$ generated by its action on the vacuum vector $|0\rangle$. It is then natural to label the currents $V(a, z)$ of a VOA by vectors $a \in \mathcal{A}$. In the axiomatic definition of VOA given in the mathematical literature (see e.g. [7, 15, 16] and references therein), one considers the currents $V(a, z)$ as formal power series[6]

$$V(a, z) = \sum_{n \in \mathbb{Z}} \mathsf{a}_n z^{-n-1}, \qquad (2.53)$$

with coefficients a_n being linear operators on $\mathcal{A}$. These data satisfy certain axioms, including the conditions $\lim_{z \to 0} V(a, z)|0\rangle = a$, $V(|0\rangle, z) = \mathrm{id}$ and $[L_{-1}, V(a, z)] = \partial_z V(a, z)$. The most important axiom is the locality axiom stating that the two formal power series $V(a, z)V(b, w)$ and $V(b, w)V(a, z)$ coincide after multiplying them with a large enough power of $z - w$. This is equivalent to the condition that the commutator $[V(a, z), V(b, w)]$ can be expanded into a finite sum of derivatives of the delta distribution supported on $z = w$.

It is possible to show that the vector space spanned by the modes a_n has a natural Lie algebra structure [7, Chapter 4]. There may, however, be additional relations among the

---

[6] Note that the conventions for mode expansions in the VOA literature often differ from those used in (2.48).

modes a_n, expressing some of them as composites of others. Examples are the relations expressing Λ_n in terms of the L_k in the case of the W_3-algebra defined above.

A simple generalisation of the notion of a VOA is a super VOA. The simplest example of a super VOA is generated from N species of *free fermions*, generated by the modes of fields $\psi_s(z)$, $\bar{\psi}_s(z)$, $s = 1, \ldots, N$,

$$\psi_s(z) = \sum_{n \in \mathbb{Z}} \psi_{s,n} z^{-n-1}, \quad \bar{\psi}_s(z) = \sum_{n \in \mathbb{Z}} \bar{\psi}_{s,n} z^{-n}, \qquad (2.54)$$

having modes satisfying the anti-commutation relations

$$\{\psi_{s,n}, \bar{\psi}_{t,m}\} = \delta_{s,t} \delta_{n,m}, \quad \{\psi_{s,n}, \psi_{t,m}\} = 0, \quad \{\bar{\psi}_{s,n}, \bar{\psi}_{t,m}\} = 0. \qquad (2.55)$$

It is possible to generalize the notion of conformal blocks to more general VOA. This is rather straightforward in genus 0. If a current $W^{(i)}(z)$ has dimension $\Delta^{(i)}$, one simply needs to replace the vector fields ξ in (2.14) by holomorphic $(1 - \Delta^{(i)})$-differentials on $C = C_{0,n}$. The definition of conformal blocks on higher genus surfaces for general VOA can be found in [7].

Connections between the theory of VOA and the operator algebraic approach to conformal field theory have recently been described in [17, 18].

2.2 Bootstrap

We will now describe a construction of large classes of conformal blocks from simpler building blocks, called gluing construction. The conformal blocks that can be constructed in this way will be argued to coincide with the conformal blocks of interest for physics. These are the conformal blocks which can appear in factorized representations of the form (2.10) for physical correlation functions. The problem of constructing the correlation functions of a CFT is thereby disentangled into a kinematic part, the construction of the conformal blocks, solved completely by exploiting only the symmetry constraints, and the problem of assembling the conformal blocks into single-valued Euclidean correlation functions. The coefficients $C_{\beta'\beta''\mathbf{m}}$ appearing in the expansion (2.10) carry the main dynamical information on a CFT, and are therefore of central interest in physical applications. We will identify general consistency constraints on these coefficients, and briefly discuss a family of cases where explicit solutions to these constraints are known.

2.2.1 Gluing construction

Our next goal will be to describe a recursive construction of large families of conformal blocks. For some cases it is known that the resulting families of conformal blocks generate bases for the relevant (sub-)spaces of the spaces of conformal blocks.

The gluing construction we are going to present is based on a geometric operation producing an n-punctured sphere $C = \mathbb{P}^1 \setminus \{x_1, \ldots, x_n\}$ by gluing two spheres C_1 and C_2 with smaller numbers of punctures. For convenient notation let us split $\{x_1, \ldots, x_n\} = I_1' \cup I_2 \cup I_1''$, where $I_1' = \{x_1, \ldots, x_l\}$, $I_2 = \{x_{l+1}, \ldots, x_m\}$, and $I_1'' = \{x_{m+1}, \ldots, x_n\}$. If $l = 0$ we set $I_1' = \emptyset$, and similarly $I_1'' = \emptyset$ for $m = n$. We then consider the following two punctured spheres,

$$C_1 = \mathbb{P}^1 \setminus (\{x_i, i \in I_1\} \cup \{x\}), \qquad C_2 = \mathbb{P}^1 \setminus (\{\infty\} \cup \{y_i, i \in I_2\}), \qquad (2.56)$$

with $I_1 = I_1' \cup I_1''$, and the positions of the punctures y_i for $i \in I_2$ are related to x_i via

$$y_i = q^{-1}(x_i - x). \qquad (2.57)$$

We want to describe how the n-punctured sphere $C_{0,n}$ can be represented as the result of a gluing operation applied to C_1 and C_2.

Let us cut (sufficiently small) discs out of C_1 and C_2, giving us the open surfaces

$$D_1^\rho = \{z \in C_1; |z - x| \geq \rho\}, \qquad D_2^\rho = \{z \in C_2; |z| \leq q^{-1}\rho\}, \qquad (2.58)$$

with q being a parameter we'll play with below. We are assuming that D_1^ρ contains x_i, $i \in I_1$, and that D_2^ρ contains y_i, $i \in I_2$. After scaling the disc D_2^ρ by a factor of q one may glue it into the hole we've cut out of C_1 to get D_1^ρ. The result of this operation is the Riemann surface

$$C = \mathbb{P}^1 \setminus \{x_n, \ldots, x_1\}, \quad \text{where } x_i = q y_i + x, \text{ if } i \in I_2. \qquad (2.59)$$

The following operation produces a conformal block associated to C from any two given conformal blocks f_{C_1} and f_{C_2} associated to $\left(C_1, \left(\bigotimes_{i \in I_1'} R_i\right) \otimes R \otimes \left(\bigotimes_{i \in I_1''} R_i\right)\right)$, where $I_1 = I_1' \cup I_1''$, and $(C_2, R \otimes \bigotimes_{i \in I_2} R_i)$, respectively.[7] Let us introduce a non-degenerate invariant bilinear form $\langle .,. \rangle_R$ on $R \otimes R$. Such a bilinear form can be defined uniquely by the properties

$$\langle L_{-n} v, w \rangle_R = \langle v, L_n w \rangle_R, \qquad \langle e_R, e_R \rangle_R = 1. \qquad (2.60)$$

We may then use f_{C_2} to define a map $V_{C_2} : \bigotimes_{i \in I_2} R_i \to R$ which satisfies

$$f_{C_2}(v \otimes w) = \langle v, V_{C_2}(w) \rangle_R. \qquad (2.61)$$

---

[7] We adopt the convention that for $I = \emptyset$ we let $R \otimes \left(\bigotimes_{i \in I} R_i\right) = R$ and $\left(\bigotimes_{i \in I} R_i\right) \otimes R = R$.

The conformal block f_C is now defined by setting

$$f_C(w' \otimes v \otimes w'') := f_{C_1}\left(w' \otimes q^{L_0} V_{C_2}(v) \otimes w''\right), \qquad (2.62)$$

for all $v \in \bigotimes_{i \in I_2}^m R_i$, $w' \in \bigotimes_{i \in I'_1} R_i$ and $w'' \in \bigotimes_{i \in I''_1} R_i$ The right-hand side of (2.62) is a series in powers of q which is believed[8] to be convergent.

By using the gluing construction recursively, one may build conformal blocks for any n-puntured Riemann sphere $C_{0,n}$ from the conformal blocks associated to $C_{0,3}$. We had noted already that we have $\dim(\mathrm{CB}(C_{0,3}, \mathcal{R})) \leq 1$ in the case of the Virasoro algebra. The choices made in the gluing construction therefore consist of two types of data:

- The different ways of gluing $C_{0,n}$ from three-punctured spheres are called *pants decompositions*. In order to distinguish pants decompositions related by monodromies,[9] let us also introduce a trivalent graph ending in the points z_r, $r = 1, \ldots, n$, having exactly one vertex v in each $C_{0,3}^v$ obtained in the pants decomposition. These data will be called *gluing patterns*, and denoted by the letter Γ. One may naturally equip the internal edges of Γ with an orientation represented by an arrow ending at the vertex v in the three-punctured sphere $C_{0,3}^v$ taking the role of C_1 in the construction above. Two examples for gluing patterns on $C_{0,4}$ are depicted in Figure 2.1.

- For each edge ϵ of Γ we need to specify the representation R_ϵ to be used in the gluing construction. These data will be collectively denoted by the letter β. The set of all possible assignments of data β to graphs Γ will be denoted by $\mathcal{C}_\Gamma$.

We will use the notation $f_{C,\Gamma}^\beta(v) \equiv \mathcal{F}_{\beta,\mathbf{v}}^\Gamma(\mathbf{z})$ for the conformal blocks constructed in this way using the notation $v = \bigotimes_{r=1}^n v_i$ if $\mathbf{v} = (v_1, \ldots, v_n)$ and $\mathbf{z} = (z_1, \ldots, z_n)$, and suppressing the choice of the representations R_i attached to the punctures z_i in the notations. β only collects the labels of the representations used 'internally' in the gluing construction.

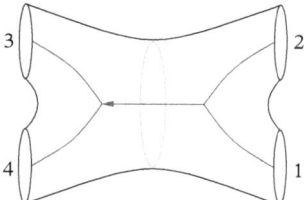 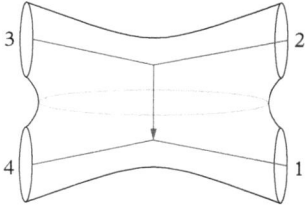

Figure 2.1 *Two gluing patterns on the four-holed sphere obtained from $C_{0,4}$ by removing discs around the four punctures. The gluing patterns on the left and on the right will be referred to as Γ_s and Γ_u (for s- and u-channel), respectively.*

---

[8] This is known to be true in several cases. A general proof has not been given yet.
[9] Variation of the position z_r along a path encircling other points, returning to the same position.

In Section 2.1.6 we had observed that conformal blocks define chiral vertex operators and vice versa. It is not hard to see that the gluing patterns correspond to the different ways of composing chiral vertex operators. Let us consider the case $n = 4$ as an example.

Example. *In the case $n = 4$ one may consider two basic cases. In the first we take $C_2 = \mathbb{P}^1 \setminus \{\infty, y_2, 0\}$ and $C_1 = \mathbb{P}^1 \setminus \{\infty, x_3, 0\}$. It is easy to see that the gluing construction yields a result which may be represented in vertex operator notation as*

$$f_C(v_4 \otimes \cdots \otimes v_1) = \left\langle v_4, V_{\rho'_s(R)}(v_3, x_3) q^{L_0} V_{\rho_s(R)}(v_2, y_2) v_1 \right\rangle, \quad (2.63)$$

where $\rho'_s(R) = \left[{}_{R_4}{}^{R_3}{}_R\right]$, $\rho_s(R) = \left[{}_R{}^{R_2}{}_{R_1}\right]$. We will associate the resulting conformal blocks with the gluing pattern on the left of Figure 2.1.

In the second case we shall take $C_2 = \mathbb{P}^1 \setminus \{\infty, y_3, 0\}$, $x_4 = \infty$ and $x = x_2$. We then have

$$f_C(v_4 \otimes \cdots \otimes v_1) = \left\langle v_4, V_{\rho'_u(R)}\left(q^{L_0} V_{\rho_u(R)}(v_3, y_3) v_2, x_2\right) v_1 \right\rangle, \quad (2.64)$$

where $\rho'_u(R) = \left[{}_{R_4}{}^{R}{}_{R_1}\right], \rho_u(R) = \left[{}_R{}^{R_3}{}_{R_2}\right]$. The conformal blocks defined in this way will be associated with the gluing pattern on the right of Figure 2.1.

We may, of course, further specialize to $x_3 = 1$ in the construction above. Choosing $y_2 = 1$ in the first case above we'll get $q = x_2$, while in the second case $y_3 = 1$ gives $q = 1 - x_2$.

2.2.2 Computing conformal blocks[+]

While it is straightforward to compute explicitly the very first few orders in the series expansions for chiral partition functions defined using the gluing construction, this will quickly become unmanageable for higher orders. There are a few tools for more efficient calculations available. The free field representation, briefly introduced in Section 2.2.2.1 yields the most explicit formulae, but is in this form only applicable for special families of conformal blocks. We then quickly mention two further representations which are sometimes useful, with pointers to the relevant literature.

2.2.2.1 Free field representation[+]

A subset of the conformal blocks can be elegantly represented using the free field representation for the Virasoro algebra introduced in Section 2.1.9. The basic building blocks are the normal ordered exponential fields constructed from the generators a_n, $n \in \mathbb{Z}$, of the free boson algebra

$$e^{2\alpha \varphi(z)} = T_\alpha \exp\left(2i\alpha \sum_{k<0} \frac{a_k}{k} z^{-k}\right) e^{-2i\alpha a_0 \log(z)} \exp\left(2i\alpha \sum_{k>0} \frac{a_k}{k} z^{-k}\right), \quad (2.65)$$

where the operator T_α maps a Fock module $\mathcal{F}_\beta$ to $\mathcal{F}_{\alpha+\beta}$. The left-hand side is a short notation for the right-hand side frequently used in the literature. It is straightforward to show that

$$[L_n, e^{2\alpha\varphi(z)}] = z^n(z\partial_z + (n+1)\Delta_\alpha)e^{2\alpha\varphi(z)}, \qquad \Delta_\alpha = \alpha(Q-\alpha), \qquad (2.66)$$

L_n being the generators of the Virasoro algebra defined in (2.50). Equation (2.66) identifies $e^{2\alpha\varphi(z)}$ as an intertwining operator between the Fock modules $\mathcal{F}_\beta$ and $\mathcal{F}_{\alpha+\beta}$. We had seen that such intertwining operators allow us to define conformal blocks on the three-punctured sphere with representations assigned to the three punctures having highest weights Δ_α, Δ_β, and $\Delta_{\alpha+\beta}$.

It is straightforward to calculate the expectation values of products of normal ordered exponentials, defined as

$$\left\langle e_0^*, e^{2\alpha_n(z_n)} \ldots e^{2\alpha_1(z_1)} e_0 \right\rangle = \prod_{s>r}(z_s - z_r)^{-2\alpha_r \alpha_s}, \qquad (2.67)$$

where $\sum_{r=1}^n \alpha_r = 0$, and e_0^* is the highest weight vector in the dual of the Fock space representation $\mathcal{F}_0$. In (2.67) we are assuming that $|z_s| > |z_r|$ for $s > r$, and the right-hand side of (2.67) is defined by the principal value of the logarithm when z_r is real and positive for $r = 1, \ldots, n$.

A generalization of these vertex operators with one discrete parameter can be defined using the so-called screening operators, defined as

$$Q_\gamma := \int_\gamma dz\, S(z), \qquad S(z) := e^{2b\varphi(z)}. \qquad (2.68)$$

Note that $\Delta_b = 1$, which implies that the commutator $[L_n, e^{2b\varphi(z)}]$ is a total derivative. This implies that products of normal ordered exponential with screening operators like

$$\mathsf{h}_s^\alpha(z) = e^{2\alpha\varphi(z)} \int_{\Gamma_s} du_1 \ldots du_s\, S(u_1) \ldots S(u_s) \qquad (2.69)$$

will define more general intertwining operators mapping $\mathcal{F}_\beta$ to $\mathcal{F}_{\beta+\alpha+bs}$, *provided* that the contour Γ_s of integration over $u_1, \ldots, u_s$ is closed. The number s of integration variables is also called the screening number. When b is real, and the real part of the parameter α is sufficiently negative, one can replace the contour Γ_s by a product of open contours $\gamma_1 \times \cdots \times \gamma_s$ starting and ending at z in order to ensure absence of boundary terms.

In this way one can obtain integral representations for the chiral partition functions associated to the special class of conformal blocks defined by the gluing construction satisfying the condition that the triple $(\Delta_1, \Delta_2, \Delta_3)$ of highest weights associated to each pair of pants is of the form $(\Delta_\alpha, \Delta_\beta, \Delta_{\alpha+\beta+bs})$. Different choices of contours Γ_s in (2.69) will define different bases for the subspace of the space of conformal blocks that can be represented in this way.

2.2.2.2 *Recursion relations and AGT representation*[++]

As previously noted, one can not use the free field representation introduced above for general conformal blocks. They define new types of special functions which will not have an integral representation with explicit integrand in general.[10]

To get more explicit information on general conformal blocks one may use two alternative types of representations. Firstly, note that the developments initiated by the discovery of relations between Virasoro conformal blocks and instanton partition functions in four-dimensional gauge theories [19] led to explicit formulae for the expansion coefficients in the series expansions for conformal blocks, see [20] for a proof of the relations conjectured in [19], and the resulting formulae for the expansion coefficients.

Secondly, there exist recursion relations for the conformal blocks that are very useful for efficient numerical calculation of conformal blocks. Such recursion relations were first obtained in [21] for the conformal blocks on the four-punctured sphere, and recently generalized in [22] to conformal blocks associated to more general Riemann surfaces.

2.2.3 Crossing symmetry

We are now going to explain why the conformal blocks constructed by the gluing construction are the ones of interest for applications in theoretical physics.

To see this we are first going to derive a more precise version of the holomorphically factorized form (2.10) for the correlation functions of a CFT. Assuming, as before, that the CFT is unitary one may decompose the Hilbert space $\mathcal{H}$ in the following form:

$$\mathcal{H} = \bigoplus_{R',R''} M_{R',R''} \otimes R' \otimes R'', \qquad (2.70)$$

where R' and R'' are unitary highest weight representations of the Virasoro algebras with generators L_n and $\bar{L}_n$, respectively, and $M_{R',R''}$ is a multiplicity space on which the Virasoro algebras act trivially. Let us then study a correlation function $\langle \prod_{r=1}^{n} \Phi_{v_r}(z_r, \bar{z}_r) \rangle$ of n vertex operators $\Phi_{v_r}(z_r, \bar{z}_r)$ associated to states $v_r \in \mathcal{H}$ of the form $v_r = m_r \otimes v'_r \otimes v''_r$, $v'_r \in R'_r, v''_r \in R''_r, m_r \in M_{R'_r,R''_r}, r = 1, \ldots, n$. We will see that such correlation functions can be represented in the following form:

$$\mathcal{Z}_{\mathbf{v}}(\mathbf{z}, \bar{\mathbf{z}}) = \sum_{\beta',\beta''} C^{\Gamma}_{\beta'\beta''\mathbf{m}} \mathcal{F}^{\Gamma}_{\beta',\mathbf{v}'}(\mathbf{z}) \mathcal{F}^{\Gamma}_{\beta'',\mathbf{v}''}(\bar{\mathbf{z}}), \qquad (2.71)$$

where $\mathcal{F}^{\Gamma}_{\beta,\mathbf{v}}(\mathbf{z})$ are the conformal blocks constructed by the gluing construction as described in Section 2.2.1. The expansion (2.71) is a more precise version of the holomorphically factorized representation (2.10) postulated previously.

---

[10] The functions defined using the free field representation as previously described have a finite-dimensional monodromy representation. This is not the case for generic conformal blocks, as we will see later.

Indeed, let us observe that the operator–state correspondence implies existence of an operator product expansion (OPE) of the following form:

$$\Phi_{w_2}(x,\bar{x})\Phi_{w_1}(y,\bar{y}) = \Phi_{\Phi_{w_2}(x-y,\bar{x}-\bar{y})w_1}(y,\bar{y})$$
$$= \sum_{\iota \in \mathcal{I}} C^{v_\iota}_{w_2,w_1}(x-y)^{\Delta_{v_\iota}-\Delta_{w_2}-\Delta_{w_1}}(\bar{x}-\bar{y})^{\bar{\Delta}_{v_\iota}-\bar{\Delta}_{w_2}-\bar{\Delta}_{w_1}}\Phi_{v_\iota}(y,\bar{y}). \qquad (2.72)$$

The first line is a consequence of the operator–state correspondence, identifying the composite field on the left of (2.72) as the field associated to the state $\Phi_{w_2}(x-y, \bar{x}-\bar{y})w_1$. The second line is then obtained by picking a basis $\{v_\iota; \iota \in \mathcal{I}\}$ for $\mathcal{H}$ consisting of eigenvectors v_ι of L_0 and $\bar{L}_0$ with eigenvalues Δ_{v_ι} and $\bar{\Delta}_{v_\iota}$, respectively, and expanding $\Phi_{w_2}(x-y,\bar{x}-\bar{y})w_1$ with respect to this basis.

Exercise 8 Consider the case of a four-point function $\langle \prod_{r=1}^{4} \Phi_{v_r}(z_r, \bar{z}_r)\rangle$. Demonstrate that applying the OPE (2.72) to the pair $\Phi_{v_2}(z_2,\bar{z}_2)\Phi_{v_1}(z_1,\bar{z}_1)$ yields an expansion of the form (2.71), with conformal blocks $\mathcal{F}^\beta_{\Gamma,\mathbf{v}}(\mathbf{z})$ constructed using the gluing pattern on the left of Figure 2.1, while application of (2.72) to the pair $\Phi_{v_3}(z_3,\bar{z}_3)\Phi_{v_2}(z_2,\bar{z}_2)$ yields an expansion of the form (2.71), with conformal blocks $\mathcal{F}^\beta_{\Gamma,\mathbf{v}}(\mathbf{z})$ constructed using the gluing pattern on the right of Figure 2.1. Hint: Use the conformal Wigner–Eckart theorem from Section 2.1.6.

It is a basic physical requirement that the correlation functions $\mathcal{Z}_\mathbf{v}(\mathbf{z},\bar{\mathbf{z}})$ are single-valued and real analytic in the variables z_r, $r = 1,\ldots,n$ away from the diagonals $z_i = z_j$. This implies that $\mathcal{Z}_\mathbf{v}(\mathbf{z},\bar{\mathbf{z}})$ defines a function on $\mathcal{M}_{0,n}$. The expansions (2.71) yield (presumably convergent) series expansions around the component of the boundary of $\mathcal{M}_{0,n}$ specified by the gluing pattern Γ, corresponding to an ordering prescription for successively performing OPEs. In order for $\mathcal{Z}_\mathbf{v}(\mathbf{z},\bar{\mathbf{z}})$ to be analytic away from the diagonals $z_i = z_j$, we need the conformal blocks $\mathcal{F}^\Gamma_{\beta,\mathbf{v}}(\mathbf{z})$ to admit an analytic continuation to the universal cover $\widetilde{\mathcal{M}}_{0,n}$ of the configuration space $\mathcal{M}_{0,n}$. We will see that this is typically the case for the conformal blocks coming from the gluing construction.

The correlation function $\mathcal{Z}_\mathbf{v}(\mathbf{z},\bar{\mathbf{z}})$ will admit many equivalent representations of the form (2.71), obtained by performing OPEs in different orders and represented by different choices of gluing patterns. The equivalence of the different representations is expressed in relations of the form

$$\sum_{\beta',\beta''} C^{\Gamma_1}_{\beta'\beta''\mathbf{m}} \mathcal{F}^{\Gamma_1}_{\beta',\mathbf{v}'}(\mathbf{z})\mathcal{F}^{\Gamma_1}_{\beta'',\mathbf{v}''}(\bar{\mathbf{z}}) = \sum_{\beta',\beta''} C^{\Gamma_2}_{\beta'\beta''\mathbf{m}} \mathcal{F}^{\Gamma_2}_{\beta',\mathbf{v}'}(\mathbf{z})\mathcal{F}^{\Gamma_2}_{\beta'',\mathbf{v}''}(\bar{\mathbf{z}}); \qquad (2.73)$$

it may be necessary to analytically continue $\mathcal{F}^{\Gamma_1}_{\beta',\mathbf{v}'}(\mathbf{z})$ and $\mathcal{F}^{\Gamma_1}_{\beta'',\mathbf{v}''}(\bar{\mathbf{z}})$ w.r.t. $z_1,\ldots,z_n$ in order to define the left-hand side in a domain of $\mathcal{M}_{0,n}$ where the right-hand side of (2.73) is defined through convergent power series expansions. These relations, often referred to as the crossing symmetry conditions, can be regarded as a system of equations constraining the remaining unknowns, the coefficients $C^{\Gamma}_{\beta'\beta''\mathbf{m}}$ in (2.71), with coefficients $\mathcal{F}^{\Gamma}_{\beta',\mathbf{v}'}(\mathbf{z})\mathcal{F}^{\Gamma}_{\beta'',\mathbf{v}''}(\bar{\mathbf{z}})$ fully determined by conformal symmetry. The next step in the bootstrap programme will be to observe that there often exist linear relations between the conformal blocks $\mathcal{F}^{\Gamma_1}_{\beta,\mathbf{v}}(\mathbf{z})$ and $\mathcal{F}^{\Gamma_2}_{\beta,\mathbf{v}}(\mathbf{z})$ associated to different gluing patterns Γ_1 and Γ_2, allowing us to exhibit the mathematical content of the crossing symmetry conditions more clearly.

2.2.4 Fusion and braiding

We are now going to observe that there often exist relations of the form

$$\mathcal{F}^{\Gamma_1}_{\beta_1,\mathbf{v}}(\mathbf{z}) = \sum_{\beta_2 \in \mathcal{S}} F^{\Gamma_1\Gamma_2}_{\beta_1\beta_2} \mathcal{F}^{\Gamma_2}_{\beta_2,\mathbf{v}}(\mathbf{z}), \tag{2.74}$$

allowing us to turn the relations (2.73) into more tractable problems. The set $\mathcal{S}$ is a subset of the set of all possible ways to colour a gluing pattern Γ with choices of intermediate representations.

The derivation of relations of the form (2.74) is a difficult problem in general, of central importance for the mathematics of conformal field theories. In order to demonstrate validity of relations of the form (2.74) it is first of all useful to observe that it suffices to derive such relations in the cases $n = 3$ and $n = 4$. This follows from the known fact [8, 23] that the transition between any two gluing patterns Γ_1 and Γ_2 can be broken up into a sequence of elementary operations localized in subsurfaces isomorphic to $C_{0,3}$ and $C_{0,4}$. These elementary operations are called braiding and fusion, braiding being depicted in Figure 2.2 while fusion is the passage from the gluing pattern on the left to that on the right of Figure 2.1. It therefore suffices to establish (2.74) in these two cases.

We shall begin by discussing the braiding operation.

Exercise 9 Let Γ_1 and Γ_2 be the conformal blocks depicted in the left and right parts of Figure 2.2, respectively. Recall that there is no internal label β needed in this case, but the conformal blocks depend on the choices of three representations R_1, R_2, R_3 associated to the three punctures. Demonstrate that

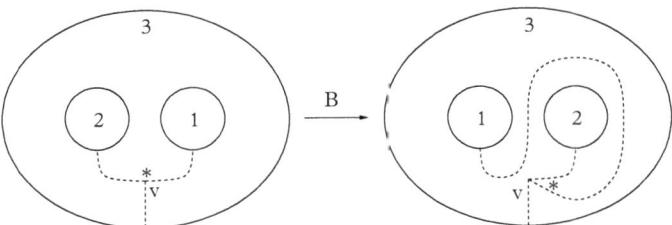

Figure 2.2 *The braiding operation.*

$$\mathcal{F}_v^{\Gamma_2} = e^{\pi i(\Delta_{R_3}-\Delta_{R_2}-\Delta_{R_1})}\mathcal{F}_v^{\Gamma_1}, \tag{2.75}$$

in this case, where Δ_R is the L_0-eigenvalue of the highest weight vector of the Virasoro representation R.

The main difficulty is to derive relations of the form (2.74) for $n = 4$. The following example illustrates how such relations can be calculated in a simple case.

Example 3 (continued) *Let us return to the case $C = C_{0,4}$ with $\alpha_2 = -b/2$, considered previously. It is instructive to check that the conformal blocks defined using the gluing pattern Γ_s introduced on the left of Figure 2.1 have chiral partition functions $\mathcal{Z}_\beta^s(z) \equiv \mathcal{Z}_{\pm 1/2}^s(z)$ given by the following linearly independent solutions of the differential equation $\mathcal{D}_{\mathrm{BPZ}}\mathcal{Z}(z) = 0$:*

$$\mathcal{Z}_{+\frac{1}{2}}^s(z) = z^{b\alpha_1}(1-z)^{b\alpha_3}F(A,B,C;z), \tag{2.76}$$

where $F(A,B,C;z)$ is the hypergeometric function, with arguments given as

$$\begin{aligned} A &= b(\alpha_1 + \alpha_3 + \alpha_4 - 3b/2) - 1, \\ B &= b(\alpha_1 + \alpha_3 - \alpha_4 - b/2), \end{aligned} \qquad C = b(2\alpha_1 - b). \tag{2.77}$$

The second linearly independent solution $\mathcal{Z}_{-1/2}^s(z)$ is given by the expressions obtained by replacing $\alpha_i \to Q - \alpha_i$ for $i = 1, 3, 4$ throughout. Similar formulae will of course represent the chiral partition functions $\mathcal{Z}_{\pm 1/2}^u(z)$ representing the conformal blocks defined from the gluing pattern Γ_u introduced on the right of Figure 2.1. Note that the representations associated to the edges marked with an arrow in Figure 2.1 are given as

$$\beta = \alpha_1 \mp \frac{b}{2} \;\text{for}\; \mathcal{Z}_{\pm\frac{1}{2}}^s, \qquad \beta = \alpha_3 \mp \frac{b}{2} \;\text{for}\; \mathcal{Z}_{\pm\frac{1}{2}}^u, \tag{2.78}$$

respectively. The restrictions on the set of representations that may be used in the gluing construction following from the presence of null vectors are often called fusion rules.

Well-known relations satisfied by the hypergeometric functions then give us the following relations:

$$\begin{pmatrix} \mathcal{Z}_{+\frac{1}{2}}^s(z) \\ \mathcal{Z}_{-\frac{1}{2}}^s(z) \end{pmatrix} = \begin{pmatrix} \frac{\Gamma(C)\Gamma(C-A-B)}{\Gamma(C-A)\Gamma(C-B)} & \frac{\Gamma(C)\Gamma(A+B-C)}{\Gamma(A)\Gamma(B)} \\ \frac{\Gamma(2-C)\Gamma(C-A-B)}{\Gamma(1-A)\Gamma(1-B)} & \frac{\Gamma(2-C)\Gamma(A+B-C)}{\Gamma(1+A-C)\Gamma(1+B-C)} \end{pmatrix} \begin{pmatrix} \mathcal{Z}_{+\frac{1}{2}}^u(z) \\ \mathcal{Z}_{-\frac{1}{2}}^u(z) \end{pmatrix}, \tag{2.79}$$

from which one may easily read off the explicit formulae for the coefficients $F_{\beta\beta'}^{\Gamma_1\Gamma_2}$ in this case.

A first family of examples for which relations of the form (2.74) are known to hold in general is found by considering the cases where $c = 1 - 6(\beta - \beta^{-1})^2$, $\beta = \sqrt{p/p'}$ with p,

p' being positive coprime integers satisfying $p < p'$. The corresponding CFTs are called minimal models. In order to parameterize the relevant set of representations of Vir_c let us introduce the set

$$\mathbb{A}_{\mathrm{KT}} = \left\{ a_{mn}; \; a_{mn} = \frac{\beta}{2}(1-m) - \frac{1}{2\beta}(1-n), \; m=1,\ldots,p', \; n=1,\ldots,p \right\}. \qquad (2.80)$$

Representations having highest weights $\Delta_a = a(a-q)$, $q = \beta^{-1} - \beta$, with $a \in \mathbb{A}_{\mathrm{KT}}$ are referred to as representations contained in the Kac-table. For conformal blocks having external representations in the Kac-table the existence of the relations (2.74) was established in [24].

There is another family of cases, for which relations of the form (2.74) have been established. This is the case when $c \geq 25$, $C = C_{C,4}$ with $\alpha_r = Q/2 + iP_r$, $P_r \in \mathbb{R}$. For this case it was found in [4] that there exist relations of the form

$$\mathcal{F}_{P,\mathbf{v}}^{\Gamma_s}(q) = \int_{\mathbb{R}_+} dP' \, F^{\Gamma_s \Gamma_u}(P, P') \mathcal{F}_{P',\mathbf{v}}^{\Gamma_u}(q). \qquad (2.81)$$

The relations (2.81) were derived in [4] using a generalization of the free field representation described in Section 2.2.2.1 to the case of non-integer screening numbers s. A similar result for $c = 1$ was obtained in [25], see also Section 2.3.7.

Even more general cases can be deduced from (2.81) using analytic continuation in α_r, $r = 1,\ldots,4$, or in the central charge c, which exist as long as $c > 1$ [26]. This gives the complete answer for the fusion transformations of the conformal blocks of arbitrary irreducible highest weight representations of the Virasoro algebra with $c > 1$.

The existence of fusion relations of the form (2.81) for generic representations is a remarkable and highly non-trivial fact that deserves to be better understood. It implies that the conformal blocks obtained from the gluing construction generate a subspace that is closed under the representation of the braid group defined by composing braiding and fusion operations. These results offer a starting point for the development of a harmonic analysis on spaces of conformal blocks, related to the harmonic analysis on the Teichmüller spaces themselves [4, 26].

Combining (2.73) and (2.74) yields a system of equations for the remaining undetermined quantities $C_{\beta'\beta''}^{\Gamma_1}$ in (2.71),

$$\sum_{\beta_1',\beta_1''} C_{\beta_1'\beta_1''\mathbf{m}}^{\Gamma_1} F_{\beta_1'\beta_2'}^{\Gamma_1 \Gamma_2} F_{\beta_1''\beta_2''}^{\Gamma_1 \Gamma_2} = C_{\beta_2'\beta_2''\mathbf{m}}^{\Gamma_2}. \qquad (2.82)$$

According to the above discussion, one may consider the coefficients $F_{\beta_1 \beta_2}^{\Gamma_1 \Gamma_2}$ as known data, fully determined by conformal symmetry alone. The system of equations (2.82) expresses the constraints on the as yet unknown data $C_{\beta'\beta''\mathbf{m}}^{\Gamma}$ following from crossing symmetry.

Finding the most general solution of (2.82) would be equivalent to a full classification and solution of all CFTs, which seems pretty hopeless in general. For special CFTs one may exploit additional constraints on the form of the coefficients $C^\Gamma_{\beta'\beta''\mathbf{m}}$, allowing us to determine them completely. One class of theories where this has been realized fairly completely will be described in Section 2.2.5.

One may hope that the powerful numerical techniques that have recently been developed for getting constraints on the coefficients $C^\Gamma_{\beta'\beta''\mathbf{m}}$ can lead to some progress in this direction. We feel unable to offer a good survey of the literature on this line of research here. As one possible starting point containing further references, we would like to mention the talk [27] and the paper [28] describing numerical evidence for the uniqueness of the Liouville CFT discussed in Section 2.2.5 within a certain class of CFTs. It would be very interesting if such techniques can be used to explore what comes beyond this class.

2.2.5 Rational and non-rational minimal models[+]

We will now discuss a few examples where the bootstrap strategy has been fully realized. The examples are known under the names of (generalized) minimal models and Liouville theory, respectively. The main simplifying features of these CFTs are (i) the absence of multiplicities, and (ii) a diagonal pairing of the representations of the two copies of Vir_c generating the Hilbert space of the theory, leading to the form

$$\mathcal{H}_{\mathrm{MM}} = \bigoplus_{R \in \mathsf{KT}} R \otimes R, \qquad \mathcal{H}_{\mathrm{LT}} = \int_{\mathbb{S}} d\alpha\, R_\alpha \otimes R_\alpha, \qquad (2.83)$$

for the Hilbert spaces $\mathcal{H}_{\mathrm{MM}}$ of the minimal models, and $\mathcal{H}_{\mathrm{LT}}$ of Liouville theory, respectively. The sets of representations appearing in $\mathcal{H}_{\mathrm{MM}}$ and in $\mathcal{H}_{\mathrm{LT}}$ are denoted as KT for 'Kac table' and $\mathbb{S}$, respectively. The representations appearing in $\mathcal{H}_{\mathrm{MM}}$ and $\mathcal{H}_{\mathrm{LT}}$ will be unitary highest weight representations of the Virasoro algebra with $c < 1$ and $c > 25$, respectively.

2.2.5.1 Factorization

Owing to the absence of multiplicities we now have $C^\Gamma_{\beta'\beta''\mathbf{m}} = C^\Gamma_{\beta'\beta''}$. In order to deduce the resulting restrictions on the coefficients $C^\Gamma_{\beta'\beta''}$ in the holomorphic factorization (2.71) let us recall the link between the OPE and an expansion over a basis for the Hilbert space $\mathcal{H}$ noted after (2.72). We may assume that $w_i \in R_{\alpha_i} \otimes R_{\alpha_i}$ for $i = 1, 2$, and the summation over a basis for $\mathcal{H}$ may be split into a summation or integration over a representation label denoted α_3, and a summation over vectors w_3 forming a basis for $R_{\alpha_3} \otimes R_{\alpha_3}$. The coefficients $C^{w_3}_{w_2,w_1}$ in (2.72) can be expressed in terms of the coefficient $C(\alpha_3|\alpha_2,\alpha_1)$ associated to $w_i = e_i$, the products of highest weight vectors in $R_{\alpha_i} \otimes R_{\alpha_i}$ for $i = 1,2,3$, respectively.

Exercise 10

(a) Use the conformal Ward identities to demonstrate that two-and three-point functions of primary fields in Liouville field theory have the form

$$\langle V_{\alpha_2}(z_2, \bar{z}_2) V_{\alpha_1}(z_1, \bar{z}_1) \rangle = \delta_{\mathbb{S}}(\alpha_2, \alpha_1) B(\alpha_1) |z_2 - z_1|^{-4\Delta_{\alpha_1}}, \tag{2.84}$$

$$\langle V_{\alpha_3}(z_3, \bar{z}_3) V_{\alpha_2}(z_2, \bar{z}_2) V_{\alpha_1}(z_1, \bar{z}_1) \rangle = |z_2 - z_1|^{2(\Delta_{\alpha_3} - \Delta_{\alpha_2} - \Delta_{\alpha_1})} |z_3 - z_2|^{2(\Delta_{\alpha_1} - \Delta_{\alpha_3} - \Delta_{\alpha_2})}$$
$$\times |z_1 - z_3|^{2(\Delta_{\alpha_2} - \Delta_{\alpha_3} - \Delta_{\alpha_1})} C(\alpha_3, \alpha_2, \alpha_1), \tag{2.85}$$

where $\delta_{\mathbb{S}}(\alpha_2, \alpha_1)$ is the delta-distribution on $\mathbb{S}$.

(b) Use the results from part (a) together with the OPE (2.72) to demonstrate that the as yet undetermined coefficients $C(\alpha_3, \alpha_2, \alpha_1)$ in three-point functions (2.85), $B(\alpha)$ in the two-point functions (2.84), and the OPE coefficients $C(\alpha_3|\alpha_2, \alpha_1)$ are related as

$$C(\alpha_3, \alpha_2, \alpha_1) = C(\alpha_3|\alpha_2, \alpha_1) B(\alpha_3). \tag{2.86}$$

Using these observations it becomes easy to see that the coefficients $C^{\Gamma}_{\beta'\beta''}$ in the holomorphic factorization (2.71) are supported on the diagonal $\beta' = \beta''$, and factorize as

$$C^{\Gamma}_{\beta'\beta''} = \delta(\beta', \beta'') \prod_{v \in \mathsf{vert}(\Gamma)} C(\alpha_1^v, \alpha_2^v, \alpha_3^v) \prod_{e \in \mathsf{i\text{-}edg}(\Gamma)} B(\beta_e), \tag{2.87}$$

where $\mathsf{vert}(\Gamma)$ and $\mathsf{i\text{-}edg}(\Gamma)$ are the sets of vertices and internal edges of Γ, respectively, with triples of representation labels $(\alpha_1^v, \alpha_2^v, \alpha_3^v)$ assigned to the punctures of the three-punctured sphere $C_{0,3}^v$ appearing in the pants decomposition associated to Γ, and representation labels β_e associated to the internal edges of Γ.

2.2.5.2 Functional equations

It turns out that the conditions following from crossing symmetry (2.73) can be written more explicitly in the special case where $n = 4$ with $\alpha_2 = -b/2$ considered above, not imposing further restrictions on α_1, α_3 and α_4 for the time being. Using (2.87) one may write the expansion (2.71) more explicitly as

$$\mathcal{Z}(z, \bar{z}) = \sum_{m=\pm\frac{1}{2}} C(\alpha_4, \alpha_3, \alpha_1 - mb) C(\alpha_1 - mb|\alpha_2, \alpha_1) \mathcal{Z}^s_m(z) \mathcal{Z}^s_m(\bar{z}) \tag{2.88}$$

$$= \sum_{m=\pm\frac{1}{2}} C(\alpha_4, \alpha_3 - mb, \alpha_1) C(\alpha_3 - mb|\alpha_3, \alpha_2) \mathcal{Z}^u_m(z) \mathcal{Z}^u_m(\bar{z}), \tag{2.89}$$

where $\mathcal{Z}^s_m(z)$ and $\mathcal{Z}^u_m(z)$, $m = \pm 1/2$, were introduced in Example 1c. Inserting (2.79) into (2.88) one obtains an expression apparently containing terms proportional to $\mathcal{Z}^{\pm 1/2}_s(z) \mathcal{Z}^{\mp 1/2}_s(\bar{z})$. No such terms occur in (2.89). Consistency therefore requires the coefficients in front of the terms $\mathcal{Z}^s_{\pm 1/2}(z) \mathcal{Z}^s_{\mp 1/2}(\bar{z})$ to vanish. This is easily seen to be the case if the coefficients $C(\alpha_3, \alpha_2, \alpha_1)$ satisfy a functional relation of the following form:

$$\frac{C(\alpha_1, \alpha_2, \alpha_3)}{C(\alpha_1, \alpha_2, \alpha_3 + b)} = d(\alpha_3) D(\alpha_1, \alpha_2, \alpha_3), \tag{2.90}$$

where $D(\alpha_1, \alpha_2, \alpha_3)$ can be assembled from the elements of the matrix (2.79) leading to the expression

$$D(\alpha_1, \alpha_2, \alpha_3) = \prod_{s_1, s_2 = \pm} \gamma\left(b\alpha_3 + s_1 b\left(\alpha_1 - \frac{Q}{2}\right) + s_2 b\left(\alpha_2 - \frac{Q}{2}\right)\right), \tag{2.91}$$

written in terms of the function $\gamma(x) = \Gamma(x)/\Gamma(1-x)$. The function $d(\alpha_3)$ in (2.90) is the product of some known and some unknown functions depending on α_3 only. We will later fix the resulting indeterminacy. A similar functional equation is obtained by noting that $0 = (b^2 L^2_{-1} + L_{-2}) e_{-1/2b}$ holds in $V_{-b/2}$. It leads to another functional equation obtained from (2.90) by replacing $b \to b^{-1}$ everywhere.

2.2.5.3 Explicit solutions

The functional relations (2.90) can be solved for $\text{Re}(b) > 0$ corresponding to $c > 1$ by an expression of the form

$$C(\alpha_1, \alpha_2, \alpha_3) = \frac{C_0 \prod_{r=1}^3 N(\alpha_r)}{\prod_{s_1, s_2 = \pm} \Upsilon_b\left(\alpha_3 + s_1\left(\alpha_1 - \frac{Q}{2}\right) + s_2\left(\alpha_2 - \frac{Q}{2}\right)\right)}, \tag{2.92}$$

using the fact that the special function

$$\log \Upsilon_b(x) = \int_0^\infty \frac{dt}{t} \left[\left(\frac{Q}{2} - x\right)^2 e^{-2t} - \frac{\sinh\left(\frac{t}{2}(Q - 2x)\right)}{\sinh(bt)\sinh(b^{-1}t)}\right], \tag{2.93}$$

satisfies the functional equations

$$\Upsilon_b(x + b) = b^{1-2bx} \gamma(bx) \Upsilon_b(x), \qquad \Upsilon_b(x) = \Upsilon_{1/b}(x). \tag{2.94}$$

As we have not yet determined the function $d(\alpha_3)$ in (2.90), we are not yet able to fix the functions $N(\alpha)$ in (2.92). One may notice, however, that the function $N(\alpha)$ can be changed by changing the normalization of the primary fields $\Phi_{e_\alpha}(z, \bar{z})$.

The bootstrap equations (2.90) remain valid for $c \leq 1$. However, it turns out that the formula (2.92) representing a solution to the equations (2.90) for $c \geq 1$ can not be used in this case. The special function Υ_b can only be used for $\operatorname{Re}(b) > 1$ covering the cases with $c \geq 1$. In order to solve the functional equations (2.90) in the case $c \leq 1$ it is convenient to use the parameters $a = -i\alpha$, $\beta = ib$. Instead of (2.92) one may now use an expression of the following form [29]:

$$C_M(a_1, a_2, a_3) = \frac{\prod_{s_1, s_2 = \pm} \Upsilon_\beta \left(\beta + a_3 + s_1\left(a_1 - \frac{q}{2}\right) + s_2\left(a_2 - \frac{q}{2}\right)\right)}{C_0 \prod_{r=1}^3 N_M(\alpha_r)}, \qquad (2.95)$$

using the notation $q = \beta^{-1} - \beta$. It can be shown that (2.95) can not be obtained from (2.92) by analytic continuation [29].

2.2.5.4 Solutions for $c \leq 1$: (generalized) minimal models

If all representations used in the definition of the conformal blocks appearing in (2.71) are highest weight representations with highest weights $\Delta_a = a(a - q)$, $a \in \mathbb{A}_{KT}$, one may show that crossing symmetry is satisfied with $C(a_3, a_2, a_1)$ chosen as the restriction of $C_M(a_1, a_2, a_3)$ to $a_i \in KT$, $i = 1, 2, 3$. It can furthermore be verified that this restriction reproduces the expressions for the three-point functions of the minimal models first calculated in [30] for a suitable choice of field normalization function $N_M(\alpha_r)$ and normalization factor C_0 in (2.95).

However, this turns out not to be the end of the story yet. It was verified numerically in [31] for various cases that there exists a *continuous* set of intermediate dimensions to be used to construct the four-point functions in the form (2.71) such that choosing the coefficients to be given in terms of (2.95) indeed solves the crossing symmetry conditions (2.73). Following [29] we shall refer to the theory defined in terms of these three-point functions as generalized minimal models. Even if the CFT defined thereby is non-unitary in general, it appears to define an interesting model of statistical mechanics [32].

For rational values of $c \leq 1$ there exist further solutions $\tilde{C}_M(a_1, a_2, a_3)$ to the crossing symmetry conditions [31]. The functions $\tilde{C}_M(a_1, a_2, a_3)$ differ from $C_M(a_1, a_2, a_3)$ by a non-analytic factor containing step functions. In the case $c = 1$ one may identify $\tilde{C}_M(a_1, a_2, a_3)$ as a limit of the expressions for the minimal model three-point functions [33], or as a limit of the Liouville three-point functions [34] to be discussed shortly. This implies crossing symmetry of the correlation functions build using $\tilde{C}_M(a_1, a_2, a_3)$ for $c = 1$. Crossing symmetry has been verified numerically for other rational values of c in [31].

The non-uniqueness of the solutions to the crossing symmetry conditions for $c \leq 1$ following from the results previously summarized is a remarkable phenomenon which deserves to be better understood.

2.2.5.5 Solution for $c > 1$: *Liouville theory*

In order to round off the bootstrap solution of first examples for interesting CFTs let us observe, on the one hand, that it has been verified analytically for the case $\alpha_r \in \frac{Q}{2} + i\mathbb{R}$, $r = 1, \ldots, 4$ in [4] that the three-point functions defined in (2.92), when used as coefficients in (2.71), will satisfy the crossing conditions (2.73). Key ingredients in this verification are the fusion relations (2.81) identifying the set of intermediate representations that are summed (in fact integrated) over in (2.71) to be $\{V_\alpha; \alpha \in \frac{Q}{2} + i\mathbb{R}_+\}$. By a suitable choice of normalization of the conformal blocks one can identify the crossing conditions as an expression of the unitarity of the fusion transformations (2.81) with respect to a natural scalar product.

It is furthermore remarkable that the correlation functions $\mathcal{Z}_V(\mathbf{z}, \bar{\mathbf{z}})$ turn out to be entirely analytic in the parameters α_r, $r = 1, \ldots, n$, allowing us to obtain representations of arbitrary correlation functions $\mathcal{Z}_V(\mathbf{z}, \bar{\mathbf{z}})$ for $c > 1$ in the form (2.71) by analytic continuation [4, 26].

It remains to notice [4], on the other hand, that there exists a choice of the normalization-dependent function $N(\alpha)$ in (2.92) such that the fields $\phi(z, \bar{z})$ and $e^{2b\phi(z,\bar{z})}$ defined as

$$\phi(z,\bar{z}) := \frac{1}{2} \frac{\partial}{\partial \alpha} \Phi_{e_\alpha}(z,\bar{z}) \bigg|_{\alpha=0}, \qquad e^{2b\phi(z,\bar{z})} := \Phi_{e_\alpha}(z,\bar{z}) \bigg|_{\alpha=b}, \qquad (2.96)$$

satisfy the equation

$$\partial_z \partial_{\bar{z}} \phi(z,\bar{z}) = \pi \mu \, e^{2b\phi(z,\bar{z})}. \qquad (2.97)$$

The choices of C_0 and $N(\alpha)$ which will do the job are

$$N(\alpha) = \left(\pi \mu \gamma(b^2) b^{2-2b^2}\right)^{-\frac{\alpha}{b}} \Upsilon(2\alpha), \qquad C_0 = \left(\pi \mu \gamma(b^2) b^{2-2b^2}\right)^{\frac{Q}{b}} \Upsilon(b). \qquad (2.98)$$

This observation may be used to identify the CFT characterized by the three-point function (2.92) as Liouville theory. It follows from (2.96) and (2.97) that the classical limit of the field $\phi(z, \bar{z})$ will satisfy the classical Liouville equation of motion.

Remark 2. It had been conjectured in [35, 36] that the function defined in (2.92) with $N(\alpha)$, C_0 given in (2.98) represents the three-point function of Liouville theory. The paper [36] describes several highly non-trivial checks of this proposal. The method previously described for finding this result was introduced in [37]. Using the probabilistic framework for the construction of Liouville theory introduced in [38], the method from [37] was recently used in the proof of (2.92), (2.98) proposed in [39]. A different method for deriving (2.92), (2.98) had previously been outlined in [4, 40]. The conformal bootstrap program for Liouville field theory was completed in [4] for genus zero, and in [26] for Riemann surfaces C of higher genus.

2.2.6 Higher genus[+]

2.2.6.1 Gluing construction

In order to construct conformal blocks associated to Riemann surfaces of genus larger than zero one needs a generalization of the gluing construction. Geometrically, one needs an operation that creates a surface C with n punctures and $g > 0$ handles by identifying annular neighbourhoods of a pair of punctures (P, P') on a single surface $\hat{C}$ of genus $g - 1$ and $n + 2$ punctures. Let t and t' be coordinates in discs D_P and $D_{P'}$ around P and P', respectively, such that $t(P) = 0 = t'(P')$. By identifying points Q and Q' in annular neighbourhoods of P and P' which satisfy $t(Q)t'(Q') = q$, one may define a Riemann surface C of genus g with n punctures. The parameter q represents one of the coordinates for the moduli space $\mathcal{M}(C)$ of complex structures on the surface of interest.

This operation has a counterpart on the level of conformal blocks defined as follows. Let $\hat{f}$ be a conformal block associated to $\hat{C}$ with representations $R, R_1, \ldots, R_n, R$ assigned to the punctures $P, P_1, \ldots, P_n, P'$, respectively. One may then define

$$f(v_1 \otimes \cdots \otimes v_n) := \sum_{\mu,\nu \in \mathcal{I}_R} B^{\mu\nu} \hat{f}(v_\mu \otimes v_1 \otimes \cdots \otimes v_n \otimes q^{L_0} v_\nu), \qquad (2.99)$$

where $\{v_\mu; \mu \in \mathcal{I}_R\}$ is a basis for the representation R, and $B^{\mu\nu}$ are the matrix elements of the inverse of the matrix formed out of $B_{\mu\nu} = \langle v_\mu, v_\nu \rangle_R$, $\mu, \nu \in \mathcal{I}_R$, with $\langle .,. \rangle_R$ being the invariant bilinear form on R. It can be checked that f defines a conformal block on C. This operation can be interpreted as taking the trace $\text{tr}_R(q^{L_0} V(w_{[n]}, \hat{C}))$ of the generalized chiral vertex operators $V(w_{[n]}, \hat{C}): R \to R$, $w_{[n]} \in R_1 \otimes \cdots \otimes R_n$, defined such that

$$\left\langle v, V(w_{[n]}, \hat{C}) v' \right\rangle_R = \hat{f}(v \otimes w_{[n]} \otimes v'), \qquad (2.100)$$

holds for all $v, v' \in R$, $w_{[n]} \in R_1 \otimes \cdots \otimes R_n$.

By using this generalized form of the gluing construction one can define, in much the same way as before, families of conformal blocks associated to gluing patterns Γ coloured by the choices of intermediate representations.

2.2.6.2 Modular transformation

Of particular importance is the case where $g = 1$, where the above definition reduces to taking traces of composites of chiral vertex operators. Writing the parameter q as $q = e^{2\pi i \tau}$, $\Im(\tau) > 0$, one may represent $C = C_{1,1}^\tau$ as a cylinder $\{w \in \mathbb{C}; w \sim w + 2\pi\}$ with additional identification $w \sim w + 2\pi\tau$. We will use the notation

$$\chi_R^{R_1}(\tau) := \operatorname{tr}_R\left(q^{L_0} V\left[{}_R{}^{R_1}{}_R\right](e_1, 0)\right), \qquad (2.101)$$

for the one-point torus conformal blocks with external representation R_1.

The conformal transformation $w \to w' = -w/\tau$ maps $C_{1,1}^\tau$ to a torus $C_{1,1}^{-1/\tau}$ obtained from the cylinder by the additional identification $w \sim w - 2\pi/\tau$. This transformation maps the a-cycle $\Im w = 0$ and b-cycle $\{w \in \mathbb{C}; w = 2\pi \tau \vartheta; \vartheta \in [0,1]\}$ on $C_{1,1}^\tau$ to the b-cycle and a-cycle on $C_{1,1}^{-1/\tau}$, respectively.

It is quite remarkable that there exist linear relations between the conformal blocks $\chi_R^{R_1}(\tau)$ of the following general form:

$$\chi_R^{R_1}(\tau) = (-i\tau)^{-\Delta_{R_1}} e^{\pi i \frac{c}{12}(\tau + \frac{1}{\tau})} \sum_{R'} S_{RR'}^{R_1} \chi_{R'}^{R_1}(-1/\tau). \qquad (2.102)$$

These relations represent the modular transformation $w \to w' = -w/\tau$ on the level of the conformal blocks. The existence of such transformations is known in the case of minimal models (see [41] for a guide to the relevant mathematical literature), and in the case of Liouville theory [42].

2.2.6.3 Modular invariance of physical correlation functions

The physical correlation functions on higher genus surfaces may be represented in a form generalizing the structure (2.71) in an obvious way. One thereby defines (presumably convergent) series expansions for the generalizations of the Schwinger functions associated with higher genus Riemann surfaces. Basic physical consistency requirements are the existence of (real) analytic continuations over the Teichmüller space of deformations of C which is essentially[11] independent of the gluing pattern used in (2.71).

It can be shown [8] that these requirements are satisfied if (i) crossing symmetry holds for genus zero correlation functions and (ii) if modular invariance holds for the partition functions $\mathcal{Z}_v^{1,1}$ associated with the one-punctured torus,

$$\mathcal{Z}_v^{1,1}(\tau, \bar\tau) = |\tau|^{-2\Delta_{R_1}} e^{-\frac{\pi c}{12} \Im(\tau + \frac{1}{\tau})} \mathcal{Z}_v^{1,1}(-1/\tau, -1/\bar\tau), \qquad (2.103)$$

where here $v \in R_1$. This condition is interesting already in the case where R_1 is the vacuum representation, corresponding to the torus $C_{1,0}$ without insertions. For the partition functions constructed from the representations in the Kac table there exists a complete classification of all modular invariant combinations of the form (2.71) with finite multiplicities [43].

---

[11] The relation between partition functions defined from different gluing patterns may involve overall factors represented by the absolute value of the transition functions of the c-th power of the Hodge line bundle.

2.2.6.4 Chiral modular bootstrap++

It can be shown that any two gluing patterns Γ_1 and Γ_2 can be related by sequences of fusion, braiding, and modular transformations [8, 23]. A transformation of the mapping class group, the fundamental group of the moduli space of Riemann surfaces $\mathcal{M}(C)$, is obtained from these transformation by noting that the elements μ of the mapping class group can be realized as diffeomorphisms of the surface to itself, mapping one gluing pattern Γ to another $\mu.\Gamma$. As one can always realize the transition from Γ to $\mu.\Gamma$ in terms of fusion, braiding, and modular transformations, one may define a representation of the mapping class group on the spaces of conformal blocks involved. The matrices or integration kernels representing these transformations therefore represent data of considerable interest for CFT.

Comparing with the discussion of higher genus conformal blocks in Section 2.1.8 one may identify the conformal blocks defined by the gluing construction as sections of the bundle $\mathcal{V}_c$ which are horizontal with respect to the canonical connection defined by the energy–momentum tensor. Fusion coefficients $F_{\beta_1\beta_2}^{\Gamma_1\Gamma_2}$, the braiding factors appearing in (2.75), and modular transformation coefficients $S_{R_1R_2}^{R_0}$ represent the constant transition functions defining the projectively flat bundle $\mathcal{W}_c$. The prefactor $(-i\tau)^{-\Delta_{R_1}}e^{\pi i \frac{c}{12}(\tau+\frac{1}{\tau})}$ in (2.102), on the other hand, represents a transition function of the particular projective line bundle $\mathcal{E}_c$ relating $\mathcal{W}_c$ to the vector bundle $\mathcal{V}_c$ in the case of conformal blocks defined using the gluing construction, as was explictly shown in [26].

Having defined the flat vector bundle $\mathcal{V}_c$ in terms of its transition functions allows us to characterise the conformal blocks as horizontal sections of $\mathcal{V}_c$. To find multivalued analytic functions on the moduli space $\mathcal{M}(C)$ with given monodromies is a mathematical problem of Riemann-Hilbert type. This Riemann-Hilbert problem provides an alternative characterisation of the relevant spaces of conformal blocks [26].

2.2.7 Verlinde loop operators+

The spaces of conformal blocks on Riemann surfaces C carry another important algebraic structure, the structure as a module over a non-commutative algebra $\mathcal{A}$ which can be identified as a quantum deformation of the algebra of functions on the moduli space of flat $SL(2,\mathbb{C})$-connections on C.

Using the fact that fusion relations simplify when one of the representations involved is a degenerate representation $V_{-b/2}$ or $V_{-1/2b}$, we will define certain operators acting on the spaces of conformal blocks defined by the gluing construction. The resulting additional algebraic structure on spaces of conformal blocks has turned out to be relevant in several applications, and it offers a useful alternative way of characterizing the conformal blocks [26].

The definition of this structure is based on the comparison of the spaces of conformal blocks on a Riemann surface C to the spaces of conformal blocks on a modified surface $\hat{C} = C \setminus \{y_0, y\}$ obtained by cutting out a disc D from C and re-gluing a twice punctured disc $\hat{D} = D \setminus \{y_0, y\}$ into $C \setminus D$. The representations associated with the punctures at

y_0 and y will both be degenerate representations $V_{-b/2}$. The gluing construction of conformal blocks on C may easily be modified to provide a construction of conformal blocks on $\hat{C}$ such that one of the three-punctured spheres appearing in this construction contains $\hat{D}$. We had seen that in this case there exist only two possibilities for the choice of representation on the boundary $\partial \hat{D}$ of $\hat{D}$, namely V_0 and V_{-b}. Fixing it to be the vacuum representation V_0, one obtains a space of conformal blocks from the gluing construction which is isomorphic to the space of conformal blocks on C thanks to the propagation of vacua.

It follows from the differential equations expressing the null vector decoupling that the conformal blocks are analytic in y at least as long as $y \in \hat{D}$. An analytic continuation of these conformal blocks to arbitrary $y \in C$ can be defined by a sequence of changes of pants decomposition, all described in terms of the elementary fusion and braiding moves explicitly computed previously. This allows us to consider the monodromies defined by analytic continuation of the conformal blocks introduced above along closed curves on C, starting and ending in $\hat{D}$. It is not hard to see, and illustrated by the examples explicitly calculated in [44, 45], that the conformal blocks obtained by this analytic continuation are linear combinations of the conformal blocks associated with $\hat{C}$ one had started from. The choices of intermediate representations may, however, get modified. If the representation V_{β_e} is assigned to the edge e of the gluing pattern $\hat{\Gamma}$ used to define conformal blocks on $\hat{C}$, one will get a linear combination of the conformal blocks with representations $V_{\beta_e + mb}$, $m \in \mathbb{Z}$ assigned to e as the result of the analytic continuation. This means in particular that the result of the analytic continuation will be a linear combination of conformal blocks defined by assigning representations V_0 and V_{-b} to $\partial \hat{D}$, in general. An operator from the spaces of conformal blocks on C to itself is obtained by projecting the result of the analytic continuation to the contribution having V_0 assigned to $\partial \hat{D}$. The operator associated with a closed curve γ on C is called the Verlinde loop operator L_γ.

The explicit computations of Verlinde loop operators described in [44, 45] identify the algebra $\mathcal{A}$ generated by these operators as a non-commutative deformation of the algebra of functions on the moduli space of flat $\mathrm{SL}(2,\mathbb{C})$-connections on C, with L_γ representing the quantized counterpart of the trace of the holonomy of a flat connection along γ.

It can be shown that this action characterizes the conformal blocks of the Virasoro algebra essentially uniquely as solutions to a natural Riemann–Hilbert problem defined in terms of the $\mathcal{A}$-module structure [26]. Indeed, the construction described defines Verlinde loop operators $\mathsf{L}_{\Gamma,\gamma}$ for each pants decomposition Γ. Different choices of Γ yield different representatives $\mathsf{L}_{\Gamma,\gamma}$ for the generators $\mathcal{L}_\gamma$ of one and the same abstract algebra $\mathcal{A}$. Fusion, braiding, and the modular transformations define the intertwining operators $\mathsf{U}_{\Gamma_2 \Gamma_1}$ between the representations generated by the operators $\mathsf{L}_{\Gamma_1,\gamma}$ and $\mathsf{L}_{\Gamma_2,\gamma}$, satisfying $\mathsf{L}_{\Gamma_2,\gamma} \mathsf{U}_{\Gamma_2 \Gamma_1} = \mathsf{U}_{\Gamma_2 \Gamma_1} \mathsf{L}_{\Gamma_1,\gamma}$. Under certain conditions one can argue that the fusion, braiding, and the modular transformations are completely characterized by this intertwining property. This is the case e.g. for Virasoro representation with $c = 1 + 6Q^2 \geq 25$, $\Delta = Q^2/4 + P^2$, $P \in \mathbb{R}$. In this case one can define a scalar product making the Verlinde operators self-adjoint. The fusion and the modular transformation may then be identified with the unitary operators mapping some of the operators $L_{\Gamma,\gamma}$ to diagonal form, explaining why they are essentially uniquely defined by this property.

Together with the discussion in Section 2.2.6.4 one may conclude that the spaces of physically relevant conformal blocks are essentially determined by the representation theory of the algebra $\mathcal{A}$ of quantized functions on the moduli space of flat $SL(2,\mathbb{C})$-connections on C. This can be regarded as a dual topological characterization of the spaces of conformal blocks that are relevant for physical applications. More details on this point of view can be found in [26]. A variant of this story will be discussed in Section 2.3 in connection with integrability.

2.2.8 Extended chiral symmetry[++]

As previously remarked, there is little hope that we can solve CFTs with only Virasoro symmetry, in general. If, for example, one is considering a CFT with a Lagrangian formulation with many fundamental fields, the best that one could generically hope for would be partial control over certain composites of the fundamental fields thanks to Virasoro symmetry. If, however, Virasoro symmetry is extended to a symmetry under a larger chiral algebra, one regains hope that exact results may be within reach. The 'size' of the symmetry must be in a reasonable proportion to the 'complexity' of the field content. Out goal in this section will be to indicate some directions of research aiming at the solution of CFTs with higher chiral symmetries.

It is fairly straightforward to generalize the definition and the gluing construction of conformal blocks to cases with higher symmetries. This amounts to forming useful combinations of Virasoro conformal blocks having intermediate states related by the higher symmetries. It is furthermore not hard to formulate suitable generalizations of the crossing symmetry and modular invariance conditions. An additional complication that will generically arise is due to the fact that the spaces of conformal blocks for higher chiral symmetries may have dimensions higher than one as is already the case in $C = C_{0,3}$. A first example where this happens is found as follows:

Exercise 11 Let $\mathcal{A} = W_3$, the extension of Vir_c by generators W_n with $\Delta = 3$, introduced in Section 2.1.9. Show that $\dim(\mathrm{CB}(C_{0,3}, \mathcal{R})) = \infty$ for generic representations $\mathcal{R}$.

Sektch of solution Use (generalized) Ward identities to compute the values of a conformal block $f_{C_{0,3}}$ on arbitrary vectors $v \in \mathcal{R}$ in terms of $f_{C_{0,3}}(e_1 \otimes e_2 \otimes (W_{-1})^k e_3)$. One may therefore define conformal blocks $f^l_{C_{0,3}}$ satisfying $f^l_{C_{0,3}}(e_1 \otimes e_2 \otimes (W_{-1})^k e_3) = \delta_{k,l}$. The conformal blocks $f^l_{C_{0,3}}$ will define a basis for generic $\mathcal{R}$. □

Additional features of the representations assigned to the punctures may make the dimension of the space of conformal blocks finite. Whenever $\dim(\mathrm{CB}(C_{0,3}, \mathcal{R}))$ is not smaller or equal to one, as is the case for the Virasoro algebra, one has to face additional complications in the bootstrap programme. For certain VOA one may nevertheless establish a picture reasonably close to the case of the Virasoro VOA. We shall now offer a few pointers to the available literature.

2.2.8.1 Relation to braided tensor categories

As pointed out in Section 2.1.6 for the case of the Virasoro VOA, one may use the Ward identities defining the conformal blocks to define a generalized notion of a tensor product of representations called a fusion product. Generalizations of this notion can be defined for many VOAs. Viewing the chiral vertex operators as a generalization of the Clebsch–Gordon maps intertwining the representation on tensor products with irreducible representations, one is naturally led to identifying the coefficients of fusion transformations as analogues of the Racah–Wigner $6j$-symbols, relating two natural ways of decomposing triple tensor products into irreducible representations by first projecting tensor products of pairs of representations onto irreducible ones.

A more abstract point of view has turned out to be useful. Given a VOA $\mathcal{A}$ representing an extension of conformal symmetry one may attempt to find categories of representations on which the fusion product defined using the $\mathcal{A}$-Ward identities is closed. One such category of representations of affine Lie algebras $\hat{\mathfrak{g}}_k$ at negative levels k was identified in the work of Kazhdan and Lusztig [10] as the category denoted $\mathcal{O}_k$ defined by certain finiteness conditions.

One needs to note that commutativity and associativity of the fusion product are not manifest, in general. However, braiding and fusion transformation of conformal blocks can be used to define natural isomorphisms between (iterated) fusion products with different orders of tensor factors, or different orders in which tensor products are iterated. The isomorphisms defined in this way express the sense in which the fusion products are commutative and associative. The resulting structures can be formalized using the mathematical notion of a braided tensor category. This perspective has first been developed for the case of VOAs associated to affine algebras $\hat{\mathfrak{g}}_k$ at negative level in the work [10]. It was furthermore shown in [10] that the braided tensor category defined in this way is equivalent to a certain category of quantum group representations. As previously indicated, this amounts to a complete characterization of the fusion and braiding transformations for the conformal blocks defined from the category $\mathcal{O}_k$ of representations of $\hat{\mathfrak{g}}_k$.

2.2.8.2 Rationality and modular tensor categories

Under certain finiteness conditions on the considered category of representations one can prove that there exist modular transformations similar to (2.102) among the conformal blocks of a VOA associated to surfaces of genus one [46]. For such VOAs it may furthermore be shown that the spaces of conformal blocks associated to surfaces of *arbitrary* genus are finite-dimensional. Such CFTs are called rational CFTs (RCFTs). Categories of representations of VOAs having this property are examples of what is called modular tensor categories. A guide to the literature on the mathematical foundations of this theory and some non-rational generalizations can be found in [41]. The book [23] develops a similar perspective, together with mathematical background and relations to the theory of three-manifold invariants.

This theory provides the basis for the powerful characterization of the set of solutions to the conditions of crossing symmetry and modular invariance characterizing physical correlation functions in terms of the theory of modular tensor categories developed in [47] and references therein.

2.2.8.3 Seriously non-rational cases

From the point of view of theoretical physics, it seems unlikely that the finiteness conditions characterizing RCFTs are realized for generic CFTs. Extending the existing theory for RCFTs to seriously non-rational CFTs like Liouville theory is an important mathematical problem. The results on Liouville theory previously outlined indicate that the development of such an extension is not hopeless, after all. Indeed, one may view the mathematical structure defined by the known fusion and modular transformations for unitary representations of the Virasoro algebra with $c > 1$ as an example for a natural non-rational generalization of modular tensor categories [26]. It seems very likely that there exists a vast class of conformal field theories of non-rational type which is insufficiently understood at the time of writing.

2.3 Relations to integrable models[+]

There are various relations between CFT and the theory of integrable models, the main subject of this Les Houches summer school. It won't be possible to discuss them all, so we will pick a particular one, with a short guide to other relations between CFT and integrable models at the end. We have chosen to discuss the relation between CFT and the theory of isomonodromic deformations of ordinary differential equation on Riemann surfaces for the following reasons.

- The corresponding classically integrable equations, especially the Painlevé VI equation, appear in many different problems of mathematical physics. The relation to CFT appears to be beneficial for the study of the isomonodromic deformation problem [48].
- There are interesting connections to $\mathcal{N} = 2$ supersymmetric gauge theories in the context of the AGT-correspondence [48, 49].
- These connections feed back into the conformal bootstrap for $c = 1$ [25]. They may furthermore serve to illustrate many elements of the CFT formalism previously presented in an interesting example.
- This relation is part of a family of relations between CFT and the quantization of moduli spaces of flat connections on Riemann surfaces, which reveal important aspects of the mathematical nature of CFT.

After a review of the relation between the isomonodromic deformation equations and the Riemann–Hilbert problem, we will in the following explain why solving the

Riemann–Hilbert problem is equivalent to constructing particular conformal blocks for the free fermion super VOA. The isomonodromic tau-functions are thereby identified as chiral partition functions. This relation is a particular instance of the general relations between infinite Grassmannians, free fermions, and integrable hierarchies reviewed and elaborated e.g. in [50], and in the physics papers [11, 51]. It is then shown how the representation theory of the Virasoro algebra can be used to give a *constructive* solution of the Riemann–Hilbert problem, basically by constructing the relevant conformal blocks for the free fermion VOA in terms of the conformal blocks for the Virasoro algebra. Other relations between CFT and the theory of integrable models are discussed at the end of Section 2.3.

2.3.1 Isomonodromic deformations and tau-function

The isomonodromic deformation problem is one of the most important classically integrable systems in mathematical physics. It arises naturally in the study of flat connections on $C_{0,n}$. Any flat connection on $C_{0,n}$ is gauge equivalent to a holomorphic connection of the form $\partial_y - A(y)$, with matrix-valued functions $A(y)$ of the form

$$A(y) = \sum_{r=1}^{n} \frac{A_r}{y - z_r}. \tag{2.104}$$

We will assume that $A_1, \ldots, A_n$ satisfy $\sum_{k=1}^{n} A_k = 0$. Let us then consider the equation for parallel transport with respect to the connection $A(y)$,

$$\frac{\partial}{\partial y} \Psi(y) = A(y) \Psi(y). \tag{2.105}$$

In order to define a unique solution one may impose the initial condition $\Psi(y_0) = \text{id}$ at a point $y_0 \in C_{0,n}$. Equation (2.105) determines the analytic continuation of a solution $\Psi(y)$ along closed paths γ_r on $C_{0,n}$, denoted as $\Psi(\gamma_r.y)$, to be of the form

$$\Psi(\gamma_r.y) = \Psi(y) M_r, \tag{2.106}$$

with y-independent matrices M_r called monodromy matrices.

The fundamental group π_1 of $C_{0,n} := \mathbb{P}^1 \setminus \{z_1, \ldots, z_n\}$ has n generators $\gamma_1, \ldots, \gamma_n$ subject to one relation $\gamma_1 \circ \gamma_2 \circ \cdots \circ \gamma_n = 1$. Having fixed an initial point y_0 allows us to represent the generators γ_r of π_1 by closed paths starting and ending at y_0. The monodromies M_r define representations[12] ρ of $\pi_1(C_{0,n})$ in $G = \text{GL}(N, \mathbb{C})$ if $A(y)$ is a family of $N \times N$-matrices. This means that $M_k := \rho(\gamma_r) \in G$, $r = 1, \ldots, n$ satisfy the relation $M_n \cdot M_{n-1} \cdots M_1 = 1$. A change of initial point y_0 results in an overall conjugation with elements of G.

---

[12] Here understood as *anti*-homomorphisms $\rho : \pi_1(C_{0,n}) \to G$.

Variation of the positions z_r of the punctures on $C_{0,n}$, which appear as parameters in the differential equation (2.105), will generically modify the monodromies. It is, however, always possible to compensate the resulting changes by variations of the matrix residues A_r of $A(y)$. We will see that variations of the positions z_r will not change the monodromies of the connection $A(y)$ provided that the matrix residues $A_k = A_k(z)$ satisfy the following equations:

$$\partial_{z_k} A_k = -\sum_{l \neq k} \frac{[A_k, A_l]}{z_k - z_l},$$
$$\partial_{z_l} A_k = \frac{y_0 - z_k}{y_0 - z_l} \frac{[A_k, A_l]}{z_k - z_l}, \quad k \neq l, \qquad \partial_{y_0} A_k = -\sum_{l \neq k} \frac{[A_l, A_k]}{y_0 - z_l}. \qquad (2.107)$$

In the limit $y_0 \to \infty$ one finds the Schlesinger equations,

$$\partial_{z_k} A_k = -\sum_{l \neq k} \frac{[A_k, A_l]}{z_k - z_l},$$
$$\partial_{z_l} A_k = \frac{[A_k, A_l]}{z_k - z_l}, \quad k \neq l. \qquad (2.108)$$

The Schlesinger equations are non-linear partial differential equations for the matrices A_r. In special cases $n = 4$ it is known that one may reduce these equations to the Painlevé VI-equation.

The Schlesinger equations define Hamiltonian flows, generated by the Hamiltonians

$$H_r := \frac{1}{2} \operatorname*{Res}_{y=z_r} \operatorname{tr} A^2(y) = \sum_{s \neq r} \frac{\operatorname{tr}(A_r A_s)}{z_r - z_s}, \qquad (2.109)$$

using the Poisson structure

$$\{A(y) \overset{\otimes}{,} A(y')\} = \left[\frac{\mathcal{P}}{y - y'}, A(y) \otimes 1 + 1 \otimes A(y') \right], \qquad (2.110)$$

where $\mathcal{P}$ denotes the permutation matrix. The tau-function $\tau(\mathbf{z})$ is defined as the generating function for the Hamiltonians H_k,

$$H_k = \partial_{z_k} \log \tau(\mathbf{z}). \qquad (2.111)$$

Integrability of (2.111) is ensured by the Schlesinger equations (2.108). The concept of a tau-function plays a key role in the theory of many classically integrable models.

2.3.2 The Riemann–Hilbert problem

Solving the Schlesinger equations is closely related to the Riemann–Hilbert problem, as we shall now explain.

We consider the cases where the matrices M_r are diagonalizable, $M_r = C_r^{-1} e^{2\pi i D_r} C_r$, for a fixed choice of diagonal matrices D_r. The Riemann–Hilbert problem is to find a multivalued analytic matrix function $\Psi(y)$ on $C_{0,n}$ such that the monodromy along γ_r is represented by (2.106). The solution to this problem is unique up to left multiplication with single-valued matrix functions. In order to fix this ambiguity we need to specify the singular behaviour of $\Psi(y)$ at $y = z_r$, leading to the following refined version of the Riemann–Hilbert problem:

> Find a matrix function $\Psi(y)$ such that the following conditions are satisfied.
>
> (i) $\Psi(y)$ is multivalued, analytic, and invertible on $C_{0,n}$, and satisfies $\Psi(y_0) = 1$.
> (ii) There exist neighbourhoods of z_k, $k = 1, \ldots, n$ where $\Psi(y)$ can be represented as
>
> $$\Psi(y) = \hat{Y}^{(k)}(y) \cdot (y - z_k)^{D_k} \cdot C_k, \qquad (2.112)$$
>
> with $\hat{Y}^{(k)}(y)$ holomorphic and invertible at $y = z_k$, $C_k \in G$, and D_k being diagonal matrices for $k = 1, \ldots, n$.

If such a function $\Psi(y)$ exists, it is uniquely determined by the monodromy data $\mathbf{C} = (C_1, \ldots, C_n)$ and the diagonal matrices $\mathbf{D} = (D_1, \ldots, D_n)$.

It is known that generic representations $\rho: \pi_1(C_{0,n}) \to G$ can be realized as a monodromy representation of such a Fuchsian system, which means that a solution to the Riemann–Hilbert problem formulated above will generically exist.

We shall now briefly indicate how the Riemann–Hilbert problem is related to the isomonodromic deformation problem. Given a solution $\Psi(y)$ to the Riemann–Hilbert problem we may define a connection $A_y(y)$ as

$$A_y(y) := (\partial_y \Psi(y)) \cdot (\Psi(y))^{-1}. \qquad (2.113)$$

It follows from (ii) that $A_y(y)$ is a rational function of y which has the form

$$A_y(y) = \sum_{r=1}^n \frac{A_r(z)}{y - z_r}. \qquad (2.114)$$

One may similarly deduce from (ii) that the variations of $\Psi(y)$ with respect to z_r can be represented in the form

$$A_r(y) := (\partial_{z_r}\Psi(y))\cdot(\Psi(y))^{-1} = -\frac{A_r(z)}{y-z_r}. \qquad (2.115)$$

Commutativity of the partial derivatives ∂_y and ∂_{z_r} acting on $\Psi(y)$ implies differential equations for the matrices A_r which turn out to be the Schlesinger equations. This is how we get solutions to the isomonodromic deformation equations from solutions to the Riemann–Hilbert problem.

Given a solution to the isomonodromic deformation equations, one may, on the other hand, construct the connection $A(y)$ via (2.103) and study the fundamental matrix solution $\Psi(y)$ of the differential (2.104) normalized by $\Psi(y_0) = 1$. If the eigenvalues of A_k do not differ by integers, the resulting function $\Psi(y)$ will satisfy the previous conditions (i) and (ii) for certain matrices $C_1,\ldots,C_n$ and diagonal matrices $D_1,\ldots,D_n$.

The monodromies M_r of $\Psi(y)$ play the role of conserved quantities the existence of which is ensuring the integrability of the Schlesinger system.

2.3.3 Relation to free fermion CFT and infinite Grassmannians

It has been known for a while that there are deep relations between classical integrable models, the geometry of infinite Grassmannians, and the conformal field theory of free fermions. The relation to infinite Grassmannians is discussed in the classic reference [50], reviews of these relations together with the link to free fermions can be found e.g in the physics papers [11, 51]. We are now going to explain how the CFT formalism introduced previously can be used to get a unified picture of these relations.

To this aim we are first going to show how to encode a solution $\Psi(y)$ to the Riemann–Hilbert problem into the definition of conformal blocks for the free fermion VOA. Such conformal blocks will be represented by states w_ψ satisfying a set of invariance conditions analogous to the conformal Ward identities.

2.3.3.1 Free fermions

The free fermion super VOA is generated by fields $\psi_s(z), \bar{\psi}_s(z), s=1,\ldots,N$. The fields $\psi_s(z)$ will be arranged into a row vector $\psi(z) = (\psi_1(z),\ldots,\psi_N(z))$, while $\bar{\psi}(z)$ will be our notation for the column vector with components $\bar{\psi}_s(z)$. The modes of $\psi(z)$ and $\bar{\psi}(z)$, introduced as

$$\psi(z) = \sum_{n\in\mathbb{Z}} \psi_n z^{-n-1}, \quad \bar{\psi}(z) = \sum_{n\in\mathbb{Z}} \bar{\psi}_n z^{-n}, \qquad (2.116)$$

are row and column vectors with components $\psi_{s,n}$ and $\bar{\psi}_{s,n}$ satisfying the commutation relations

$$\{\psi_{s,n},\bar{\psi}_{t,m}\} = \delta_{s,t}\delta_{n,-m}, \quad \{\psi_{s,n},\psi_{t,m}\} = 0, \quad \{\bar{\psi}_{s,n},\bar{\psi}_{t,m}\} = 0. \qquad (2.117)$$

We will here consider a representation generated from a highest weight vector e_0 satisfying

$$\psi_{s,n} e_0 = 0, \quad n \geq 0, \qquad \bar{\psi}_{s,n} e_0 = 0, \quad n > 0. \tag{2.118}$$

The Fock space $\mathcal{F}$ is generated from e_0 by the action of the modes $\psi_{s,n}$, $n < 0$, and $\bar{\psi}_{s,m}$, $m \leq 0$.

2.3.3.2 Free fermion conformal blocks from solutions to the Riemann–Hilbert problem

We are going to construct states w_ψ in the dual $\mathcal{F}'$ of the fermionic Fock space $\mathcal{F}$, characterized by a set of Ward identities defined from a solution $\Psi(y)$ of the RH problem. Let us define the following infinite-dimensional spaces of multi-valued vector functions on $\tilde{C}_{0,n}$:

$$\begin{aligned}\mathcal{R} &= \left\{ v(y) \cdot \Psi(y); \; v(y) \in \mathbb{C}^N \otimes \mathbb{C}[\mathbb{P}^1 \setminus \{\infty\}] \right\}, \\ \bar{\mathcal{R}} &= \left\{ \Psi^{-1}(y) \cdot \bar{v}(y); \; \bar{v}(y) \in \mathbb{C}^N \otimes \mathbb{C}[\mathbb{P}^1 \setminus \{\infty\}] \right\},\end{aligned} \tag{2.119}$$

where v and $\bar{v}$ are row and column vectors with N components, respectively, and $\mathbb{C}[\mathbb{P}^1 \setminus \{\infty\}]$ is the space of meromorphic functions on $\mathbb{P}^1$ having poles at ∞ only. The elements of the space $\mathcal{R}$ represent solutions of a generalization of the RH problem where the condition of regularity at infinity has been dropped.

The vectors w_ψ we are about to define are required to satisfy the conditions

$$\psi[\bar{g}] w_\psi = 0, \qquad \bar{\psi}[g] w_\psi = 0, \tag{2.120}$$

where $g \in \mathcal{R}, \bar{g} \in \bar{\mathcal{R}}$, and the operators $\psi[\bar{g}]$ are constructed as

$$\psi[\bar{g}] = \frac{1}{2\pi i} \int_C dz \, \psi(z) \cdot \bar{g}(z), \qquad \bar{\psi}[g] = \frac{1}{2\pi i} \int_C dz \, g(z) \cdot \bar{\psi}(z), \tag{2.121}$$

with C being a circle separating ∞ from $z_1, \ldots, z_n$. One may recognize the identities (2.120) as analogues of the conformal Ward identities defining Virasoro conformal blocks where the role of the Virasoro algebra is taken by the free fermion VOA, and the role of the space of vector fields on C is taken by the sets $\mathcal{R}$ and $\bar{\mathcal{R}}$. The definition of the state w_ψ by means of the identities (2.120) can be considered as a fermionic analogue of the one-point localization discussed in Section 2.1.7 for the case of the Virasoro algebra.

It can easily be shown that the vector w_ψ is defined uniquely up to normalization by the identities (2.120). Using the notation $\langle v, w \rangle_\mathcal{F}$ for the pairing between a vector $v \in \mathcal{F}$ and a vector w in the dual $\mathcal{F}'$ of $\mathcal{F}$, this means that the identities (2.120) can be used to calculate the values of $\langle v, w_\psi \rangle_\mathcal{F}$ for $w_\psi \in \mathcal{F}'$ satisfying (2.120) and arbitrary $v \in \mathcal{F}$ in terms of $\langle e_0, w_\psi \rangle_\mathcal{F}$. It is not hard to check that this is the case. This implies that the space of conformal blocks for the free fermionic VOA is one dimensional.

2.3.3.3 Explicit construction of conformal blocks for the free fermion VOA

Associated with a solution $\Psi(y) \equiv \Psi(y_0, y)$ to the Riemann–Hilbert problem as previously formulated, we have constructed a free fermion conformal block w_Ψ. Let $G(x, y)$ be the matrix which has the two-point function

$$\langle \bar{\psi}_s(x) \psi_t(y) \rangle_\Psi \equiv \frac{\langle e_0, \bar{\psi}_s(x) \psi_t(y) w_\Psi \rangle}{\langle e_0, w_\Psi \rangle}, \tag{2.122}$$

as its matrix element in row s and column t. We are now going to show that $G(x, y)$ has a very simple relation to the solution $\Psi(x, y)$ of the Riemann–Hilbert problem, namely

$$G(x, y) = \frac{\mathbb{I}}{x - y} \Psi(x, y) = \frac{\mathbb{I}}{x - y} + R(x, y). \tag{2.123}$$

In order to see this, let us expand $G(x, y)$ around $x = \infty$ and $y = \infty$, respectively:

$$\begin{aligned} G(x, y) &= \sum_{l \geq 0} y^{-l-1} \bar{G}_l(x) & \bar{G}_l(x) &= -x^l \mathbb{I} + \sum_{k > 0} x^{-k} R_{kl}, \\ G(x, y) &= \sum_{k > 0} x^{-k} G_k(y), & G_k(y) &= y^{k-1} \mathbb{I} + \sum_{l \geq 0} y^{-l-1} R_{kl}. \end{aligned} \tag{2.124}$$

Note that rows of the matrix-valued functions $G_k(y)$, $k > 0$, and the columns of $\bar{G}_l(x)$, $l \geq 0$, defined in this way generate a basis for the spaces $\mathcal{R}$ and $\bar{\mathcal{R}}$ introduced in (2.119), respectively. We claim that a vector w_Ψ satisfying (2.120) can be represented in terms of the expansion coefficients R_{kl} introduced in (2.124) explicitly as

$$w_\Psi = \mathcal{N} \exp\left(-\sum_{k > 0} \sum_{l \geq 0} \psi_{-k} \cdot R_{kl} \cdot \bar{\psi}_{-l} \right) e_0. \tag{2.125}$$

This can be verified by a straightforward computation. In a similar way one may furthermore check that the two-point function defined by this conformal block is exactly the function $G(x, y)$ introduced in (2.123) above.

2.3.3.4 Chiral partition functions as tau-functions

Out of the free fermion VOA one may define a representation of the Virasoro algebra by introducing the energy–momentum tensor as

$$T(z) = \frac{1}{2} \lim_{w \to z} \sum_{s=1}^{N} \left(\partial_z \psi_s(w) \bar{\psi}_s(z) + \partial_z \bar{\psi}_s(w) \psi_s(z) + \frac{1}{(w - z)^2} \right). \tag{2.126}$$

Conformal blocks for the free fermion VOA represent conformal blocks for the Virasoro algebra defined via (2.126). We will see that the conformal Ward identities determine the dependence of the normalisation factor $\mathcal{N}$ in (2.125) on $\mathbf{z} = (z_1,\ldots,z_n)$. It turns out that $\mathcal{N}$ can be chosen in such a way that the isomonodromic tau-functions coincide with the chiral partition functions $\langle e_0, w_\Psi \rangle$.

To this aim let us compare the Taylor expansion of matrix functions $\Psi(y)$ solving the Riemann–Hilbert problem to the one of the two-point function (2.122). We have, on the one hand,

$$\Psi(y) = \mathbb{I} + (y - y_0)\mathcal{J}(y_0) + \frac{1}{2}(y - y_0)^2\left[\mathcal{J}^2(y_0) + \partial_y\mathcal{J}(y_0)\right] + \mathcal{O}((y-y_0)^3), \quad (2.127)$$

where $\mathcal{J}(y) = (\Psi(y))^{-1}\partial_y\Psi(y) = (\Psi(y))^{-1}A(y)\Psi(y)$. The residues of the trace part of $\mathcal{J}^2(y)$ are the Schlesinger Hamiltonians,

$$\frac{1}{2}\operatorname*{Res}_{y=z_r}(\operatorname{tr}\mathcal{J}^2(y)) = \frac{1}{2}\operatorname*{Res}_{y=z_r}(\operatorname{tr}A^2(y)) = \sum_{s\neq r}\frac{\operatorname{tr}(A_rA_s)}{z_r - z_s}. \quad (2.128)$$

Note, on the other hand, that the same expansion of $\Psi(y)$ around y_0 can be calculated using the OPE of the two fermionic fields in (2.122). At second order in the expansion around y_0 one finds the energy–momentum tensor $T(y)$ in the trace part of the expansion of $G(x,y)$ around $x = y$. Using only the Ward identities one can show[13] that

$$\langle e_0, T(y) w_\Psi \rangle_{\mathcal{F}} = \sum_{r=1}^{n}\left(\frac{\Delta_r}{(y-z_r)^2} + \frac{1}{y-z_r}\partial_{z_r}\right)\langle e_0, w_\Psi \rangle_{\mathcal{F}}, \quad (2.129)$$

where $\Delta_r = \operatorname{tr}(D_r^2)$, with D_r being the diagonal matrices introduced in (2.112).

The definition of the tau-function $\tau(\mathbf{z})$ relates the Schlesinger Hamiltonians H_r to the derivatives of $\log\tau(\mathbf{z})$ with respect to z_r, $r = 1,\ldots,n$, while the residues of the poles of the expectation value of $T(y)$ near $y = z_r$ are identified with the derivatives of the chiral partition functions via the definition of the conformal blocks. Comparison shows that we must have $H_r = \partial_{z_r}\log\langle e_0, w_\Psi\rangle$ implying that we can choose $\mathcal{N} = \mathcal{N}(\mathbf{z})$ in (2.125) in such a way that $\tau(\mathbf{z}) = \langle e_0, w_\Psi \rangle$, as claimed.

In the one-point localization discussed in Section 2.1.7, for the case of the Virasoro algebra one identifies the derivatives with respect to z_r with operators $L_{-1}^{(r)}$ acting on the vacuum representation which here gets replaced by the Fock space representation of the free fermion VOA. The isomonodromic deformation flows are thereby identified with

---

[13] In order to derive (2.129) one may start by noting that it is built into the definitions that the fermion two-point function permits an analytic continuation to the universal cover of $C_{0,n}$ with singularities of the form specified by (2.112) only at $x = z_r$ or $y = z_r$, $r = 1,\ldots,n$. Using the fact that the energy–momentum tensor (2.126) appears in the trace part of the free fermion OPE, one may define the expectation value $t(y)$ of the energy–momentum tensor, and show that it is holomorphic on $C_{0,n}$ with singular behaviour near z_r of the form (2.129). This implies that the free fermion conformal block is a particular conformal block on $C_{0,n}$ for the Virasoro algebra defined via (2.126), represented in terms of the one-point localization discussed in Section 2.1.7. It follows from (2.34) that the residues of $t(y)$ get represented in terms of the derivatives ∂_{z_r}. Working out the details of this argument is a good exercise.

commuting flows in the fermionic Fock space generated by commuting subalgebras of the Virasoro algebra.

2.3.3.5 Isomonodromic flows in infinite Grassmannians

The space $\mathcal{R}$ may be recognized as an element of the Grassmannian of the Hilbert space $\mathcal{H} = L^2(S^1) \otimes \mathbb{C}^N$, consisting of all closed subspaces W of $\mathcal{H}$ such that

(i) the orthogonal projection $W \to \mathcal{H}_+$ is a Fredholm operator, and

(ii) the orthogonal projection $W \to \mathcal{H}_-$ is a compact operator.

A subspace W can be defined by specifying a basis for W. The set of all subspaces W satisfying the conditions above forms an infinite-dimensional manifold called the infinite Grassmannian.

Generalizing the construction in Section 2.3.3.2 slightly, one gets a one-to-one correspondence between subspaces W in the infinite Grassmannian and states w_W in the fermionic Fock space. Abelian subalgebras in the free fermion VOA define commuting flows in the fermionic Fock space. The matrix elements $\langle e_0, w_W \rangle_\mathcal{F}$ can be used to represent the tau-functions of various integrable hierarchies, see e.g. [11, 51] for reviews.

In order to describe the isomonodromic deformation flows as flows in infinite Grassmannians one should choose as subspaces $W \subset \mathcal{H}$ the spaces $\mathcal{R}$ considered above

2.3.3.6 Discussion

Relations between the Riemann–Hilbert problem and the theory of free fermions were first established in [52] based on an explicit construction of fermionic twist fields. The relation between the vertex operator constructions of [52] and conformal field theory was discussed in [53], establishing the relation between tau-functions and expectation values of fermionic twist fields with the help of the fermionic construction of the energy–momentum tensor.

The approach as described follows a somewhat different route. Assuming that a solution to the Riemann–Hilbert problem exists, we have outlined a simple way to represent the isomonodromic tau-function as a fermionic matrix element using only the defining Ward identities. This explains in particular how the isomonodromic deformation problem fits into the general framework for integrable hierarchies based on the infinite Grassmannians [50]. Our approach is also related to the work [54] identifying the isomonodromic tau-functions as determinants of certain Cauchy–Riemann operators.

It remains to find more explicit and computable descriptions of the state w_Ψ. More recent versions of the fermionic twist field construction [49, 55, 56] fulfil this task, leading to explicit representations for the tau-functions in the case $N = 2$. We will in the following describe the approach of [49], followed by a brief guide to other vertex operator constructions.

2.3.4 Trace coordinates for moduli spaces of flat connections

As previously indicated, we will temporarily restrict attention to the case $N = 2$, and furthermore assume that the connection $A(y)$ is traceless, which implies that its holonomies are elements of $G = \mathrm{SL}(2,\mathbb{C})$. The goal of this section is to introduce useful coordinates for the space of monodromy data in this case which will be used in the construction of a solution to the Riemann–Hilbert problem, in Section 2.3.5. It will be fairly easy to remove the condition of vanishing trace later, while the generalization of the construction described in Section 2.3.5 to $N > 2$ represents an interesting open problem.

The Riemann–Hilbert correspondence between flat connections $\partial_y - A(y)$ and representations $\rho : \pi_1(C_{0,n}) \to G$ discussed previously relates the moduli space $\mathcal{M}_{\mathrm{flat}}(C_{0,n})$ of flat connections on $C_{0,n}$ to the so-called character variety $\mathcal{M}_{\mathrm{char}}(C_{0,n}) = \mathrm{Hom}(\pi_1(C_{0,n}), \mathrm{SL}(2,\mathbb{C}))/\mathrm{SL}(2,\mathbb{C})$. Useful sets of coordinates for $\mathcal{M}_{\mathrm{flat}}(C_{0,n})$ are given by the trace functions $L_\gamma := \mathrm{tr}\,\rho(\gamma)$ associated with simple closed curves γ on $C_{0,n}$. Minimal sets of trace functions that can be used to parameterize $\mathcal{M}_{\mathrm{flat}}(C_{0,n})$ can be identified using pants decompositions.

To simplify the exposition, we shall restrict attention to the case $n = 4$ in the following. By using pants decompositions one may easily generalize the following definitions to the cases with $n > 4$. Conjugacy classes of irreducible representations of $\pi_1(C_{0,4})$ are uniquely specified by seven invariants,

$$L_k = \mathrm{Tr}\,M_k = 2\cos 2\pi m_k, \qquad k = 1,\ldots,4, \tag{2.130a}$$

$$L_s = \mathrm{Tr}\,M_1 M_2, \qquad L_t = \mathrm{Tr}\,M_1 M_3, \qquad L_u = \mathrm{Tr}\,M_2 M_3, \tag{2.130b}$$

generating the algebra of invariant polynomial functions on $\mathcal{M}_{\mathrm{char}}(C_{0,n})$. The monodromies M_r are associated with the curves γ_r encircling the punctures z_r for $r = 1,2,3,4$, respectively. These trace functions satisfy the quartic equation

$$L_1 L_2 L_3 L_4 + L_s L_t L_u + L_s^2 + L_t^2 + L_u^2 + L_1^2 + L_2^2 + L_3^2 + L_4^2 =$$
$$= (L_1 L_2 + L_3 L_4) L_s + (L_1 L_3 + L_2 L_4) L_t + (L_2 L_3 + L_1 L_4) L_u + 4. \tag{2.131}$$

The affine algebraic variety defined by (2.131) is a concrete representation for the character variety of $C_{0,4}$. For fixed choices of $m_1,\ldots,m_4$ in (2.130a) one may use equation (2.131) to describe the character variety as a cubic surface in $\mathbb{C}^3$. This surface admits a parameterization in terms of coordinates (λ, κ) of the form

$$L_s = 2\cos 2\pi\lambda, \quad \begin{aligned} (\sin(2\pi\lambda))^2 L_t &= C_t^+(\lambda) e^{i\kappa} + C_t^0(\lambda) + C_t^-(\lambda) e^{-i\kappa}, \\ (\sin(2\pi\lambda))^2 L_u &= C_u^+(\lambda) e^{i\kappa} + C_u^0(\lambda) + C_u^-(\lambda) e^{-i\kappa}, \end{aligned} \tag{2.132}$$

where

$$C_u^\pm(\lambda) = -4 \prod_{s=\pm 1} \sin\pi(\lambda + s(m_1 \mp m_2)) \sin\pi(\lambda + s(m_3 \mp m_4)), \tag{2.133a}$$

$$C_u^0(\lambda) = 2\,[\cos 2\pi m_2 \cos 2\pi m_3 + \cos 2\pi m_1 \cos 2\pi m_4] \tag{2.134}$$
$$- 2\cos 2\pi\lambda\,[\cos 2\pi m_1 \cos 2\pi m_3 + \cos 2\pi m_2 \cos 2\pi m_4],$$

together with similar formulae for C_t^k, $k=\pm,0$. Explicit formulae expressing the monodromy matrices $M_r = M_r(\lambda,\kappa)$, $r=1,\ldots,4$, in terms of close relatives[14] of our coordinates (λ,κ) as defined can be found in [57]. These coordinates are also closely related to the coordinates used in [58].

2.3.5 Solving the Riemann–Hilbert problem using CFT

We will now describe how to solve the Riemann–Hilbert problem in terms of conformal blocks for the Virasoro algebra Vir_c at $c=1$. For the case $c=1$ we shall replace the parameters α_r and β used above by variables m_r and p giving the conformal dimensions as $\Delta_{m_r} = m_r^2$, for $r=1,\ldots,4$, and $\Delta_p = p^2$. The chiral vertex operator $V_{p_2,p_1}^m(z)$ maps $\mathcal{V}_{p_1}$ to $\mathcal{V}_{p_2}$. We shall use the notation $\mathsf{h}_s(y)$ for the chiral vertex operator associated to the degenerate representation $\mathcal{V}_{\frac{1}{2}}$ which maps $\mathcal{V}_p$ to $\mathcal{V}_{p-s/2}$, $s=\pm 1$, for all p.

We will find it convenient to assume that the vertex operators $V_{p_2,p_1}^m(z)$ are normalized by

$$V_{p_2,p_1}^m(z)\, v_{p_1} = N(p_2,m,p_1)\, z^{\Delta_{p_2}-\Delta_{p_1}-\Delta_m}\left[v_{p_2}+\mathcal{O}(z)\right], \qquad (2.135)$$

with $N(p_3,p_2,p_1)$ being chosen as

$$N(p_3,p_2,p_1) = \qquad (2.136)$$
$$= \frac{G(1+p_3-p_2-p_1)G(1+p_1-p_3-p_2)G(1+p_2-p_1-p_3)G(1+p_3+p_2+p_1)}{G(1+2p_3)G(1-2p_2)G(1-2p_1)},$$

where $G(p)$ is the Barnes G-function that satisfies $G(p+1)=\Gamma(p)G(p)$.

Definition: Out of the vector

$$v_\epsilon^{\mathrm{D}}(\lambda,\kappa) := \sum_{n\in\mathbb{Z}} e^{in\kappa}\, v_\epsilon(\lambda+n), \qquad v_\epsilon(p) := V_{m_4+\frac{\epsilon}{2},p}^{m_3}(z_3)\, V_{p,m_1}^{m_2}(z_2)\, V_{m_1,0}^{m_1}(z_1)\, e_0. \qquad (2.137\mathrm{a})$$

let us define (a) the matrix $\Psi(y;y_0)$ which has elements

$$\Psi_{s's}(y;y_0) := \frac{\pi s'(y_0-y)^{\frac{1}{2}}}{\sin 2\pi m_4}\, \frac{\left\langle v_{m_4},\, \mathsf{h}_{-s'}(y_0)\mathsf{h}_s(y)\, v_{s-s'}^{\mathrm{D}}(\lambda,\kappa)\right\rangle}{\left\langle v_{m_4},\, v_0^{\mathrm{D}}(\lambda,\kappa)\right\rangle}, \qquad (2.137\mathrm{b})$$

and (b) the function

$$\tau(z) = \left\langle v_{m_4},\, v_0^{\Gamma}(\lambda,\kappa)\right\rangle. \qquad (2.137\mathrm{c})$$

---

[14] The relation between our coordinates (λ,κ) and the coordinates used in [54] is given in [49, Eqn. (6.69)].

We claim that $\Psi_{s's}(y;y_0)$ represents the solution to the Riemann–Hilbert problem which has monodromy matrices $M_r(\lambda,\kappa)$ parameterized by the complex numbers (λ,κ), while $\tau(\mathbf{z})$ is the associated tau-function. The proof of this statement is given in [49]. This result yields in particular a proof of the relation between the tau-function for Painlevé VI and Virasoro conformal blocks discovered in [48]. At this point we only remark that the prefactor in (2.137b) ensures the normalization $\Psi(y_0;y_0) = 1$.

In order to show that (2.137) represents a solution to the Riemann–Hilbert problem we mainly need to compute the monodromies along the closed curves γ_r. It is straightforward to check that the analytic continuation with respect to the variable y can be represented in terms of a composition of the fusion and braiding operations introduced in Section 2.2.4. The main ingredient for carrying out this computation used in [49] are the fusion relations (2.79). The value of the normalization (2.135) adopted above is that it turns all the gamma functions appearing in the relations (2.79) into trigonometric functions.

While the details of this calculation are somewhat technical, it is fairly easy to understand the role of the Fourier transformation in the definition (2.137a). The relations between conformal blocks constructed from different gluing patterns imply, in particular, relations of the form

$$\mathsf{h}_{-s_1}(y)\, V^m_{p_2-\frac{s_1}{2},p_1}(z_2) = \sum_{s_2=\pm} B_{s_1 s_2} V^m_{p_2,p_1-\frac{s_2}{2}}(z_2) \mathsf{h}_{s_2}(y). \qquad (2.138)$$

Using such relations to compute the monodromy of the vector $\mathsf{h}_{s_2}(y)v_\epsilon(p)$ along the contour γ_u encircling z_2 and z_3 one finds a result of the form

$$\mathsf{h}_{s_1}(\gamma_u.y)\, v_{\epsilon+s_1\frac{1}{2}}(p) = \sum_{s_2=\pm} M_{s_1 s_2}(p,\mathsf{T}_p)\, \mathsf{h}_{s_2}(y)\, v_{\epsilon+s_2\frac{1}{2}}(p),$$

where T_p is the shift operator acting as $\mathsf{T}_p v_\epsilon(p) = v_\epsilon(p+1)$. This means that the monodromy matrices of degenerate fields are *operator valued*, acting non-trivially on the spaces of conformal blocks. The Fourier transformation in (2.137a) diagonalizes T_p. It may furthermore be checked that the traces of the monodromy matrices depend on λ only via $e^{2\pi i \lambda}$, which is unaffected by the Fourier transformation. We conclude that the Fourier transformation in (2.137a) maps the operatorial realization of the trace functions on spaces of conformal blocks to the classical expressions (2.133) used to define the coordinates (λ,κ).

By means of an argument very similar to that described in Section 2.3.3.4, one may then show that the function $\tau(\mathbf{z})$ defined in (2.137c) represents the isomonodromic tau-function. A completely different method to prove this result was presented in [59].

2.3.6 Non-abelian fermionization

We will now see how the free fermionic representation of tau-functions described in Section 2.3.3 is related to a simple generalization of the bosonic construction in Section 2.3.5.

To this aim let us introduce an additional free field φ_0,

$$\varphi_0(w)\varphi_0(z) \sim -\frac{1}{2}\log(w-z).$$

Note furthermore that we have

$$\Delta_{-b/2}\Big|_{b=i} = \frac{1}{4}, \qquad \Delta_{-b}\Big|_{b=i} = 1. \tag{2.139}$$

Let us construct the fields

$$\psi_s(z) := e^{i\varphi_0(z)}\mathsf{h}'_s(z), \qquad \bar{\psi}_s(z) := s e^{-i\varphi_0(z)}\mathsf{h}'_{-s}(z), \qquad s = \pm 1, \tag{2.140}$$

where $\mathsf{h}'_s(z)$ are related to the fields $\mathsf{h}_s(z)$ by a change of normalization

$$\mathsf{h}'_+(z) = \mathsf{h}_+(z), \qquad \mathsf{h}'_-(z) = \frac{\pi}{\sin 2\pi \mathsf{p}}\mathsf{h}_-(z), \tag{2.141}$$

with p satisfying $\mathsf{p} v_p = p v_p$ for all $v_p \in \mathcal{V}_p$. The fields $\psi_s(z)$, $\bar{\psi}_s(z)$ have the free fermion OPE,

$$\psi_s(w)\psi_{s'}(z) \sim \text{regular}, \tag{2.142a}$$

$$\psi_s(w)\bar{\psi}_{s'}(z) \sim \frac{\delta_{s,s'}}{w-z}. \tag{2.142b}$$

This means that the fields $\psi_s(w)$, $\bar{\psi}_s(w)$ generate a representation of the fermionic VOA.

A straightforward generalization of the construction described in Section 2.3.5 will allow us to represent the solution of the Riemann–Hilbert problem for monodromies in GL(2) in the form

$$\Psi_{s's}(y;y_0) := (y_0 - y)\frac{\langle v_0, \bar{\psi}_{s'}(y_0)\psi_s(y) w_\Psi \rangle_{\mathcal{F}}}{\langle v_0, w_\Psi \rangle_{\mathcal{F}}}, \tag{2.143}$$

where w_Ψ is a state in the fermionic Fock space which can be represented in bosonized form as the tensor product of the state defined in (2.137a) with a state in the Fock space generated by the modes of φ_0 created by a product of exponential vertex operators.

This furnishes a more explicit description of the state w_Ψ we had associated in Section 2.3.3 with a solution $\Psi(y)$ of the Riemann–Hilbert problem. One may, in particular, use the explicit formulae for the series expansions of Virasoro conformal blocks provided by the AGT-correspondence [19, 20] to obtain very detailed information on the isomonodromic tau-functions that had previously been unknown, as illustrated by the examples studied in [48].

Another approach was recently described in [55]. A generalization of the constructions presented in Section 2.3.3 can be used to construct fermionic twist fields from the solutions of the Riemann–Hilbert problem on the three-punctured sphere. The state w_Ψ can then be created from the vacuum by a composition of these fermionic twist fields. A similar construction was used in [56] to prove the formulae for the isomonodromic tau-functions furnished by the AGT-correspondence without explicit reference to conformal field theory.

2.3.7 Implications for bosonic CFT

Going back to the Schlesinger equations (2.108), note that the three points z_1, z_3, z_4 can be mapped to 0, 1, and ∞ using Möbius transformations. The Schlesinger system then reduces to the Painlevé VI equation,

$$-\frac{1}{2}\left(z(z-1)\zeta''\right)^2 = \qquad (2.144)$$

$$= \det\begin{pmatrix} 2m_1^2 & z\zeta' - \zeta & \zeta' + m_1^2 + m_2^2 + m_3^2 - m_4^2 \\ z\zeta' - \zeta & 2m_2^2 & (z-1)\zeta' - \zeta \\ \zeta' + m_1^2 + m_2^2 + m_3^2 - m_4^2 & (z-1)\zeta' - \zeta & 2m_3^2 \end{pmatrix},$$

satisfied by the logarithmic derivative of the tau-function,

$$\zeta(z) = z(z-1)\frac{d}{dz}\ln\tau. \qquad (2.145)$$

The representation (2.137c) of $\tau(\mathbf{z}) \equiv \tau_s(z)$ as a Fourier transform of the $c=1$ Virasoro conformal block can be written as

$$\tau_s(z) = \sum_{n\in\mathbb{Z}} \langle v_{m_4}, V^{m_3}_{m_4,\lambda_s+n}(1) V^{m_2}_{\lambda_s+n,m_1}(z) e_{m_1}\rangle e^{in\kappa_s}. \qquad (2.146)$$

We have renamed (λ,κ) to (λ_s,κ_s) in order to indicate the gluing pattern on which the definition of these coordinates was based. Assuming without loss of generality that $-\frac{1}{2} < \mathrm{Re}(\lambda) < \frac{1}{2}$, and taking into account the normalization (2.136) of the chiral vertex operators, we may deduce the asymptotic behaviour,

$$\tau_s(z) = \sum_{n=0,\pm 1} N(m_4,m_3,\lambda_s+n)N(\lambda_s+n,m_2,m_1) e^{in\kappa_s} z^{(\lambda_s+n)^2-m_1^2-m_2^2}$$
$$+ O\left(z^{\lambda_s^2-m_1^2-m_2^2+1}\right). \qquad (2.147)$$

This is equivalent to Jimbo's asymptotic formula [57, Theorem 1.1] expressing the asymptotic behavior of the Painlevé VI tau-function in terms of monodromy data.

It is perfectly possible, of course, to replace the construction (2.137b) based on the gluing pattern Γ_s by a similar construction using the gluing pattern Γ_u. The definition of the tau-function $\tau_u(z)$ associated with the gluing pattern Γ_u has asymptotics of the form

$$\tau_u(z) = \sum_{n=0,\pm 1} N(m_4, \lambda_u + n, m_1) N(\lambda_u + n, m_3, m_2) e^{in\kappa_u} (1-z)^{(\lambda_u+n)^2 - m_2^2 - m_3^2}$$
$$+ O\left((1-z)^{\lambda_u^2 - m_2^2 - m_3^2 + 1}\right), \tag{2.148}$$

expressed in terms of coordinates (λ_u, κ_u) defined by expressions similar to (2.132), but with trace function L_u now represented simply as $L_u = 2\cos(2\pi\lambda_u)$. Higher-order terms in the expansion (2.148) are uniquely determined by the Painlevé equation (2.144).

The pairs (λ_s, κ_s) and (λ_u, κ_u) are two different sets of coordinates for the character variety, and must therefore be related by a change of coordinates.[15] For given monodromy data μ one may use the differential equation (2.144) to determine the tau-function uniquely up to a multiplicative constant. It follows that the tau-functions $\tau_s(z;\mu)$ and $\tau_u(z;\mu)$ associated with the same monodromy data μ must coincide up to a multiplicative constant,

$$\tau_s(z,\mu) = F(\mu)\tau_u(z,\mu). \tag{2.149}$$

An explanation for this observation is offered by the relation to the conformal blocks of the free fermion VOA previously described. As the spaces of free fermion and free boson conformal blocks are one-dimensional, the fusion relations for these conformal blocks simplify to the form (2.149). The factor of proportionality $F(\mu)$ in (2.149) is independent of z, but depends in a highly non-trivial way on the monodromy data μ. An explicit formula for $F(\mu)$ has been proposed in [25] and a proof was given in [60].

This result has interesting consequences for the theory of conformal blocks of the Virasoro algebra at $c = 1$. Note that the Fourier transformations used to construct $\tau_s(z;\mu)$ and $\tau_u(z;\mu)$ out of conformal blocks can be regarded as changes of basis in the space of conformal blocks. Equation (2.149) expresses the fact that the fusion transformation becomes *diagonal* in the basis defined by the Fourier transformations. It is possible to invert the Fourier transformation in (2.146), allowing the authors of [25] to compute the fusion transformations of Virasoro conformal blocks at $c = 1$ explicitly.

It would be very interesting if these remarkable results could be used to demonstrate the crossing symmetry of the correlation functions of the $c = 1$ generalized minimal model defined by the three point functions (2.95) analytically.

2.3.8 Further relations to integrable models++

Apart from the relations between CFT and the isomondromic deformation problem described previously, there exist further important relations to quantum integrable models. We want to indicate some of them here.

---

[15] Beautiful descriptions of this change of coordinates can be found in [25, 58].

2.3.8.1 Semiclassical limit of Virasoro conformal blocks and the Garnier system

It seems interesting to note that conformal field theory is related to the isomonodromic deformation problem in the classical limit $c \to \infty$.

The Schlesinger system has an alternative description, known as the Garnier system, describing the isomonodromic deformations of the second-order ODE naturally associated with the matrix differential equation $(\partial_y - A(y))s = 0$. We refer to [61] for a review and further references.

As shown in [62], one may describe the leading asymptotics of Virasoro conformal blocks on $C_{0,n}$ with $n-3$ insertions of degenerate representations in terms of the generating function for a change of coordinates between two natural sets of Darboux coordinates for the Garnier system. One set of coordinates is natural for the Hamiltonian formulation of the Garnier system [61], the other coordinates are the complex Fenchel–Nielsen coordinates for $\mathcal{M}_{\text{flat}}(C_{0,n})$ introduced in Section 2.3.4. The result of [62] characterizes the leading classical asymptotics of Virasoro conformal blocks completely and clarifies in which sense conformal field theory represents a quantization of the isomonodromic deformation problem.

2.3.8.2 Yang's functions for the Hitchin systems

The Yang's function was introduced in [63] as a useful potential for the Bethe Ansatz equations in a certain integrable system. More recently it was realized in [64] that the quantization conditions in much larger classes of integrable systems can be described in terms of potentials generalizing the Yang's function. An large class of classically integrable systems is provided by the Hitchin systems, introduced in [65], associated with the choice of a Riemann surface C and a Lie algebra $\mathfrak{g}$. The Hamiltonians of the Hitchin system have been quantized in the work of Beilinson and Drinfeld on the geometric Langlands programme using methods from conformal field theory. A review can be found in [66].

More recently it was realized in [67] that the Yang's functions for the Hitchin systems associated with $\mathfrak{g} = \mathfrak{sl}_2$ are given by the $c \to \infty$ limits of the Virasoro conformal blocks associated with the Riemann surface C. This leads to a geometric characterization of the Yang's function as generating function for the variety of opers within $\mathcal{M}_{\text{flat}}(C)$, as independently proposed in [58]. There exist natural quantization conditions for the quantum Hitchin system which can be re-expressed in terms of the Yang's function [68].

2.3.8.3 Integrable structure of conformal field theories

Important relations between conformal field theory and quantum integrable models follow from the observation originally made in [69, 70] and later generalized in [71] that large abelian subalgebras can be defined as the subspaces of various VOAs generated by the elements commuting with a suitable set of screening operators. These screening operators can be identified with the interaction terms of perturbed CFTs in the light-cone representation. The elements commuting with the screening charges are thereby related to the conserved quantities in integrable perturbed CFTs. One may in this sense regard the infinite-dimensional abelian algebras generated by the commuting charges of integrable perturbed CFT as the remnant of the conformal symmetry surviving integrable perturbations.

This point of view was developed much further in the series of papers [72] by using techniques from the theory of quantum integrable models to construct the so-called T- and Q-operators, generating functions for the local and non-local integrals of motion, combined with profound studies of their analytic properties.

We see that conformal field theory has various relations to both classical and quantum integrable models both at finite central charge c and in the limit $c \to \infty$. These relations continue to represent an attractive topic for research in mathematical physics.

Acknowledgements

The author thanks the organizers of the Les Houches summer school for the invitation, and for setting up an inspiring event.

Special thanks to M. Alim, A. Balasubramanian, and O. Lisovyy for comments on the draft.

This work was supported by the Deutsche Forschungsgemeinschaft (DFG) through the collaborative Research Centre SFB 676 'Particles, Strings and the Early Universe', project A10.

References

[1] A. Recknagel, V. Schomerus, *Boundary Conformal Field Theory and the Worldsheet Approach to D-Branes*, Cambridge Monographs on Mathematical Physics Cambridge University Press, 2013.

[2] Y. Kawahigashi, *Conformal Field Theory, Tensor Categories and Operator Algebras*, J. Phys. **A48** (2015) 303001.

[3] V.G. Kac, A.K. Raina, *Bombay lectures highest weight representations of infinite-dimensional Lie algebras*, Advanced Series in Mathematical Physics 2. World Scientific, 1987.

[4] J. Teschner, *Liouville theory revisited*, Class. Quant. Grav. 18 (2001) R153–R222.

[5] G. Felder, J. Fröhlich, G. Keller, *On the structure of unitary conformal field theory. I. Existence of conformal blocks*, Comm. Math. Phys. 124 (1989) 417–463.

[6] A.A. Belavin, A.M. Polyakov, A.B. Zamolodchikov, *Infinite conformal symmetry in two-dimensional quantum field theory.* Nucl. Phys. **B241** (1984) 333–380.

[7] E. Frenkel, D. Ben-Zvi, *Vertex algebras and algebraic curves.* Second edition. Mathematical Surveys and Monographs, 88. American Mathematical Society, Providence, RI, 2004.

[8] G. Moore, N. Seiberg, *Classical and quantum conformal field theory*, Comm. Math. Phys. 123 (1989) 177–254.

[9] G. Felder, J. Fröhlich, G. Keller, *On the structure of unitary conformal field theory. II. Representation-theoretic approach*, Comm. Math. Phys. 130 (1990) 1–49.

[10] D. Kazhdan, G. Lusztig *Tensor structures arising from affine Lie algebras. I,II:* J. Amer. Math. Soc. 6 (1993) 905-947, 949-1011; *IV:* J. Amer. Math. Soc. 7 (1994) 383–453.

[11] L. Alvarez-Gaume, C. Gomez, G. Moore, C. Vafa, *Strings in the Operator Formalism*, Nucl. Phys. **B303** (1988) 455–521.

[12] J. Dubedat, *SLE and Virasoro Representations: Localization*, Comm. Math. Phys. **336** (2015) 695–760.
[13] D. Friedan, S. Shenker, *The Analytic Geometry of Two-Dimensional Conformal Field Theory*, Nucl. Phys. **B281** (1987) 509–545.
[14] A. Zamolodchikov, *Infinite additional symmetries in two-dimensional conformal field theory*, Theor. Math. Phys. **65** (1985) 1205–1213.
[15] R. E. Borcherds, *Vertex algebras, Kac-Moody algebras, and the Monster*, Proc. Natl. Acad. Sci. U.S.A. **83** (1986), 3068–3071.
[16] I. Frenkel, J. Lepowsky, A. Meurman, *Vertex Operator Algebras and the Monster*, Academic Press, Boston, 1989.
[17] S. Carpi, Y. Kawahigashi, R. Longo, M. Weiner, *From vertex operator algebras to conformal nets and back*. Memoirs of the American Mathematical Society **254** (2018) no. 1213, vi + 85.
[18] J.E. Tener, *Geometric realization of algebraic conformal field theories*, Preprint arXiv:1611.01176.
[19] L.F. Alday, D. Gaiotto, and Y. Tachikawa, *Liouville Correlation Functions from Four-dimensional Gauge Theories*, Lett. Math. Phys. **91** (2010) 167–197.
[20] V.A. Alba, V.A. Fateev, A.V. Litvinov, G.M. Tarnopolsky, *On combinatorial expansion of the conformal blocks arising from AGT conjecture*, Lett. Math. Phys. **98** (2011) 33–64.
[21] Al. Zamolodchikov, *Conformal symmetry in two-dimensional space: Recursion representation of conformal block*, Theor. Math. Phys. **73** (1987) 1088–1093.
[22] M. Cho, S. Collier, X. Yin, *Recursive Representations of Arbitrary Virasoro Conformal Blocks*. JHEP **1904** (2019) 018.
[23] B. Bakalov, A. Kirillov, Jr., *On the Lego-Teichmüller game*. Transform. Groups **5** (2000) 207–244.
[24] Y.-Z. Huang, *Virasoro Vertex Operator Algebras, the (Nonmeromorphic) Operator Product Expansion and the Tensor Product Theory*, Journal of Algebra **182** (1996) 201–234.
[25] N. Iorgov, O. Lisovyy, Yu. Tykhyy, *Painlevé VI connection problem and monodromy of $c=1$ conformal blocks*, JHEP **12** (2013) 029.
[26] J. Teschner, G. S. Vartanov, *Supersymmetric gauge theories, quantization of moduli spaces of flat connections, and conformal field theory*. Adv. Theor. Math. Phys. **19** (2015) 1–135.
[27] Xi Yin, *Conformal Bootstrap in Two Dimensions*, Talk at String-Math 2017, available at https://stringmath2017.desy.de/e45470/
[28] S. Collier, P. Kravchuk, Y.-H. Lin, X. Yin, *Bootstrapping the Spectral Function: On the Uniqueness of Liouville and the Universality of BTZ*. JHEP **1809** (2018) 150
[29] Al.Zamolodchikov *On the Three-point Function in Minimal Liouville Gravity*, Preprint arXiv:hep-th/0505063.
[30] V. Dotsenko, V. Fateev. *Four point correlation functions and the operator algebra in the two-dimensional conformal invariant theories with the central charge $c < 1$*, Nucl.Phys. **B251** (1985) 691–734.
[31] S. Ribault, R. Santachiara, *Liouville theory with a central charge less than one*, JHEP **1508** (2015) 109.
[32] Y. Ikhlef, J.L. Jacobsen, H. Saleur, *Three-point functions in $c \leq 1$ Liouville theory and conformal loop ensembles*. Phys. Rev. Lett. **116** (2016) 130601.
[33] I. Runkel, G.M.T. Watts, *A Nonrational CFT with $c = 1$ as a limit of minimal models*. JHEP **0109** (2001) 006.
[34] Volker Schomerus, *Rolling tachyons from Liouville theory*, JHEP **0311** (2003) 043.
[35] H. Dorn, H.-J. Otto, *Two and three-point functions in Liouville theory*. Nucl. Phys. B **429** (1994) 375–388.

[36] A.B. Zamolodchikov, Al.B. Zamolodchikov, *Structure Constants and Conformal Bootstrap in Liouville Field Theory*. Nucl. Phys. **B477** (1996) 577–605.
[37] J.Teschner, *On the Liouville three-point function*. Phys.Lett., **B363** (1995) 63.
[38] F. David, A. Kupiainen, R. Rhodes, V. Vargas, *Liouville Quantum Gravity on the Riemann sphere*, Comm. Math. Phys. **342** (2016) 869–907.
[39] A. Kupiainen, R. Rhodes, V. Vargas, *Integrability of Liouville theory: proof of the DOZZ Formula*. Preprint arXiv:1707.08785.
[40] J. Teschner, *A lecture on the Liouville vertex operators*, Int. J. Mod. Phys. **A19S2** (2004) 436–458.
[41] Y.-Z. Huang, J. Lepowsky, *Tensor categories and the mathematics of rational and logarithmic conformal field theory*, J.Phys. **A46** (2013) 494009.
[42] L. Hadasz, P. Jaskolski, P. Suchanek, *Modular bootstrap in Liouville field theory*, Phys. Lett **B685** (2010) 79–85.
[43] A. Cappelli, C. Itzykson, J.-B. Zuber, *The A-D-E classification of minimal and $A_1^{(1)}$ conformal invariant theories*, Comm. Math. Phys. **113** (1987) 1–26.
[44] L.F. Alday, D. Gaiotto, S. Gukov, Y. Tachikawa, H. Verlinde, *Loop and surface operators in $\mathcal{N}=2$ gauge theory and Liouville modular geometry*, J. High Energy Phys. **1001** (2010) 113.
[45] N. Drukker, J. Gomis, T. Okuda, J. Teschner, *Gauge Theory Loop Operators and Liouville Theory*, J. High Energy Phys. **1002** (2010) 057.
[46] Y. Zhu, *Modular invariance of characters of vertex operator algebras*, J. AMS **9** (1996) 237–302.
[47] J. Fuchs, I. Runkel, C. Schweigert, *TFT construction of RCFT correlators IV: Structure constants and correlation functions*, Nucl. Phys. **B715** (2005) 539–638.
[48] O. Gamayun, N. Iorgov, O. Lisovyy, *Conformal field theory of Painlevé VI*, JHEP **10** (2012) 038.
[49] N. Iorgov, O. Lisovyy, J. Teschner, *Isomonodromic Tau-Functions from Liouville Conformal Blocks*, Comm. Math. Phys. **336** (2015) 671–694.
[50] G. Segal, G. Wilson, *Loop groups and equations of KdV type*, Publ. Math. IHES **61** (1985) 5–65.
[51] R. Dijkgraaf, *Intersection theory, integrable hierarchies and topological field theory*, NATO Sci. Ser. B **295** (1992) 95–158.
[52] M. Sato, T. Miwa, M. Jimbo, *Holonomic quantum fields. II – The Riemann-Hilbert Problem*, Publ. RIMS, Kyoto Univ. **15** (1979) 201–278.
[53] G.W. Moore, *Geometry of the string equations*, Comm. Math. Phys. **133** (1990) 261–304.
[54] J. Palmer, *Determinants of Cauchy-Riemann operators as τ-functions*, Acta Appl. Math. **18** (1990) 199–223.
[55] P. Gavrylenko, A. Marshakov, *Free fermions, W-algebras and isomonodromic deformations*, Theor. Math. Phys. **187** (2016) 649–677.
[56] P. Gavrylenko, O. Lisovyy, *Fredholm determinant and Nekrasov sum representations of isomonodromic tau functions*. Commun. Math. Phys. **363** (2018) 1–58.
[57] M. Jimbo, *Monodromy problem and the boundary condition for some Painlevé equations*. Publ. Res. Inst. Math. Sci. **18** (1982), no. 3, 1137–1161.
[58] N. Nekrasov, A. Rosly, S. Shatashvili, *Darboux coordinates, Yang-Yang functional, and gauge theory*, Nucl. Phys. Proc. Suppl. **216** (2011) 69–93.
[59] M.A. Bershtein, A.I. Shchechkin, *Bilinear Equations on Painlevé τ Functions from CFT*, Comm. Math. Phys. **339** (2015) 1021–1061.
[60] A. Its, O. Lisovyy, A. Prokhorov, *Monodromy dependence and connection formulae for isomonodromic tau functions*. Duke Math. J. **167** (2018) no.7, 1347–1432.

[61] K. Iwasaki, H. Kimura, S. Shimomura, and M. Yoshida. *From Gauss to Painlevé, a Modern Theory of Special Functions*, Volume **E 16**. Aspects of Mathematics, 1991.
[62] J. Teschner, *Semiclassical Limit of Virasoro conformal blocks and the isomonodromic deformation problem*, Preprint arXiv:1707.07968.
[63] C.N. Yang, C.P. Yang, *Thermodynamics of a one-dimensional system of bosons with repulsive delta-function interaction*, J. Math. Phys. **10** (1969) 1115.
[64] N. Nekrasov, S. Shatashvili, *Quantization of Integrable Systems and Four Dimensional Gauge Theories*, In: Proceedings of the 16th International Congress on Mathematical Physics, Prague, August 2009, P. Exner, Editor, pp. 265–289, World Scientific 2010, p. 265–289.
[65] N. Hitchin, *Stable bundles and integrable systems*, Duke Math. J. **54** (1987) 91–114.
[66] E. Frenkel, *Lectures on the Langlands Program and Conformal Field Theory*, in: Frontiers in Number Theory, Physics, and Geometry II (P. Cartier, P. Moussa, B. Julia, and P. Vanhove, eds.), pp. 387–533. Springer, 2007.
[67] J. Teschner, *Quantization of the Hitchin moduli spaces, Liouville theory, and the geometric Langlands correspondence I*. Adv. Theor. Math. Phys. **15** (2011) 471–564.
[68] J. Teschner, *Quantisation conditions of the quantum Hitchin system and the real geometric Langlands correspondence*. In: Geometry and Physics: A Festschrift in Honour of Nigel Hitchin, J.E. Andersen, A. Dancer, O. Garcia-Prada (Eds): Oxford University Press (2018).
[69] R. Sasaki, I. Yamanaka, *Virasoro algebra, vertex operators, quantum Sine- Gordon and solvable Quantum Field theories*, Adv. Stud. in Pure Math. **16** (1988) 271–296.
[70] T. Eguchi, S.K. Yang, *Deformation of conformal field theories and soliton equations*, Phys. Lett. **B224** (1989) 373–378.
[71] B. Feigin, E. Frenkel, *Free field resolutions in affine Toda field theories*, Phys. Lett. **B276** (1992) 79–86.
[72] V.V. Bazhanov, S.L. Lukyanov, A.B. Zamolodchikov, *Integrable structure of conformal field theory*, Part 1: Comm. Math. Phys. **177** (1996) 381–398, Part 2: Comm. Math. Phys. **190** (1997) 247–278, Part 3: Comm. Math. Phys. **200** (1999) 297–324.

3
Lectures on the holographic duality of gauge fields and strings

Gordon W. Semenoff

Department of Physics and Astronomy, University of British Columbia, 6224 Agricultural Road, Vancouver, British Columbia, Canada V6T 1Z1

Chapter Contents

3 Lectures on the holographic duality of gauge fields and strings   121
Gordon W. SEMENOFF

    Preface   123
  3.1 Preamble about gauge fields   123
  3.2 The Wilson loop   129
  3.3 The large N expansion   134
  3.4 D branes, black branes, and the Maldacena conjecture   142
  3.5 $\mathcal{N}=4$ superconformal quantum field theory   150
  3.6 Holographic Wilson loops   160
  3.7 Epilogue   179
    References   179

Preface

This is a summary of lecture notes that I have presented as a pedagogical review of some of the basics of the holographic duality between string theory and gauge field theory. I presented these lectures to a mixed audience of students and early career researchers who were all theoretical physicists but with diverse backgrounds. I have made no attempt to be complete, or to give a comprehensive review of the subject or even a comprehensive introduction. There are plenty of reviews [30] [31] [32] [33] [41] [44] [47] [52] [53] [54] and even a few excellent books [57] [3] [62] [63] [64] [65] available and they more than adequately fill the gap. What I have prepared is a discussion of some aspects of the AdS/CFT correspondence as I best understand them, with the hope that it will help students of the school where the lectures are presented to appreciate the more advanced courses on more specialized topics which will come later.

3.1 Preamble about gauge fields

In this section we shall begin by sketching some of the basics of Yang–Mills theory [1]. I assume that students have seen such an introduction already. What I have written here is intended to define some of the ideas and to fix some of the notation and conventions for the material that will follow.

3.1.1 Basics

Throughout these lectures, we will use the natural system of units where the physical constants $\hbar = 1$ and $c = 1$. We will consider quantum field theories in either Minkowski space or Euclidean space. In four space-time dimensions, the Cartesian coordinates of Minkowski space are the time, x^0, and the three space dimensions, x^1, x^2, x^3. They combine to form the position four-vector, denoted by x^μ. Lorentz indices are lowered and raised by the Minkowski metric,

$$\eta_{\mu\nu} = \begin{bmatrix} -1 & 0 & 0 & 0 \\ 0 & 1 & 0 & 0 \\ 0 & 0 & 1 & 0 \\ 0 & 0 & 0 & 1 \end{bmatrix} \quad (3.1)$$

and its inverse, $\eta^{\nu\lambda}$ (so that $\eta_{\mu\nu}\eta^{\nu\lambda} \equiv \sum_\nu \eta_{\mu\nu}\eta^{\nu\lambda} = \delta_\mu^\lambda$). In this expression and hereafter, we use the Einstein summation convention where repeated up and down indices are assumed to be summed. In Euclidean space, the metric is simply the unit matrix, $\delta_{\mu\nu}$. An incremental translation dx^μ typically has an up index and a derivative by a four-vector has down indices, $\partial_\mu = \frac{\partial}{\partial x^\mu}$. These indices can be lowered and raised by the metric, $dx_\mu \equiv \eta_{\mu\nu} dx^\nu$ or $\partial^\mu = \eta^{\mu\nu}\partial_\nu$.

The basic dynamical variable of Yang–Mills theory is a connection field

$$A_\mu(x)$$

with a lower Lorentz index, so that the combination $A \equiv A_\mu(x)dx^\mu$ is a one-form. At the same time as being a one-form, the connection takes values in the Lie algebra of the gauge group. We will practically exclusively use the Lie group $U(N)$ for the gauge group. In that case, for each value of the index μ and for each value of coordinates, x, $A_\mu(x)$ is a Hermitian matrix. It obeys $A_\mu^\dagger(x) = A_\mu(x)$, where

$$[A_\mu^\dagger(x)]_{ab} = [A_\mu^*(x)]_{ba}. \tag{3.2}$$

Sometimes, we will also be interested in the Lie algebra for the Lie group $SU(N)$. In that case $A_\mu(x)$ are traceless as well as Hermitian matrices. It is possible and straightforward to generalize our discussion to other Lie algebras if it is needed.

The connection field is used to form covariant derivatives. Covariant derivatives are designed to act on wave functions. The wave function of a quark is a space-time dependent object with N complex components, which we shall denote by

$$\Psi(x) = \begin{bmatrix} \psi_1(x) \\ \psi_2(x) \\ \dots \\ \psi_N(x) \end{bmatrix}. \tag{3.3}$$

A gauge transformation maps it to

$$\Psi(x) \to g(x)\Psi(x) \tag{3.4}$$

where, for each value of the space-time coordinates, x, $g(x)$ is an $N \times N$ unitary matrix, that is, an $N \times N$ matrix which obeys $g^\dagger(x)g(x) = 1$ and $g(x)g^\dagger(x) = 1$. The covariant derivative of the wave function is defined as

$$D_\mu \Psi(x) \equiv \left(\partial_\mu - iA_\mu(x)\right)\Psi(x). \tag{3.5}$$

The purpose of a covariant derivative is to define a derivative in such a way that it is compatible with gauge transformations. Accordingly, the covariant derivative of the wave function must gauge transform in the same way as the wave function, that is,

$$D_\mu \Psi(x) \to g(x) D_\mu \Psi(x). \tag{3.6}$$

This fixes the gauge transformation of the connection so that the gauge transformations are

$$A_\mu(x) \to g(x)A_\mu(x)g^\dagger(x) - i\partial_\mu g(x)g^\dagger(x) \tag{3.7}$$
$$\Psi(x) \to g(x)\Psi(x). \tag{3.8}$$

A basic notion of geometry is that of parallel transport. It is a way of transporting a vector along a curve in such a way that it remains 'parallel' to its original orientation. Here, the wave function $\Psi(x)$ in (3.3) is the N component vector with complex valued components and a change in orientation of this vector is the multiplication by an $N \times N$ unitary matrix, as in (3.8). We can think of these gauge transformations (3.7) and (3.8) as the analogue of general coordinate transformations in Riemannian geometry. The data which defines what is meant by 'parallel' is stored in the connection $A_\mu(x)$. The result of parallel transport of the wave function $\Psi(x_1)$ along a curve C to the point x_2 is $\Psi(x_2) = U_C(x_2, x_1)\Psi(x_1)$, where $U_C(x_2, x_1)$ is a unitary matrix which we will construct in the paragraphs to follow.

Consider the parametric representation of a curve, C, in space-time given by four functions $x^\mu(\tau)$ of the curve parameter τ. As τ runs over its range, the functions $x^\mu(\tau)$ trace the curve C in space-time. For the purposes of the following argument, C is an open curve (that is, a curve with distinct beginning and end points).

The wave function is constant along the curve C if, at each point τ of that curve,

$$\dot{x}^\mu(\tau) D_\mu \Psi(x(\tau)) = 0. \tag{3.9}$$

We have used the notation $\dot{x}^\mu(\tau) = \frac{d}{d\tau} x^\mu(\tau)$. Equation (3.9) can be integrated to show that $\Psi(x)$ is constant along the curve C if, for any two points $x_1^\mu = x^\mu(\tau_1)$ and $x_2^\mu = x^\mu(\tau_2)$ lying on C,

$$\Psi(x_2) = U_C(x_2, x_1)\Psi(x_1),$$

where $U_C(x_2, x_1)$ is the path-ordered exponential of the line integral of the connection field along C,

$$U_C(x_2, x_1) = \mathcal{P} e^{i \int_{\tau_1}^{\tau_2} d\tau \dot{x}^\mu(\tau) A_\mu(x(\tau))} \tag{3.10}$$

$$\equiv \sum_{n=0}^{\infty} \frac{i^n}{n!} \int_{\tau_1}^{\tau_2} d\tau_1 \dot{x}_{\mu_1}(\tau_1) \ldots \int_{\tau_1}^{\tau_2} d\tau_n \dot{x}_{\mu_n}(\tau_n) \mathcal{P} A_{\mu_1}(x(\tau_1)) \ldots A_{\mu_n}(x(\tau_n)). \tag{3.11}$$

The symbol $\mathcal{P}$ orders the matrices so that those with later arguments are to the left of those with earlier arguments. $U_C(x_2, x_1)$ is itself a unitary matrix which tells us how the wave function $\Psi(x)$ is re-oriented as it is parallel transported along C. Moreover,

$$U_C^{-1}(x_2, x_1) = U_C^\dagger(x_2, x_1) = U_C(x_1, x_2).$$

Also, under the gauge transformation (3.7) and (3.8),

$$U_C(x_2, x_1) \to g(x_2) U_C(x_2, x_1) g^{-1}(x_1), \tag{3.12}$$

so the parallel transported wave function has the correct gauge transformation property,

$$U_C(x_2, x_1)\Psi(x_1) \to g(x_2) U_C(x_2, x_1)\Psi(x_1) \quad \text{when} \quad \Psi(x_1) \to g(x_1)\Psi(x_1),$$
$$A_\mu(x) \to g(x) A_\mu(x) g^\dagger(x) - i\partial_\mu g(x) g^\dagger(x). \tag{3.13}$$

Generally, parallel transport depends on the path, C. A measure of its path dependence is called the curvature of the connection, defined as a commutator of covariant derivatives,

$$F_{\mu\nu}(x) = i[D_\mu, D_\nu] = \partial_\mu A_\nu - \partial_\nu A_\mu - i[A_\mu, A_\nu]. \tag{3.14}$$

Each component of this curvature tensor is itself an $N \times N$ Hermitian matrix. Under a gauge transformation

$$F_{\mu\nu}(x) \to g(x) F_{\mu\nu}(x) g^\dagger(x). \tag{3.15}$$

In any region of space-time where $F_{\mu\nu} \neq 0$ the parallel transport of a wave function depends on the curve along which it is transported. $F_{\mu\nu}$ cannot be any arbitrary set of six Hermitian matrices. It is constrained by the fact that it is derived from a connection, as in equation (3.14). The fact that the curvature is a commutator of covariant derivatives plus the Jacobi identity for commutators,

$$[D_\mu [D_\nu, D_\lambda]] + [D_\mu [D_\lambda, D_\mu]] + [D_\lambda [D_\mu, D_\nu]] = 0,$$

requires that $F_{\mu\nu}(x)$ satisfies the Bianchi identity,

$$D_\mu F_{\nu\lambda}(x) + D_\nu F_{\lambda\mu}(x) + D_\lambda F_{\mu\nu}(x) = 0. \tag{3.16}$$

Note that, in the covariant derivative of $F_{\mu\nu}(x)$, the connection appears in a commutator,

$$D_\mu F_{\nu\lambda}(x) \equiv \partial_\mu F_{\nu\lambda}(x) - i[A_\mu(x), F_{\nu\lambda}(x)].$$

The quark wave function that we have considered here is an N-component complex vector and it transforms in the fundamental representation of the gauge group, that is, like $\Psi(x) \to g(x)\Psi(x)$. There are other representations that are of interest to us. One which will appear a lot in the following is the *adjoint representation*. In that case, the wave function is an $N \times N$ complex matrix and its gauge transformation is

$$\psi(x) \to g(x)\psi(x)g^{-1}(x). \tag{3.17}$$

The covariant derivative acts on such a wave function as

$$D_\mu \Psi(x) = \partial_\mu \Psi(x) - i[A_\mu(x), \Psi(x)].$$

Under parallel transport along C, it obeys

$$\Psi(x_1) \to U_C(x_2, x_1)\Psi(x_1)U_C^\dagger(x_2, x_1) = U_C(x_2, x_1)\Psi(x_1)U_C(x_1, x_2). \tag{3.18}$$

This consideration can easily be generalized to wave functions which transform in higher representations of the gauge group.

3.1.2 Yang–Mills theory

In Yang–Mills theory, the connection $A_\mu(x)$ becomes a dynamical variable. The connection itself is a redundant description of this variable; more properly said, the gauge orbit of a connection is the dynamical variable. A gauge orbit is an equivalence class of connections where the equivalence relation is $A \sim A'$ if A and A' are related by a gauge transformation. The principle of gauge invariance requires that all physical quantities in Yang–Mills theory should be gauge invariant. This principle is particularly important when the Yang–Mills theory is quantized, since the logical and mathematical consistency of the theory depends on it.

The dynamics of a classical or quantum mechanical theory must be fixed by specifying the equations of motion for the dynamical variables. The equations of motion can be obtained from an action by using a variational principle. The usual action is one which is quadratic in the curvature, so that the dynamics tend to favour field configurations with vanishing curvature,[1]

$$S_{\text{YM}} = \int dx\, \mathcal{L}_{\text{YM}}, \qquad \mathcal{L}_{\text{YM}} = -\frac{1}{2g_{\text{YM}}^2} \text{Tr}\left(F_{\mu\nu} F^{\mu\nu}\right). \tag{3.23}$$

Here, the integration is over space-time. The symbol Tr means taking the trace of the matrix

$$\text{Tr}\, M \equiv \sum_{a=1}^{N} M_{aa}. \tag{3.24}$$

---

[1] We will at times assume that the four-dimensional space-time is Euclidean, rather than Lorentzian. In the Euclidean case, $-\text{Tr} F_{\mu\nu} F^{\mu\nu}$ is replaced by $\text{Tr} F^{\mu\nu} F_{\mu\nu}$ which we sometimes write as $\text{Tr} F_{\mu\nu} F_{\mu\nu}$. The factor of $\frac{1}{2}$ in the Lagrangian density is conventional. There is another way of writing gauge fields by expanding the connection in a complete basis for $N \times N$ Hermitian matrices, T^a, $a = 1, 2, ..., N^2$,

$$A_\mu(x) = \sum_{a=1}^{N^2} T^a A_\mu^a(x). \tag{3.19}$$

Also,

$$F_{\mu\nu}(x) = \sum_{a=1}^{N^2} T^a F_{\mu\nu}^a(x). \tag{3.20}$$

Typically, the basis matrices have a commutation relation

$$[T^a, T^b] = if^{abc} T^c \tag{3.21}$$

and they are normalized so that $\text{Tr}\, T^a T^b = \frac{1}{2} \delta^{ab}$. With this notation, the Lagrangian density has the conventional normalization

$$\mathcal{L}_{\text{YM}} = -\frac{1}{4g_{\text{YM}}^2} \sum_{a=1}^{N^2} F_{\mu\nu}^a F^{a\mu\nu}, \tag{3.22}$$

with $F_{\mu\nu}^a = \partial_\mu A_\nu^a - \partial_\nu A_\mu^a - f^{abc} A_\mu^b A_\nu^c$.

Matrices appearing in a trace can be permuted cyclically, $\text{Tr}ABC = \text{Tr}BCA = \text{Tr}CAB$. One can use this cyclicity of the trace to show that the Lagrangian density $\mathcal{L}$ in (3.23) is gauge invariant.

The parameter g_{YM} is the coupling constant. Dimensional analysis shows that, in four-dimensional Yang–Mills theory, it is a dimensionless constant and four-dimensional classical Yang–Mills theory therefore has no parameters with non-zero scaling dimensions. It is thus scale invariant and it turns out to be conformally invariant. The scale invariance generally does not survive quantization, except in some special cases, such as the maximally supersymmetric Yang–Mills theory which will be of central interest to us later on in this chapter. In that case, the content of the theory in quark fields is carefully tuned to result in maximal supersymmetry, and at the same time it is scale invariant in both the classical and quantum theories. We will return to a detailed discussion of this theory in later sections.

The equation of motion that results from applying the variational principle to the action of Yang–Mills theory is

$$D_\mu F^{\mu\nu}(x) = 0. \tag{3.25}$$

These are the classical Yang–Mills field equations. Together with the Bianchi identity (3.16), they are the analogue of Maxwell's equations for Yang–Mills theory or, rather, the Yang–Mills theory equations with gauge group $U(1)$ are the source-free Maxwell equations of classical electrodynamics.

Of most interest to us will be quantized Yang–Mills theory. We shall take the quantization as being defined by the functional integral where, for example, time-ordered correlation functions of gauge invariant operators are obtained as

$$\langle \mathcal{T} O_1(x_1) \ldots O_n(x_n) \rangle = \frac{\int dA_\mu(x) e^{iS[A]} O_1(x_1) \ldots O_n(x_n)}{\int dA_\mu(x) e^{iS[A]}},$$

where S is the action given in equation (3.23) and $O_i(x_i)$ are gauge invariant operators constructed from the connection field, $A_\mu(x)$ and its derivatives. Here, we have assumed that the space-time is Minkowski space. If it were Euclidean, rather than Minkowski, the functional integral would be

$$\langle \mathcal{T} O_1(x_1) \ldots O_n(x_n) \rangle = \frac{\int dA_\mu(x) e^{-S[A]} O_1(x_1) \ldots O_n(x_n)}{\int dA_\mu(x) e^{-S[A]}},$$

where $\mathcal{T}$ now denotes Euclidean time ordering. This functional integral is not computable in space-time dimensions greater than two. Even getting an understanding of its salient properties has been a difficult problem which has been intensively researched over the past fifty years and this research continues today. The best understood analytic tool is perturbation theory, which is an asymptotic expansion in powers of the coupling g_{YM}. The limit at $g_{\text{YM}} = 0$ is exactly solvable and perturbation theory gives a systematic technique for computing corrections to that limit when g_{YM} is small enough. However, this expansion has many subtleties and, in generic Yang–Mills theory, it can only be applied in a certain kinematic regime. There is an interesting alternative expansion of the theory about the limit where N is taken to be large, where N is the rank of the $U(N)$ gauge group. However, again, in dimensions greater than two, even the leading term in the large N expansion, that is, the solution of the theory when N is taken to infinity, is thus far absent. What we will learn in the following is that, for certain versions of Yang–Mills theory, the duality with certain solutions of string theory will make accessible the leading order of a double expansion, one where a large N limit is taken while holding the combination $g_{YM}^2 N$ fixed, and then a limit of large $g_{YM}^2 N$ is taken. Moreover, the technology for correcting the semi-classical limit of string theory should make this limit correctable in a systematic expansion in $1/N$ and in $1/g_{YM}^2 N$. Developing an appreciation of this duality is the goal of the remaining sections of these lectures.

3.2 The Wilson loop

The matrix $U_C(x_2, x_1)$ is not gauge invariant, it transforms as in equation (3.13). This is so even when C is a closed curve, so that the final and initial points are the same ($x_2 = x_1$). However, in that case, it transforms by conjugation $U_C(x,x) \to g(x) U_C(x,x) g^{-1}(x)$ and, even though the matrix itself is not gauge invariant, we can get a gauge invariant object by taking its trace,[2]

$$\mathrm{Tr}\,[U_C(x,x)] \to \mathrm{Tr}\left[g(x) U_C(x,x) g^{-1}(x)\right] = \mathrm{Tr}\,[U_C(x,x)].$$

This trace is an important gauge invariant quantity, called the *Wilson loop*,

$$w[C] = \mathrm{Tr}\,U_C(x,x) = \mathrm{Tr}\mathcal{P} e^{i \oint_C d\tau \dot{x}^\mu(\tau) A_\mu(x(\tau))}. \tag{3.26}$$

The result of the trace, although it generally depends on the curve C, does not depend on which point on the curve we would choose as the initial and final point. It is one of the gauge invariant quantities that it is sometimes useful to analyse and it, as well as a slight generalization of it, will play an important role in our study of Yang–Mills theory.

For example, in the quantized Yang–Mills theory, the Wilson loop has the expectation value given by the functional integral (in Euclidean space)

$$W[C] = \langle w[C] \rangle = \frac{\int [dA_\mu(x)] e^{-\int d^4 x \frac{1}{2g_{YM}^2} \mathrm{Tr} F_{\mu\nu} F_{\mu\nu}} \mathrm{Tr}\mathcal{P} e^{i \oint_C d\tau \dot{x}^\mu(\tau) A_\mu(x(\tau))}}{\int [dA_\mu(x)] e^{-\int d^4 x \frac{1}{2g_{YM}^2} \mathrm{Tr} F_{\mu\nu} F_{\mu\nu}}}. \tag{3.27}$$

---

[2] In fact, when $U_C(x,x)$ transforms as $U_C(x,x) \to g(x) U_C(x,x) g^{-1}(x)$, its eigenvalues are gauge invariant. Its trace is the sum of its eigenvalues. We could easily imagine studying more elaborate gauge invariant combinations of $U_C(x,x)$ which give other characteristics of its eigenvalues.

Later on, we will be interested in understanding the dependence of the Wilson loop on the coupling constant g_{YM} and the rank of the gauge group, N, in a sypersymmetric and conformally symmetric version of Yang–Mills theory and in some circumstances where these can be computed. This will give us a way of studying the duality between the Yang–Mills theory and string theory. Historically, and beyond the scope of what we will discuss in the following, the behaviour of the Wilson loop for large contours C has been important in the study of the dynamical behaviour of quantized Yang–Mills theory [6]. For example, a diagnostic of confinement in Yang–Mills theory is the area law for the Wilson loop. For a large contour, C, the area law is the behaviour of the expectation value,

$$W[C] \sim \exp(-\sigma \cdot \text{area}[C]). \tag{3.28}$$

Here, area[C] is the area of a minimal two-dimensional surface where the boundary of that surface is the closed curve, C. The coefficient of the area, σ, is a constant with dimensions of inverse area or energy per unit length. It is called the string tension.

An alternative behaviour, which is characteristic of a deconfined phase of gauge theory, is the perimeter law,

$$W[C] \sim \exp(-\delta M \cdot \text{perimeter}[C]), \tag{3.29}$$

where the perimeter is the length of the curve C, Here, the coefficient, δM, when $\hbar = c = 1$, has dimensions of inverse distance, or of energy or mass. δM can be interpreted as contributing to the renormalization of the quark mass due to interaction of the quark with quantized gauge fields.

In space-times with dimensions greater than two, the primary tool for calculating the right-hand side of equation (3.27) is perturbation theory using Feynman diagrams. This perturbation theory is super-renormalizable in three dimensions and renormalizable in four dimensions. Three dimensions is an interesting case, as once a few ultraviolet divergences are dealt with, it is ultraviolet finite. No dimensionful scales are generated by renormalization and the only parameter with non-zero scaling dimension is the coupling constant. The theory is very strongly coupled in the low energy regime and it is expected to be confining. It must therefore have a string tension and dimensional analysis then tells us that the string tension must be proportional to g_{YM}^4 times a dimensionless number.

In four dimensions, the coupling constant of classical Yang–Mills theory is dimensionless, but it becomes scale dependent in the quantized theory. Perturbative calculations can only be reliable in regimes where perturbation theory can be trusted, that is, where corrections are controlled by a small parameter. In four-dimensional Yang–Mills theory, the coupling constant has a non-zero, negative beta-function, so that at leading order it becomes a scale-dependent running coupling,

$$\frac{1}{g_{YM}^2(\mu_1)} - \frac{1}{g_{YM}^2(\mu_2)} = \frac{11}{16\pi^2} \ln(\mu_1^2/\mu_2^2). \tag{3.30}$$

From this formula, we can see that the coupling $g_{YM}(\mu)$ is small and the coupling is weak, so that perturbation theory is applicable, when the energy scale μ is large. This is the regime where the energy and momentum exchanged in an interaction are large, or, according to quantum mechanical uncertainty, the time and distance scales involved are small. The fact that $g_{YM}(\mu)$ gets smaller as μ gets larger is called *asymptotic freedom*.

On the other hand, when the scale μ is decreased, the running coupling gets larger. This fact, that the running coupling is large when the momentum scale is small, is called *infrared slavery*. The latter phenomenon is thought to be at the root of confinement. However, confining behaviour is outside of the regime where perturbation theory is reliable. The only other computational tool is numerics, for attempting to perform the integral in equation (3.27) numerically. Considerable progress has been made in this direction, to the point where confinement is a well-established behaviour of three and four space-time dimensional Yang–Mills theories. The distance scale where the theory crosses over from weak coupling to strong coupling is called the mass scale. This scale, rather than the coupling constant, can be thought of as a parameter of the quantized Yang–Mills theory. The other is the N of the $U(N)$ gauge group.

Quantum chromodynamics (QCD) is the currently widely accepted theory of the strong interactions. This model is a Yang–Mills theory with $SU(3)$ gauge group coupled to some fundamental representation massive fermions, which are the quarks. The mass scale of QCD is around 250 *Mev*. Much of the interesting physics of QCD occurs in the strong coupling regime, at mass scales below 250 *Mev*, including the formation of the hadron spectrum, and practically all of nuclear physics, for example.

The running coupling constant in Yang–Mills theory has implications for the Wilson loop. Asymptotic freedom implies that Wilson loops for curves C whose spatial extent and structure are smaller than the inverse mass scale should be computable by perturbation theory. Those which are larger are not computable. As we have already discussed, usually, the larger C regime is the interesting one—for example, the area versus perimeter law test for confinement is implemented there—and perturbative Yang–Mills theory cannot be used to compute the Wilson loop in that regime. Aside from using computers and numerics to compute the right-hand side, an endeavour which has seen great advances and found a lot of interesting results, there are no reliable analytic techniques.

It has long been conjectured that the quantized Yang–Mills theory that we have been discussing has a dual description as a string theory [10] [11] [12] [23]. In that description, the right-hand side of (3.27) could be replaced by a string model, where the Wilson loop, for example, would be computed by (in Euclidean space)

$$W[C] = \sum_\chi g_s^{-\chi} \int_{\xi:\delta\xi=C} [d\xi]\, \mu[\xi,\chi]\, e^{-\sigma \cdot \text{Area}[\xi]}. \tag{3.31}$$

Here, the objects ξ, which are being integrated over, are two-dimensional surfaces with the property that their boundary is the curve C. Possible measure factors, which we denote by $\mu[\xi,\chi]$, could arise, for example, from integrating out degrees of freedom which live on the surfaces or ghost fields. The entity χ is an integer. It is the topological Euler number ($\chi = 2 - 2 \cdot$ #handles$-$#boundaries ≤ 2) of the two-dimensional worldsheet of the string, ξ. The summand has a Boltzmann factor, $e^{-\sigma \cdot \text{Area}[\xi]}$, which favours surfaces with minimal area. The parameter σ has the dimensions of inverse area and it is called the *string tension*. In fundamental string theory, it is usually denoted as $1/2\pi\alpha'$. The dimensionless parameter g_s which controls the Euler number is called the *string coupling*. If g_s is small, ξ with the simplest topologies are emphasized.

The appealing feature of such a model is that it could be computable in an opposite regime to the perturbative regime where (3.27) is tractable. If g_s is small, therefore emphasizing ξs with disc-like geometry, and if C is a large loop, compared with the scale that is set by the inverse of the string tension, then the area of a disc which has C as boundary should also be large compared to the inverse string tension, $\sigma \text{Area}[\xi] >> 1$, and the integral would be computable by semi-classical techniques. The area law and confinement would then be obtained from this model at the semi-classical limit,

$$W[C] \sim \frac{1}{g_s} \mu[\xi_0, \chi = -1] \, e^{-\sigma \cdot \text{Area}[\xi_0]}, \qquad (3.32)$$

where ξ_0 is the minimal surface. In a confining Yang–Mills theory, one might imagine a duality where the quantum field theory representation (3.27) is useful for small Wilson loops, with spatial size smaller than the confinement scale, and where the string model (3.31) is useful for large Wilson loops, with size and features much larger in spatial extent than the confinement scale.

The existence of such a string representation of the gauge theory is hinted at by the strong coupling expansion of lattice gauge theory, as well as the behaviour of some lower-dimensional solvable models, like Yang–Mills theory in two dimensions. Another piece of evidence in its favour is the large N expansion of Yang–Mills theory, which we will review shortly. It organizes the Feynman diagram expansion of the right-hand side of (3.27) into a summation over surfaces with given Euler number and then a summation over Euler numbers, analogous to the right-hand side of (3.31). What one learns from that latter expansion is that

$$g_s \sim \frac{1}{N} \qquad (3.33)$$

and that the infinite N limit projects one to the case where ξ are discs. In that case, the disc amplitude,

$$\int_{\xi : \delta \xi = C} [d\xi] \, \mu[\xi, \chi = -1] \, e^{-\sigma \cdot \text{Area}[\xi]},$$

is identified with the sum of planar Feynman diagrams, that is, those diagrams that can be drawn on a plane without crossing lines. In the following we will show that this planar limit of Yang–Mills theory is indeed the result of the appropriate large N limit.

The subject of following sections, the *AdS/CFT correspondence*, contains an explicit example of at least one gauge field theory where a string model of the sort outlined in equation (3.31) is indeed realized, and where some computations using this duality between gauge fields and strings can be performed [24] [25] [26] [27] [28]. Interestingly, the gauge theory in question is not a confining Yang–Mills theory of the type that we have been discussing. It is the maximally supersymmetric and conformal invariant $\mathcal{N}=4$ Yang–Mills theory in four-dimensional space-time. The Wilson loop is a slight modification of the one which we have described previously, adapted to the $\mathcal{N}=4$ theory. In this explicit example, the measure $\mu[\xi,\chi]$ is rather complicated. However, it has a simple and remarkable description. If we imagine that the discs ξ are not restricted to be embedded in our four-dimensional space-time, but they are allowed to wander into extra dimensions so that they live in the ten-dimensional space-time $AdS_5 \times S^5$, and their boundary is the curve C which resides in the four-space-time dimensional boundary of AdS_5, the right-hand side of (3.31) is thought to be an exact representation of (3.27). The space $AdS_5 \times S^5$ with a constant dilaton and a Ramond–Ramond 4-form field with N units of five-form flux piercing the S^5 is a solution of type IIB super-gravity and also IIB string theory. ξ are the world-sheets of the critical type IIB superstring with Dirichlet boundary conditions, so that ξ has boundary C at the boundary of AdS_5. One could, in principle, by integrating out the higher dimensional degrees of freedom, present the theory as four-dimensional, with a measure $\mu[\xi,\chi]$.

This duality, rather than being one between long- and short-distance physics, is for a scale invariant theory which does not exhibit quark confinement. Rather, the coupling constant has vanishing beta function and it is tuneable. The duality is then between the weak and strong coupling limits of the theory. The Yang–Mills quantum field theory is computable in perturbation theory when the coupling constant is small. As we shall discuss in Section 3.3, this is so when $g_{YM}^2 N \ll 1$. The factor of N arises from the multiplicity of particles, or from index sums in Feynman diagrams. The string theory is computable when the string coupling constant is weak, which turns out to be achieved by putting $g_{YM} \to 0$ and $N \to \infty$ so that $g_{YM}^2 N$ is held finite, and when the string tension is large, that turns out to be when $g_{YM}^2 N$ is tuned to large values.

In those limits, the semi-classical limit is an accurate description of the string theory. Both limits are, in principle, solvable in their leading orders and they are also, in principle, systematically correctable, one as an expansion in $g_{YM}^2 N$ and an asymptotic expansion in $1/N$, the other, also as an asymptotic expansion in $1/N$, and then in $1/\sqrt{g_{YM}^2 N}$. This gives the possibility of using the string theory to study the gauge theory in its strong coupling regime, and using the gauge theory to study a certain strongly coupled limit of the string theory. This feature is what makes the duality very interesting.

Equation (3.31) which describes the string dual of the Wilson loop also has implications for local operators. In the pure Yang–Mills theory that we have been discussing a Wilson loop for a small circle of radius a centred on the origin has an expansion in local operators,

$$W[C] = N + C_4 a^4 \text{Tr} F_{\mu\nu}(0) F^{\mu\nu}(0) + \ldots,$$

where C_4 is a coefficient. We could imagine using this expansion to compute a correlation functions of gauge invariant operators by considering the expectation values of several small Wilson loops. Then, the string theory dual calculation would consider a thin minimal surface which connects the locations of the Wilson loops. This minimal surface is the world-sheet of a closed string, which in the low energy limit is the propagator of one of the particles in the string spectrum, typically a graviton. The general duality statement is as follows. Consider local operators $O_j(x)$ in $\mathcal{N}=4$ Yang–Mills theory. We assume that these are operators with good conformal dimensions, that is, they obey commutation relation

$$-i\bigl[D, O_j(0)\bigr] = \Delta_j O_j(0)$$

with the dilatation generator D, where their conformal dimensions are the numbers Δ_j. (We will supply more details as to the meaning of these in later sections.) These Yang–Mills theory operators are dual to degrees of freedom, $\phi_j(r,x)$ in the IIB superstring theory on $AdS_5 \times S^5$ background. Then, the generating functional for correlations of $O_i(x)$ in the Yang–Mills theory is expressed in terms of the string theory partition function [25] [26],

$$\left\langle \mathcal{T} e^{i \int d^4 x \varphi_j(x) O_j(x)} \right\rangle = Z\left[\lim_{r \to \infty} \phi_j(r,x) = r^{\Delta_j} \varphi_j(x) \right],$$

where the string theory partition function on the right-hand side is for the string theory with boundary conditions such that each $\phi_j(r,x)$ has the boundary condition that it approaches the classical field $\varphi_i(x)$ as r approaches the boundary of AdS_5.

3.3 The large N expansion

The large N expansion is a way of reorganizing the perturbative expansion of a gauge field theory into contributions that can be characterized by the topology of the two-dimensional surfaces on which the Feynman diagrams can be drawn without crossing any lines. In pure Yang–Mills theory, any connected Feynman diagram can be seen to be proportional to

$$\sim \lambda^{L+\varepsilon/2-1} N^{\chi-\varepsilon/2}. \tag{3.34}$$

Here, N is the N of the $U(N)$ gauge group of the Yang–Mills theory[3] and

$$\lambda \equiv g_{YM}^2 N \tag{3.35}$$

---

[3] The Yang–Mills theories of interest to us later actually have an $SU(N)$, rather than $U(N)$ gauge group. If it were $SU(N)$, formula (3.34) would be modified in a significant way. We shall not worry about this difference here, as we will either always consider the infinite N limit, where $U(N)$ and $SU(N)$ are indeed indistinguishable in the leading order, or the pure Yang–Mills theory where the difference between $U(N)$ and $SU(N)$ is a $U(1)$ subgroup whose fields are free fields which decouple from the other Yang–Mills fields and whose contributions in various cases can be understood and taken into account if necessary.

is the combination of the Yang–Mills coupling and N called the 't Hooft coupling, The parameter χ is the topological Euler character of the graph (and the Euler character of the two-dimensional surface on which the graph can be drawn without crossing any lines). L is the number of loops in the graph and ε is the number of external lines. We shall derive expression (3.34) shortly.

When we take the large N limit, where we put $N \to \infty$ while holding the 't Hooft coupling λ and the number of loops L and the number of external lines ε constant, equation (3.34) tells us that the dominant order is the sum of all Feynman diagrams with maximum Euler character, χ.

If we consider vacuum diagrams, with no external lines, so that $\varepsilon = 0$, the maximum Euler character is $\chi = 2$, the Euler number of the 2-sphere and the Feynman diagrams of this order can be drawn on a 2-sphere without crossing lines.

If we consider a contribution with external lines, $\varepsilon > 0$, the faces of the diagram into which the the external lines are emitted are counted as holes rather than faces. The Euler character is maximal if the number of holes is minimal. The surface can generally have only one hole. (To be clear, in this case, all emissions of external lines are into a single hole.) The 2-sphere with a hole cut out of it has $\chi = 1$ and these diagrams are of order $N^{1-\varepsilon/2}$. The Riemann surface on which the graphs of this leading order can be drawn has $\chi = 1$ and the topology of a plane. The graphs are therefore called planar and the truncation of the full Yang–Mills theory to the sum over planar graphs for any multi-point function is called the planar limit of Yang–Mills theory.

The contribution to this leading order, planar, contains Feynman diagrams to all orders in the usual perturbative loop expansion in the Yang–Mills coupling constant. The planar limit of Yang–Mills theory therefore appears to still be a highly non-trivial dynamical system in its own right. The diagrams are weighted by the 't Hooft coupling λ to the power of the number of loops. Thus, the weak coupling limit of this planar theory occurs where λ is small, in that small λ favours diagrams with smaller numbers of loops, whereas strong coupling is large λ in that it favours diagrams with large numbers of loops. Before AdS/CFT, even this leading, planar limit of Yang–Mills theory was not solvable for Yang–Mills theory in space-time dimensions greater than two. With the advent of AdS/CFT, the strong coupling limit of the planar limit of the four-dimensional $\mathcal{N} = 4$ supersymmetric Yang–Mills theory and some field theories which are related to it have become solvable in many circumstances. In the following, we shall define what we mean by 'strong coupling limit' in this context.

The analogue of the large N expansion in statistical models with matrix-valued degrees of freedom had been studied for a long time, particularly in combinatorics [2] [3] and the statistical mechanical theories of triangulated random surfaces [18]. Application of the idea to Yang–Mills theory is due to t'Hooft and dates back to the 1970's [12] [9]. The motivation was to understand in what way Yang–Mills theory could look like a string theory. The large N expansion is indeed one way: reorganize the perturbation theory into one that resembles the string perturbative expansion of quantities in the closed string sigma model in powers of the string coupling constant g_s. In the string sigma model, the expansion is in the powers $g_s^{-\chi}$, where χ is the Euler character of the world-sheets of the strings. There, the leading order contains strings with world-sheets which are topologically equivalent to the two-sphere.

We now want to find a basic derivation of some of the previous statements about the large N limit. In order to discuss perturbation theory, we should consider the gauge-fixed Lagrangian density

$$\mathcal{L}_{\text{gf}} = \text{Tr}\left\{-\partial_\mu A_\nu \partial^\mu A^\nu + \left(1 - \frac{1}{\xi}\right)(\partial_\mu A^\mu)^2 + \partial_\mu \bar{c}\partial^\mu c - ig_{\text{YM}}\partial_\mu \bar{c}[A^\mu, c] - \right.$$
$$\left. -ig_{\text{YM}}\partial_\mu A_\nu [A^\mu, A^\nu] + \frac{1}{2}g_{\text{YM}}^2 [A_\mu, A_\nu][A^\mu, A^\nu]\right\}, \tag{3.36}$$

where ξ is the gauge fixing parameter and c and $\bar{c}$ are Faddeev–Popov ghosts, which are also matrix-valued fields. Here, we have scaled the fields so that the Yang–Mills coupling constant appears in front of the Lagrangian density.

To get the large N expansion, it is convenient to adopt a fat graph notation for the propagators of matrix-valued fields. The fat graph for the propagator is depicted in Figure 3.1 and for the three- and four-point vertices is depicted in Figure 3.2. The propagators of matrix-valued fields carry the matrix indices. For example, the gluon propagator in the Feynman gauge ($\xi = 1$) and the Fadeev–Popov ghost propagator are

$$\langle A^\mu_{ab}(x) A^\nu_{cd}(y)\rangle\big|_{g_{\text{YM}}=0} = \delta_{ad}\delta_{bc}\int\frac{d^4k}{(2\pi)^4}e^{ik_\mu(x-y)^\mu}\frac{-i\eta^{\mu\nu}}{k^2 - i\epsilon} \tag{3.37}$$

$$\langle c_{ab}(x)\bar{c}_{cd}(y)\rangle\big|_{g_{\text{YM}}=0} = \delta_{ad}\delta_{bc}\int\frac{d^4k}{(2\pi)^4}e^{ik_\mu(x-y)^\mu}\frac{i}{k^2 - i\epsilon}. \tag{3.38}$$

The features of the propagators that are important to us are the delta-functions with matrix indices on the right-hand side. They are a result of global $U(N)$ symmetry that is left over after gauge fixing and they are appropriate to $U(N)$ gauge theory. For $SU(N)$, $\delta_{ad}\delta_{bc}$ would be replaced by $\delta_{ad}\delta_{bc} - \frac{1}{N}\delta_{ab}\delta_{cd}$ and the large N expansion would be more complicated, although the differences would not be seen in the leading order. In the

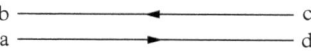

Figure 3.1 *The propagator of the Yang–Mills field in fat graph notation has two lines with indices which are connected by the lines identified as depicted.*

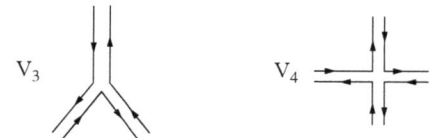

Figure 3.2 *The three- and four-point vertices of Yang–Mills theory in fat graph notation.*

following, for simplicity, we shall mostly be interested in $U(N)$. The solid lines in the propagator in Figure 3.1 connect the indices that are identified by the delta-functions in equations (3.37) and (3.38).

Vertices are obtained by using these propagators in Wick contractions with the interaction terms in (i times) the action

$$g_{YM}\text{Tr}\partial_\mu \bar{c}[A^\mu, c] \tag{3.39}$$

$$g_{YM}\text{Tr}\partial_\mu A_\nu [A^\mu, A^\nu] \tag{3.40}$$

$$g_{YM}^2 \text{Tr}\frac{i}{2}[A_\mu, A_\nu][A^\mu, A^\nu]. \tag{3.41}$$

The fat graph notation for the vertices is depicted in Figure 3.2.

We are interested in studying the dependence of generic Feynman diagrams on N. The counting of powers of N in the large N expansion is best seen by using the fat graph notation that we have outlined previously.

Consider a connected Feynman diagram which contributes to an ε-point correlation function. We shall call it a *graph*. It is an assembly of vertices and of propagators connecting vertices to each other as well as to the external points.

Let us say that there are V_3 three-point vertices and V_4 four-point vertices. The internal double lines of the graph form 'edges'. Let us say that there are E edges. As well, there must be ε external lines, that is, lines which connect each of the ε external points to vertices which are in the graph.

Now, let us find the factors of g_{YM} and of N which must go with a given graph.

1. Each three-point vertex of a graph is a vertex from which three lines emanate. Each three-point vertex is accompanied by a factor of g_{YM} and the three-point vertices therefore contribute the factor

$$(g_{YM})^{V_3}.$$

2. Each four-point vertex is a point from which four lines emanate. Each is accompanied by a factor of g_{YM}^2. The four-point vertices contribute the factor

$$(g_{YM})^{2V_4}.$$

3. Each face yields a factor of N coming from the summation over indices in a closed line, resulting in the factor

$$(N)^F.$$

4. A summary of the factors so far is the product of the above three quantities,

$$(g_{YM})^{V_3+2V_4}(N)^F.$$

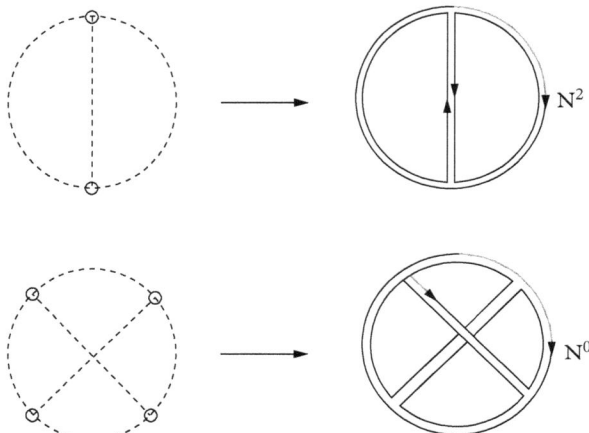

Figure 3.3 *Some examples of Feynman diagrams which would be used to calculate the vacuum amplitude. The graphs on the right are equivalent to the graphs on the left, but they are written in fat-graph notation. Once the graph is depicted that way, the number of closed index lines is equal to the number of faces of the graph. The sum over indices in each closed index line produces a factor of N. The top graph has two loops and it is* planar *in that it has been drawn on the plane without crossing any lines. It has three faces which contribute a factor of N^3. (The exterior of the graph counts as a face.) It also has two vertices, which together give it a factor of g_{YM}^2. The total factor with the graph is therefore $g_{YM}^2 N^3 = \lambda N^2$. Since the graph is planar, and it has no external lines, its Euler character is that of the 2-sphere, $\chi = 2$. The second graph has four loops and it is not planar. It has two index loops, so it has the factor $g_{YM}^4 N^2 = \lambda^2$. Since it can be drawn on a torus without crossing lines, its Euler character is $\chi = 0$. The factors of both graphs agree with the general formula in equation (3.34), $\lambda^{L+\varepsilon/2-1} N^{\chi-\varepsilon/2}$, when we put the number of external lines ε to zero.*

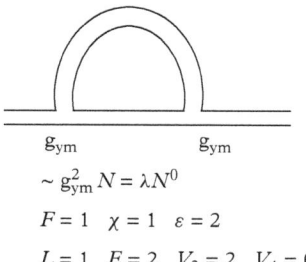

Figure 3.4 *A one-loop correction to the Yang–Mills field propagator is depicted. The graph has two internal lines, thus two edges, $E = 2$. It has two three-point vertices and no four-point vertices, thus, $V_3 = 2$, $V_4 = 0$. It has one face and thus $F = 1$. The index sum corresponding to the face produces a factor of N. The Euler character of this graph is that of the plane, $\chi = 1$. To see this, one can consider the entire exterior of the graph as a hole into which the two external lines are emitted. The number of loops is $\ell = 1$. The counting displayed in the figure yields λN^0, which agrees with equation (3.34), $\lambda^{L+\varepsilon/2-1} N^{\chi-\varepsilon/2}$.*

The large N expansion 139

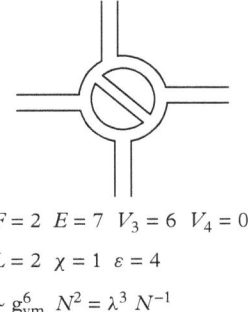

$$F = 2 \quad E = 7 \quad V_3 = 6 \quad V_4 = 0$$
$$L = 2 \quad \chi = 1 \quad \varepsilon = 4$$
$$\sim g_{ym}^6 \quad N^2 = \lambda^3 \, N^{-1}$$

Figure 3.5 *A two-loop correction to the Yang–Mills field four-point function is depicted. The graph has seven internal lines, thus seven edges, $E = 7$. It has six three-point vertices and no four-point vertices, thus $V_3 = 6$, $V_4 = 0$. It has two faces and thus $F = 2$. The two index sums produce a factor of N^2. The Euler character of this graph is that of the plane, $\chi = 1$. To see this, one can consider the entire exterior of the graph as a hole into which the four external lines are emitted. The number of external lines is $\varepsilon = 4$ and the number of loops is $\ell = 2$. The counting displayed in the figure yields a factor of λ/N which agrees with what we would obtain from equation (3.34), $\lambda^{L+\varepsilon/2-1} N^{\chi-\varepsilon/2}$.*

Now, there are some identities that the graph must obey.

1. Each three-point vertex is a source of three lines and each four-point vertex is a source of four lines. Each internal line must attach to two vertices, each external line to one vertex. This relates the numbers of vertices and lines as

$$\varepsilon + 2E = 3V_3 + 4V_4.$$

2. The number of loops is equal to the number of momentum space volume integrals which need to be done in order to evaluate the Feynman graph. Each external line has a predetermined momentum, so they are not counted. Each internal line has a momentum and each vertex has a momentum conserving delta-function. When all momentum conserving delta-functions are taken into account, one must be left over to conserve momentum in the entire graph. Thus, the net number of momentum space volume integrals is

$$L = E - V_3 - V_4 + 1.$$

3. The Euler character χ is a topological characteristic of the graph. It is equal to the number of faces minus the number of edges plus the total number of vertices,

$$\chi = F - E + (V_3 + V_4).$$

This is also a topological characteristic of the two-dimensional surface on which the graph can be drawn without crossing any lines. For the surface, this is

$$\chi = 2 - 2g - h,$$

where g is the number of handles on the surface and h is the number of holes in the surface. Both of these are non-negative integers, so the maximum Euler number is two, that of the sphere, $\chi_{\text{sphere}} = 2$. A torus has one handle and no holes and it has Euler character $\chi_{\text{torus}} = 0$. A disc has no handles and one hole and it has $\chi_{\text{disc}} = 1$. An annulus has $\chi_{\text{annulus}} = 0$. In summary,

	sphere	disc	torus	cylinder	...
$-\chi$	-2	-1	0	0	...

Putting these together, some simple algebra tells us that a Feynman graph is weighted by the overall factor

$$\lambda^{L+\varepsilon/2-1} N^{\chi-\varepsilon/2} \cdot \text{Feynman integral.} \qquad (3.42)$$

which is the expression that we used in equation (3.34).

3.3.1 Large N factorization

There is one rather remarkable result of the large N expansion, which is called large N factorization. If, in the Yang–Mills theory, we compute the correlation function of any number of $U(N)$ invariant operators,

$$\langle \mathcal{O}_1(x_1)\mathcal{O}_2(x_2)\ldots\mathcal{O}_k(x_k)\rangle = \langle \mathcal{O}_1(x_1)\rangle \langle \mathcal{O}_2(x_2)\rangle \ldots \langle \mathcal{O}_k(x_k)\rangle \left\{1 + \mathcal{O}\left(\frac{1}{N^2}\right)\right\}. \qquad (3.43)$$

This property is called large N factorization. In the leading order, $SU(N)$ invariant operators behave as though they are classical fields, they are uncorrelated. Of course, in the Ads/CFT correspondence, which we shall describe shortly, the classical variables are classical string theory degrees of freedom. The large N limit is the one which turns off the string interactions, that is, amplitudes where strings split or join. However, even classical, or more accurately, tree-level string theory is technically very complicated and we shall need another limit to make it solvable. We will come back to this subject in later sections.

It is easy to prove factorization if we make some assumptions about the operators. Let us assume that they are gauge invariant products of the matrix-valued Yang–Mills fields and the derivatives of Yang–Mills fields, all evaluated at the same point, and that they are formed by taking a single trace of a product of such matrices. An example would be

$\text{Tr}F_{\mu\nu}(x)F^{\mu\nu}(x)$ or $\text{Tr}F_{\mu\nu}(x)F^{\nu\rho}(x)F_\rho{}^\mu(x)$. We could then add them to the Lagrangian density of Yang–Mills theory with position-dependent coefficients,

$$\mathcal{L} \to \mathcal{L} + \sum_i g_i(x)\mathcal{O}_i(x)$$

and treat them as vertices. Then, a generalization of the above arguments which includes the new vertices (and does not depend on the coupling constants actually being constants) would tell us that the dominant diagrams contributing to $\langle \mathcal{O}_i(x_i)\rangle$ are the connected planar diagrams, which are of order N^2. These are just the leading part of the connected vacuum amplitude of the Yang–Mills theory with the generalized vertices included, and then Taylor expanded to leading oder in $g_i(x_i)$, and with the other g_is set to zero. Thus, the leading order connected diagrams in $\langle \mathcal{O}_i(x_i)\rangle$ are of order N^2.[4]

Moreover, the leading order of $\langle \mathcal{O}_1(x_1)\rangle\langle \mathcal{O}_2(x_2)\rangle$ would be N^4. On the other hand, the connected two-point function,

$$\langle \mathcal{O}_1(x_1)\mathcal{O}_2(x_2)\rangle_C = \langle \mathcal{O}_1(x_1)\mathcal{O}_2(x_2)\rangle - \langle \mathcal{O}_1(x_1)\rangle\langle \mathcal{O}_2(x_2)\rangle,$$

would be the coefficient of $g_1(x_1)g_2(x_2)$ in the functional Taylor expansion in the g_is of the sum of connected planar vacuum diagrams for the Yang–Mills theory with enhanced Lagrangian density, $\mathcal{L} + \sum_i g_i(x)\mathcal{O}_i(x)$. As we have already argued, the leading order contributions to this quantity must be of order N^2. Thus, in summary,

$$\langle \mathcal{O}_1(x_1)\rangle\langle \mathcal{O}_2(x_2)\rangle \sim N^4$$

and

$$\langle \mathcal{O}_1(x_1)\mathcal{O}_2(x_2)\rangle - \langle \mathcal{O}_1(x_1)\rangle\langle \mathcal{O}_2(x_2)\rangle \sim N^2$$

and

$$\langle \mathcal{O}_1(x_1)\mathcal{O}_2(x_2)\rangle = \langle \mathcal{O}_1(x_1)\rangle\langle \mathcal{O}_2(x_2)\rangle\left\{1 + \mathcal{O}(1/N^2)\right\}.$$

It is easy to generalize this argument to higher-order correlation functions. The upshot is that, at large N, single-trace gauge invariant local operators are uncorrelated. It is also easy to generalize to other operators, like correlators of Wilson loops.

There are several other important facts that are related to factorization that I will leave to the reader to derive, or find in the literature. An important one is what happens when we couple the Yang–Mills theory to a quark which transforms in the fundamental

---

[4] Of course, non-connected diagrams cancel with the expansion of the denominator in the expression

$$\langle \mathcal{O}_i(x_i)\rangle = \frac{\int [dA] e^{-S} \mathcal{O}_i(x_i)}{\int [dA] e^{-S}}.$$

The diagrams which do not cancel are those which are connected to the point x_i. This is one of Goldstone's theorems.

representation of the gauge group, that is, a quark field which, instead of being matrix-valued like the Yang–Mills field, is a complex vector, with only one rather than two of the gauge group indices. In the fat-graph notation, the propagator of such a field contains only a single line. When they occur in Feynman diagrams, these propagators must form boundaries of the graph. For example, a closed internal loop with a single line propagator should be interpreted as a hole in the graph. It therefore has smaller Euler character than the graphs without the internal loop and it is therefore subdominant in the large N expansion. For this reason, fundamental representation quarks do not contribute to the large N limit of the vacuum energy of a Yang–Mills theory which is coupled to them.

3.4 D branes, black branes, and the Maldacena conjecture

3.4.1 D3 branes and black D3 branes

The AdS/CFT correspondence is a duality between string theory on asymptotically anti-de Sitter background space-times and conformal field theories defined on flat space of dimensions one smaller than the anti-de Sitter space. These theories are *holographic* in the sense that all of the information that is needed to specify a state of a string theory, which contains a theory of quantum gravity, is encoded in the quantum state of a quantum field theory on flat space-time and in a lower dimension. The quantum field theory state is the hologram. This is a generalization of ideas which stemmed from the use of black hole entropy to count the degrees of freedom in quantum gravity [20] [21] [45].

This duality can hold at various levels, and there are several examples. Here, we will consider the duality in its best established example, the mapping between $\mathcal{N}=4$ supersymmetric Yang–Mills theory[5] in four flat space-time dimensions and type IIB superstring theory on the ten-dimensional $AdS_5 \times S^5$ background. The $\mathcal{N}=4$ theory is a superconformal field theory, thus the CFT in AdS/CFT. We will also consider the duality in its strongest form, where it is a one-to-one mapping of all of the quantum states, operators, observables, processes, etc. between the two theories. For example, the vacuum state of the $\mathcal{N}=4$ theory corresponds to the state of the string theory which is the empty $AdS_5 \otimes S^5$ geometry, without the excitation of gravitons or other modes in the string theory spectrum.

The $\mathcal{N}=4$ theory has two parameters, the dimensionless coupling constant g_{YM} which governs the strength of interactions, and the N of the $SU(N)$ gauge group. These are combined to form the 't Hooft coupling,

$$\lambda \equiv g_{YM}^2 N, \qquad (3.44)$$

which plays an important role in the large N limit.

The string theory also has two parameters. One of them is the string coupling constant, g_s which regulates the strength of string interactions, that is the propensity of strings

---

[5] We will call it $\mathcal{N}=4$ theory for short.

to split and join. The other parameter is the string tension, which is effectively the radius of curvature of the geometry in which the string is embedded, written in units of fundamental string tension $2\pi\alpha'$. The AdS_5 and S^5 are constant negative and positive curvature spaces respectively, which have the same radii of curvature, which we call L. Fundamental strings also have a string tension, α', which has the units of length squared and the dimensionless parameter is $L/\sqrt{\alpha'}$.

The parameters of $\mathcal{N} = 4$ theory and the IIB string are mapped onto each other as

$$4\pi g_s = g_{YM}^2 \qquad (3.45)$$

$$\frac{L^4}{\alpha'^2} = \lambda \equiv g_{YM}^2 N. \qquad (3.46)$$

With these two identifications, we are supposed to be able to find a one-to-one mapping of all of the attributes of the gauge field theory to the string theory and vice versa. Of course, the precise nature of this mapping in all of its detail is still a subject that is undergoing development. One of the fascinating frontiers of this subject is the project of filling in the details of exactly what is dual to what.

What makes this mapping interesting is the fact that it is a weak coupling to strong coupling duality. The gauge field theory is tractable in the limit where both g_{YM} and λ are small, so that perturbation theory, which is practically the only analytic approach, is accurate. The string theory, on the other hand, is tractable where it is semi-classical, in the limit where g_s is small, and the limit of weak curvature, $\frac{L}{\sqrt{\alpha'}} \to \infty$.

Of course, the duality in its strongest form, and even in the weaker forms where it is only approximate, is only conjectured to exist. There is no proof of it at the mathematical level. It is not even clear that there can be such a proof, as it would require a mathematically precise definition of the theories on both sides of the mapping. A weaker version of the duality would have it holding only in the large N limit or only in the weak curvature $\frac{L}{\sqrt{\alpha'}} \to \infty$ limit. Both of these are still very interesting as even an understanding of non-trivial gauge field theories in the large N limit and tree-level superstring theory is far from complete thus far. The weak curvature limit would still allow us to use the duality to compute the strong coupling limit of the quantum field theory.

3.4.2 Weakly coupled string theory with D3 branes

Let us first begin with weakly coupled IIB string theory on ten-dimensional Minkowski space with N flat, infinite D3 branes occupying the Cartesian space-time dimensions x^0, x^1, x^2, x^3 as

dimension	0	1	2	3	4	5	6	7	8	9
N D3	X	X	X	X	O	O	O	O	O	O

In the above table, the D3 branes fill the dimensions marked by X and they sit at points in the dimensions marked 0. They are distributed this way, parallel, flat, and infinitely extended, because this configuration should be a local minimum of the energy. Also, when they have coinciding positions, they violate as few symmetries as possible. They are $\frac{1}{2}$-BPS objects which reduce the number of supersymmetries of the IIB string theory from 32 to 16. They also have Poincare symmetry in the dimensions that they occupy, x^0, x^1, x^2, x^3. The transverse directions $x^4,...,x^9$ have an $SO(6)$ ($=SU(4)$) symmetry under rotating these coordinates into each other. These symmetries will be enhanced to superconformal symmetry in the low energy limit. This leads to a doubling of the supersymmetries, back to 32, which is needed for the equivalence to full IIB string theory on $AdS_5 \times S^5$.

This system contains closed strings moving in ten-dimensional space-time surrounding the D3 branes and open strings which begin and end on the D3 branes. The string coupling constant g_s governs the propensity of strings to interact by splitting and joining. This interaction will be suppressed if this coupling constant is small, $g_s << 1$. However, this is not quite enough to guarantee that the interactions are weak. The open strings can begin or end on any one of the N D3 branes. Summing over these possibilities can produce factors of N. These factors will occur in such a way that the effective coupling constant is $g_s N$. Thus, the open string sector of this system will be weakly coupled and perturbative string theory will give an accurate description of this system only when $g_s N << 1$. Of course, this also guarantees that $g_s << 1$.

Thus, when $g_s N << 1$, the system is accurately described by these weakly interacting strings on flat ten-dimensional space-time. In this weak coupling limit, both the open and closed strings have states which behave as massless relativistic particles, as well as infinite towers of massive relativistic particles with masses $M^2 \sim \frac{1}{\alpha'}$.

Now, let us take the low energy limit. In this limit, the massive states of both the open and closed strings should decouple. Intuitively, 'taking the low energy limit' is simply agreeing that we restrict our ability to probe the system so that we are only allowed low frequency, large wavelength probes, to the extent that we can never disturb it violently enough to excite a massive state of a string. The low energy limit of the closed string sector of IIB string theory is IIB super-gravity and, in this background of flat ten-dimensional Minkowski space, the low energy excitations of super-gravity are the fields of the graviton super-multiplet, which contains the graviton, dilaton, antisymmetric tensor field, and their superpartners. Thus, the closed string degrees of freedom appear as weakly interacting super-gravitons propagating in the bulk of the flat ten-dimensional space-time surrounding the D3 branes. Their interactions with each other and with the open strings are weak since the strength of the interactions scales like powers of the ten-dimensional Planck length times the energies of the strings,

$$g_s^2 (\alpha')^4 \cdot (\text{energy})^8,$$

which is small because g_s is small and the energy is much smaller than $1/\sqrt{\alpha'}$.

As well as the closed strings, there are open strings which begin and end on the D3 branes. The massless modes of these open strings are the quantum fields of the $\mathcal{N}=4$ theory, living on the world-volume of the D3 brane. In the low energy limit, they decouple from the massive degrees of freedom of the open and closed strings. However, they do not become free field theory in the low energy limit. The $\mathcal{N}=4$ theory is conformally invariant and scaling to low energies does not change the strength of the interactions. It remains a non-trivial interacting quantum field theory in the low energy limit. The coupling constant of the gauge field theory is directly related to the coupling constant of the string theory,

$$g_{YM}^2 = 4\pi g_s. \tag{3.47}$$

Our conclusion is that, in the limits $g_s N << 1$ and where we keep states of the theory where energies are much smaller than $\frac{1}{\sqrt{\alpha'}}$, the remaining dynamical system consists of free super-gravitons propagating in flat ten-dimensional space-time and fully interacting four space-time-dimensional $\mathcal{N}=4$ theory. What is more, the $\mathcal{N}=4$ theory is weakly coupled and one should be able to do accurate calculations of quantities like correlation functions using perturbation theory, that is, in an asymptotic expansion in $g_{YM}^2 N$ about free field theory.

3.4.3 The black D3 brane solution of IIB super-gravity

Now, let us turn to another description of the stack of N D3 branes, that of a Ramond–Ramond-charged state of IIB super-gravity which has as a classical solution a black D3 brane geometry with N units of Ramond–Ramond four-form charge. It is actually a conjecture that the D3 branes can be described in this way. The description is accurate in a regime that is different from that of weak coupling which we have discussed previously. Since it is in a different regime, and there is no accurate description of the system in the intermediate range, it is not possible to follow the system adiabatically from one limit to the other, and therefore no rigorous proof that they describe the same system. On the other hand, the equivalence, which was first realized by Polchinski [22], fits so well with so much of what is known of the states of string theory there is little doubt that it is correct.

The low energy effective action for the type IIB super-string is the action of type IIB super-gravity, whose bosonic part is

$$S_{\text{SUGRA}} = \frac{1}{(2\pi)^7 \alpha'^4} \int d^{10}x \sqrt{-g} \left\{ e^{-2\varphi} \left(R + 4(\nabla\varphi)^2 \right) - \sum_p \frac{2}{(8-p)!} F_{p+2}^2 \right\}. \tag{3.48}$$

This theory contains the metric tensor, $g_{AB}(x)$, the dilaton, $\varphi(x)$, and (p+2)-form fields $F_{p+2}(x) = dA_{p+1}(x)$, the field strengths for the Ramond–Ramond (p+1)-form fields $A_{p+1}(x)$. Type IIB super-gravity which we are interested in here has only p-odd forms, whereas the IIA super-gravity would have only p-even forms. Moreover, in the IIB theory the five-form F_5 is constrained to be self-dual, $F_5 = *F_5$, where the star denotes the Hodge dual. This constraint is usually imposed at the level of the equations of motion which are derived from (3.48) using a variational principle.[6]

The extremal black D3 brane solution of the super-gravity theory (3.48) in the string frame is given by the metric, dilaton, and Ramond–Ramond 5-form fields

$$ds^2 = H(r)^{-\frac{1}{2}} \eta_{\mu\nu} dx^\mu dx^\nu + H(r)^{\frac{1}{2}} dx^m dx^m, \tag{3.49}$$

$$e^{2\varphi(r)} = g_s^2 \tag{3.50}$$

$$F_5 = (1 + *) dx^0 \wedge dx^1 \wedge dx^2 \wedge dx^3 \wedge d\left(1 - H^{-1}(r)\right) \tag{3.51}$$

$$\mu, \nu = 0, 1, 2, 3; \quad m, n = 4, \ldots, 9; \quad r^2 = x^m x^m$$

$$H(r) = 1 + \frac{L^4}{r^4}, \quad L^4 = 4\pi g_s N \alpha'^2, \tag{3.52}$$

respectively. The element $\eta_{\mu\nu} dx^\mu dx^\nu$ is that of four-dimensional Minkowski space. $F_5 = dA_4$ has N units of flux through the 5-sphere, that is the five-dimensional space with coordinates $x^4, \ldots, x^9$ constrained by $(x^4)^2 + \ldots + (x^9)^2 = 1$. This 5-sphere links the position of the black brane and the flux of the 5-form integrated over the 5-sphere is N,

$$\int_{S^5} F_5 = N. \tag{3.53}$$

This is an extremal charged black hole. The event horizon is located at $r = 0$.

The function $H(r)$ is singular near the horizon at $r = 0$. However, the geometry is regular there, and when $r \ll L$, it approaches

$$\begin{aligned}ds^2 &= \frac{r^2}{L^2} \eta_{\mu\nu} dx^\mu dx^\nu + \frac{L^2}{r^2} dx^m dx^m, \\ &= \frac{r^2}{L^2} \eta_{\mu\nu} dx^\mu dx^\nu + \frac{L^2}{r^2} dr^2 + L^2 d\Omega_5^2,\end{aligned} \tag{3.54}$$

where $d\Omega_5^2$ is the metric of the unit five-sphere. We shall sometimes write the 5-sphere coordinates more explicitly as six-component unit vectors $\hat{\theta}$ where $\hat{\theta}^2 = 1$ and the line element is denoted $d\hat{\theta} \cdot d\hat{\theta}$.

What we obtain is the metric of the direct product space-time $AdS_5 \times S^5$. AdS_5 is a space of constant negative curvature, whereas S^5 is a space of positive constant curvature and the radius of curvature of both spaces is L. The coordinate r is the AdS_5 radius. It

---

[6] The action (3.48) is presented in the *string frame*. This terminology has to do with how the dilaton is separated from the metric. The *Einstein frame* is obtained by absorbing a factor $e^{-\varphi(x)}$ into $g_{\mu\nu}(x)$. We shall not need to do this here.

has the units of length. It is common to rescale r so that it has energy units. This is done by redefining it as $r \to L^2 r$, whence

$$ds^2 = L^2 \left[r^2 \eta_{\mu\nu} dx^\mu dx^\nu + \frac{dr^2}{r^2} + d\hat{\theta} \cdot d\hat{\theta} \right]. \tag{3.55}$$

Now, we must ask where the solution (3.49)–(3.52) is reliable as a background for the IIB string theory. First of all, it is derived from the super-gavity equations of motion, without the corrections that should occur if these equations were embedded in string theory. This procedure will be self-consistent when those corrections are small. They would be suppressed by powers of $\frac{\sqrt{\alpha'}}{L}$, and negligibly small when the scale of the curvature of the space-time is much greater than the string scale,

$$L \gg \sqrt{\alpha'} \to 4\pi g_s N \gg 1, \tag{3.56}$$

which is opposite to the limit of weakly interacting strings that we discussed in Section 3.42. This is the limit where the massive modes of the string decouple and the physics is governed by the lowest lying modes, the fields of ten-dimensional type IIB super-gravity.

Once we have turned off the massive modes of the string, we have ten-dimensional super-gravity. In principle, this is a quantum field theory including quantum gravity. It is described by a classical solution only when the quantum fluctuations are small. If the classical solution is to be accurate we should therefore shut off the quantum fluctuations of gravity. This is done by making the characteristic scale of the geometry, L, much larger than the Planck length,

$$L \gg \ell_P, \tag{3.57}$$

where the Planck length can be deduced from the coefficient of the action (3.48),

$$\ell_P \sim \left(\alpha'^4 g_s^2 \right)^{\frac{1}{8}}. \tag{3.58}$$

The action when evaluated on the classical black D3 brane is proportional to $\frac{L^8}{\ell_P^8}$ and it is large in this limit, justifying the semi-classical approximation. The condition (3.57) requires that

$$N \gg 1. \tag{3.59}$$

Thus, to be guaranteed to be a good description of the string theory, we need N to be large and the coupling to be strong, $4\pi g_s N \gg 1$.

Now, let us consider the low energy excitations of the IIB string about the background in equations (3.49)–(3.52). There are two ways in which a super-gravity excitation can have low energy. Firstly, it can have very long wavelength. Long wavelength gravitons, that is, gravitons with wavelengths much larger than L, and therefore with the low energies $E << \frac{1}{L}$, occupy the asymptotic region of the black D3 brane, where $r >> L$. Their wavelengths are sufficiently long that they do not fit into the throat of the geometry at smaller values of $r \sim L$. What is more, their coupling to the other degrees of freedom scale to zero with energy, so that they are decoupled from those other degrees of freedom, and from each other. They are effectively free gravitons propagating in flat ten-dimensional Minkowski space asymptotically far from the black hole.

The other way that an excitation of the background can have low energy is to occupy the region near the horizon, $r \sim 0$, that is, to have fallen down into the gravitational potential well which is the black hole.[7] The energy of a string theory state in that region, that is the energy that must be injected in order to excite a typical string sitting at radius r, is $E_r \sim \frac{1}{\sqrt{\alpha'}}$, which is a high energy. However, as seen from an observer at infinity, this energy is red-shifted to $\sqrt{-g_{00}}E_r = H(r)^{-\frac{1}{4}}E_r \sim \frac{r}{L}\frac{1}{\sqrt{\alpha'}}$ and it can be arbitrarily small if r is small. Thus, within a distance $\sqrt{\alpha'}$ of the horizon, we can find all of the excitations of the string theory and they can all be viewed by an observer at infinity as having low energy.

The last statement, that we can find the full spectrum of the IIB string at small r might worry the reader, since we have said that the solution of super-gravity can only be trusted when the massive modes of the string decouple. This worry would be justified in that such a high energy excitation should back-react on the geometry. The answer to this question lies in the fact that the near-horizon geometry, with $AdS_5 \times S^5$, metric given in (3.55), is thought to be an exact solution of not only super-gravity, but the IIB string theory itself. This means that all corrections that would have arisen from integrating out massive string modes, which are proportional to $\frac{\alpha'}{L^2}$, should vanish in this region anyway. It also means that the back-reaction even from locally high energy string states vanishes.

We conclude that the very low energy limit of the super-gravity degrees of freedom is of these two kinds, which are decoupled from each other, free super-gravitons propagating in ten-dimensional Minkowski space, and the full IIB string theory in the near horizon geometry, $AdS_5 \times S^5$.

3.4.4 The Maldacena conjecture

Now, we return to the idea that in the two scenarios described in Sections 3.4.2 and 3.4.3, we are simply describing the same dynamical system in two different limits, both where $N \to \infty$ holding $4\pi g_s N$ fixed and then, in the first where $4\pi g_s N << 1$ and the second where $4\pi g_s N >> 1$ and in both cases at low energy $<< 1/\sqrt{\alpha'}$. In both cases there was a subsystem of almost free super-gravitons propagating on ten-dimensional Minkowski space and we can identify these subsystems. The great insight of Maldacena was to then identify the other two subsystems, the $\mathcal{N} = 4$ theory of low energy open strings and the closed IIB string on the $AdS_5 \times S^5$ near-horizon geometry. The idea is that these are descriptions of one and the same system, each description being analytically tractable

---

[7] For an extremal black hole, the horizon is a boundary of the space-time, rather than an event horizon.

in a different limit. The parameters are identified as $g_{YM}^2 = 4\pi g_s$, as in equation (3.47) Moreover, defining $\lambda \equiv 4\pi g_s N = g_{YM}^2 N$, we have $L^2 = \sqrt{\lambda}\alpha'$, as in equation (3.52). It is in principle a one-to-one identification of the two, including the identification of all of the quantum states, observables, operators, and amplitudes. As such, it is rather remarkable. It says that one observer whose observational tools are, for example, particle accelerators, might see this world as four space-time dimensional and containing $\mathcal{N} = 4$ theory with gauge fields and adjoint representation quarks, whereas a differently equipped observer would see exactly the same world as being ten-dimensional and the elementary objects as strings. Of course, λ would be the same in both worlds, and if it were large, for example, the second, string observer would be able to do accurate calculations, whereas the first, gauge field observer, although they might have guessed that gauge fields were the degrees of freedom, would be frustrated to find that this theory to describe their world is strongly coupled.

Given the duality between a gauge and string theory that we have been discussing, to make it useful, we must identify what is dual to what. One can go a long way in such an identification by comparing the symmetries of the theories and identifying objects with the same quantum numbers. Given a set of gauge invariant local operators in the gauge theory, $O_i(x)$, the generating functional

$$Z[\phi_0] = \left\langle e^{i \int d^4 x \phi_0^i(x) O_i(x)} \right\rangle \tag{3.60}$$

can be used to find correlation functions. One must take functional derivatives by the source fields $\phi_0^i(x)$ and then set them to zero. In order to be gauge invariant, the operators $O_i(x)$ must be composite. Moreover, it is convenient to organize them according to their scaling dimensions Δ_i, that is they should obey

$$-i[D, O_i(x)] = i(x^\mu \partial_\mu + \Delta_i) O_i(x).$$

On the string side of the correspondence the procedure is as follows. Firstly, we identify the degree of freedom of the IIB string which is dual to the operator $O_i(x)$ of the gauge theory. Generally, the way that degrees of freedom transform under symmetries as well as the quantum number Δ_i is a useful guide in this identification. In practical terms, the degrees of freedom are often fields which are in the string spectrum. Then, for that degree of freedom, we require a boundary condition in the string theory on $AdS_5 \times S^5$ that the field $\phi^i(x)$ approaches a constant times the classical source function $r^{\Delta_i} \phi_0^i(x)$ at the boundary of AdS_5.

The theory is then solved and the on-shell partition function of the string theory is found, call it $Z_{\text{string}}[\phi^i \to r^{\Delta_i} \phi_0^i]$, which is a functional of $\phi_0^i(x)$. The prescription for the generating functional is then

$$\left\langle e^{i \int d^4 x \phi_0^i(x) O_i(x)} \right\rangle = Z_{\text{string}}[\phi^i \to r^{\Delta_i} \phi_0^i]. \tag{3.61}$$

In the parameter regime $\lambda \gg 1$ where the string theory is semi-classical and at large N where it is super-gravity, the latter partition function is obtained from the on-shell action in classical super-gravity and

$$\left\langle e^{i\int d^4x \phi_0^i(x) O_i(x)} \right\rangle \approx e^{iS_{sugra}[\phi^i \to r^{\Delta_i}\phi_0^i]}. \tag{3.62}$$

Here, $O^i(x)$ must the Yang–Mills theory operators which are dual to the fields of the super-gravity theory.

3.5 $\mathcal{N} = 4$ superconformal quantum field theory

In this section we shall outline some of the details of $\mathcal{N} = 4$ supersymmetric Yang–Mills theory. Much of this section is quite technical and it could be skipped on first reading of these notes. Very few details beyond the bosonic part of the Lagrangian density will be needed in the following sections. I include these details mainly for completeness.

The $\mathcal{N} = 4$ theory which appears on one side of the AdS/CFT duality is of interest for a number of reasons. Of course, it is the quantum field theory side of the example of the AdS/CFT correspondence that we have discussed in Section 3.4. This duality has an enormous number of potential applications in many contexts. In addition, $\mathcal{N} = 4$ is a simple, albeit non-trivial example of a supersymmetric and conformally symmetric quantum field theory in four dimensions, where the high degree of supersymmetry simplifies a number of properties but leaves the detailed dynamics of the theory highly non-trivial. The conjectured integrability of the planar limit of the theory, and then also the IIB string theory in $AdS_5 \times S^5$ background in the limit of vanishing string coupling, have also been of great interest. For example, integrability can be used to do computations of the conformal dimensions of large classes of local operators with amazing efficiency.

The $\mathcal{N} = 4$ theory is a highly non-trivial four-dimensional quantum field theory which is thought to exhbit an exact non-perturbative $SL(2,\mathcal{Z})$ duality, which interchanges electric and magnetic fields and charges. If we combine the theta-angle and coupling constant to form the complex parameter

$$\tau = \frac{\theta}{2\pi} + \frac{4\pi i}{g_{YM}^2}, \tag{3.63}$$

the $SL(2,Z)$ has the generators

$$\tau \to -\frac{1}{\tau}, \quad \tau \to \tau + 1. \tag{3.64}$$

Finally, in the study of scattering amplitudes using modern techniques, infrared cut-off $\mathcal{N} = 4$ theory has been claimed to be the simplest interacting four-dimensional Yang–Mills theory.

Historically, $\mathcal{N} = 4$ theory first appeared in the work of Brink, Scherk, and Schwarz [2], who constructed the four–dimensional field theory by using dimensional reduction of $\mathcal{N} = 1$ supersymmetric Yang–Mills theory in ten dimensions. The field content is

1. Six scalar fields $\phi^i(x)$, $i = 1, \ldots, 6$
2. A vector gauge field $A_\mu(x)$
3. Four Weyl spinor fields $\lambda_\alpha^a(x)$, $a = 1, 2, 3, 4$ and their conjugates $\bar{\lambda}_{\dot\alpha a}(x)$

The scalar and vector fields are $N \times N$ Hermitian matrices and λ and $\bar{\lambda}$ are Hermitian conjugates of each other. Lorentz indices are raised and lowered by the Minkowski metric (3.1). The space-time vector indices are $\mu = 0, 1, 2, 3$, spinor indices are $\alpha, \dot\alpha = 1, 2$ and $i = 1, 2, \ldots, 6$ and $a = 1, 2, 3, 4$ are $O(6) = SU(4)$ R-symmetry indices. The un-dotted and dotted spinors transform under the $SL(2, C)$ universal cover of the Lorentz group in the usual way. The spinor indices are contracted using the $SL(2, C)$ antisymmetric invariant tensors $\epsilon_{\alpha\beta}$ or $\epsilon^{\alpha\beta}$ and $\epsilon_{\dot\alpha\dot\beta}$ or $\epsilon^{\dot\alpha\dot\beta}$ where $-\epsilon_{21} = \epsilon_{12} = 1 = -\epsilon^{12} = \epsilon^{21}$. Ergo $\lambda^\alpha \equiv \epsilon^{\alpha\beta}\lambda_\beta$, for example. The sets of 2×2 matrices $\sigma^\mu_{\alpha\dot\beta}$ and $\bar{\sigma}^\mu_{\dot\alpha\beta}$ are defined by the four-dimensional Dirac matrices in a representation where γ^5 is diagonal,

$$\gamma^\mu = \begin{bmatrix} 0 & \sigma^\mu \\ \bar{\sigma}^\mu & 0 \end{bmatrix} \tag{3.65}$$

and Dirac matrices obey the algebra

$$[\gamma^\mu, \gamma^\nu] = -2\eta^{\mu\nu}. \tag{3.66}$$

In a particular basis, $\sigma^\mu_{\alpha\dot\beta} = (-\mathcal{I}, \vec{\sigma})$ and $\bar{\sigma}^\mu = (-\mathcal{I}, -\vec{\sigma})$ where $\vec{\sigma}$ are the Pauli matrices,

$$\sigma^1 = \begin{bmatrix} 0 & 1 \\ 1 & 0 \end{bmatrix}, \ \sigma^2 = \begin{bmatrix} 0 & -i \\ i & 0 \end{bmatrix}, \ \sigma^3 = \begin{bmatrix} 1 & 0 \\ 0 & -1 \end{bmatrix}. \tag{3.67}$$

Also,

$$\sigma^{\mu\nu}_{\alpha\beta} = \sigma^\mu_{\alpha\dot\beta}\epsilon^{\dot\beta\dot\gamma}\bar{\sigma}^\nu_{\dot\gamma\beta} - \sigma^\nu_{\alpha\dot\beta}\epsilon^{\dot\beta\dot\gamma}\bar{\sigma}^\mu_{\dot\gamma\beta} \tag{3.68}$$

$$\bar{\sigma}^{\mu\nu}_{\dot\alpha\dot\beta} = \bar{\sigma}^\mu_{\dot\alpha\beta}\epsilon^{\beta\gamma}\sigma^\nu_{\gamma\dot\beta} - \bar{\sigma}^\nu_{\dot\alpha\beta}\epsilon^{\beta\gamma}\sigma^\mu_{\gamma\dot\beta} \tag{3.69}$$

For more details on these conventions, see [4].

In component fields (this is the superfield action written in component fields in the *Wess–Zumino gauge*), the Lagrangian density of $\mathcal{N}=4$ is

$$\mathcal{L} = \mathrm{Tr}\left[-\frac{1}{2}F_{\mu\nu}F^{\mu\nu} - \sum_{i=1}^{6} D_\mu \phi^i D^\mu \phi^i + g_{YM}^2 \frac{1}{2}\sum_{i,j}\left[\phi^i,\phi^j\right]^2 \right.$$
$$\left. -i\sum_{a=1}^{4}\bar{\lambda}^a \bar{\sigma}^\mu D_\mu \lambda_a + g_{YM}\sum_{iab} C_i^{ab}\epsilon_{\alpha\beta}\lambda_a^\alpha\left[\phi^i,\lambda_b^\beta\right] + g_{YM}\sum_{iab}\bar{C}_{iab}\epsilon^{\dot{\alpha}\dot{\beta}}\bar{\lambda}_{\dot{\alpha}}^a\left[\phi^i,\bar{\lambda}_{\dot{\beta}}^b\right]\right], \tag{3.70}$$

where the trace is over the $N\times N$ matrix indices. The first term is the usual Yang–Mills action and the second and third terms contain the six adjoint representation scalar fields. The matrices C_{iab} and $\bar{C}_i^{ab}$ are obtained from the 8×8 Euclidean Dirac matrices of six-dimensional Euclidean space, in a chiral representation where $\gamma^1\gamma^2\ldots\gamma^6$ is diagonal,

$$\gamma_i = \begin{bmatrix} 0 & \bar{C}_{i\bar{a}\bar{b}} \\ C_i^{ab} & 0 \end{bmatrix}, \; i=1,\ldots,6, \; a,b,\bar{a},\bar{b}=1,2,3,4 \tag{3.71}$$

normalized by the anti-commutator

$$\{\gamma_i,\gamma_j\} = 2\delta_{ij}. \tag{3.72}$$

Shortly, we shall also need the spin tensors

$$\bar{C}_{ijb}^{a} = C_i^{ac}\bar{C}_{jcb} - C_j^{ac}\bar{C}_{icb} \tag{3.73}$$
$$C_{ija}{}^{b} = \bar{C}_{iac}C_j^{cb} - \bar{C}_{jac}C_i^{cb}. \tag{3.74}$$

The dual of the field strength tensor is

$$\tilde{F}_{\mu\nu} = \frac{1}{2}\epsilon_{\mu\nu\rho\sigma}F^{\rho\sigma}, \tag{3.75}$$

where $\epsilon_{\mu\nu\rho\sigma}$ is the completely antisymmetric tensor with $\epsilon_{0123}=1$. The self-dual and anti-self-dual field strength tensors are defined by

$$F_{\mu\nu}^{\pm} = \frac{1}{\sqrt{2}}\left(F_{\mu\nu}\pm\tilde{F}_{\mu\nu}\right). \tag{3.76}$$

The supersymmetry transformations are generated by a Weyl spinor supercharge

$$\delta_\alpha^a X = \frac{1}{i}\left[Q_\alpha^a,X\right]_\mp, \tag{3.77}$$

where the $\mp$ means that we should use a commutator or anti-commutator if X is bosonic or fermionic, respectively. The transformations are

$$\delta_\alpha^a \phi^i = C^{iab} \lambda_{\alpha b} \tag{3.78}$$

$$\delta_\alpha^a \lambda_{\beta b} = F_{\mu\nu}^+ \sigma_{\alpha\beta}^{\mu\nu} \delta_b^a + g_{\text{YM}} \left[\phi^i, \phi^j\right] (\bar{C}_{ij})^a_{\ b} \epsilon_{\alpha\beta} \tag{3.79}$$

$$\delta_\alpha^a \bar{\lambda}^b_{\dot\beta} = C_i^{ab} \sigma^\mu_{\alpha\dot\beta} D_\mu \phi^i \tag{3.80}$$

$$\delta_\alpha^a A^\mu = \sigma^\mu_{\alpha\dot\gamma} \epsilon^{\dot\gamma\dot\beta} \bar{\lambda}^a_{\dot\beta}. \tag{3.81}$$

The quantum field theory described by the Lagrangian (3.70) contains no dimensional parameters and at the classical level it has exact scale and therefore conformal invariance. It also has the remarkable feature that the conformal invariance survives at the quantum level in that the beta-function for its only coupling constant g_{YM} vanishes. This is easily seen at the one-loop level by the old formula for the 1-loop beta-function in Yang–Mills theory,

$$\beta(g_{\text{YM}}, N) = -\frac{g_{\text{YM}}^3}{16\pi^2} \left(\frac{11}{3}N - \frac{1}{6} \sum_k C_2(\text{scalars}) - \frac{1}{3} \sum_k C_2(\text{Weyl spinors}) \right), \tag{3.82}$$

where C_2 is the quadratic Casimir invariant of the representation under which the scalar and spinor fields transform. $C_2 = N$ for the adjoint representation. $\mathcal{N} = 4$ has six real scalars and eight Weyl spinors, so one can see that the right-hand side of (3.82) vanishes. The one-loop beta-function is zero. This was extended to two [8] and three loops [13] [14]. There is a general algebraic proof in the light-cone gauge [17] as well as an algebraic argument which we shall review here.

3.5.1 *PSU(2, 2|4)*

The symmetries of $\mathcal{N} = 4$ Yang–Mills theory are described by Lie superalgebra $PSU(2,2|4)$. In the following we shall describe the three components of this algebra, the Poincaré algebra, its conformal extension, its supersymmetric extension, and finally its superconformal extension.

3.5.1.1 *Poincaré algebra*

The Poincaré symmetry of the field theory contains symmetry under space-time translations, rotations, and Lorentz boosts. These transformations form a non-compact Lie group called the Poincaré group. The Lie algebra corresponding to the Poincaré group is generated by four momentum operators P_μ which generate the space-time translations and six (antisymmetric) $L_{\mu\nu}$ which generate the rotations and Lorentz boosts. In any relativistic quantum field theory, this Lie algebra should be realized by a set of Hermitian operators. Moreover, in a local quantum field theory the appropriate

Hermitian operators are integrals of moments of components of a conserved, symmetric stress–energy tensor.

The commutation relations of the Lie algebra basis elements are

$$[P_\mu, P_\nu] = 0 \tag{3.83}$$

$$[L_{\mu\nu}, P_\lambda] = -i\left(\eta_{\mu\lambda} P_\nu - \eta_{\nu\lambda} P_\mu\right) \tag{3.84}$$

$$[L_{\mu\nu}, L_{\rho\sigma}] = -i\left(\eta_{\mu\rho} L_{\nu\sigma} + \eta_{\nu\sigma} L_{\mu\rho} - \eta_{\mu\sigma} L_{\nu\rho} - \eta_{\nu\rho} L_{\mu\sigma}\right). \tag{3.85}$$

The quantum field theory can be thought of as a set of infinite-dimensional unitary representations of this algebra.

3.5.1.2 Conformal symmetry

According to the Coleman–Mandula theorem [4], only certain extensions of the Poincaré algebra are allowed as the symmetry algebra of a quantum field theory that has unitary time evolution. One interesting possible extension of the Poincaré algebra is the conformal algebra in four dimensions. Field theories with this symmetry do not have asymptotic states (in fact the spectrum of the mass operator $P_\mu P^\mu$ is continuous) and they do not have an S-matrix. However, they are of interest, as generic renormalizable quantum field theories must approach theories with conformal symmetry in their high and low energy limits. The $\mathcal{N}=4$ Yang–Mills theory will also have conformal symmetry. The appropriate Lie algebra is that of the non-compact Lie group $SO(2,3)$, which has fifteen generators. These include the Poincaré generators with the algebra (3.83)–(3.85), as well as the generator of dilations, D, and a generator of special conformal transformations, K_μ, so that, in addition to (3.83–3.85), the remaining algebra is

$$[D, P_\mu] = -iP_\mu, \quad [D, L_{\mu\nu}] = 0, \quad [D, K_\mu] = iK_\mu \tag{3.86}$$

$$[L_{\mu\nu}, K_\lambda] = -i\left(\eta_{\mu\lambda} K_\nu - \eta_{\nu\lambda} K_\mu\right) \tag{3.87}$$

$$[P_\mu, K_\nu] = 2i\left(L_{\mu\nu} - \eta_{\mu\nu} D\right) \tag{3.88}$$

$$[K_\mu, K_\nu] = 0. \tag{3.89}$$

An operator $O_\Delta(x)$ is said to have dimension Δ if it has the property

$$i[D, O_\Delta(x)] = \left(x^\mu \partial_\mu + \Delta\right) O_\Delta(x). \tag{3.90}$$

Under a finite transformation,

$$s^{iD} O_\Delta(x) s^{-iD} = s^\Delta O_\Delta(sx). \tag{3.91}$$

Then, the algebraic relations in (3.86) imply that K_μ is a lowering operator for the dimension. To see this, consider the action of the special conformal generator $[K_\mu, O_\Delta(x)]$. The dimension of the resulting operator is obtained by taking the commutator

$$[D, [K_\mu, O_\Delta(x)]] = -[K_\mu, [O_\Delta(x), D]] - [O_\Delta(x), [D, K_\mu]]$$
$$= -i(\Delta - 1)[K_\mu . O_\Delta(x)], \quad (3.92)$$

where we have used the Jacobi identity for commutators and the algebra in equations (3.90) and (3.86). We see that the commutator with K_μ has lowered the dimension of $O_\Delta(x)$ by one unit. Similarly, P_μ is a raising operator,

$$[D, [P_\mu, O_\Delta(x)]] = -[P_\mu, [O_\Delta(x), D]] - [O_\Delta(x), [D, P_\mu]]$$
$$= -i(\Delta + 1)[P_\mu . O_\Delta(x)]. \quad (3.93)$$

In a relativistic quantum field theory, the dimensions of operators other than the identity must be positive and they have a lower bound. If four dimensions, the lower bound for bosonic operators is $\Delta_{\min} = 1$, the dimension of a free scalar field, and for fermionic operators it is $\Delta_{\min} = \frac{3}{2}$, the dimension of a free fermion field. This means that for any operator $O_\Delta(0)$ which has a given dimension Δ, the dimension cannot be lowered beyond a certain positive minimum value, that is, there must be operators such that

$$[K_\mu, \mathcal{O}_{\Delta_0}(0)] = 0. \quad (3.94)$$

Such an operator is called a *primary operator*. It has dimension Δ_0. Commutators of a primary operator with P_μ,

$$[P_{\mu_1}, [P_{\mu_2}, \ldots [P_{\mu_k}, \mathcal{O}_{\Delta_0}(0)] \ldots]] = (-i)^k \partial_{\mu_1} \partial_{\mu_2} \ldots \partial_{\mu_k} \mathcal{O}_{\Delta_0}(x)\Big|_{x=0}, \quad (3.95)$$

raise the dimension of $\mathcal{O}_{\Delta_0}(0)$ from Δ_0 to $\Delta_0 + k$. If $O_p(x)$ has dimension Δ_p, then $[P_\mu, O_p(x)]$ has dimension $\Delta_p + 1$, $[P_\mu, [P_\nu, O_p(x)]]$ has dimension $\Delta_p + 2$, and so on. The resulting operators are called *descendants*.

Note that we are evaluating $O_\Delta(x)$ at $x = 0$, the fixed point of the conformal group. $O_\Delta(x)$ at other points is in principle obtained by considering the Taylor expansion using the descendants of $O_\Delta(0)$. The set of all commutators of the generators of the conformal group with a primary operator forms a representation of the conformal algebra which is infinite-dimensional, as it must be for a non-compact Lie algebra. Also, note that, since $L_{\mu\nu}$ commutes with D, it does not change the dimension of an operator. It must therefore either commute with $O_\Delta(0)$, as would be the case for a scalar, or there must be several $O_{\Delta s}(0)$ with the same dimension, so that

$$[L_{\mu\nu}, O_{\Delta s}(0)] = iO_{\Delta s'}(0)[\Sigma_{\mu\nu}]^{s'}_{s}, \quad (3.96)$$

that is, the operators $O_{\Delta s}$ carry a finite-dimensional representation of the algebra with Lorentz spin matrices $\Sigma_{\mu\nu}$.

Conformal symmetry determines the coordinate dependence of two-point functions up to normalization, which can be chosen so that

$$\langle O_{\Delta_1}(x) O_{\Delta_2}(0) \rangle = \frac{\delta_{\Delta_1, \Delta_2}}{|x|^{\Delta_1 + \Delta_2}}. \tag{3.97}$$

The three-point functions are also determined up to a constant,

$$\langle O_{\Delta_1}(x) O_{\Delta_2}(y) O_{\Delta_3}(z) \rangle = \frac{c(\Delta_1, \Delta_2, \Delta_3)}{|x-y|^{\Delta_1+\Delta_2-\Delta_3}|y-z|^{\Delta_2+\Delta_3-\Delta_1}|z-x|^{\Delta_3+\Delta_1-\Delta_2}}. \tag{3.98}$$

The fact that these are determined follows from the ability to use finite conformal transformations to move the two or three points to some reference values. This no longer works for higher-point functions since the coordinates have some invariant ombinations. For example for the four-point functions one can construct two combinations of coordinates, called harmonic ratios, which are invariant under conformal transformations.

3.5.1.3 Supersymmetry

Another allowed extension of the Poincaré algebra is obtained by adding some anti-commuting generators which transform like spinors under Lorentz transformations and rotations. Consider the un-dotted spinor supercharges Q_α^a and their conjugate dotted spinors $\bar{Q}_{\dot{\alpha} a}$, where $a = 1, ..., \mathcal{N}$ in theories with different degrees of supersymmetry, Generally, $\mathcal{N} = 1, 2, 4$. Here we will consider the case specific to $\mathcal{N} = 4$. In addition to the Poincaré algebra (3.83)–(3.85), a supersymmetric relativistic quantum field theory contains Q_α^a and $\bar{Q}_{\dot{\alpha} a}$ which have the algebra

$$[P_\mu, Q_\alpha^a] = 0, \quad [P_\mu, \bar{Q}_{\dot{\alpha} a}] = 0 \tag{3.99}$$

$$[L^{\mu\nu}, Q_\alpha^a] = i\sigma^{\mu\nu}_{\alpha\beta} \epsilon^{\beta\gamma} Q_\gamma^a, \quad [L^{\mu\nu}, \bar{Q}_{\dot{\alpha} a}] = i\bar{\sigma}^{\mu\nu}_{\dot{\alpha}\dot{\beta}} \epsilon^{\dot{\beta}\dot{\gamma}} \bar{Q}^a_{\dot{\gamma}} \tag{3.100}$$

$$\{Q_\alpha^a, Q_\beta^b\} = 0, \quad \{Q_\alpha^a, \bar{Q}_{\dot{\beta} b}\} = 2\delta^a_b \sigma^\mu_{\alpha\dot{\beta}} P_\mu, \quad \{\bar{Q}_{\dot{\alpha} a}, \bar{Q}_{\dot{\beta} b}\} = 0. \tag{3.101}$$

The indices $a, b = 1, 2, 3, 4$ transform under $SU(4)$ R-symmetry. To distinguish the supercharges Q_α^a and $\bar{Q}_{\dot{\alpha} a}$ from the conformal supercharges which we shall now introduce, these are called the Poincaré supercharges.

3.5.1.4 Superconformal algebra

Now, let us consider the situation where a supersymmetric theory also happens to be a conformal field theory. It is clear from the algebra (3.101) that the supercharges have dimension $\frac{1}{2}$,

$$[D, Q_\alpha^a] = -\frac{i}{2} Q_\alpha^a, \quad [D, \bar{Q}_{\dot{\alpha} a}] = -\frac{i}{2} \bar{Q}_{\dot{\alpha} a}. \tag{3.102}$$

To close the algebra, more anti-commuting generators are needed. These are the conformal supercharges, $S_{\alpha a}$ and $\bar{S}^a_{\dot\alpha}$, which are commutators of the Poincaré supercharges with K_μ,

$$[K^\mu, Q^a_\alpha] = \sigma^\mu_{\alpha\dot\alpha}\epsilon^{\dot\alpha\dot\beta}\bar{S}^a_{\dot\beta}, \quad [K^\mu, \bar{Q}_{\dot\alpha a}] = \bar{\sigma}^\mu_{\dot\alpha\alpha}\epsilon^{\alpha\beta}S_{\beta a} \tag{3.103}$$

$$[D, S_{\alpha a}] = \frac{i}{2}S_{\alpha a}, \quad [D, \bar{S}^a_{\dot\alpha}] = \frac{i}{2}\bar{S}^a_{\dot\alpha}. \tag{3.104}$$

The remainder of the algebra is

$$\{S_{\alpha a}, S_{\beta b}\} = 0, \quad \{S_{\alpha a}, \bar{S}^b_{\dot\beta}\} = 2\delta^b_a \sigma^\mu_{\alpha\dot\beta} K_\mu, \quad \{\bar{S}^a_{\dot\alpha}, \bar{S}^b_{\dot\beta}\} = 0 \tag{3.105}$$

$$\{Q^a_\alpha, S_{\beta b}\} = \epsilon_{\alpha\beta}\left[\bar{C}_{ij}{}^a{}_b T^{ij} + \delta^a_b D\right] + \frac{1}{2}\delta^a_b \sigma^{\mu\nu}_{\alpha\beta} L_{\mu\nu} \tag{3.106}$$

$$\{\bar{Q}_{\dot\alpha a}, \bar{S}^b_{\dot\beta}\} = \epsilon_{\dot\alpha\dot\beta}\left[\bar{C}_{ija}{}^b T^{ij} + \delta^b_a D\right] + \frac{1}{2}\delta^b_a \bar{\sigma}^{\mu\nu}_{\dot\alpha\dot\beta} L_{\mu\nu} \tag{3.107}$$

$$\{Q^a_\alpha, \bar{S}^b_{\dot\beta}\} = 0, \quad \{\bar{Q}_{\dot\alpha a}, S_{\beta b}\} = 0. \tag{3.108}$$

where T^{ij} is a generator of the $SO(6) \sim SU(4)$ Lie algebra. In $SO(6)$ it generates a rotation in the ij-plane and the spin matrices C and $\bar{C}$ convert them to the 4 and $\bar{4}$ representations of $SU(4)$, so that

$$[\bar{C}_{ija}{}^b T^{ij}, Q^c_\alpha] = i\delta^c_a Q^b_\alpha, \quad [\bar{C}_{ija}{}^b T^{ij}, \bar{Q}_{\dot\alpha c}] = -i\bar{Q}_{\dot\alpha a}\delta^b_c.$$

A primary operator $\mathcal{O}_{\Delta_0}(0)$ is an operator which has a conformal dimension Δ_0 and which obeys $[K_\mu, \mathcal{O}_{\Delta_0}(0)] = 0$.

We note that (anti-)commutators of an operator of dimension Δ with S and $\bar{S}$, if they are non-zero, must yield operators with dimensions $\Delta - \frac{1}{2}$. We can therefore use S and $\bar{S}$ to lower the dimension of an operator. This process of lowering the dimension using S and $\bar{S}$ must truncate at operators of a minimal dimension. There must therefore exist operators which obey

$$\left[S^a_\alpha, \tilde{\mathcal{O}}_{\Delta_0 s}(0)\right] = 0, \quad \left[\bar{S}_{\dot\alpha a}, \tilde{\mathcal{O}}_{\Delta_0 s}(0)\right] = 0 \tag{3.109}$$

(these are anticommutators if $\tilde{\mathcal{O}}_{\Delta_0 s}(0)$ is fermionic). An operator $\tilde{\mathcal{O}}_{\Delta_0 s}(0)$ with this property is called a *superconformal primary operator*. Here, we have anticipated that there can be more than one such operator and the label s runs over the set of operators.

It is easy to see that superconformal primary operators are always primary operators. However, the converse does not have to be true, it is possible that a primary is not a superconformal primary.

Commutator brackets of the other algebra elements with a superconformal primary generate a representation of $PSU(2,2|4)$ (modulo the usual need to eliminate null

158 *Lectures on the holographic duality of gauge fields and strings*

vectors). Generally, representations of the superconformal algebra can split into several representations of the conformal subalgebra.

3.5.1.5 *Short representations*

Though all non-trivial unitary representations of the superconformal algebra are infinite-dimensional, there is a sense in which some have fewer states than others. Such representations are called *short representations*. They occur when the superconformal primary operator, as well as obeying $[S_{\alpha a}, \chi_{\Delta sk}(0)] = 0$ and $[\bar{S}^a_{\dot\alpha}, \chi_{\Delta sk}(0)] = 0$, also commutes with some of the Poincaré supercharges,

$$[Q^a_\alpha, \chi_{\Delta sk}(0)] = 0 \text{ for some of the } a. \tag{3.110}$$

These are called *chiral primary operators* and we have denoted them by the symbol $\chi_{\Delta sk}(0)$. Such an operator has conformal dimension Δ and, for clarity, we have displayed its indices s and k explicitly. These indices transform under spin and internal symmetry, respectively.

Assume that a chiral primary operator exists and consider the values of a for which (3.110) holds. Then, trivially, the following double commutator vanishes identically:

$$\{S_{\beta b}, [Q^a_\alpha, \chi_{\Delta sk}(0)]\} = 0. \tag{3.111}$$

Using the Jacobi identity, we find

$$0 = \{S_{\beta b}, [Q^a_\alpha, \chi_{\Delta sk}(0)]\} = -\{Q_{\alpha a}, [\chi_{\Delta sk}(0), S^b_\beta]\} - [\chi_{\Delta sk}(0), \{S^b_\beta, Q_{\alpha a}\}]$$
$$= [\chi_{\Delta sk}(0), \epsilon_{\alpha\beta}\left[C^{ija}_b T_{ij} + \delta^a_b D\right] + 2\delta^a_b \sigma^{\mu\nu}_{\alpha\beta} M_{\mu\nu}], \tag{3.112}$$

where we have used the fact that $\chi_{\Delta sk}(0)$ is a superconformal primary (3.109) and the superconformal algebra equation (3.106). Now, the fact that both $L_{\mu\nu}$ and T_{ij} commute with D tells us that the set of operators degenerate with $\chi_{\Delta sk}(0)$ must carry a representation of both $SO(1,3)$ and $SO(6)$. This implies that

$$[T_{ij}, \chi_{\Delta sk}(0)] = i\chi_{\Delta sk'}(0)[t_{ij}]^{k'}_k, \quad [L_{\mu\nu}, \chi_{\Delta sk}(0)] = i\chi_{\Delta s'k}(0)[\Sigma_{\mu\nu}]^{s'}_s, \tag{3.113}$$

where t_{ij} and $\Sigma_{\mu\nu}$ are representations of $SU(4) \sim SO(6)$ and $SO(3,1)$ respectively. Equation (3.112) then implies

$$\epsilon_{\alpha\beta}\bar{C}^{ija}{}_b [t_{ij}]^{k'}_k \chi_{\Delta sk'}(0) - \epsilon_{\alpha\beta}\delta^b_a \Delta\chi_{\Delta sk}(0) - \delta^a_b \sigma^{\mu\nu}_{\alpha\beta}[\Sigma_{\mu\nu}]^{s'}_s \chi_{\Delta s'k}(0) = 0. \tag{3.114}$$

For a Lorentz scalar, $\Sigma_{\mu\nu} = 0$. Let us consider that case. We must understand the implication of (3.114). C^{ij} is a set of matrices in the fundamental representation of $SU(4)$ which is a rank 3 algebra. Assuming a certain normalization (and remembering that rotations in orthogonal planes in R^6 should commute), we can write the Cartan subalgebra as

$$C^{12} = \begin{bmatrix} 1 & 0 & 0 & 0 \\ 0 & 1 & 0 & 0 \\ 0 & 0 & -1 & 0 \\ 0 & 0 & 0 & -1 \end{bmatrix}, \ C^{34} = \begin{bmatrix} 1 & 0 & 0 & 0 \\ 0 & -1 & 0 & 0 \\ 0 & 0 & 1 & 0 \\ 0 & 0 & 0 & -1 \end{bmatrix}, \ C^{56} = \begin{bmatrix} 1 & 0 & 0 & 0 \\ 0 & -1 & 0 & 0 \\ 0 & 0 & -1 & 0 \\ 0 & 0 & 0 & 1 \end{bmatrix}.$$
(3.115)

The remaining twelve generators have vanishing diagonal elements. The eigenvalues of these matrices are the weights of the fundamental representation. If the eigenvalues of t_{12}, t_{34}, t_{56} are $(\mathcal{J}_1, \mathcal{J}_2, \mathcal{J}_3)$, respectively, the 1-1 component of (3.114) is then $\mathcal{J}_1 + \mathcal{J}_2 + \mathcal{J}_3 = \Delta$, and the 2-2 component is $\mathcal{J}_1 - \mathcal{J}_2 - \mathcal{J}_3 = \Delta$. The equation is obeyed when $\mathcal{J}_2 = \mathcal{J}_3 = 0$ and when $\mathcal{J}_1 = \Delta$. The chiral primary operator has $\Delta = \mathcal{J}_1$ which, due to the Lie algebra which quantizes $\mathcal{J}_1$, must be an integer. More importantly, it is independent of the coupling constant g_{YM} or the number N. Since it is not expected to change discontinuously as g_{YM} is varied, we expect that it is independent of the coupling and it is given by its value at $g_{YM} = 0$, that is, by the free field value, which is its engineering dimension.

An operator of this form is found by taking the complex combination of scalar fields $z = \phi^1 + i\phi^2$, and the generator whose eigenvalue is $\mathcal{J}_1$ generates a phase transform of $z = \phi^1 + i\phi^2$. The composite

$$O_k(0) = \text{Tr } z(0)^k \tag{3.116}$$

$[T_{12}, O_k] = kO_k$, $[T_{34}, O_k] = 0$, $[T_{56}, O_k] = 0$. This operator has tree-level dimension k. However, its dimension is determined by the algebra, so it must also have dimension k, unaffected by radiative corrections in the interacting theory.

This, a superconformal primary operator which commutes with half of the supercharges, is called a *chiral primary operator*. A single trace operator of this form is

$$O_k = \text{Tr } \phi^{(i_1)}(0)\phi^{i_2}(0)\ldots\phi^{i_k)}(0) - \text{Traces}, \tag{3.117}$$

which is made completely symmetric and traceless in its indices.

This is the first no-renormalization theorem—the dimension of chiral primary operators does not renormalize. It implies that, suitably normalized, their two-point functions are known exactly,

$$\langle O_{\Delta_1}(x) O_{\Delta_2}(0) \rangle = \frac{\delta_{\Delta_1, \Delta_2}}{|x|^{\Delta_1 + \Delta_2}} \tag{3.118}$$

(where Δ are positive integers). There is actually a stronger statement that can be made. The three-point functions of these operators is also thought not to renormalize:

$$\langle O_{\Delta_1}(x)O_{\Delta_2}(y)O_{\Delta_3}(z)\rangle = \frac{c(\Delta_1,\Delta_2,\Delta_3)}{|x-y|^{\Delta_1+\Delta_2-\Delta_3}|y-z|^{\Delta_2+\Delta_3-\Delta_1}|z-x|^{\Delta_3+\Delta_1-\Delta_2}} \quad (3.119)$$

and the coefficients $c(\Delta_1,\Delta_2,\Delta_3)$ are given by their tree-level values.

It is also an important fact that all of the operators in the entire representation of the superconformal algebra that is generated from a chiral primary operator have the same anomalous dimension. Such an operator is, for example, $\text{Tr}F_{\mu\nu}F^{\mu\nu}$ itself which can be obtained from $\text{Tr}\phi^i\phi^j - \frac{1}{6}\delta^{ij}\text{Tr}(\phi^k)^2$ by taking four commutators with supercharges. This implies that the operator has dimension four. Since the Lagrangian also must have dimension 4, the coupling constant g_{YM} must be dimensionless and therefore scale independent. A similar argument for conformal invariance was originally given in [13].

3.6 Holographic Wilson loops

In this section, we will give a simple review of the computation of the Wilson loop, both in the Yang–Mills theory and on the gravity side of the AdS/CFT duality. We will not give a complete review of this subject, but instead concentrate on explaining some of the basic ideas.

As we have already discussed, the Wilson loop is the quantum expectation value of the holonomy of a heavy quark wave function when we would drag the quark around a closed curve C in space-time (which we shall take to be Euclidean). The loops that are readily accessible to us in AdS/CFT are rather special ones which contain the scalar fields in the exponential

$$W[C] = <\text{Tr}\mathcal{P}e^{\int_0^1 d\tau\left(i\dot{x}^\mu(\tau)A_\mu(x(\tau))+|\dot{x}(\tau)|\hat{\theta}\cdot\phi(x(\tau))\right)}>. \quad (3.120)$$

Here, the expectation value is in the vacuum state of the $\mathcal{N}=4$ superconformal Yang–Mills theory with $SU(N)$ gauge group. The functions $x^\mu(\tau)$ sweep out the curve C as τ goes from 0 to 1. The scalar is multiplied by $\hat{\theta}$ which is a unit vector residing on the S^5. This expectation value is the measure of the holonomy of the wave function of a heavy quark when it is parallel transported along the curve C.

A weak coupling expansion of the Wilson loop begins with a Taylor expansion of the path ordered exponential

$$W[C] = \left[N + \frac{g_{YM}^2}{2}\int d\tau \int d\tau' \left[|\dot{x}(\tau)||\dot{x}(\tau')|\hat{\theta}^i\hat{\theta}^j <\mathcal{P}\phi_i^{ab}(x(\tau))\phi_j^{ba}(x(\tau'))>_0 \right.\right.$$
$$\left.\left. -\dot{x}^\mu(\tau)\dot{x}^\nu(\tau') <\mathcal{P}A_\mu^{ab}(x(\tau))A_\nu^{ba}(x(\tau'))>_0\right]+\ldots\right] \quad (3.121)$$

where, in the Feynman gauge, the free-field propagators are proportional to Green functions for the four-dimensional Laplacian,

$$<A_\mu^{ab}(x)A_\nu^{cd}(y)>_0 = \frac{\delta_{\mu\nu}\delta^{ad}\delta^{bc}}{8\pi^2(x-y)^2}, \quad <\phi_i^{ab}(x)\phi_j^{cd}(y)>_0 = \frac{\delta_{ij}\delta^{ad}\delta^{bc}}{8\pi^2(x-y)^2}.$$

We have taken the gauge group to be $U(N)$.

We get the expression

$$W[C] = \text{Tr}\left[1 + \frac{\lambda}{16\pi^2}\int d\tau d\tau' \frac{|\dot{x}(\tau)||\dot{x}(\tau')| - \dot{x}(\tau)\cdot\dot{x}(\tau')}{(x(\tau)-x(\tau'))^2} + \ldots\right]. \tag{3.122}$$

The contributions from the vector and scalar fields in this expression have the interesting consequence that the singularity at $\tau \to \tau'$ that would normally lead to a linearly ultraviolet divergence is absent here. For smooth trajectories, the integrals in (3.122) are finite.

3.6.1 Straight line and circle Wilson loop

The result (3.122) gives the leading terms in an expansion of the Wilson loop in the coupling constant λ. The integrals over the curve parameter can easily be done for some simple examples. For example, for a straight line Wilson loop, where $x^\mu(\tau) = (\tau, 0, 0, 0)$, the correction term vanishes and

$$W[\text{straight line}] = N. \tag{3.123}$$

For a circle on the other hand, $x^\mu(\tau) = a(\cos\tau, \sin\tau, 0, 0)$, the result is

$$W[\text{circle}] = N(1 + \lambda/8 + \ldots), \tag{3.124}$$

Scale invariance tells us that it does not depend on the radius of the circle. This is an expansion in small λ about $\lambda = 0$.

The prescription for computing this Wilson loop at strong coupling is to seek a minimal surface D whose boundary traces the curve C placed on the boundary of AdS_5 (at $r \to \infty$), while sitting at a point on S_5. The Wilson loop expectation value is the classical limit of the disc amplitude, which is given by

$$W[C] = \frac{1}{g_s}\exp\left(-\frac{1}{2\pi\alpha'}\inf_{D:\partial D=C} A[D] + ML[C]\right), \tag{3.125}$$

where $A[D]$ is the area of D. The counter-term, $ML[C]$, where $L[C]$ is the length of C and M is the infinite mass of the heavy quark, cancels a linear divergence that always occurs in the first term in the exponent. The factor of $\frac{1}{g_s} = \frac{4\pi}{\lambda}N$ arises from the fact that this is a disc amplitude. The factor of N is expected, on the gauge theory side, it comes from the trace which is taken to find the Wilson loop. The remaining factor $\frac{4\pi}{\lambda}$ should be regarded as sub-leading in the semi-classical limit.

As an example, consider the Wilson loop which is a straight line, $x^\mu(\tau) = (\tau,0,0,0)$. By symmetry, one might expect that the world-sheet which extremizes the area is simply the surface which extends along the x^1-direction and the r-direction, with the embedding equations

$$x^\mu(\tau,\sigma) = (\tau,0,0,0)$$
$$r(\tau,\sigma) = \sigma$$

and the position on the 5-sphere is given by a unit vector in 6-dimensional space $\hat{\theta}$, so that the intrinsic metric is

$$ds^2 = L^2\left[\frac{d\sigma^2}{\sigma^2} + \sigma^2 d\tau^2\right],$$

which is the metric of AdS_2 with the same radius of curvature, L, as the AdS_5 that it is embedded in. The area of this world-sheet is given by

$$\frac{1}{2\pi\alpha'}A[D] = \frac{1}{2\pi\alpha'}\int d\tau \int_0^\infty d\sigma \sqrt{\det g} = \frac{1}{2\pi\alpha'}\int d\tau \int_0^\infty d\sigma L^2.$$

This expression is infinite. To make sense of it, we could call the factor

$$\int d\tau = L[C],$$

the length of the curve. The other factor is

$$\frac{1}{2\pi\alpha'}\int_0^\infty d\sigma L^2 = \frac{\sqrt{\lambda}}{2\pi}\int_0^\infty dr \equiv M, \qquad (3.126)$$

which we shall show can be interpreted as the mass of the heavy quark. This also coincides with quark masses in probe brane constructions of heavy quarks [43]. Then, the area term in equation (3.125) and the length term cancel each other

$$\left[\frac{1}{2\pi\alpha'}A[D] - ML[C]\right]_{\text{straight line}} = 0.$$

We find that for a straight line $W = \frac{1}{g_s}$ which, to within the accuracy of our computation, is the same as $W = N$ (where we recall that $N = \frac{1}{g_s} \cdot \frac{\lambda}{4\pi}$). This is identical to the result (3.124) which we found at weak coupling, but of course here it is good in the strong coupling regime. We might conjecture that this is so in between these two limits, that when C is an infinite straight line, $W[C] = N$ to all orders in perturbation theory. This has indeed been confirmed at weak coupling to a few orders at small λ. Improving the string theory calculation to check the higher orders or the strong coupling expansion in $1/\sqrt{\lambda}$, and in particular, to check that the coefficient is indeed changed from $\frac{1}{g_s}$ to N seems to be more difficult. It would require solving for the fluctuations of the full type IIB superstring sigma model about the classical solution. At the time of writing, this has not been done in full detail. Some discussion can be found in references [66] and [67].

A reason for this triviality of the straight line would be supersymmetry. The straight-line Wilson loop operator of interest commutes with half of the Poincaré supercharges of the $\mathcal{N}=4$ theory.

For the circle Wilson loop, $x^\mu(\tau)$ is, for example, the locus of $(x^1)^2 + (x^2)^2 = a^2$. The minimal surface in AdS_5 with this boundary, and with maximal symmetry, turns out to be the locus of $(x^1)^2 + (x^2)^2 + \frac{1}{r^2} = a^2$, which is a Euclidean AdS_2 embedded in AdS_5. Using a particular world-sheet coordinate system, the embedding equations are

$$x^1(\sigma,\tau) = \frac{a}{\cosh a\sigma}\cos a\tau, \quad x^2(\sigma,\tau) = \frac{a}{\cosh a\sigma}\sin a\tau, \quad r(\sigma,\tau) = \frac{1}{a}\coth a\sigma, \quad \hat{\theta}\subset S^5. \tag{3.127}$$

Here, $\sigma = 0$ is located at $r=\infty$ and $\sigma=\infty$ is located at $r=\frac{1}{a}$. The induced metric of the world-sheet is that of a Euclidean AdS_2 black hole,

$$ds^2 = L^2\left(\frac{a}{\sinh a\sigma}\right)^2\left[d\sigma^2 + d\tau^2\right], \tag{3.128}$$

where the horizon is located at $\sigma = \infty$. The absence of a conical singularity at the point $\sigma = \infty$ requires peridicity of the Euclidean time with a particular period. In that region,

$$ds^2 \approx L^2\left(4a^2 e^{-2a\sigma}d\sigma^2 + 4a^2 e^{-2a\sigma}d\tau^2\right) = 4L^2\left[dy^2 + a^2 y^2 d\tau^2\right]$$

and there is no conical singularity at $y=0$ if the time has the correct periodic identification, $\tau \sim \tau + 2\pi/a$ (which we already knew from equation (3.127)). Of course, in Euclidean quantum field theory, periodic time means finite temperature, which, in this case, is interpreted as a Hawking temperature of the black hole. The Hawking temperature is thus

$$T_H = \frac{a}{2\pi}. \tag{3.129}$$

For the surface which we have been discussing, we can evaluate the exponent in the contribution to the Wilson loop in (3.125),

$$\frac{1}{2\pi\alpha'}A[D] - ML[C] = \frac{1}{2\pi\alpha'}\int_0^{2\pi/a}d\tau\int_0^\infty d\sigma L^2\left(\frac{a}{\sinh a\sigma}\right)^2 - \left(\frac{\sqrt{\lambda}}{2\pi}\int_0^\infty dr\right)(2\pi a)$$

$$= \frac{\sqrt{\lambda}}{2\pi}(2\pi a)\left[\int_0^\infty d\sigma\frac{dr}{d\sigma} - \int_0^\infty dr\right] = -\sqrt{\lambda}a\int_0^{\frac{1}{a}}dr = -\sqrt{\lambda}$$

where, in the last term in the first line, we have used the length of the curve, $L[C] = 2\pi a$ and the heavy quark mass given in equation (3.126). We see that the circle Wilson loop, in the limit of large λ, is then given by

$$W[\text{circle}] = Ne^{\sqrt{\lambda}}. \tag{3.130}$$

Generally, because of conformal symmetry, we expect the circle Wilson loop to be independent of the radius of the circle. Indeed, our results indicate that, in the large N limit,

$$W[\text{circle}] = Ne^{f(\lambda)}, \quad f(\lambda) = \begin{cases} \frac{\lambda}{8} & \lambda \ll 1 \\ \sqrt{\lambda} & \lambda \gg 1 \end{cases}. \tag{3.131}$$

In fact, this function is known to all orders in λ. It was conjectured using re-summed perturbation theory [18] and confirmed using supersymmetric localization [51] that the circle Wilson loop is given by the matrix model

$$W[\text{circle}] = \frac{\int dM \, e^{-\frac{8\pi^2 N}{\lambda} \text{Tr} \, M^2} \, \text{Tr} \, e^M}{\int dM \, e^{-\frac{8\pi^2 N}{\lambda} \text{Tr} \, M^2}}. \tag{3.132}$$

This integral can be computed exactly [37],

$$W[\text{circle}] = L^1_{N-1}[-\lambda/4N] e^{\lambda/8N}, \tag{3.133}$$

where

$$L^n_m = \frac{1}{n!} e^x x^{-m} \left(\frac{d}{dx}\right)^n x^{m+n} e^{-x} \tag{3.134}$$

is the Laguerre polynomial. The large N limit is

$$\lim_{N \to \infty} W[\text{circle}] = N \frac{2}{\sqrt{\lambda}} I_1(\sqrt{\lambda}), \tag{3.135}$$

where I_k is the modified Bessel function. Asymptotic expansions of the Bessel function can be used to show that the small and large λ limits agree with our computations summarized in (3.131).

3.6.2 Wilson loops and heavy quarks

It is enlightening to put the discussion of the Wilson loop into a more physical context. To do this, we begin with $\mathcal{N}=4$ superconformal Yang–Mills theory with gauge group $SU(N+1)$. Consider the scenario where the scalar field gets a non-zero vacuum expectation value,

$$\left\langle \phi_6^{N+1\,N+1}(x)\right\rangle = \varphi.$$

The expectation values of all other components of that and the other scalar fields vanish,

$$\left\langle \phi_6^{ab}(x)\right\rangle = 0 \text{ either } a \neq N+1 \text{ or } b \neq N+1$$
$$\left\langle \phi_i^{a,b}(x)\right\rangle = 0, \ i = 1,2,...,5.$$

With such a condensate, the gauge symmetry will be reduced by the Higgs mechanism, from $SU(N+1)$ to $SU(N) \times U(1)$ and the SO(6) R-symmetry is reduced to SO(5). Once the symmetry is broken, there are three types of fields. There is still a full $\mathcal{N}=4$ super-multiplet of massless fields with gauge group $SU(N)$ and which, in isolation, would be a superconformal quantum field theory. These massless fields are singlets under the $U(1)$ gauge symmetry. Then there are massless fields which constitute a full $\mathcal{N}=4$ super-multiplet with gauge group $U(1)$ and which are singlets under the SU(N) gauge symmetry. Then, there is a short $\frac{1}{2}$-BPS super-multiplet of massive W-fields. The latter multiplet contains vector, scalar, and spinor fields (which we generically refer to as W-fields). These transform under the fundamental representations of both the residual $SU(N)$ and $U(1)$. The bosonic fields in the super-multiplet are vectors

$$W_\mu^a = A_\mu^{aN+1}, \ \bar{W}_\mu^a = A_\mu^{N+1a}$$

and scalars

$$\Phi_i = \phi_i^{aN+1}, \ \bar{\Phi}_i = \phi_i^{N+1a}, \ I = 1,2,...,5.$$

These fields get masses from the commutator terms in the Yang–Mills action, for example, if we write

$$\phi_i^{ab} \to \varphi \delta_{i6} \delta^{aN+1} \delta^{bN+1} + \phi_i^{ab}$$

then,

$$-g_{YM}^2 \sum_{i<j=1}^{6} \left[\phi_i(x),\phi_j(x)\right]^2 \to g_{YM}^2 \varphi^2 \sum_{i=1}^{5}\sum_{a=1}^{N} \bar{\Phi}_i^a \Phi_i^a + \text{ cubic } + \text{quartic},$$

and we would identify the mass as $M_W = g_{YM}|\varphi|$. Also, the SO(6) R-symmetry is reduced to SO(5) symmetry by the choice of which scalar gets a vacuum expectation value. The orientation of this expectation value is generally a point on the coset $SO(6)/SO(5) = S^5$ which can be represented by a six-dimensional unit vector $\hat{\theta}_0$. In the above discussion, we have taken the simple case where $\hat{\theta}_0$ is in the 6-direction.

Let us consider the quantum amplitude whose modulus-squared is the probability that one of the scalar fields in the W-boson super-multiplet will propagate from position

166   *Lectures on the holographic duality of gauge fields and strings*

x_i^μ with colour index a to position x_f^μ with colour index b. (The $\mathcal{N} = 4$ theory is not in a confining phase, so global colour should be a good quantum number and we can ascribe 'colours' to particles.) Let us denote the amplitude by $\mathcal{A}_{ab}(x_f, x_i)$. We discuss $\mathcal{A}_{ab}(x_f, x_i)$ in the large N limit, at both the weak coupling and the strong coupling limits, in both cases at the limit where the W-boson mass is large. For this purpose, it is instructive to use a world-line functional integral representation of the propagator on Euclidean space,

$$\mathcal{A}_{ab}(x_f, x_i) = \int_0^\infty [dT] \int [dx^\mu(\tau)] e^{-S[x]} \left\langle \left[\mathcal{P} e^{\oint d\tau (i\dot{x}^\mu(\tau) A_\mu(x(\tau)) + |\dot{x}(\tau)| \hat{\theta}_0 \cdot \vec{\Phi}(x(\tau)))} \right]_{ab} \right\rangle. \tag{3.136}$$

Of course, this propagator is only non-zero in the gauge-fixed theory and even then it can depend on the way in which the gauge is fixed. Nevertheless, it should contain some physical information, such as the value of the physical mass of the W-boson. This could be deduced from the asymptotic behaviour for large proper time, where

$$\mathcal{A}_{ab}(x_f, x_i) \sim \delta_{ab} \exp\left(-M\sqrt{(x_f - x_i)^2}\right). \tag{3.137}$$

(Remember that we are in Euclidean space. If we were in Minkowski space, the decaying exponential would be replaced by a phase $e^{-iM\sqrt{-(x_f - x_i)^2}}$.) That the coefficient M of the proper time in the exponent in equation (3.137) is gauge invariant is equivalent to the statement that the pole in the propagator is gauge invariant.

The residual global colour symmetry of the gauge fixed theory tells us that equation (3.137) is equivalent to

$$\mathcal{A}_{ab}(x_f, x_i) = \frac{\delta_{ab}}{N} \int_0^\infty [dT] \int [dx^\mu(\tau)] e^{-S[x]} \left\langle \mathrm{Tr} \mathcal{P} e^{i\oint (\hat{A} + \Phi^1)} \right\rangle \tag{3.138}$$

$$= \frac{\delta_{ab}}{N} \int_0^\infty [dT] \int [dx^\mu(\tau)] e^{-\tilde{S}[x]} \tag{3.139}$$

$$\tilde{S} = \int_0^1 d\tau \left[\frac{1}{4T} \dot{x}^\mu(\tau) \dot{x}^\mu(\tau) + M^2 T \right] - \ln W[C], \tag{3.140}$$

where $W[C] = \left\langle \mathrm{Tr} \mathcal{P} e^{i\oint (\hat{A} + \Phi^1)} \right\rangle$ is the open Wilson loop for the curve C which is parameterized by $x^\mu(\tau)$. The potential term in the action is given by the logarithm of this Wilson loop, and the functional integration has boundary conditions, $x(1) = x_f$ and $x(0) = x_i$.

The Wilson loop computes the interaction of the particle with the remaining massless fields of the $\mathcal{N} = 4$ supersymmetric Yang–Mills theory, which are by themselves an $\mathcal{N} = 4$ theory with residual $SU(N)$ gauge group. The influence on the amplitude (3.138) of the W-bosons as well as the residual $U(1)$ gauge fields and other fields in their supermultiplets can be neglected in the large N limit. This is due to the fact that they transform in the fundamental or singlet representations of $SU(N)$, respectively, and their loops therefore contribute only to sub-leading orders in the large N limit. The one open W-boson line which is allowed in the amplitude is that which is computed by the world-line functional integral in (3.138).

We have written a gauge fixed version of the world-line functional integral where T is not the einbein, as it would be in the completely time reparameterization invariant version of the integral, but here it is a constant, which plays the role of Schwinger's proper time. We have used $[dT]$ to denote the measure whose details depend on the regularization of the functional integral over the $x^\mu(\tau)$. The T-dependence of the measure can be found by comparing with the Schwinger representation for the propagator of a free particle where

$$\mathcal{A}_0(x_f, x_i) = \int_0^\infty \frac{dT}{16\pi^2 T^2} e^{-\frac{(x_f-x_i)^2}{4T} - M^2 T} = \frac{M\sqrt{(x_f-x_i)^2} K_1(M\sqrt{(x_f-x_i)^2})}{4\pi^2(x_f-x_i)^2}. \quad (3.141)$$

A derivation of the world-line path integral for the scalar field propagator using zeta-function regularization can be found in the appendices of [58].

When the mass of the W-boson is large compared to any other quantity with dimensions (here, these could only be functions of the trajectory itself), we can study this propagator in the semiclassical approximation. The dimensionless number that controls this approximation is $\frac{1}{M\sqrt{(x_f-x_i)^2}}$. To study the semi-classical approximation, we begin with the classical equations of motion

$$-\frac{\ddot{x}^\mu(\tau)}{2T} - \frac{\delta}{\delta \tilde{x}^\mu(\tau)} \ln W[\tilde{x}]\bigg|_{\tilde{x}^\mu(\tau) = x^\mu(\tau)} = 0, \quad T^2 = \frac{1}{4M^2} \int_0^1 d\tau (\dot{x}^\mu)^2, \quad (3.142)$$

where we see that the presence of the Wilson loop creates a force term, the analogue of the Lorentz force on a relativistic charged particle. This term should be evaluated on the trajectory $x^\mu(\tau)$. The solution of equations (3.142) is a straight-line trajectory,

$$x_0^\mu(\tau) = (x_f^\mu - x_i^\mu)\tau + x_i^\mu, \quad T = \frac{1}{2M}\sqrt{(x_f-x_i)^2}. \quad (3.143)$$

To see this, we note that

$$-\frac{1}{2T}\ddot{x}_0^\mu(\tau) = 0.$$

Moreover, we can conjecture that the functional derivative of the Wilson loop vanishes when it is evaluated on an infinite straight line,

$$\frac{\delta}{\delta x^\nu(\tau)} \ln W[x^\mu]\bigg|_{x^\mu = \text{straight line}} = 0. \quad (3.144)$$

This is indeed true of the perturbative expression given in equation (3.122), even for a finite straight line. Indeed, for the infinite straight line, it is known that the first derivative by the contour, evaluated on the straight line, can be expressed as an anticommutator of a supercharge with another fermionic operator [49]. Therefore, as long as the superconformal symmetry of the $SU(N)$ $\mathcal{N}=4$ theory remains intact, equation (3.144) must hold, independent of the size of λ.

We conclude that the straight line is a legitimate saddle point of the functional integral in equation (3.138). We can proceed to evaluate the action on that saddle point. The world-line action contributes the term

$$\int_0^1 d\tau \left[\frac{1}{4T} \dot{x}^\mu(\tau) \dot{x}^\mu(\tau) + M^2 T \right] = M\sqrt{(x_f - x_i)^2}.$$

The semi-classical limit of the amplitude is thus given by plugging the classical solutions (3.143) into the action, to get

$$\mathcal{A}_{ab}(x_f, x_i) \sim e^{-M\sqrt{(x_f-x_i)^2}} \delta_{ab}. \tag{3.145}$$

Let us review what we have learned from this computation. Our only inputs were large N, which allowed us to ignore loops of the W-bosons, and the large mass limit for the W-boson, which allows us to do a semi-classical computation. Then, in addition, we know that (3.144) is independent of the coupling constant, and, in fact, independent of the large N expansion. This tells us that the straight line is a solution of the classical equation (3.142). Moreover, since the Wilson line cannot correct the exponential behaviour—it does not depend on M—and, because of conformal symmetry it does not depend on $(x_f - x_i)^2$ either. This tells us that M is not renormalized and that the semi-classical large N limit must always be of the form (3.145).

Now, let us move to the strong coupling side, which is described by the IIB string theory on $AdS_5 \times S^5$ background. On that side, we must first find the appropriate state of the string theory—the one that is dual to the state of $\mathcal{N}=4$ theory with $SU(N+1)$ gauge symmetry spontaneously broken to $SU(N) \times U(1)$ by the Higgs mechanism.

The $AdS^5 \times S^5$ background of the string theory is the gravitational field (in the near horizon region) due to a stack of $N+1$ D3 branes. We create the state with $SU(N+1)$ Higgsed to $SU(N) \times U(1)$ by separating one of the D3 branes from the stack of $N+1$ D3 branes and putting it at some distance from the stack. There are six directions which they can go in, being taken away from the stack and remaining parallel to the branes that are in the stack. We will call the direction that is chosen the unit 6-vector $\hat{\theta}_0$.

There are then three types of open strings. There are those which begin and end on the remaining stack of N D3 branes, those which begin and end on the separated D3 brane, and those which stretch from the stack of N D3 branes to the separated D3 brane.

The strings which connect the stack of N D3 branes to the separated D3 brane have a mass gap which is proportional to the distance of separation times the string tension. The lowest energy states of these strings are the massive W-boson super-multiplet.

The lowest modes of open strings with both ends attached to the stack of N D3 branes or both ends attached to the separated D3 brane are two $\mathcal{N}=4$ super-multiplets, the first with $SU(N)$ gauge symmetry and the second with $U(1)$ gauge symmetry. In the limits which create the $AdS_5 \times S^5$ background, the separated brane is located at a constant value of the AdS radius, $r = r_M$, and it is extended in the x^μ directions, parallel to the horizon at $r=0$ and the boundary at $r=\infty$ and it sits at a point on S^5 given by the unit 6-vector $\hat{\theta}_0$, the direction in which the D3 brane was separated from the stack.

It is interesting that the separated D3 brane will float at a fixed radius in AdS^5. This is a consequence of the fact that it is a $\frac{1}{2}$-BPS state of the string theory where the energy of the state does not depend on the distance of separation. Moving the D3 brane along the radius is a flat direction in the energy landscape. The D3 brane is attracted to the stack of D3 branes by the gravitational interaction. This attraction is exactly balanced by a repulsive interaction due to their Ramond–Ramond charges so that the net force is equal to zero. Another way to see this is to consider the separated D3 brane a probe brane. Then, in the leading order in string perturbation theory, the disc amplitude, the free energy of the separated D3 brane, is given by the Dirac–Born–Infeld action augmented by a Chern–Simons term for the Ramond–Ramond field

$$S_{\text{DBI}} = T_3 \int d^4x \left[-\sqrt{-\det(g_{\mu\nu} - 2\pi\alpha' F_{\mu\nu})} + \omega^{(4)} \right].$$

For a flat brane located at radius r_M the embedding metric obtained by setting $r = r_M$ and leaving only dx^μ non-zero in the AdS_5 metric to get the induced metric of the D3 brane is

$$g^{D3}_{\mu\nu} dx^\mu dx^\nu = L^2 r_M^2 dx_\mu dx^\mu.$$

The Ramond–Ramond 4-form $\omega^{(4)}$ is given by

$$\omega^{(4)} = L^4 r_M^4 dx^0 dx^1 dx^2 dx^3.$$

If (for future reference) the brane world-volume gauge field has a constant electric field, $F_{01} = E$,

$$S_{\text{DBI}} = \frac{L^4 r_M^4}{(2\pi)^3 \alpha'^2 g_s} \int d^4x \left[-\sqrt{1 - \frac{(2\pi\alpha')^2}{L^4 r_M^4} E^2} + 1 \right]. \quad (3.146)$$

We see that, when $E = 0$, the action (3.146) is zero, independent of the radius r_M. When $E \neq 0$ but small, and using $4\pi g_s = g_{\text{YM}}^2$, the energy of the electric field is

$$S_{\text{DBI}} \approx \frac{1}{g_{\text{YM}}^2} E^2,$$

which is what we expect for the energy of the $U(1)$ field (up to a factor of 4 which has been absorbed into the normalization of g_{YM}^2).

Thus we describe the appropriate state of the string theory as a probe D3 brane floating in AdS^5, parallel to both the boundary and the horizon, at radius r_M. Since in this strong coupling limit $\sqrt{\lambda} \gg 1$, we can treat the string theory semi-classically. We also work at weak string coupling g_s where we only need to compute the disc amplitude. We therefore take the semi-classical limit of the disc amplitude which is the semi-classical limit of

$$\mathcal{A}_{ab}(x_f, x_i) = \frac{\delta_{ab}}{N} \frac{1}{g_s} \int [dXdrd\hat{\theta}\ldots] e^{-\frac{L^2}{4\pi\alpha'} \int \left(r^2 \partial X \cdot \bar{\partial} X + \frac{1}{r^2} \partial r \bar{\partial} r + \partial \hat{\theta} \bar{\partial} \hat{\theta}\right) + \ldots}, \qquad (3.147)$$

where $\partial = \frac{\partial}{\partial(\sigma+i\tau)}$ and we have written only the bosonic part of the sigma-model action as the Polyakov action in the conformal gauge. The other fields in the supersymmetric world-sheet theory do not contribute to the leading order in the large $\sqrt{\lambda}$ limit. The factor $\frac{\delta_{ab}}{N}$ needs some explanation. The initial state in the amplitude is an open string which is suspended between the separated D3 brane and the D3 brane labelled 'a' in the stack of N D3 branes and located at $x^\mu = x_i^\mu$ and at the point $\hat{\theta} = \hat{\theta}_0$ on the 5-sphere which coincides with the direction in the six-dimensional space in which the separated D3 brane was pulled away from the stack. The final state in the amplitude is an open string which is suspended between the separated D3 brane and the D3 brane labelled 'b' in the stack and located at $x^\mu = x_f^\mu$ and $\hat{\theta} = \hat{\theta}_0$. The disc-like world-sheet should interpolate between these initial and final strings. This is only possible if the endpoint of the string follows the same D3 brane, therefore the amplitude must vanish unless $a = b$, thus the δ_{ab}. Then, by symmetry, the amplitude is independent of the index a so that we can average over index, thus the $\frac{1}{N}$. Then the stack of D3 branes is replaced by the $AdS_5 \times S^5$ geometry, and the information about individual D3 branes in the stack is lost, the open string must simply go to the AdS_5 horizon at $r = 0$.

In the large $\sqrt{\lambda}$ limit, to implement the semi-classical approximation, we look for a solution of the classical field equations for the embedding functions of the string

$$\left(r(\sigma,\tau), x^\mu(\sigma,\tau), \hat{\theta}(\sigma,\tau)\right)$$

into the $AdS_5 \times S^5$ space-time. These equations are obtained from the action in the functional integral in equation (3.147) by a variational method. They are

$$-\partial_a \left(\frac{1}{r^2} \partial_a r\right) + r\partial_a X \partial_a X - \frac{1}{r^3} \partial_a r \partial_a r = 0$$

$$\partial_a(r^2 \partial_a X_\mu) = 0$$

$$\partial_a^2 \hat{\theta} - \hat{\theta}(\hat{\theta} \cdot \partial_a^2 \hat{\theta}) = 0, \quad \hat{\theta}^2 = 1$$

with the appropriate boundary conditions. These are solved by

$$r(\sigma,\tau) = \frac{1}{\sqrt{(x_f - x_i)^2}\,\sigma}, \quad X(\sigma,\tau) = (x_f - x_i)\tau + x_i, \quad \hat{\theta}(\sigma,\tau) = \hat{\theta}_0.$$

This ansatz also solves the Virasoro constraints,

$$r^2 \partial_\sigma X \cdot \partial_\tau X + \frac{1}{r^2}\partial_\sigma r \partial_\tau r + \partial_\sigma \hat{\theta} \cdot \partial_\tau \hat{\theta} = 0$$

$$r^2 \partial_\sigma X \cdot \partial_\sigma X - r^2 \partial X_\tau \cdot \partial_\tau X - \frac{1}{r^2}\partial_\tau r \partial_\tau r + \frac{1}{r^2}\partial_\sigma r \partial_\sigma r + \partial_\sigma \hat{\theta} \cdot \partial_\sigma \hat{\theta} - \partial_\tau \hat{\theta} \cdot \partial_\tau \hat{\theta} = 0,$$

which are associated with fixing the conformal gauge for the world-sheet coordinates in the functional integral in equation (3.147). In order to satisfy the boundary conditions, σ should be integrated from $\frac{1}{\sqrt{(x_f-x_i)^2}\,r_M}$ to ∞ and τ from 0 to 1. The on-shell action is

$$\frac{\sqrt{\lambda}\alpha'}{4\pi\alpha'}\int\left(r^2\partial X \cdot \bar{\partial}X + \frac{1}{r^2}\partial r \bar{\partial} r + \partial\hat{\theta} \cdot \bar{\partial}\hat{\theta}\right) = \frac{\sqrt{\lambda}}{2\pi}\sqrt{(x_f-x_i)^2}\,r_M,$$

where we have used $L^2 = \sqrt{\lambda}\alpha'$ and the amplitude is

$$\mathcal{A}(x_f, x_i)_{ab} \sim \frac{\delta_{ab}}{N}\frac{1}{g_s}e^{-\frac{\sqrt{\lambda}\,r_M}{2\pi}\sqrt{(x_f-x_i)^2}}. \tag{3.148}$$

This result is interesting. It has the form given in equation (3.145), however, unlike (3.145) which is valid where $M\sqrt{(x_f - x_i)^2} \gg 1$, (3.148) is valid when $\sqrt{\lambda} \gg 1$. If we assume that they are simultaneously valid, we can make the identification

$$M = \frac{\sqrt{\lambda}\,r_M}{2\pi}, \quad r_M = \frac{2\pi M}{\sqrt{\lambda}}. \tag{3.149}$$

The result is also compatible with the Wilson loop at strong coupling being equal to one. This is notwithstanding the factor $\frac{1}{N}\frac{1}{g_s} = \frac{4\pi}{\lambda}$ which should be computed by fluctuations about the classical solution.

The other Wilson loop which is sometimes computable is the circle and we can develop something similar to the above reasoning for that case also. This has been done in an application to the study of the Schwinger pair production by a strong electric field where circle-like instantons dominate the semi-classical limit of the Euclidean world-line path integral that is used to compute the imaginary part of the vacuum energy [55]. In the following we will review another development which is related to the circle, but it is in Lorentzian space-time. As the reader will see, this subject is incomplete and there are several interesting open directions.

Consider the following thought experiment. We consider the same scenario as above, Yang–Mills theory with $SU(N)$ gauge group broken by the Higgs mechanism to $SU(N) \times U(1)$. We will work in Lorentzian space-time with real, rather than Euclidean time. We will use the residual $U(1)$ symmetry as one which leads to what we will call a conserved electric charge. The only objects in the theory which carry this electric charge are the W-bosons. We will consider the situation where a constant electric field which couples to this charge permeates the whole of space-time. This will be a weak field, sufficiently weak that we can neglect Schwinger pair production of charged particle–antiparticle pairs which should always occur at a slow rate in a constant electric field. The criterion for neglecting pair production of particles of mass M in an electric field of strength E is $\frac{E}{M^2} \ll 1$. In the string theory, it was argued in [55] that, to avoid Schwinger pair production, we must also require that $\frac{\sqrt{\lambda}}{2\pi}\frac{E}{M^2} \ll 1$. Then, we will consider a scalar particle in the W-boson super-multiplet and we will imagine that we prepare a state where we inject the particle from spatial infinity, travelling at almost the speed of light, and in a direction which is precisely anti-parallel to the electric field. The particle, when injected, has a particular colour state with label a. We will ask what the quantum amplitude is for the particle to return to precisely the same position, with precisely oppositely directed velocity after a time which is precisely equal to its classical travel time. We will consider the limiting case where the initial velocity of the particle approaches the speed of light, the injection position is infinitely far from the place where the particle stops and turns around, and the travel time approaches infinity. Some of the ideas which we discuss here appeared in [59] and [60].

The classical trajectory of a relativistic particle in a constant electric field has constant proper acceleration. It is therefore a Rindler trajectory. Our first aim is to do a semi-classical computation of the amplitude,

$$\mathcal{A}_{ab} = \frac{\delta_{ab}}{N} \int [dx][dT] e^{iS} \left\langle \text{Tr} \mathcal{P} e^{i \oint d\tau \left(\dot{x}^\mu(\tau) A_\mu(x(\tau)) + \Phi_i(x(\tau))\hat{\theta}_0^i\right)} \right\rangle, \qquad (3.150)$$

in the context of the gauge theory, where the world-line action of the particle is

$$S = \int d\tau \left[\frac{1}{4T} \dot{x}_\mu(\tau) \dot{x}^\mu(\tau) - M^2 T + \frac{E}{2}(x_0 \dot{x}_1 - x_1 \dot{x}_0) \right].$$

The first terms are the same as we used for the relativistic scalar particle in the previous discussion. The terms with coefficient E are the coupling to the electric field, $\int d\tau \dot{x}^\mu(\tau) A_\mu(x(\tau))$ where we use the gauge $A_\mu = -\frac{1}{2} F_{\mu\nu} x^\nu$ for a constant electromagnetic field $F^{\mu\nu}$, and then specialize to constant electric field. The equations of motion for the integration variables $(x_\mu(\tau), T)$ are

$$-\frac{1}{2T} \ddot{x}_0(\tau) + E\dot{x}_1(\tau) + \frac{1}{i}\frac{\delta}{\delta x^0(\tau)} \ln W[C] = 0 \qquad (3.151)$$

$$-\frac{1}{2T} \ddot{x}_1(\tau) + E\dot{x}_0(\tau) + \frac{1}{i}\frac{\delta}{\delta x_1(\tau)} \ln W[C] = 0 \qquad (3.152)$$

$$-\frac{1}{2T} \ddot{x}_{2,3}(\tau) - \frac{\delta}{\delta x_{2,3}(\tau)} \ln W[C] = 0 \qquad (3.153)$$

$$T^2 = -\frac{1}{4M^2} \int_{-\tau_P/2}^{\tau_P/2} d\tau \dot{x}_\mu(\tau) \dot{x}^\mu(\tau). \tag{3.154}$$

Then, we can use symmetry to argue that, when evaluated on the Rindler trajectory, $\frac{\delta}{\delta x^\mu(\tau)} \ln W[C]$ vanishes. As a consequence, equations (3.151)–(3.154) are solved by the Rindler trajectory itself, with constant proper acceleration $\frac{E}{M}$ due to the electric field, and in which we use the boundary conditions of our thought experiment to write the solution of equations (3.151)–(3.154) as

$$x_\mu(\tau) = \left(\frac{M}{E} \sinh \frac{E}{M}\tau, \frac{M}{E} \cosh \frac{E}{M}\tau, 0, 0\right) \tag{3.155}$$

$$T = \frac{\tau_P}{2M}, \quad -\tau_P/2 \leq \tau \leq \tau_P/2. \tag{3.156}$$

Here, τ_P is the total proper time accumulated by the W-boson during its flight and the initial and final points are

$$(x_f)_\mu = \left(\frac{M}{E} \sinh \frac{E}{2M}\tau_P, \frac{M}{E} \cosh \frac{E}{2M}\tau_P, 0, 0\right)$$

$$(x_i)_\mu = \left(-\frac{M}{E} \sinh \frac{E}{2M}\tau_P, \frac{M}{E} \cosh \frac{E}{2M}\tau_P, 0, 0\right).$$

The initial and final speeds, $v_{f,i} = \tanh \frac{E}{2M}\tau_P$, should be close to the speed of light, $v_{f,i} \sim 1$. Together with our previous weak-field assumption, we need to be in the regime

$$1 << \frac{M^2}{E} << M\tau_P,$$

which we shall assume from now on. Because of gauge invariance issues, we shall consider only the part of the phase of the amplitude $\mathcal{A}$ which grows linearly in τ_P. The coefficient of τ_P should be related to the energy of the particle in its rest frame. In fact, let us evaluate the world-line action on the classical trajectory that we have found. The result is

$$S = -M\tau_P + \frac{M}{2}\tau_P \tag{3.157}$$

and, in this limit,

$$\mathcal{A}_{ab} \approx \frac{\delta_{ab}}{N} e^{-iM\tau_P + iM\tau_P/2} W[\text{Rindler}]. \tag{3.158}$$

We have separated the on-shell action (3.157) into two parts in order to remind ourselves that the first term is the energy of the W boson in its rest frame, equal to its rest mass, M, times τ_P, as we would expect. The second term in (3.157), $\frac{M}{2}\tau_P$, is from the interaction energy of the W-boson with the electric field, which cancels exactly half of its rest energy.

The on-shell action becomes the phase in the amplitude (3.158). The remaining factor is the open Wilson loop evaluated on the Rindler trajectory (3.155). In the case of the straight line we argued that, as a consequence of symmetry, the Wilson loop could not contribute to the phase. We shall see that this is not the case here. This is due to the presence of two dimensionful parameters, the proper time τ_P and the acceleration $\frac{E}{M}$.

The Rindler trajectory in equation (3.155) is a hyperbola, the locus of the equation

$$(x_0)^2 - (x_1)^2 + \frac{M^2}{E^2} = 0,$$

where $\frac{E}{M}$ is the proper acceleration. This is the Lorentzian analogue of the circle which would be obtained by putting $x_0 \to -ix_0$. We therefore expect that some of the simplifications that characterize the circle Wilson loop should also apply to this trajectory. Indeed, this is the case. For example, the sum of the vector and the scalar particle propagators connecting any two points on the trajectory is a constant. Using

$$\left\langle \mathcal{T} A^{ab}_\mu(x) A^{cd}_\nu(y) \right\rangle_0 = \frac{\delta^{ad}\delta^{bc}\eta_{\mu\nu}}{8\pi^2 (x-y)^\mu (x-y)_\mu + i\epsilon}$$

$$\left\langle \mathcal{T} \Phi^{ab}_i(x) \Phi^{cd}_j(y) \right\rangle_0 = \frac{\delta^{ad}\delta^{bc}\delta_{ij}}{8\pi^2 (x-y)^\mu (x-y)_\mu + i\epsilon}$$

(where the subscript 0 on the bracket indicates that it should be evaluated in the gauge fixed $\mathcal{N} = 4$ theory with coupling g_{YM} set to zero), we find, upon plugging in the Rindler trajectory (3.155),

$$\dot{x}^\mu(\tau)\dot{x}^\nu(\tau')\left\langle \mathcal{T} A^{ab}_\mu(x(\tau)) A^{cd}_\nu(x(\tau')) \right\rangle_0 + \sqrt{-\dot{x}(\tau)^2}\sqrt{-\dot{x}(\tau')^2}\left\langle \mathcal{T} \Phi^{ab}(x(\tau)) \cdot \hat{\theta}_0 \Phi^{cd}(x(\tau')) \cdot \hat{\theta}_0 \right\rangle_0$$

$$= \frac{1 - \cosh\frac{E}{M}(\tau - \tau')}{8\pi^2 \frac{M^2}{E^2}(2 - 2\cosh(\frac{E}{M}(\tau - \tau'))) - i\epsilon}\delta^{ad}\delta^{bc} = \frac{E^2}{16\pi^2 M^2}\delta^{ad}\delta^{bc}.$$

This is similar to the circle Wilson loop, however, in the case of the circle, the constant propagator connects points on a finite circle contour, whereas here the contour is open and of proper length τ_P, which we shall eventually take as being large. Each integral over the contour parameter produces a factor of τ_P. Let us consider the leading order which is similar to the Euclidean case, given in equations (3.122) and (3.124),

$$W[\text{Rindler}] = N\left[1 - \frac{\lambda}{8}\left(\frac{E}{2\pi M}\right)^2 \tau_P^2 + \cdots\right]. \quad (3.159)$$

The first thing that we learn from equation (3.159) is that the correction is large and it diverges as τ_P becomes large. Perturbation theory only makes sense if this correction is small. We shall therefore keep τ_P finite and assume that λ is small enough that the corrections are indeed small. We will resum some of the perturbation theory and afterwards we will examine the result by relaxing the assumption of small corrections. To sum all of the Feynman diagrams which contain only lines whose endpoints are on the trajectory, so that their propagators are constants, all we need to do is to solve the combinatorics of the matrix indices. As in the case of the Euclidean circle, this is easiest done using a matrix model. If these are the only contributions that we keep, we compute

$$W[\text{Rindler}] = \frac{\int dA\, e^{-\frac{8\pi^2 M^2 N}{\lambda E^2} \text{Tr} A^2}\, \text{Tr} e^{i\tau_P A}}{\int dA\, e^{-\frac{8\pi^2 M^2 N}{\lambda E^2} \text{Tr} A^2}}, \tag{3.160}$$

which can be integrated exactly to give

$$W[\text{Rindler}] = L_{N-1}^1\left[\frac{1}{N}\frac{E^2 \tau_P^2}{16\pi^2 M^2}\right] \exp\left[-\frac{1}{N}\frac{E^2 \tau_P^2}{32\pi^2 M^2}\right], \tag{3.161}$$

where L_{N-1}^1 is the modified Laguerre polynomial given in equation (3.134). The first observation that we can make about this result is that since, for any finite value of N, $L_{N-1}^1(x)$ is a polynomial in its argument, for large τ_P, when N is finite, the expression in equation (3.161) decays like a Gaussian. This is a more severe dependence than the linear in τ_P phase that is expected in the amplitude for a particle. We must interpret this result as telling us that the amplitude is actually zero when τ_P is taken to be large.

In retrospect, the amplitude being zero is not surprising, as the W-boson is accelerating and therefore it is expected to emit bremsstrahlung in the way of soft massless vector particles. Because of infrared singularities, the amplitude for emitting any finite number of vector particles is zero. This is the reason for the damping, and it can also be seen as coming directly from an infrared singularity. In spite of this, we can still extract something from this discussion. For this, we note that the graphs where bremsstrahlung external lines are attached are non-planar. These amplitudes are suppressed in the large N limit. This means that the damping would be expected to go away in the large N limit. (Here we are taking the large N limit before the large τ_P limit.) To see that this solves our immediate problem, we take the large N limit of equation (3.161) to get

$$W[\text{Rindler}] \sim N \frac{1}{\sqrt{\frac{E\tau_P}{4\pi M}}}\, \mathcal{J}_1\left(\frac{E\tau_P}{2\pi M}\right), \tag{3.162}$$

where $\mathcal{J}_n(x)$ is a Bessel function. Then, we take the large τ_P limit and we find

$$W[\text{Rindler}] \sim N \sqrt{\frac{2}{\pi}\frac{4\pi M}{E\tau_P}}\, \exp\left(i\frac{E\tau_P}{2\pi M}\right), \tag{3.163}$$

where we have used the asmyptotic limit $\mathcal{J}_1(x) \sim \sqrt{\frac{2}{\pi x}}\cos(x)$ and we have assumed that an $i\epsilon$ prescription faithfully carried through the above reasoning (it should replace M by $M - i\epsilon$) picks out the appropriate phase. In equation (3.163), we have found the oscillating behaviour that is expected for a correction to the propagator. Moreover, we can take the coefficient of the linear growth in τ_P in the phase seriously. Including the contribution of the Wilson loop, the amplitude in the large N limit is then

$$\mathcal{A}_{ab} \approx \delta_{ab} e^{-i\left(M - \frac{1}{2}M - \frac{\sqrt{\lambda}E}{2\pi M}\right)\tau_P}. \tag{3.164}$$

This result is interesting as it suggests a shift of the rest energy of the accelerated particle by $M \to M - \sqrt{\lambda}\frac{E}{2\pi M}$. The Unruh temperature of the acceleration is

$$T_U = \frac{1}{2\pi}\frac{E}{M}, \tag{3.165}$$

where we have used the fact that the proper acceleration is $\frac{E}{M}$. Then, the energy shift could perhaps be interpreted as $M \to M - \sqrt{\lambda}T_U$ which could perhaps be interpreted as a free energy (H=E-TS) with $E = M$ the internal energy, T the Unruh temperature (which coincides with the Hawking temperature on the world-sheet), and $S = \sqrt{\lambda}$.

Let us examine this system in the strong coupling limit. Again, we must consider the semi-classical limit of the string sigma model which describes an open string ending on a D3 brane which is suspended at radius r_M in AdS_5, and located at a point $\hat{\theta}_0$ on the 5-sphere. The Polyakov action for the bosonic coordinates of the string, in the conformal gauge, is

$$S = -\frac{L^2}{4\pi\alpha'}\int d\tau d\sigma \eta^{ab}\left\{\frac{1}{r^2}\partial_a r\partial_b r + r^2 \partial_a x_\mu \partial_b x^\mu + \partial_a \hat{\theta} \cdot \partial_b \hat{\theta}\right\} + \int_{r=r_M}\frac{E}{2}(x_0 \dot{x}_1 - x_1 \dot{x}_0),$$

where $\eta^{ab} = \begin{bmatrix} -1 & 0 \\ 0 & 1 \end{bmatrix}$ and $\partial_a = \begin{bmatrix} \partial_\tau \\ \partial_\sigma \end{bmatrix}$. This leads to the same equations of motion,

$$-\partial_a\left(\frac{1}{r^2}\partial_a r\right) + r\partial_a x^\mu \partial_a x_\mu - \frac{1}{r^3}\partial_a r \partial_a r = 0$$

$$\partial_a(r^2 \partial_a x_\mu) = 0$$

$$\partial_a^2 \hat{\theta} - \hat{\theta}(\hat{\theta} \cdot \partial_a^2 \hat{\theta}) = 0, \quad \hat{\theta}^2 = 1$$

and Virasoro constraints,

$$r^2 \partial_\sigma x_\mu \partial_\tau x^\mu + \frac{1}{r^2}\partial_\sigma r \partial_\tau r + \partial_\sigma \hat{\theta} \cdot \partial_\tau \hat{\theta} = 0$$

$$r^2 \partial_\sigma x_\mu \partial_\sigma x^\mu + r^2 \partial_\tau x_\mu \partial_\tau x^\mu + \frac{1}{r^2}\partial_\tau r \partial_\tau r + \frac{1}{r^2}\partial_\sigma r \partial_\sigma r + \partial_\sigma \hat{\theta} \cdot \partial_\sigma \hat{\theta} + \partial_\tau \hat{\theta} \cdot \partial_\tau \hat{\theta} = 0$$

that we found before. We have copied them here for the reader's convenience. In addition, there are boundary conditions to be imposed at the end of the string which intersects the D3 brane,

$$\frac{\sqrt{\lambda}}{2\pi} r_M^2 \partial_\sigma x_0(\sigma_0, \tau) + E \partial_\tau x_1(\sigma_0, \tau) = 0 \tag{3.166}$$

$$\frac{\sqrt{\lambda}}{2\pi} r_M^2 \partial_\sigma x_1(\sigma_0, \tau) + E \partial_\tau x_0(\sigma_0, \tau) = 0 \tag{3.167}$$

$$\frac{\sqrt{\lambda}}{2\pi} r_M^2 \partial_\sigma x_{2,3}(\sigma_0, \tau) = 0, \ r(\sigma_0, \tau) = r_M. \tag{3.168}$$

It is through these boundary conditions that the information about the presence of the electric field enters the dynamics of the string. We shall solve these equations with $x_2 = x_3 = 0$ and $\hat{\theta} = \hat{\theta}_0$, a constant. The equations of motion for the remaining variables have a solution given by the replacement of Euclidean by real time in equation (3.127):

$$x_0(\sigma, \tau) = b \frac{\sinh a\tau}{\cosh a\sigma}, \ x_1(\sigma, \tau) = b \frac{\cosh a\tau}{\cosh a\sigma}, \ r(\sigma, \tau) = \frac{1}{b} \coth a\sigma, \tag{3.169}$$

These are a solution of the equations of motion and the Virasoro constraints. They must be adjusted to fit the boundary conditions. They obey the boundary conditions (3.166) and (3.167) when

$$\tanh a\sigma_0 = \frac{2\pi E}{\sqrt{\lambda} r_M^2} = \frac{\sqrt{\lambda}}{2\pi} \frac{E}{M^2}$$

or

$$\cosh a\sigma_0 = \frac{1}{\sqrt{1 - \left(\frac{\sqrt{\lambda}}{2\pi} \frac{E}{M^2}\right)^2}}, \ \sinh a\sigma_0 = \frac{\frac{\sqrt{\lambda}}{2\pi} \frac{E}{M^2}}{\sqrt{1 - \left(\frac{\sqrt{\lambda}}{2\pi} \frac{E}{M^2}\right)^2}}$$

and the condition (3.168) when

$$r(\sigma_0) = r_M = \frac{1}{b} \coth a\sigma_0 = \frac{1}{b} \frac{\sqrt{\lambda} r_M^2}{2\pi E} \to b = \frac{M}{E}.$$

We can also adjust a so that τ records the proper time of the trajectory,

$$a = \frac{\frac{E}{M}}{\sqrt{1 - \left(\frac{\sqrt{\lambda}}{2\pi} \frac{E}{M^2}\right)^2}}. \tag{3.170}$$

Here, the world-sheet coordinates lie in the ranges $-\tau_P/2 < \tau < \tau_P/2$ and $\sigma_0 \leq \sigma < \infty$.

The solution that we have found is the locus of the curve

$$(x_0)^2 - (x_1)^2 - \frac{1}{r^2} + \frac{M^2}{E^2} = 0$$

where, at radius $r = r_M$ the endpoint of the string traces a Rindler trajectory, but with acceleration given in equation (3.170) which is enhanced from what it would be for a particle with the same charge and mass. This acceleration diverges at what was argued in [55] to be an upper critical electric field. In fact, remembering that $M = \frac{\sqrt{\lambda} r_M}{2\pi}$ we can see that this is the same upper critical field that we would deduce from the Born–Infeld action in equation (3.146). This means that, when the electric field is strong enough, the combination $\frac{\sqrt{\lambda}}{2\pi} \frac{E}{M^2}$ is of order one, we expect that Schwinger pair production becomes important, and that it competes with the process that we are considering. Since we did not take it into account in our quantum field theory computation, the best that we can do for comparison with that computation is to assume that $\frac{\sqrt{\lambda}}{2\pi} \frac{E}{M^2} \ll 1$. In the following, we shall make this assumption by keeping only terms which are linear in $\frac{\sqrt{\lambda}}{2\pi} \frac{E}{M^2}$.

The induced metric of the world-sheet is

$$ds^2 = L^2 \left(\frac{a}{\sinh a\sigma}\right)^2 \left[d\sigma^2 - d\tau^2\right], \tag{3.171}$$

which is just the analytic continuation of the Euclidean expression that we found in the context of the circle Wilson loop (3.128) where the parameter a is now the acceleration. This is a Lorentzian signature AdS_2 black hole metric. The general solution for the world-sheet of a string with any time-like boundary placed at the boundary of AdS_5 was found by Mikhailov [46]. Our solution agrees with it in the appropriate limit.

It is very interesting that the Hawking temperature of this classical world-sheet, which is given in equation (3.129), coincides with the Unruh temperature of the space-time acceleration which was given in equation (3.165),

$$T_H = \frac{a}{2\pi} = \frac{1}{2\pi} \frac{\frac{E}{M}}{\sqrt{1 - \left(\frac{\sqrt{\lambda}}{2\pi} \frac{E}{M^2}\right)^2}} = T_U.$$

It is also interesting that both of these temperatures diverge, like the acceleration does, at the critical electric field.

The event horizon of the metric (3.171) is located at $\sigma \to \infty$ which is at $r = \frac{E}{M}$. If we compute the action that is due only to the world-sheet which is located above the event horizon, we obtain

$$S = \left(1 - \frac{\sqrt{\lambda}}{2\pi} \frac{E}{M^2}\right) M\tau_P - \frac{M}{2}\tau_P + \dots \tag{3.172}$$

and the string theory prediction for the amplitude is

$$\mathcal{A}_{ab} = \delta_{ab} \, e^{iM\tau_P - i\frac{1}{2}M\tau_P - i\frac{\sqrt{\lambda}}{2\pi}\frac{E}{M}\tau_P + \ldots}, \quad (3.173)$$

where the ellipses indicate that, following our discussion above, we have ignored terms of higher than linear order in $\frac{\sqrt{\lambda}}{2\pi}\frac{E}{M^2}$. This agrees precisely with our quantum field theory result in equation (3.164).

This rather remarkable agreement suggests that the correct result from the string theory side is, as we have done here, to take the area of the world-sheet which is above the event horizon. This brings up a paradox as the world-sheet does not just end at the horizon, it can be continued smoothly beyond the horizon and, in fact, as pointed out in [60], it curls back and reaches the boundary region of AdS_5 at a second Rindler trajectory for the W-boson's anti-particle. So, why not the full world-sheet? The answer could lie in causality. As the endpoint of the open string moves with a constant acceleration, the rest of the string should lag behind it. As we have seen, this string should sweep out the section of AdS_2 which is above the event horizon. However, the string does not have time to make the connection through the wormhole to the other parts of the extended would-be world-sheet. Instead, as advocated in [56], a shock-front forms at the event horizon. The world-sheet is not analytic there, the event horizon is a line where the exterior AdS_2 is joined to a null surface which descends from the event horizon to the Poincaré horizon of AdS_5. Being null, the latter surface has zero proper area and it does not contribute to the on-shell action.

3.7 Epilogue

In this chapter I have attempted to give a simple introduction to the idea of duality between gauge fields and string theory in its best understood form. I have made no attempt to be complete or even to address what some would regard as the most important topics. To my mind, there is little need for that. There is already a lot of excellent literature about this subject including comprehensive introductory textbooks which I encourage you to look at. What I hope that I have accomplished is to give some appreciation of the beauty of the ideas which led us to this point.

References

[1] C. N. Yang and R. L. Mills, 'Conservation of Isotopic Spin and Isotopic Gauge Invariance,' Phys. Rev. **96**, 191 (1954).
[2] W. T. Tutte, A census of planar triangulations. Canad. J. Math. 14(1962), 21–38.
[3] W. T. Tutte, A new branch of enumerative graph theory, Bull. Amer. Math. Soc. 68 (1962), 500.
[4] S. R. Coleman and J. Mandula, 'All Possible Symmetries of the S-Matrix,' Phys. Rev. **159**, 1251 (1967).

[5] G. 't Hooft, 'A Planar Diagram Theory for Strong Interactions,' Nucl. Phys. B **72**, 461 (1974).
[6] K. G. Wilson, 'Confinement of Quarks,' Phys. Rev. D **10**, 2445 (1974).
[7] L. Brink, J. H. Schwarz, and J. Scherk, 'Supersymmetric Yang-Mills Theories,' Nucl. Phys. B **121**, 77 (1977).
[8] O. V. Tarasov and A. A. Vladimirov, 'Two Loop Renormalization Of The Yang-Mills Theory In An Arbitrary Gauge,' Sov. J. Nucl. Phys. **25**, 585 (1977) [Yad. Fiz. **25**, 1104 (1977)].
[9] E. Brezin, C. Itzykson, G. Parisi, and J. B. Zuber, 'Planar Diagrams,' Commun. Math. Phys. **59**, 35 (1978).
[10] Y. Nambu, 'QCD and the String Model,' Phys. Lett. **B 80** (1979) 372.
[11] Y. M. Makeenko and A. A. Migdal, 'Exact Equation for the Loop Average in Multicolor QCD,' Phys. Lett. **88B**, 135 (1979) Erratum: [Phys. Lett. **89B**, 437 (1980)].
[12] A. M. Polyakov, 'Gauge Fields as Rings of Glue,' Nucl. Phys. B **164**, 171 (1980).
[13] L. V. Avdeev, O. V. Tarasov and A. A. Vladimirov, 'Vanishing Of The Three Loop Charge Renormalization Function In A Supersymmetric Gauge Theory,' Phys. Lett. B **96**, 94 (1980).
[14] M. T. Grisaru, M. Rocek, and W. Siegel, 'Zero Three Loop Beta Function In N=4 Superyang-Mills Theory,' Phys. Rev. Lett. **45**, 1063 (1980).
[15] M. F. Sohnius and P. C. West, 'Conformal Invariance In N=4 Supersymmetric Yang-Mills Theory,' Phys. Lett. B **100**, 245 (1981).
[16] M. B. Green, J. H. Schwarz, and L. Brink, 'N=4 Yang-Mills And N=8 Supergravity As Limits Of String Theories,' Nucl. Phys. B **198**, 474 (1982).
[17] S. Mandelstam, 'Light Cone Superspace And The Ultraviolet Finiteness Of The N=4 Model,' Nucl. Phys. B **213**, 149 (1983).
[18] V. A. Kazakov, A. A. Migdal, and I. K. Kostov, Phys. Lett. **157B**, 295 (1985). doi:10.1016/0370-2693(85)90669-0
[19] J. Wess and J. Bagger, 'Supersymmetry and supergravity,' *Princeton, USA: Univ. Pr. (1992) 259 p*
[20] G. 't Hooft, 'Dimensional reduction in quantum gravity,' Conf. Proc. C **930308**, 284 (1993) [gr-qc/9310026].
[21] L. Susskind, 'The World as a hologram,' J. Math. Phys. **36**, 6377 (1995) [hep-th/9409089].
[22] J. Polchinski, 'Dirichlet Branes and Ramond-Ramond charges,' Phys. Rev. Lett. **75**, 4724 (1995) [hep-th/9510017].
[23] A. M. Polyakov, 'String theory and quark confinement,' Nucl. Phys. Proc. Suppl. **68**, 1 (1998) [hep-th/9711002].
[24] J. M. Maldacena, 'The Large N limit of superconformal field theories and supergravity,' Int. J. Theor. Phys. **38**, 1113 (1999) [Adv. Theor. Math. Phys. **2**, 231 (1998)] [hep-th/9711200].
[25] S. S. Gubser, I. R. Klebanov, and A. M. Polyakov, 'Gauge theory correlators from noncritical string theory,' Phys. Lett. B **428**, 105 (1998) [hep-th/9802109].
[26] E. Witten, 'Anti-de Sitter space and holography,' Adv. Theor. Math. Phys. **2**, 253 (1998) [hep-th/9802150].
[27] S. J. Rey and J. T. Yee, 'Macroscopic strings as heavy quarks in large N gauge theory and anti-de Sitter supergravity,' Eur. Phys. J. C **22**, 379 (2001) [hep-th/9803001].
[28] J. M. Maldacena, 'Wilson loops in large N field theories,' Phys. Rev. Lett. **80**, 4859 (1998) [hep-th/9803002].
[29] E. Witten, 'Anti-de Sitter space, thermal phase transition, and confinement in gauge theories,' Adv. Theor. Math. Phys. **2**, 505 (1998) [hep-th/9803131].
[30] J. L. Petersen, Int. J. Mod. Phys. A **14**, 3597 (1999) doi:10.1142/S0217751X99001676 [hep-th/9902131].

References

[31] O. Aharony, S. S. Gubser, J. M. Maldacena, H. Ooguri, and Y. Oz, 'Large N field theories, string theory and gravity,' Phys. Rept. **323**, 183 (2000) [hep-th/9905111].

[32] J. Erickson, G. W. Semenoff, and K. Zarembo, 'BPS versus nonBPS Wilson loops in N=4 supersymmetric Yang-Mills theory,' Phys. Lett. B **466**, 239 (1999) [hep-th/9906211].

[33] P. Di Vecchia, 'Large N gauge theories and AdS / CFT correspondence,' hep-th/9908148.

[34] J. K. Erickson, G. W. Semenoff, R. J. Szabo, and K. Zarembo, 'Static potential in N=4 supersymmetric Yang-Mills theory,' Phys. Rev. D **61**, 105006 (2000) [hep-th/9911088].

[35] J. K. Erickson, G. W. Semenoff, and K. Zarembo, 'Wilson loops in N=4 supersymmetric Yang-Mills theory,' Nucl. Phys. B **582**, 155 (2000) [hep-th/0003055].

[36] J. McGreevy, L. Susskind, and N. Toumbas, 'Invasion of the giant gravitons from anti-de Sitter space,' JHEP **0006**, 008 (2000) [arXiv:hep-th/0003075].

[37] N. Drukker and D. J. Gross, 'An Exact prediction of N=4 SUSYM theory for string theory,' J. Math. Phys. **42**, 2896 (2001) [hep-th/0010274].

[38] G. Policastro, D. T. Son, and A. O. Starinets, Phys. Rev. Lett. **87**, 081601 (2001) [hep-th/0104066].

[39] G. W. Semenoff and K. Zarembo, 'More exact predictions of SUSYM for string theory,' Nucl. Phys. B **616**, 34 (2001) [hep-th/0106015].

[40] J. Plefka and M. Staudacher, 'Two loops to two loops in N=4 supersymmetric Yang-Mills theory,' JHEP **0109**, 031 (2001) [hep-th/0108182].

[41] G. W. Semenoff and K. Zarembo, 'Wilson loops in SYM theory: From weak to strong coupling,' Nucl. Phys. Proc. Suppl. **108**, 106 (2002) [hep-th/0202156].

[42] S. S. Gubser, I. R. Klebanov, and A. M. Polyakov, 'A semi-classical limit of the gauge/string correspondence,' Nucl. Phys. B **636**, 99 (2002) [arXiv:hep-th/0204051].

[43] A. Karch and E. Katz, 'Adding flavor to AdS/CFT,' **JHEP** 0206 (2002) 043 [arXiv: hep-th/0205236].

[44] F. Bigazzi, A. L. Cotrone, M. Petrini, and A. Zaffaroni, 'Supergravity duals of supersymmetric four-dimensional gauge theories,' Riv. Nuovo Cim. **25N12**, 1 (2002) [hep-th/0303191].

[45] R. Bousso, 'The Holographic principle,' Rev. Mod. Phys. 74, 825 (2002) black hole [hep-th/0203101].

[46] A. Mikhailov, 'Nonlinear waves in AdS/CFT correspondence,' arXiv:hep-th/0305196.

[47] J. M. Maldacena, 'TASI 2003 lectures on AdS / CFT,' hep-th/0309246.

[48] P. Kovtun, D. T. Son, and A. O. Starinets, 'Viscosity in strongly interacting quantum field theories from black hole physics,' Phys. Rev. Lett. **94**, 111601 (2005) [hep-th/0405231].

[49] G. W. Semenoff and D. Young, 'Wavy Wilson line and AdS / CFT,' Int. J. Mod. Phys. A **20**, 2833 (2005) [hep-th/0405288].

[50] L. F. Alday and J. M. Maldacena, 'Comments on operators with large spin,' JHEP **0711**, 019 (2007) [arXiv:0708.0672 [hep-th]].

[51] V. Pestun, 'Localization of gauge theory on a four-sphere and supersymmetric Wilson loops,' Commun. Math. Phys. **313**, 71 (2012) [arXiv:0712.2824 [hep-th]].

[52] J. McGreevy, 'Holographic duality with a view toward many-body physics,' Adv. High Energy Phys. **2010**, 723105 (2010) [arXiv:0909.0518 [hep-th]].

[53] J. Polchinski, 'Introduction to Gauge/Gravity Duality,' arXiv:1010.6134 [hep-th].

[54] V. E. Hubeny, S. Minwalla, and M. Rangamani, 'The fluid/gravity correspondence,' arXiv:1107.5780 [hep-th].

[55] G. W. Semenoff and K. Zarembo, 'Holographic Schwinger Effect,' Phys. Rev. Lett. **107**, 171601 (2011) [arXiv:1109.2920 [hep-th]].

[56] J. A. Garcia, A. A. Guijosa, and E. J. Pulido, JHEP 1301, 096 (2013) [JHEP 1301, 096 (2013)] [arXiv:1210.4175 [hep-th]].
[57] A. V. Ramallo, 'Introduction to the AdS/CFT correspondence,' Springer Proc. Phys. **161**, 411 (2015) [arXiv:1310.4319 [hep-th]].
[58] J. Gordon and G. W. Semenoff, 'World-line instantons and the Schwinger effect as a Wentzel-Kramers-Brillouin exact path integral,' J. Math. Phys. **56**, 022111 (2015) Erratum: [J. Math. Phys. **59**, no. 1, 019901 (2018)] [arXiv:1407.0987 [hep-th]].
[59] V. E. Hubeny and G. W. Semenoff, 'Holographic Accelerated Heavy Quark-Anti-Quark Pair,' arXiv:1410.1172 [hep-th].
[60] V. E. Hubeny and G. W. Semenoff, 'String worldsheet for accelerating quark,' JHEP **1510**, 071 (2015) [arXiv:1410.1171 [hep-th]].
[61] M. Ammon and J. Erdmenger, 'Gauge/gravity duality : Foundations and applications,' Cambridge University Press, 2015.
[62] J. Zaanen, Y. Liu, Y.W. Sun, and K. Schalm, 'Holographic Duality in Condensed Matter Physics' Cambridge University Press, 2015.
[63] M. Natsuume, 'AdS/CFT Duality User Guide', Springer, 2016.
[64] H. Nastase, 'String Theory Methods for Condensed Matter Physics' Cambridge University Press, 2017.
[65] S. A. Hartnoll, A. Lucas, and S. Sachdev, 'Holographic Quantum Matter', MIT Press, 2018.
[66] J. Aguilera-Damia, A. Faraggi, L. A. Pando Zayas, V. Rathee, and G. A. Silva, 'Zeta-function Regularization of Holographic Wilson Loops,' arXiv:1802.03016 [hep-th].
[67] J. Aguilera-Damia, A. Faraggi, L. A. Pando Zayas, V. Rathee, and G. A. Silva, 'Toward Precision Holography in Type IIA with Wilson Loops,' arXiv:1805.00859 [hep-th].

4
Introduction to scattering amplitudes

David A. Kosower

Institut de Physique Théorique, Université Paris Saclay, CEA, CNRS,
F-91191 Gif-sur-Yvette, France

Chapter Contents

4 Introduction to scattering amplitudes — 183
David A. KOSOWER

Preface — 185

4.1 Kinematics — 186
4.2 Amplitudes — 190
4.3 Factorization — 194
4.4 On-shell (BCFW) recursion relations — 196

References — 204

Preface

The subject of this chapter is *scattering amplitudes*, and the new techniques that have been developed over the past two decades to compute them: on-shell methods.

You can think of amplitudes, rather than quantum fields, as the central objects in this approach to quantum field theory; to borrow a phrase from Lance Dixon, we'll be doing quantum field theory without the quantum fields.

The techniques cover a broad range of applications:

- Computing standard-model backgrounds, and quantum chromodynamics (QCD) backgrounds in particular, for LHC physics.
- Solving the $\mathcal{N}=4$ supersymmetric Yang–Mills theory for all values of the coupling.
- Determining whether $\mathcal{N}=8$ supergravity is ultraviolet-finite.
- Finding a new set of principles upon which to base a formulation of quantum field theory.

There are several key ideas underlying on-shell methods:

- Use only information from physical, on-shell states. Trees and loops still contain off-shell ('virtual') states; we just don't use them directly in computations.
- Use general properties of scattering amplitudes as tools for calculations: factorization, which leads to the Britto–Cachazo–Feng–Witten (BCFW) on-shell recursion relations; unitarity, which leads to the generalized unitarity method; and the existence of a path-integral formulation, which leads to a representation in terms of Feynman integrals, even if no Feynman diagrams are used.

This chapter will cover a subset of the lecture topics: kinematics and calculations of tree amplitudes.

Two specialized books have appeared in recent years, and offer the reader a more comprehensive and deeper look at the subject: that of Elvang and Huang [1], and that of Henn and Plefka [2]. In addition, some of these ideas are starting to appear in 'conventional' quantum field theory textbooks: for example, that of Srednicki [3], the second edition of Zee [4], and the recent book by Schwartz [5].

Why might we expect there to exist better alternatives to the standard techniques of Feynman diagrams? The latter, after all, have been enormously successful for over six decades, and provide not only a means of calculating amplitudes but a language for describing effects qualitatively, as well as a framework for making general arguments about perturbation theory. Even with new methods, traditional diagrammatic approaches conserve their utility for proving general statements, as long as one doesn't have to use them for calculations.

If we look at Yang–Mills theories, which are of course of prime importance in particle physics, we find a factorial proliferation of Feynman diagrams as the loop order or the number of external legs grows, and an even faster growth in the total number of terms.

Yet results of computations are much simpler than this proliferation would suggest; and some answers are *remarkably* simple. One celebrated example is the Parke–Taylor amplitude [6, 7], the color-ordered amplitude for a maximally helicity-violating configuration of n gluons,

$$A_n(1^+, \overbrace{\ldots}^{+}, m_1^-, \overbrace{\ldots}^{+}, m_2^-, \overbrace{\ldots}^{+}, n^+) = i \frac{\langle m_1 m_2 \rangle^4}{\langle 1\,2 \rangle \langle 2\,3 \rangle \cdots \langle (n-1)n \rangle \langle n\,1 \rangle}. \quad (4.1)$$

This one line summarizes the results of calculating an infinite number of Feynman diagrams. Even for small n, say $n = 4$ or 5, it is much simpler than the raw expression obtained from writing down the expressions produced by the diagrams.

If we believe that a simple result should have a simple derivation,[1] then there should be a simple way of obtaining this result. In fact, there are several, and we'll see one in this chapter.

In the above equation, $\langle 1\,2 \rangle$ is a *spinor product*, a kind of square root of a Lorentz invariant. The use of spinor quantities is critical to the simplicity of methods and results, so our first task is to introduce them.

4.1 Kinematics

Our focus will be primarily on particles and external states characterized by massless four-momenta k_μ, that is with $k^2 = 0$. At sufficiently high energies, particles with fixed mass will look more and more massless, so this is clearly relevant to high-energy scattering. In fact, the techniques will work for massive particles too, but we will need to build them up differently.

Our first task is to learn how to take the square root of k_μ. Introduce the four-vector of Pauli matrices, consisting of the classical triplet along with the 2×2 identity matrix for the time component,

$$\sigma^\mu = (1, \vec{\sigma}) : \sigma^0 = \begin{pmatrix} 1 & 0 \\ 0 & 1 \end{pmatrix}, \quad \sigma^1 = \begin{pmatrix} 0 & 1 \\ 1 & 0 \end{pmatrix}, \quad \sigma^2 = \begin{pmatrix} 0 & -i \\ i & 0 \end{pmatrix}, \quad \sigma^3 = \begin{pmatrix} 1 & 0 \\ 0 & -1 \end{pmatrix}. \quad (4.2)$$

We can use these matrices to recast the real four-momentum k^μ as a complex 2×2 matrix,

$$\underline{k}_{\alpha\dot\alpha} \equiv k \cdot \sigma = \begin{pmatrix} k^0 - k^3 & -k^1 + ik^2 \\ -k^1 - ik^2 & k^0 + k^3 \end{pmatrix}. \quad (4.3)$$

In this equation, the α and $\dot\alpha$ indices run over the values 1 and 2. Notice that $k^2 = \det(\underline{k})$. Also, Lorentz transformations on k_μ represented by 4×4 matrices in $SO(1,3)$ are replaced by 2×2 matrices in $SL(2, \mathbb{C})$, that is 2×2 matrices with unit determinant,

---

[1] A saying I'm told is due to Feynman.

$$k'_\mu = \Lambda_\mu{}^\nu k_\nu \longrightarrow \underline{k}' = u_\Lambda \underline{k} u_\Lambda^\dagger. \tag{4.4}$$

The α and $\dot{\alpha}$ indices are $SL(2,\mathbb{C})$ indices.

For massless particles, $k^2 = 0$; this means that our matrix has rank 1 (or less), and we can therefore write it as an outer product of two vectors,

$$\underline{k}_{\alpha\dot{\alpha}} = \lambda_\alpha \tilde{\lambda}_{\dot{\alpha}}. \tag{4.5}$$

The objects λ and $\tilde{\lambda}$ are complex two-vectors (more precisely, elements of a complex projective space $\mathbb{CP}^1$). They are called *spinors*, as they transform under a spin-½ representation. It is important to remember that they are not Grassmann objects, however. They are still bosonic. They are the basic variables in terms of which we will express amplitudes.

It's natural to form invariant quantities using the $SL(2,\mathbb{C})$ invariant tensor, the two-dimensional ε symbol:

$$\varepsilon^{\alpha\beta}\lambda_\alpha \lambda'_\beta, \quad \text{and} \quad \varepsilon^{\dot{\alpha}\dot{\beta}}\tilde{\lambda}_{\dot{\alpha}}\tilde{\lambda}'_{\dot{\beta}}. \tag{4.6}$$

These are called *spinor products*.

It's convenient to introduce additional notation,

$$\begin{aligned}
|j^+\rangle &\equiv \lambda_j^\alpha \equiv |j\rangle, \\
|j^-\rangle &\equiv \tilde{\lambda}_{j\dot{\alpha}} \equiv |j], \\
\langle j| &\equiv \langle j^-| \leftrightarrow \varepsilon_{\alpha\beta}\lambda_j^\beta, \\
[j| &\equiv \langle j^+| \leftrightarrow \varepsilon^{\dot{\alpha}\dot{\beta}}\tilde{\lambda}_{j\dot{\beta}}.
\end{aligned} \tag{4.7}$$

Here j is an index into a set of external momenta. With this simplified notation, the spinor products are,

$$\begin{aligned}
\langle ij\rangle &\equiv \langle i^-|j^+\rangle = \varepsilon^{\alpha\beta}\lambda_{i\alpha}\lambda_{j\beta}, \\
[ij] &\equiv \langle i^+|j^-\rangle = \varepsilon^{\dot{\alpha}\dot{\beta}}\tilde{\lambda}_{j\dot{\alpha}}\tilde{\lambda}_{i\dot{\beta}}.
\end{aligned} \tag{4.8}$$

In the literature, the sign conventions for the angle product, $\langle ij\rangle$, are uniform. There are, unfortunately, two opposite sign conventions for the bracket product, $[ij]$. The convention used in this chapter is the 'QCD' convention; the other one is the 'SUSY' convention.

Spinor products are not merely abstract or formal objects; like invariants, they can—and *should!*—be used for numerical calculations as well. To see how that would go, we can give explicit representations for λ and $\tilde{\lambda}$; let us define,

$$k_\pm \equiv k^0 \pm k^3,$$
$$e^{\pm i\phi_k} \equiv \frac{k^1 \pm ik^2}{\sqrt{k_+ k_-}}, \tag{4.9}$$

where the latter is a phase because $k^2 = 0$. (The second form uses $\hat{z}$ as a reference axis; of course this is an arbitrary choice.) In terms of these variables, we can write,

$$\lambda_\alpha = \begin{pmatrix} \sqrt{k_+} \\ e^{i\phi_k}\sqrt{k_-} \end{pmatrix}, \quad \tilde{\lambda}_{\dot{\alpha}} = \begin{pmatrix} \sqrt{k_+} & e^{-i\phi_k}\sqrt{k_-} \end{pmatrix} \tag{4.10}$$

In the rest of the chapter, I will drop the underline on the matrix representation of k, allowing context to determine whether it represents a four-vector or a 2×2 complex matrix.

We get explicit formulæ for the spinor products,

$$\begin{aligned} \langle ij \rangle &= e^{i\phi_{k_i}}\sqrt{k_{i-}k_{j+}} - e^{i\phi_{k_j}}\sqrt{k_{i+}k_{j-}}, \\ [ij] &= e^{-i\phi_{k_j}}\sqrt{k_{i+}k_{j-}} - e^{-i\phi_{k_i}}\sqrt{k_{i-}k_{j+}}. \end{aligned} \tag{4.11}$$

Now if $k^0_{i,j} > 0$, then

$$[ij] = \langle ji \rangle^*; \tag{4.12}$$

in general,

$$[ij] = \operatorname{sign}(k^0_i k^0_j) \langle ji \rangle^* \tag{4.13}$$

for real momenta. (This corresponds to multiplying each definition in equation (4.11) by $\sqrt{\operatorname{sign}(k^0_i)} \times \sqrt{\operatorname{sign}(k^0_j)}$, and being careful with the branch cut of the square root.) As a result,

$$\langle ij \rangle [ji] = 2 k_i \cdot k_j. \tag{4.14}$$

It's in this sense that spinor products are square roots of Lorentz products. The normalizations and signs used here match those of the S@M package [8] (though not all aspects of the notation).

The following are key properties of the spinor product:

- Antisymmetry,
  $\langle ji \rangle = -\langle ij \rangle$, $[ji] = -[ij]$,
  $\langle ii \rangle = 0 = [ii]$.

- *Gordon identity*: reconstruct momenta by dotting into $(\sigma^\mu)^{\alpha\dot\alpha}$,
  $2k_j^\mu = (\sigma^\mu)^{\alpha\dot\alpha}\lambda_{j\alpha}\tilde\lambda_{j\dot\alpha} = \langle j|\sigma^\mu|j] \equiv \langle j|\mu|j]$.
- Fierz identity,
  $\langle i|\mu|j]\langle q|\mu|r] = 2\langle iq\rangle[rj]$.

Now, any two-dimensional vector can be expressed in terms of two basis vectors (here, with complex coefficients),

$$\lambda_q = c_i \lambda_i + c_j \lambda_j. \tag{4.15}$$

If we now contract respectively with λ_j and λ_i, we obtain the following expressions for the coefficients:

$$c_i = \frac{\langle jq\rangle}{\langle ji\rangle}, \qquad c_j = \frac{\langle iq\rangle}{\langle ij\rangle}. \tag{4.16}$$

If we use these expressions, and contract with λ_r, we obtain the

- Schouten identity,
  $\langle ij\rangle\langle rq\rangle = \langle iq\rangle\langle rj\rangle + \langle ir\rangle\langle jq\rangle$.

For real momenta, $\tilde\lambda_j = \text{sign}(k_j^0)\lambda_j^*$. We can redefine λ while keeping the momentum fixed,

$$\begin{aligned}\lambda_j &\mapsto \tau\lambda_j, \\ k_j &\mapsto |\tau|^2 k_j,\end{aligned} \tag{4.17}$$

as long as τ is a pure phase.

Nothing in our constructions above depends on the relation between λ and $\tilde\lambda$, however. This is crucial; we can relax the connection, thinking of λ and $\tilde\lambda$ as independent. The momenta we obtain are then *complex*.

The phase τ corresponds to helicity under little-group transformations: λ has weight $-\frac{1}{2}$, $\tilde\lambda$ weight $+\frac{1}{2}$. We can define a helicity operator,

$$\mathbb{h} = -\frac{1}{2}\left[\sum_i \lambda_i \frac{\partial}{\partial \lambda_i} - \tilde\lambda_i \frac{\partial}{\partial \tilde\lambda_i}\right]. \tag{4.18}$$

What about massive momenta? Introduce a null reference momentum q, that is $q^2 = 0$. Define,

$$k_\mu^\flat = k_\mu - \frac{k^2}{2k\cdot q}q_\mu. \tag{4.19}$$

Then $(k^\flat)^2 = 0$, and we can build spinors for it. (If k is null, then of course $k^\flat = k$.) Intermediate quantities will depend—perhaps only implicitly—on q, but physical quantities such as differential cross sections shouldn't depend on it.

4.2 Amplitudes

Let's now compute some amplitudes in Yang–Mills theory, that is, the theory of spin-1 particles. Massless vectors have two physical polarizations, and it's convenient to use a basis of circular polarizations or helicities.

We'll start by computing the simplest possible amplitude, the three-point amplitude. In fact, this will turn out to be the only building block we need. All other amplitudes can be derived from it. I'll follow here the argument given in [9].

In three-particle kinematics,

$$k_3^2 = 0 = (k_1 + k_2)^2 = 2 k_1 \cdot k_2, \tag{4.20}$$

so that all invariants vanish. At first glance, this implies that all spinor products vanish too,

$$\begin{aligned}\langle 1\,2\rangle = 0, \langle 2\,3\rangle = 0, \langle 3\,1\rangle = 0,\\ [1\,2] = 0, [2\,3] = 0, [3\,1] = 0.\end{aligned} \tag{4.21}$$

This implies that the three-point amplitude vanishes (or at best, might be ill-defined),

$$A_3(1,2,3) = 0. \tag{4.22}$$

This is actually the correct result for on-shell momenta, because an on-shell gluon cannot split into two on-shell gluons. It's not very useful, however, as the promised building block for all amplitudes.

If, however, we decouple $\tilde\lambda$ from λ, taking $\tilde\lambda \neq \pm\lambda^*$—that is if we allow the external momenta to be complex, then the momentum invariants can vanish without forcing all spinor products to vanish,

$$k_i \cdot k_j \Rightarrow \langle ij \rangle = 0 \quad \text{or} \quad [ij] = 0, \tag{4.23}$$

without both necessarily vanishing.

Let's first examine what happens if we satisfy the on-shell conditions by setting the bracket products to zero,

$$[1\,2] = 0, [2\,3] = 0, [3\,1] = 0. \tag{4.24}$$

We then have three invariants to play with: $\langle 1\,2\rangle$, $\langle 2\,3\rangle$, and $\langle 3\,1\rangle$.

Write an ansatz for A_3,

$$A_3 = \text{const} \, \langle 1\,2 \rangle^{e_{12}} \langle 2\,3 \rangle^{e_{23}} \langle 3\,1 \rangle^{e_{31}}. \tag{4.25}$$

Our ansatz has to be homogeneous because it can only change by a phase when we redefine the spinors by a phase. That phase maps to the helicity of the spinor, so we obtain three equations,

$$\begin{aligned} h_1 &= -\frac{1}{2}(e_{12} + e_{31}), \\ h_2 &= -\frac{1}{2}(e_{12} + e_{23}), \\ h_3 &= -\frac{1}{2}(e_{23} + e_{31}). \end{aligned} \tag{4.26}$$

We can solve them to obtain,

$$\begin{aligned} e_{12} &= h_3 - h_1 - h_2, \\ e_{23} &= h_1 - h_2 - h_3, \\ e_{31} &= h_2 - h_1 - h_3. \end{aligned} \tag{4.27}$$

For a three-gluon amplitude with helicities $(--+)$, we find,

$$e_{12} = 3, \quad e_{23} = e_{31} = -1. \tag{4.28}$$

That is,

$$A_3 = \text{const} \times \frac{\langle 1\,2 \rangle^3}{\langle 2\,3 \rangle \langle 3\,1 \rangle}. \tag{4.29}$$

An n-point amplitude should have dimension $4 - n$, and that's exactly what we get when 'const' is dimensionless. We'll identify the constant later. As we go back to real momenta, all spinor products will vanish; but the numerator has a higher power, so the amplitude will go smoothly to zero, as it should.

Repeating the exercise for gravity, with gravitons of helicity ± 2, we obtain,

$$M_3 = \text{const} \times \frac{\langle 1\,2 \rangle^6}{\langle 2\,3 \rangle^2 \langle 3\,1 \rangle^2}, \tag{4.30}$$

though here the constant has dimensions of $1/M$. Up to the constants, we see that $M_3 = A_3^2$. This is the simplest example of an idea that has had powerful consequences in recent years, that 'Gravity = (Yang–Mills)^2'.

192 *Introduction to scattering amplitudes*

What about other helicities? $(++-)$ gives

$$e_{12} = -3, \quad e_{23} = e_{31} = 1. \tag{4.31}$$

If we just used these values, we would get

$$A_3 = \text{const} \times \frac{\langle 2\,3\rangle \langle 3\,1\rangle}{\langle 1\,2\rangle^3}. \tag{4.32}$$

This is bad. It has the wrong dimensions unless 'const' is dimensionful; but there is no dimensionful constant in pure Yang–Mills theory. Even worse, it's also pathologically singular as we go back to real momenta.

Notice, though, that this helicity configuration is just the parity conjugate of the first one we considered. So instead, we can obtain its amplitude by starting with the other possible solution to the on-shell equations,

$$[1\,2], [2\,3], [3\,1] \neq 0, \tag{4.33}$$

along with

$$\langle 1\,2\rangle = 0, \langle 2\,3\rangle = 0, \langle 3\,1\rangle = 0. \tag{4.34}$$

We then get

$$A_3 = \text{const} \times \frac{[1\,2]^3}{[2\,3][3\,1]}, \tag{4.35}$$

with the constant again being dimensionless. We get similar results for the cyclic permutations of the $(--+)$ and $(++-)$ helicities we've considered thus far.

This leaves us with two additional helicity configurations to consider: $(---)$ and its complex conjugate $(+++)$. Consider the $(---)$ helicity configuration, with the first solution (4.24) to the on-shell equations. Using our solutions for the exponents (4.27), we would get

$$A_3 = \text{const} \times \langle 1\,2\rangle \langle 2\,3\rangle \langle 3\,1\rangle, \tag{4.36}$$

which again would require a dimensionful constant. Here, use of the other on-shell solution (4.34) can't save the day, and this amplitude truly vanishes in pure Yang–Mills theory. It can however be non-vanishing if we add the right higher-dimension operator to our theory, along with its dimensionful coupling.

What is the 'const' that appeared above? Looking at our expression for the three-point amplitude $A_3(--+)$, we see that it is antisymmetric in $1 \leftrightarrow 2$. We expect it to be symmetric, because of Bose symmetry of gluons. This can be achieved if the constant carries indices, and contains a factor of a fully antisymmetric tensor. The correct one is precisely the Lie structure constant $f^{a_1 a_2 a_3}$. In addition, it can be multiplied by an arbitrary real constant, otherwise known as the Yang–Mills coupling g (along with a factor of i ultimately due to tree-level unitarity).

In general, the full amplitude for n-gluon scattering can be written as follows:

$$\mathcal{A}_n = \sum_{\rho \in S_n/Z_n} \mathrm{Tr}(T^{a_{\rho(1)}} T^{a_{\rho(2)}} \cdots T^{a_{\rho(n)}}) \qquad (4.37)$$
$$\times A_n(\lambda_{\rho(1)}, \tilde{\lambda}_{\rho(1)}, h_{\rho(1)}; \lambda_{\rho(2)}, \tilde{\lambda}_{\rho(2)}, h_{\rho(2)}; \ldots; \lambda_{\rho(n)}, \tilde{\lambda}_{\rho(n)}, h_{\rho(n)}).$$

In this expression, the sum is over the non-cyclic permutations of the n labels $1, \ldots, n$ of the external momenta. The trace factor depends on the color indicies a_i of the external gluons. The last factor is the so-called color-ordered amplitude, which depends on the spinors and helicities of the external gluons (but not on their color indices). It is the color-ordered three-point amplitude that we derived previously.

How are the amplitudes we computed previously related to what we would obtain from a more conventional approach, using a Lagrangian? We could also compute the three-point amplitude from the Yang–Mills Langrangian, by simply evaluating the Feynman three-point vertex. I won't do that calculation here—I leave it as an exercise for the student—but I will discuss the one missing ingredient in that exercise. This is a spinorial representation for the gluon polarization vectors. The Feynman vertex, after all, we will have in terms of gluon momenta k_i and polarization vectors ε_i; how do we rewrite it in terms of spinors?

To provide a suitable representation, we again need a null reference vector q^μ, that is with $q^2 = 0$, along with its associated spinors $|q\rangle$ and $|q]$. Then, define

$$\varepsilon_\mu^{(-)}(k_j, q) \equiv \frac{\langle j|\mu|q]}{\sqrt{2}[jq]}. \qquad (4.38)$$

When we rotate the phase of $|j\rangle$ (and that of $|j]$ in the opposite direction, as appropriate for real k_j), then $\varepsilon^{(-)}$ acquires twice the phase angle; that is, it has helicity -1. It is in contrast unchanged if we rotate the phase of $|q]$, as that cancels between numerator and denominator. Similarly,

$$\varepsilon_\mu^{(+)} \equiv \frac{\langle q|\mu|j]}{\sqrt{2}\langle qj\rangle}. \qquad (4.39)$$

These polarization vectors, along with some rules for choosing q so as to minimize the number of non-vanishing terms in a calculation, are known as *Chinese magic* because they were first written down by a trio (Xu, Zhang, Chang) then at Qinghua University in Beijing, in a 1984 preprint, when Chinese universities were just emerging from the darkness of the Cultural Revolution.

A simple exercise to test your understanding of spinor algebra and the spinor-helicity representation of polarization vectors is to compute the four-point amplitude for the helicity configuration $(--++)$. Clever choices of q will greatly simplify that calculation. I'll just record the result for the color-ordered amplitude stripped of the coupling,

$$A_4(1^-,2^-,3^+,4^+) = i\frac{\langle 1\,2\rangle^3}{\langle 2\,3\rangle\langle 3\,4\rangle\langle 4\,1\rangle}, \qquad (4.40)$$

which is just the $n=4$, $m_1=1$, $m_2=2$ case of equation (4.1).

4.3 Factorization

To go beyond the three-point amplitude and compute additional amplitudes, we will rely on the physical property of factorization of scattering amplitudes. It's a generic property of amplitudes in any sensible quantum field theory.

In its simplest form, as a sum of two external momenta approaches the momentum of a physical state, the amplitude factorizes on the corresponding pole. That is, as $p^2 = (k_1+k_2)^2 \to m_X^2$, we will find

$$A(1+2 \to 3+\cdots+n) \to A_L(1+2 \to X)\frac{i}{p^2 - m_X^2}A_R(X \to 3+\cdots+n)+\cdots, \qquad (4.41)$$

where the remaining terms are non-singular in the limit.

More generally, as a sum of several momenta approaches an on-shell physical state, $p^2 = (k_1+\cdots+k_r)^2 \to m_X^2$, then

$$A(1+\cdots+r \to (r+1)+\cdots+n) \to A_L(1+\cdots+r \to X)\frac{i}{p^2 - m_X^2}A_R(X \to (r+1)+\cdots+n). \qquad (4.42)$$

Strictly speaking, this isn't quite right in Yang–Mills theories; it needs modification for $n=2$ or beyond tree level.

Here we're focusing on pure Yang–Mills theories—where $m_X=0$. For $n \geq 3$ everything works straightforwardly,

$$A_n(\ldots,a,b,c,\ldots) \to A_4(a,b,c,X^\pm)\frac{i}{(k_a+k_b+k_c)^2}A_{n-2}((-X)^\mp,\ldots) \qquad (4.43)$$

as $(k_a+k_b+k_c)^2 \to 0$ but $s_{ab}, s_{bc}, s_{ac} \not\to 0$. The factorization for $n=3$ is depicted in Figure 4.1. The only addition is the requirement to sum over the helicities of the intermediate leg. We use the convention that all momenta are outgoing in an amplitude, which explains the sign flip in the last factor ($-X$ denotes $-k_X$). It also explains the helicity flip; when you flip the sign of the momentum, the sign of the helicity also flips.

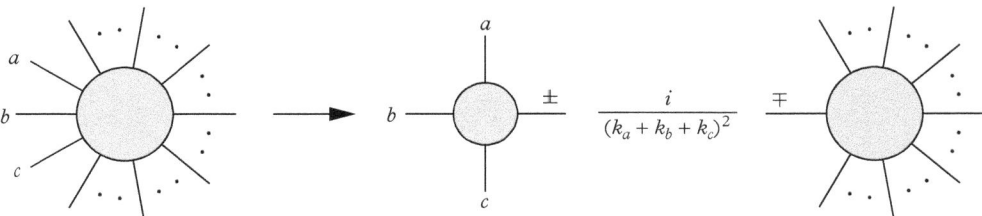

Figure 4.1 *General factorization of an amplitude.*

As an example, consider three-particle factorizations of the six-point amplitude $A_6(1^-,2^-,3^-,4^+,5^+,6^+)$. This amplitude has a remarkably simple form,

$$\frac{(\langle 2\,3\rangle[5\,6]\langle 1|1+2+3|4])^2}{s_{234}s_{23}s_{34}s_{56}s_{61}} + \frac{(\langle 1\,2\rangle[4\,5]\langle 3|1+2+3|6])^2}{s_{345}s_{34}s_{45}s_{61}s_{12}} \\ + \frac{s_{123}\langle 1\,2\rangle[5\,6]\langle 1|1+2+3|4]\langle 1\,2\rangle[4\,5]\langle 3|1+2+3|6]}{s_{12}s_{23}s_{34}s_{45}s_{56}s_{61}} \quad (4.44)$$

(where I've removed an overall factor of i). In this expression, the notation '$\langle 1|1+2+3|4]$' stands for '$\langle 1|k_1+k_2+k_3|4]$'. This expression is finite as $s_{123} \to 0$ (as long as s_{12}, s_{23}, and $s_{31} \not\to 0$). That reflects the vanishing of $A_4(1^-,2^-,3^-,X^\pm)$, which we'll see a little later.

As $s_{234} \to 0$, we pick up the first term in equation (4.44); using the vanishing of spinor products $\langle 1\,1\rangle$ and $[4\,4]$ and introducing $K = k_2 + k_3 + k_4$, we can rewrite

$$\langle 1|1+2+3|4] = \langle 1|2+3+4|4], \quad (4.45)$$

which factorizes in the limit when $K^2 = 0$ to

$$\langle 1\,K\rangle[K\,4]. \quad (4.46)$$

The first term in equation (4.44) then factorizes as follows:

$$\frac{\langle 2\,3\rangle[K\,4]^2}{[2\,3]s_{34}} \times \frac{1}{s_{234}} \times \frac{[5\,6]\langle 1\,K\rangle^2}{\langle 5\,6\rangle s_{61}}. \quad (4.47)$$

The first factor only involves the momenta k_2, k_3, k_4, and K; the second factor is the required propagator; and the last factor only involves the momenta k_1, k_5, k_6, and K. We can perform a few algebraic manipulations to make the structure of the first and last factors a little clearer. Using momentum conservation in the limit, we can introduce factors of

196 *Introduction to scattering amplitudes*

$$\frac{[K\,4]}{[K\,4]} = -\frac{[K\,4]\langle 4\,3\rangle}{[K\,2]\langle 2\,3\rangle},$$
$$\frac{\langle 1\,K\rangle}{\langle 1\,K\rangle} = \frac{\langle 1\,K\rangle[1\,6]}{\langle K\,5\rangle[5\,6]}.$$
(4.48)

Equation (4.47) then becomes

$$\frac{[4K]^3}{[23][34][K2]} \times \frac{1}{s_{234}} \times \frac{\langle 1K\rangle^3}{\langle 56\rangle\langle 61\rangle\langle K5\rangle}$$
$$= A_L(2^-, 3^-, 4^+, K^+) \times \text{propagator} \times A_R(1^-, (-K)^-, 5^+, 6^+),$$
(4.49)

where I've put back the factor of i removed in equation (4.44). In going to the last line, we must note that when we flip the sign of a momentum, each of the associated spinors must change by a factor of i,

$$|-K\rangle = i|K\rangle.$$
(4.50)

What about the two-particle case? If we're taking the momenta to be real, then as we've seen, $A_3 = 0$, so the above simple factorization can't work. Combining the amplitude with the propagator, we would get $0/0$, which is ambiguous. If we look at our result for A_3 with complex momenta, however, and take the limit of the momenta becoming real, we see that it vanishes linearly:

$$A_3 \sim \frac{E^3}{E^2} \sim E \to 0.$$
(4.51)

This tells us that combining $A_3(1, 2, X)$ with the propagator gives an object with well-defined scaling behavior as $s_{12} \to 0$ (but $s_{12} \neq 0$), namely $\sim 1/\sqrt{s_{12}}$. These objects are called *splitting amplitudes*, and they play an important role in understanding and capturing the infrared behavior of amplitudes. The limit where the sum of two real momenta goes on shell is called a *collinear* limit, because in the limit the two momenta become proportional: $k_1 \cdot k_2 = 0$ implies that $k_2 = \alpha k_1$.

If we take the momenta to be complex, however, the two-particle case works just like the multiparticle case: $A_3 \neq 0$, and the propagator diverges as $s_{12} \to 0$, because either $\langle 1\,2\rangle \to 0$ or $[1\,2] \to 0$.

4.4 On-shell (BCFW) recursion relations

Our next subject is the construction of all scattering amplitudes, using the three-point amplitudes we determined earlier as building blocks, and using only on-shell quantities in our construction. We'll re-derive the recursion relations, based on factorization, that were first written down by Britto, Cachazo, Feng, and Witten [11].

We start by defining a shift operation on the spinors associated with two momenta k_j and k_l. It depends on a complex parameter z; we'll denote the operation $([j,l))$:

$$\tilde{\lambda}_j = |j] \to |j] - z|l], \tag{4.52}$$
$$\lambda_l = |l\rangle \to |l\rangle + z|j\rangle.$$

This induces a shift of the corresponding external momenta,

$$k_j^\mu \to k_j^\mu(z) \equiv k_j^\mu - \frac{z}{2}\langle j|\mu|l],$$
$$k_l^\mu \to k_l^\mu(z) \equiv k_l^\mu + \frac{z}{2}\langle j|\mu|l], \tag{4.53}$$

while all momenta other than $k_{j,l}$ remain unchanged. The shift operation also defines a z-dependent continuation of all n-point amplitudes, $A_n \to A_n(z)$.

The shifted momenta are still on shell,

$$k_j^2(z) = k_j^2 - z\langle j|j|l] + \frac{z^2}{4}(\langle j|\mu|l])^2 = 0, \tag{4.54}$$

because each term vanishes separately: k_j^2 because the original momentum is on shell; the second because the spinor product $\langle jj\rangle$ vanishes; and the last one, for the same reason after using the Fierz identity. Similarly, $k_l^2(z) = 0$.

Also,

$$k_j^\mu(z) + k_l^\mu(z) = k_j^\mu + k_l^\mu, \tag{4.55}$$

because the same shift vector appears in both momenta, but with opposite sign. This means that all momenta are still on shell, and overall momentum conservation is left undisturbed. The shifted amplitude is then a scattering amplitude for a shifted, complex, momentum configuration.

We'll make one assumption, that $A(z) \to 0$ as $z \to \infty$. This assumption can in fact be proven correct for some classes of amplitudes and shift choices; and can also be relaxed in some ways, but we won't discuss that.

Let's consider the contour integral,

$$\frac{1}{2\pi i}\oint_C \frac{dz}{z} A(z), \tag{4.56}$$

taken around a circle C at infinity. The integration is shown in Figure 4.2. We can evaluate this integral in two different ways, and derive a useful equation by equating the two evaluations. The first evaluation is direct: the integral vanishes because the measure dz/z is proportional to $d\theta$, and $A(z) \to 0$ as the circle is taken to infinity.

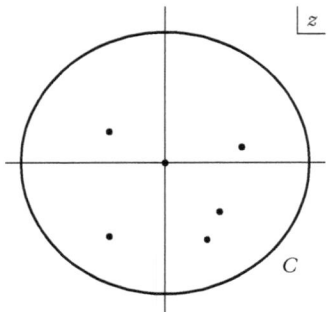

Figure 4.2 *Contour integral for deriving the BCFW on-shell recursion relation.*

We can also evaluate the integral by residues. To do this, we need to find the poles of the integrand. One pole is at $z = 0$; the residue there is simply $A(0)$, which is our original amplitude before the shift (4.52). Equating the evaluation by residues to the direct result already shows us that we can obtain an equation for $A(0)$ in terms of the other residues,

$$A(0) = -\sum_{\text{poles } a} \operatorname*{Res}_{z=z_a} \frac{A(z)}{z}. \tag{4.57}$$

Where do the other poles come from? Factorization!

A generic propagator denominator won't vanish if it depends solely on external momenta,

$$(K_{i\cdots r} \equiv k_i + \cdots k_r)^2 \neq 0, \tag{4.58}$$

where we assume a generic configuration of the momenta.

On the other hand, if the denominator depends on z, as z wanders all over the complex plane, such an expression is guaranteed to vanish somewhere. The squared momentum sum will depend on z when k_j is a member of the set $\{k_i \cdots k_r\}$, and k_l isn't (or vice versa),

$$K^{\mu}_{i\cdots r}(z) = K^{\mu}_{i\cdots r} - \frac{z}{2} \langle j| \mu |l], \tag{4.59}$$

so that

$$K^2_{i\cdots r}(z) = K^2_{i\cdots r} - z \langle j| K_{i\cdots r} |l], \tag{4.60}$$

which vanishes when

$$z = z_{ir} \equiv \frac{K^2_{i\cdots r}}{\langle j| K_{i\cdots r} |l]}. \tag{4.61}$$

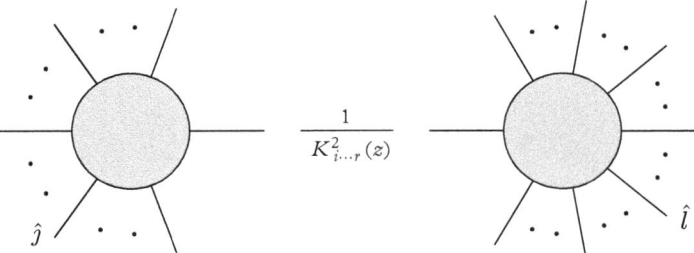

Figure 4.3 *A factorization giving rise to one term in the BCFW on-shell recursion.*

This corresponds to an *exact* factorization of $A(z)$ with k_j on one side and k_l on the other side. Unlike factorization in real momenta, discussed earlier, where we only approach the pole, here we want to approach it arbitrarily closely in order to take its residue. An example term is shown in Figure 4.3.

The sum over poles then becomes a sum over partitions of the external legs $1, \ldots, n$, into two subsets, where each of the two shifted legs is in a different subset,

$$\sum_{\text{poles}\,a} = \sum_{\substack{\text{partitions}\,P \\ \text{s.t. } j\in P,\, l\in \overline{P}}}. \qquad (4.62)$$

P denotes one of the subsets (say $\{i, \ldots, r\}$), and $\overline{P}$ its complement. To obtain a non-trivial factorization each of P and $\overline{P}$ must contain at least two external momenta. Each partition corresponds to a pair of residues at $z = z_{ir}$, one for each intermediate helicity in the factorization. What is the residue at $z = z_{ir}$? In the factorization limit, we can rewrite

$$\begin{aligned}\frac{A(z)}{z} &= i\frac{A_L(z)A_R(z)}{zK^2_{i\cdots r}(z)} \\ &= -i\frac{A_L(z)A_R(z)}{z(z-z_{ir})\langle j|K_{i\cdots r}|l]}.\end{aligned} \qquad (4.63)$$

Taking the residue, we obtain

$$\begin{aligned}\operatorname*{Res}_{z=z_{ir}} \frac{A(z)}{z} &= -i\frac{A_L(z_{ir})A_R(z_{ir})}{z_{ir}\langle j|K_{i\cdots r}|l]} \\ &= -i\frac{A_L(z_{ir})A_R(z_{ir})}{K^2_{i\cdots r}}.\end{aligned} \qquad (4.64)$$

Plugging this back into equation (4.57), we obtain the BCFW recursion,

$$A_n(1,\ldots,n) = \sum_{\text{partitions}\,P}\sum_{h=\pm} A_{\#P+1}(\ldots,\hat{j},\ldots,-\hat{P}^h)\times \frac{i}{P^2}\times A_{\#\overline{P}+1}(\ldots,\hat{l},\ldots,\hat{P}^{-h}). \qquad (4.65)$$

At each stage of the recursion, the sum is taken over partitions P into two subsets where $j \in P$ (one of the two subsets), and $l \in \overline{P}$, and the hatted momenta and spinors are evaluated at $z = z_P$:

$$\hat{k}_j = k_j(z_P), \qquad |\hat{j}] = |j] - z_P|l]. \tag{4.66}$$

The intermediate momentum P is determined by momentum conservation from the momenta in the subset P, and is of course dependent on z as well. The pole value z_P is given by equation (4.61) with $K_{i...r}$ replaced by P. Other momenta and spinors are independent of z. The 'propagator' factor in each term has the square of the naive, unshifted momentum P in the denominator. This recursion relation extends in a straightforward way to amplitudes with fermions, and can be extended to massive particles as well. In a helicity framework, the spin projectors that would appear in the numerator of the propagator are absorbed into the amplitudes $A_{L,R}$. Having non-trivial amplitudes on either side of each factorization requires each subset to contain two or more momenta, so that each must also contain no more than $n-2$ momenta, and thus have fewer legs than the original amplitude. They are on-shell amplitudes, but for complex momenta. Because the number of momenta in each amplitude on the right-hand side is smaller than on the left-hand side, the recursion will terminate—at the three-point amplitudes we obtained earlier.

How should we choose j and l? The number of terms in the recursion will go as $|l-j| \times (n-3)$, so it's best to choose them close in the cyclic ordering. The best choice, when possible, is generally to have them nearest neighbors.

What shifts are legitimate? To get an idea, let's look at the Parke–Taylor amplitude (4.1). Strictly speaking, we haven't computed it yet (though the three- and four-point amplitudes we've seen are examples of it); and it's also a very special amplitude, and might not tell us how things work in general. As we saw in the Introduction, this amplitude is the one for a special configuration of gluon helicities, where two gluons have negative helicities, and the rest are of positive helicity. It's also special because only λ spinors appear, and it's free of the $\tilde{\lambda}$ spinors. (The form given in eq. (4.1), incidentally, is not the form you will find in the Parke–Taylor paper, but rather was first written down by Mangano, Parke, and Xu.)

Let's take $m_1 = 1$, $m_2 = 4$, and study its behavior under a few different shifts. Under a $[1^-, 2^+\rangle$ shift,

$$|2\rangle \to |2\rangle + z|1\rangle,$$
$$\langle 1\,2\rangle \to \langle 1\,2\rangle, \tag{4.67}$$
$$\langle 2\,3\rangle \to \langle 2\,3\rangle + z\langle 1\,3\rangle,$$

so only the denominator is affected, and $A_n \sim 1/z$ as $z \to \infty$, satisfying the assumption we made in deriving the recursion relation.

Under a $\langle 1^-, 3^+\rangle$ shift,

$$|3\rangle \to |3\rangle + z|1\rangle,$$
$$\langle 2\,3\rangle \to \langle 2\,3\rangle - z\langle 1\,3\rangle, \tag{4.68}$$
$$\langle 3\,4\rangle \to \langle 3\,4\rangle + z\langle 1\,4\rangle.$$

Once again, only the denominator is affected, but here $A_n \sim 1/z^2$ as $z \to \infty$. For suitable helicities, this is in fact a generic behavior for a wide class of non-nearest neighbor shifts, in which the large-z behavior is better than the minimum needed for our assumption.

Under a $[2^+, 3^+\rangle$ shift,

$$|3\rangle \to |3\rangle + z|2\rangle,$$
$$\langle 2\,3\rangle \to \langle 2\,3\rangle,$$
$$\langle 3\,4\rangle \to \langle 3\,4\rangle + z\langle 2\,4\rangle,$$
(4.69)

once again $A_n \sim 1/z$; under a $[1^-, 4^-\rangle$ shift,

$$|4\rangle \to |4\rangle + z|1\rangle,$$
$$\langle 1\,4\rangle \to \langle 1\,4\rangle,$$
$$\langle 3\,4\rangle \to \langle 3\,4\rangle - z\langle 1\,3\rangle,$$
$$\langle 4\,5\rangle \to \langle 4\,5\rangle + z\langle 1\,5\rangle,$$
(4.70)

$A_n \sim 1/z^2$.

However, under a $[2^+, 1^-\rangle$ shift,

$$|1\rangle \to |1\rangle + z|2\rangle,$$
$$\langle 1\,4\rangle \to \langle 1\,4\rangle + z\langle 2\,4\rangle,$$
$$\langle 1\,2\rangle \to \langle 1\,2\rangle,$$
$$\langle n\,1\rangle \to \langle n\,1\rangle + z\langle n\,2\rangle,$$
(4.71)

the numerator is affected too, and $A_n \sim z^3$ as $z \to \infty$. This shift is a bad one, and fails to satisfy the assumption made in deriving the recursion relation. A similar problem afflicts a $[3^+, 1^-\rangle$ shift, under which,

$$|1\rangle \to |1\rangle + z|3\rangle,$$
$$\langle 1\,4\rangle \to \langle 1\,4\rangle + z\langle 3\,4\rangle,$$
$$\langle 1\,2\rangle \to \langle 1\,2\rangle - z\langle 2\,3\rangle,$$
$$\langle n\,1\rangle \to \langle n\,1\rangle + z\langle n\,3\rangle,$$
(4.72)

where $A_n \sim z^2$.

Even though the details in these results are special to the Parke–Taylor amplitudes, the behavior of shifts with the given gluon helicity patterns holds much more generally: $[j^-, l^+\rangle$, $[j^+, l^+\rangle$, and $[j^-, l^-\rangle$ have good large-z behavior and are legitimate shifts, while $[j^+, l^-\rangle$ has bad behavior and isn't a legitimate shift.

We can give a power-counting argument for the adequacy of the first of these shifts $[j^-, l^+\rangle$ based on Feynman diagrams—we won't do any explicit calculations, but only a gedanken calculation! We'll use the polarization vectors in the spinor-helicity representation, along with a couple of basic facts about the vertices from the Lagrangian:

- the three-point vertex is linear in momenta
- the four-point vertex is independent of momenta

Consider all possible diagrams that can arise in the calculation of an n-point amplitude. Trace a path in the each diagram from leg j to leg l. The momentum flowing at any point in that path is z-dependent, so a diagram where a pair of three-point vertices along with a propagator are replaced by a four-point vertex will lose a power of z, and have better large-z behavior. The worst behavior will come from diagrams with only three-point vertices along that path. Every such diagram has one more vertex than propagators (whatever that number may be), and so the vertices and propagators yield an overall factor of z. Next, we must tie on the polarization vectors, and this is where the helicity dependence enters. The two polarization vectors are

$$\varepsilon_j^{(-)} = \frac{\langle j|\mu|q]}{\sqrt{2}([jq] - z[lq])} \sim \frac{1}{z},$$
$$\varepsilon_l^{(+)} = \frac{\langle q|\mu|l]}{\sqrt{2}(\langle ql\rangle + z[qj])} \sim \frac{1}{z}. \tag{4.73}$$

The two polarization vectors together improve the behavior of A to at worst $1/z$ as $z \to \infty$. Of course, as we have seen previously, the actual behavior could be even better; but this is good enough to use the recursion.

The proof for $[j^+, l^+\rangle$ and $[j^-, l^-\rangle$ shifts is more involved; switching the roles of j and l in equation (4.73) suggests (but does not prove!) that $[j^+, l^-\rangle$ is an unsuitable shift.

Let's do an example, and calculate $A_4(1^-, 2^-, 3^+, 4^+)$. Use the $[2, 3\rangle$ shift,

$$|2] \to |2] - z|3],$$
$$|3\rangle \to |3\rangle + z|2\rangle. \tag{4.74}$$

The other conjugate spinors are untouched: $|2\rangle \to |2\rangle$ and $|3] \to |3]$.

Only one partition is possible with this shift: leg 2 is on one side, and leg 3 on the other. The term is shown diagrammatically in Figure 4.4. Because the $(---)$ three-point amplitude vanishes, the only non-vanishing amplitude has positive helicity assigned to the internal line coming from the left-side amplitude. The sum over internal helicities reduces to one term; writing down the factors from left to right, we have

$$i\frac{\langle 1\,2\rangle^3}{\langle 2(-\hat{P})\rangle\langle(-\hat{P})1\rangle} \times \frac{i}{s_{34}} \times (-i)\frac{[3\,4]^3}{[\hat{P}3][4\hat{P}]}. \tag{4.75}$$

Recall that when we flip the sign of a momentum, each of the associated spinors must change by a factor of i,

$$|-\hat{P}\rangle = i|\hat{P}\rangle. \tag{4.76}$$

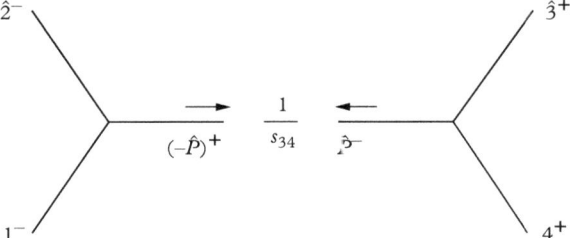

Figure 4.4 *Diagrammatic representation for the ECFW term for the four-point amplitude $A_4(1^-,2^-,3^+,4^+)$.*

We can then pair up angle products containing $\hat{P}$ with corresponding bracket products to form spinor sandwiches. There are two ways to do this; the better way, because of the choice of shift is to pair $\langle 2\,\hat{P}\rangle$ with $[4\,\hat{P}]$, but the final answer won't depend on the pairing. Our amplitude then becomes

$$A_4 = -i\frac{\langle 1\,2\rangle^3 [3\,4]^3}{\langle 4\,3\rangle \langle 1|\hat{P}|3]\langle 2|\hat{P}|4]}. \tag{4.77}$$

We can replace $\hat{P}$ using the shift vector

$$\hat{P} = P - \frac{z_{12}}{2}\langle 2|\mu|3], \tag{4.78}$$

where $P = k_1 + k_2 = -(k_3 + k_4)$. The choice of pairing ensures that the term with the shift vector vanishes after use of the Fierz identity,

$$\langle 1|\hat{P}|3] = \langle 1|P|3] - \frac{z_{12}}{2}\langle 1|\mu\,3]\langle 2|\mu|3] = \langle 1|P|3]. \tag{4.79}$$

This simplifies the expression for the amplitude to

$$-i\frac{\langle 1\,2\rangle^3 [3\,4]^2}{\langle 4\,3\rangle \langle 1|P|3]\langle 2|P|4]}. \tag{4.80}$$

We can simplify $\langle 1|P|3] = -\langle 1|4|3]$ and $\langle 2|P|4] = -\langle 2|3|4]$, and cancel factors to obtain the Parke–Taylor form,

$$i\frac{\langle 1\,2\rangle^3}{\langle 2\,3\rangle \langle 3\,4\rangle \langle 4\,1\rangle}. \tag{4.81}$$

The n-point generalization of this formula earlier is of course the Parke–Taylor amplitude, which can be derived in a very similar way. This amplitude is also called the maximally helicity violating amplitude or MHV amplitude, because amplitudes with even larger net helicity simply vanish,

$$\begin{aligned} A_n(++\cdots+) &= 0, \\ A_n(-+\cdots+) &= 0. \end{aligned} \tag{4.82}$$

Both of these results can be proven by induction using the recursion relations.

References

[1] H. Elvang and Y. t. Huang, *Scattering Amplitudes in Gauge Theory and Gravity*, Cambridge University Press, 2015; ISBN: 978–1–107–06925–1 [arXiv:1308.1697 [hep-th]].
[2] J. M. Henn and J. C. Plefka, *Scattering Amplitudes in Gauge Theories*, Springer-Verlag Lecture Notes in Physics (volume 883), 2014; ISBNs 978–3–642–54021–9 and 978–3–642–54022–6.
[3] M. Srednicki, *Quantum Field Theory*, Cambridge University Press, 2007; ISBN 978–0–521–86449–7.
[4] A. Zee, *Quantum Field Theory in a Nutshell (2nd ed.)*, Princeton University Press, 2010; ISBN 978–1–400–83532–4.
[5] M. Schwartz, *Quantum Field Theory and the Standard Model*, Cambridge University Press, 2013; ISBN 978–1–107–03473–0.
[6] S. J. Parke and T. R. Taylor, Phys. Rev. Lett. **56**, 2459 (1986). [doi:10.1103/PhysRevLett.56.2459]
[7] M. L. Mangano, S. J. Parke, and Z. Xu, Nucl. Phys. B **298**, 653 (1988). [doi:10.1016/0550-3213(88)90001-6]
[8] D. Maître and P. Mastrolia, Comput. Phys. Commun. **179**, 501 (2008). [doi:10.1016/j.cpc.2008.05.002][arXiv:0710.5559 [hep-ph]].
[9] P. Benincasa and F. Cachazo, arXiv:0705.4305 [hep-th].
[10] Z. Xu, D. H. Zhang, and L. Chang, Nucl. Phys. B **291**, 392 (1987). [doi:10.1016/0550-3213(87)90479-2]
[11] R. Britto, F. Cachazo, B. Feng, and E. Witten, Phys. Rev. Lett. **94**, 181602 (2005) [doi:10.1103/PhysRevLett.94.181602] [arXiv:hep-th/0501052].

5
Integrability in sigma-models

K. ZAREMBO

Nordita, Stockholm University and KTH Royal Institute of Technology,
Roslagstullsbacken 23, SE-106 91 Stockholm, Sweden
Department of Physics and Astronomy, Uppsala University
SE-751 08 Uppsala, Sweden

Chapter Contents

5 Integrability in sigma-models   205
K. ZAREMBO

	Preface	207
5.1	Introduction	207
5.2	Geometry	208
5.3	Principal chiral field	212
5.4	Symmetric cosets	216
5.5	B-field and topology	221
5.6	Quantum sigma-models	224
5.7	$O(N)$ model: large N	228
5.8	$O(N)$ model: exact solution	230
5.9	Crash course in superalgebras	238
5.10	Supercoset sigma-models	241
	Acknowledgements	245
	References	246

Preface

The following topics are covered in this chapter: (1) homogeneous spaces, (2) classical integrability of sigma-models in two dimensions, (3) topological terms, (4) background-field method and beta-function, (5) S-matrix bootstrap in the $O(N)$ model, (6) supersymmetric cosets and strings on $AdS_d \times X$.

5.1 Introduction

A sigma-model is a field theory wherein fields take values in a curved manifold $\mathcal{M}$. In other words, a field configuration is a map[1]

$$X^M(\sigma): \Sigma \to \mathcal{M}. \tag{5.1}$$

Sigma-models arise as effective field theories in many physical contexts that range from low-energy QCD [1] to condensed-matter systems [2].

Another vast area of applications of sigma-models is string theory, where they govern string propagation in curved backgrounds. The ultimate goal of this chapter is to introduce string sigma-models relevant for holographic duality whose key feature is complete integrability. Quantization of integrable sigma-models will then be exemplified by the S-matrix bootstrap in the $O(N)$ model [3]. Although sigma-models that arise in the holographic duality are in many respects different, methods for their exact solution are conceptually the same.

The canonical example of holography, the AdS/CFT correspondence, deals with string theory on $AdS_5 \times S^5$, a non-linear manifold whose curvature is controlled by the 't Hooft coupling of the dual gauge theory. The sigma-model on $AdS_5 \times S^5$ [4] is completely integrable [5], a remarkable property not only of this particular model, but of a whole class of holographic string backgrounds, which underlies proliferation of integrability methods in the gauge–string duality. The goal of this chapter is a step-by-step introduction to integrable sigma-models.

Given local coordinates X^M on $\mathcal{M}$, the most general two-derivative Lagrangian of a sigma-model is

$$\mathcal{L} = \frac{1}{2}\left(\sqrt{|h|}h^{ab}G_{MN}(X)\partial_a X^M \partial_b X^N + \varepsilon^{ab}B_{MN}(X)\partial_a X^M \partial_b X^N\right). \tag{5.2}$$

Transformations that leave the metric and B-field invariant translate into global symmetries of the sigma-model. Symmetries are neither necessary nor sufficient for integrability, but they allow one to build a large class of integrable models. The sigma-models arising in the holographic duality are precisely of this type. For this reason we concentrate on the cases when the target space $\mathcal{M}$ admits an action of a (simple) Lie group (or supergroup) G, and in Section 5.2 recollect basic facts about action of Lie groups on manifolds,

---

[1] The space-time manifold Σ, for the purpose of this chapter, will almost exclusively be the 2d Minkowski space with $(+-)$ signature or the Euclidean $\mathbb{R}^2$.

following [6]. We then describe classical integrability of sigma-models on such manifolds, following [7].

5.2 Geometry

We start by collecting basic geometric facts about homogeneous spaces, closely following [6].

Def 1 *A map*
$$T_g(x): G \times \mathcal{M} \to \mathcal{M}$$
defines left (right) action of group G on manifold $\mathcal{M}$, if
$$T_g T_h = T_{gh} \qquad (T_g T_h = T_{hg})$$
and $T_1 = \mathrm{id}$.

The stability group—or the little group—of a point $x \in \mathcal{M}$ is defined as the set of all elements in G that leave x intact:
$$H_x = \{g \in G | T_g(x) = x\}.$$

Def 2 $\mathcal{M}$ *is a left (right) homogeneous space of group G, if the action of G on $\mathcal{M}$ is transitive, namely if*
$$\forall x, y \in \mathcal{M} \;\; \exists g \in G: \;\; T_g(x) = y.$$

Stability groups of any two points in a homogeneous space are isomorphic to one another. To see this we notice that any two points in $\mathcal{M}$, x and y, are related by the group action: $T_g(x) = y$ and $T_{g^{-1}}(y) = x$. Consider now an element $h \in H_x$ ($T_h(x) = x$), then
$$y = T_g(x) = T_g T_h(x) = T_g T_h T_{g^{-1}}(y) = T_{ghg^{-1}}(y).$$

Therefore $ghg^{-1} \in H_y$ and the stability groups H_y and H_x are related by conjugation:
$$H_y = g H_x g^{-1}.$$

In particular, they have the same multiplication table.

Given a subgroup $H \subset G$, one can define a (right) coset G/H as a set of equivalence classes with respect to right multiplication by H:
$$G/H = \{g \sim gh | g \in G, h \in H\}.$$

A set gH, obtained by multiplying all elements of H by g, constitutes one point in G/H (Figure 5.1). One can show that for a closed Lie subgroup $H \subset G$, the coset space G/H is a smooth manifold. The left coset $H \backslash G$ is defined in a similar way.

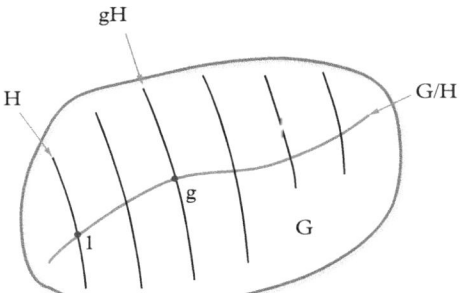

Figure 5.1 *A group manifold foliated over equivalence classes gH. In the coset G/H, each equivalence class is projected to one point.*

The coset G/H is a left homogeneous space of G, with the group action defined by left multiplication:

$$T_k(gH) = kgH. \tag{5.3}$$

Interestingly, the converse is also true, by virtue of the following theorem.

Th 1. *Homogeneous space $\mathcal{M}$ is isomorphic to the coset of its symmetry group G by the stability group H_{x_0} of any point $x_0 \in \mathcal{M}$: $\mathcal{M} = G/H_{x_0}$.*

Proof. Define a map $f : G/H_{x_0} \to \mathcal{M}$ by

$$f(gH) = T_g(x_0).$$

This map
- does not depend on the representative in the equivalence class gH:

$$T_{gh}(x_0) = T_g T_h(x_0) = T_g(x_0),$$

and thus maps the whole equivalence class to a single point in $\mathcal{M}$.
- covers all $\mathcal{M}$. This follows from transitivity of the group action.
- is one-to-one. Indeed, assuming that $f(g_1 H) = f(g_2 H)$, we have $T_{g_1}(x_0) = T_{g_2}(x_0)$ and $T_{g_2^{-1} g_1}(x_0) = x_0$. Consequently, $g_2^{-1} g_1 \in H_{x_0}$. Denoting $g_2^{-1} g_1 = h$ we can thus write $g_1 = g_2 h$ with $h \in H_{x_0}$, which proves that g_1, g_2 belong to the same equivalence class. Different equivalence classes, therefore, map to different points on $\mathcal{M}$.

The construction does not really depend on the base point x_0, because the stability groups H_x for different $x \in \mathcal{M}$ are all isomorphic. □

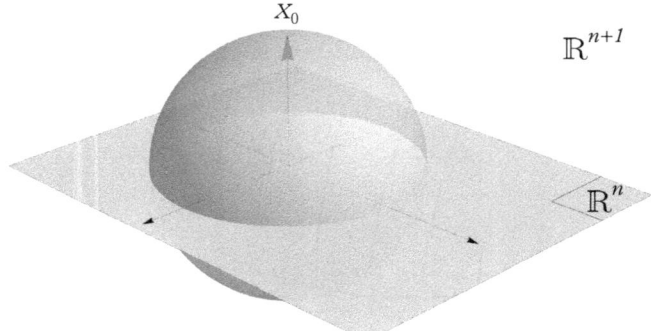

Figure 5.2 *n-dimensional sphere is $SO(n+1)/SO(n)$.*

An algebraic characterization of homogeneous spaces as cosets of a group by its subgroup is very convenient for field-theory applications. Since the construction may look rather abstract in the beginning, we illustrate it on a number of examples.

Ex 1. *Group manifold itself is both left and right homogeneous space.*

Proof. The group action is defined by left or right multiplication:
$$T_g^L(x) = gx$$
$$T_g^R(x) = xg. \tag{5.4}$$

□

Ex 2. *n-dimensional sphere S^n*

Proof. The $SO(n+1)$ rotation group acts transitively on the unit sphere in $\mathbb{R}^{n+1}$. The stability group of the north pole is the $SO(n)$ rotation group of the lateral hyperplane (Figure 5.2). Consequently,
$$S^n = SO(n+1)/SO(n). \tag{5.5}$$

□

Ex 3. *Complex projective space $\mathbb{CP}^n = \{\mathbf{z} \sim \lambda\mathbf{z} | \lambda \in \mathbb{C}^*, \mathbf{z} \in \mathbb{C}^{n+1}\}$*

Proof. $SU(n+1)$ acts transitively on $\mathbb{CP}^n$. The invariance subgroup of the point $(1,0) \sim (\lambda,0)$ is
$$\begin{pmatrix} e^{i\varphi} & \\ & SU(n) \times e^{-\frac{i\varphi}{n}} \end{pmatrix} \begin{pmatrix} 1 \\ 0 \\ \vdots \\ 0 \end{pmatrix}$$

Consequently, $\mathbb{CP}^n = SU(n+1)/SU(n) \times U(1)$.

□

Ex 4. *Anti-de-Sitter space*

Proof. The $(d+1)$-dimensional anti-de-Sitter space has coordinates (x^μ, z) and the line element

$$ds^2 = \frac{dx_\mu dx^\mu + dz^2}{z^2}, \quad (5.6)$$

where $\mu = 0 \ldots d-1$ and the indices are contracted with the $-+\ldots+$ Minkowski metric (for AdS_{d+1}) or with a Euclidean metric (for $EAdS_{d+1}$). The AdS space can be realized as a hypersurface in $\mathbb{R}^{d,2}$ (or $\mathbb{R}^{d+1,1}$) with embedding coordinates,

$$X^\mu = \frac{x^\mu}{z}$$
$$X^{-1} = \frac{x^2 + z^2 + 1}{2z}$$
$$X^d = \frac{x^2 + z^2 - 1}{2z}, \quad (5.7)$$

that satisfy

$$\eta_{MN} X^M X^N + 1 = 0, \quad (5.8)$$

where $M, N = -1 \ldots d$ and η_{MN} is the pseudo-Euclidean metric with signature $-\mp +\ldots+$ (the upper sign is for Minkowskian AdS).

Euclidean anti-de-Sitter (Lobachevsky) space $EAdS_{d+1}$ is the upper sheet of the two-sheeted hyperboloid imbedded into the $(d+2)$-dimensional Minkowski space, Figure 5.3(a). The hyperboloid asymptotes to the future light-cone, and the boundary of Euclidean AdS, the sphere S^d at infinity shown in Figure 5.3(a) as a circle, can be thought of as a set of all light rays emitted from the origin.

Minkowski AdS, as defined by equation (5.8), is a one-sheeted hyperboloid, Figure 5.3(b). The relationship between the space with the Poincaré metric (5.6) and the surface (5.8) is more intricate in this case. The coordinates (x^μ, z) cover only part of the hyperboloid, the Poincaré patch with $X^{-1} > X^d$ (which corresponds to

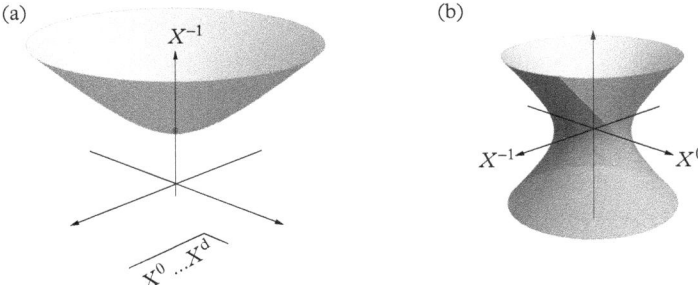

Figure 5.3 (a) $EAdS_{d+1}$ (b) AdS_{d+1}.

$z > 0$). The two patches, shaded differently in Figure 5.3(b), are glued together along the horizon at $z = \infty$. The lateral sections of the hyperboloid are actually time-like, and on the boundary that has the $S^{d-1} \times S^1$ geometry, the time direction is cyclic. The space which is usually called global AdS_{d+1} is the universal cover of the hyperboloid (5.8) obtained by unwinding the time direction.

The embedding (5.8) defines a transitive action of $SO(d,2)$ on AdS_{d+1} and of $SO(d+1,1)$ on $EAdS_{d+1}$. The stability group of the point shown as a dot in Figure 5.3 is $SO(d+1)$ for $EAdS_{d+1}$ and $SO(d,1)$ for AdS_{d+1}. Hence,

$$EAdS_{d+1} = SO(d+1,1)/SO(d+1), \qquad AdS_{d+1} = SO(d,2)/SO(d,1). \tag{5.9}$$

□

5.3 Principal chiral field

The principal chiral field is a non-linear field that takes values in a group manifold:

$$g(x) : \Sigma \to G.$$

One can define a Lie-algebra-valued current

$$j_a = g^{-1} \partial_a g \in \mathfrak{g}, \tag{5.10}$$

and construct the Lagrangian as

$$\mathcal{L} = -\operatorname{tr} j_a j^a = -\operatorname{tr} j \wedge *j. \tag{5.11}$$

Here 'tr' denotes the invariant quadratic form on $\mathfrak{g}$. The overall normalization and sign are chosen so as to give a conventional kinetic term to X^A in the parameterization $g = e^{iX^A T_A}$, where T_A are Lie algebra generators normalized such that $\operatorname{tr} T_A T_B = \delta_{AB}/2$. Another commonly used normalization includes an extra factor of $1/4$.

The Lagrangian is invariant under global $G_L \times G_R$ transformations

$$g(x) \to g_L g(x), \qquad g(x) \to g(x) g_R. \tag{5.12}$$

The current is left-invariant and transforms in the adjoint under right multiplications: $j_a \to g_R^{-1} j_a g_R$. The invariance of the Lagrangian in the latter case follows from the invariance of the quadratic form, denoted by tr, under the group action.

The equations of motion of the principal chiral field follow from an infinitesimal variation

$$\delta g = g\xi, \qquad \xi \in \mathfrak{g}.$$

The variation of the current is

$$\delta j_a = \partial_a \xi + [j_a, \xi]. \tag{5.13}$$

The commutator term does not contribute to the variation of the Lagrangian, because of the invariance of the quadratic form under the adjoint action of the Lie algebra. Partial integration in the remaining derivative term yields

$$\delta S = 2 \int d^2 x \, \text{tr} \, \xi \, \partial_a j^a.$$

The equations of motion therefore are

$$\partial_a j^a = 0, \tag{5.14}$$

and are equivalent to the conservation of current j_a. The current j_a is nothing but the Noether current of the right group multiplication and the equations of motion can be alternatively derived from the Noether theorem.

When written in terms of $g(x)$, the conservation of current is a second-order non-linear differential equation. Boundary conditions, for instance,

$$g(x^0, \pm\infty) = 1, \tag{5.15}$$

and initial conditions at some fixed time slice lead to a well-defined Cauchy problem.

However, it is far more convenient to regard the current itself as a dynamical variable. To this end, we notice that (5.10) defines a pure-gauge potential, a flat connection, whose curvature is equal to zero:

$$\partial_a j_b - \partial_b j_a + [j_a, j_b] = 0. \tag{5.16}$$

Unlike (5.14), this equation is an identity, a consequence of definition (5.10). Nevertheless we can treat equations (5.14) and (5.16) on equal footing, as two equations of motion for the two components of the current. Once the current is known, $g(x)$ can be reconstructed from (5.10) after imposing the boundary conditions (5.15).

In the index-free notations,

$$d * j = 0 \tag{5.17}$$
$$dj + j \wedge j = 0. \tag{5.18}$$

These two equations can be combined into one by the following simple trick. Let us multiply the first equation by z and add to the second equation: $z(5.17) + (5.18)$. We do not lose any information by doing so, because the two expressions are null as soon as their arbitrary linear combination is null. Explicitly,

$$d(j+z*j)+j\wedge j=0. \tag{5.19}$$

Here z is just a dummy variable, an arbitrary complex number for instance.

The last equation suggest redefining the current to $j+z*j$. And indeed, using the identities

$$*a\wedge b = -a\wedge *b, \qquad *^2 = 1,$$

valid for one-forms in two dimensions, the last term in the equation can be expressed through the new current as well, up to a rescaling factor

$$(j+z*j)\wedge(j+z*j) = \left(1-z^2\right)j\wedge j.$$

A simple rescaling factor

$$L = \frac{j+z*j}{1-z^2}, \tag{5.20}$$

makes the current so defined flat:

$$dL+L\wedge L = 0. \tag{5.21}$$

The current L is called the Lax connection. Explicitly,

$$L_a = \frac{j_a + z\varepsilon_{ab}j^b}{1-z^2}. \tag{5.22}$$

The equations of motion for the principal chiral field are equivalent to the condition that the Lax connection is flat [8]:

$$\partial_a L_b - \partial_b L_a + [L_a, L_b] = 0 \quad \forall z. \tag{5.23}$$

The zero curvature, or Lax representation of the equations of motion is a hallmark of integrability.

By virtue of the zero curvature condition, the holonomy (the Wilson line) of L is invariant under continuous deformations of the contour. The holonomy along a constant time slice defines the monodromy matrix

$$M(t;z) = P\exp\int_{-\infty}^{+\infty}dx L_1(t,x;z), \tag{5.24}$$

which for generic complex z is an element of the complexified group $G^{\mathbb{C}}$. Translation of the contour along the time direction is a continuous deformation and therefore leaves the monodromy matrix intact. Hence, the monodromy matrix is time-independent:

$$\partial_t M(t;z) = 0. \tag{5.25}$$

A quantity that does not change with time, once the equations of motion are imposed, is called the conserved charge. The monodromy matrix defines an infinite number of those because it contains an auxiliary parameter z (called the spectral parameter). The individual conserved charges are generated by expanding the monodromy matrix in z, for instance by the Laurent expansion at infinity:

$$M(z) = \exp\left(\sum_{n=1}^{\infty} \frac{Q_n}{z^n}\right). \tag{5.26}$$

Written explicitly in terms of currents, the monodromy matrix is

$$M(z) = P\exp\int_{-\infty}^{+\infty} dx \frac{j_1 - zj_0}{1 - z^2}, \tag{5.27}$$

and to the first order in $1/z$,

$$M(z) = 1 + \frac{1}{z}\int_{-\infty}^{+\infty} dx j_0 + \mathcal{O}\left(\frac{1}{z^2}\right). \tag{5.28}$$

The first charge in the hierarchy, Q_1, therefore, is just the Noether charge of right group multiplications.

Exercise 1 *Compute the second (Yangian) charge Q_2 by expanding the monodromy matrix to the second order in $1/z$.*

Exercise 2 *Derive the Noether current of the left group multiplications. Show that the corresponding charge appears at the first order in the Taylor expansion of the monodromy matrix at $z = 0$.*

Higher conserved charges obtained by expanding the monodromy matrix at $z = \infty$ are non-local, they cannot be represented by integrals of local densities. Local conserved quantities are generated by a Taylor expansion at $z = \pm 1$. The conventional energy and momentum of the sigma-model, however, cannot be obtained from the monodromy matrix. Integrability of the principal chiral field is more consistent with an alternative canonical structure that does not follow from the Lagrangian (5.11), but whose Hamiltonian and momentum appear as first terms in the expansion of the monodromy matrix at $z = \pm 1$ [9].

The zero-curvature representation has far-reaching consequences. It can be used to explicitly construct solutions of the equations of motion, or to find the action-angle variables, in which the dynamics is trivial. The counterpart of the Lax representation in quantum theory is the Algebraic Bethe Ansatz [10], which defines the building blocks of

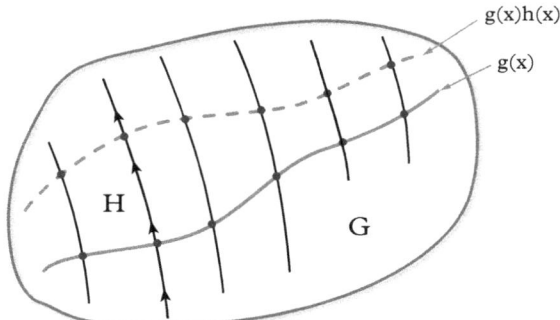

Figure 5.4 *Locally, a coset can be viewed as a section of the group manifold transverse to the orbits of H. Two sections that differ by multiplication with elements in H are equivalent.*

the quantum spectrum. The usual issues of UV regularization and definition of the true ground state are very non-trivial in integrable QFTs, and it is often easier to reconstruct the final answer from the constraints imposed by integrability on the dynamics, rather than to solve the model directly. Unfortunately, many steps of this bootstrap procedure are model dependent and may also depend on the boundary conditions. We exemplify the integrability bootstrap in Section 5.8 by the exact solution of the $O(N)$ sigma-model.

5.4 Symmetric cosets

Sigma-models on homogeneous spaces are also integrable, provided that the target space is *symmetric*, an algebraic property described later. The action of a sigma-model can be always written as (5.2) in an explicit coordinate system, but for homogeneous spaces an elegant invariant construction based on the coset representation [11] is in many ways more convenient. Points in a coset are orbits of H. Picking one representative per orbit introduces an explicit coordinate system. This defines a section of the group manifold that is transverse to the orbits of H at any point.[2] A field configuration then is a map from space-time to the group manifold which is restricted to this section (Figure 5.4). But any other choice differing by right multiplication with elements in H is equivalent, and it is desirable to have a formulation covariant with respect to the action of H. In other words, any map $g(x)$ should be allowed, but the field configurations $g(x)$ and $g(x)h(x)$ with $h(x) \in H$ should be physically indistinguishable. This can be achieved by realizing the right action of H as a gauge symmetry:

$$g(x) \to g(x)h(x), \qquad h(x) \in H. \tag{5.29}$$

For this construction to work the action of the sigma-model has to be gauge invariant.

---

[2] Such a section may not be globally defined.

The Lagrangian (5.11) fails to satisfy this condition. Indeed, the current (5.10) transforms as

$$j_a \to h^{-1}j_a h + h^{-1}\partial_a h, \qquad (5.30)$$

and the inhomogeneous term does not cancel in the gauge variation of the action.

The inhomogeneous term in the variation of the current can be projected out by decomposing the Lie algebra $\mathfrak{g}$ of G into the denominator subalgebra $\mathfrak{h}$ and its orthogonal complement $\mathfrak{f}$ with respect to the invariant quadratic form:

$$\mathfrak{g} = \mathfrak{h} \oplus \mathfrak{f}. \qquad (5.31)$$

The current then expands in two components,

$$j_a = A_a + K_a, \qquad A_a \in \mathfrak{h},\ K_a \in \mathfrak{f}. \qquad (5.32)$$

Since $h \in H$, and $h^{-1}\partial_a h \in \mathfrak{h}$, it is A_a that transforms as a gauge field,

$$A_a \to h^{-1}A_a h + h^{-1}\partial_a h, \qquad (5.33)$$

while the coset component transforms as a matter field in the adjoint,

$$K_a \to h^{-1}K_a h. \qquad (5.34)$$

The simplest gauge-invariant Lagrangian built from these fields is

$$\mathcal{L} = -\operatorname{tr} K_a K^a. \qquad (5.35)$$

It is manifestly invariant under the local gauge transformations (5.29) and is also symmetric under global left multiplications $g(x) \to g_L g(x), g_L \in G$, because the current is left invariant from the very beginning. The model possesses all the requisite symmetries of the coset space G/H.

In order to make the discussion less abstract let us illustrate the coset construction by a simple example of the Hopf fibration $S^3 \to S^2$:

Ex 5. $S^2 = SU(2)/U(1)$.

Proof. The generators of $\mathfrak{su}(2)$ (the Pauli matrices) are split into the coset generators and the denominator subalgebra as

$$\mathfrak{h} = <\sigma_1>, \qquad \mathfrak{f} = <\sigma_2, \sigma_3>. \qquad (5.36)$$

218 Integrability in sigma-models

The generic element of $SU(2)$ can be parameterized by three Euler angles:

$$g = e^{\frac{i\varphi\sigma_1}{2}} e^{\frac{i\theta\sigma_3}{2}} e^{\frac{i\psi\sigma_1}{2}}$$

The last factor belongs to the coset denominator and can be removed by a gauge transformation. We can just drop it altogether, which is equivalent to fixing a gauge. For the current we then have

$$g^{-1}dg = e^{-\frac{i\theta\sigma_3}{2}}\left(d + \frac{i}{2}d\varphi\sigma_1\right)e^{\frac{i\theta\sigma_3}{2}} = \frac{i}{2}\Big(\underbrace{d\theta\sigma_3 + d\varphi\sin\theta\sigma_2}_{K} + \underbrace{d\varphi\cos\theta\sigma_1}_{A}\Big).$$

The resulting line element is

$$ds^2 = -2\operatorname{tr} K^2 = \frac{1}{2}\operatorname{tr}(d\theta\sigma_3 + d\varphi\sin\theta\sigma_2)^2 = d\theta^2 + \sin^2\theta\, d\varphi^2,$$

which is the familiar metric of the sphere. □

Returning to the general case, we can re-write the variation of the current (5.13) as

$$\delta j_a = D_a\xi + [K_a, \xi], \tag{5.37}$$

where D_a is the standard covariant derivative,

$$D_a = \partial_a + [A_a, \cdot]. \tag{5.38}$$

The ensuing variation of the Lagrangian is

$$\delta\mathcal{L} = -2\operatorname{tr}\delta K_a K^a = -2\operatorname{tr}\delta j_a K^a = -2\operatorname{tr} D_a\xi K^a,$$

where the first equality follows from orthogonality of $\delta A_a \in \mathfrak{h}$ and $K^a \in \mathfrak{f}$, and the commutator term in (5.37) was dropped because of the invariance of the quadratic form. The resulting equations of motion are

$$D_a K^a = 0. \tag{5.39}$$

Together with the flatness condition (5.16), they form a complete set of equations for the current components. The flatness condition, when expressed in terms of K_a and A_a, takes the form

$$F_{ab} + D_a K_b - D_b K_a + [K_a, K_b] = 0, \tag{5.40}$$

where $F_{ab} = \partial_a A_b - \partial_b A_a + [A_a, A_b]$ is the field strength of the gauge connection.

Before proceeding further let us pause for a few general remarks on the coset decomposition (5.31). It is orthogonal with respect to the invariant scalar product on $\mathfrak{g}$, but what about the commutation relations? How are they affected by the coset decomposition? Since $\mathfrak{h}$ is a subalgebra, the commutator of two elements of $\mathfrak{h}$ is again an element of $\mathfrak{h}$,

$$[\mathfrak{h},\mathfrak{h}] \subset \mathfrak{h}. \tag{5.41}$$

In terms of the structure constraints this means that $f_{\mathfrak{h}\mathfrak{h}\mathfrak{f}} = 0$. Antisymmetry of the structure constants (a very general property that follows from the existence of a non-degenerate invariant quadratic form) implies that $f_{\mathfrak{h}\mathfrak{f}\mathfrak{h}} = 0$. Hence,

$$[\mathfrak{h},\mathfrak{f}] \subset \mathfrak{f}. \tag{5.42}$$

What can be said about $[\mathfrak{f},\mathfrak{f}]$? Since $f_{\mathfrak{f}\mathfrak{f}\mathfrak{h}}$ are related to $f_{\mathfrak{h}\mathfrak{f}\mathfrak{f}}$, they are different from zero. The structure constants $f_{\mathfrak{f}\mathfrak{f}\mathfrak{f}}$ are not restricted by any general principles. An interesting special cases arises if they also vanish. Then,

$$[\mathfrak{f},\mathfrak{f}] \subset \mathfrak{h}. \tag{5.43}$$

If the Lie algebra decomposition obeys this constraint, the coset space is called symmetric. The commutation relations (5.41), (5.42), and (5.43) follow a systematic pattern in this case. One can formalize it by introducing a $\mathbb{Z}_2$ transformation acting on the Lie algebra $\mathfrak{g}$ as

$$\Omega(\mathfrak{h}) = \mathfrak{h}, \qquad \Omega(\mathfrak{f}) = -\mathfrak{f}. \tag{5.44}$$

This operation constitutes a symmetry of $\mathfrak{g}$ once (5.43) holds. Compatibility of the $\mathbb{Z}_2$ symmetry (5.44) with the commutation relations of $\mathfrak{g}$ can be taken as a definition of symmetric homogeneous space.

One can invert the logic and think of a symmetric space as being defined by the symmetry group G and a $\mathbb{Z}_2$ automorphism of its Lie algebra (a linear map $\Omega : \mathfrak{g} \to \mathfrak{g}$ that preserves the commutation relations and squares to the identity: $\Omega^2 = \text{id}$). The denominator of the coset $\mathfrak{h}$ is then identified with the invariant subalgebra of Ω. The $\mathbb{Z}_2$ automorphism of the Lie algebra translates into a geometric reflection symmetry of the homogeneous space, as illustrated by the following example.

Ex 6. $S^2 = SU(2)/U(1)$ *is a symmetric space.*

Proof. The coset decomposition (5.36) is $\mathbb{Z}_2$ invariant because $[\sigma_2,\sigma_3] = 2i\sigma_1 \in \mathfrak{h}$. Since σ^3 pairs with θ in the Euler parameterization, the $\mathbb{Z}_2$ symmetry of the algebra translates to the invariance of the sphere under $\theta \to -\theta$. □

Exercise 3 Check that $S^n, \mathbb{CP}^n$, and $(E)AdS_{d+1}$ are symmetric spaces.

Exercise 4 *The conformal algebra* $\mathfrak{so}(d,2)$ *is generated by Lorentz transformations* $L_{\mu\nu}$, *translations* P_μ, *dilatation* D, *and special conformal transformations* K_μ. *Its commutation relations are*

$$[L_{\mu\nu}, L_{\lambda\rho}] = \eta_{\mu\lambda} L_{\nu\rho} + \eta_{\nu\rho} L_{\mu\lambda} - \eta_{\mu\rho} L_{\nu\lambda} - \eta_{\nu\lambda} L_{\mu\rho},$$
$$[L_{\mu\nu}, P_\lambda] = \eta_{\mu\lambda} P_\nu - \eta_{\nu\lambda} P_\mu, \qquad [L_{\mu\nu}, K_\lambda] = \eta_{\mu\lambda} K_\nu - \eta_{\nu\lambda} K_\mu,$$
$$[D, P_\mu] = P_\mu, \qquad [D, K_\mu] = -K_\mu, \qquad [P_\mu, K_\nu] = 2L_{\mu\nu} - 2\eta_{\mu\nu} D. \qquad (5.45)$$

The Killing metric on $\mathfrak{so}(d,2)$ *is*

$$\operatorname{tr} L_{\mu\nu} L_{\lambda\rho} = \eta_{\mu[\lambda} \eta_{\nu\rho]}, \qquad \operatorname{tr} D^2 = -1, \qquad \operatorname{tr} P_\mu K_\nu = -2\eta_{\mu\nu}. \qquad (5.46)$$

Consider a $\mathbb{Z}_2$ *transformation,*

$$\Omega(L_{\mu\nu}) = L_{\mu\nu}, \qquad \Omega(D) = -D, \qquad \Omega(K_\mu) = P_\mu, \qquad \Omega(P_\mu) = K_\mu. \qquad (5.47)$$

Check that Ω *is consistent with the commutation relations of* $\mathfrak{so}(4,2)$, *and therefore constitutes an automorphism of the algebra. Show that the invariant subalgebra of* Ω *is* $\mathfrak{so}(d,1)$, *and the symmetric coset defined by* Ω *is* AdS_{d+1}.

Take the coset representative to be

$$g(x,z) = e^{iP_\mu x^\mu} z^{-D}, \qquad (5.48)$$

and derive the metric in the (x^μ, z) *coordinates from the coset construction.*

A remarkable fact about symmetric-space sigma-models is their complete integrability [12]. While coset sigma-models, which are not symmetric, can still be integrable in exceptional cases [13], integrability of symmetric cosets can be demonstrated in a uniform, essentially algebraic way.

Here is the key point, that singles out symmetric spaces. The first two terms in (5.40) belong to $\mathfrak{h}$ and $\mathfrak{f}$, respectively, because of (5.41) and (5.42), while the third term in general has both $\mathfrak{f}$ and $\mathfrak{h}$ components. But for symmetric cosets all four terms in the flatness condition (5.40) have definite $\mathbb{Z}_2$ grading and the equation can therefore be neatly projected onto $\mathfrak{h}$ and $\mathfrak{f}$ subspaces of the coset decomposition:

$$F_{ab} + [K_a, K_b] = 0$$
$$D_a K_b - D_b K_a = 0. \qquad (5.49)$$

These equations, together with the equation of motion (5.39) follow from the flatness condition for the Lax connection:

$$L_a = A_a + \frac{z^2+1}{z^2-1} K_a - \frac{2z}{z^2-1} \varepsilon_{ab} K^b. \qquad (5.50)$$

Exercise 5 *Show that the zero-curvature condition (5.23) for (5.50) is equivalent to the equations of motion (5.39), (5.49).*

5.5 B-field and topology

So far we have discussed the metric term in the sigma-model Lagrangian (5.2), and showed how to construct it for manifolds that admit an action of a symmetry group G. Is it also possible to construct an invariant B-field? To better formulate the question, it is convenient to switch to the form notations:

$$\frac{1}{2}\int_\Sigma d^2 x \varepsilon^{ab} B_{MN} \partial_a X^M \partial_b X^N = \int_\Sigma B = \int_D H, \tag{5.51}$$

where

$$B = B_{MN} dX^M \wedge dX^N, \quad H = dB = H_{MNL} dX^M \wedge dX^N \wedge dX^L, \quad H_{MNL} = \partial_{[M} B_{NL]}, \tag{5.52}$$

and D is a fiducial three-dimensional space whose boundary is Σ: $\partial D = \Sigma$.

Re-writing a 2d integral in a 3d form may look superficial, but it helps to uncover interesting possibilities that would be overlooked otherwise. The three-form H, as defined in (5.52), is exact. Interestingly, this requirement can be relaxed and replaced by a weaker condition that the form is closed ($dH = 0$). The three-dimensional form of the B-field coupling cannot be then transformed back to a two-dimensional integral of an invariant two-form. Yet it defines a consistent local coupling in two-dimensional field theory [14, 15].

If $\Sigma = S^2$, then D is the three-dimensional ball. Field configurations on S^2 can be unfolded to $\mathbb{R}^2$ by stereographic projection and, conversely, field configurations on $\mathbb{R}^2$ sufficiently well behaved at infinity can be compactified on S^2, so the ensuing construction applies to conventional field theories on the flat 2d space under assumption of regularity conditions at infinity.

The B-field coupling in the sigma-model action can thus be constructed from an invariant two-form on the target space or from an invariant closed three-form. Both cases lead to topological terms in the sigma-model action. We consider them in turn. A more comprehensive introduction to topological terms in QFT can be found in [16].

5.5.1 Theta-term

If the coset denominator contains an Abelian factor, $H \ni U(1)$, the corresponding field strength F_{ab} is a gauge-invariant two-form. The coset sigma-model on G/H then admits a theta-term:

$$S_\theta = \frac{\theta}{4\pi}\int d^2 x \varepsilon^{ab} F_{ab} = \frac{\theta}{2\pi}\int F. \tag{5.53}$$

The theta-term is topological, it only depends on the values of the fields at infinity,

$$\frac{1}{2\pi}\int F = \frac{1}{2\pi}\oint_\infty A.$$

Since the field must approach pure gauge at $x \to \infty$: $A = d\varphi$,

$$\frac{1}{2\pi}\oint_\infty A = \frac{1}{2\pi}\oint_\infty d\varphi = \frac{\Delta\varphi}{2\pi} = n.$$

The theta-term is non-zero only on topologically non-trivial field configurations (instantons) and measures their topological charge. As such it does not contribute to the equations of motion. In the path integral the theta-term produces a phase factor $e^{iS_\theta} = e^{i\theta n}$. Since n is an integer, the theta-angle is a periodic variable: the path integral does not change under $\theta \to \theta + 2\pi$.

Examples of sigma-models that admit a theta-term, and have instantons, are the $\mathbb{CP}^n$ models, for any n. These include the $O(3)$ model (because $S^2 = \mathbb{CP}^1$), while $O(n)$ models with $n > 3$ do not admit a theta-term and do not have instantons.

The theta-term measures the winding of pure gauge field at infinity and is associated with the homotopy group[3] $\pi_1(H) = \mathbb{Z}$. If $\pi_2(G/H) = \mathbb{Z}$, the degree of mapping $\Sigma \to G/H$ also defines a topological charge, but one can show on very general grounds that $\pi_2(G/H) = \pi_1(H)$ and the degree of mapping actually coincides with the winding of the pure gauge at infinity, so we do not get anything new. It is instructive to explicitly work out the equivalence of the two definitions of the topological charge for the $O(3)$ model.

Exercise 6 *The $O(3)$ sigma model can be realized as the $SU(2)/U(1) = S^2$ coset, or as a field theory of three-dimensional unit vector $\mathbf{n}$, $\mathbf{n}^2 = 1$. The degree of mapping defined by $\mathbf{n}$ (the number of times the target-space S^2 wraps Σ) is given by the integral*

$$n = \frac{1}{8\pi}\int d^2x\, \varepsilon^{ab}\mathbf{n}\cdot \partial_a\mathbf{n}\times \partial_b\mathbf{n}.$$

Prove that this number coincides with the instanton charge in the coset formulation, perhaps up to a sign. (Hint: use $F = -K\wedge K$, which is an identity, not an equation of motion).

Instanton solutions of arbitrary topological charge in the $O(3)$ model have been constructed in [17]. An interplay between instantons and integrability in this and other two-dimensional models remains an interesting open problem.

5.5.2 Wess–Zumino term

An invariant three-form on a group manifold can be obtained by taking the wedge product of three currents. This defines the Wess–Zumino (WZ) term that can be added to the action of the principal chiral field sigma-model [18]:

$$S_{\text{WZ}} = \frac{1}{3}\int_D \mathrm{tr}\, j\wedge j\wedge j. \tag{5.54}$$

---

[3] The definition and properties of homotopy groups can be found, for instance, in [6].

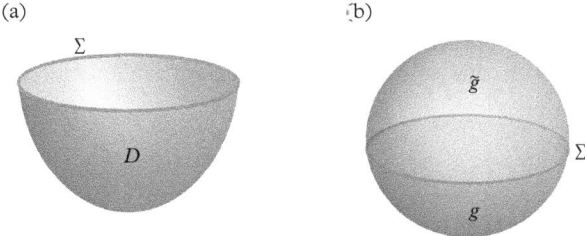

Figure 5.5 *(a) The space-time manifold Σ is the boundary of the integration domain D in the WZ term. (b) Two different continuations of the field variable in the interior of D can be combined into a map $\hat{g}: S^3 \to G$.*

The boundary of the integration domain is the physical 2d space-time: $\partial D = \Sigma$, Figure 5.5(a). The definition tacitly assumes that the field variable $g(x)$ is continued into the interior of D in some way. This is always possible to do, since $\pi_2(G) = 0$ for any Lie group G and consequently any map $g: S^2 \to G$ is contractible. The process of contraction defines a map $D \to G$.

If considered as part of the action functional in a 2d field theory, the WZ term should be independent of the way the field variable is continued inside D. Let us see if this is indeed the case [15, 18], by considering the WZ term evaluated on two different maps, $g: D \to G$ and $\tilde{g}: D \to G$, which coincide on the boundary, $g|_\Sigma = \tilde{g}|_\Sigma$ and thus correspond to one and the same field configuration in space-time. Consider first infinitesimally close maps: $\tilde{g} = g(1 + \xi)$, where $\xi \in \mathfrak{g}$ is small. Substituting the variation of the current from (5.13) in the WZ functional, we find

$$\delta S_{\rm WZ} = \int_D d\xi \wedge j \wedge j = -\int_D \xi\, d(j \wedge j). \tag{5.55}$$

The boundary term can be omitted upon integration by parts because ξ vanishes on Σ. The flatness condition (5.18) then gives

$$d(j \wedge j) = dj \wedge j - j \wedge dj = -j \wedge j \wedge j + j \wedge j \wedge j = 0,$$

so the variation of the WZ action indeed vanishes.

Continuous deformations of the map g do not change the WZ term, but can any two maps from D to the group manifold be continuously connected? Consider two such maps g and $\tilde{g}$ and combine them into a map $\hat{g}: S^3 \to G$ defined as shown in Figure 5.5(b):

$$\hat{g}(x) = \begin{cases} \tilde{g}(x) & x \in \text{upper hemisphere} \\ g(x) & x \in \text{lower hemisphere}. \end{cases}$$

Then,

$$S_{\rm WZ}[\tilde{g}] - S_{\rm WZ}[g] = \frac{1}{3} \int_{S^3} \text{tr}\, \hat{j} \wedge \hat{j} \wedge \hat{j} = 16\pi^2 in. \tag{5.56}$$

The number n does not change under continuous deformations of the map $\hat{g}$ and therefore is a topological invariant. Under appropriate normalization of the quadratic form on the Lie algebra this number is an integer that characterizes an element of $\pi_3(G) = \mathbb{Z}$.

Exercise 7 *For $G = SU(2) = S^3$, $\hat{g}$ defines a map $S^3 \to S^3$. Show that n given by the integral (5.56) is the degree of this map, the number of times $\hat{g}(x)$ wraps the target S^3 while x goes around the sphere.*

The WZ action thus is a multi-valued functional. Depending on how the field variable g is continued inside the three-dimensional domain of integration, the result may change by an integer:

$$S_{\mathrm{WZ}} \sim S_{\mathrm{WZ}} + 16\pi^2 i n. \tag{5.57}$$

This is not a problem in classical field theory, because the equations of motion only depend on the variation of the action, and the variation of the WZ action is single-valued. But in quantum theory the action enters the path integral through the phase factor,

$$e^{\frac{k}{8\pi} S_{\mathrm{WZ}}} \in \text{path integral},$$

where k is the coupling constant (called level). This factor will be single valued provided that the coupling is quantized:

$$k \in \mathbb{Z}. \tag{5.58}$$

The principal chiral field with the WZ term, the Wess–Zumino–Witten (WZW) model, is defined by the action [18]

$$S_{\mathrm{WZW}} = -\frac{1}{\kappa^2} \int_\Sigma \mathrm{tr}\, j \wedge *j + \frac{ik}{24\pi^2} \int_D \mathrm{tr}\, j \wedge j \wedge j. \tag{5.59}$$

Exercise 8 *Derive the equations of motion for the WZW model and show that they admit a Lax representation [19]. What happens at $\kappa^2 = 2\pi/k$?*

5.6 Quantum sigma-models

A quantum coset sigma-model is defined by the path integral (in Euclidean signature)

$$Z = \int \mathcal{D}g\, e^{\frac{1}{\kappa^2} \int d^2x\, \mathrm{tr}\, K_a K^a}. \tag{5.60}$$

This is a highly non-linear quantum field theory with a single dimensionless coupling. To get a first glance at the properties of this QFT we consider its response to a macroscopic external force, assuming momentarily that residual quantum effects can be treated perturbatively.

To this end we separate the integration variable in the path integral into the classical background field $\bar{g}$, which models the external force and which we assume to be a solution of the classical equations of motion, and quantum field X whose fluctuations are assumed to be small:

$$g = \bar{g} e^X. \tag{5.61}$$

The field X belongs to the Lie algebra of G. An appropriate gauge transformation (5.29) can eliminate the $\mathfrak{h}$ component of X:

$$X \in \mathfrak{f}. \tag{5.62}$$

The restriction to the coset subspace is a convenient gauge-fixing condition.

To develop perturbation theory we need to expand the current (5.10) in powers of X. We start by deriving a general formula for such an expansion. Substituting (5.61) into the definition of the current, we get

$$j_a = e^{-X} \left(\partial_a + \bar{j}_a \right) e^X = \bar{j}_a + \int_0^1 ds \; e^{-sX} \mathcal{D}_a X e^{sX}, \tag{5.63}$$

where $\bar{j}_a = \bar{g}^{-1} \partial_a \bar{g}$ is the background current and $\mathcal{D}_a = \partial_a + [\bar{j}_a, \cdot]$. The last term expands in multiple commutators of X. To write the series in a compact form, we commute both sides of (5.63) with X, which eliminates the integral over s,

$$[X, j_a - \bar{j}_a] = -\int_0^1 ds \frac{d}{ds} e^{-sX} \mathcal{D}_a X e^{sX} = \mathcal{D}_a X - e^{-X} \mathcal{D}_a X e^X. \tag{5.64}$$

Introducing the notation

$$\mathrm{ad} X \cdot Y = [X, Y], \tag{5.65}$$

and taking into account that

$$e^{-X} Y e^X = e^{-\mathrm{ad} X} Y,$$

we find

$$j_a = \bar{j}_a + \frac{1 - e^{-\mathrm{ad} X}}{\mathrm{ad} X} \mathcal{D}_a X. \tag{5.66}$$

Using this formula the current can be expanded to any desired order in X.

After rescaling $X \to \kappa X$, the first two orders become

$$j_a \simeq \bar{j}_a + \kappa \mathcal{D}_a X - \frac{\kappa^2}{2}[X, \mathcal{D}_a X]$$
$$= \bar{A}_a + \bar{K}_a + \kappa \left(D_a X + [\bar{K}_a, X]\right) - \frac{\kappa^2}{2}\left([X, D_a X] + [X, [\bar{K}_a, X]]\right), \quad (5.67)$$

where D_a is the covariant derivative with respect to the background gauge field. We can use $\mathbb{Z}_2$ grading to assign commutators to either $\mathfrak{h}$ or $\mathfrak{f}$, and thus already here assume that the coset is a symmetric space.

The action of the sigma-model, to the leading order in κ, is

$$S = S_{\text{cl}} - \int d^2 x \, \text{tr}\left(D_a X D^a X - [\bar{K}_a, X][\bar{K}^a, X]\right) + \mathcal{O}(\kappa). \quad (5.68)$$

Gaussian integration over X then yields

$$Z \simeq e^{-S_{\text{cl}}} \det^{-\frac{1}{2}}\left(-D^2 + \text{ad}\,\bar{K}^2\right). \quad (5.69)$$

The result of these manipulations is a one-loop effective action for the background field,

$$S_{\text{eff}} = S_{\text{cl}} + \frac{1}{2}\ln\det\left(-D^2 + \text{ad}\,\bar{K}^2\right). \quad (5.70)$$

The one-loop determinant diverges and requires regularization. Since in two dimensions the only scalar diagram that diverges is the bubble, the divergent part of the effective action is

$$\bigcirc = \frac{1}{2}\int d^2 x \, \text{tr}_{\mathfrak{f}} \, \text{ad}\,\bar{K}^2 \times \int^{\Lambda} \frac{d^2 p}{(2\pi)^2} \frac{1}{p^2} = \frac{1}{4\pi} \ln \frac{\Lambda}{\mu} \int d^2 x \, \text{tr}_{\mathfrak{f}} \, \text{ad}\,\bar{K}^2. \quad (5.71)$$

As a matrix in the adjoint representation, $\text{ad}\,\bar{K}_a$ has $\mathbb{Z}_2$ grading -1 and thus the following matrix structure:

$$\text{ad}\,\bar{K}_a : \begin{pmatrix} 0 & * \\ * & 0 \end{pmatrix} \begin{pmatrix} \mathfrak{h} \\ \mathfrak{f} \end{pmatrix}. \quad (5.72)$$

So,

$$\text{tr}_{\mathfrak{f}}\,\text{ad}\,\bar{K}^2 = \langle \mathfrak{f}|\,\text{ad}\,\bar{K}_a\,|\mathfrak{h}\rangle \langle \mathfrak{h}|\,\text{ad}\,\bar{K}^a\,|\mathfrak{f}\rangle = \text{tr}_{\mathfrak{h}}\,\text{ad}\,\bar{K}^2 = \frac{1}{2}\text{tr}_{\mathfrak{g}}\,\text{ad}\,\bar{K}^2.$$

The divergence renormalizes the sigma-model coupling constant,

$$S_{\text{eff}} = -\int d^2x \left(\frac{1}{\kappa^2}\text{tr} - \frac{1}{8\pi}\ln\frac{\Lambda}{\mu}\text{tr}_{\text{adj}}\right)\bar{K}_a\bar{K}^a + \text{finite}. \quad (5.73)$$

Taking into account that

$$\text{tr}_{\text{adj}} XY = 2C_G \text{tr} XY, \quad (5.74)$$

where C_G is the dual Coxeter number[4] of the group G, we find that the coupling renormalized at scale μ is related to the bare coupling by

$$\frac{1}{\kappa^2(\mu)} = \frac{1}{\kappa^2(\Lambda)} - \frac{C_G}{4\pi}\ln\frac{\Lambda}{\mu}. \quad (5.75)$$

The running of sigma-model coupling with the energy scale is governed by the beta-function, which can be read off from the above calculation:

$$\beta(\kappa) = \frac{d\kappa(\mu)}{d\ln\mu} = -\frac{C_G}{8\pi}\kappa^3. \quad (5.76)$$

The one-loop beta-function is negative, which means that two-dimensional coset sigma-models are asymptotically free quantum field theories [20].

Exercise 9 *Calculate the beta-function of the WZW model (5.59)* [15].

The RG flow in two-dimensional sigma-models is often interpreted geometrically as shrinking of the target-space in the IR [21]. The coupling constant rescales the metric in the sigma-model action (5.2). Its inverse thus plays the role of an overall radius of the target space. The theory flows to stronger coupling in the IR, with the radius of the target space becoming smaller and smaller along the flow, Figure 5.6.

The RG flow will eventually stop and generate a dynamical mass scale through dimensional transmutation:[5]

$$M = \Lambda e^{-\frac{4\pi}{C_G\kappa^2}}. \quad (5.77)$$

This means that coset sigma-models are massive field theories, with particles transforming under some representation of the symmetry group G. The dynamics of these particles is non-perturbative at low energies, but is highly constrained by integrability, of course under the assumption that integrability is not broken by quantization.

---

[4] $C_{SU(N)} = N$, $C_{SO(N)} = N - 2$.
[5] In the presence of a theta-term or a WZ term this is not necessarily true. The flow may then end in a non-trivial CFT in the IR.

228 *Integrability in sigma-models*

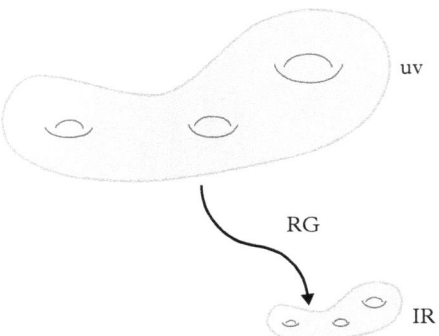

Figure 5.6 *The RG flow in a sigma-model.*

Table 5.1 Examples of classically integrable sigma-models, which can be either integrable or not at the quantum level.

QFT	Coset	Integrability
Principal chiral field	G	✓
$O(N)$ model	$SO(N)/SO(N-1)$	✓
$\mathbb{CP}^N$ model	$SU(N+1)/SU(N) \times U(1)$	✗
Superstring on $AdS_5 \times S^5$	$PSU(2,2|4)/SO(4,1) \times SO(5)$	✓
$O(3)$ model with $\theta \neq 0, \pi$	$SO(3)/SO(2)$	✗

Conservation laws are often affected by quantum anomalies, and integrable QFTs are no exception. Whether or not higher conserved charges are anomalous is a dynamical question, with a model-dependent answer. There are cases where integrability is preserved at the quantum level, and cases where integrability is broken by anomalies. Some representative examples are listed in Table 5.1.

The dynamical information of a massive QFT is contained in its S-matrix. The S-matrix of an integrable QFT is so constrained by symmetries and analyticity that it can be reconstructed from the kinematical requirements alone, with the help of a minimal amount of explicit perturbative data. We illustrate how such integrable bootstrap works, taking the $O(N)$ model as an example. The departure point for constructing the exact scattering theory for the $O(N)$ model is its solution at large-N, which can be obtained by elementary manipulations of the path integral described in Section 5.7.

5.7   *O(N)* model: large *N*

The $O(N)$ model is a quantum field theory of an N-component vector field **n** subject to the constraint $\mathbf{n}^2 = 1$. The constraint can be imposed by a Lagrange multiplier in the path integral:

$$Z = \int D\mathbf{n} D\sigma \, e^{\frac{i}{\kappa^2} \int d^2 x [(\partial \mathbf{n})^2 - \sigma(\mathbf{n}^2 - 1)]}. \tag{5.78}$$

Integration over σ yields the requisite delta-function, so $D\mathbf{n}$ is the linear, unconstrained measure in the field space. Integrating over $\mathbf{n}$ first we get an effective action for σ:

$$S_{\text{eff}} = \frac{1}{\kappa^2} \int d^2 x \sigma + \frac{iN}{2} \ln \det(-\partial^2 - \sigma). \tag{5.79}$$

The two terms are of the same order of magnitude in the 't Hooft limit:

$$N \to \infty, \qquad \lambda = \frac{\kappa^2 N}{2} - \text{fixed}. \tag{5.80}$$

In the 't Hooft limit the effective action is of an overall order N. The semiclassical approximation becomes exact at $N \to \infty$, and it suffices to solve the classical equations of motion that follow from (5.79). Denoting

$$\langle \sigma \rangle = m^2, \tag{5.81}$$

and taking the variation of (5.79), we get

$$\frac{1}{\lambda} = i \langle x | \frac{1}{-\partial^2 - m^2} | x \rangle \quad \Longleftrightarrow \quad \frac{1}{\lambda} = \int \frac{d^2 p}{(2\pi)^2} \frac{i}{p^2 - m^2}. \tag{5.82}$$

This is known as the gap equation[6], which after the Wick rotation, $p_0 \to ip_0$, becomes

$$\frac{1}{\lambda} = \int \frac{d^2 p}{(2\pi)^2} \frac{1}{p^2 + m^2} = \frac{1}{2\pi} \ln \frac{\Lambda}{m}. \tag{5.83}$$

The UV cut-off Λ is necessary for regulating the divergent momentum integral. The gap equation determines the mass scale of the theory:

$$m = \Lambda e^{-\frac{2\pi}{\lambda}}. \tag{5.84}$$

This is consistent with (5.77) at large N, when the difference between N and $C_{SO(N)} = N - 2$ is immaterial.

If we return to (5.78), the expectation value of the Lagrange multiplier $\langle \sigma \rangle = m^2$ gives mass to the field $\mathbf{n}$, which thus describes N massive particles in the vector representation of $SO(N)$. The $N - 1$ Goldstone modes of the naive perturbation theory rearrange

---

[6] The ground state of the system is of course homogeneous, but the gap equation also has space-time dependent soliton solutions that describe excited states; see [22] for an example.

themselves into N massive states at low energies. This rearrangement is obviously non-perturbative.

Exercise 10 *Calculate the $2 \to 2$ scattering amplitude in the $O(N)$ model to the first non-vanishing order in $1/N$ [3].*

5.8 O(N) model: exact solution

Integrability strongly constrains the scattering processes. The constraints follow from no more than simple kinematics. Consider, to begin with, the scattering of two identical particles $A(p_1^{\text{in}}) + A(p_2^{\text{in}}) \to A(p_1^{\text{out}}) + A(p_2^{\text{out}})$ in two dimensions, where p_i^{in}, p_i^{out} are incoming and outgoing spacial momenta. If the initial momenta are given, energy and momentum conservation constitute two equations for two final-state momenta:

$$p_1^{\text{out}} + p_2^{\text{out}} = p_1^{\text{in}} + p_2^{\text{in}}$$
$$E(p_1^{\text{out}}) + E(p_2^{\text{out}}) = E(p_1^{\text{in}}) + E(p_2^{\text{in}}).$$

Two equations for two unknowns have a discrete set of solutions—two-body scattering has no phase space in two dimensions.

It is easy to convince oneself that the conservation laws admit but two obvious solutions: $p_1^{\text{out}} = p_1^{\text{in}}, p_2^{\text{out}} = p_2^{\text{in}}$ (transmission) and $p_1^{\text{out}} = p_2^{\text{in}}, p_2^{\text{out}} = p_1^{\text{in}}$ (reflection). This is true in any translationally invariant theory in two dimensions, but imagine now that there is an infinite number of additional conservation laws. In any interaction process, even if it involves more than two particles, exchange of momenta would still be possible, but any other outcome of the scattering process would be kinematically forbidden, because conservation laws constitute a hugely overconstrained system of equations. In integrable theories, which feature an infinite number of conservation laws, particle production is forbidden, scattering is diagonal in particle number, and the final-state momenta of any $n \to n$ process constitute a permutation of the initial-state momenta: $p_i^{\text{out}} = p_{\sigma(i)}^{\text{in}}$. Any permutation can be written as a product of transpositions, likewise $n \to n$ scattering in an integrable theory factorizes into a sequence of elementary $2 \to 2$ processes. The dynamics of an integrable QFT is thus encoded in the two-body scattering S-matrix. This is a huge simplification that puts integrable models in between trivial (free) theories and generic QFTs, where the intricate phase space of intermediate states makes the analytic structure of the S-matrix enormously complicated.

Taking into account that the $O(N)$ model is integrable, and drawing intuition from the large-N approximation, we conclude that the model is characterized by:

- N massive particles transforming in the vector representation of $SO(N)$
- no particle production
- factorized scattering.

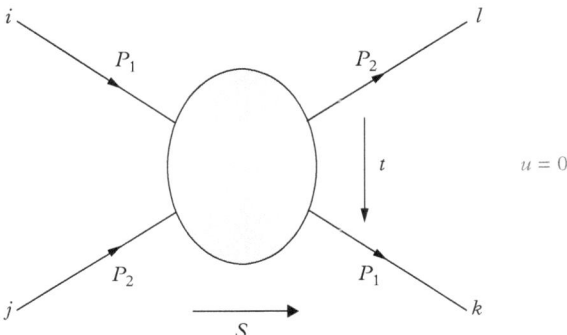

Figure 5.7 *The 2 → 2 scattering amplitude.*

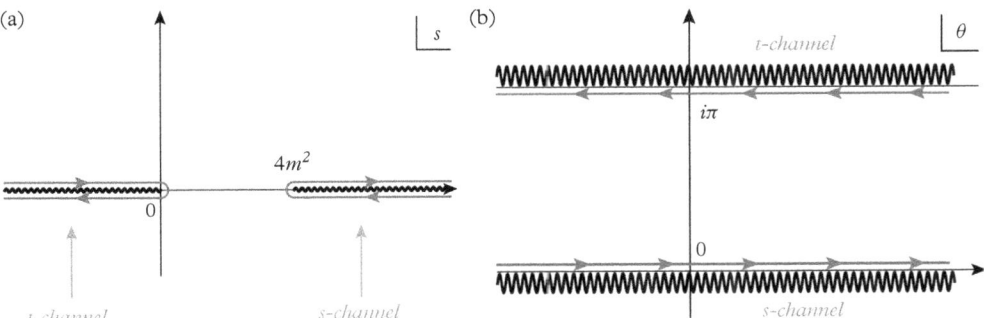

Figure 5.8 *Analytic structure of the S-matrix: (a) in the s-plane; (b) in the rapidity plane.*

These basic data are sufficient to completely reconstruct the S-matrix of the theory. Our exposition of this bootstrap solution of the model closely follows the original derivation [3].

Consider 2 → 2 scattering, illustrated in Figure 5.7. The Mandelstam invariant $u = 0$ in two dimensions, while s and t are related by

$$s + t = 4m^2. \tag{5.85}$$

Hence the 2 → 2 S-matrix is a function of only one kinematic variable: $S_{ij}^{kl}(s)$. Moreover, the S-matrix is analytic in s almost everywhere on the complex plane. All the singularities of the S-matrix are associated with intermediate virtual states going on-shell. And here integrability brings in enormous simplifications. The absence of particle production implies that the only allowed singularities are two-particle cuts in the s- and t-channels, Figure 5.8(a). The paucity of analytic functions on the complex plane with two cuts is the ultimate reason for the success of integrability bootstrap.

Two-dimensional relativistic kinematics drastically simplifies in the rapidity variables:

$$p_{1,2} = m \sinh \theta_{1,2}$$
$$E_{1,2} = m \cosh \theta_{1,2}. \tag{5.86}$$

The centre-of-mass energy squared is then

$$s = 4m^2 \cosh^2 \frac{\theta_1 - \theta_2}{2}. \tag{5.87}$$

The S-matrix is thus a function of the rapidity difference between the two colliding particles, $S = S(\theta_1 - \theta_2)$. The entire s-plane maps onto the strip of the theta-plane between the lines $\text{Im}\,\theta = 0$ and $\text{Im}\,\theta = \pi$ (called the physical strip), as shown in Figure 5.8.

In spite of what the name might suggest, most of the physical strip corresponds to unphysical kinematics. The S-matrix is a probability amplitude of a physical scattering process only for real positive $s > 4m^2$, cf. (5.87). The rest is analytic continuation. More precisely, the scattering amplitude is defined for s real positive $+i\epsilon$, according to the usual Feynman prescription. The upper side of the s-channel cut maps to the positive real semi-axis on the θ-plane (Figure 5.8). Analytic continuation to the entire s-plane leads to a set of S-matrix axioms [3, 23]:

- Unitarity (+ real analyticity):

$$S(-\theta)S(\theta) = 1 \tag{5.88}$$

- Crossing symmetry:[7]

$$S_{ij}^{kl}(\theta) = S_{il}^{kj}(i\pi - \theta) \tag{5.89}$$

- Yang–Baxter equation (YBE), Figure 5.9:

$$S_{12}(\theta_1 - \theta_2)S_{13}(\theta_1 - \theta_3)S_{23}(\theta_2 - \theta_3) = S_{23}(\theta_2 - \theta_3)S_{13}(\theta_1 - \theta_3)S_{12}(\theta_1 - \theta_2) \tag{5.90}$$

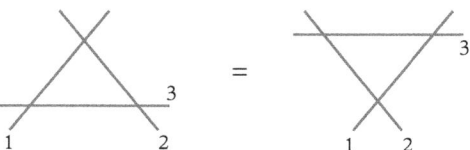

Figure 5.9 *The Yang–Baxter equation.*

---

[7] Physical particles in the $O(N)$ model are their own anti-particles. More generally, crossing involves charge conjugation.

The YBE is a consistency condition on the factorization of the $3 \to 3$ scattering into elemental $2 \to 2$ processes. The outcome should not depend on the order in which the three particles collide pairwise. The labels on S_{ab} indicate that its acts on the $SO(N)$ indices of particles a and b.

Exercise 11 *Write down the YBE in the explicit index notations.*

Symmetries and the YBE usually produce a skeleton of the S-matrix, a fixed matrix structure without the detailed dependence on the rapidity. Indeed, if $S_{ij}^{kl}(\theta)$ solves the YBE, then $f(\theta) S_{ij}^{kl}(\lambda\theta)$ does so also. To bootstrap the remaining ambiguity one needs to invoke crossing symmetry, unitarity, and analyticity. Here we show how this program works for the $O(N)$ model.

The general form of the S-matrix consistent with the $O(N)$ symmetry is

$$S_{ij}^{kl}(\theta) = \sigma(\theta) \left(\delta_i^k \delta_j^l + \frac{1}{p(\theta)} \delta_i^l \delta_j^k + \frac{1}{k(\theta)} \delta_{ij} \delta^{kl} \right). \tag{5.91}$$

Firstly, we impose the YBE. The overall 'dressing' factor $\sigma(\theta)$ then drops out. It is in principle straightforward to determine what the constraints on the two remaining functions are, albeit the calculations are rather lengthy. The algebra can be facilitated by introducing permutation and trace operators, that act on the tensor product of two $SO(N)$ vectors as

$$P a \otimes b = b \otimes a, \qquad K a \otimes b = (a \cdot b) \mathbf{1}. \tag{5.92}$$

In these notations,

$$S = \sigma(\theta) \left(1 + \frac{1}{p(\theta)} P + \frac{1}{k(\theta)} K \right). \tag{5.93}$$

The permutation and trace operators satisfy the following braiding relations:

$$\begin{aligned} P_{ab} P_{bc} &= P_{ac} P_{ab}, & K_{ab}^2 &= N K_{ab}, & K_{ab} P_{ab} &= P_{ab} K_{ab} = K_{ab}, \\ K_{ab} P_{cb} &= P_{cb} K_{ac}, & K_{ab} K_{cb} &= P_{bc} P_{ab} K_{bc}, & P_{ab} P_{bc} K_{ab} &= P_{ac} K_{ab}. \end{aligned} \tag{5.94}$$

Algebraic manipulations based on (5.94)—the details of which we omit[8]—reduce the YBE to a set of three independent equations on functions $k(\theta)$ and $p(\theta)$:

$$\begin{aligned} & p(u) - p(v) = p(u-v) \\ & k(u) - k(v) = -p(u-v) \\ & k(u)\left(k(u-v) + k(v) + p(u) + N\right) = k(v) + k(u-v) - p(u). \end{aligned} \tag{5.95}$$

---

[8] When dealing with the YBE, one really appreciates the power of *Mathematica*.

234   *Integrability in sigma-models*

The first two equations simply tell us that $k(\theta)$ and $p(\theta)$ are linear functions,

$$p(\theta) = -\lambda\theta, \qquad k(\theta) = \lambda\theta - \frac{1}{2\Delta}, \tag{5.96}$$

where λ and Δ are arbitrary constants. The last equation is then also satisfied, provided

$$\Delta = \frac{1}{N-2}. \tag{5.97}$$

As anticipated, the YBE fixes the functional structure of the S-matrix up to rescaling $\theta \to \lambda\theta$, and up to an overall dressing factor. The permutation and trace terms in the S-matrix map to one another under the crossing transformation, which interchanges the $SO(N)$ indices j and l in (5.91). The crossing-symmetry condition (5.89) then requires that

$$p(i\pi - \theta) = k(\theta), \tag{5.98}$$

which fixes

$$\lambda = \frac{1}{2\pi i \Delta}. \tag{5.99}$$

This determines the S-matrix up to the dressing factor:

$$S_{ij}^{kl}(\theta) = \sigma(\theta)\left(\delta_i^k \delta_j^l - \frac{2\pi i \Delta}{\theta}\delta_i^l \delta_j^k + \frac{2\pi i \Delta}{\theta - i\pi}\delta_{ij}\delta^{kl}\right). \tag{5.100}$$

The dressing factor is constrained by the unitarity and crossing conditions (5.88) and (5.89). Substitution of (5.93) into (5.88), and the use of the identities $P^2 = 1$, $K^2 = NK$, $PK = K = KP$, gives

$$S(-\theta)S(\theta) = \sigma(-\theta)\sigma(\theta)\left[1 + \frac{1}{p(-\theta)p(\theta)} + \left(\frac{1}{p(-\theta)} + \frac{1}{p(\theta)}\right)P\right.$$
$$\left. + \left(\frac{1}{k(-\theta)} + \frac{1}{k(\theta)} + \frac{1}{k(-\theta)p(\theta)} + \frac{1}{k(-\theta)p(\theta)} + \frac{N}{k(-\theta)k(\theta)}\right)K\right].$$

The coefficients of P and K in this equation identically vanish, and the unitarity condition boils down to

$$\sigma(-\theta)\sigma(\theta) = \frac{\theta^2}{\theta^2 + 4\pi^2\Delta^2}. \tag{5.101}$$

The crossing symmetry simply requires that

$$\sigma(i\pi - \theta) = \sigma(\theta). \tag{5.102}$$

We need to solve these two functional equations on $\sigma(\theta)$.

The dressing factor $\sigma(\theta)$ by itself is not an amplitude of any physical process. It is much more instructive to reformulate the crossing/unitarity equations in terms of a physical S-matrix element for some particular scattering channel. General features of integrability bootstrap, which are important, for example, in the context of thermodynamic Bethe ansatz (TBA), become more transparent if the equations are written in terms of a physical amplitude.

To construct these we concentrate on a $U(1)$ subgroup of $SO(N)$ generated by rotations in the (12) plane. An $SO(N)$ vector has two components with $U(1)$ charges $+1$ and -1 and $N-2$ neutral components. In terms of the fields in the path integral (5.78), the charge $+1$ state corresponds to

$$\phi = n_1 + in_2.$$

Because of the charge conservation and the absence of particle production the $\phi\phi \to \phi\phi$ amplitude is purely elastic, and the S-matrix element

$$s(\theta) \equiv S_{\phi\phi \to \phi\phi}(\theta) \tag{5.103}$$

is a pure phase,

$$s(\theta) = e^{i\Phi(\theta)}. \tag{5.104}$$

From (5.100) we find

$$s(\theta) = \sigma(\theta) \frac{\theta - 2\pi i \Delta}{\theta}. \tag{5.105}$$

The crossing/unitarity equations for $s(\theta)$ are

$$s(i\pi - \theta) = s(\theta) \frac{\theta (\theta + 2\pi i \Delta - i\pi)}{(\theta - 2\pi i \Delta)(\theta - i\pi)}$$

$$s(-\theta)s(\theta) = 1. \tag{5.106}$$

From these we infer that

$$s\left(\theta + \frac{i\pi}{2}\right) s\left(\theta - \frac{i\pi}{2}\right) = \frac{\left(\theta - \frac{i\pi}{2}\right)\left(\theta + \frac{i\pi}{2} - 2\pi i \Delta\right)}{\left(\theta + \frac{i\pi}{2}\right)\left(\theta - \frac{i\pi}{2} + 2\pi i \Delta\right)}. \tag{5.107}$$

Written in terms of the phase, the equation becomes

$$\Phi\left(\theta+\frac{i\pi}{2}\right)+\Phi\left(\theta-\frac{i\pi}{2}\right)=2\arctan\frac{2\theta}{\pi}-2\arctan\frac{2\theta}{\pi(1-4\Delta)}+\text{p.a.} \tag{5.108}$$

The last term (to be specified shortly) reflects an ambiguity in choosing the branch of the logarithm when reformulating the problem in terms of the scattering phase, which is defined up to shifts by integer multiples of 2π. The branch of the logarithm can be chosen differently for different θ, and we can parameterize the phase ambiguity by writing

$$\text{p.a.}=2\pi\sum_k(\vartheta(\theta-i\alpha_k)+\vartheta(\theta+i\alpha_k)), \tag{5.109}$$

where $\vartheta(x)$ is the step function, and $\{\alpha_k\}$ is a potentially arbitrary set of parameters.

Another function, that plays an important role in all types of Bethe ansatz equations, is the derivative of the scattering phase,

$$K(\theta)=\frac{d\Phi(\theta)}{d\theta}. \tag{5.110}$$

For this function we get a difference equation

$$K\left(\theta+\frac{i\pi}{2}\right)+K\left(\theta-\frac{i\pi}{2}\right)=\frac{\pi}{\theta^2+\frac{\pi^2}{4}}-\frac{\pi(1-4\Delta)}{\theta^2+\frac{\pi^2}{4}(1-4\Delta)^2}$$
$$+2\pi\sum_k(\delta(\theta-i\alpha_k)+\delta(\theta+i\alpha_k)), \tag{5.111}$$

which can be also written as

$$2\cos\left(\frac{\pi}{2}\frac{d}{d\theta}\right)K(\theta)=\frac{\pi}{\theta^2+\frac{\pi^2}{4}}-\frac{\pi(1-4\Delta)}{\theta^2+\frac{\pi^2}{4}(1-4\Delta)^2}$$
$$+2\pi\sum_k(\delta(\theta-i\alpha_k)+\delta(\theta+i\alpha_k)). \tag{5.112}$$

The equation becomes algebraic in the Fourier space,

$$\cosh\frac{\pi\omega}{2}K(\omega)=\pi e^{-\frac{\pi}{2}|\omega|}-\pi e^{-\frac{\pi}{2}(1-4\Delta)|\omega|}+2\pi\sum_k\cosh\alpha_k\omega. \tag{5.113}$$

Fourier transforming back to the θ-representation, we get

$$2\pi K(\theta) = \psi\left(1 + \frac{i\theta}{2\pi}\right) + \psi\left(1 - \frac{i\theta}{2\pi}\right) - \psi\left(\frac{1}{2} + \frac{i\theta}{2\pi}\right) - \psi\left(\frac{1}{2} - \frac{i\theta}{2\pi}\right)$$
$$- \psi\left(1 - \Delta + \frac{i\theta}{2\pi}\right) - \psi\left(1 - \Delta - \frac{i\theta}{2\pi}\right)$$
$$+ \psi\left(\frac{1}{2} - \Delta + \frac{i\theta}{2\pi}\right) + \psi\left(\frac{1}{2} - \Delta - \frac{i\theta}{2\pi}\right) + \sum_k \frac{4\pi \cos\alpha_k \cosh\theta}{\sinh^2\theta + \cos^2\alpha_k}. \quad (5.114)$$

Integrating and exponentiating the result, we find

$$s(\theta) = S_{\text{CDD}}(\theta) \frac{\Gamma\left(1 + \frac{i\theta}{2\pi}\right)\Gamma\left(\frac{1}{2} - \frac{i\theta}{2\pi}\right)\Gamma\left(1 - \Delta - \frac{i\theta}{2\pi}\right)\Gamma\left(\frac{1}{2} - \Delta + \frac{i\theta}{2\pi}\right)}{\Gamma\left(1 - \frac{i\theta}{2\pi}\right)\Gamma\left(\frac{1}{2} + \frac{i\theta}{2\pi}\right)\Gamma\left(1 - \Delta + \frac{i\theta}{2\pi}\right)\Gamma\left(\frac{1}{2} - \Delta - \frac{i\theta}{2\pi}\right)}, \quad (5.115)$$

where

$$S_{\text{CDD}}(\theta) = \prod_k \frac{\sinh\theta - i\cos\alpha_k}{\sinh\theta + i\cos\alpha_k} \quad (5.116)$$

is known as the CDD (Castillejo–Dalitz–Dyson) factor.

The CDD factor parameterizes fundamental ambiguity in the crossing + unitarity conditions, which do not fix the S-matrix completely. However, each CDD factor introduces extra poles in the rapidity plane. Any pole on the physical strip is associated with a bound state, and no new poles can be added by hand without changing the particle content of the theory. Usually this means that the minimal solution of the crossing equation is the physical one, but sometimes the minimal solution has spurious poles that have to be cancelled by CDD factors. The $O(N)$ model is a good example of this phenomenon.

The minimal solution without CDD factors would have a pole on the physical strip, at $\theta = i\pi(1 - 2\Delta)$. The pole has to be cancelled by a single CDD factor with

$$\alpha = \frac{\pi}{2}(1 - 4\Delta).$$

After which (5.115) becomes

$$s(\theta) = -\frac{\Gamma\left(1 + \frac{i\theta}{2\pi}\right)\Gamma\left(\frac{1}{2} - \frac{i\theta}{2\pi}\right)\Gamma\left(\Delta - \frac{i\theta}{2\pi}\right)\Gamma\left(\frac{1}{2} + \Delta + \frac{i\theta}{2\pi}\right)}{\Gamma\left(1 - \frac{i\theta}{2\pi}\right)\Gamma\left(\frac{1}{2} + \frac{i\theta}{2\pi}\right)\Gamma\left(\Delta + \frac{i\theta}{2\pi}\right)\Gamma\left(\frac{1}{2} + \Delta - \frac{i\theta}{2\pi}\right)}. \quad (5.117)$$

This is the exact scattering amplitude of the ϕ quanta.

The dressing factor can be extracted from (5.105):

$$\sigma(\theta) = Q(\theta)Q(i\pi - \theta), \qquad Q(\theta) = \frac{\Gamma\left(\frac{1}{2} - \frac{i\theta}{2\pi}\right)\Gamma\left(\Delta - \frac{i\theta}{2\pi}\right)}{\Gamma\left(-\frac{i\theta}{2\pi}\right)\Gamma\left(\frac{1}{2} + \Delta - \frac{i\theta}{2\pi}\right)}. \tag{5.118}$$

Crossing symmetry is manifest in (5.118), while unitarity is manifest in (5.117).

Exercise 12 Expand the exact S-matrix of the $O(N)$ model to the first non-vanishing order in $1/N$, and compare with the result of the explicit calculation from Exercise 10.

5.9 Crash course in superalgebras

String theory requires supersymmetry for internal consistency. If a sigma-model is to describe critical ten-dimensional string it has to be supersymmetric. Supersymmetric counterparts of the coset sigma-models that we have studied so far are ubiquitous in holographic duality, and many of them are integrable, just like in the purely bosonic case. Fields in these models take values in a supermanifold, and are invariant under the action of a supergroup. Any detailed introduction to supergroups and superalgebras (which can be found in [24]) would occupy too large a space-time volume, but the most common examples can be introduced very easily, as algebras of matrices of a certain type, just like most common examples of ordinary Lie algebras are algebras of matrices constrained to be traceless, anti-symmetric, and so on.

In addition to bosonic generators T_M, a superalgebra contains supercharges Q_α. An element of a superalgebra is a linear combination,

$$X = X^M T_M + \theta^\alpha Q_\alpha,$$

where θ^α are Grassmann-odd variables. The commutator on a superalgebra must satisfy the usual requirements of linearity, anti-symmetry, and obeying the Jacobi identity. A group element is defined by an exponential map $g = e^X$, which is actually polynomial in supercharges because a sufficiently large power of Grassmann variables vanishes identically.

In any finite-dimensional representation the generators T_M and Q_α are ordinary numeric matrices. An element of the Lie superalgebra X then forms a supermatrix— a matrix some of whose elements are Grassmann-odd numbers.

Complex supermatrices can be defined as linear transformations of $\mathbb{C}^{(n|m)}$, a space of vectors whose first n components are Grassmann-even and the last m of which are Grassmann-odd:

$$\left(\begin{array}{c|c} B & \Theta \\ \hline \Theta' & B' \end{array}\right) \left(\begin{array}{c} b \\ \hline f \end{array}\right).$$

To preserve the Grassmann-parity structure, the diagonal blocks of a supermatrix must be even and the off-diagonal blocks must be odd.

Supermatrices can be multiplied and commuted, with the commutator obeying all the usual axioms of the Lie bracket. The algebra of all $(n|m) \times (n|m)$ supermatrices is denoted by $\mathfrak{gl}(n|m)$. This superalgebra is not simple. An obvious way to get a simple algebra is to impose the traceless condition, and here the first deviation from the usual Lie algebras occurs.

Consider this simple example:

$$\left[\begin{pmatrix} 0 & \theta_1 \\ \psi_1 & 0 \end{pmatrix}, \begin{pmatrix} 0 & \theta_2 \\ \psi_2 & 0 \end{pmatrix}\right] = \begin{pmatrix} \theta_1\psi_2 - \theta_2\psi_1 & 0 \\ 0 & \psi_1\theta_2 - \psi_2\theta_1 \end{pmatrix}.$$

When θ_i, ψ_i are ordinary numbers the sum of diagonal matrix elements on the right-hand side vanishes; the trace of a commutator is always zero. But if θ's and ψ's are Grassmann numbers, the diagonal matrix elements differ by a sign and the trace does not vanish. Since the trace of a commutator of supermatrices is not zero, we conclude that traceless supermatrices do not form a subalgebra in $\mathfrak{gl}(1|1)$.

This can be fixed in the following way. In the above example the difference, rather then the sum, of the diagonal matrix elements is zero. This is know as supertrace, in general defined as

$$\mathrm{Str}\left(\begin{array}{c|c} B & \Theta \\ \hline \Theta' & B' \end{array}\right) = \mathrm{tr}\, B - \mathrm{tr}\, B'. \tag{5.119}$$

Exercise 13 *Show that for any supermatrices,*

$$\mathrm{Str}[M_1, M_2] = 0. \tag{5.120}$$

Because of the last property, restriction to *supertraceless* supermatrices gives a closed algebra, known as the special linear superalgebra $\mathfrak{sl}(n|m)$. The $\mathfrak{sl}(n|m)$ superalgebras are simple if $m \neq n$, but the case of $m = n$ is special. Since

$$\mathrm{Str}\left(\begin{array}{c|c} 1 & 0 \\ \hline 0 & 1 \end{array}\right) = n - m$$

for $m = n$ the trace condition does not eliminate the unit matrix! The unit matrix commutes with anything and thus represents a central element of the algebra. The factor-algebra, where this central element is set to zero is denoted by

$$\mathfrak{psl}(n|n) = \mathfrak{sl}(n|n)/C. \tag{5.121}$$

In any irreducible representation the central element takes a constant numeric value, and $\mathfrak{psl}(n|n)$ is defined by keeping only those representations of $\mathfrak{sl}(n|n)$ in which the central

240 *Integrability in sigma-models*

element is equal to zero. It is interesting to notice that the defining $(n|n)$-dimensional representation of $\mathfrak{sl}(n|n)$ is not a representation of $\mathfrak{psl}(n|n)$.

The $\mathfrak{psl}(n|n)$ superalgebras are remarkable in many respects. Being finite-dimensional they admit a central extension (never happens for ordinary Lie algebras). They also have zero dual Coxeter number, where the dual Coxeter number is defined as in the bosonic case through equation (5.74). To see this, consider a diagonal supermatrix $K = \mathrm{diag}(a_1 \ldots a_n | b_1 \ldots b_n)$, which represents an element of $\mathfrak{psl}(n|n)$ if the sums of a_i's and b_I's are separately zero. As for its action in the adjoint representation, the eigenvalues of $\mathrm{ad}\,K$ are $a_i - a_j$, $b_I - b_{\tilde{J}}$, $a_i - b_{\tilde{J}}$, and $b_I - a_j$ for all unordered pairs of indices ij, $I\tilde{J}$, $i\tilde{J}$, and Ij. Therefore,

$$\mathrm{Str}_{\mathrm{adj}} K^2 = \sum_{ij}(a_i - a_j)^2 + \sum_{I\tilde{J}}(b_I - b_{\tilde{J}})^2 - \sum_{i\tilde{J}}(a_i - b_{\tilde{J}})^2 - \sum_{Ij}(b_I - a_j)^2 = 0.$$

The unitary superalgebras $(\mathfrak{p})\mathfrak{su}(p,q|r,s)$ are real forms of $(\mathfrak{p})\mathfrak{sl}(n|m)$ singled out by the Hermiticity condition

$$M^\dagger = -B^{-1}MB, \qquad B = \mathrm{diag}\left(+^p - ^q | +^r - ^s\right). \tag{5.122}$$

Another important class of superalgebras is the supersymmetric counterpart of $\mathfrak{so}(n)$. Ordinary orthogonal algebra is singled out by the anti-symmetry condition imposed on elements of $\mathfrak{gl}(n)$. For supermatrices this does not work because transposition is not compatible with commutators. In general, $(M_1 M_2)^t \neq M_2^t M_1^t$ for supermatrices, since their off-diagonal Grassmann entries do not commute but anti-commute.

The supersymmetric generalization of transposition, called supertransposition, is defined as

$$\begin{pmatrix} A & \Theta \\ \Psi & B \end{pmatrix}^{st} = \begin{pmatrix} A^t & -\Psi^t \\ \Theta^t & B^t \end{pmatrix}. \tag{5.123}$$

It follows that

$$(M_1 M_2)^{st} = M_2^{st} M_1^{st}, \tag{5.124}$$

and consequently supertransposition preserves commutators up to a sign.

An orthosymplectic algebra $\mathfrak{osp}(n|2m)$ is defined as a subset of $\mathfrak{gl}(n|2m)$ singled out by the condition

$$M^{st} = -H^{-1}MH, \tag{5.125}$$

where

$$H = \begin{pmatrix} 1_{n\times n} & 0 \\ 0 & \mathcal{J} \end{pmatrix}, \tag{5.126}$$

and $\mathcal{J}$ is the $2m \times 2m$ symplectic matrix,

$$\mathcal{J} = \begin{pmatrix} 0 & 1_{m \times m} \\ -1_{m \times m} & 0 \end{pmatrix}. \tag{5.127}$$

Explicitly,

$$\mathfrak{osp}(n|2m) = \left\{ \begin{pmatrix} A & \Theta \\ \Psi & B \end{pmatrix} \in \mathfrak{gl}(n|2m) \;\middle|\; A^t = -A, B^t = \mathcal{J}B\mathcal{J}, \Theta^t = \mathcal{J}\Psi^t \right\}. \tag{5.128}$$

The bosonic subalgebra of $\mathfrak{osp}(n|2m)$ is $\mathfrak{so}(n) \oplus \mathfrak{sp}(2m)$.

Exercise 14 *Check that superalgebras $\mathfrak{osp}(2n+2|2n)$ have zero dual Coxeter number.*

5.10 Supercoset sigma-models

A natural supersymmetric generalization of a symmetric space is based on an automorphism $\Omega : \mathfrak{g} \to \mathfrak{g}$ that squares to the fermion parity,

$$\Omega^2 = (-1)^F. \tag{5.129}$$

Such automorphism has order four: $\Omega^4 = \mathrm{id}$, and the superalgebra has a $\mathbb{Z}_4$ grading,

$$\mathfrak{g} = \mathfrak{h}_0 \oplus \mathfrak{h}_1 \oplus \mathfrak{h}_2 \oplus \mathfrak{h}_3, \tag{5.130}$$

where

$$\Omega(\mathfrak{h}_n) = i^n \mathfrak{h}_n. \tag{5.131}$$

The $\mathbb{Z}_4$ decomposition is consistent with the Grassmann parity: $\mathfrak{h}_0 \oplus \mathfrak{h}_2$ form the bosonic subalgebra of $\mathfrak{g}$ and all the supercharges belong to $\mathfrak{h}_1 \oplus \mathfrak{h}_3$.

If H_0 is the subgroup of G whose Lie algebra $\mathfrak{h}_0$ is $\mathbb{Z}_4$-invariant, the coset G/H_0 is called semi-symmetric superspace [25]. Anti-de-Sitter geometries in various dimensions arise as bosonic sections of $\mathbb{Z}_4$ cosets, which play an important role in the holographic duality. The key example is string theory on $AdS_5 \times S^5$ whose sigma-model [4] is a semi-symmetric coset [26]. Just as for ordinary symmetric-space sigma-models, integrability of semi-symmetric cosets can be established at the algebraic level, using constraints imposed on the equations of motion by the $\mathbb{Z}_4$-symmetry.

The current (5.10) decomposes into four components according to their $\mathbb{Z}_4$ grading,

$$j = j_0 + j_1 + j_2 + j_3. \tag{5.132}$$

242 *Integrability in sigma-models*

The current components j_0, j_2 expand in bosonic generators of the superalgebra, while the components j_1, j_3 are fermionic—they are linear combinations of supercharges with Grassmann-odd coefficients.

The action of the $\mathbb{Z}_4$ coset sigma-model is

$$S = \int d^2x \, \text{Str}\left(\sqrt{-h}h^{ab}j_{2a}j_{2b} + \varepsilon^{ab}j_{1a}j_{3b}\right). \tag{5.133}$$

The first term is the usual sigma-model Lagrangian, while the second term descends from the WZ action, which for supercosets is not topological and can be written in a manifestly 2d form.

Exercise 15 *Derive the equations of motion for a $\mathbb{Z}_4$ coset and check that the combined system of equations of motion and the flatness condition of the left-invariant current admits zero-curvature representation with the Lax connection* [5]

$$L = j_0 + \frac{z^2+1}{z^2-1}j_2 - \frac{2z}{z^2-1}*j_2 + \sqrt{\frac{z+1}{z-1}}j_1 + \sqrt{\frac{z-1}{z+1}}j_3. \tag{5.134}$$

Therefore the model is completely integrable.

The action (5.133), taken at face value, is quadratic in derivatives, which is fine for bosonic degrees of freedom, but a fermion Lagrangian is normally of Dirac type and is expected to be linear in derivatives. It is instructive to see how this apparent contradiction is resolved in $\mathbb{Z}_4$ cosets.

To get some intuition about the structure of the sigma-model action we can take the coset representative in the standard exponential form,

$$g = e^X, \quad X = X_1 + X_2 + X_3 = \theta_1^\alpha Q_{1\alpha} + X^M T_M + \theta_3^{\dot\alpha} Q_{3\dot\alpha}, \tag{5.135}$$

and expand the currents in X to the linear order. Then $j_{na} \simeq \partial_a X_n$. The sigma-model term becomes $(\partial X^M)^2$, while the WZ term gives $\varepsilon^{ab}\partial_a X_1 \partial_b X_3$, which looks like a second-order kinetic term for fermions, but in fact is a total derivative that integrates to zero,

$$\varepsilon^{ab}\partial_a X_1 \partial_b X_3 = \partial_a\left(\varepsilon^{ab} X_1 \partial_b X_3\right).$$

The quadratic term for fermions appears only at the next order in the expansion.

The cubic part of the action, one can show, contains[9]

$$\mathcal{L}_F^{(3)} = \text{Str}\left(\Pi_-^{ab} X_1 [\partial_a X_2, \partial_b X_1] + \Pi_+^{ab} X_3 [\partial_a X_2, \partial_b X_3]\right)$$
$$= \Pi_-^{ab} \theta_{1\alpha} f_{M\beta}^\alpha \partial_a X^M \partial_b \theta_1^\beta + \Pi_+^{ab} \theta_{3\dot\alpha} f_{M\dot\beta}^{\dot\alpha} \partial_a X^M \partial_b \theta_3^{\dot\beta}, \quad (5.136)$$

where f_{AB}^C are structure constants of the superalgebra $\mathfrak{g}$, and $\Pi_\pm$ are the world-sheet chirality projectors,

$$\Pi_\pm^{ab} = \eta^{ab} \pm \varepsilon^{ab}. \quad (5.137)$$

This action is exactly the same as the quadratic part of the Green–Schwarz (GS) action for the superstring [27], with structure constants $f_{M\alpha}^\beta$ replacing the Dirac matrices.

The GS action is notoriously difficult to quantize covariantly, one manifestation of which is absence of kinetic terms for fermions. Those can be induced if bosonic currents have a non-zero expectation value: $\langle \partial X^M \rangle \neq 0$. The most familiar instance when this happens is the light-cone gauge:

$$X^+ = \tau. \quad (5.138)$$

Taking into account that

$$\Pi_\pm^{0b} \partial_b = \partial_\pm, \quad (5.139)$$

where $\partial_\pm$ are the world-sheet light-cone derivatives, we see that the GS action in the light-cone gauge describes the standard left- and right-moving Majorana fermions,

$$\mathcal{L}_{F,\text{lc}}^{(3)} = \theta_1 f_+ \partial_- \theta_1 + \theta_3 f_+ \partial_+ \theta_3. \quad (5.140)$$

Exercise 16 *Consider a slightly more general setup. As in Section 5.6, take the coset representative in the form (5.61) where $\bar{g}$ is the background field that belongs to the bosonic subgroup of G and satisfies the equations of motion, and expand the action to the quadratic order in fluctuations. Reproduce the trilinear terms in the action (5.136) from this more general result. Show that the light-cone gauge-fixed action in addition to (5.140) contains a mass term that couples θ_1 and θ_3.*

Exercise 17 *Use the results from Exercise 16 to compute the one-loop beta-function for generic semi-symmetric coset [28]. Show that the beta-function vanishes as soon as the superalgebra $\mathfrak{g}$ has zero dual Coxeter number ($\text{Str}_{\text{adj}} K^2 = 0 \ \forall \ K \in \mathfrak{h}_2$).*

---

[9] To simplify the notations we consider a flat Minkowski metric on the world-sheet from now on.

Semi-symmetric superspaces are all classified [25]. As we have seen each of them gives rise to an integrable GS-type sigma-model. Not all of these sigma-models are consistent as string backgrounds. There are a number of additional constraints that need to be satisfied. The sigma-model must be conformally invariant and must have zero beta-function, to begin with. The one-loop beta-function of semi-symmetric cosets is proportional to the supertrace of the background current in the adjoint, and only those cosets whose dual Coxeter number vanishes have a chance to be consistent as string sigma-models. Two infinite series of superalgebras with vanishing adjoint supertrace were identified in Section 5.9: $\mathfrak{psl}(n|n)$ and $\mathfrak{osp}(2n+2|2n)$.

To define a critical superstring in ten dimensions a sigma-model should also have central charge 26, which is more or less equivalent to the condition that the physical degrees of freedom in the light-cone gauge are $8_b + 8_f$ transverse fluctuation modes of the string. We are not going to present a comprehensive analysis of all possible cases, but restrict exposition to two representative examples.

The $\mathfrak{psl}(m|m)$ superalgebras admit a number of $\mathbb{Z}_4$ automorphisms. One of them, defined for even $m = 2n$, is a combination of supertransposition (5.123) and a conjugation:

$$\Omega = -st \circ \text{Ad}\,\text{diag}(\mathcal{J}, \mathcal{J}), \qquad (5.141)$$

where $\mathcal{J}$ is the symplectic matrix (5.127). The invariant subalgebra of Ω is

$$\mathfrak{h}_0 = \{\text{diag}(A, B) | \, A = \mathcal{J} A^t \mathcal{J}, B = \mathcal{J} B^t \mathcal{J}\}, \qquad (5.142)$$

namely,

$$\mathfrak{h}_0 = \mathfrak{sp}(2n) \oplus \mathfrak{sp}(2n). \qquad (5.143)$$

The bosonic dimension of the coset (the number of generators in $\mathfrak{h}_2$) is $2[(4n^2 - 1) - (2n^2 + n)] = 4n^2 - 2n - 2$. For $n = 2$, this equals 10.

Choosing the real form to be $\mathfrak{psu}(2,2|4)$, and taking into account that $SU(4) = Spin(6)$, $SU(2,2) = Spin(4,2)$, $USp(4) = Spin(5)$ and $USp(2,2) = Spin(4,1)$, we get the coset $PSU(2,2|4)/Spin(4,1) \times Spin(5)$ whose bosonic section is

$$Spin(4,2)/Spin(4,1) \times Spin(6)/Spin(5) = AdS_5 \times S^5.$$

The coset therefore supersymmetrizes $AdS_5 \times S^5$, the background that plays so prominent a rôle in the AdS/CFT correspondence.

The off-diagonal blocks of a $\mathfrak{psu}(2,2|4)$ element supply $2 \times 4^2 = 32$ real supercharges. In the coset parameterization (5.135) those give rise to 32 fermion fields in the sigma-model. However, only half of them correspond to physical, propagating degrees of freedom. This is readily seen in the light-cone gauge associated with $T_+ = (+ + - - | + + - -) \sim D - \mathcal{J}$, where D is the dilatation generator of the conformal algebra and $\mathcal{J}$ is the $\mathfrak{so}(6)$ angular momentum. So chosen T_+ is indeed a null element of $\mathfrak{psu}(2,2|4)$:

Str $T_+^2 = 0$. The light-cone coordinate X^+ is a combination of global AdS time and an angle on S^5, so the classical solution (5.138) describes a point-like string moving along the big circle of S^5 at the speed of light [29]. The kinetic terms in the gauge-fixed fermion action contain $f_+ = \text{ad}\, T_+$. When restricted to the fermionic subspace of $\mathfrak{psu}(2,2|4)$, $\text{ad}\, T_+$ has eight eigenvalues $+2$, eight eigenvalues -2, and sixteen zero eigenvalues. Half of the fermions (those corresponding to the zero modes of $\text{ad}\, T_+$) do not appear in the action at all. This is a manifestation of the kappa-symmetry, a Grassmann-odd gauge symmetry that at the linearized level acts as a shift symmetry: $\theta \to \theta_\alpha + \varepsilon\kappa_\alpha$, where κ_α are zero eigenvectors of $\text{ad}\, T_+$. Gauge-fixing the kappa-symmetry leaves sixteen Majorana fermions, eight of which are left- and eight are right-moving. This is the spectrum of critical superstring in ten dimensions.

Another example is the sigma-model based on the $\mathfrak{osp}(2n+2|2n)$ series of superalgebras. These superalgebras admit a $\mathbb{Z}_4$ automorphism

$$\Omega = \text{Ad}\,\text{diag}(\mathcal{J}|+^p -^{n-p} +^p -^{n-p}). \tag{5.144}$$

The invariant subalgebra is $\mathfrak{h}_0 = \mathfrak{u}(n+1) \oplus \mathfrak{sp}(2p) \oplus \mathfrak{sp}(2n-2p)$, and the bosonic dimension of the coset is $(n+1)(2n+1) - (n+1)^2 + (2n^2+n) - (2p^2+p) - [2(n-p)^2 + n - p] = n^2 + n + 4np - 4p^2$, which gives ten for $n=2$ and $p=1$ (for larger n the dimension is too big). Taking into account that the appropriate real form of $\mathfrak{sp}(4)$ is $\mathfrak{so}(3,2)$, that $\mathfrak{so}(6) = \mathfrak{su}(3)$, and that the denominator subalgebra is $\mathfrak{u}(3) \oplus \mathfrak{so}(3,1)$, we get the coset $OSp(6|4)/U(3) \times SO(3,1)$ whose bosonic section is

$$SU(4)/U(3) \times SO(3,2)/SO(3,1) = \mathbb{CP}^3 \times AdS_4, \tag{5.145}$$

another representative holographic background.

Exercise 18 *Show that the number of kappa symmetries in the $OSp(6|4)/U(3) \times SO(3,1)$ coset is such that the light-cone spectrum contains $8_L + 8_R$ Majorana fermions.*

Acknowledgements

I would like to thank the organizers of the Les Houches summer school 'Integrability: From Statistical Systems To Gauge Theory' for the opportunity to present these lectures. I am grateful to X. Chen-Lin, D. Medina-Rincon, and E. Widén for carefully reading the notes and for many suggestions for improvements. This work was supported by the Marie Curie network GATIS of the European Union's FP7 Programme under REA Grant Agreement No. 317089, by the ERC advanced grant No 341222, by the Swedish Research Council (VR) grant 2013-4329, by the grant 'Exact Results in Gauge and String Theories' from the Knut and Alice Wallenberg foundation, and by RFBR grant 15-01-99504.

References

[1] S. Weinberg, 'Dynamical approach to current algebra', Phys. Rev. Lett. 18, 188 (1967).
S. Weinberg, 'Nonlinear realizations of chiral symmetry', Phys. Rev. 166, 1568 (1968).

[2] F. D. M. Haldane, 'Continuum dynamics of the 1-D Heisenberg antiferromagnetic identification with the O(3) nonlinear sigma model', Phys. Lett. A93, 464 (1983).
F. D. M. Haldane, 'Nonlinear field theory of large spin Heisenberg antiferromagnets. Semiclassically quantized solitons of the one-dimensional easy Axis Neel state', Phys. Rev. Lett. 50, 1153 (1983).
A. M. Tsvelik, 'Quantum field theory in condensed matter physics', Cambridge Univ. Press (2003).

[3] A. B. Zamolodchikov and A. B. Zamolodchikov, 'Factorized S-matrices in two dimensions as the exact solutions of certain relativistic quantum field models', Annals Phys. 120, 253 (1979).

[4] R. R. Metsaev and A. A. Tseytlin, 'Type IIB superstring action in $AdS_5 \times S^5$ background', Nucl. Phys. B533, 109 (1998), hep-th/9805028.

[5] I. Bena, J. Polchinski, and R. Roiban, 'Hidden symmetries of the $AdS_5 \times S^5$ superstring', Phys. Rev. D69, 046002 (2004), hep-th/0305116.

[6] B. A. Dubrovin, A. T. Fomenko, and S. P. Novikov, 'Modern geometry – methods and applications: Part ii: The geometry and topology of manifolds', Springer (1985).

[7] L. D. Faddeev and L. A. Takhtajan, 'Hamiltonian methods in the theory of solitons', Springer (1987).

[8] V. E. Zakharov and A. V. Mikhailov, 'Relativistically Invariant Two-Dimensional Models in Field Theory Integrable by the Inverse Problem Technique', Sov. Phys. JETP 47, 1017 (1978).

[9] L. D. Faddeev and N. Y. Reshetikhin, 'Integrability of the principal chiral field model in (1+1)-dimension', Ann. Phys. 167, 227 (1986).

[10] L. D. Faddeev, 'How Algebraic Bethe Ansatz works for integrable model', hep-th/9605187.

[11] C. G. Callan, Jr., S. R. Coleman, J. Wess, and B. Zumino, 'Structure of phenomenological Lagrangians. 2.', Phys. Rev. 177, 2247 (1969).

[12] H. Eichenherr and M. Forger, 'On the Dual Symmetry of the Nonlinear Sigma Models', Nucl. Phys. B155, 381 (1979).

[13] D. Bykov, 'Complex structures and zero-curvature equations for σ-models', Phys. Lett. B760, 341 (2016), 1605.01093.

[14] S. Novikov, 'The Hamiltonian formalism and a many valued analog of Morse theory', Usp.Mat.Nauk 37N5, 3 (1982).

[15] E. Witten, 'Global Aspects of Current Algebra', Nucl.Phys. B223, 422 (1983).

[16] A. G. Abanov, 'Topology, geometry and quantum interference in condensed matter physics', 1708.07192.

[17] A. M. Polyakov and A. A. Belavin, 'Metastable States of Two-Dimensional Isotropic Ferromagnets', JETP Lett. 22, 245 (1975).

[18] E. Witten, 'Nonabelian Bosonization in Two-Dimensions', Commun.Math.Phys. 92, 455 (1984).

[19] A. P. Veselov and L. A. Takhtajan, 'Integrability of the Novikov equations for principal chiral fields with a multivalued Lagrangian', Sov. Phys. Dokl. 29, 994 (1984).

[20] A. M. Polyakov, 'Interaction of Goldstone Particles in Two-Dimensions. Applications to Ferromagnets and Massive Yang-Mills Fields', Phys.Lett. B59, 79 (1975).

[21] D. Friedan, 'Nonlinear Models in Two Epsilon Dimensions', Phys. Rev. Lett. 45, 1057 (1980).

[22] K. Zarembo, 'Quantum Giant Magnons', JHEP 0805, 047 (2008), 0802.3681.

[23] P. Dorey, 'Exact S matrices', hep-th/9810026.

[24] V. G. Kac, 'A Sketch of Lie Superalgebra Theory', Commun. Math. Phys. 53, 31 (1977).
L. Frappat, P. Sorba and A. Sciarrino, 'Dictionary on Lie superalgebras', hep-th/9607161.
[25] V. V. Serganova, 'Classification of real simple Lie superalgebras and symmetric superspaces', Funct. Anal. Appl. 17, 200 (1983).
[26] N. Berkovits, M. Bershadsky, T. Hauer, S. Zhukov, and B. Zwiebach, 'Superstring theory on $AdS_2 \times S^2$ as a coset supermanifold', Nucl. Phys. E567, 61 (2000), hep-th/9907200.
[27] M. B. Green and J. H. Schwarz, 'Covariant Description of Superstrings', Phys. Lett. B136, 367 (1984).
[28] K. Zarembo, 'Strings on Semisymmetric Superspaces', JHEP 1005, 002 (2010), 1003.0465.
[29] D. E. Berenstein, J. M. Maldacena, and H. S. Nastase, 'Strings in flat space and pp waves from $N = 4$ super Yang Mills', JHEP 0204, 013 (2002), hep-th/0202021.

6
Integrability in 2D fields theory/sigma-models

Sergei L. Lukyanov and
Alexander B. Zamolodchikov

NHETC, Department of Physics and Astronomy Rutgers University Piscataway, NJ 08855-0849, USA

Chapter Contents

6 Integrability in 2D fields theory/sigma-models   248
    Sergei L. LUKYANOV and Alexander B. ZAMOLODCHIKOV

	Preface	250
6.1	Liouville theorem	250
6.2	Free particle on a group manifold	253
6.3	Non-linear sigma model	255
6.4	Principal chiral field	258
6.5	Integrable examples of NLSM	260
6.6	The cigar	265
6.7	B-field	275
6.8	Integrable examples of NLSM with B-field	278
6.9	Wick rotation	279
6.10	Free scalar field on a cylinder	281
6.11	General structure of the Hilbert space of 2D CFT	285
6.12	Dimensional transmutation in PCF	287
6.13	RG flow equations	290
6.14	String equations	293
6.15	The quantum cigar	294
6.16	The quantum sausage	308
6.17	Other integrable structures in the cigar NLSM	314
	Acknowledgements	317
	References	317

Preface

This is an extended version of two lectures given by the first author at the 2016 Les Houches summer school.

6.1 Liouville theorem

Consider a dynamical Hamiltonian system with phase space of dimension $2n$. The system is Liouville integrable if it possesses n functionally independent *integral of motions* (IM) F_i, $i=1,\ldots,n$, $\{H,F_j\}=0$, in involution $\{F_i,F_j\}=0$. The independence means that at a generic point, the dF_i are linearly independent. Also, there cannot be more than n independent quantities in involution otherwise the *Poisson bracket* (PB) would be degenerate. Therefore the Hamiltonian H must be a function of the F_i.

Liouville theorem [1, 2]: The solution of the equations of a Liouville integrable system can be obtained in *quadratures*.

6.1.1 Free particle on an ellipsoid (Jacobi problem)

Exercise 1[1] Consider a particle of mass m constrained to move on the two-dimensional ellipsoid,

$$\left(\frac{x}{a}\right)^2 + \left(\frac{y}{b}\right)^2 + \left(\frac{z}{c}\right)^2 = 1.$$

Show that

$$F = \left(\frac{x^2}{a^4} + \frac{y^2}{b^4} + \frac{z^2}{c^4}\right)\left[\left(\frac{\dot{x}}{a}\right)^2 + \left(\frac{\dot{y}}{b}\right)^2 + \left(\frac{\dot{z}}{c}\right)^2\right]$$

is an IM (*Joachimsthal–Jacobi* IM).

Exercise 2 Let $\lambda > -c^2 > \mu > -b^2 > \nu > -a^2$ be three roots of the equation $f(u) = \frac{x^2}{a^2+u} + \frac{y^2}{b^2+u} + \frac{z^2}{a^2+u} = 1$. The triple (λ,μ,ν) turns out to be a set of orthogonal coordinates in the first octant $x,y,z > 0$. Eq. $\lambda = 0$ defines an ellipsoid, so that μ and ν can serve as generalized coordinates for a particle on the ellipsoid. Show that the Lagrangian of the problem has the form

$$L = \frac{1}{2}\left(G_{\mu\mu}\dot{\mu}^2 + G_{\nu\nu}\dot{\nu}^2\right), \tag{6.1}$$

---

[1] This is a classic problem of geodesics on an ellipsoid. For a two-dimensional ellipsoid, its solution was announced by Jacobi on 28 December 1838. Using the remarkable substitution, he reduced this problem to quadratures. This result was published in the paper, C. Jacobi, JRAM 19, (1839), 309–313.

where

$$G_{\mu\mu} = \frac{m}{4} \frac{(\mu-\nu)\mu}{(a^2+\mu)(b^2+\mu)(c^2+\mu)}, \quad G_{\nu\nu} = \frac{m}{4} \frac{(\nu-\mu)\nu}{(a^2+\nu)(b^2+\nu)(c^2+\nu)}.$$

Exercise 3 Let p_μ, p_ν be momenta canonically conjugated to μ and ν, respectively (i.e. $\{\mu, p_\mu\} = \{\nu, p_\nu\} = 1$, $\{\mu, \nu\} = \{p_\mu, p_\nu\} = 0$). Show that the Hamiltonian and the Joachimsthal–Jacobi IM are expressed through the canonical variables as follows:

$$H = \frac{p_\mu^2}{2G_{\mu\mu}} + \frac{p_\nu^2}{2G_{\nu\nu}}, \quad F = -\frac{2}{m(abc)^2}\left(\frac{\nu p_\mu^2}{2G_{\mu\mu}} + \frac{\mu p_\nu^2}{2G_{\nu\nu}}\right).$$

Check that $\{H, F\} = 0$ and using the Hamilton–Jacobi approach integrate the equations of motion in quadrature (see Ch.2.13 in [2]).

6.1.2 The Euler top

The configuration space of a rigid body is a six-dimensional manifold $\mathbb{R}^3 \times SO(3)$ (see Figure 6.1). For a free moving top, it is convenient to work in the C-system whose origin coincides with the centre of mass and the axes of the moving frame coincide with the principal axes of the top [1–3]. Ignoring the trivial motion of the centre of mass, the configuration of the rigid body at each moment of time is specified by a 3×3 special orthogonal matrix g which describes the rotation of a fixed frame $\{e_a\}_{a=1}^3$ w.r.t. the moving one $\{e'_a\}_{a=1}^3$:

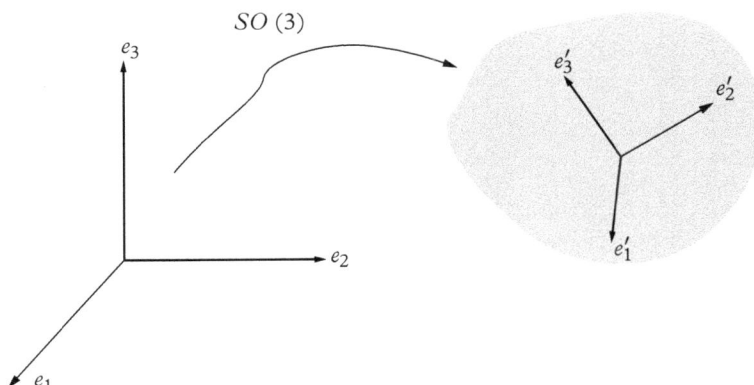

Figure 6.1 *The configuration of a rigid body is uniquely specified by the orientation of the moving axes $\{e'_a\}_{a=1}^3$ w.r.t. the fixed ones $\{e_a\}_{a=1}^3$ together with a three-dimensional vector that specifies the position of the origin of the moving frame. Reproduced with permission from John Wiley & Sons.*

$$e'_a(t) = \sum_a g_{ab}(t) e_b.$$

Let $\omega = \omega(t)$ be an instantaneous angular velocity. The Lagrangian of the free moving top coincides with its kinetic energy and is given by

$$L = \sum_a \frac{I_a \omega_a^2}{2},$$

where $\omega_a = \omega \cdot e'_a$ are projections of ω on the principal axes.

Exercise 4 Let t_a be 3×3 matrices,

$$t_1 = i \begin{pmatrix} 0 & 0 & 0 \\ 0 & 0 & -1 \\ 0 & 1 & 0 \end{pmatrix}, \quad t_2 = i \begin{pmatrix} 0 & 0 & 1 \\ 0 & 0 & 0 \\ -1 & 0 & 0 \end{pmatrix}, \quad t_3 = i \begin{pmatrix} 0 & -1 & 0 \\ 1 & 0 & 0 \\ 0 & 0 & 0 \end{pmatrix},$$

satisfying the commutation relations $[t_a, t_b] = i\varepsilon_{abc} t_c$. Show that

$$\omega \cdot e'_a = -i \left\langle t_a, \dot{g} g^{-1} \right\rangle, \quad \omega \cdot e_a = -i \left\langle t_a, g^{-1} \dot{g} \right\rangle,$$

where the angular brackets are defined as

$$\langle t_a, t_b \rangle \equiv \frac{1}{2} \operatorname{Tr}(t_a t_b) = \delta_{ab}.$$

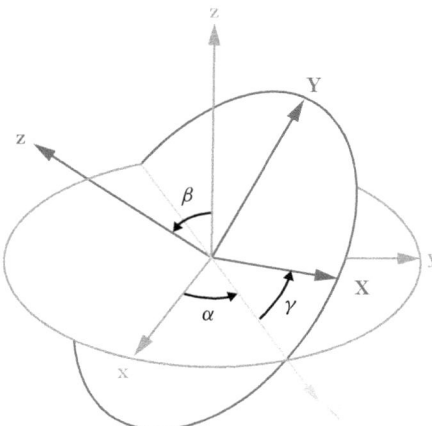

Figure 6.2 *The definition of the Euler angles (α, β, γ) that specify the orientation of the coordinate system (X, Y, Z) w.r.t (x, y, z). Reproduced with permission from John Wiley & Sons.*

Exercise 5 Let the Euler angles (α, β, γ) (see Figure 6.2),

$$g = e^{i\gamma t_3} e^{i\beta t_1} e^{i\alpha t_3} \qquad (0 \leq \alpha, \gamma < 2\pi, \quad 0 \leq \beta \leq \pi), \tag{6.2}$$

be chosen as the generalized coordinates on $SO(3)$. Consider a set of six functions $\{L_a \equiv \mathbf{L} \cdot \mathbf{e}'_a; N_a \equiv \mathbf{e}'_a \cdot \mathbf{e}_3 = g_{a,3}\}_{a=1,2,3}$ on the phase space of the top. Express these functions in terms of (α, β, γ) and their canonically conjugate momenta $(p_\alpha, p_\beta, p_\gamma)$ and derive the following PB relations:

$$\{L_a, L_b\} = -\varepsilon_{abc} L_c, \quad \{L_a, N_b\} = -\varepsilon_{abc} N_c, \quad \{N_a, N_b\} = 0. \tag{6.3}$$

Show that

$$H = \sum_a \frac{L_a^2}{2I_a}, \quad L^2 = \sum_a L_a^2, \quad \tilde{L}_3 = \mathbf{L} \cdot \mathbf{e}_3 = \sum_a N_a L_a \tag{6.4}$$

constitute a set of involutive and functionally independent IM. Integrate the equations of motion in quadrature (see Ch. 37 of [3]).

6.2 Free particle on a group manifold

6.2.1 Spherical top. Free particle on the 3-sphere

The rigid body with equal moments of inertia, $I_1 = I_2 = I_3 = \frac{1}{f^2}$, is called a spherical top. Its free motion is described by the Lagrangian

$$L = -\frac{1}{2f^2} \left\langle \dot{g} g^{-1}, \dot{g} g^{-1} \right\rangle. \tag{6.5}$$

It is well known that $SO(3)$ is not a simply connected Lie group. The universal cover of $SO(3)$ coincides with $SU(2)$—the group of 2×2 unitary matrices:

$$g g^\dagger = 1, \quad \det(g) = 1 \tag{6.6}$$

(here we use the same notation g for elements of $SU(2)$). Any such matrix can be written in the form

$$g = \begin{pmatrix} n_4 + in_3 & n_2 + in_1 \\ -n_2 + in_1 & n_4 - in_3 \end{pmatrix}, \quad n_1^2 + n_2^2 + n_3^2 + n_4^2 = 1, \tag{6.7}$$

so that, regarded as a manifold, $SU(2)$ is a 3-sphere. Since (6.5) contains only the elements of the Lie algebra, the Lagrangian is identical for both groups $SO(3)$ and $SU(2)$. The difference appears in the global aspects of the motion.

Exercise 6 Let σ_a be the standard set of Pauli matrices. The Euler decomposition for the $SU(2)$ matrices is given by

$$g = e^{i\gamma t_3} e^{i\beta t_1} e^{i\alpha t_3} \ : \ \ t_a = \tfrac{1}{2}\sigma_a, \quad 0 \le \tfrac{1}{2}(\alpha \pm \gamma) < 2\pi, \ \ 0 \le \beta \le \pi. \tag{6.8}$$

It is convenient to substitute the Euler angles by the Hopf coordinates (θ, χ_1, χ_2): $\theta = \frac{\pi - \beta}{2}$, $\chi_1 = \frac{1}{2}(\alpha - \gamma)$, $\chi_2 = \frac{1}{2}(\alpha + \gamma)$ (for a review of the Hopf coordinates, see e.g. [41]). Show that

$$n_1 \pm i n_2 = e^{\pm i \chi_1} \cos(\theta), \quad n_4 \pm i n_3 = e^{\pm i \chi_2} \sin(\theta) \quad \left(0 \le \theta \le \tfrac{\pi}{2}, \ \chi_i \sim \chi_i + 2\pi\right) \tag{6.9}$$

and the metric of the round sphere of radius $\frac{2}{f}$ is given by

$$G_{ab} dX^a dX^b = \frac{4}{f^2} \left((d\theta)^2 + \cos^2(\theta)(d\chi_1)^2 + \sin^2(\theta)(d\chi_2)^2 \right). \tag{6.10}$$

As has already been mentioned, the Lagrangian stays the same for the $SO(3)$ and $SU(2)$ group. It can be written as

$$L = \frac{G_{ab}}{2} \dot{X}^a \dot{X}^b, \tag{6.11}$$

where $X^a = (\theta, \chi_1, \chi_2)$. However, the spherical top problem and the problem of the free moving particle on the round 3-sphere correspond to different compactification conditions, namely:[2]

Spherical top : $\alpha \equiv \chi_2 + \chi_1 \sim \alpha + 2\pi$, $\gamma \equiv \chi_2 - \chi_1 \sim \gamma + 2\pi$
Particle on 3-sphere : $\chi_1 \sim \chi_1 + 2\pi$, $\chi_2 \sim \chi_2 + 2\pi$.

Notice that $\theta = 0, \frac{\pi}{2}$ look like singular points of the metric. In fact, they do not correspond to any geometric singularities of the round 3-sphere. Indeed, consider the metric in the vicinity $\theta = 0$ where it can be approximated by

$$G_{ab} dX^a dX^b \approx \frac{4}{f^2} \left((d\theta)^2 + \theta^2 (d\chi_2)^2 + (d\chi_1)^2 \right). \tag{6.12}$$

Since $\chi_2 \sim \chi_2 + 2\pi$ is an angle type variable, we can change the coordinate frame by introducing $X = \theta \cos(\chi_2)$, $Y = \theta \sin(\chi_2)$. In the new coordinates the metric reads as

$$G_{ab} dX^a dX^b \approx \frac{4}{f^2} \left((dX)^2 + (dY)^2 + (d\chi_1)^2 \right). \tag{6.13}$$

---

[2] The manifold $SO(3) = SU(2)/Z_2$ can be identified with the real projective space RP^3, so that the spherical top can be thought of as a free particle on RP^3.

This is a flat Euclidean metric on $\mathbb{R}^2 \times S^1$ and $\theta = 0$ corresponds to the circle at $X = Y = 0$, parameterized by the $\chi_1 \sim \chi_1 + 2\pi$. This shows that $\theta = 0$ is just a coordinate singularity rather than a geometric one. Notice that the compactification condition $\chi_2 \to \chi_2 + const$ with $const \neq 2\pi$ yields a conical type singularity on the surface. Of course, the same analyses can be applied to $\theta = \pi$. The Hopf coordinate (θ, χ_1, χ_2) turn out to be a convenient parameterization of the 3-sphere, since a single chart $\{0 < \theta < \frac{\pi}{2}, \; 0 \leq \chi_i < 2\pi\}$ covers almost the whole sphere except two circles of co-dimension 2 at $\theta = 0$ and $\theta = \frac{\pi}{2}$. In what follows we will trade the Hopf coordinate θ for

$$\zeta = n_1^2 + n_2^2 - n_3^2 - n_4^2 = \cos(2\theta) = -\cos(\beta) \; : \; -1 < \zeta < 1. \tag{6.14}$$

6.2.2 General case

The Lagrangian

$$L = -\frac{1}{2f^2} \left\langle \dot{g}g^{-1}, \dot{g}g^{-1} \right\rangle \tag{6.15}$$

makes sense for any simple Lie Group $\mathfrak{G}$. It governs the free motion of a particle in the principal homogeneous space of $\mathfrak{G}$. With this generalization $\dot{g}g^{-1}$ is an element of the Lie algebra $\mathfrak{g}$, which is spanned by infinitesimal group elements t_a satisfying the commutation relations

$$[t_a, t_b] = i f_{ab}{}^c \, t_c.$$

The bracket $\langle \ldots \rangle$ in (6.15) stands for the Killing form (one-half of the trace over the adjoint representation of $\mathfrak{g}$). The generators t_a can be chosen to satisfy the condition

$$\langle t_a, t_b \rangle = \delta_{ab}.$$

As well as in the $SU(2)$ case, the Lagrangian (6.15) takes the general form (6.11) in any coordinate frame X^a, $a = 1, \ldots, \dim \mathfrak{G}$.

6.3 Non-linear sigma model

The Lagrangian $L = \frac{G_{ab}}{2} \dot{X}^a \dot{X}^b$ admits an immediate field theory generalization which is usually referred to as the *non-linear sigma model* (NLSM).[3] In the simplest setup

---

[3] The origin of the term 'sigma model' for a field theory where the scalar values are on a manifold is from Gell-Mann and Levy's 1960 paper 'The Axial Vector Current in β-Decay' which introduced two models. The first of these is called the *linear sigma model*, and it is a renormalized Heisenberg-inspired Mexican hat model for a pion condensate. The model has four fields, ϕ^i where $i = 0, 1, 2, 3$, which have a regular Mexican hat potential,

the model can be formulated as follows. Let $\mathcal{M}_D$ be a D-manifold (the *target space*) equipped with a Riemannian metric G. Let also $\mathcal{W}_2$ be a two-dimensional manifold (the *world sheet*) with a Minkowski type metric η. Consider a map from the world sheet to the target space, which in some coordinate frame is given by a set of functions $X^a(x^0, x^1)$ with $a = 1, \ldots, D$ and $(x^0, x^1) \in \mathcal{W}_2$. The NLSM action is given by

$$S = \frac{1}{2} \int_{\mathcal{W}_2} \mathrm{d}^2 x \sqrt{-\eta}\, \eta^{\mu\nu}\, G_{ab}(X)\, \partial_\mu X^a \partial_\nu X^b. \tag{6.16}$$

Exercise 7 Show that the Euler–Lagrange equations corresponding to (6.16) are given by

$$\eta^{\mu\nu} \left(\partial_\mu \partial_\nu X^a + \Gamma^a{}_{bc} \partial_\mu X^b \partial_\nu X^c \right) = 0 \qquad (a = 1, \ldots, D), \tag{6.17}$$

where $\Gamma^a{}_{bc} = \Gamma^a{}_{cb}$ stands for the symmetric Christoffel symbols built from the metric G.

The NLSM possesses a curious property. Calculating the energy–momentum tensor, one finds

$$T_{\mu\nu} \equiv \frac{\pi}{\sqrt{-\eta}} \frac{\delta S}{\delta \eta^{\mu\nu}} = \frac{\pi}{2} \left(G_{ab} \partial_\mu X^a \partial_\nu X^b - \tfrac{1}{2} \eta_{\mu\nu}\, \eta^{\lambda\sigma} G_{ab} \partial_\lambda X^a \partial_\sigma X^b \right). \tag{6.18}$$

It turns out that the trace of the energy–momentum tensor vanishes, $\eta^{\mu\nu} T_{\mu\nu} = 0$. In fact this is a manifestation of the fact that the NLSM does not contains any length scale and hence, it possesses a scale invariance. More formally, the action (6.16) is invariant w.r.t. the conformal transformations of the the world sheet metric, $\eta_{\mu\nu} \mapsto \lambda \eta_{\mu\nu}$ (Weyl symmetry). In what follows we will assume that the world sheet metric is flat and the theory is Lorentz invariant. The Lorentz group is very special in $1+1$ Minkowski space. Let $x^\pm = x^0 \pm x^1$, then under the Lorentz boost with rapidity θ, $x^\pm \mapsto \mathrm{e}^{\mp\theta}\, x^\pm$, whereas linear combinations

$$T_{\pm\pm} = T_{00} + T_{11} \pm 2 T_{01} \qquad \Theta = T_{00} - T_{11}, \tag{6.19}$$

so that the vacuum values are on a sphere S^3. This makes three field directions light, and these modes are the three pions, and one field direction heavy, and this mode was called the 'sigma'. It was a predicted particle, and it was identified with the $\sigma(600)$ except that this resonance is very strange and was delisted, and is too broad to be a real sigma, so the model is not good. Ignoring renormalizability, the mass of the σ is adjusted by making the wall of the Mexican hat potential tighter oscillating, and in the limit of infinitely fast oscillations, you just end up restricting the π fields to a sphere, and there is no finite energy σ. This limit is the non-renormalizable non-linear sigma model in the paper. It was called that, because it is the non-linear version of the renormalizable sigma model Gell-Mann and Levy believed, but it is a misnomer, because the non-linear theory doesn't have a sigma, that's the whole point of going to the non-linear version.

transform irreducibly and their Lorentz spins equal ± 2 and 0, respectively:

$$x^{\pm} \mapsto e^{\mp\theta} x^{\pm} \; : \quad T_{\pm\pm} \mapsto e^{\pm 2\theta} T_{\pm}, \quad \Theta \mapsto \Theta. \tag{6.20}$$

The continuity equations $\partial^\nu T_{\mu\nu} = 0$ can be written in the form

$$\partial_- T_{++} + \partial_+ \Theta = 0, \quad \partial_+ T_{--} + \partial_- \Theta = 0, \tag{6.21}$$

where we use the notations $\partial_\pm = \frac{1}{2}(\partial_0 \pm \partial_1)$. Supplementing the equation with the condition $\Theta = 0$, one finds

$$\partial_- T_{++} = \partial_+ T_{--} = 0 \quad \text{where} \quad T_{\pm\pm} = 2\pi\, G_{ab}(X) \partial_\pm X^a \partial_\pm X^b. \tag{6.22}$$

This implies that $T_{++}(T_{--})$ is a chiral field which depends on the variable x^+ (x^-) only. Let us consider the model in a finite size geometry, with the spatial coordinate x^1 compactified on a circle of circumference R, so that $T_{\mu\nu}(x^0, x^1+R) = T_{\mu\nu}(x^0, x^1)$. The non-vanishing components of the energy–momentum tensor can be expanded in Fourier series,

$$T_{++}(x^0, x^1) = \left(\frac{2\pi}{R}\right)^2 \sum_{n=-\infty}^{\infty} L_n^{(+)}(x^0)\, e^{-\frac{2\pi i n}{R} x^1} \tag{6.23}$$

$$T_{--}(x^0, x^1) = \left(\frac{2\pi}{R}\right)^2 \sum_{n=-\infty}^{\infty} L_n^{(-)}(x^0)\, e^{+\frac{2\pi i n}{R} x^1},$$

and equations (6.22) imply

$$L_n^{(+)}(x^0) = L_n\, e^{-\frac{2\pi i n}{R} x^0}, \quad L_n^{(-)}(x^0) = \bar{L}_n\, e^{-\frac{2\pi i n}{R} x^0}. \tag{6.24}$$

Thus we conclude that any monomial of the form

$$L_{n_1}^{(+)} \cdots L_{n_N}^{(+)} L_{m_1}^{(-)} \cdots L_{m_M}^{(-)} \quad \text{with} \quad n_1 + \ldots + n_N + m_1 + \ldots + m_M = 0, \tag{6.25}$$

is an IM.

Exercise 8 Show that $T_{\pm\pm}(x) \equiv T_{\pm\pm}(0, x)$ satisfy the following equal time PB relations:

$$\{T_{\pm\pm}(x), T_{\pm\pm}(y)\} = \pm 2\pi\, (T_{\pm\pm}(x) + T_{\pm\pm}(y))\, \delta'(x-y), \quad \{T_{++}(x), T_{--}(y)\} = 0. \tag{6.26}$$

Here the prime stands for the derivative of the δ-function. Formulae (6.26) can be equivalently written in terms of the Fourier expansion coefficients L_n and $\bar{L}_n$ as

$$\{L_n, L_m\} = -\mathrm{i}\,(n-m)\, L_{n+m}, \quad \{\bar{L}_n, \bar{L}_m\} = -\mathrm{i}\,(n-m)\, \bar{L}_{n+m}, \quad \{L_n, \bar{L}_m\} = 0. \tag{6.27}$$

The monomials (6.25) are conserved quantities, i.e. they Poisson commute with the Hamiltonian $\mathbb{H}_R$ and the total momentum $\mathbb{P}_R$,

$$\mathbb{H}_R = \frac{2}{\pi} \int_0^R dx\, T_{00} = \frac{2\pi}{R} (L_0 + \bar{L}_0)$$

$$\mathbb{P}_R = \frac{2}{\pi} \int_0^R dx\, T_{01} = \frac{2\pi}{R} (L_0 - \bar{L}_0). \qquad (6.28)$$

However they do not mutually commute in general. Nevertheless, it is not difficult to construct an infinite set of involutive IM. Let G be a positive definite metric, i.e. $T_{\pm\pm} = 2\pi G_{ab} \partial_\pm X^a \partial_\pm X^b > 0$, then the energy–momentum tensor components can be represented in the form

$$T_{\pm\pm}(x_\pm) = j_\pm^2(x)|_{x=x_\pm} \quad \text{with} \quad j_\pm(x+R) = j_\pm(x). \qquad (6.29)$$

Exercise 9 Show that

$$\{j_\pm(x), j_\pm(y)\} = \pi\, \delta'(x-y), \quad \{j_+(x), j_-(y)\} = 0, \qquad (6.30)$$

imply equations (6.26). In terms of the Fourier modes

$$j_\pm(x) = \frac{2\pi}{R} \sum_{n=-\infty}^{\infty} a_n^{(\pm)} e^{-\frac{2\pi i n}{R} x}, \qquad (6.31)$$

the PB relations (6.30) are equivalent to

$$\{a_n^{(\pm)}, a_m^{(\pm)}\} = -i\, \frac{n}{2}\, \delta_{n+m,0}, \quad \{a_n^{(+)}, a_m^{(-)}\} = 0. \qquad (6.32)$$

The above exercises lead us to the conclusion that $a_n^{(+)} a_{-n}^{(+)}$ and $a_m^{(-)} a_{-m}^{(-)}$ are involutive IM for any integers n and m.

6.4 Principal chiral field

In this section we consider an immediate generalization of a classical mechanics particle on a group manifold—the field theory whose Lagrangian density is given by

$$\mathcal{L} = -\frac{1}{2f^2}\, \eta^{\mu\nu} \langle J_\mu, J_\nu \rangle, \quad J_\mu = \partial_\mu g\, g^{-1}. \qquad (6.33)$$

As before, g takes values in the principal homogeneous space of the simple Lie group, while the currents J_μ belong to the corresponding Lie algebra. The model can be defined in $d+1$-dimensional space-time, $x^\mu = (x^0, x^1, \ldots x^d) \in R^{1,d}$, equipped with the Minkowski metric $\eta^{\mu\nu}$ ($\eta^{00} = -\eta^{11} = \ldots = 1$). In fact, it was originally introduced by

Feza Gürsey in 1960 for $d=3$, to describe the effective interactions of mesons in the chiral limit (where the masses of the quarks go to zero) but without necessarily mentioning quarks at all. For this reason the model is usually referred to as the *principal chiral field* (*PCF*) [2, 4, 5]. It is also of great interest in mathematics [6].

We will focus on the $1+1$-dimensional case only, i.e. $\boldsymbol{g} = \boldsymbol{g}(x^0, x^1)$. Then (6.33) can be rewritten as

$$\mathcal{L} = -\frac{2}{f^2} \langle \boldsymbol{J}_+ \boldsymbol{J}_- \rangle. \tag{6.34}$$

It is easy to check that the currents $\boldsymbol{J}_\pm = \partial_\pm \boldsymbol{g} \boldsymbol{g}^{-1}$ satisfy the Bianchi type identity:

$$\partial_- \boldsymbol{J}_+ - \partial_+ \boldsymbol{J}_- + [\boldsymbol{J}_+, \boldsymbol{J}_-] = 0. \tag{6.35}$$

Combining this with the Lagrange–Euler equations $\partial^\mu \boldsymbol{J}_\mu = 0$, or, equivalently,

$$\partial_- \boldsymbol{J}_+ + \partial_+ \boldsymbol{J}_- = 0, \tag{6.36}$$

one obtains

$$\partial_- \boldsymbol{J}_+ + \tfrac{1}{2}[\boldsymbol{J}_+, \boldsymbol{J}_-] = 0, \quad \partial_+ \boldsymbol{J}_- - \tfrac{1}{2}[\boldsymbol{J}_+, \boldsymbol{J}_-] = 0. \tag{6.37}$$

Let us consider a pair of differential operators

$$\boldsymbol{D}_\pm = \partial_\pm - \boldsymbol{A}_\pm, \quad \boldsymbol{A}_\pm = \lambda_\pm \boldsymbol{J}_\pm. \tag{6.38}$$

It is easy to see that (6.37) yields

$$[\boldsymbol{D}_+, \boldsymbol{D}_-] = \left(\lambda_+ \lambda_- - \frac{\lambda_-}{2} - \frac{\lambda_+}{2}\right)[\boldsymbol{J}_+, \boldsymbol{J}_-]. \tag{6.39}$$

If we choose the arbitrary parameters λ_+ and λ_- to satisfy the condition

$$\frac{1}{\lambda_+} + \frac{1}{\lambda_-} = 2, \quad \text{i.e.,} \quad \lambda_\pm = \frac{1}{1 \pm \lambda}, \tag{6.40}$$

then the commutator (6.39) turns out to be zero:

$$[\boldsymbol{D}_+, \boldsymbol{D}_-] = 0. \tag{6.41}$$

A condition of this type is called a *zero-curvature representation* (*ZCR*). Notice that (6.38) can be understood as a world sheet connection (gauge field $A_\mu(x^0, x^1)$) with values in a Lie algebra $\mathfrak{g}$. Then (6.39) means that it has zero curvature, or equivalently, the gauge field strength $\boldsymbol{F}_{\mu\nu}(x^0, x^1) = [\boldsymbol{D}_\mu, \boldsymbol{D}_\nu] = \partial_\mu \boldsymbol{A}_\nu - \partial_\nu \boldsymbol{A}_\mu + [\boldsymbol{A}_\mu, \boldsymbol{A}_\nu]$ vanishes everywhere.

So far, there was no question of *boundary conditions* (*BC*) so that the above discussion was of a formal nature. There are several ways to impose BC. For definiteness, let us set

$$J_\mu(x^0, x^1 + R) = J_\mu(x^0, x^1). \tag{6.42}$$

We may now choose some matrix representation $\mathcal{R}$ for the Lie group and introduce the *monodromy matrix*:

$$M_\mathcal{R}(x^0) = \pi_\mathcal{R} \left[\overleftarrow{\mathcal{P}} \exp\left(\int_0^R dx^1 A_1(x^0, x^1) \right) \right]. \tag{6.43}$$

Exercise 10 Using the ZCR show that

$$\partial_0 M_\mathcal{R} = \left[\pi_\mathcal{R}\big(A_0(x^0, 0)\big), M_\mathcal{R} \right], \tag{6.44}$$

and, hence, the matrix trace $T_\mathcal{R} = \text{Tr}[M_\mathcal{R}]$ is an IM.

It is crucial that the flat connection depends on an arbitrary complex variable λ—the so-called spectral parameter, so that $T_\mathcal{R} = T_\mathcal{R}(\lambda)$ constitutes a continuous family of IM, rather than a single conserved charge. It is possible to show (see e.g. Ch.3.9 in [2] for details) that $T_\mathcal{R}(\lambda)$ are involutive families of IM, Poisson commuting for different choices of the spectral parameter and the representation:

$$\{T_\mathcal{R}(\lambda), T_{\mathcal{R}'}(\lambda')\} = 0. \tag{6.45}$$

Motivated by the above consideration we accept the (somewhat) technical definition of classical integrability:

A classical field theory in two space-time dimensions is said to be integrable if it admits the ZCR.

6.5 Integrable examples of NLSM

6.5.1 Anisotropic $SU(2)$ PCF

Consider the field theory counterpart of the so-called symmetrical top, such that $I_1 = I_2 \neq I_3$. The Lagrangian of the model, which is usually referred to as the anisotropic $SU(2)$ PCF, can be written in the form

$$\mathcal{L} = \frac{2}{f_\perp^2} \left(j_+^1 j_-^1 + j_+^2 j_-^2 \right) + \frac{2}{f_\parallel^2} j_+^3 j_-^3. \tag{6.46}$$

Here we use the real fields $j_a = j_a(x^0, x^1)$ defined by means of the relation

$$\partial_{\pm} \mathbf{g}\mathbf{g}^{-1} = i \sum_a j_{\pm}^a t_a. \tag{6.47}$$

The model can be interpreted as a NLSM on the 3-sphere with a one-parameter deformed round metric. As a deformation parameter one can choose

$$z = \operatorname{arcsinh}\left(\frac{f_{\|}}{\sqrt{f_{\perp}^2 - f_{\|}^2}}\right). \tag{6.48}$$

Exercise 11 Show that the anisotropic NLSM possesses the ZCR of the form

$$D_+ = \partial_+ - i\left[\lambda_+^{\perp}\left(j_+^1 t_1 + j_+^2 t_2\right) + \lambda_+^{\|} j_+^3 t_3\right] \tag{6.49}$$
$$D_- = \partial_- - i\left[\lambda_-^{\perp}\left(j_-^1 t_1 + j_-^2 t_2\right) + \lambda_-^{\|} j_-^3 t_3\right],$$

where

$$\lambda_{\pm}^{\perp} = \frac{\cosh(\mu)}{\sinh(\mu) \mp \sinh(z)}, \quad \lambda_{\pm}^{\|} = \frac{\sinh(\mu) \pm \frac{1}{\sinh(z)}}{\sinh(\mu) \mp \sinh(z)} \tag{6.50}$$

and the unconstrained complex variable μ can be thought of as the spectral parameter.

6.5.2 $O(N)$ NLSM

Suppose the moment of inertia w.r.t. the symmetry axis of the top becomes very large, $I_1 = I_2 < I_3 \to \infty$. In this case the configuration of the top is completely specified by the unit vector $\mathbf{n} = \mathbf{e}_3' = (\cos(\alpha)\sin(\beta), \sin(\alpha)\sin(\beta), \cos(\beta)) \in \mathbb{R}^3$ along the symmetry axis. For the anisotropic $SU(2)$ PCF, this implies that as $f_{\|} \to 0$, the second term in the r.h.s. of (6.46) can be neglected and the Lagrangian takes the form

$$\mathcal{L} = \frac{2}{g^2}\left(\partial_+\beta\partial_-\beta + \sin^2(\beta)\partial_+\alpha\partial_-\alpha\right) \tag{6.51}$$

(here we substitute $f_{\perp}^2$ by g^2), or, equivalently,

$$\mathcal{L} = \frac{1}{2g^2} \eta^{\mu\nu} \partial_\mu \mathbf{n} \cdot \partial_\nu \mathbf{n} \quad \text{where} \quad \mathbf{n} \cdot \mathbf{n} = 1. \tag{6.52}$$

The Lagrangian corresponds to the NLSM on the round 2-sphere S^2 with radius g^{-1}. The model is also known as the $O(3)$ NLSM [7].

Exercise 12 Construct the ZCR for the O(3) NLSM.

As has been already discussed, the Lie Group $SU(2)$ can be identified with the 3-sphere equipped with the round metric. Hence, the Lagrangian of the PCF in this case can be written in the form (6.52) where the unit vector $\boldsymbol{n} \in R^4$ is expressed in terms of the Hopf coordinates as $\boldsymbol{n} = (\cos(u)\cos(\chi_1), \cos(u)\sin(\chi_1), \sin(u)\cos(\chi_2), \sin(u)\sin(\chi_2))$. One can also consider the NLSM (6.52) with $\boldsymbol{n} = (n_1, \ldots, n_N) \in \mathbb{R}^N$. The corresponding model is known as the $O(N)$-sigma model. Notice that g^{-1} can be interpreted as a radius of the round sphere S^{N-1} which is the target manifold of the model. It turn out that the $O(N)$-sigma model is an integrable model for any $N \geq 2$, i.e. the corresponding equations of motion admit the ZCR.

6.5.3 Fateev NLSM

Let us focus on the $SU(2)$-case. The PCF is invariant w.r.t. the global transformation

$$g \mapsto U_L g U_R. \tag{6.53}$$

The current $J_\mu = \partial_\mu g g^{-1}$ is not affected by the right action of the group, while under the left action it transforms as

$$J_\mu \mapsto U_L J_\mu U_L^{-1}. \tag{6.54}$$

The anisotropic $SU(2)$ PCF can be interpreted as a perturbation of the $SU(2)$ PCF by the Lorentz scalar field $\eta^{\mu\nu} j_\mu^3 j_\nu^3$ which breaks the global symmetry of the original action down to $U_L(1) \otimes SU_R(2)$. Similarly to the 'left' current J_μ, one can introduce the right currents

$$\tilde{J}_\mu = g^{-1} \partial_\mu g = \mathrm{i} \sum_a \tilde{j}_\pm^a \mathbf{t}_a. \tag{6.55}$$

Notice that $\eta^{\mu\nu} j_\mu^a j_\nu^a = \eta^{\mu\nu} \tilde{j}_\mu^a \tilde{j}_\nu^a = 2\eta^{\mu\nu} \mathrm{Tr}(\partial_\mu g \partial_\nu g^{-1})$, where g is understood as a 2×2 unitary matrix. Therefore one can also consider the perturbation in the form $\eta^{\mu\nu} \tilde{j}_\mu^3 \tilde{j}_\nu^3$ which breaks the original global symmetry down to $SU_L(2) \otimes U_R(1)$. This is of course another integrable model, in a sense equivalent to the anisotropic $SU(2)$ PCF. An immediate question arises: what happens for the perturbation of the form $\eta^{\mu\nu}(l j_\mu^3 j_\nu^3 + r \tilde{j}_\mu^3 \tilde{j}_\nu^3)$ which breaks the $SU_L(2) \times SU_R(2)$ symmetry down to $U_L(1) \otimes U_R(1)$? It turns out that this two-parameter deformation spoils the integrability. However, the NLSM

$$\mathcal{L} = \eta^{\mu\nu} \frac{u \mathrm{Tr}(\partial_\mu g \partial_\nu g^{-1}) + \frac{l}{2} j_\mu^3 j_\nu^3 + \frac{r}{2} \tilde{j}_\mu^3 \tilde{j}_\nu^3}{4(u+r)(u+l) - rl \left(\mathrm{Tr}(g\sigma_3 g^{-1}\sigma_3)\right)^2} \qquad (g \text{ is } 2 \times 2\ SU(2) \text{ matrix}) \tag{6.56}$$

which involves three parameters (u,l,r), turns out to be integrable. This model was introduced by Fateev in 1996 [8], and the corresponding ZCR was found in [9].[4]

Exercise 13 Let β, $\alpha = \chi_2 + \chi_1$, $\gamma = \chi_2 - \chi_1$ be the Euler angles. Show that in the coordinate frame $(\zeta, X^R, X^L) = (-\cos(\beta), \alpha, \gamma)$, the target space metric for the Fateev NLSM has the form

$$G = G_{\zeta\zeta}(\zeta)(d\zeta)^2 + \sum_{A,B \in \{R,L\}} G_{AB}(\zeta) dX^A dX^B \qquad (6.57)$$

and its non-vanishing components depend on the coordinate ζ only:

$$G_{\zeta\zeta} = \frac{1}{4u} \frac{(1+p^{-2}\kappa)(1+p^2\kappa)}{(1-\zeta^2)(1-\kappa^2\zeta^2)}$$

$$G_{RR} = \frac{1}{4u} \frac{1+p^2\kappa - (p^2+\kappa)\kappa\zeta^2}{1-\kappa^2\zeta^2} \qquad (6.58)$$

$$G_{LL} = \frac{1}{4u} \frac{1+p^{-2}\kappa - (p^{-2}+\kappa)\kappa\zeta^2}{1-\kappa^2\zeta^2}$$

$$G_{RL} = -\frac{1}{4u} \frac{(1-\kappa^2)\zeta}{1-\kappa^2\zeta^2},$$

where we use the parameters

$$\kappa = \sqrt{\frac{rl}{(u+r)(u+l)}}, \qquad p^2 = \sqrt{\frac{l(u+r)}{r(u+l)}}.$$

Exercise 14 For $\kappa = 0$, equations (6.58) describe the round metric on the 3-sphere of radius $\frac{1}{\sqrt{u}}$. Show that the scalar curvature of the Fateev metric is given by

$$R_G = \frac{\pi^2}{u} \left[\frac{2(1-\kappa^2)}{a_1 a_2 \kappa} \frac{1+\kappa^2\zeta^2}{1-\kappa^2\zeta^2} - \frac{1}{2\kappa} \frac{(1-\kappa^2)^2}{(a_1+a_2\kappa)(a_2+a_1\kappa)} \right], \qquad (6.59)$$

where we use the parameters

$$a_1 = \frac{\pi}{2\sqrt{-r(u+l)}}, \qquad a_2 = \frac{\pi}{2\sqrt{-l(u+r)}}. \qquad (6.60)$$

Equation (6.59) implies that, for $\kappa \in (0,1)$, $a_1, a_2 > 0$ the curvature remains finite and takes maximum value at $\zeta = \pm 1$. The metric is degenerate at $\zeta = \pm 1$. The degeneration points form two submanifolds of codimension two (circles). Show that for $\chi_i \sim \chi_i + 2\pi$ these are coordinate singularities which can be removed by choosing suitable local coordinate frames.

---

[4] A different, but gauge equivalent ZCR for the Fateev model was independently constructed in [10] (for detailed explanation see [11])

6.5.4 The sausage

Let us consider the limit $u, p^2 \to \infty$ of the metric (6.58) keeping $g^{-2} \equiv \frac{p^2 \kappa}{4u}$ fixed [12]. Then $G_{ab} dX^a dX^b = G_{ab}^{(\text{saus})} dX^a dX^b + O(u^{-1})$, where

$$G_{ab}^{(\text{saus})} dX^a dX^b = \frac{1}{g^2} \left[\frac{(d\zeta)^2}{(1-\zeta^2)(1-\kappa^2\zeta^2)} + \frac{(1-\zeta^2)(d\alpha)^2}{1-\kappa^2\zeta^2} \right]. \tag{6.61}$$

As $\kappa = 0$ it becomes the metric of the round 2-sphere of radius g^{-1}. (Recall that $\zeta = -\cos(\beta)$ and β and α should be identified with the polar and azimuthal angles, respectively.) Thus (6.61) can be interpreted as a deformation of the round metric on the 2-sphere of radius g^{-1} and $0 \leq \kappa < 1$ plays the rôle of a deformation parameter. It is useful to look at the asymptotic form of the target space background as $\kappa \to 1$. Let us cover the manifold by two charts $\mathcal{C}_+$ and $\mathcal{C}_-$ depending on the value of the 'global' coordinate ζ: $-\kappa < \zeta \leq 1$ and $-1 \leq \zeta < \kappa$, respectively. In the domain covered by the chart $\mathcal{C}_\pm$, we introduce a complex coordinate $Z_\pm$:

$$|Z_\pm|^2 = \frac{1+\kappa}{1-\kappa} \left(\frac{1-\zeta}{1+\zeta} \right)^{\pm 1}, \qquad \arg(Z_\pm) = \alpha \qquad \left(|Z_\pm| \leq \tfrac{1+\kappa}{1-\kappa} \right). \tag{6.62}$$

As $\kappa \to 1$, the target space metric asymptotically approaches

$$\mathcal{C}_\sigma : \quad G_{ab}^{(\text{saus})} dX^a dX^b \approx \frac{1}{g^2} \frac{|dZ_\sigma|^2}{1+|Z_\sigma|^2} \qquad (\sigma = \pm). \tag{6.63}$$

It is the metric of a semi-infinite (Hamilton) 'cigar' [13], which is also known as the metric of a 2D Euclidean black hole [14, 15].

Exercise 15 Show that $\frac{|dZ|^2}{1+|Z|^2}$ coincides with the induced metric of the surface of revolution

$$z = \log\left(\frac{\sqrt{2-x^2-y^2}+1}{\sqrt{1-x^2-y^2}} \right) - \sqrt{2-x^2-y^2}$$

in 3D Euclidean space (see Figure 6.3).

For sufficiently large Z_σ, $1 < |Z_\sigma| < \frac{1+\kappa}{1-\kappa}$ the metric (6.58) is almost flat and it is well approximated by the Euclidean metric on a finite cylinder—product of S^1 and a line segment $|\phi| < \frac{1}{2}\log(\frac{1+\kappa}{1-\kappa})$ (see Figure 6.4):

$$\mathcal{C}_+ \cap \mathcal{C}_- : \quad G_{ab}^{(\text{saus})} dX^a dX^b \approx \frac{1}{g^2} \left[(d\phi)^2 + (d\alpha)^2 \right]. \tag{6.64}$$

Here we use the coordinate $\phi = -\frac{1}{2}\log(\frac{1-\zeta}{1+\zeta})$.

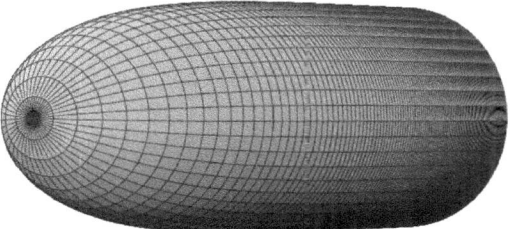

Figure 6.3 *A depiction of the 'cigar' manifold embedded in three-dimensional space. Reproduced with permission from John Wiley & Sons.*

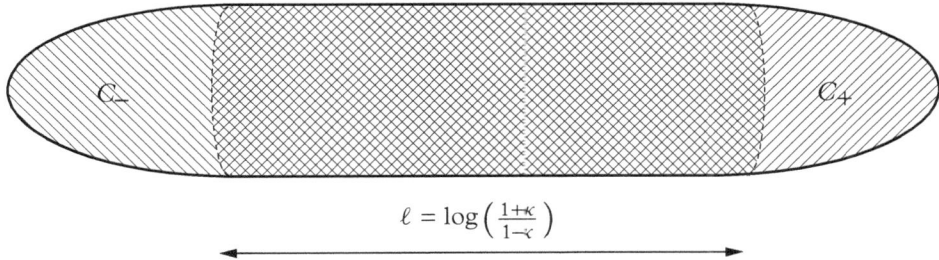

Figure 6.4 *The target space of the sausage NLSM as $\kappa \to 1_-$. The charts C_+ and C_- used in equation (6.63) cover the right and left portions of the sausage respectively. Reproduced with permission from John Wiley & Sons.*

Finally, let us note that (to the best of our knowledge) the NLSM with the target space of the ellipsoid (field theory counterpart of the Jacobi problem) does not possesses a ZCR, i.e. it is a not integrable field theory.

6.6 The cigar

6.6.1 General solution of the Cauchy problem

The cigar NLSM is a somewhat special example of an integrable field theory. The target space in this case coincides with $\mathbb{R}^2$ and it is convenient to use the complex coordinate $Z = X + iY$. Then the cigar metric is given by

$$G_{ab}^{(\text{cigar})} dX^a dX^b = \frac{1}{2\pi} \frac{|dZ|^2}{1 + |Z|^2}. \tag{6.65}$$

Notice that the classical equations of motion for the NLSM do not depend on the overall factor for the metric. In order to make the link between quantum and classical relations simpler, we choose g^2 in the cigar metric (6.63) as 2π. For calculations it is sometimes convenient to use the coordinate frame (ϕ, α) such that

$$Z = \sinh(\phi)\, e^{i\alpha} \qquad (\phi > 0,\ \alpha \sim \alpha + 2\pi). \tag{6.66}$$

With these variables the Lagrangian of the model reads explicitly as

$$\mathcal{L} = \frac{1}{\pi}\left(\partial_+\phi\partial_-\phi + \tanh^2(\phi)\,\partial_+\alpha\partial_-\alpha\right). \tag{6.67}$$

The model is invariant under the $U(1)$-rotations, $Z \to e^{i\chi}Z$ ($\alpha \to \alpha + \chi$), so that the corresponding current

$$\mathcal{J}_\mu = \tanh^2(\phi)\,\partial_\mu\alpha, \tag{6.68}$$

obeys the continuity equation

$$\partial_-\mathcal{J}_+ + \partial_+\mathcal{J}_- = 0. \tag{6.69}$$

These equations can be resolved in terms of the 'dual' field $\tilde{\alpha}$ such that

$$\mathcal{J}_+ = \partial_+\tilde{\alpha},\quad \mathcal{J}_- = -\partial_-\tilde{\alpha}, \tag{6.70}$$

or, equivalently, $\mathcal{J}_\mu = \epsilon_{\mu\nu}\,\partial_\nu\tilde{\alpha}$ with $\epsilon_{01} = -\epsilon_{10} = 1$, $\epsilon_{11} = \epsilon_{00} = 0$.

Let us consider the model with quasiperiodic BC such that

$$Z(x^0, x^1 + R) = e^{2\pi i k} Z(x^0, x^1). \tag{6.71}$$

The real parameter k is defined modulo integer, so it can be chosen to be in the interval $-\frac{1}{2} < k \leq \frac{1}{2}$. It is usually referred to as a twist parameter. Together with the twist parameter, one should consider the conserved charge m associated with the conserved $U(1)$-current,

$$m = \int_0^R \frac{dx^1}{2\pi}\,\mathcal{J}_0 = \int_0^R \frac{dx^1}{2\pi}\,\partial_1\tilde{\alpha} = \frac{1}{2\pi}\left[\tilde{\alpha}\left(x^0, R\right) - \tilde{\alpha}\left(x^0, 0\right)\right]. \tag{6.72}$$

Exercise 16 Consider the fields

$$\begin{aligned}
\psi_+ &= (\partial_+\phi + i\tanh(\phi)\partial_+\alpha)\, e^{+i(\alpha+\tilde{\alpha})},\\
\psi_+^* &= (\partial_+\phi - i\tanh(\phi)\partial_+\alpha)\, e^{-i(\alpha+\tilde{\alpha})},\\
\psi_- &= (\partial_-\phi + i\tanh(\phi)\partial_-\alpha)\, e^{+i(\alpha-\tilde{\alpha})},\\
\psi_-^* &= (\partial_-\phi - i\tanh(\phi)\partial_-\alpha)\, e^{-i(\alpha-\tilde{\alpha})},
\end{aligned} \tag{6.73}$$

satisfying the quasiperiodic BC

$$\psi_\pm(x^0, x^1+R) = e^{+2\pi i(k\pm m)}\,\psi_\pm(x^0,x^1), \quad \psi_\pm^*(x^0,x^1+R) = e^{-2\pi i(k\pm m)}\,\psi_\pm^*(x^0,x^1) \tag{6.74}$$

($\psi_\pm^*$ is just the complex conjugate of $\psi_\pm$). Using the Euler–Lagrange equations for (6.67) show that

$$\partial_\mp \psi_\pm = \partial_\mp \psi_\pm^* = 0. \tag{6.75}$$

Exercise 17 Let e, f, h be formal operators satisfying the $sl(2)$ commutation relations,

$$[h,e] = 2e, \quad [h,f] = -2f, \quad [e,f] = h, \tag{6.76}$$

and

$$\omega = e^{-\frac{1}{2}(\alpha+\tilde{\alpha})h}\, e^{\phi(e+f)}\, e^{\frac{1}{2}(\alpha-\tilde{\alpha})h}\;:\;\omega(x_0,x_1+R) = e^{-i(k+m)h}\,\omega(x_0,x_1)\,e^{i(k-m)h}. \tag{6.77}$$

Show that

$$\psi_+ f + \psi_+^* e = \partial_+ \omega \omega^{-1}, \quad \psi_- f + \psi_-^* e = \omega^{-1}\partial_-\omega. \tag{6.78}$$

Now consider the Cauchy problem for the Euler–Lagrange equations. Suppose we are given the initial data including Z and $\partial_0 Z$ at $x^0 = 0$. The initial value of the dual field $\tilde{\alpha}$ is given by the relation $\tilde{\alpha} = \tilde{\alpha}_0 + \int_0^{x^1} dx^1 \tanh^2(\phi)\partial_0\alpha$, where $\tilde{\alpha}_0$ is an arbitrary constant. Then equations (6.73) allow one to find the fields $\psi_\pm$ at $x_0 = 0$. As follows from equations (6.75), $\psi_\pm(x^0,x^1) = \psi_\pm(x^0 \pm x^1)$, i.e. these fields can be found at any moment of time up to a constant phase factor $e^{i\tilde{\alpha}_0}$. Assuming that $\psi_\pm, \psi_\pm^*$ in (6.78) are given, we may now consider these relations as equations for determining ω for any value (x_0,x_1). It is not difficult to see that the solution of (6.78) can be expressed in terms of the right and left path-ordered exponentials (see Figure 6.5):

$$\omega(x^0,x^1) = \overleftarrow{\mathcal{P}}\exp\left(\int_0^{x^0+x^1} dx^+ \left(\psi_+ f + \psi_+^* e\right)\right) \omega_0\, \overrightarrow{\mathcal{P}}\exp\left(\int_0^{x^0-x^1} dx^- \left(\psi_- f + \psi_-^* e\right)\right), \tag{6.79}$$

where $\omega_0 = \omega(0,0)$. This formal solution can be used to reconstruct $\omega(x^0,x^1)$ in any matrix representation of the $sl(2)$ commutation relations (6.76). In particular, for the two-dimensional representation $\pi_{\frac{1}{2}}(e) = \sigma_+ \equiv \begin{pmatrix} 0 & 1 \\ 0 & 0 \end{pmatrix}, \pi_{\frac{1}{2}}(f) = \sigma_- \equiv \begin{pmatrix} 0 & 0 \\ 1 & 0 \end{pmatrix}, \pi_{\frac{1}{2}}(h) = \sigma_3 \equiv \begin{pmatrix} 1 & 0 \\ 0 & -1 \end{pmatrix}$, then

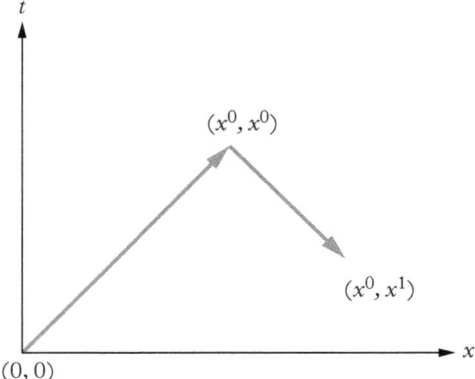

Figure 6.5 *The integration path from (0, 0) to (x_0, x_1) can be broken up into two light-cone segments. In terms of the light cone coordinates $x_\pm = x_0 \pm x_1$, these segments can be described by $0 < x_+ < x_0 + x_1$ with $x_- = 0$ and $0 < x_- < x_0 - x_1$ with $x_+ = x_0 + x_1$. Reproduced with permission from John Wiley & Sons.*

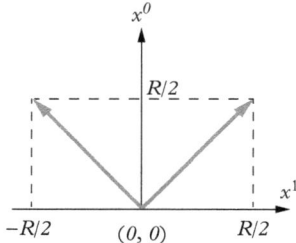

Figure 6.6 *The values of $\omega(\frac{R}{2}, \frac{R}{2})$ and $\omega(\frac{R}{2}, -\frac{R}{2})$, represented by the tips of the right and left arrow respectively, should be related according to the periodicity condition (6.77). Reproduced with permission from John Wiley & Sons.*

$$\pi_{\frac{1}{2}}(\omega) = \begin{pmatrix} e^{-i\tilde{\alpha}}\sqrt{1+|Z|^2} & Z^* \\ Z & e^{+i\tilde{\alpha}}\sqrt{1+|Z|^2} \end{pmatrix}, \quad \pi_{\frac{1}{2}}(\omega_0) = \begin{pmatrix} e^{-i\tilde{\alpha}_0}\cosh(\phi_0) & e^{-i\alpha_0}\sinh(\phi_0) \\ e^{+i\alpha_0}\sinh(\phi_0) & e^{+i\tilde{\alpha}_0}\cosh(\phi_0) \end{pmatrix}$$
(6.80)

and, hence, (6.79) allows one to reconstruct Z for $x^0 \neq 0$. At first glance the solution depends on the arbitrary constant ϕ_0. In fact, one should take into account the quasiperiodicity condition $Z(\frac{R}{2}, \frac{R}{2}) = e^{2\pi i k} Z(\frac{R}{2}, -\frac{R}{2})$, or equivalently (see Figure 6.6),

$$\text{Tr}_{\frac{1}{2}}\left[\sigma_\pm \omega\left(\frac{R}{2}, \frac{R}{2}\right)\right] = e^{\pm 2\pi i k}\, \text{Tr}_{\frac{1}{2}}\left[\sigma_\pm \omega\left(\frac{R}{2}, -\frac{R}{2}\right)\right],$$
(6.81)

which sets the value of ϕ_0.

Also, it is not difficult to see that $Z(x^0, x^1)$ does not depend on the arbitrary chosen constant $\tilde{\alpha}_0$, whereas the dependence on α_0 occurs through the overall phase factor $e^{i\alpha_0}$, which is just a manifestation of the global $U(1)$ invariance of the model.

6.6.2 Poisson structure

Let us discuss now the Poisson structure underlying the model under consideration.

Exercise 18 Show that the following equal time PB relations hold:

$$\{\psi_\pm(x), \psi_\pm(y)\} = \pm 2\pi \epsilon(x-y) \, \psi_\pm(x) \psi_\pm(y)$$
$$\{\psi_\pm^*(x), \psi_\pm^*(y)\} = \pm 2\pi \epsilon(x-y) \, \psi_\pm^*(x) \psi_\pm^*(y)$$
$$\{\psi_\pm^*(x), \psi_\pm(y)\} = \pm 2\pi \left(\delta'(x-y) - \epsilon(x-y) \, \psi_\pm^*(x)\psi_\pm(y)\right) \quad (6.82)$$
$$\{\psi_\pm(x), \psi_\mp(y)\} = \{\psi_\pm(x), \psi_\mp^*(y)\} = \{\psi_\pm^*(x), \psi_\mp^*(y)\} = 0.$$

Here, $\psi_\pm(x) \equiv \psi_\pm(0,x)$, $\psi_\pm^*(x) \equiv \psi_\pm^*(0,x)$, and

$$\epsilon(x) = 2n+1, \quad \text{for} \quad nR < x < (n+1)R; \quad n = 0, \pm 1, \pm 2 \ldots. \quad (6.83)$$

Exercise 19 Let $\phi'_\pm = \phi'_\pm(x), \alpha_\pm = \alpha_\pm(x)$ such that

$$\phi'_\pm(x+R) = \phi'_\pm(x), \quad \alpha_\pm(x+R) = \alpha_\pm(x) + \pi(k \pm m), \quad (6.84)$$

and satisfying the PB relations

$$\{\phi'_\pm(x), \phi'_\pm(y)\} = \pi \delta'(x-y), \quad \{\alpha_\pm(x), \alpha_\pm(y)\} = -\tfrac{\pi}{2} \, \epsilon(x-y) \quad (6.85)$$
$$\{\phi'_\pm(x), \phi'_\mp(y)\} = \{\alpha_\pm(x), \alpha_\mp(y)\} = \{\phi'_\pm(x), \alpha_\mp(y)\} = \{\phi'_\mp(x), \alpha_\mp(y)\} = 0.$$

Show that

$$\psi_\pm(x_\pm) = \pm\left[(\phi'_\pm + i\alpha'_\pm) \, e^{+2i\alpha_\pm}\right]_{x=x_\pm}, \quad \psi_\pm^*(x_\pm) = \pm\left[(\phi'_\pm - i\alpha'_\pm) \, e^{-2i\alpha_\pm}\right]_{x=x_\pm} \quad (6.86)$$

satisfy the equal time PB relations (6.82) and the boundary conditions (6.74) (the prime here stands for the derivative w.r.t. the argument).

The result of Exercise 19 can be obtained without brute force calculations. Indeed, consider the cigar equations of motion in the case when the field ϕ becomes large, i.e. in the asymptotically flat domain of the target manifold. In this case they asymptotically approach the pair of D'Alembert equations, $\partial_+\partial_-\phi \approx 0$ and $\partial_+\partial_-\alpha \approx 0$, whose general solution can be written as $\phi \approx \phi_+(x_+) + \phi_-(x_-)$ and $\alpha \approx \alpha_+(x_+) + \alpha_-(x_-)$. The dual field $\tilde{\alpha}$ takes the form $\tilde{\alpha} \approx \alpha_+(x_+) - \alpha_-(x_-)$. Substituting these asymptotic formulae to the definition (6.73) of $\psi_\pm$ one obtains the relations (6.86). On the other hand, formulae (6.82) give the equal time PB relations which are the same independently of the choice of x^0. Taking x^0 such that the field ϕ becomes large, it is easy to arrive at the statement quoted in Exercise 19.[5]

The non-vanishing components of the energy–momentum tensor $T_{\pm\pm} = 2\pi G_{ab}\partial X^a_\pm \partial X^b_\pm$ can be written as

$$T_{\pm\pm} = \psi^*_\pm \psi_\pm = (\phi'_\pm)^2 + (\alpha'_\pm)^2. \tag{6.87}$$

Notice that the general relations $\partial_\mp T_{\pm\pm} = 0$ discussed previously, follow immediately from (6.75). Also, the quasiperiodic boundary conditions imply that $T_{\pm\pm}(x_0, x_1) = T_{\pm\pm}(x_0, x_1 + R)$.

Since the function $\phi'_\pm = \phi'_\pm(x), \alpha_\pm = \alpha_\pm(x)$ (6.84) can be expanded in a Fourier series of the form

$$\phi'_\pm(x) = \frac{2\pi}{R} \sum_{n=-\infty}^{\infty} a_n^{(\pm)} e^{-\frac{2\pi i n}{R}x} \tag{6.88}$$

$$\alpha_\pm(x) = \alpha_0^{(\pm)} + \frac{2\pi x}{R} b_0^{(\pm)} + i \sum_{n\neq 0}^{\infty} \frac{b_n^{(\pm)}}{n} e^{-\frac{2\pi i n}{R}x},$$

where $b_0^{(\pm)} = \frac{1}{2}(m \pm k)$. The Hamiltonian $\mathbb{H}_R$ and the total momentum $\mathbb{P}_R$ of the model are given by

$$\mathbb{H}_R = \int_0^R \frac{dx^1}{2\pi}(T_{++} + T_{--}) = \frac{2\pi}{R} \sum_{\sigma=\pm} \sum_{n=-\infty}^{\infty} \left(a_{-n}^{(\sigma)} a_n^{(\sigma)} + b_{-n}^{(\sigma)} b_n^{(\sigma)}\right)$$

$$\mathbb{P}_R = \int_0^R \frac{dx^1}{2\pi}(T_{++} - T_{--}) = \frac{2\pi}{R} \sum_{\sigma=\pm} \sigma \sum_{n=-\infty}^{\infty} \left(a_{-n}^{(\sigma)} a_n^{(\sigma)} + b_{-n}^{(\sigma)} b_n^{(\sigma)}\right). \tag{6.89}$$

Exercise 20 Show that

$$\{a_n^{(\pm)}, a_l^{(\pm)}\} = \{b_n^{(\pm)}, b_l^{(\pm)}\} = -i\frac{n}{2}\delta_{n+l,0}, \quad \{\alpha_0^{(\pm)}, b_0^{(\pm)}\} = \frac{1}{2}, \tag{6.90}$$

whereas all other PB involving the expansion coefficients vanish. Thus $\left\{a_{-n}^{(\sigma)} a_n^{(\sigma)}, b_{-n}^{(\sigma)} b_n^{(\sigma)}\right\}_{n=0,\sigma=\pm}^{\infty}$ is an involutive set of IM.

Exercise 21 Consider a solution of the equations of motion corresponding to $a_0^{(+)} = -a_0^{(-)} = \frac{v}{2} < 0$ and $a_n^{(\pm)} = b_n^{(\pm)} = 0$ for $n \neq 0$. Using general relations (6.79) and (6.80) find $Z = Z(x^0, x^1)$. Show that, with a proper choice of the constants $\alpha_0^{(\pm)}$, and using the invariance w.r.t. the overall shift of the time $x^0 \mapsto x^0 + \text{const}$, the solution can be chosen to satisfy the following asymptotic conditions as $x^0 \to \pm\infty$:

---

[5] Notice that the overall sign factor in (6.86) does not affect the PB relations.

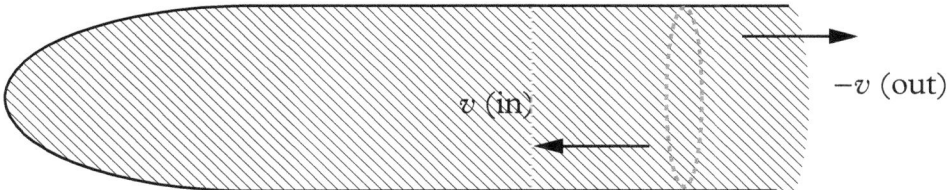

Figure 6.7 *The classical scattering problem for the cigar NLSM. Reproduced with permission from John Wiley & Sons.*

$$Z \to \begin{cases} e^{\frac{2\pi}{R}(vx^0 + kx^1)} \, (1 + o(1)) & \text{as } x^0 \to -\infty \\ e^{\frac{2\pi}{R}(-vx^0 + kx^1)} \, e^{\Delta z} \, (1 + o(1)) & \text{as } x^0 \to +\infty \end{cases}, \quad (6.91)$$

where the time delay Δ_Z in this classical scattering problem (see Figure 6.7) is given by the equation

$$e^{\Delta_Z} = \frac{(m+k-iv)(m-k-iv)}{4v^2}. \quad (6.92)$$

Exercise 22 Show that equation (6.81) yields to the following conditions:

$$a_0^{(+)} + a_0^{(-)} = 0 \quad (6.93)$$

$$e^{2\pi a_0^{(+)} - \phi_0} \int_0^R dx \, \alpha'_+ e^{-2\Delta\phi_+} + e^{2\pi a_0^{(-)} + \phi_0} \int_0^R dx \, \alpha'_- e^{-2\Delta\phi_-} = 0,$$

where $\Delta\phi_\pm(x) = \int_0^x dx \, \phi'_\pm$ and $a_0^{(\pm)} = \frac{1}{2\pi} \Delta(\phi_\pm(R))$.

Let us introduce the notation $\phi_\pm(x) = C - \frac{1}{2}(2\pi a_0^{(\pm)} \mp \phi_0) + \int_0^x dx \, \phi'_\pm$, where C is an arbitrary constant and

$$\mathcal{X}_\pm = \int_0^R dx \, \alpha'_\pm e^{-2\phi_\pm}. \quad (6.94)$$

Then the conditions in (6.93) take the form

$$a_0^{(+)} + a_0^{(-)} = 0, \qquad \mathcal{X}_+ + \mathcal{X}_- = 0. \quad (6.95)$$

Formulae (6.95) have a simple Hamiltonian interpretation. Namely, the phase space of the model under consideration admits an extension such that the antiderivatives $\phi_\pm$ are regarded as dynamical variables. Apparently, $\phi_\pm$ are quasiperiodic fields, i.e. in the interval $x \in [0, R]$ they have Fourier expansions similar to that for $\alpha_\pm$ (6.88),

$$\phi_\pm(x) = \phi_0^{(\pm)} + \frac{2\pi x}{R} a_0^{(\pm)} + i \sum_{n \neq 0}^{\infty} \frac{a_n^{(\pm)}}{n} e^{-\frac{2\pi i n}{R} x}. \tag{6.96}$$

Assuming that $\phi_0^{(\pm)}$ is canonically conjugate to $a_0^{(\pm)}$,

$$\{\phi_0^{(\pm)}, a_0^{(\pm)}\} = \frac{1}{2}, \tag{6.97}$$

it is easy to see that the PB $\{\phi'_\pm(x), \phi'_\pm(y)\} = \pi\, \delta'(x-y)$ are replaced by

$$\{\phi_\pm(x), \phi_\pm(y)\} = -\frac{\pi}{2}\, \epsilon(x-y), \tag{6.98}$$

where $\epsilon(-x) = -\epsilon(x), \epsilon'(x) = 2\delta(x)$, whereas

$$\{\phi_\pm(x), \phi_\mp(y)\} = 0. \tag{6.99}$$

In the extended phase space both $a_0^{(\pm)}$ and $\phi_0^{(\pm)}$ are regarded as dynamical variables and $a_0^{(+)} + a_0^{(-)} = 0$ and $\mathcal{X}_+ + \mathcal{X}_- = 0$ are first class constraints. Then the conditions $\phi_0^{(+)} + \phi_0^{(-)} = c$ and $a_0^{(+)} - a_0^{(-)} = v$, where c and v are some non-dynamical constants, can be interpreted as the gauge fixing conditions imposed on top of the first class constraints. Notice that v turns out to be an important parameter labelling the gauge orbits generated by the first class constraints. At the same time a value of the non-dynamical constant c is not essential—it can be set to be any number, say zero, by a proper constant shift of x^0. The relations $b_0^{(\pm)} = \frac{1}{2}(m \pm k)$, where m and k are non-dynamical variables, can be also regarded as first class constraints. In this case the gauge fixing can be achieved by assigning α and $\tilde{\alpha}$ certain values α_0 and $\tilde{\alpha}_0$ at some point of the world sheet (it is easy to see that, similar to the constant c, $(\alpha_0, \tilde{\alpha}_0)$ are not essential parameters in the theory). This way the phase space of the model with quasiperiodic BC is obtained through Hamiltonian reduction from the extended phase space by means of eight second class constraints.

In dealing with a system with constraints, a special rôle belongs to the gauge invariant observables—dynamical variables which Poisson commute (in a certain weak sense) with the first class constraints. It is clear that the components of the energy–momentum tensor should be gauge invariant observables. Let us consider the component $T_{++} = (\phi'_+)^2 + (\alpha'_+)^2$. Then a simple calculation shows that

$$\{T_{++}(x), \alpha'_+ e^{-2\phi_+}(y)\} = 2\phi'_+ \alpha'_+ e^{-2\phi_+}(y) 2\pi \delta(x-y) + \alpha'_+(x) e^{-2\phi_+}(y) 2\pi \delta'(x-y)$$
$$= -2\pi \partial_y \left(\alpha_+(y) e^{-2\phi_+}(y) \delta(x-y)\right). \tag{6.100}$$

Integrating both sides of (6.100) over the segment $(0, R)$, one finds

$$\{T_{++}(x), \mathcal{X}_+\} = \alpha_+ e^{-2\phi_+}(0) 2\pi \delta(x) - \alpha_+ e^{-2\phi_+}(R) 2\pi \delta(x-R). \tag{6.101}$$

As x belongs to the open segment $(0,R)$, the contact terms can be ignored and, therefore,

$$\{T_{++}(x), \mathcal{X}_+\} = 0, \qquad x \in (0, R). \tag{6.102}$$

Also, it is easy to see that $\{T_{++}(x), \mathcal{X}_-\} = \{T_{++}(x), a_0^{(\pm)}\} = 0$. Thus we arrive at the expected conclusion that T_{++} and, of course, T_{--} are gauge invariant observables.

Let us consider now the fields

$$\begin{aligned}
W_{\pm 2} &= \psi_\pm^* \psi_\pm \\
W_{\pm 3} &= \frac{1}{2i} \left(\psi_\pm^* \partial_\pm \psi_\pm - \psi_\pm \partial_\pm \psi_\pm^* \right) \\
W_{\pm 4} &= -\frac{1}{2} \left(\psi_\pm^* \partial_\pm^2 \psi_\pm + \psi_\pm \partial_\pm^2 \psi_\pm^* \right) + (\psi_\pm^* \psi_\pm)^2
\end{aligned} \tag{6.103}$$

with the Lorentz spins ± 2, ± 3, and ± 4, respectively (notice that $W_{\pm 2} \equiv T_{\pm\pm}$). It turns out that they possess properties similar to those of $T_{\pm\pm}$: they are *local* w.r.t. the original complex field $Z(x^0, x^1)$,

$$\partial_\mp W_{\pm s} = 0, \quad W_{\pm s}\left(x^0, x^1\right) = W_{\pm s}\left(x^0, x^1 + R\right) \qquad (s \geq 2), \tag{6.104}$$

and

$$\{W_s(x), \mathcal{X}_\pm\} = 0 \qquad x \in (0, R). \tag{6.105}$$

In what follows, W_s will be referred to as W-currents. When considering their properties, it is sufficient to focus on the case with positive Lorentz spin.

Exercise 23 Show that

$$\begin{aligned}
W_2 &= (\phi'_+)^2 + (\alpha'_+)^2 \\
W_3 &= 2(\alpha'_+)^3 + 2(\phi'_+)^2 \alpha'_+ - \phi''_+ \alpha'_+ + \phi'_+ \alpha''_+ \\
W_4 &= (\phi'_+)^4 + 5(\alpha'_+)^4 + 6(\phi'_+)^2 (\alpha'_+)^2 + 4\phi'_+ \alpha''_+ \alpha'_+ - 4\phi''_+ (\alpha'_+)^2 - \phi'''_+ \phi'_+ - \alpha'''_+ \alpha'_+
\end{aligned} \tag{6.106}$$

and then prove formula (6.105).

6.6.3 Integrable structures

The purpose of the next exercise is to give a rough idea about the hidden integrable structures [16] in the cigar NLSM.

Exercise 24 Let W_s be the W-currents defined in eq. (6.103). Introduce the set $\left\{I_{s-1}^{(I)}\right\}_{s=2}^{4}$, where

$$I_{s-1}^{(I)} = \int_0^R \frac{dx^1}{2\pi} W_s(x^0, x^1). \tag{6.107}$$

Show that $\left\{I_{s-1}^{(I)}\right\}_{s=2}^{4}$ is an involutive set of IM and check that

$$\left\{I_2^{(I)}, \psi\right\} = -i\left(-\psi'' + 2(\psi^*\psi)\psi\right) \tag{6.108}$$

(here and below we drop the subscript '+' in the notation ψ_+).

The Hamiltonian flow $\partial_\tau \psi = \left\{I_2^{(I)}, \psi\right\}$ coincides with the so-called non-linear Schrödinger equation—the famous integrable equation in $(1+1)$-dimensions [5]. Notice that in the theory of classical integrable systems the PB relations

$$\{\psi(x), \psi(y)\} = 2\pi\epsilon(x-y)\,\psi(x)\psi(y) \tag{6.109}$$
$$\{\psi^*(x), \psi(y)\} = 2\pi\left(\delta'(x-y) - \epsilon(x-y)\,\psi^*(x)\psi(y)\right)$$

are known as the second Poisson structure for the non-linear Schrödinger equation.

Results of Exercises 23, 24 can be generalized as follows. It is possible to show that for any $s = 2, 3, 4, 5, \ldots$ there exists a W-current of the Lorentz spin s satisfying the conditions (6.104) and (6.105), such that $I_{s-1}^{(I)}$ of the form (6.107) is a member of an infinite involutive set $\left\{I_{s-1}^{(I)}\right\}_{s=2}^{\infty}$ of local IM. The term 'local' here means that, as a density of $I_{s-1}^{(I)}$, the field W_s is a *real* polynomial with constant coefficients involving ψ and ψ^* and their derivatives similar, to (6.103). When W_s is expressed in terms of ϕ_+ and α_+, it is a *homogeneous* polynomial of order s with constant coefficients w.r.t. the fields ϕ'_+, α'_+ and their higher derivatives (in other words, any monomials appearing within W_s contains exactly s derivative symbols; see (6.106) for an illustration). Notice that for given spin s there exists only one (up to an overall constant factor) local IM from the set whose first representatives are $I_1^{(I)} = \int_0^R \frac{dx}{2\pi}\psi^*\psi$ and $I_2^{(I)} = \int_0^R \frac{dx}{2\pi i}\psi^*\partial\psi$. The corresponding density W_s is defined up to the total derivative ∂O, where O is a differential polynomial involving the W-currents with spin lower then s.

It deserves mentioning that the integrable structure described above is not unique. There exists another involutive set of local IM $\{I_{2j-1}^{(II)}\}_{j=1}^{\infty}$, where $I_{2j-1}^{(II)} = \int_0^R \frac{dx}{2\pi} W_{2j}^{(II)}(x^0, x^1)$ and the local densities of the first representatives are given by

$$W_2^{(II)} = \psi^*\psi \tag{6.110}$$
$$W_4^{(II)} = -\frac{1}{2}\left(\psi^*\partial^2\psi + \psi\partial^2\psi^*\right) + 4(\psi^*\psi)^2$$

or, equivalently,

$$W_2^{(II)} = (\phi'_+)^2 + (\alpha'_+)^2 \tag{6.111}$$
$$W_4^{(II)} = 4(\phi'_+)^4 + 8(\alpha'_+)^4 + 12(\phi'_+)^2(\alpha'_+)^2 + 4\phi'_+\alpha''_+\alpha'_+ - 4\phi''_+(\alpha'_+)^2 - \phi'''_+\phi'_+ - \alpha'''_+\alpha'_+.$$

6.7 B-field

6.7.1 WZWN model

We now return to the PCF. The Lagrange–Euler equation for this model has the form of the continuity equation $\partial_-J_+ + \partial_+J_- = 0$. Let us slightly modify this equation by adding an extra term with an arbitrary parameter ν,

$$\partial_-J_+ + \partial_+J_- - \nu[J_+,J_-] = 0. \tag{6.112}$$

Combining this with the Bianchi identity one obtains two equations,

$$\partial_-J_+ + \tfrac{1-\nu}{2}[J_+,J_-] = 0 \tag{6.113}$$
$$\partial_+J_- - \tfrac{1+\nu}{2}[J_+,J_-] = 0.$$

It is now easy to see that the zero-curvature condition for the connection

$$D_+ = \partial_+ - \lambda_+ J_+, \qquad D_- = \partial_- - \lambda_- J_-, \tag{6.114}$$

can still be preserved if λ_+ and λ_- satisfy the constraint

$$\frac{1+\nu}{\lambda_+} + \frac{1-\nu}{\lambda_-} = 2, \tag{6.115}$$

i.e.

$$\lambda_\pm = \frac{1 \pm \nu}{1 \pm \lambda}. \tag{6.116}$$

Given this, an immediate question arises: Does the Lagrangian for the modified equations of motions exist? The following exercise answers this question in the case of the $SU(2)$ group.

Exercise 25 Show that equations (6.112) for $SU(2)$ follows from the Lagrangian

$$\mathcal{L} = \frac{8}{f^2}\Big(\partial_+\theta\partial_-\theta + \cos^2(\theta)\partial_+\chi_1\partial_-\chi_1 + \sin^2(\theta)\partial_+\chi_2\partial_-\chi_2 \\ - \frac{\nu}{2}\cos(2\theta)\left(\partial_+\chi_1\partial_-\chi_2 - \partial_-\chi_1\partial_+\chi_2\right)\Big), \tag{6.117}$$

where the Hopf coordinates (θ, χ_1, χ_2) are used.

There is a nice geometrical interpretation of the additional term in (6.117). The volume form for the round metric of the 3-sphere of unit radius is given by

$$\Omega = \sin(\theta)\cos(\theta)\,d\chi_1 \wedge d\theta \wedge d\chi_2 \,:\quad \int_{S^3} \Omega = 2\pi^2. \qquad (6.118)$$

Since this 3-form is closed, there is a local 2-form $d^{-1}\Omega$. With the Hopf coordinates it reads explicitly as

$$d^{-1}\Omega = \tfrac{1}{4}\cos(2\theta)\,d\chi_1 \wedge d\chi_2. \qquad (6.119)$$

Using this observation, the Lagrangian (6.117) can be written in the form

$$\mathcal{L} = 2\left(G_{ab}\partial_+ X^a \partial_- X^b - \tfrac{1}{2} B_{ab}\left(\partial_+ X^a \partial_- X^b - \partial_- X^a \partial_+ X^b\right)\right), \qquad (6.120)$$

where

$$\frac{1}{2!} B_{ab} dX^a \wedge dX^b = \frac{2\nu}{f^2}\cos(2\theta)\,d\chi_1 \wedge d\chi_2 = \frac{8\nu}{f^2}\,d^{-1}\Omega. \qquad (6.121)$$

The result of Exercise 25 can be generalized for any simple Lie group $\mathfrak{G}$. For this purpose, let us note that the 3-form Ω can be expressed in terms of the Maurer–Cartan 1-form $d\mathbf{g}\mathbf{g}^{-1}$:

$$\Omega = \frac{1}{48} \langle \left[d\mathbf{g}\mathbf{g}^{-1} \stackrel{\wedge}{,} d\mathbf{g}\mathbf{g}^{-1}\right] \stackrel{\wedge}{,} d\mathbf{g}\mathbf{g}^{-1} \rangle. \qquad (6.122)$$

It is not difficult to see that this formula defines a bi-invariant and closed 3-form for $\mathfrak{G}$. Since $d\Omega = 0$, there exists a local 2-form B such that[6]

$$dB = \frac{k}{\pi}\,\Omega, \quad \text{where} \quad k = \frac{8\pi\nu}{f^2}, \qquad (6.123)$$

and therefore the Lagrangian (6.120) can be introduced for any simple Lie group. It is now straightforward to show that this Lagrangian indeed yields the equations of motion (6.112), which can be equivalently rewritten in terms of the the right currents $\tilde{J}_\pm = \mathbf{g}^{-1}\partial_\pm \mathbf{g}$:

---

[6] For the case of $SU(N)$, the Killing form can be written as $\langle \mathfrak{a},\mathfrak{b}\rangle = 2\mathrm{Tr}(\mathfrak{a}\mathfrak{b})$, where the trace is taken over the N-dimensional fundamental representation, and equation (6.123) takes the form

$$dB = \frac{k}{12\pi}\,\mathrm{Tr}\left(d\mathbf{g}\mathbf{g}^{-1} \wedge d\mathbf{g}\mathbf{g}^{-1} \wedge d\mathbf{g}\mathbf{g}^{-1}\right).$$

$$\partial_+ \tilde{J}_- + \tfrac{1-\nu}{2}\, [\tilde{J}_+, \tilde{J}_-] = 0 \qquad (6.124)$$
$$\partial_- \tilde{J}_+ - \tfrac{1+\nu}{2}\, [\tilde{J}_+, \tilde{J}_-] = 0.$$

The constructed integrable model is known as the Wess–Zumino–Witten–Novikov (WZWN) model and is usually represented by the action

$$\mathcal{S}_{\text{WZWN}} = \frac{1}{f^2} \int d^2 x \sqrt{-\eta}\, \eta^{\mu\nu}\, \text{Tr}\left(\partial_\mu g \partial_\nu g^{-1}\right) + \mathrm{k}\, W[g], \qquad (6.125)$$

where $W[g]$ is called the Wess–Zumino term. It deserves mentioning that $\nu = \frac{\mathrm{k}f^2}{8\pi} = \pm 1$ are very special cases of the WZWN model. Indeed, if $\nu = 1$ then, as follows from equations (6.112), (6.124),

$$\partial_- J_+ = 0, \qquad \partial_+ \tilde{J}_- = 0. \qquad (6.126)$$

The general solution of these equations is

$$g(x^0, x^1) = g_+(x^+) g_-(x^-), \qquad (6.127)$$

where $g_+(x^+)$ and $g_-(x^-)$ are arbitrary $\mathfrak{G}$-valued functions of one coordinate. At $\nu = -1$ the factorization is instead $g(x^0, x^1) = g_-(x^-) g_+(x^+)$.

6.7.2 NLSM with *B*-field

The WZWN model leads to an important generalization of NLSM.

Let the target space be a *D*-manifold equipped with an affine connection Γ and a covariantly constant Riemannian metric G. Consider the harmonic map problem which describes a map of the two-dimensional world sheet $\mathcal{W}_2$ into the affine-metric manifold

$$\partial_+ \partial_- X^a + \Gamma^a{}_{bc}\, \partial_+ X^b \partial_- X^c = 0, \qquad (6.128)$$

where $\Gamma^a{}_{bc}$ stands for the Christoffel symbol of the affine connection. For a general target space background, equations (6.128) cannot be derived from the variational principal. However, as was observed in [18], if the connection is compatible with the metric and the covariant torsion tensor

$$H_{abc} = G_{ad}\left(\Gamma^d{}_{bc} - \Gamma^d{}_{cb}\right) \qquad (6.129)$$

is a closed three-form,

$$H_{abc} = \partial_a B_{bc} + \partial_b B_{ca} + \partial_c B_{ab}, \qquad (6.130)$$

then (6.128) follows from the NLSM action with an additional (Kalb–Ramond) term[7]

$$S = \tfrac{1}{2} \int_{W_2} d^2x \left(\sqrt{-\eta}\, \eta^{\mu\nu} G_{ab} \partial_\mu X^a \partial_\nu X^b + \epsilon^{\mu\nu} B_{ab}\, \partial_\mu X^a \partial_\nu X^b \right). \tag{6.131}$$

The 2-form B_{ab} provides an anti-symmetric component to the affine connection and is sometimes known as the *torsion potential*.

6.8 Integrable examples of NLSM with B-field

6.8.1 Modified anisotropic $SU(2)$ PCF

The Wess–Zumino term does not spoil the integrability of the anisotropic $SU(2)$ PCF.

Exercise 26 Show that the NLSM

$$S = \int d^2x \left(\frac{2}{f_\perp^2} \left(j_+^1 j_-^1 + j_+^2 j_-^2 \right) + \frac{2}{f_\parallel^2}\, j_+^3 j_-^3 \right) + \mathrm{k} W[g] \tag{6.132}$$

possesses the ZCR of the form

$$D_+ = \partial_+ - i \left[\lambda_+^\perp \left(j_+^1\, \mathbf{t}_1 + j_+^2\, \mathbf{t}_2 \right) + \lambda_+^\parallel\, j_+^3 \mathbf{t}_3 \right] \tag{6.133}$$
$$D_- = \partial_- - i \left[\lambda_-^\perp \left(j_-^1\, \mathbf{t}_1 + j_-^2\, \mathbf{t}_2 \right) + \lambda_-^\parallel\, j_-^3 \mathbf{t}_3 \right],$$

where

$$\lambda_+^\perp = \frac{\cosh(\mu)}{\sinh(\mu) - \sinh(z_1)}$$
$$\lambda_+^\parallel = 1 + \coth\left(\tfrac{z_1+z_2}{2}\right) \frac{\cosh(z_1)}{\sinh(\mu) - \sinh(z_1)}$$
$$\lambda_-^\perp = \frac{\cosh(\mu)}{\sinh(\mu) + \sinh(z_2)} \tag{6.134}$$
$$\lambda_-^\parallel = 1 - \coth\left(\tfrac{z_1+z_2}{2}\right) \frac{\cosh(z_2)}{\sinh(\mu) + \sinh(z_2)}$$

and z_1 and z_2 are related to the couplings in (6.132) by

$$\frac{f_\parallel}{f_\perp} = \tanh\left(\tfrac{z_1+z_2}{2}\right), \qquad \tfrac{\mathrm{k}}{8\pi}\, f_\perp f_\parallel = \tanh\left(\tfrac{z_2-z_1}{2}\right). \tag{6.135}$$

---

[7] Recall that the antisymmetric Levi-Civita symbol $\epsilon^{\mu\nu}$ is a tensor density rather than a tensor, so we do not need a factor of $\sqrt{-\eta}$ in the B-term.

6.8.2 Modified Fateev model

The Fateev NLSM admits an integrable deformation which involves the B-field [9]. Components of the modified metric and B-field,

$$G = G_{\zeta\zeta}(\zeta)(d\zeta)^2 + \sum_{A,B\in\{R,L\}} G_{AB}(\zeta) dX^A dX^B$$

$$B = \frac{1}{2} \sum_{A,B\in\{R,L\}} B_{AB} \, dX^A \wedge dX^B, \tag{6.136}$$

depend on additional parameters $-\infty < h, \bar{h} < +\infty$ and read explicitly as

$$G_{\zeta\zeta} = \frac{1}{4u} \frac{(1+p^{-2}\kappa)(1+p^2\kappa)}{(1-\zeta^2)(1-\kappa^2\zeta^2)}$$

$$G_{RR} = \frac{1}{4u} \frac{(c+1)(\bar{c}-1)}{(1-\kappa^2)(c+\zeta)(\bar{c}-\zeta)} \left(1 + p^2\kappa - (p^2+\kappa)\kappa\zeta^2\right) \tag{6.137}$$

$$G_{LL} = \frac{1}{4u} \frac{(c+1)(\bar{c}-1)}{(1-\kappa^2)(c+\zeta)(\bar{c}-\zeta)} \left(1 + p^{-2}\kappa - (p^{-2}+\kappa)\kappa\zeta^2\right)$$

$$G_{RL} = -\frac{1}{4u} \frac{(c+1)(\bar{c}-1)}{(1-\kappa^2)(c+\zeta)(\bar{c}-\zeta)} (1-\kappa^2)\zeta,$$

and

$$B_{RL} = -\frac{\sqrt{(1+p^{-2}\kappa)(1+p^2\kappa)}}{4u} \frac{(c+1)(\bar{c}-1)}{(1-\kappa^2)(\bar{c}+c)} (1-\zeta) \left(h \frac{c-1}{c+\zeta} + \bar{h} \frac{\bar{c}+1}{\bar{c}-\zeta}\right), \tag{6.138}$$

where

$$c = +\sqrt{\frac{1+h^2}{\kappa^2+h^2}}, \quad \bar{c} = +\sqrt{\frac{1+\bar{h}^2}{\kappa^2+\bar{h}^2}}. \tag{6.139}$$

Exercise 27 Consider the limiting form of the modified Fateev NLSM as $\kappa, p^{-2}, h, \bar{h} \to 0$, whereas κp^2, u, and $\bar{h}/h$ are kept fixed. Show that the resulting model is equivalent to (6.132) upon the following identification of the parameters:

$$\kappa p^2 = \frac{f_\parallel^2}{f_\perp^2} - 1, \quad 4u = f_\parallel^2, \quad \frac{\bar{h}}{h} = -e^{z_1-z_2}. \tag{6.140}$$

6.9 Wick rotation

We now turn to the quantization of the $1+1$ integrable models described earlier. Working with QFT it is usually convenient to perform the Wick rotation [19] and consider the model in the Euclidean picture. The Wick rotation means that we make a continuation to imaginary values of time

$$x^0 = -ix^2. \tag{6.141}$$

The variable x^2, which runs from $-\infty$ to $+\infty$, is usually referred to as the Euclidean time. The Wick rotation brings $x^\pm = x^0 \pm x^1$ to

$$x^+ \mapsto -iz, \quad x^- \mapsto -i\bar{z}, \tag{6.142}$$

where $z = x_2 + ix_1$ and $\bar{z} = x_2 - ix_1$ are a pair of complex conjugate coordinates.[8]

Let us consider the Euclidean cylinder with the compact coordinate $x_1 \sim x_1 + R$, and the coordinate x_2 along the cylinder. Obviously, in this geometry there are two natural Hamiltonian pictures to choose between. Firstly (see Figure 6.8), one can interpret x_2 as the Euclidean time and associate the space of states $\mathcal{H}_R$ to the 'equal-time' section $x_2 = const$ which is the circle of circumference R; this choice was actually implied through our earlier discussion.

The 'time' evolution of the states in this picture (which will be referred to later as the R-channel) is described by the Hamiltonian

$$\hat{\mathbb{H}}_R = -\frac{2}{\pi} \int_0^R dx_1 \, T_{22}(x_2, x_1), \tag{6.143}$$

where $T_{22}(x_2, x_1) = -T_{00}(x^0, x^1)|_{x^0 = -ix_2}$. If $E_0(R)$ is the lowest eigenvalue the partition function of the QFT on a very long but finite cylinder (of length L) is

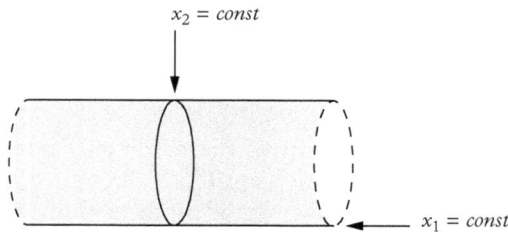

Figure 6.8 *The space-time cylinder. The system can be interpreted as a QFT in finite volume with x_2 playing the role of the Euclidean time (the 'R-channel'); or as a QFT at the finite temperature R_{-1} (the 'L-channel'). Reproduced with permission from John Wiley & Sons.*

---

[8] After the Wick rotation, there is no need to distinguish between the upper and lower indices, i.e. $x^\mu = x_\mu$ ($\mu = 1, 2$).

$$Z = e^{-LE_0(R)}. \qquad (6.144)$$

Alternatively, one could take the compact coordinate x_1 for the Euclidean time (this means considering the QFT at finite temperature R^{-1}) so that the space $\mathcal{H}_\infty$ of states would be associated with the infinite (or very long) line $x_1 = const$. The associated Hamiltonian is

$$\hat{\mathbb{H}}_\infty = \frac{2}{\pi} \int dx_2 \; T_{11}(x_2, x_1) \qquad (6.145)$$

and the partition function can be calculated in this Hamiltonian picture ('the L-channel') as

$$Z = \mathrm{Tr}_{\mathcal{H}_\infty}\left[e^{-R\hat{\mathbb{H}}_\infty} \right]. \qquad (6.146)$$

6.10 Free scalar field on a cylinder

As an illustration of a free scalar field on a cylinder [20], [21], we consider the theory described by the Euclidean action

$$\mathcal{A}[\phi(x)] = \frac{1}{4\pi} \int dx_2 \int_0^R dx_1 \left((\partial_\mu \phi)^2 + m^2 \phi^2 \right) \qquad (6.147)$$

with periodic BC $\phi(x_2, x_1 + R) = \phi(x_2, x_1)$ imposed.

Exercise 28 Show that the Matsubara propagator $D(x) = D(x_1, x_2)$ for the model (6.147) can be written in the form of an absolutely convergent series

$$D(x) = \sum_{n=-\infty}^{\infty} K_0(m|z - inR|) \qquad (z = x_2 + ix_1). \qquad (6.148)$$

Here we use the conventional modified Bessel function: $K_\nu(r) = \frac{1}{2} \int_{-\infty}^{\infty} d\theta \; e^{\nu\theta - r\cosh\theta}$.

Hint: $D(x)$ is defined by the set of conditions:

$$(m^2 - \partial_\mu \partial^\mu) D(x) = 2\pi \delta^{(2)}(x) \qquad (6.149)$$
$$D(x_2, x_1 + R) = D(x_2, x_1), \quad \lim_{x_2 \to \pm\infty} D(x_2, x_1) = 0.$$

Let us consider the model (6.147) in the L-channel. First of all we note that as $L \to \infty$ the eigenvalues of $\hat{\mathbb{H}}_\infty$ are composed of an extensive part proportional to the length i.e. $\langle \hat{\mathbb{H}}_\infty \rangle = L\mathcal{E}_0 + o(1)$ and dimensional analysis shows that $\mathcal{E}_0 = Cm^2$, where C is some dimensionless constant. For an infinitely large L the model describes a non-interacting Bose gas of particles of mass m with the relativistic dispersion relation

$$\varepsilon(p) = \sqrt{m^2 + p^2}. \tag{6.150}$$

Using the textbook formula (see e.g. Ch.56 in [22]) for the partition function of a non-interacting Bose gas, one finds

$$\log Z = -Cm^2 RL - L \int_{-\infty}^{\infty} \frac{dp}{2\pi} \log\left(1 - e^{-R\varepsilon(p)}\right), \tag{6.151}$$

and, hence, taking into account the general relation (6.140),

$$E_0(R) = Cm^2 R + \frac{\pi}{R}\mathfrak{F}(r), \tag{6.152}$$

where $r = mR$ and we use

$$\mathfrak{F}(r) = \frac{r}{2\pi^2} \int_{-\infty}^{\infty} d\theta \, \cosh\theta \, \log\left(1 - e^{-r\cosh\theta}\right). \tag{6.153}$$

Exercise 29 Show that

$$\mathfrak{F}(r) \to \begin{cases} -\frac{1}{\pi^2}\left(\frac{r\pi}{2}\right)^{\frac{1}{2}} e^{-r} + O(r^{-\frac{1}{2}}e^{-r}) & \text{as } r \to \infty \\ -\frac{1}{6} + \frac{r}{2\pi} + \frac{r^2}{4\pi^2} \log\left(\frac{r}{4\pi}e^{\gamma_E - \frac{1}{2}}\right) + O(r^4) & \text{as } r \to 0 \end{cases} \tag{6.154}$$

(here $\gamma_E = 0.577216\ldots$ stands for the Euler constant).

Using (6.152), (6.154) it is easy to see that, at $m = 0$, the ground state energy in the R-channel is given by

$$E_0(R)|_{m=0} = -\frac{\pi}{6R}. \tag{6.155}$$

It is instructive to reproduce this result from the canonical quantization in the R-channel. At $m = 0$, the equation of motion for the action (6.147) turns out to be the Laplace equation

$$\partial\bar{\partial}\phi = 0. \tag{6.156}$$

Here and below we use the shortcut notations $\partial = \frac{\partial}{\partial z}$ and $\bar{\partial} = \frac{\partial}{\partial \bar{z}}$. Taking into account the periodic BC, $\phi(x_1 + R, x_2) = \phi(x_1, x_2)$, one has

$$\phi(x) = \phi_0 - \frac{2\pi}{R} i(z + \bar{z}) \mathcal{P} + i \sum_{n=-\infty}^{\infty} \frac{a_n^{(+)}}{n} e^{-\frac{2\pi}{R} n z} + i \sum_{n=-\infty}^{\infty} \frac{a_n^{(-)}}{n} e^{-\frac{2\pi}{R} n \bar{z}}. \quad (6.157)$$

The canonical quantization procedure implies that

$$\left[a_n^{(+)}, a_m^{(+)}\right] = \left[a_n^{(-)}, a_m^{(-)}\right] = \frac{n}{2} \delta_{n+m,0}, \quad \left[a_n^{(+)}, a_m^{(-)}\right] = 0 \quad (6.158)$$

and

$$\mathcal{P} = \frac{1}{2i} \frac{\partial}{\partial \phi_0} \quad (6.159)$$

is the momentum conjugate to the zero-mode ϕ_0. As usual, one can construct the Fock space $\mathcal{F}$ for the algebra of the creation and annihilation operators $a_n^{(\pm)}$, $n = \pm 1, \pm 2, \ldots$. The Fock space consists of the Fock vacuum $|0\rangle$:

$$a_n^{(\pm)} |0\rangle = 0 \quad \text{for} \quad n = 1, 2, 3 \ldots \quad (6.160)$$

and the states generated by the action of $a_n^{(\pm)}$ with $n < 0$. The Hamiltonian of the model takes the form

$$\hat{\mathbb{H}}_R = \frac{2\pi}{R} \left(-\frac{c}{12} + 2\mathcal{P}^2 + 2 \sum_{n=1}^{\infty} \left(a_{-n}^{(+)} a_n^{(+)} + a_{-n}^{(-)} a_n^{(-)} \right) \right), \quad (6.161)$$

where c is some constant which appears from the normal ordering of the creation–annihilation. It acts in the space $\mathcal{H}_R = L^2(-\infty < \varphi_0 < \infty) \otimes \mathcal{F}$, where L^2 stands for the space of square integrable functions. This Hilbert space admits a spectral decomposition in the form of a direct integral

$$\mathcal{H}_R = \int_{-\infty < P < \infty}^{\oplus} \mathcal{F}_P, \quad (6.162)$$

and the energy spectrum in the eigensubspaces $\mathcal{F}_P = e^{2iP\phi_0} \mathcal{F}$ is given by

$$E = \frac{2\pi}{R} \left(-\frac{c}{12} + 2P^2 + \mathcal{N} \right), \quad \mathcal{N} = 0, 1, 2 \ldots. \quad (6.163)$$

The ground state, i.e. the state with the minimum energy, corresponds to $P = 0$ and $\mathcal{N} = 0$, and hence

$$E_0(R) = -\frac{\pi c}{6R}. \tag{6.164}$$

To calculate the constant c, let us note that it is related to the vacuum expectation value of the energy–momentum tensor. At $m = 0$, the equation of motion $\partial\bar{\partial}\phi = 0$ possesses the conformal invariance, so that the model is a *conformal field theory* (*CFT*). For the Euclidean 2D CFT, the trace of the energy–momentum tensor vanishes $T_{11} + T_{22} = 0$ and its non-trivial components

$$T = T_{11} - T_{22} + 2iT_{12}, \qquad \bar{T} = T_{11} - T_{22} - 2iT_{12}, \tag{6.165}$$

satisfy the Cauchy–Riemann conditions:

$$\bar{\partial} T = \partial \bar{T} = 0. \tag{6.166}$$

Also, as follows from the general relation (6.143),

$$E_0(R) = \frac{R}{2\pi} \langle (T + \bar{T}) \rangle = \frac{R}{\pi} \langle T \rangle. \tag{6.167}$$

For the theory described by the action (6.147) with $m = 0$, the holomorphic and antiholomorphic components of the energy–momentum tensor are given by

$$T(z) = -(\partial\phi)^2, \qquad \bar{T}(\bar{z}) = -(\bar{\partial}\phi)^2. \tag{6.168}$$

The product of two fields $\partial\phi$ ($\bar{\partial}\phi$) taken at the same point is singular.

Exercise 30 Show that the singular part of the *operator product expansion* (*OPE*) of the fields $\partial\phi$ is given by

$$\partial\phi(z_1)\partial\phi(z_2) = -\frac{1}{2} \frac{1}{(z_1 - z_2)^2} + O(1) \quad \text{as} \quad z_1 \to z_2. \tag{6.169}$$

In writing the formulae (6.168) a certain regularization procedure was assumed, namely

$$T(z) = -(\partial\phi)^2 \equiv -\lim_{\epsilon \to 0} \left(\frac{1}{2\epsilon^2} + \partial\phi(z+\epsilon)\partial\phi(z) \right), \tag{6.170}$$

and similar for $\bar{T}$.

Exercise 31 Using the Matsubara propagator obtained in Exercise 28 show that

$$\langle \phi(x)\phi(0)\rangle = \frac{\pi}{mR} - \frac{1}{2}\log\left[4\sinh\left(\tfrac{\pi z}{R}\right)\sinh\left(\tfrac{\pi \bar{z}}{R}\right)\right] + O(m) \quad \text{as } m \to 0 \quad (6.171)$$

and hence

$$\lim_{m\to 0}\langle \partial\phi(z)\partial\phi(0)\rangle = -\frac{\pi^2}{2R^2\sinh^2\left(\tfrac{\pi z}{R}\right)}. \quad (6.172)$$

Up to a sign factor, the constant term in the short distance expansion of the massless two-point correlation function (6.172),

$$\langle \partial\phi(z)\partial\phi(0)\rangle_{m=0} = -\frac{1}{2z^2} + \frac{\pi^2}{6R^2} + O(z^2), \quad (6.173)$$

coincides with the vacuum expectation value $\langle T\rangle$, i.e.

$$\langle T\rangle = -\frac{\pi^2}{6R^2}. \quad (6.174)$$

Combining this with (6.167), (6.170) one finds that the constant c from equation (6.164) equals one. Thus we come to equation (6.155), which had been already obtained using the L-channel picture.

Exercise 32 Similar to $\langle (\partial\phi)^2\rangle = \frac{\pi^2}{6R^2}$, show that

$$\langle (\partial\phi)^4\rangle = \frac{\pi^4}{12R^4} \quad (6.175)$$

$$\langle (\partial^2\phi)^2\rangle = \frac{\pi^4}{15R^4}.$$

(The composite holomorphic fields $(\partial\phi)^n$ can be defined recursively as the regular part of the OPE $\partial\phi(z+\epsilon)(\partial\phi)^{n-1}(z)$ at $\epsilon = 0$. Also $(\partial^2\phi)^2 = \text{reg}\left(\partial^2\phi(z+\epsilon)\partial^2\phi(z)\right)|_{\epsilon=0}$.)

6.11 General structure of the Hilbert space of 2D CFT

The model discussed in Section 6.10 provides a simple illustration of the general relations for 2D CFT. For such class of theories, the energy–momentum tensor is traceless, $T_{11} = -T_{22}$, and its spin $+2$ and -2 components (6.165) satisfy the Cauchy–Riemann conditions (6.166). In the world sheet geometry of an infinite flat cylinder, T and $\bar{T}$ admit expansions [23], [20] of the form[9]

$$T(z) = \left(\frac{2\pi}{R}\right)^2 \left[-\frac{c}{24} + \sum_{n=-\infty}^{\infty} L_n \, e^{-\frac{2\pi}{R}nz}\right], \quad \bar{T}(\bar{z}) = \left(\frac{2\pi}{R}\right)^2 \left[-\frac{c}{24} + \sum_{n=-\infty}^{\infty} \bar{L}_n \, e^{-\frac{2\pi}{R}n\bar{z}}\right],$$
(6.176)

where the expansion coefficients generate two copies of the Virasoro algebra,

$$[L_n, L_m] = (n-m)L_{n+m} + \frac{c}{12} n(n^2-1) \, \delta_{n+m,0}$$
$$[\bar{L}_n, \bar{L}_m] = (n-m)\bar{L}_{n+m} + \frac{c}{12} n(n^2-1) \, \delta_{n+m,0} \tag{6.177}$$
$$[L_n, \bar{L}_m] = 0.$$

Here c is the so-called central charge—an important characteristic of a CFT. The Hamiltonian of a CFT is given by

$$\hat{\mathbb{H}}_R = \frac{2\pi}{R}\left(-\frac{c}{12} + L_0 + \bar{L}_0\right). \tag{6.178}$$

The space of states of a general CFT decomposes into a sum $\mathcal{H}_R = \oplus_a \mathcal{H}_a$, where $\mathcal{H}_a = \mathcal{V}_{\Delta_a} \otimes \bar{\mathcal{V}}_{\bar{\Delta}_a}$ is an irreducible highest weight representation of $vir \oplus \overline{vir}$ with the highest weight $(\Delta_a, \bar{\Delta}_a)$. The space $\mathcal{V}_\Delta$ is spanned by the vectors

$$L_{-n_1} L_{-n_2} \ldots L_{-n_N} |\Delta\rangle, \quad 0 \leq n_1 \leq n_2 \leq \ldots \leq n_N,$$

---

[9] Recall how equations (6.176) are dictated by the conformal invariance. Let

$$T(w) = \sum_{n=-\infty}^{\infty} \frac{L_n}{w^{n+2}}$$

be the holomorphic component of the energy–momentum tensor for a CFT defined on a complex plane without the origin $w = 0$. Conformal invariance implies that under a conformal map $w = w(z)$, $T(w)$ transforms as a projective connection, i.e. according to the rule

$$T(w) \mapsto T(z) = \left[(\partial w)^2 \, T(z) + \frac{c}{12} \{w, z\}\right]_{w=w(z)},$$

where $\{w, z\}$ is the Schwarzian derivative, $\{w, z\} = \frac{\partial^3 w}{\partial w} - \frac{3}{2}\left(\frac{\partial^2 w}{\partial w}\right)^2$. The conformal map $w(z) = e^{\frac{2\pi z}{R}}$ transforms the infinite cylinder to the complex plane without the origin $w = 0$ and equations (6.176) are a specialization of the general relation for this particular map. Also, the commutation relations (6.177) follow from the singular part of the OPE:

$$T(w)T(0) = \frac{c}{2w^4} + \frac{2T(0)}{w^2} + \frac{\partial T(0)}{w} + O(1), \quad \bar{T}(w)\bar{T}(0) = \frac{c}{2\bar{w}^4} + \frac{2\bar{T}(0)}{\bar{w}^2} + \frac{\bar{\partial} \bar{T}(0)}{\bar{w}} + O(1), \quad \bar{T}(w)T(0) = O(1).$$

where $|\Delta\rangle$ is the highest weight state which satisfies

$$L_n|\Delta\rangle = 0 \quad \text{for} \quad n > 0; \quad L_0|\Delta\rangle = \Delta|\Delta\rangle. \tag{6.179}$$

Exercise 33 For the massless model (6.147) show that

$$L_n = \sum_{m \neq 0, n} a_m^{(+)} a_{n-m}^{(+)} + 2\mathcal{P} a_n^{(+)}, \quad L_0 = \mathcal{P}^2 + 2\sum_{m>0} a_{-m}^{(+)} a_m^{(+)}$$

$$\bar{L}_n = \sum_{m \neq 0, n} a_m^{(-)} a_{n-m}^{(-)} + 2\mathcal{P} a_n^{(-)}, \quad \bar{L}_0 = \mathcal{P}^2 + 2\sum_{m>0} a_{-m}^{(-)} a_m^{(-)}. \tag{6.180}$$

Using equations (6.158) show that the operators (6.180) satisfy the commutation relations (6.177) with $c = 1$. In this case $\mathcal{H}_a = \mathcal{F}_P$ and $\Delta_a = P^2$. States of the form $e^{2P\phi_0}|0\rangle$ are the conformal primary states.

6.12 Dimensional transmutation in PCF

Earlier, it was discussed that any classical NLSM possesses scale (in fact, conformal) invariance. If we assume that the model remains scale invariant at the quantum level, its ground state energy in the R-channel should be given by a simple formula $E_0(R) = -\frac{\pi c}{6R}$, where c is some constant. In fact, this naive expectation turns out to be wrong. The conformal symmetry is spoiled by quantum effects for most cases of NLSM. The corresponding mechanism is known as dimensional transmutation. Here we discuss dimensional transmutation using our favourite example the *principal chiral field*, whose Euclidean action is given by

$$\mathcal{A}[\mathbf{g}(x)] = -\int d^2x \, \frac{1}{2f_0^2} \left\langle \partial_\mu \mathbf{g} \mathbf{g}^{-1}, \partial_\mu \mathbf{g} \mathbf{g}^{-1} \right\rangle. \tag{6.181}$$

Let us study the effective action which arises in the one-loop approximation. We consider the wave functional

$$\Psi[\mathbf{g}(x_1)] = \int_{\mathbf{g}(x)|_{x_2=0}=\mathbf{g}(x_1)} \mathcal{D}\mathbf{g}(x) \, e^{-\mathcal{A}[\mathbf{g}(x)]}, \tag{6.182}$$

which is just the Euclidean analogue of the Schrödinger wave function. The classical limit would correspond to taking the minimum of the action $\mathcal{A}$ with the Dirichlet boundary conditions $\mathbf{g}(x)|_{x_2=0} = \mathbf{g}(x_1)$ (see Figure 6.9). To perform the integration we write the field $\mathbf{g}$ in the form

$$\mathbf{g}(x) = \mathbf{h}(x)\mathbf{g}_{\text{cl}}(x), \tag{6.183}$$

288 *Integrability in 2D fields theory/sigma-models*

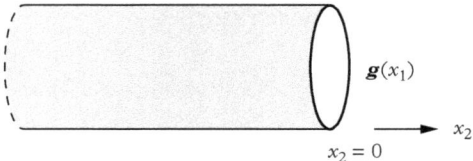

Figure 6.9 *The Dirichlet boundary condition* $g(x)|_{x_2=0} = g(x_1)$ *entering into equation (6.182) that defines the functional* $[g(x_1)]$. *Reproduced with permission from John Wiley & Sons.*

where g_{cl} is a solution of the classical equations of motion. Integration over $h(x)$ corresponds to the inclusion of quantum fluctuations. It is clear that we should fix the boundary condition

$$h(x)|_{x_2=0} = I. \tag{6.184}$$

Notice that $J_\mu \equiv \partial_\mu g g^{-1}, \tilde{J}_\mu \equiv g^{-1}\partial_\mu g = \tilde{J}_\mu^{\text{cl}} + g_{\text{cl}}^{-1}(h^{-1}\partial_\mu h)g_{\text{cl}}$ and hence

$$-\frac{1}{2f_0^2}\langle J_\mu, J_\mu\rangle = -\frac{1}{2f_0^2}\langle \tilde{J}_\mu, \tilde{J}_\mu\rangle \tag{6.185}$$

$$= -\frac{1}{2f_0^2}\langle J_\mu^{\text{cl}}, J_\mu^{\text{cl}}\rangle - \frac{1}{2f_0^2}\langle h^{-1}\partial_\mu h, h^{-1}\partial_\mu h\rangle - \frac{1}{f_0^2}\langle J_\mu^{\text{cl}}, h^{-1}\partial_\mu h\rangle.$$

If we want to examine only one-loop corrections to the classical action, we have to consider only small fluctuations of the matrix h. We can write

$$h = e^{i\phi} = I + i\phi - \frac{1}{2}\phi^2 + \ldots. \tag{6.186}$$

The substitution of (6.183), (6.186) into the Euclidean action (6.181) gives

$$\mathcal{A}[g(x)] = \mathcal{A}[g_{\text{cl}}(x)] + \frac{1}{2f_0^2}\int d^2x\langle\partial_\mu\phi, \partial_\mu\phi\rangle + \frac{1}{2f_0^2}\int d^2x\langle J_\mu^{\text{cl}}, [\phi, \partial_\mu\phi]\rangle + O(\phi^3) \tag{6.187}$$

(we have omitted the term $J_\mu^{\text{cl}}\partial_\mu\phi$ because of the classical equations of motion $\partial_\mu J_\mu^{\text{cl}} = 0$). Now the term of the effective action which is quadratic in $J^{\text{cl}} = i\sum_a j_a^{\text{cl}} t^a$ with $[t^a, t^b] = f^{abc} t^c$, will be given by the Feynman diagram shown in Figure 6.10

$$\delta\mathcal{A} = -\int \frac{d^2q}{(2\pi)^2} j_a^{\text{cl}}(q) j_b^{\text{cl}}(-q) \times \frac{1}{4} f^{acd} f^{bcd} \int \frac{d^2p}{(2\pi)^2} \frac{(2p+q)_\mu (2p+q)_\nu}{8p^2(p+q)^2}. \tag{6.188}$$

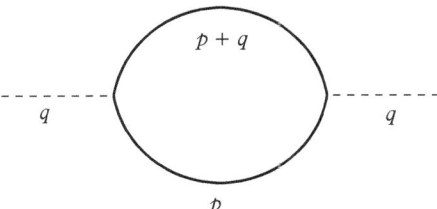

Figure 6.10 *The Feynman diagram that gives a one-loop correction to the classical action. Reproduced with permission from John Wiley & Sons.*

Notice that $f_{abc}f_{dbc} = C_2(\mathfrak{G})\,\delta^{ad}$, where $C_2(\mathfrak{G})$ is the value of the quadratic Casimir in the adjoint representation. If we extract the ultraviolet divergent contribution, the formula takes the form

$$\delta A = -\int \frac{d^2q}{(2\pi)^2}\, j_a(q)j_a(-q) \times \frac{1}{4}C_2(\mathfrak{G}) \int \frac{d^2p}{(2\pi)^2}\, \frac{p_\mu p_\nu}{p^4} + \text{finite part} \qquad (6.189)$$

$$= -\frac{C_2(\mathfrak{G})}{8\pi}\,\log(\Lambda) \int d^2x\, \frac{1}{2}\left\langle \partial_\mu g_{\rm cl}g_{\rm cl}^{-1},\partial_\mu g_{\rm cl}g_{\rm cl}^{-1}\right\rangle + \text{finite part},$$

where Λ is a momentum cut-off. If we introduce now the renormalized coupling constant by the formula

$$\frac{1}{f^2} = \frac{1}{f_0^2} - \frac{C_2(\mathfrak{G})}{8\pi}\,\log(\Lambda/E), \qquad (6.190)$$

the effective action will be finite in the given order. Notice that, to make the argument of the logarithm in (6.190) dimensionless, we introduce some typical energy scale E (say, the temperature R^{-1}). The formula (6.190) can be rewritten in the form

$$\frac{4}{f^2} = -\frac{C_2(\mathfrak{G})}{2\pi}\,\log(E_*/E), \qquad (6.191)$$

where

$$E_* = \Lambda\, e^{-\frac{8\pi}{f_0^2 C_2(\mathfrak{G})}}. \qquad (6.192)$$

The consistent removal of the UV cut-off requires that the bare coupling be given by a cutoff dependence $f_0 = f_0(\Lambda)$ such that E_* remains fixed as $\Lambda \to \infty$. The dimensionful parameter E_* sets a *renormalization group* (RG) invariant scale to the problem and can be interpreted as an inverse correlation length. This way the bare dimensionless coupling f_0 transmutes into the energy scale E_*.

Let us consider equation (6.191) for the case $SU(2)$, where $C_2(SU(2)) = j(j+1)|_{j=1} = 2$. As already pointed out, the ratio $\frac{2}{f}$ is interpreted as the radius of the 3-sphere. Thus we see that at the ultraviolet limit, $E \sim R^{-1} \to \infty$, the radius of the 3-

sphere becomes infinitely large, i.e. the sphere effectively becomes 3D flat Euclidean space. In standard terminology, the theory becomes 'asymptotically free' at the short distance limit. At the same time as E approaches to E_*, the radius of the sphere shrinks to zero. In this regime the theory becomes strongly interacting, so that the perturbative expansion cannot be applied any more.

Similar considerations can be used for any simple Lie group. This suggests that the dimensionless combination $\frac{RE_0}{\pi}$ is a function of the scaling variable $r = RE_*$, and

$$\frac{RE_0}{\pi} \asymp -\frac{1}{6}\dim(\mathfrak{G}) + \sum_{n=1}^{\infty} c_n \, [f(r)]^{2n} \quad \text{as} \quad r \to 0. \tag{6.193}$$

Here c_n are some numerical coefficients and $f(r)$ stands for the 'running' coupling constant:

$$[f(r)]^2 = \frac{8\pi}{C_2(\mathfrak{G})\log(1/r)} + O\!\left(\log(\log(1/r))/\log^2(1/r)\right). \tag{6.194}$$

In the large-R limit, similar to the model (6.147),

$$\frac{RE_0}{\pi} = Cr^2 + O(\sqrt{r}e^{-r}), \tag{6.195}$$

where C is some dimensionless constant.

6.13 RG flow equations

The Ricci tensor for the round 3-sphere of radius $\mathfrak{r}$ is given by $R_{ab} = \frac{2}{\mathfrak{r}^2} G_{ab}$. Also, as has been shown previously, $E\frac{d\mathfrak{r}}{dE} = -\frac{1}{\mathfrak{r}}$. Combining these two relations one finds

$$\partial_\tau G_{ab} = -R_{ab}, \tag{6.196}$$

where we use the 'RG time'

$$\tau = \frac{1}{2\pi}\log(E^*/E). \tag{6.197}$$

Exercise 34 [24, 28]. Show that for the $O(N)$ NLSM, the radius $\mathfrak{r}$ of the S^{N-1} sphere flows at the lowest perturbative order according to the equation

$$\mathfrak{r}^2 = -(N-2)\,\tau,$$

and therefore equation (6.196) is still satisfied. (Useful fact: For the round sphere S^D of radius $\mathfrak{r}$, the Ricci tensor is given by $R_{ab} = \frac{D-1}{\mathfrak{r}^2} G_{ab}$.)

Exercise 35 [12]. Show that the sausage metric,

$$G_{ab}dX^a dX^b = \frac{1}{g^2}\left[\frac{(d\zeta)^2}{(1-\zeta^2)(1-\kappa^2\zeta^2)} + \frac{(1-\zeta^2)(d\alpha)^2}{1-\kappa^2\zeta^2}\right], \qquad (6.198)$$

satisfies the partial differential equation (6.196) as

$$\kappa(\tau) = -\tanh\left(\frac{2\pi\tau}{n}\right), \quad g^2(\tau) = \frac{2\pi}{n\kappa(\tau)}, \qquad (6.199)$$

where n stands for a RG-invariant (i.e. τ-independent) constant.[10]

Exercise 36 Show that the ansatz

$$G_{ab}dX^a dX^b = \frac{1}{4e^2}\left(\frac{(\mu-\nu)\mu\,(d\mu)^2}{(a^2+\mu)(b^2+\mu)(c^2+\mu)} + \frac{(\nu-\mu)\nu\,(d\nu)^2}{(a^2+\nu)(b^2+\nu)(c^2+\nu)}\right), \qquad (6.200)$$

where the e^2, a, b, c are some functions of τ, does not solve the equation (6.196). Notice that (6.200) is the intrinsic metric of the ellipsoid $(\frac{x}{a})^2 + (\frac{y}{b})^2 + (\frac{z}{c})^2 = 1$ with $a > b > c$ (see Section 6.1.1).

Remarkably, the partial differential equation (6.196) describes the RG flow in the general NLSM at the lowest perturbative order [25]. The latter statement needs some clarification. To illustrate a subtlety we consider the sausage metric (6.198). Let us change variable ζ by ϕ, such that

$$\zeta = \frac{\text{cn}(\phi,\kappa)}{\text{dn}(\phi,\kappa)}, \qquad (6.201)$$

where we use the conventional Jacobi functions.[11] In the coordinate frame (ϕ,α) the sausage metric reads as[12]

$$G_{ab}dX^a dX^b = \frac{n\kappa}{2\pi}\left((d\phi)^2 + \text{sn}^2(\phi,\kappa)(d\alpha)^2\right), \qquad (6.202)$$

---

[10] The scalar curvature R_G for the sausage metric is given by

$$R_G = \frac{4\pi}{n}(\kappa^{-1}-\kappa)\frac{1+\kappa^2\zeta^2}{1-\kappa^2\zeta^2} \quad : \quad \frac{4\pi}{n}(\kappa^{-1}-\kappa) \leq R_G \leq \frac{4\pi}{n}(\kappa^{-1}+\kappa).$$

The maximum value of R_G is achieved at the tips $\zeta = \pm 1$ and can be thought of as a parameter of the perturbative expansion.

[11] Recall that

$$\text{cn}^2(\phi,\kappa) + \text{sn}^2(\phi,\kappa) = 1, \quad \text{dn}^2(\varphi,\kappa) = 1 - \kappa^2\text{sn}^2(\phi,\kappa).$$

[12] Notice that the metric (6.202) nicely interpolates between the round metric of the 2-sphere $(\lim_{\kappa\to 0}\text{sn}(\phi,\kappa) = \sin(\phi))$ and the cigar metric $\left(\lim_{\substack{\kappa\to 1\\ \phi-\text{fixed}}}\text{sn}(\phi,\kappa) = \tanh(\phi)\right)$.

where we replace g^2 by the RG invariant parameter n (6.199). Since the coordinate transformation (6.201) depends on the flowing parameter $\kappa = \kappa(\tau)$, the metric (6.202) does not satisfy the equation $\partial_\tau \dot{G}_{ab} = -R_{ab}$. This shows that the general form of the RG flow equation should admit the possibility of various coordinate transformations. Under an arbitrary infinitesimal reparameterization,

$$\delta G_{ab} = -(\nabla_a V_b + \nabla_b V_a)\, \delta\tau, \tag{6.203}$$

where V_a stands for an arbitrary covariant vector field (one form). Taking this into account, one arrives at the general form of the RG flow equation,

$$\partial_\tau G_{ab} = -(R_{ab} + \nabla_a V_b + \nabla_b V_a). \tag{6.204}$$

Exercise 37 [9]. Sometimes it is convenient to express the Jacobi functions in terms of the ϑ-functions and ϑ-constants. For example, if ϕ and κ in (6.202) are substituted by u and q:

$$\phi = \vartheta_3^2(0,q^2)u, \qquad \kappa = \frac{\vartheta_2^2(0,q^2)}{\vartheta_3^2(0,q^2)}, \tag{6.205}$$

then $\operatorname{sn}(\phi,\kappa) = \frac{\vartheta_3(0,q^2)\vartheta_1(u,q^2)}{\vartheta_2(0,q^2)\vartheta_4(u,q^2)}$. Using u and q instead ϕ and κ, the RG-flow in the sausage NLSM at the one-loop order is described by the following formulae:

$$G_{ab}\mathrm{d}X^\alpha \mathrm{d}X^b = \frac{n}{2\pi}\left(\vartheta_2^2(0,q^2)\vartheta_3^2(0,q^2)(\mathrm{d}u)^2 + \frac{\vartheta_1^2(u,q^2)}{\vartheta_4^2(u,q^2)}\,(\mathrm{d}\alpha)^2\right)$$

$$\mathrm{e}^{\frac{2\pi\tau}{n}} \equiv \left(\frac{E_*}{E}\right)^{\frac{1}{n}} = \frac{\vartheta_4(0,q)}{\vartheta_3(0,q)}. \tag{6.206}$$

Show that the metric (6.206) with the RG-invariant parameter n, satisfies the RG-flow equation (6.204) with

$$V_a = \partial_a \Psi \ : \quad \Psi(u,\tau) = \log\left(\frac{\vartheta_2(0,q^2)}{\vartheta_4(u,q^2)}\right). \tag{6.207}$$

It is usually assumed (in fact, it can be proved under certain assumptions) that for a general RG transformation, the covariant vector field V_a in (6.204) is a gradient, i.e. $V_a = \partial_a \Psi$. Taking this into account, one has

$$\partial_\tau G_{ab} = -(R_{ab} + \nabla_a \nabla_b \Psi + \nabla_b \nabla_a \Psi). \tag{6.208}$$

Also, it turns out that Ψ can be chosen to satisfy the additional relation

$$\partial_\tau \left(2\Psi - \log\sqrt{G}\right) = -\frac{1}{4}\left(-G^{ab}R_{ab} + 4\nabla_a\Psi\nabla^a\Psi - 4\nabla_a\nabla^a\Psi\right), \tag{6.209}$$

where $G = \det G_{ab}$.

Exercise 38 [9]. Show that G_{ab} and Ψ defined by equations (6.206) and (6.207) satisfy (6.209).

The RG flow equations are known to two-loop order for the general NLSM with torsion. At the lowest perturbative order the equations form a coupled system of PDEs [26, 27],

$$\begin{aligned}
\partial_\tau G_{ab} &= -\left(R_{ab} - \frac{1}{4}H_a{}^{cd}H_{cdb} + \nabla_a\nabla_b\Psi + \nabla_b\nabla_a\Psi\right) \\
\partial_\tau B_{ab} &= -\left(\frac{1}{2}\nabla_c H^c{}_{ab} - \nabla_c\Psi H^c{}_{ab}\right) \\
\partial_\tau\left(2\Psi - \log\sqrt{G}\right) &= -\frac{1}{4}\left(-G^{ab}R_{ab} + \frac{1}{12}H_{abc}H^{abc} + 4\nabla_a\Psi\nabla^a\Psi - 4\nabla_a\nabla^a\Psi\right),
\end{aligned} \tag{6.210}$$

where

$$H_{abc} = \partial_a B_{bc} + \partial_b B_{ca} + \partial_c B_{ab}. \tag{6.211}$$

6.14 String equations

Among the class of quantum NLSM, a special rôle belongs to the theories which keep the scale invariance at the quantum level. Within the one-loop approximation they correspond to the stationary solutions of the system (6.210), such that $\partial_\tau G_{ab} = \partial_\tau B_{ab} = 0$ and

$$\Psi = -C\tau + \Phi, \tag{6.212}$$

where C is some constant[13] and Φ is a τ-independent scalar field, which is usually referred to as the dilaton [26]. In the stationary case, equations (6.210) become a system of so-called *string equations* [29]:

---

[13] The constant C is expressed in terms of the central charge c of the CFT corresponding to the stationary solution of the RG-flow equation: $C = \frac{\pi(c-D)}{12}$, where D is the dimension of the target manifold.

$$R_{ab} - \frac{1}{4} H_a{}^{cd} H_{cdb} + 2\nabla_a \nabla_b \Phi = 0$$
$$\frac{1}{2} \nabla_c H^c{}_{ab} - \nabla_c \Phi H^c{}_{ab} = 0 \qquad (6.213)$$
$$-G^{ab} R_{ab} + \frac{1}{12} H_{abc} H^{abc} + 4\nabla_a \Phi \nabla^a \Phi - 4\nabla_a \nabla^a \Phi = 8C.$$

Exercise 39 [30]. Show that equations (6.213) admit a solution given by the following relations

$$G_{ab} dX^a dX^b = \frac{k}{2\pi} \left[(d\theta)^2 + \frac{\varepsilon^2 \cos^2(\theta) (d\chi_1)^2}{\cos^2(\theta) + \varepsilon^2 \sin^2(\theta)} + \frac{\sin^2(\theta) (d\chi_2)^2}{\cos^2(\theta) + \varepsilon^2 \sin^2(\theta)} \right]$$
$$\frac{1}{2!} B_{ab} dX^a \wedge dX^b = \pm \frac{k}{2\pi} \frac{\varepsilon^2 \sin^2(\theta)}{\cos^2(\theta) + \varepsilon^2 \sin^2(\theta)} \; d\chi_1 \wedge d\chi_2 \qquad (6.214)$$
$$\Phi = -\frac{1}{2} \log\left(\cos^2(\theta) + \varepsilon^2 \sin^2(\theta)\right).$$

Here k and ε are RG invariant parameters. At $\varepsilon = 1$, the corresponding NLSM coincides with the WZWN model. For general values of ε, the model is known as the deformed WZWN model.

Thus we see that the scale invariance imposes strong restrictions on both the target space metric G and the torsion potential B. Remarkably, the dilaton field is also incorporated in the action of the NLSM—it is coupled to the the scalar curvature $\mathcal{R}^{(2)}$ associated with the world sheet metric $\gamma_{\mu\nu}$:

$$\mathcal{A} = \frac{1}{2} \int d^2 x \left(\sqrt{\gamma} \gamma^{\mu\nu} \; G_{ab}(X) \partial_\mu X^a \partial_\nu X^b + i\epsilon^{\mu\nu} B_{ab}(X) \partial_\mu X^a \partial_\nu X^b + \frac{1}{2\pi} \; \Phi(X) \; \sqrt{\gamma} \; \mathcal{R}^{(2)} \right).$$
$$(6.215)$$

Here $\gamma_{\mu\nu}$ has the Euclidean signature and $\epsilon^{12} = -\epsilon^{21} = 1$. The term with the dilaton disappears for the flat world-sheet metric. However it gives a finite contribution to the energy–momentum tensor $T_{\mu\nu} = \frac{\pi}{\sqrt{\gamma}} \frac{\delta \mathcal{A}}{\delta \gamma^{\mu\nu}}$.

Exercise 40 Show that the dilaton term yields the following modification to the holomorphic and antiholomorphic components of $T_{\mu\nu}$:

$$T = -2\pi \; G_{ab}(X) \partial X^a \partial X^b + \partial^2 \Phi(X) \qquad (6.216)$$
$$\bar{T} = -2\pi \; G_{ab}(X) \bar\partial X^a \bar\partial X^b + \bar\partial^2 \Phi(X).$$

6.15 The quantum cigar

The cigar NLSM is a remarkable example of a CFT [15]. First of all one should check that the cigar metric satisfies the string equations.

Exercise 41 [14]. Show that

$$G_{ab}\mathrm{d}X^a\mathrm{d}X^b = \frac{n}{2\pi}\frac{\mathrm{d}Z\mathrm{d}Z^*}{1+|Z|^2}$$
$$\Phi = -\frac{1}{2}\log\left(1+|Z|^2\right) \tag{6.217}$$

satisfy the string equations (6.213).[14]

In the presence of the dilaton, the energy-momentum tensor should be modified according to the formulae (6.216). Thus we have

$$T = -n\frac{\partial Z \partial Z^*}{1+|Z|^2} - \frac{1}{2}\partial^2\log\left(1+|Z|^2\right) + O(1/n) \tag{6.218}$$
$$\bar{T} = -n\frac{\bar\partial Z \bar\partial Z^*}{1+|Z|^2} - \frac{1}{2}\bar\partial^2\log\left(1+|Z|^2\right) + O(1/n).$$

Notice that the parameter $\frac{8\pi}{n}$ coincides with the scalar curvature at the tip of the cigar (see footnote 10, where κ should be set to one). On the other hand, this parameter plays the rôle of the Planck constant, so that the symbol $O(1/n)$ in (6.218) stands for higher loop corrections. In the asymptotically flat domain of the target space where $|Z| \sim e^\phi \gg 1$, the field ϕ and $\alpha = \arg(Z)$ are split on the holomorphic and antiholomorphic components: $\phi \asymp \phi_+(z) + \phi_-(\bar z)$, $\alpha \asymp \alpha_+(z) + \alpha_-(\bar z)$. In this case, the non-trivial components of the energy–momentum tensor take the form

$$T \asymp -n\,(\partial\phi_+)^2 - \partial^2\phi_+ - n\,(\partial\alpha_+)^2 \tag{6.219}$$
$$\bar T \asymp -n\,(\bar\partial\phi_-)^2 - \bar\partial^2\phi_- - n\,(\bar\partial\alpha_-)^2.$$

6.15.1 W-algebra

The non-perturbative quantization of the cigar NLSM is based on the 'free field representation' of the classical W-currents discussed in Section 6.1. It goes along the following route [31], [32]. Let us introduce the quantum 'free fields' $\varphi_\pm$ and $\vartheta_\pm$:

$$\varphi_\pm(z) = \varphi_0^{(\pm)} - \frac{2\pi}{R}iza_0^{(\pm)} + i\sum_{l\neq 0}^{\infty}\frac{a_m^{(\pm)}}{m}e^{-\frac{2\pi m}{R}z}. \tag{6.220}$$

$$\vartheta_\pm(z) = \vartheta_0^{(\pm)} - \frac{2\pi}{R}izb_0^{(\pm)} + i\sum_{m\neq 0}^{\infty}\frac{b_m^{(\pm)}}{m}e^{-\frac{2\pi m}{R}z},$$

with

$$\left[a_m^{(\pm)}, a_l^{(\pm)}\right] = \left[b_m^{(\pm)}, b_l^{(\pm)}\right] = \frac{m}{2}\delta_{m+l,0}, \quad \left[\vartheta_0^{(\pm)}, b_0^{(\pm)}\right] = \frac{i}{2}, \tag{6.221}$$

---
[14] NLSM with 2D target space do not possess the B-field.

Notice that we do not include the Plank constant (i.e. $\frac{1}{n}$) in the commutation relations. Instead we will replace the classical free fields $\phi_\pm$ and $\alpha_\pm$ by the quantum fields (6.220) using the formal substitution $\phi_\pm \mapsto \frac{\varphi_\pm}{\sqrt{n}}$ and $\alpha_\pm \mapsto \frac{\vartheta_\pm}{\sqrt{n}}$.

Let us focus on one chirality and consider the W-current with positive spin. The case of negative spins is only notationally different. The generating currents $W_s(z)$ with $s = 2, 3, 4 \ldots$, can be characterized by the condition that they commute with the *screening operator*:[15]

$$\oint_z dw\, B(w) W_s(z) = 0 \qquad \text{where} \qquad B = \partial \vartheta_+ e^{-\frac{2\varphi_+}{\sqrt{n}}}, \tag{6.222}$$

and the integration is over a small contour around z. The vanishing of the integral (6.222) implies that the singular part of the OPE of $B(w) W_s(z)$ is a total derivative $\partial_w(\cdots)$. It is easy to check that this condition fixes W_2 uniquely up to overall normalization:

$$W_2(z) = -(\partial \varphi_+)^2 - (\partial \vartheta_+)^2 - \frac{1}{\sqrt{n}} \partial^2 \varphi_+. \tag{6.223}$$

We may now observe that this spin-2 current has exactly the same modification to the classical form as the holomorphic component of the energy–momentum tensor within the one-loop approximation. It seems natural to identify the quantum field W_2 with T. Notice the vertex $e^{\frac{2}{\sqrt{n}} \varphi_+}$ is a primary field[16] with zero conformal dimensions w.r.t. the modified energy–momentum tensor (6.223), and therefore the 'screening' charge density $B(w)$ is a type $(1,0)$ form (its conformal dimension Δ_+ equals 1, whereas $\Delta_- = 0$). This makes (6.222) the conformally invariant condition.

Exercise 42 Show that the central charge corresponding to the energy–momentum tensor (6.223) is given by

$$c = 2 + \frac{6}{n}. \tag{6.224}$$

---

[15] Here and below, exponential fields are always understood as the normal ordered operators, for example

$$e^{\omega \varphi_\pm} = \exp\left(\omega \varphi_0^{(\pm)} - i\omega \frac{2\pi z}{R} a_0^{(\pm)}\right) \exp\left(i\omega \sum_{l<0}^{\infty} \frac{a_l^{(\pm)}}{l} e^{-\frac{2\pi l}{R} z}\right) \exp\left(i\omega \sum_{l>0}^{\infty} \frac{a_l^{(\pm)}}{l} e^{-\frac{2\pi l}{R} z}\right).$$

[16] Recall that the field $\mathcal{O}$ is called a conformal primary with conformal dimensions (Δ_+, Δ_-) if it satisfies the following OPE:

$$T(z)\mathcal{O}(0) = \frac{\Delta_+}{z^2} \mathcal{O}(0) + \frac{1}{z} \partial \mathcal{O}(0) + O(1), \qquad \bar{T}(\bar{z})\mathcal{O}(0) = \frac{\Delta_-}{\bar{z}^2} \mathcal{O}(0) + \frac{1}{\bar{z}} \bar{\partial} \mathcal{O}(0) + O(1).$$

In the case $s = 3$, the condition (6.222) defines the spin-3 current W_3 up to normalization and the adding of the derivative ∂W_2. Up to this ambiguity, W_3 can be chosen as

$$iW_3 = \frac{6n+4}{3}(\partial\vartheta_+)^3 + 2n(\partial\varphi_+)^2\partial\vartheta_+ - n\sqrt{n}\partial^2\varphi_+\partial\vartheta_+ + (n+2)\partial\varphi_+\partial^2\vartheta_+ + \frac{n+2}{6}\partial^3\vartheta_+, \tag{6.225}$$

where the ambiguity in adding a term proportional to ∂W_2 is fixed by demanding that (6.225) is a conformal primary. One can notice that the current W_3 is antisymmetric under reflection $\vartheta_+ \mapsto -\vartheta_+$, but there is no symmetry w.r.t. $\varphi_+ \mapsto -\varphi_+$. Also it deserves to be mentioned that the classical currents $W_s^{(cl)}$ discussed in Section 6.6.2 (see equation (6.106)), appear in the classical limit $n \to \infty$:

$$W_s^{(cl)} = \lim_{n\to\infty} n^{2-\frac{3s}{2}} \, W_s|_{\substack{\varphi_+ \mapsto \sqrt{n}\phi_+ \\ \vartheta_+ \mapsto \sqrt{n}\alpha_+}}, \tag{6.226}$$

for the spin -2 and -3 currents, and similarly for the antiholomorphic W-currents with $s = -2, -3$.

The higher currents $W_4, W_5, \ldots$ can be found by a direct computation of the OPE with screening density B (6.222), or recursively, by studying the singular parts of the operator product expansions of the lower currents, starting with $W_3(w)W_3(z)$ and continuing upward. Thus the product $W_3(w)W_3(z)$ contains the singular term $\sim (w-z)^{-2}$, which involves, besides the derivative $\partial^2 T$ and the composite operator T^2, the new current W_4. Further operator products with W_4 define higher W's, etc. With the proper normalization of the quantum W_s-current the formula (6.226) can be applied for $s > 3$.

Exercise 43 The field T^2 is defined as

$$T^2(z) = \lim_{w\to z}\left(T(w)T(z) - \frac{c}{2(w-z)^4} - \frac{2T(z)}{(w-z)^2} - \frac{\partial T(z)}{w-z}\right). \tag{6.227}$$

Show that

$$T^2 = (\partial\varphi_+)^4 + (\partial\vartheta_+)^4 + 2(\partial\varphi_+)^2(\partial\vartheta_+)^2 + \frac{2}{\sqrt{n}}(\partial\vartheta_+)^2\partial^2\varphi_+ + \frac{1+n}{n}(\partial^2\varphi_+)^2 + (\partial^2\vartheta_+)^2$$

$$+ \partial\left(\frac{2}{3\sqrt{n}}(\partial\varphi_+)^3 - \partial\varphi_+\partial^2\varphi_+ - \partial\theta_+\partial^2\vartheta_+ - \frac{1}{3\sqrt{n}}\partial^3\varphi_+\right). \tag{6.228}$$

In fact the easiest way to generate the higher-spin W-currents is to observe that the screening operator commutes with the holomorphic fields

$$\Psi(z) = \left(\sqrt{n}\partial\varphi_+ + i\sqrt{n+2}\partial\vartheta_+\right) e^{+\frac{2i\vartheta_+}{\sqrt{n+2}}} \tag{6.229}$$

$$\Psi^*(z) = \left(\sqrt{n}\partial\varphi_+ - i\sqrt{n+2}\partial\vartheta_+\right) e^{-\frac{2i\vartheta_+}{\sqrt{n+2}}}$$

in the same way as the W-currents, i.e.

$$\oint_z dw\, B(w)\Psi(z) = 0, \qquad \oint_z dw\, B(w)\Psi^*(z) = 0. \qquad (6.230)$$

These fields are quantum counterparts of ψ_+ and ψ_+^* (6.86) discussed in Section 6.6.2. They turn out to be primary conformal fields with the non-integer conformal dimension $\Delta = 1 + \frac{1}{n+2}$ and, therefore, they are not local—they extend the notion of Z_k parafermions [33] to non-integer $k = -n - 2$. Nevertheless, both Ψ and Ψ^* are local w.r.t. the exponential $B(w)$, hence the integration contour in (6.230)—a small contour around z—is indeed a closed one. It follows from (6.230) that all the fields generated by the OPE of $\Psi(z)\Psi^*(0)$ satisfy the condition (6.222). Thus we have

$$\Psi(z)\Psi^*(0) = -z^{-\frac{2}{n+2}} \left(\frac{n+2}{z^2} + \frac{n}{2}(T(z) + T(0)) + \frac{z}{2\sqrt{n+2}}(W_3(z) + W_3(0)) \right.$$
$$\left. + \frac{z^2}{4(n+2)}(W_4(z) + W_4(0)) + \ldots \right). \qquad (6.231)$$

Exercise 44 Show that the current W_4 defined by equation (6.231) can be written as

$$W_4 = W_4^{(\text{sym})} + \partial \mathcal{O}_3 + aT^2 + b\partial^2 T, \qquad (6.232)$$

where

$$W_4^{(\text{sym})} = (4n^2 + 9n + 4)(\partial^2 \varphi_+)^2 + (4n^2 + 7n + 2)(\partial^2 \vartheta_+)^2 + n(3n+4)(\partial \varphi_+)^4$$
$$+ (n+2)(3n+2)(\partial \vartheta_+)^4 + 6n(n+2)(\partial \varphi_+)^2(\partial \vartheta_+)^2. \qquad (6.233)$$

and $\mathcal{O}_3$ is some local field of Lorentz spin -3.

The last two terms in (6.232) are irrelevant—they represent the ambiguity in adding composites of the lower W-currents—and exact values of the constants a,b are not important here. The explicit form of the field $\mathcal{O}_3$ is not important for the present discussion either. Important is the symmetry of $W_4^{(\text{sym})}$ with respect to the $\varphi \mapsto -\varphi$ reflection. A similar property can be observed for the next few W-currents of even spins. Using the freedom of adding derivatives and composite fields built from the lower-spin W-currents, the fields W_{2j} can be brought to the form

$$W_{2j} = W_{2j}^{(\text{sym})} + \partial \mathcal{O}_{2j-1} \qquad (j = 1, 2, \ldots), \qquad (6.234)$$

where the expressions for $W_{2j}^{(\text{sym})}$ are even with respect to the reflection $\varphi \mapsto -\varphi$. It is expected that this statement is valid for all even-spin W-currents.

6.15.2 Highest weight representations

It should be emphasized that the W-currents generate a closed algebraic structure which can be understood without any references to the free field realization. We shall call it the W-algebra and denote it $\mathfrak{W}_+$. The extra sign factor indicates that the W-algebra is built from the holomorphic currents. All the above can be repeated with the antiholomorphic currents of negative Lorentz spin. The corresponding W-algebra will be denoted by $\mathfrak{W}_-$. Of course, $\mathfrak{W}_+$ and $\mathfrak{W}_-$ are completely identical from the formal algebraic point of view, in most cases it is sufficient to focus on $\mathfrak{W}_+$.

Equations (6.223), (6.225) allow one to introduce the structure of the highest weight representation of $\mathfrak{W}_+$ in the Fock space $\mathcal{F}_+$ for the creation–annihilation operators $a_m^{(+)}$, $b_m^{(+)}$. The space $\mathcal{F}_+$ consists of the Fock vacuum $|0\rangle$ annihilated by all $a_l^{(+)}$, $b_l^{(+)}$ with $l > 0$ and the states generated by the action of $a_m^{(+)}$, $b_m^{(+)}$ with $m < 0$ on the Fock vacuum. The operators $a_0^{(+)}$, $b_0^{(+)}$, which commute with all $a_m^{(+)}$, $b_m^{(+)}$, act in $\mathcal{F}_+$ by multiplication of the constants P and Q, respectively. Since all the W-currents are periodic in the x_1-direction, one can expand them in the Fourier series with coefficients

$$\widetilde{W}_s(m) = \left(\frac{R}{2\pi}\right)^{s-1} \oint_R \frac{dz}{2\pi i}\, W_s(z)\, e^{\frac{2\pi m}{R}z} \quad (s=2,3,\ldots), \qquad (6.235)$$

where the closed integration contour wraps around the cylinder (i.e. $\oint_R \frac{dz}{2\pi i} \equiv \int_0^R \frac{dx_1}{2\pi}$, see Figure 6.11). The free field realization of the W-currents defines the action of $\widetilde{W}_s(m)$ in the Fock space $\mathcal{F}_+$.

Exercise 45 Shows that the Fock vacuum satisfies the conditions

$$\widetilde{W}_s(m)|\mathbf{w}\rangle = 0 \quad (m=1,2,\ldots)$$
$$\widetilde{W}_s(0)|\mathbf{w}\rangle = w_s|\mathbf{w}\rangle. \qquad (6.236)$$

where $w_s = w_s(P,Q)$ are certain polynomials of degree s in the variables P and Q. In particular,

$$w_2(P,Q) = P^2 + Q^2 - \frac{1}{12}$$
$$w_3(P,Q) = 2Q\left(nP^2 + \left(n+\frac{2}{3}\right)Q^2 - \frac{2n+1}{12}\right). \qquad (6.237)$$

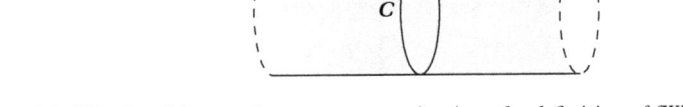

Figure 6.11 *The closed integration contour entering into the definition of $\int W_s(m)$ in equation (6.235). Reproduced with permission from John Wiley & Sons.*

The last exercise implies that the Fock vacuum is the highest vector for the W-algebra corresponding to the highest weight $(w_2, w_3, \ldots)$. In fact, since all the W-currents can be generated from spin -2 and -3 currents, it is sufficient to consider the case $s = 2, 3$ in (6.236) to unambiguously define the highest vector $|\mathbf{w}\rangle$.

The highest weight representation is generated by the action of $\widetilde{W}_s(l)$ with $l < 0$ on the highest vector. It is possible to show that for $n > 0$ this representation of the W-algebra is irreducible and coincides with the Fock space $\mathcal{F}_+$ as a linear space. Notice that the highest weight components (6.237) are even w.r.t. the reflection $P \mapsto -P$ and, hence P and $-P$ correspond to equivalent representations of $\mathfrak{W}_+$. Using this property, one can introduce the operator $\hat{s}_+$ which is uniquely defined by the conditions

$$\hat{s}_+ = \hat{s}_+(P|Q) : \mathcal{F}_+ \mapsto \mathcal{F}_+, \quad \hat{s}_+|0\rangle = |0\rangle, \quad \hat{s}_+ \widetilde{W}_s(m) = \widetilde{W}_s^{(\mathrm{ref})}(m) \hat{s}_+, \qquad (6.238)$$

where $\widetilde{W}_s^{(\mathrm{ref})}(m) = \widetilde{W}_s(m)|_{P \mapsto -P}$. Below we will use the so-called 'reflection' operator $\hat{R}_+$ whose definition is slightly different from that of $\hat{s}_+$:

$$\hat{R}_+ = \hat{c}\, \hat{s}_+. \qquad (6.239)$$

The action of the operator $\hat{c}$ in the Fock space $\mathcal{F}_+$ is defined by the conditions $\hat{c} a_m^{(+)} = -a_m^{(+)} \hat{c}$, $\hat{c} b_m^{(+)} = +b_m^{(+)} \hat{c}$ and $\hat{c}|0\rangle = |0\rangle$.

The space $\mathcal{F}_+$ naturally splits into the finite-dimensional *level subspaces*

$$\mathcal{F}_+ = \oplus_{L=0}^{\infty} \mathcal{F}_+^{(L)} : \quad \hat{\mathbb{L}}_+ \mathcal{F}_+^{(L)} = L\, \mathcal{F}_+^{(L)}, \qquad (6.240)$$

where the grading operator is given by

$$\hat{\mathbb{L}}_+ = 2 \sum_{m=1}^{\infty} \left(a_{-m}^{(+)} a_m^{(+)} + b_{-m}^{(+)} b_m^{(+)} \right). \qquad (6.241)$$

It is easy to see that both operators $\hat{s}_+$ and $\hat{R}_+$ act invariantly in each level subspace $\mathcal{F}_+^{(L)}$.

Exercise 46 (a) The states $|a\rangle = a_{-1}^{(+)}|0\rangle$ and $|b\rangle = b_{-1}^{(+)}|0\rangle$ form a basis in the first level subspace $\mathcal{F}_+^{(1)}$. Calculate the matrices of $\hat{s}_+$ and $\hat{R}_+$ in this basis. (b) Repeat the calculation for the five-dimensional second level subspace $\mathcal{F}_+^{(2)}$. Use the basis $|1\rangle = \left(a_{-1}^{(+)}\right)^2 |0\rangle, |2\rangle = a_{-2}^{(+)}|0\rangle, |3\rangle = \left(b_{-1}^{(+)}\right)^2 |0\rangle, |4\rangle = b_{-2}^{(+)}|0\rangle, |5\rangle = a_{-1}^{(+)} b_{-1}^{(+)}|0\rangle$. (c) Diagonalize the operator $\hat{R}_+$ in the first two level subspaces.

6.15.3 Involutive local IM

Let us take a closer look at the operators

$$\widetilde{W}_s(0) = \left(\frac{R}{2\pi}\right)^{s-1} \oint_R \frac{dz}{2\pi i} W_s(z) \quad (s=2,3,\ldots). \tag{6.242}$$

Using the Cauchy–Riemann equation $\bar{\partial} W_s = 0$ and the periodicity $W_s(z+iR) = W_s(z)$, it is straightforward to show that $\widetilde{W}_s(0)$ does not depends on x_2, i.e. it is an IM (see Figure 6.12) Furthermore, since $\widetilde{W}_s(0)$ is given by the integral over the local field W_s, it is a local IM [32]. As has already been mentioned, the even spin W-current can be chosen in the form

$$W_{2j} = W_{2j}^{(\text{sym})} + \partial \mathcal{O}_{2j-1}, \tag{6.243}$$

where the expressions for $W_{2j}^{(\text{sym})}$ are even w.r.t the reflection $\varphi \mapsto -\varphi$. Since the total derivative (6.243) does not contribute to the integral (6.242), the local IM[17]

$$\hat{I}_{2j-1}^{(+)} = C_{2j-1} \oint_R \frac{dz}{2\pi i} W_{2j}^{(\text{sym})}(z) \tag{6.244}$$

are invariant under the reflection $\varphi \mapsto -\varphi$. Therefore, they all commute with the reflection operator (6.239):

$$\left[\hat{I}_{2j-1}^{(+)}, \hat{R}_+\right] = 0. \tag{6.245}$$

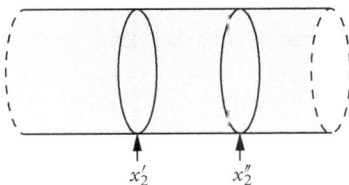

Figure 6.12 *The value of the integral (6.242) does not depend on the Euclidean-time slice. Reproduced with permission from John Wiley & Sons.*

---

[17] The overall constant C_{2j-1} can be chosen at will. Below, the constant C_1 is set to one.

Recall that $\hat{I}^{(+)}_{2j-1}$ and $\hat{R}_+$ act on the Fock space $\mathcal{F}_+$ for the creation–annihilation operators $a^{(+)}_l$, $b^{(+)}_l$. The grading operator $\mathbb{L}_+$ (6.241) essentially coincides with I_1:

$$\hat{I}^{(+)}_1 = \frac{2\pi}{R}(w_2 + \hat{\mathbb{L}}_+), \qquad (6.246)$$

where $w_2 = P^2 + Q^2 - \frac{1}{12}$. Therefore all the local IM $\hat{I}^{(+)}_{2j-1}$, as well as the reflection operator $\hat{R}_+$, act invariantly in each level subspace. It turns out that the eigenstates of $\hat{R}_+$ acting in $\mathcal{F}^{(L)}_+$ are not degenerate and form a basis in the level subspaces (see Exercise 46). Then the commutativity condition (6.245) implies that

$$\left[\hat{I}^{(+)}_{2l-1}, \hat{I}^{(+)}_{2j-1}\right] = 0. \qquad (6.247)$$

Exercise 47 For $j = 1, 2, 3$, show that,

$$\hat{I}^{(+)}_{2j-1} = \oint_R \frac{dz}{2\pi i}\left(\sum_{l+m=j} C^{(j)}_{lm}(\partial\varphi_+)^{2l}(\partial\vartheta_+)^{2m} + \ldots\right), \qquad (6.248)$$

where the dots stand for monomials which include higher derivatives of $\partial\varphi_+$ and $\partial\vartheta_+$ and the numerical coefficients $C^{(j)}_{lm}$ can be written as

$$C^{(j)}_{lm} = C_{2j-1}\frac{(-2)^{j+1}(2j-2)!}{(j+1)!}\frac{\left((n+2)(\frac{1}{2}-j)\right)_l\left((-n)(\frac{1}{2}-j)\right)_m}{l!m!}(-n)^{l-1}(n+2)^{m-1}, \qquad (6.249)$$

where $(a)_m = \prod_{i=0}^{m-1}(a+i)$ is the Pochhammer symbol.

It turns out that (6.248), (6.249) hold true for any positive integer j [32]. Also, the overall normalization constant C_{2j-1} is usually set to

$$C_{2j-1} = \frac{(-8)^{-j}(j+1)!\,n(n+2)}{\left((n+2)(\frac{1}{2}-j)\right)_j\left((-n)(\frac{1}{2}-j)\right)_j}. \qquad (6.250)$$

Exercise 48 Show that the vacuum eigenvalues of the local IM $\hat{I}^{(+)}_{2j-1}$ and $\hat{I}^{(-)}_{2j-1}$ are given by a certain polynomial in P^2 and Q^2 of degree j. In particular, using the results of Exercise 32, show that

$$I_1^{(\text{vac})}(P,Q) = \frac{2\pi}{R}\left(P^2 + Q^2 - \frac{1}{12}\right) \qquad (6.251)$$

$$I_3^{(\text{vac})}(P,Q) = C_3\left(\frac{2\pi}{R}\right)^3 \Big(n(3n+4)P^4 + 6n(n+2)P^2Q^2 + (n+2)(3n+2)Q^4$$

$$-\frac{1}{2}n(2n+3)P^2 - \frac{1}{2}(n+2)(2n+1)Q^2 + \frac{18n^2+36n+11}{240}\Big).$$

Exercise 49 Using the bases from Exercise 46, calculate the matrices of $\hat{I}_3$ in the level subspace $\mathcal{F}^{(1)}$ and $\mathcal{F}^{(2)}$. Check the commutativity condition (6.245).

Identical considerations can be applied to the local IM $\hat{I}_{2j-1}^{(-)}$ built from the antiholomorphic W-currents. Thus we conclude that $\left\{\hat{I}_{2j-1}^{(+)}, \hat{I}_{2j-1}^{(-)}\right\}_{j=1}^{\infty}$ is an involutive family of quantum local IM in the cigar NLSM.

6.15.4 The sine-Liouville model

Further exploration of the cigar W-algebra yields the following surprising fact [34].

Exercise 50 Show that the parafermionic currents, and therefore all the W-currents, commute with two additional screening charges:

$$\oint_z dw\, F^{(\pm)}(w)\Psi(z) = 0,\quad \oint_z dw\, F^{(\pm)}(w)\,\Psi^*(z) = 0,\ \text{ where }\ F^{(\pm)} = e^{-\sqrt{n}\varphi_+ \pm i\sqrt{n+2}\vartheta_+}$$

$$(6.252)$$

and, hence,

$$\oint_z dw\, F^{(\pm)}(w)\, W_c(z) = 0. \qquad (6.253)$$

The screening charges (6.252) are sometimes referred to as fermionic. It is not difficult to check that the W-algebra can be alternatively defined through the condition (6.253) instead of (6.222). Contrary to the 'bosonic' screening (6.222), the classical limit $n \to \infty$ for the fermionic screenings is not well defined. In its turn, the fermionic screenings are perfectly suitable for taking $n \to 0$, which is a singular limit for the 'bosonic' screening charge. This observation suggests that the cigar NLSM admits a dual description which is most suitable for study in the strong coupling regime. The dual action is given by [35]

$$\mathcal{A} = \int d^2x \left[\frac{1}{4\pi}\left((\partial_\sigma\varphi)^2 + (\partial_\sigma\vartheta)^2\right) - 2\mu e^{-\sqrt{n}\varphi}\cos\left(\sqrt{n+2}\vartheta\right)\right]. \qquad (6.254)$$

The arguments in support of the dual description go along the following line. Let us take the 'zero-mode' of the field φ,

$$\varphi_0 = \int_0^R \frac{dx_1}{R} \, \varphi(x), \tag{6.255}$$

and consider the region $\varphi_0 \to +\infty$ in the configuration space. Here the potential term in the action (6.254) can be neglected, so that φ and ϑ become asymptotically free massless fields satisfying the Laplace equation, i.e. $\varphi \asymp \varphi_+(z) + \varphi_-(\bar{z})$ and $\vartheta \asymp \vartheta_+(z) - \vartheta_-(\bar{z})$ (here we use the symbol $\asymp$ to emphasize the asymptotic nature of the relations). Because of the equations of motion, any homogeneous differential polynomial $W_s(\partial \varphi_+, \partial^2 \varphi_+, \ldots; \partial \vartheta_+, \partial^2 \vartheta_+ \ldots)$ of overall degree s and with constant coefficients is a holomorphic current, $\bar{\partial} W_s = 0$. We may now consider the potential term perturbatively in μ. If one adjusts the constant coefficients in W_s to satisfy the commutativity condition (6.253), then the Cauchy–Riemann condition $\bar{\partial} W_s = 0$ will be satisfied to the first order in μ. In fact μ is a somewhat fake parameter—by an additive shift $\varphi \mapsto \varphi + const$, which does not affect the W-currents the value of μ can be chosen to be an arbitrary real number. For this reason it is expected that (6.254) defines the CFT, the so-called sine-Liouville model, which possesses an infinite set of holomorphic and antiholomorphic currents with Lorentz spins $s = \pm 2, \pm 3 \ldots$.

How are the original fields in the NLSM description related to the fields from the sine-Liouville fields? Our consideration suggests that, in the asymptotic domain, $\phi \asymp \frac{\varphi}{\sqrt{n}} + const$ and $\tilde{\alpha} = \frac{\vartheta}{\sqrt{n+2}} + const$. The sine-Liouville fields satisfy the following BC:

$$\varphi(x_1 + R, x_2) = \varphi(x_1, x_2), \quad \vartheta(x_1 + R, x_2) = \vartheta(x_1, x_2) + \frac{2\pi q}{\sqrt{n+2}}, \tag{6.256}$$

where the constant q is related to the parameter m (see Section 6.6.1, equation (6.74)) as $m = \frac{q}{n+2}$. Apparently, a consistent description of the quantum theory requires that the $U(1)$ charge q takes integer values only: $q \in \mathbb{Z}$. Recall that in the classical limit we considered the quasiperiodic BC imposed on the field α:

$$\alpha(x_1 + R, x_2) = \alpha(x_1, x_2) + 2\pi k. \tag{6.257}$$

The twist parameter k admits a natural interpretation in the dual description—it can be identified with the so-called quasimomentum in the sine-Liouville model. The sine-Liouville action is invariant under the transformation $\vartheta \mapsto \vartheta + \frac{2\pi}{\sqrt{n+2}}$. Because of this periodicity, the space of states of the theory with BC (6.256) splits on the orthogonal subspaces $\mathcal{H}_{q,k}$ such that for any state $|S\rangle \in \mathcal{H}_{q,k}$, the corresponding wave functional $\Psi_S[\varphi(x), \vartheta(x)]$ transforms as

$$\Psi_S\left[\varphi(x_1), \vartheta(x_1) + \frac{2\pi}{\sqrt{n+2}}\right] = e^{2\pi i k} \, \Psi_S[\varphi(x_1), \vartheta(x_1)]. \tag{6.258}$$

6.15.5 Space of states

To get a better idea about the states $|S\rangle \in \mathcal{H}_{q,k}$, and the corresponding wave functionals Ψ_S, we note that they are organized in the highest weight representation of the tensor product of two W-algebras, $\mathfrak{W}_+ \otimes \mathfrak{W}_-$. The (formal) algebraic construction of the W-algebra highest weight representation follows the procedure discussed previously. In the case of the tensor product $\mathfrak{W}_+ \otimes \mathfrak{W}_-$ the highest vector $|\mathbf{w}_+, \mathbf{w}_-\rangle$ is characterized by the highest weight $\mathbf{w}_\pm \equiv (w_{\pm 2}, w_{\pm 3})$. The free field realization of the W-algebra suggests to parameterize $w_{\pm s}$ by $(P, Q_\pm)$ such that $w_{\pm s} = w_s(P, Q_\pm)$ with the polynomials $w_s(P, Q)$ given by equations (6.237). Formulae (6.256), (6.257) imply that $Q_\pm$ are expressed in terms of the pair (q, k) as follows:

$$Q_\pm = \frac{1}{2}\left(\frac{q}{\sqrt{n+2}} \pm k\sqrt{n+2}\right). \tag{6.259}$$

Let us denote the corresponding highest weight representation of $\mathfrak{W}_+ \otimes \mathfrak{W}_-$ by $\mathcal{V}_{P;q,k} \equiv \mathcal{V}_{-P;q,k}$, then the space of states [35], [38] $\mathcal{H}_{q,k}$ is given by the direct integral

$$\mathcal{H}_{q,k} = \int_{P>0}^{\oplus} \mathcal{V}_{P;q,k}. \tag{6.260}$$

Notice that common eigenvectors of the involutive set of integrals of motion $\{I_{2j-1}^{(+)}, I_{2j-1}^{(-)}\}_{j=1}^{\infty}$ form a basis of each component $\mathcal{V}_{P;q,k}$ in the decomposition (6.260).

The free field representation of $\mathfrak{W}_+ \otimes \mathfrak{W}_-$ allows one to naturally associate the $\varphi_0 \to +\infty$ asymptotic for the wave functional $\Psi_S[\varphi(x), \vartheta(x)]$ for the normalizable states $|S\rangle$ from $\mathcal{H}_{M,k}$, with an element of the space $L^2(-\infty < \varphi_0 < \infty) \otimes \mathcal{F}$. Here $\mathcal{F} = \mathcal{F}_+ \otimes \mathcal{F}_-$ is the Fock space of the oscillators $a_l^{(\pm)}, b_l^{(\pm)}$, with $l \in \mathbb{Z}$, $l \neq 0$, $b_0^{(\pm)} = Q_\pm$ and L^2 stands for the space of square integrable functions. Notice that both zero mode plane waves $e^{\pm 2iP\varphi_0}$ times the Fock vacuum $|0\rangle$ are highest vectors in the free field realization of $\mathfrak{W}_+ \otimes \mathfrak{W}_-$. It is clear that the correct wave functionals Ψ_P, associated with the highest vectors in the direct sum (6.260), correspond to certain linear combination of these states, i.e.

$$\Psi_P[\varphi(x_1), \vartheta(x_1)] \sim \left(e^{-2iP\varphi_0} + S(P)\, e^{+2iP\varphi_0}\right)|0\rangle \quad \text{at} \quad \varphi_0 \to +\infty, \tag{6.261}$$

with some reflection amplitude $S(P)$ dependent on more complicated dynamics in the region of φ_0 where the exponential term is important. The amplitude $S(P)$ is an important characteristic of the theory. It was originally found in the work [35]:[18]

$$S(P) \equiv S_{q,\lambda}(p) \quad \text{with} \quad p = 2P\sqrt{n}, \quad \lambda = (n+2)k, \tag{6.262}$$

---

[18] In the minisuperspace approximation the reflection amplitude was derived in [36].

and

$$S_{q,\lambda}(p) = \left(\frac{\pi\mu}{n}\right)^{-\frac{2ip}{n}} \frac{\Gamma(\frac{1}{2}+\frac{q+\lambda}{2}-\frac{ip}{2})\Gamma(\frac{1}{2}+\frac{q-\lambda}{2}-\frac{ip}{2})}{\Gamma(\frac{1}{2}+\frac{q+\lambda}{2}+\frac{ip}{2})\Gamma(\frac{1}{2}+\frac{q-\lambda}{2}+\frac{ip}{2})} \frac{\Gamma(1+ip)}{\Gamma(1-ip)} \frac{\Gamma(1+\frac{ip}{n})}{\Gamma(1-\frac{2ip}{n})}. \quad (6.263)$$

More generally, if $|S\rangle$ is some state in $\mathcal{V}_{P;q,k}$,

$$\Psi_S[\varphi(x_1),\vartheta(x_1)] \sim \left(e^{-2iP\varphi_0} + \hat{S}(P)\, e^{+2iP\varphi_0}\right)|s\rangle \quad \text{at} \quad \varphi_0 \to +\infty, \quad (6.264)$$

where $|s\rangle \in \mathcal{F}$ and $\hat{S}(P)$ is now a unitary operator in $\mathcal{F}$ which can be expressed in terms of the operators $\hat{s}_+$ (6.238), the analogous operator $\hat{s}_-$ corresponding to $\mathfrak{W}_-$ and the vacuum amplitude $S(P)$:

$$\hat{S}_{q,k}(P) = S(P)\, \hat{s}_+(P|Q_+)\hat{s}_-(P|Q_-). \quad (6.265)$$

Exercise 51 In the quantum case it is convenient to use (p,q,λ) instead of the parameters (v,m,k), which have been used in the classical description of the cigar NLSM. These two triples are related as

$$p \equiv 2P\sqrt{n} = nv, \qquad \lambda = k(n+2), \qquad q = m(n+2). \quad (6.266)$$

Consider the limit $n \to \infty$ of the reflection amplitude $S_{q\lambda}(p)$ (6.263) keeping (v,m,k) fixed. Show that

$$\lim_{\substack{n\to\infty \\ v,m,q-\text{fixed}}} \left(\frac{\partial}{\partial q} + i\frac{\partial}{\partial p}\right) \log S_{q,\lambda}(p) = \Delta_Z, \quad (6.267)$$

where Δ_Z stands for the classical time delay from Exercise 21.

6.15.6 W-primary fields

As well as the space of states, the space of operators of the cigar NLSM/sine-Liouville model can be classified w.r.t. the (adjoint) action of the $\mathfrak{W}_+ \otimes \mathfrak{W}_-$ algebra. In this classification a special rôle belongs to the W-primary fields [15], [36], i.e. the conformal primaries satisfying the extra requirements $[\tilde{W}_{\pm 3}(0), \mathcal{O}] = w_\pm \mathcal{O}$. The theory possesses W-primaries $\mathcal{O}_{q,\lambda}^{(j)}$ whose conformal dimensions $\Delta_\pm$ and the W-heights $w_\pm$ are given by[19]

$$\Delta_\pm = \frac{(q\pm\lambda)^2}{4(n+2)} - \frac{j(j+1)}{n} \quad (6.268)$$

$$w_{\pm 3} = \frac{q\pm\lambda}{\sqrt{n+2}} \left(\frac{3n+2}{12(n+2)}(q\pm\lambda)^2 - j(j+1) - \frac{n+2}{6}\right).$$

---

[19] These formulae are specializations of equations (6.237) for $P = -\frac{i}{\sqrt{n}}(j+1/2)$, $Q_\pm = \frac{q\pm\lambda}{2\sqrt{n+2}}$. Also one should take into account that $w_{\pm 2} = \Delta_\pm - \frac{c}{24}$, which follows from the relation (6.176).

Here $j \geq -\frac{1}{2}$ is a real parameter, whereas $\lambda \in \mathbb{R}$ and $M \in \mathbb{Z}$, so that $\mathcal{O}_{q,\lambda}^{(j)}$ are semilocal fields—they carry the Lorentz spin $s \equiv \Delta_+ - \Delta_- = \frac{q\lambda}{n+2}$. They become local, bosonic, or fermionic, when $\frac{\lambda}{n+2}$ is integer or half-integer. In the classical limit $n \to +\infty$,

$$\begin{pmatrix} \mathcal{O}_{0,-1}^{(1/2)} & \mathcal{O}_{-1,0}^{(1/2)} \\ \mathcal{O}_{+1,0}^{(1/2)} & \mathcal{O}_{0,+1}^{(1/2)} \end{pmatrix} \longrightarrow \pi_{\frac{1}{2}}(\omega) = \begin{pmatrix} e^{-i\tilde{\alpha}}\sqrt{1+|Z|^2} & Z^* \\ Z & e^{+i\tilde{\alpha}}\sqrt{1+|Z|^2} \end{pmatrix}, \quad (6.269)$$

where $\pi_{\frac{1}{2}}(\omega)$ stands for the matrix of $\omega = e^{-\frac{1}{2}(\alpha+\tilde{\alpha})\mathrm{h}} e^{\phi(\mathrm{e}+\mathrm{f})} e^{\frac{1}{2}(\alpha-\tilde{\alpha})\mathrm{h}}$ in the two-dimensional representation of $sl(2)$ generators $\mathrm{h}, \mathrm{e}, \mathrm{f}$ (see equations (6.73)–(6.81) in Section 6.6.1 from the first lecture). More generally, in the case of the $(2j+1)$-dimensional representation ($j = \frac{1}{2}, 1, \frac{3}{2}, \ldots$), one has the relation

$$\lim_{\substack{n \to +\infty \\ j,q,\lambda-\text{fixed}}} \mathcal{O}_{q,\lambda}^{(j)}\Big|_{\substack{\lambda=-m-m' \\ q=+m-m'}} = \mathcal{N}_{m'm}^{(j)} (m'|\pi_j(\omega)|m), \quad (6.270)$$

where $|m\rangle : \pi_j(\mathrm{h})|m\rangle = m|m\rangle$, $\langle m'|m\rangle = \delta_{m',m}$ ($m = j, j-1, \ldots, -j$) and $\mathcal{N}_{m'm}^{(j)}$ is some constant which depends on the normalization of $\mathcal{O}_{q,\lambda}^{(j)}$.

Exercise 52 Using an explicit expression for the Wigner d-matrix in terms of Jacobi polynomials (see e.g.[37]), show that the classical limit of the W-primaries can be expressed in terms of the conventional hypergeometric function:

$$\lim_{\substack{n \to +\infty \\ j,q,\lambda-\text{fixed}}} \mathcal{O}_{q,\lambda}^{(j)}\Big|_{\substack{\lambda=-m-m' \\ q=+m-m'}} = \frac{\Gamma(1+j+m)\Gamma(1+j-m')}{\Gamma(1+2j)\Gamma(m-m'+1)} \quad (6.271)$$

$$\times e^{-i(m+m')\tilde{\alpha}} Z^{m-m'} \left(1+|Z|^2\right)^{\frac{m+m'}{2}} {}_2F_1\left(m-j, m+j+1, 1+m-m'; -|Z|^2\right).$$

Here the overall factor is chosen in such a way that the r.h.s. $\sim e^{-i(m+m')\tilde{\alpha}} e^{i(m-m')\alpha} |Z|^{2j}$ as $|Z| \to +\infty$, for $j > \frac{1}{2}$. Show that for $q = \lambda = 0$ and $j = 1, 2$, equations (6.271) yields

$$\lim_{n \to +\infty} \mathcal{O}_{0,0}^{(1)} = |Z|^2 + \tfrac{1}{2}, \quad (6.272)$$

$$\lim_{n \to +\infty} \mathcal{O}_{0,0}^{(2)} = |Z|^2\left(1+|Z|^2\right) + \tfrac{1}{6}.$$

Finally, let us note that $\mathcal{O}_{q,k}^{(j)}$ can be understood as exponentials built from the sine-Liouville fields:

$$\mathcal{O}_{q,\lambda}^{(j)} = e^{\frac{2j\varphi}{\sqrt{n}}} e^{\frac{i\lambda\vartheta}{\sqrt{n+2}}} e^{\frac{iq\tilde{\vartheta}}{\sqrt{n+2}}}, \quad (6.273)$$

where $\tilde{\vartheta}$ is a T-dual to ϑ, i.e., $\partial_\mu \tilde{\vartheta} = i\epsilon_{\mu\nu}\partial_\nu\vartheta$.

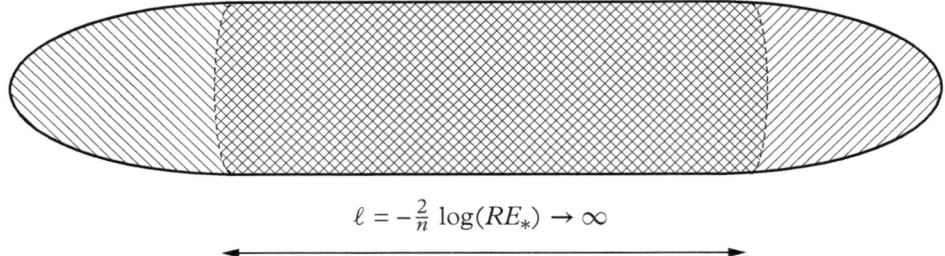

Figure 6.13 *The target space of the sausage NLSM when $\kappa \to 1_-$, which correponds to the UV limit of the model. Reproduced with permission from John Wiley & Sons.*

6.16 The quantum sausage

6.16.1 Dual form of the sausage NLSM

Let us consider the NLSM corresponding to the sausage metric

$$G_{ab}dX^a dX^b = \frac{n\kappa}{2\pi}\left[\frac{(d\zeta)^2}{(1-\zeta^2)(1-\kappa^2\zeta^2)} + \frac{(1-\zeta^2)(d\alpha)^2}{1-\kappa^2\zeta^2}\right] \tag{6.274}$$

for short distance when the RG time $\tau = \frac{1}{2\pi}\log(RE_*) \to -\infty$. Recall that under the RG flow equation n remains fixed, whereas $\kappa = -\tanh(\frac{2\pi\tau}{n})$, or equivalently,

$$(RE_*)^{\frac{1}{n}} = \sqrt{\frac{1-\kappa}{1+\kappa}}. \tag{6.275}$$

In the central region (see Figure 6.13) the sausage looks like a half infinite cylinder equipped with the flat metric. If one formally sets $\kappa = 1$ in (6.274) and ignores the presence of the two infinitely separated tips, then

$$G_{\mu\nu}dX^\mu dX^\nu \approx \frac{n}{2\pi}\left((d\phi)^2 + (d\alpha)^2\right), \tag{6.276}$$

where we use $\phi = -\log\left(\frac{1-\zeta}{1+\zeta}\right)$, such that

$$-\infty \leftarrow \tfrac{1}{n}\log(RE_*) < \phi < -\tfrac{1}{n}\log(RE_*) \to +\infty. \tag{6.277}$$

The NLSM action corresponding to (6.276) has the form

$$\mathcal{A}_0 = \frac{n}{4\pi}\int d^2x\left((\partial_\mu\phi)^2 + (\partial_\mu\alpha)^2\right). \tag{6.278}$$

We can now apply the T-duality transformation to the field α, i.e. substitute α by the T-dual field ϑ such that $\partial_\mu \alpha = \frac{1}{\sqrt{n+2}} i\epsilon_{\mu\nu}\partial_\nu \vartheta$ (we ignore the difference between n and $n+2$ here, since the parameter n is assumed to be large). The substitution of the fields (ϕ, α) by the pair (φ, ϑ), with $\phi = \frac{1}{\sqrt{n}}\varphi$, brings the action to the form

$$\tilde{\mathcal{A}}_0 = \frac{1}{4\pi} \int d^2x \left((\partial_\mu \varphi)^2 + (\partial_\mu \vartheta)^2 \right). \tag{6.279}$$

The advantage of $\tilde{\mathcal{A}}_0$ compared to the action (6.278), is that it allows one to easily incorporate the effects of the sausage tips. As was discussed in Section 6.15, the part of the sausage which includes the central region and also the left tip can be described by the sine-Liouville action

$$\tilde{\mathcal{A}}^{(\text{left})} = \tilde{\mathcal{A}}_0 - 2\mu \int d^2x \, e^{-\sqrt{n}\varphi} \cos(\sqrt{n+2}\vartheta). \tag{6.280}$$

Apparently, the right tip together with the central region is governed by the action which is related to (6.280) by the flip $\varphi \to -\varphi$, i.e.

$$\tilde{\mathcal{A}}^{(\text{right})} = \tilde{\mathcal{A}}_0 - 2\mu \int d^2x \, e^{+\sqrt{n}\varphi} \cos(\sqrt{n+2}\vartheta). \tag{6.281}$$

At this point one can guess that the sausage NLSM admits the dual description by means of the Euclidean action[20]

$$\tilde{\mathcal{A}}^{(\text{saus})} = \int d^2x \left[\frac{1}{4\pi} \left((\partial_\sigma \varphi)^2 + (\partial_\sigma \vartheta)^2 \right) - 4\mu \cosh\left(\sqrt{n}\varphi\right) \cos\left(\sqrt{n+2}\vartheta\right) \right]. \tag{6.282}$$

Remarkably, this naive guess turns out to be correct! There are many indications that (6.282) provides a dual description of the sausage NLSM. Some of them will be discussed below.

Contrary to the sine-Liouville model where dimensionless μ was a somewhat fake parameter, the coupling μ in (6.282) is an important, dimensionful characteristic of the theory. Notice that the field $\cosh(\sqrt{n}\varphi)\cos(\sqrt{n+2}\vartheta)$ has the scale dimensions $d = \Delta_+ + \Delta_- = 1$ w.r.t. the conventional energy–momentum tensor of the 'unperturbed' free theory (6.279). Therefore μ has dimensions of energy, i.e.

$$\mu = CE_*. \tag{6.283}$$

where C is some dimensionless constant which a priori depends on the RG-invariant parameter n (recall that E_* is identified with the inverse correlation length).

---

[20] The dual form of the sausage model was originally proposed by Aleosha Zamolodchikov.

6.16.2 Hilbert space of the sausage NLSM

The Hilbert space of the sine-Liouville theory is classified w.r.t. the action of the $\mathfrak{W}_+ \otimes \mathfrak{W}_-$-algebra. As has already been mentioned the W-algebras related by the reflection $\varphi \to -\varphi$ are algebraically isomorphic. Hence the Hilbert spaces corresponding to the 'left' and 'right' sine-Liouville models (6.280) and (6.281) can be identified and understood as an extended Hilbert space of the sausage NLSM.

As before, we shall consider the boundary conditions,

$$\varphi(x_1+R,x_2) = \varphi(x_1,x_2), \quad \vartheta(x_1+R,x_2) = \vartheta(x_1,x_2) + \frac{2\pi q}{\sqrt{n+2}} \quad (q \in \mathbb{Z}). \quad (6.284)$$

Since the action (6.282) is invariant under the transformation $\vartheta \mapsto \vartheta + \frac{2\pi}{\sqrt{n+2}}$, the space of states splits on the orthogonal subspaces $\mathcal{H}^{(\text{saus})}_{q,k}$ characterized by the quasimomentum k, whose values can be restricted to the first Brillouin zone, $-\frac{1}{2} \le k \le \frac{1}{2}$. Similarly to the case of a quantum particle in a periodic potential (see Figure 6.14), each $\mathcal{H}^{(\text{saus})}_{q,k}$ is split into invariant subspaces labelled by a positive integer $N = 1, 2, \ldots$, which corresponds to the splitting of the energy spectrum into bands:

$$\mathcal{H}^{(\text{saus})}_{q,k} = \oplus_{N=1}^{\infty} \mathcal{H}^{(N)}_{q,k}. \quad (6.285)$$

Notice that in the cigar NLSM/sine-Liouville model, the forbidden bands have zero width, and therefore, the splitting into bands appears to be somewhat artificial in that

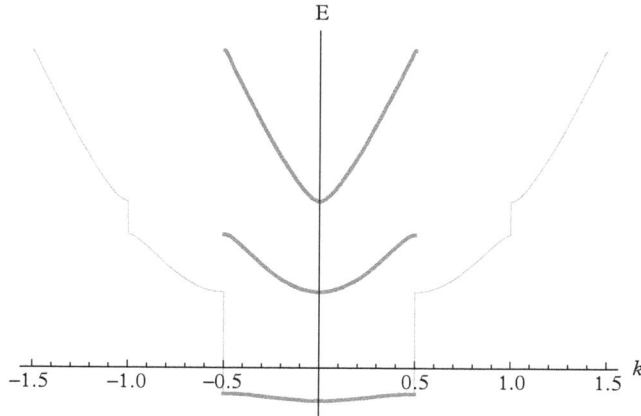

Figure 6.14 *The energy spectrum of the quantum mechanical particle in the periodic potential $V = -V_0 \cos(X)$. The quasimomentum $-\frac{1}{2} \le k \le \frac{1}{2}$ labels the Floquet solutions of the Schrödinger equation inside the N-th conducting band: $\Psi^{(N)}_k(X+2\pi) = e^{2\pi i k} \Psi^{(N)}_k(X)$. Reproduced with permission from John Wiley & Sons.*

case. In the sausage NLSM the widths of the forbidden bands are non-zero and depend on the dimensionless parameter μR. As has been mentioned previously, $\mathcal{H}_{q,k}^{(\text{saus})}$ can be classified w.r.t. the action of $\mathfrak{W}_+ \otimes \mathfrak{W}_-$-algebra and admits the decomposition similar to (6.260):

$$\mathcal{H}_{q,k}^{(\text{saus})} = \oplus_{\substack{P_i > 0 \\ K \in \mathbb{Z}}} \mathcal{V}_{P_i; q, k+K}. \tag{6.286}$$

However, contrary to the case of the cigar NLSM, the individual components $\mathcal{V}_{P_i; q, k+K}$ are not invariant subspaces of the Hamiltonian any more. Also, as we will see, the zero-mode momentum P is quantized, i.e. takes a certain discrete set of admissible values $\{P_i\}$ for given μR.

Let us recall the sine-Liouville model possesses an infinite set of involutive local IM. Since $\{\hat{I}_{2j-1}^{(+)}, \hat{I}_{2j-1}^{(-)}\}_{j=1}^{\infty}$ are invariant under the reflection $\varphi \mapsto -\varphi$, they are IM for both theories (6.280) and (6.281). This simple observation suggests that the quantum sausage NLSM is an integrable QFT, as well as its classical counterpart. In particular, the quantum theory possesses the local IM $\{\hat{\mathbb{I}}_{2j-1}^{(+)}, \hat{\mathbb{I}}_{2j-1}^{(-)}\}_{j=1}^{\infty}$ which can be thought of as a certain deformation of the conformal one. In particular (see Exercise 46)

$$\hat{\mathbb{I}}_{2j-1}^{(\pm)} = \int_0^R \frac{dx^1}{2\pi} \left(\sum_{l+m=j} C_{lm}^{(j)} (\partial_\pm \varphi)^{2l} (\partial_\pm \vartheta)^{2m} + \ldots \right), \tag{6.287}$$

where we return to the Minkowski light cone variables $x^\pm = x^0 \pm x^1$ and

$$C_{lm}^{(j)} = (-1)^j \, 2^{1-2j} \, \frac{(2j-2)!}{l!m!} \, \frac{\big((n+2)(\tfrac{1}{2}-j)\big)_l \big((-n)(\tfrac{1}{2}-j)\big)_m}{\big((n+2)(\tfrac{1}{2}-j)\big)_j \big((-n)(\tfrac{1}{2}-j)\big)_j} \, (-n)^l (n+2)^m \tag{6.288}$$

and the dots stand for monomials which include higher derivatives of φ and ϑ, as well as terms proportional to powers of μ. Notice that the Hamiltonian $\mathbb{H}_R$ and the total momentum $\mathbb{P}_R$ in the R-channel are given by $\hat{\mathbb{H}}_R = \hat{\mathbb{I}}_1^{(+)} + \hat{\mathbb{I}}_1^{(-)}$ and $\hat{\mathbb{P}}_R = \hat{\mathbb{I}}_1^{(+)} - \hat{\mathbb{I}}_1^{(-)}$, with

$$\hat{\mathbb{I}}_1^{(\pm)} = \int_0^R \frac{dx^1}{2\pi} \left((\partial_\pm \varphi)^2 + (\partial_\pm \vartheta)^2 - 4\mu \cosh\left(\sqrt{n}\varphi\right) \cos\left(\sqrt{n+2}\,\vartheta\right) \right). \tag{6.289}$$

In all likelihood the common eigenvectors of $\{\hat{\mathbb{I}}_{2j-1}^{(+)}, \hat{\mathbb{I}}_{2j-1}^{(-)}\}_{j=1}^{\infty}$ form a basis in each invariant subspace $\mathcal{H}_{M,k}^{(N)}$ of the Hilbert space of the sausage NLSM.

Finally, let us note that the sausage NLSM possesses the usual QFT discrete symmetries, such as $P-$ and $C-$ invariance. The parity transformation, $x_1 \mapsto -x_1$, interchanges $\hat{\mathbb{I}}_{2j-1}^{(+)}$ and $\hat{\mathbb{I}}_{2j-1}^{(-)}$ and acts inside $\mathcal{H}_{q,k}^{(N)}$. The C-conjugation can be defined as a transformation $\varphi \mapsto \varphi$ and $\vartheta \mapsto -\vartheta$, which preserves the form of the action $\tilde{\mathcal{A}}^{(\text{saus})}$ (6.282). All the local IM are C-invariant, while $\mathcal{H}_{q,k}^{(N)} \mapsto \mathcal{H}_{q,-k}^{(N)}$ under C-conjugation.

6.16.3 Small-R limit in the sausage NLSM

The Hamiltonian of the sausage model is bounded from below in each $\mathcal{H}_{M,k}^{(\text{saus})}$. The ground states in the neutral sectors $\mathcal{H}_{0,k}^{(\text{saus})}$ are referred to below as the k-vacuums, and the corresponding energies are denoted as E_k. Of course, the k-vacuum belongs to the first band: $|\text{vac}\rangle_k \in \mathcal{H}_{0,k}^{(1)}$.

Let us consider the k-vacuum energy as $R \to 0$ [35], [38]. Because of the scaling properties of the interaction $\cosh(\sqrt{n}\varphi)\cos(\sqrt{n+2}\vartheta)$, one can rescale the problem to the circle of circumference 2π, substituting (6.282) by

$$\mathcal{A} = \int dx_2 \int_0^{2\pi} dx_1 \left[\frac{1}{4\pi} \left((\partial_\sigma \varphi)^2 + (\partial_\sigma \vartheta)^2 \right) - 2\mu \frac{R}{2\pi} \left(e^{-\sqrt{n}\varphi} + e^{+\sqrt{n}\varphi} \right) \cos\left(\sqrt{n+2}\vartheta\right) \right]. \tag{6.290}$$

Let $\Psi_k^{(\text{vac})}[\varphi(x_1), \vartheta(x_1)]$ be the ground state wave functional and define again the zero-mode φ_0 as in equation (6.255):

$$\varphi_0 = \int_0^R \frac{dx_1}{R} \varphi(x). \tag{6.291}$$

As $R \to 0$ there is a large region $\frac{1}{n}\log\left(\frac{\mu R}{2n}\right) < \frac{\varphi_0}{\sqrt{n}} < -\frac{1}{n}\log\left(\frac{\mu R}{2n}\right)$ (to be compared with equation (6.277)) in the configuration space where one can neglect the interaction term in (6.290) and consider $\varphi(x)$ and $\vartheta(x)$ as free massless fields $\varphi(x) \asymp \varphi_+(z) + \varphi_-(\bar{z})$, $\vartheta(x) \asymp \vartheta_+(z) - \vartheta_-(\bar{z})$. In this region the ground state wave functional is expected to be a superposition,

$$\Psi_k^{(\text{vac})}[\varphi(x_1), \vartheta(x_1)] \sim \left(c_1 e^{-2iP\varphi_0} + c_2 e^{+2iP\varphi_0} \right) |0\rangle, \tag{6.292}$$

of two zero-mode plane waves with some R-dependent zero mode momentum P times the Fock vacuum $|0\rangle$ of the oscillators $a_l^{(\pm)}$ and $b_l^{(\pm)}$ ($l \neq 0$) (6.221). The corresponding k-vacuum energy is determined at $R \to 0$ mainly by $P(R)$:

$$\frac{RE_k}{\pi} = -\frac{1}{3} + (n+2)k^2 + 4P^2(R) + O\left(R^{\frac{4}{n}}, R^2 \log R\right) \tag{6.293}$$

up to power corrections in R.[21] The momentum $P(R)$ is quantized due to the the right and left potential walls at $\frac{\varphi_0}{\sqrt{n}} \sim \pm \frac{1}{n} \log\left(\frac{\mu R}{2n}\right)$ in the action (6.290), which correspond to the sausage tips in the NLSM description. Consider say the left wall at $\frac{\varphi_0}{\sqrt{n}} \sim \frac{1}{n} \log\left(\frac{\mu R}{2n}\right)$. If R is small enough, the second exponent $e^{+\sqrt{n}\varphi}$ in the potential term of (6.290) is small in this region and does not affect the dynamics which therefore are expected to be the same as in the sine-Liouville theory. Thus for the reflected wave we have (see equation (6.261))

$$\Psi_k^{(\text{vac})}[\varphi(x_1), \vartheta(x_1)] \sim \left(e^{-2iP\varphi_0} + \left(\frac{R}{2\pi}\right)^{-\frac{4iP}{\sqrt{n}}} S(P)\, e^{+2iP\varphi_0}\right)|0\rangle \quad \text{at} \quad \varphi_0 \to +\infty, \tag{6.294}$$

where $S_{0,k}(P)$ is the same reflection amplitude (6.263) as in the sine-Liouville theory. The factor $\left(\frac{R}{2\pi}\right)^{-\frac{4iP}{\sqrt{n}}}$ appears due to the rescaling of μ in (6.290). A similar consideration about the right wall reflection leads to the following quantization condition:

$$\left(\frac{R}{2\pi}\right)^{-\frac{8iP}{\sqrt{n}}} (S(P))^2 = 1. \tag{6.295}$$

For the k-ground state momentum $P_0 \equiv \frac{p_0}{2\sqrt{n}}$ this equation reads

$$-\frac{4p_0}{n} \log\left(\frac{\mu R}{2n}\right) + 2\delta_\lambda(p_0) = 2\pi, \tag{6.296}$$

where $\lambda = k(n+2)$ and

$$\delta_\lambda(p) = -i\log\left[\frac{\Gamma(\frac{1}{2}+\frac{\lambda}{2}-\frac{ip}{2})\Gamma(\frac{1}{2}-\frac{\lambda}{2}-\frac{ip}{2})}{\Gamma(\frac{1}{2}+\frac{\lambda}{2}+\frac{ip}{2})\Gamma(\frac{1}{2}-\frac{\lambda}{2}+\frac{ip}{2})} \frac{\Gamma(1+ip)}{\Gamma(1-ip)} \frac{\Gamma(1+\frac{ip}{n})}{\Gamma(1-\frac{ip}{n})}\right], \tag{6.297}$$

where the branch of the logarithm is chosen so that $\delta_\lambda(0) = 0$. Together with (6.293) equation (6.296) determines the most important part of the $R \to 0$ asymptotic of the k-vacuum energy E_k. For example, using the regular expansion of the reflection phase in odd powers of p,

$$\delta_\lambda(p) = \delta'_\lambda(0)p + \frac{1}{3!}\delta'''_\lambda(0)p^3 + \ldots, \tag{6.298}$$

---

[21] The presence of the term $\propto R^2 \log(R)$ (6.293) is a peculiar instanton effect.

one develops RE_k/π systematically in $1/\log(R)$:

$$\frac{RE_k}{\pi} = -\frac{1}{3} + (n+2)k^2 + \frac{\pi^2}{n\left(\ell + \delta'_\lambda(0)\right)^2} + O(\ell^{-5}), \qquad (6.299)$$

where

$$\ell = -\frac{2}{n}\log\left(\frac{\mu R}{2n}\right) = \log\left(\frac{1+\kappa}{1-\kappa}\right). \qquad (6.300)$$

Notice that ℓ can be interpreted as an effective length of the sausage (see Figure 6.13). Remarkably, essentially the same result for the small R-asymptotic can be obtained within the so-called minisuperspace approximation for the cigar NLSM.

All the above considerations can be applied to all the k-vacuum eigenvalues of local IM:

$$\hat{\mathbb{I}}^{(+)}_{2j-1}|\mathrm{vac}\rangle_k = \hat{\mathbb{I}}^{(-)}_{2j-1}|\mathrm{vac}\rangle_k = \mathbb{I}^{(\mathrm{vac})}_{2j-1}|\mathrm{vac}\rangle_k \qquad (6.301)$$

(here we take into account that unbroken P-invariance implies that $\hat{\mathbb{I}}^{(+)}_{2j-1}$ and $\hat{\mathbb{I}}^{(-)}_{2j-1}$ have the same vacuum eigenvalues). Similar to the k-vacuum energy, the $R \to 0$ asymptotic of higher spin local IM $\mathbb{I}^{(\mathrm{vac})}_{2j-1}$ are given by

$$\mathbb{I}^{(\mathrm{vac})}_{2j-1} = I^{(\mathrm{vac})}_{2j-1}\left(\tfrac{p_0(R)}{2\sqrt{n}}, \tfrac{k}{2}\sqrt{n+2}\right) + O\left(R^{\frac{4}{n}+1-2j}, R^{3-2j}\right), \qquad (6.302)$$

where $I^{(\mathrm{vac})}_{2j-1}(P,Q)$ are vacuum eigenvalues of the sine-Liouville local IM, which are certain polynomials of P^2 and Q^2 of degree j (see Exercise 44).

6.17 Other integrable structures in the cigar NLSM

Let us return to the cigar NLSM and take a closer look at the local IM $I^{(+)}_{2j-1}$. As follows from the above consideration, they can be introduced through two requirements: (1) Each I_{s-1} is an integral over a local density P_s, which is a homogeneous differential polynomial in $\partial\varphi_+, \partial^2\varphi_+,\ldots\partial\vartheta_+, \partial^2\vartheta_+,\ldots$ of overall degree s with constant coefficients; (2) $\hat{I}_s$ commute with a combined set of four fermionic screening charges for both the sine-Liouville models (6.281) and (6.282), i.e.

$$\left[\hat{I}_{s-1}, X\right] = 0, \quad \text{where} \quad X = \oint dw \, e^{\sigma\sqrt{n}\varphi_+ + \sigma'\sqrt{n+2}\vartheta_+}(w) \quad (\sigma,\sigma' = \pm). \qquad (6.303)$$

For given degree $s = 1,2,3\ldots$, the homogeneous differential polynomials form a finite-dimensional linear space. Starting with the most general linear combination of the differential monomials, one can adjust all the coefficients in P_s modulo an overall factor by means of the conditions (6.303). Brute force calculations show that for $s = 2j = 2, 4, 6$ there exists only one (up to the normalization) operator $\hat{I}_{2j-1}$ satisfying (6.303), whereas there are no solutions of (6.303) for odd s. Then one can check that the constructed operators $\hat{I}_{2j-1}$ mutually commute.

In spite of the formal algorithm described above requiring a considerable amount of computational effort, it provides a systematic way to generate the involutive set of local IM. Remarkably, similar procedures allow one to construct different involutive sets of local IM. Namely, the condition $[\hat{I}_{s-1}, X] = 0$ with

$$X = \left\{ \oint dw\, e^{-\sqrt{n}\varphi_+ \pm \sqrt{n+2}\vartheta_+}(w) \quad \text{and} \quad \oint dw\, e^{+\frac{2\varphi_+}{\sqrt{n}}}(w) \right\} \tag{6.304}$$

yields the commuting set with exactly one IM for any $s = 1, 2, 3, \ldots$. The first representatives from this set are given by [16],

$$\hat{I}_1^{(I)} = \oint_R \frac{dz}{2\pi i}\, T(z)$$
$$\hat{I}_2^{(I)} = C_2^{(I)} \oint_R \frac{dz}{2\pi i}\, W_3(z) \tag{6.305}$$
$$\hat{I}_3^{(I)} = C_3^{(I)} \oint_R \frac{dz}{2\pi i}\, \left(W_4^{(\text{sym})} - 3n(n+1)\, T^2 \right)(z),$$

where $W_4^{(\text{sym})}$ and T^2 are given by equations (6.233) and (6.228), respectively.

Yet another involutive set $\left\{ \hat{I}_{2j-1}^{(II)} \right\}_{j=1}^{\infty}$ can be obtained from the system of screening charges

$$X = \left\{ \oint dw\, e^{-\sqrt{n}\varphi_+ \pm \sqrt{n+2}\vartheta_+}(w) \quad \text{and} \quad \oint dw\, e^{+\frac{4\varphi_+}{\sqrt{n}}}(w) \right\}. \tag{6.306}$$

The lowest spin IM from this set are given by

$$\hat{I}_1^{(II)} = \oint_R \frac{dz}{2\pi i}\, T(z) \tag{6.307}$$
$$\hat{I}_3^{(II)} = C_3^{(II)} \oint_R \frac{dz}{2\pi i}\, \left(W_4^{(\text{sym})} - \frac{3n(n+2)(n+4)}{n+6}\, T^2 \right)(z).$$

Notice that the classical limit of the commuting families $\left\{ \hat{I}_s^{(I)} \right\}_{s=1}^{\infty}$ and $\left\{ \hat{I}_{2j-1}^{(II)} \right\}_{j=1}^{\infty}$ was discussed briefly in Section 6.6.3 from the first lecture.

Following the same logic as for the sausage NLSM we can guess at this point, that the following Euclidean actions define integrable QFT:

$$\tilde{\mathcal{A}}^{(I)} = \int d^2x \left[\frac{1}{4\pi} \left((\partial_\sigma \varphi)^2 + (\partial_\sigma \vartheta)^2 \right) - 2\mu \, e^{-\sqrt{n}\varphi} \cos(\sqrt{n+2}\vartheta) + \mu' \, e^{\frac{2\varphi}{\sqrt{n}}} \right] \quad (6.308)$$

$$\tilde{\mathcal{A}}^{(II)} = \int d^2x \left[\frac{1}{4\pi} \left((\partial_\sigma \varphi)^2 + (\partial_\sigma \vartheta)^2 \right) - 2\mu \, e^{-\sqrt{n}\varphi} \cos(\sqrt{n+2}\vartheta) + \mu' \, e^{\frac{4\varphi}{\sqrt{n}}} \right],$$

where μ' is a dimensionful coupling. Recall that, in the cigar NLSM, the classical limit of the fields $e^{\frac{2j\varphi}{\sqrt{n}}}$ for $j = 1, 2$ is described by the equations (6.272). This gives a hint that (6.308) provide the dual description for the theories whose classical ($n \to \infty$) limit is governed by the Lagrangians

$$\mathcal{L}^{(I)} = \frac{1}{4\pi} \left(\frac{\partial_\mu Z \partial^\mu Z^*}{1 + |Z|^2} - m^2 |Z|^2 \right) \quad (6.309)$$

$$\mathcal{L}^{(II)} = \frac{1}{4\pi} \left(\frac{\partial_\mu Z \partial^\mu Z^*}{1 + |Z|^2} - m^2 |Z|^2 \left(1 + |Z|^2 \right) \right).$$

The models (6.309) are well known in the theory of classical integrable systems under the name of the complex sinh-Gordon I (Pohlmeyer–Lund–Regge) model [7, 39] and the complex sinh-Gordon II (Getmanov) model [40].

Exercise 53 Show that the Lagrange–Euler equation corresponding to both the Lagrangians (6.309) admit the zero curvature representation of the form

$$\left[D_+, \omega D_- \omega^{-1} \right] = 0 \quad (6.310)$$

with

$$D_\pm = \partial_\pm \mp (\psi_\pm f + \psi_\pm^* e) - \frac{im}{4} \lambda^{\pm 1} \, a, \quad (6.311)$$

where $a = h$ for the Lund–Regge model, whereas for the Getmanov model h, e, f, and a should be understood as 3×3 matrixes[22]

$$h = \begin{pmatrix} 2 & 0 & 0 \\ 0 & 0 & 0 \\ 0 & 0 & -2 \end{pmatrix}, \quad e = -i\sqrt{2} \begin{pmatrix} 0 & 1 & 0 \\ 0 & 0 & 1 \\ 0 & 0 & 0 \end{pmatrix}, \quad f = i\sqrt{2} \begin{pmatrix} 0 & 0 & 0 \\ 1 & 0 & 0 \\ 0 & 1 & 0 \end{pmatrix},$$

$$a = \frac{2}{3} \begin{pmatrix} 1 & 0 & 0 \\ 0 & -2 & 0 \\ 0 & 0 & 1 \end{pmatrix}. \quad (6.312)$$

For other notations see equations (6.66)–(6.78) in Section 6.6.1.

---

[22] From the formal algebraic point of view h, e, f, and a satisfy the following set of commutation relations:
$[h, e] = +2e, \quad [h, f] = -2f, \quad [e, f] = h, \quad [h, a] = 0, \quad [a, [a, e]] = 4e, \quad [a, [a, f]] = 4f.$

Acknowledgements

The authors acknowledge the help of Gleb Kotcusov in the preparation of the lectures. SL would like to thank the organizers of the school for their hospitality and for creating, together with the participants, a pleasant and stimulating atmosphere.

This work is partially supported by the NSF under grant number NSF-PHY-1404056.

References

[1] V.I Arnold, Mathematical Methods of Classical Mechanics, Springer.
[2] O. Babelon, D. Bernard, and M. Talon, Introduction to classical integrable systems, Cambridge Univ. Press 2003.
[3] L.D. Landau and E.M. Lifshitz, Mechanics, Pergamon Press.
[4] V. E. Zakharov and A. V. Mikhailov, Sov. Phys. JETP **47**, 1017 (1978) [Zh. Eksp. Teor. Fiz. **74**, 1953 (1978)].
[5] L. D. Faddeev and L. A. Takhtajan, Hamiltonian Methods in the Theory of Solitons, Springer Series in Soviet Mathematics 1987.
[6] K. Uhlenbeck, 'Harmonic maps into Lie groups: classical solutions of the chiral model,' J. Differential Geom. **30**, 1 (1989).
[7] K. Pohlmeyer, Commun. Math. Phys. **46**, 207 (1976).
[8] V. A. Fateev, Nucl. Phys. B **473**, 509 (1996).
[9] S. L. Lukyanov, Nucl. Phys. B **865**, 308 (2012) [arXiv:hep-th/1205.3201].
[10] C. Klimcik, Lett. in Math. Phys. 50, 043508 (2009) [arXiv:math-ph/1402.2105].
[11] B. Hoare, R. Roiban, and A. A. Tseytlin, JHEP **1406**, 002 (2014) [arXiv:hep-th/1403.5517].
[12] V. A. Fateev, E. Onofri, and A. B. Zamolodchikov, Nucl. Phys. B **406**, 521 (1993).
[13] R. S. Hamilton, Mathematics and general relativity (Santa Cruz, CA, 1986) Contemp. Math., vol. 71, Amer. Math. Soc., Providence, RI, 1988. pp. 237-262.
[14] S. Elitzur, A. Forge, and E. Rabinovici, Nucl. Phys. B **359**, 581 (1991).
[15] E. Witten, Phys. Rev. D **44**, 314 (1991).
[16] V. A. Fateev and S. L. Lukyanov, J. Phys. A **39**, 12889 (2006) [arXiv:hep-th/0510271].
[17] E. Witten, Commun. Math. Phys. **92**, 455 (1984).
[18] E. Braaten, T. L. Curtright, and C. K. Zachos, Nucl. Phys. B **260**, 630 (1985).
[19] A. B. Zamolodchikov, Nucl. Phys. B **342**, 695 (1990).
[20] P. H. Ginsparg, 'Applied Conformal Field Theory', Lectures given at Les Houches summer session, Jun 28 – Aug 5, 1988 [arXiv:hep-th/9108028].
[21] P. Di Francesco, P. Mathieu, D. Sénéchal, Conformal Field Theory, Springer-Verlag New York 1997.
[22] L.D. Landau and E.M. Lifshitz, Statistical Physics, Pergamon Press.
[23] A. A. Belavin, A. M. Polyakov, and A. B. Zamolodchikov, Nucl. Phys. B **241**, 333 (1984).
[24] A. M. Polyakov, Gauge Fields and Strings, Harwood (1987).
[25] D. Friedan, Phys. Rev. Lett. **45**, 1057 (1980); Annals Phys. **163**, 318 (1985).
[26] E. S. Fradkin and A. A. Tseytlin, Nucl. Phys. B **261**, 1 (1985).
[27] C. G. Callan, Jr., E. J. Martinec, M. J. Perry, and D. Friedan, Nucl. Phys. B **262**, 593 (1985).
[28] M. E. Peskin and D. V. Schroeder, An Introduction to Quantum Field Theory, Westview Press, 1995.

[29] M. B. Green, J. H. Schwarz, E. Witten, Superstring Theory. Vol. 1: Introduction. Cambridge etc., Cambridge University Press, 1987.
[30] S. F. Hassan and A. Sen, Nucl. Phys. B **405**, 143 (1993) [arXiv:hep-th/9210121].
[31] I. Bakas and E. Kiritsis, Int. J. Mod. Phys. A **7S1A**, 55 (1992) [arXiv:hep-th/9109029].
[32] S. L. Lukyanov, E. S. Vitchev, and A. B. Zamolodchikov, Nucl. Phys. B **683**, 423 (2004) [arXiv:hep-th/0312168].
[33] V. A. Fateev and A. B. Zamolodchikov, Sov. Phys. JETP **62**, 215 (1985) [Zh. Eksp. Teor. Fiz. **89**, 380 (1985)].
[34] V. A. Fateev and A. B. Zamolodchikov, Nucl. Phys. B **280**, 644 (1987).
[35] Our presentation here follows unpublished notes by Al. and A. Zamolodchikov, 1998.
[36] R. Dijkgraaf, H. L. Verlinde, and E. P. Verlinde, Nucl. Phys. B **371**, 269 (1992).
[37] L.D. Landau and E.M. Lifshitz, Quantum Mechanics, Pergamon Press.
[38] A. B. Zamolodchikov and A. B. Zamolodchikov, Nucl. Phys. B **477**, 577 (1996) [arXiv:hep-th/9506136].
[39] F. Lund and T, Regge, Phys. Rev. D, 14, 1524 (1976).
[40] B. S. Getmanov, Theor. Math. Phys. **48**, 572 (1982) [Teor. Mat. Fiz. **48**, 13 (1981)].
[41] I. Bakas, S. Kong, and L. Ni, Journal für die reine und angewandte Math. **663**, 209 (2012) [arXiv:math.DG/0906.0589].

7
Applications of integrable models in condensed matter and cold atom physics

Fabian H.L. Essler

The Rudolf Peierls Centre for Theoretical Physics,
Oxford University, Oxford OX1 3NP, UK

Chapter Contents

7 **Applications of integrable models in condensed matter and cold atom physics**   319
  Fabian H.L. ESSLER

    Preface   321
7.1   Physical realizations of integrable models   321
7.2   Eigenstates of integrable and generic local Hamiltonians   331
7.3   Macro states in the spin-1/2 Heisenberg chain   334
7.4   Quantum quenches   346
    References   349

Preface

These are lecture notes for a short course given at the 2016 Les Houches Summer School on *Integrability: From Statistical Systems to Gauge Theory*. I start with a discussion of approximate physical realizations of $1+1$-dimensional integrable models in solids and systems of ultra-cold trapped atoms. I then turn to local properties of energy eigenstates away from the edges of the spectrum. In generic models these states are thermal, while in integrable models non-thermal states with finite entropy densities coexist with thermal ones. I discuss how to construct these atypical states by means of the Bethe ansatz. Finally I outline the crucial role these states play in describing the stationary states reached at late times after quantum quenches to integrable theories.

7.1 Physical realizations of integrable models

7.1.1 Anisotropic crystals

One realization of integrable models is in *anisotropic solids*. An example is shown in Figure 7.1(a). The relevant degrees of freedom ultimately arise from the outer electrons of the Cu atoms. They tunnel predominantly between neighbouring Cu atoms and interact by a screened Coulomb repulsion. A crude model for this situation is the one-dimensional *Hubbard model*,

$$H_{\text{Hub}} = -t \sum_{j=1}^{L} \sum_{\sigma=\uparrow,\downarrow} c_{j,\sigma}^{\dagger} c_{j+1,\sigma} - \text{h.c.} + U \sum_{j=1}^{L} n_{j,\uparrow} n_{j,\downarrow}. \quad (7.1)$$

Here $n_{j,\sigma} = c_{j,\sigma}^{\dagger} c_{j,\sigma}$ are number operators for spin-σ electrons at site j, and $c_{j,\sigma}$ are fermionic annihilation operators fulfilling canonical anticommutation relations $\{c_{j,\sigma}, c_{l,\tau}^{\dagger}\} = \delta_{j,l} \delta_{\sigma,\tau}$. We are interested in the half-filled band case, which corresponds to one electron per site on average. In the kind of material shown in the figure one typically is in the regime

$$t \ll U. \quad (7.2)$$

The low-energy sector of the Hubbard model in this limit is described by the spin-$\frac{1}{2}$ *Heisenberg antiferromagnet*,

$$H_{\text{Heis}} = \mathcal{J} \sum_{j=1}^{L} S_j^x S_{j+1}^x + S_j^y S_{j+1}^y + S_j^z S_{j+1}^z, \quad (7.3)$$

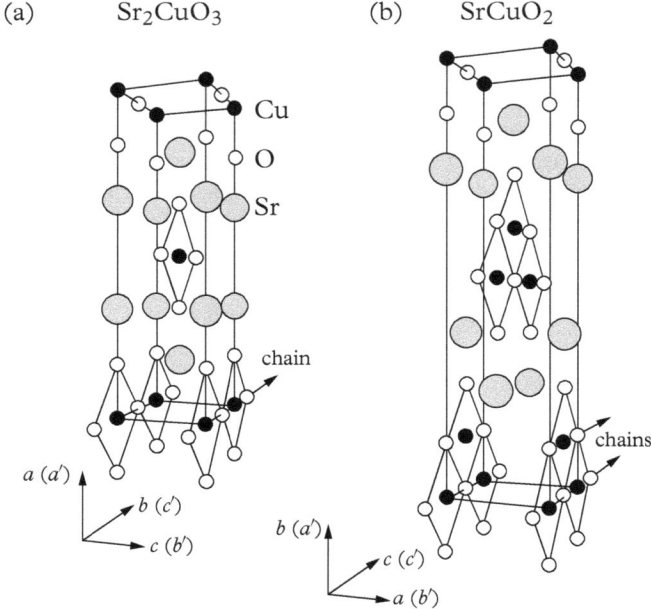

Figure 7.1 *Crystal structure of the quasi-one-dimensional spin chain materials (a) Sr_2CuO_3 and (b) $SrCuO_2$. Figure taken from A.V. Sologubenko et al., Phys. Rev. B**64**, 054412 (2001). Reprinted with permission.*

where $\mathcal{J} = 4t^2/U$ and S_j^α are spin operators

$$[S_j^\alpha, S_k^\beta] = i\delta_{j,k}\epsilon_{\alpha\beta\gamma}S_k^\gamma \tag{7.4}$$

acting on a spin-$\frac{1}{2}$ representation of SU(2) at every site. While the Hubbard model only provides a qualitative description of the electronic degrees of freedom in Sr_2CuO_3 and $SrCuO_2$, their magnetic properties are fairly well described by H_{Heis}, see e.g. [1].

The way the Heisenberg Hamiltonian arises from the Hubbard model is a particular example of a general method for carrying out strong coupling expansions [2]. In case you haven't encountered this before, the main steps are summarized in the Section 7.1.1.1.

7.1.1.1 *Effective Hamiltonians from strong coupling expansions*

Our starting point is the *atomic limit* limit $t \to 0$. Here the Hamiltonian becomes

$$U\hat{D} = U\sum_{j=1}^{L} n_{j,\uparrow}n_{j,\downarrow}, \tag{7.5}$$

and simply counts the number of doubly occupied sites. For any $U > 0$ there are 2^L zero-energy ground states, corresponding to singly occupied sites with either a spin-up or a spin-down electron. The lowest excited states involve precisely one doubly occupied site and have energy U. There are $L(L-1)2^{L-2}$ such states (as we have precisely L electrons). The next lowest excited states involve two doubly occupied sites etc.

On general grounds we expect that switching on a hopping $t \neq 0$ will lift the macroscopic degeneracies in the sectors with $0,1,2,\ldots$ double occupancies, but retain the structure of the spectrum (at sufficiently low energies). In other words, the low-energy sector involves 2^L states that are adiabatically connected with states that have no doubly occupied sites. In order to work out how the Hamiltonian acts in the low-energy subspace, we first define a basis of states at site j by $|0\rangle_j$, $|\uparrow\rangle_j = c_{j,\uparrow}^\dagger |0\rangle_j$, $|\downarrow\rangle_j = c_{j,\downarrow}^\dagger |0\rangle_j$, $|2\rangle_j = c_{j,\uparrow}^\dagger c_{j,\downarrow}^\dagger |0\rangle_j$. Clearly $|0\rangle_j$ and $|2\rangle_j$ are bosonic, while $|\uparrow\rangle$ and $|\downarrow\rangle$ are fermionic. We now introduce *Hubbard operators* by

$$X_j^{ab} = |a\rangle_j \, _j\langle b|, \quad a,b \in \{0,\uparrow,\downarrow,2\}. \tag{7.6}$$

It is easily verified that they fulfil the algebra

$$X_j^{ab} X_n^{cd} = (1 - \delta_{j,n}) X_n^{cd} X_j^{ab} (-1)^{(\epsilon_a + \epsilon_b)(\epsilon_c + \epsilon_d)} + \delta_{j,n} X_j^{ad} \delta_{b,c}, \tag{7.7}$$

where we have defined $\epsilon_0 = \epsilon_2 = 0$, $\epsilon_\uparrow = \epsilon_\downarrow = 1$. The Hubbard Hamiltonian is expressed in terms of these operators as

$$H_{\text{Hub}} = U\hat{D} + t\hat{T}_0 + t\hat{T}_1 + t\hat{T}_{-1}, \tag{7.8}$$

where $\hat{D} = \sum_j X_j^{22}$ and $\hat{T}_n$ changes the number of double occupancies by n,

$$\hat{T}_0 = -\sum_{j,\sigma} X_j^{\sigma 0} X_{j+1}^{0\sigma} + X_j^{2\sigma} X_{j+1}^{\sigma 2} + \text{h.c.},$$

$$\hat{T}_1 = -\sum_{j,\sigma} \varepsilon_\sigma \left[X_j^{2\bar\sigma} X_{j+1}^{0\sigma} + X_j^{0\bar\sigma} X_{j+1}^{2\sigma} \right] = \left(\hat{T}_{-1}\right)^\dagger. \tag{7.9}$$

Here we have defined $\varepsilon_\uparrow = 1 = -\varepsilon_\downarrow$. The Hamiltonian (7.8) contains terms that act within the subspaces of a conserved number of double occupancies, and terms that connect these subspaces. The idea is to eliminate such 'off-diagonal' terms order by order in perturbation theory in t/U by means of a unitary transformation. In other words, consider the unitary transformation

$$H' = e^{iS} H_{\text{Hub}} e^{-iS} = H_{\text{Hub}} + [iS, H_{\text{Hub}}] + \frac{1}{2}[iS, [iS, H_{\text{Hub}}]] + \ldots, \tag{7.10}$$

where S has a power series expansion of the form

$$S = \frac{t}{U}S^{(0)} + \frac{t^2}{U^2}S^{(1)} + \frac{t^3}{U^3}S^{(2)} + \ldots. \tag{7.11}$$

By choosing $S^{(n)}$ appropriately, we can remove off-diagonal terms order by order in t/U. You can check that

$$iS^{(0)} = \hat{T}_1 - \hat{T}_{-1}, \qquad iS^{(1)} = [T_1 + T_{-1}, T_0] \tag{7.12}$$

gives a transformed Hamiltonian of the form

$$H' = U\hat{D} + t\hat{T}_0 + \frac{t^2}{U}[\hat{T}_1, \hat{T}_{-1}] + \mathcal{O}(t^3/U^2). \tag{7.13}$$

Now consider the action of H' on the low-lying states (i.e. those without doubly occupied sites) $|\psi\rangle$. We have

$$H'|\psi\rangle \simeq \frac{t^2}{U}[\hat{T}_1, \hat{T}_{-1}]|\psi\rangle = \frac{t^2}{U}\left[\sum_j 2X_j^{-\sigma\sigma}X_{j+1}^{\sigma-\sigma} - 2X_j^{\sigma\sigma}X_{j+1}^{-\sigma-\sigma}\right]|\psi\rangle$$

$$\equiv \frac{4t^2}{U}\sum_j\left[\mathbf{S}_j \cdot \mathbf{S}_{j+1} - \frac{1}{4}\right]|\psi\rangle, \tag{7.14}$$

where S_j^α are the usual spin-$\frac{1}{2}$ operators

$$S_j^z = \frac{1}{2}\left[c_{j,\uparrow}^\dagger c_{j,\uparrow} - c_{j,\downarrow}^\dagger c_{j,\downarrow}\right], \quad S^+ = c_{j,\uparrow}^\dagger c_{j,\downarrow}, \quad S^- = c_{j,\downarrow}^\dagger c_{j,\uparrow}. \tag{7.15}$$

The emergent spin interaction is known as *Anderson superexchange*. As we have seen, it arises as a result of strong repulsive interactions between electrons. We note that the sign of the exchange interaction is positive, leading to *antiferromagnetic* correlations. The analogous derivation for strongly repulsive bosons leads to a ferromagnetic interaction instead. Finally, we note that the procedure we have been following can be extended to higher orders in t/U. This will generate *ring-exchange* interactions in addition to the Heisenberg term.

7.1.1.2 Field theory description of the low energy limit

At low energies and large distances the physics of the Heisenberg chain can be described by a relativistic QFT, the compactified free boson [3]

$$\boxed{H = \frac{v}{16\pi}\int dx\left[(\partial_x\Phi)^2 + (\partial_x\Theta)^2\right].} \tag{7.16}$$

Here $v = \pi \mathcal{J} a_0/2$ (where a_0 is the lattice spacing), $\Phi = \varphi + \bar{\varphi} = \Phi + 4\pi$ is a Bose field, $\Theta = \varphi - \bar{\varphi}$ the dual field

$$[\Phi(x), \Theta(y)] = 4\pi i\, \mathrm{sgn}(x-y). \tag{7.17}$$

The models (7.1), (7.3), and (7.16) are all integrable. There is however an important variation of the theme we have just discussed. As we have seen, the Heisenberg model is an effective Hamiltonian for the low-energy degrees of freedom of electronic models of solids. In many materials the effective exchange interaction obtained by projecting to the low-energy subspace is anisotropic. We also may apply an external magnetic field to the system, which leads us to the integrable spin-$\frac{1}{2}$ XXZ chain

$$\boxed{H(\Delta, h) = \mathcal{J} \sum_j S_j^x S_{j+1}^x + S_j^y S_{j+1}^y + \Delta S_j^z S_{j+1}^z + h \sum_j S_j^z.} \tag{7.18}$$

In a class of materials with low crystal symmetry the application of a magnetic field has the interesting effect of inducing a staggered field in the perpendicular direction as well (see [4] and references therein),

$$H = H(\Delta, h) + h' \sum_j (-1)^j S_j^x. \tag{7.19}$$

This model is not integrable. However, in the field theory limit it maps onto

$$\boxed{H = \frac{v}{16\pi} \int dx \left[(\partial_x \Phi)^2 + (\partial_x \Theta)^2 \right] + \mu \int dx\, \cos(\beta \Phi).} \tag{7.20}$$

This is the Hamiltonian of the quantum sine-Gordon QFT, which is integrable. This is one of many examples, in which the low-energy limit of a non-integrable lattice model is described by an integrable QFT.

7.1.2 Cold atoms in optical traps

In the last decade or so a second class of (approximate) realizations of integrable models has become available. These are in systems of ultra-cold trapped atoms.

- The quantum mechanical particles are entire atoms. The physics of interest is related to the motion of these atoms and short-range interactions between them.
- A large number of atoms $\sim 10^6$–10^9 is spatially confined by electromagnetic trapping fields that couple to the dipole moments of the atoms.

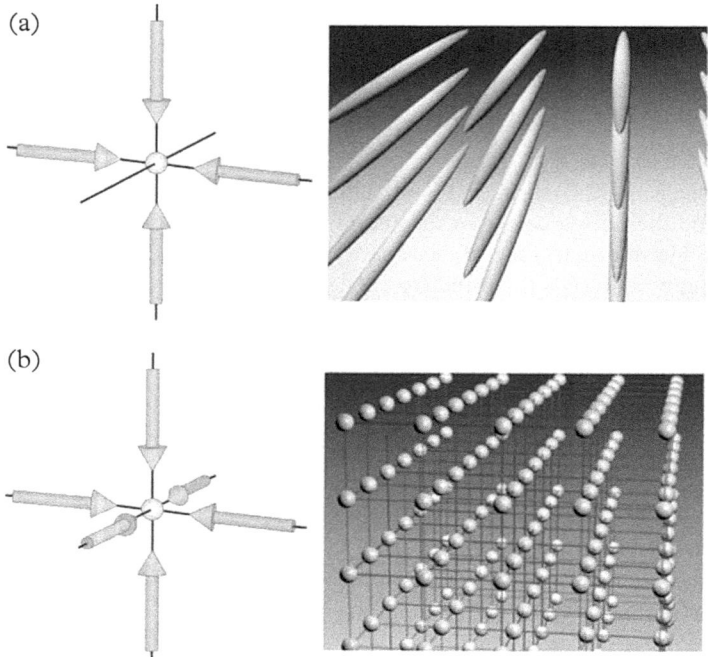

Figure 7.2 *Optically trapped atoms. (a) For a 2D optical lattice, the atoms are confined to an array of tightly confining 1D potential tubes. (b) In the 3D case, the optical lattice can be approximated by a 3D simple cubic array of tightly confining harmonic oscillator potentials at each lattice site. Figure taken from I. Bloch, Nature Phys. 1, 23 (2005). Reprinted with permission from Springer Nature.*

- The trapping fields can be shaped to realize ensembles of 1D tubes as shown in Figure 7.2. The tunnelling amplitude between tubes can be made extremely small, so that over the time scale of the experiment no tunnelling events occur.

The Hamiltonian governing the dynamics of N ultra-cold bosonic or fermionic atoms trapped in 1D tubes is of the form

$$H = H_{\text{LL}} + \underbrace{\lambda \sum_j x_j^2}_{V}, \qquad (7.21)$$

where V is a harmonic potential that arises as a result of the trap and

$$H_{\text{LL}} = -\frac{\hbar^2}{2m} \sum_{j=1}^{N} \frac{\partial^2}{\partial x_j^2} + 2c \sum_{j<k} \delta(x_j - x_k). \qquad (7.22)$$

Here m is the mass of the bosons, and c is the tuneable interaction strength (that can be either positive or negative). The Hamiltonian for a single species of bosons is the integrable Lieb–Liniger model [5] and has the following second-quantized form:

$$H_{\text{LL}} = \int dx \left\{ \frac{\hbar^2}{2m} \partial_x \Phi^\dagger(x) \partial_x \Phi(x) + c \Phi^\dagger(x) \Phi^\dagger(x) \Phi(x) \Phi(x) \right\}, \quad (7.23)$$

where $\Phi^\dagger$, Φ are complex bosonic fields satisfying $[\Phi(x), \Phi^\dagger(y)] = \delta(x-y)$. This is the appropriate description for e.g. Rb atoms [6–8]. Fermionic ^{173}Yt atoms that have six degenerate nuclear spin states have been trapped recently [9]. The first quantized Hamiltonian has the same form as (7.22), while in second quantization we get an SU(6) Yang–Gaudin model

$$H_{\text{YG}} = \int dx \left\{ \frac{\hbar^2}{2m} \sum_a \partial_x \Psi_a^\dagger(x) \partial_x \Psi_a(x) + c \sum_{a,b} \Psi_a^\dagger(x) \Psi_b^\dagger(x) \Psi_b(x) \Psi_a(x) \right\}, \quad (7.24)$$

where $\{\Psi_a(x), \Psi_b^\dagger(y)\} = \delta_{a,b} \delta(x-y)$.

All of these are *continuum models*. Using lasers it is possible to impose a lattice potential, which allows the realization of bosonic and fermionic Hubbard models, cf. [11], [12].

7.1.3 Deviations from integrability

Physical realizations of integrable models are necessarily approximate. In all cases the Hamiltonian describing the actual physical system has the form

$$H_{\text{phys}} = H_0 + \lambda V. \quad (7.25)$$

In cold atom realizations the main contribution to V arises from the harmonic trap potential, which breaks translational invariance as well as integrability. Integrable models can still be used to obtain a good description of many physical properties in the centre of the trap. The situation in solids is somewhat simpler. There the parameter λ (which we take to have dimensions of energy) may be roughly a factor of 10^2–10^3 smaller than the natural energy scale $\mathcal{J}$ of H_0. In many cases the effects of V turn out to be negligible in the following sense.

1. Many (but not all) physical properties are smooth functions of $\lambda/\mathcal{J} \ll 1$. This is typically the case for thermodynamic properties.
2. Actual experiments are not infinitely precise, but have a finite instrumental resolution. If the effects of $\lambda \neq 0$ are smaller than this resolution, we are in business.
3. How important the V term is depends very much on what questions we are asking. Let us consider for example the effects of three-dimensional couplings in a

quantum magnet: crystals are three dimensional and hence there are always small 'interchain' exchange interactions. Let us denote their characteristic energy scale by $\mathcal{J}_\perp$,

$$H = \underbrace{\mathcal{J}\sum_{j,k,l}\mathbf{S}_{j,k,l}\cdot\mathbf{S}_{j+1,k,l}}_{\text{uncoupled chains}} + \mathcal{J}_\perp \sum_{j,k,l}\mathbf{S}_{j,k,l}\cdot\mathbf{S}_{j,k+1,l} + \mathbf{S}_{j,k,l}\cdot\mathbf{S}_{j,k,l+1}. \qquad (7.26)$$

The effects of the interchain coupling terms depend on the temperature and the energy/frequency scale ω at which we probe the system e.g. in a scattering experiment. At very low temperatures the continuous global SU(2) symmetry of (7.26) gets broken spontaneously and magnetic long-range order develops

$$\lim_{\epsilon\to 0}\lim_{V\to\infty}\frac{1}{Z}\mathrm{Tr}\left[e^{-H_\epsilon/k_B T}S^z_{j,k,l}\right] = me^{i\pi(j+k+l)} \neq 0, \quad Z = \mathrm{Tr}\left[e^{-H/k_B T}\right], \qquad (7.27)$$

where ϵ is an infinitesimal symmetry breaking field. The emergence of order is a pure interchain coupling effect, because continuous symmetries cannot be broken spontaneously in one-dimensional models with short-range interactions (Mermin–Wagner–Hohenberg theorem). So at $T \lesssim \mathcal{J}_\perp$ the extra contributions to our integrable model have a big effect on static ($\omega = 0$) properties. There also will be strong 3D effects related to the presence of gapless Goldstone modes (which must be present as a result of the spontaneous breaking of a continuous symmetry) at low frequencies $\omega \lesssim \mathcal{J}_\perp$. On the other hand, if we probe the system at finite frequencies $\omega \gg \mathcal{J}_\perp$ the effects of the interchain coupling will again be small. Hence in the window

$$\mathcal{J}_\perp \ll T, \omega \lesssim \mathcal{J} \qquad (7.28)$$

our system will look to a good approximation to be one-dimensional and integrable (at frequencies/temperatures large compared to $\mathcal{J}$, physics beyond that of the spin degrees of freedom will eventually become important).

7.1.4 Some standard experimental probes

We conclude this section with a brief discussion of some standard experimental probes that have been used to study realizations of quantum magnets in both solids and trapped cold atoms.

7.1.4.1 Equilibrium thermodynamics

Information about different phases and phase transitions in solids can be obtained by means of thermodynamic probes. Let us consider the example of a spin-$\frac{1}{2}$ magnet described by a Hamiltonian $H = H_0 + \mathbf{h}\cdot\mathbf{S}$, where $\mathbf{h}$ is an external magnetic field.

The *partition function* Z and Helmholtz free energy F are defined as

$$Z = \text{Tr}\left(e^{-\beta H}\right), \quad \beta = \frac{1}{k_B T}, \quad F = -k_B T \ln(Z). \tag{7.29}$$

The *magnetization per site* is

$$m^\alpha(\mathbf{h}, T) = \langle S_j^\alpha \rangle = \frac{1}{Z} \text{Tr}\left(S_j^\alpha \, e^{-\beta H}\right) = \frac{1}{L} \frac{\partial F}{\partial h^\alpha}. \tag{7.30}$$

The *magnetic susceptibilities* quantify how susceptible the magnetization is to changes in the magnetic field,

$$\chi^{\alpha\beta}(\mathbf{h}, T) = \frac{\partial m^c(\mathbf{h}, T)}{\partial h^\beta}. \tag{7.31}$$

Finally, the *specific heat* quantifies the change in internal energy U under a change in temperature,

$$C = \frac{\partial U}{\partial T} = -T \frac{\partial^2 F}{\partial T^2}. \tag{7.32}$$

Thermodynamic quantities can readily be determined in integrable models, and we will see later how this can be done.

7.1.4.2 Dynamical response functions in equilibrium

Extremely detailed information on physical properties of quantum magnets is provided by *inelastic neutron scattering* (Figure 7.3). The neutron scattering cross section is proportional to a particular retarded correlation function in the sample, the so-called *dynamical structure factor* (DSF) $S^{\alpha\beta}(\omega, \mathbf{q})$,

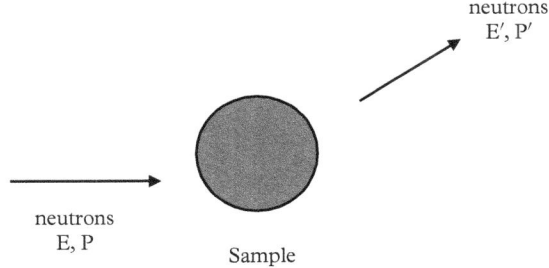

Figure 7.3 *Inelastic neutron scattering.*

$$\frac{d^2\sigma}{d\Omega d\omega} \propto \frac{\delta_{\alpha\beta} - q_\alpha q_\beta}{\mathbf{q}\cdot\mathbf{q}} \underbrace{\int_{-\infty}^{\infty} \frac{dt}{2\pi} e^{i\omega t} \frac{1}{L}\sum_{j,\ell} e^{-i\mathbf{q}\cdot(\mathbf{R}_j - \mathbf{R}_\ell)} \frac{1}{Z}\text{Tr}\left(e^{-H/k_B T} S_j^\alpha(t) S_\ell^\beta(0)\right)}_{S^{\alpha\beta}(\omega,\mathbf{q})}. \qquad (7.33)$$

It is easy to see that the DSF contains information on the structure and the kinematics of excitations by considering $T = 0$ and going to a Lehmann representation in terms of eigenstates $|n\rangle$ of H with corresponding energies E_n. Assuming that we have translational invariance and denoting the momentum eigenvalues of $|n\rangle$ by $\mathbf{P}_n$, we obtain

$$S^{\alpha\alpha}(\omega,\mathbf{q}) = \sum_n |\langle\text{GS}|S_0^\alpha|n\rangle|^2 \delta(\omega - (E_n - E_{\text{GS}})) \delta_{\mathbf{q},\mathbf{P}_n}. \qquad (7.34)$$

This shows that the transverse components $\alpha = x, y$ probe excitations over the ground state with $\delta S^z = \pm 1$, while the longitudinal component is susceptible to excited states with $\delta S^z = 0$. Single-particle excitations in general, and spinwaves in particular, give rise to δ-function lines in the DSF at zero temperature that follow the single-particle dispersion $\epsilon(\mathbf{q})$, i.e.

$$S^{xx}(\omega,\mathbf{q}) \propto \delta(\omega - \epsilon(\mathbf{q})). \qquad (7.35)$$

The dynamical structure factor in the spin-1/2 Heisenberg chain (7.3) looks very different; it exhibits a scattering continuum rather than single-particle excitations. This is because the elementary excitations are spin-1/2 objects called *spinons*, and all excited states in (7.34) involve an even number of these [13].

7.1.4.3 Non-equilibrium protocols

Approximate realizations of integrable models in cold atom systems are very interesting because experimental probes are different from those in solids, and because it is possible to investigate the non-equilibrium evolution in almost isolated quantum systems. The latter is very difficult in solids, because there are always other degrees of freedom present (lattice vibrations, impurities, other bands), and this 'environment' is felt after extremely short times (picoseconds). In contrast, in cold atom systems the main coupling to the environment is through heating by the lasers that generate the confining potential, but because the characteristic energy scale is tiny, it takes on the order of seconds for these effects to become important. This leaves a lot of time for conducting experiments! An example of the kind of things that can be done is shown in Figure 7.4. Initially a cloud of Rb atoms sits in the centre of a 1D tube. At time $t = 0$ the cloud is split through a laser beam, which bestows momentum p on half the atoms, and momentum $-p$ on the other half. The atoms then start moving up the 'walls' of the confining potential. It is possible to extract the density of atoms as a function of position and time from measurements by repeating the experiment many times. The result is shown in Figure 7.4.

One of the amazing things about cold atom experiments is that it is possible to measure not only expectation values of observables, but the full QM probability distributions [14].

Eigenstates of integrable and generic local Hamiltonians 331

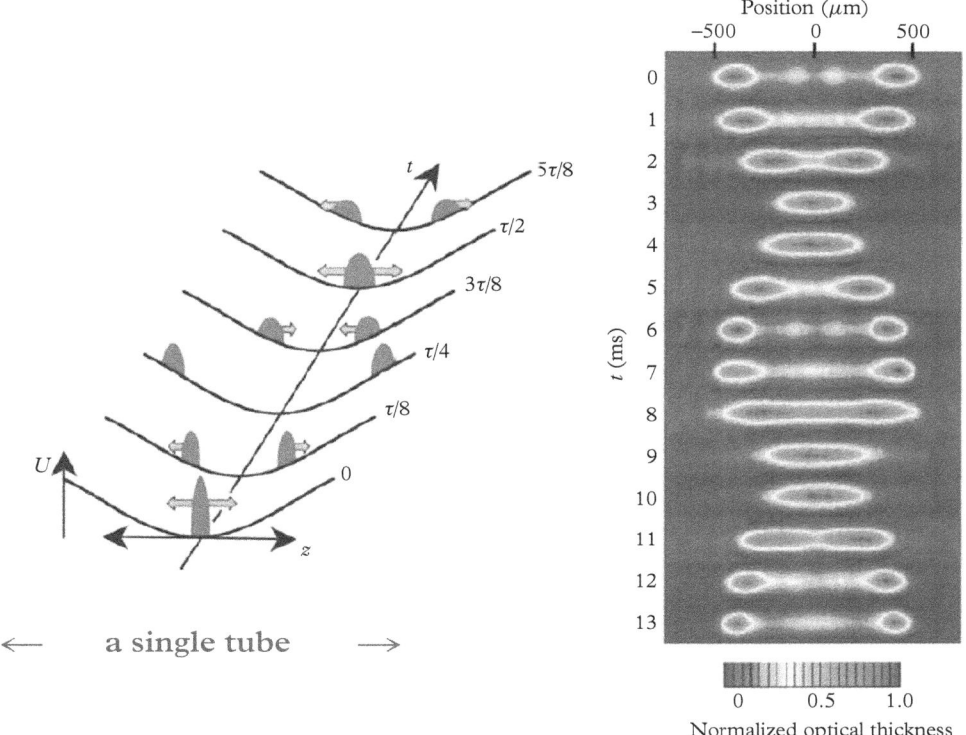

Figure 7.4 *Quantum Newton's Cradle experiments, T. Kinoshita, T. Wenger, D. S. Weiss, A quantum Newton's cradle,* Nature **440**, *900 (2006). Reprinted with permission from Springer Nature.*

7.2 Eigenstates of integrable and generic local Hamiltonians

In Section 7.1 we have seen that many 1+1-dimensional integrable models are approximately realized in solids and systems of ultra-cold atoms. We have argued that many properties will look very similar for integrable and non-integrable models. However, integrable models are in some ways fundamentally different from generic ones, and we now want to focus on some of these differences at the level of energy eigenstates.

Let us consider a lattice model with a translationally invariant Hamiltonian of the form

$$H = \sum_j H_j, \qquad (7.36)$$

where H_j acts non-trivially only on sites within a fixed distance d of site j. Let us further assume for simplicity that the Hilbert space at site j is finite dimensional. We call Hamiltonians of this form *local* (the terminology refers to the fact that the density of

H is a local operator). We term H *generic* if it has no local (in the same sense) conservation laws.

Non-local conservation laws

Any Hamiltonian has a huge number of *non-local* conservation laws as can be seen in the following. Consider the set of one-dimensional projectors on energy eigenstates

$$H|\psi_n\rangle = E_n|\psi_n\rangle, \qquad I_n = |\psi_n\rangle\langle\psi_n|, n = 1, \ldots \dim(\mathcal{H}), \tag{7.37}$$

where $\dim(\mathcal{H})$ is the dimension of the Hilbert space of our model. By construction these form a mutually commuting set of conservation laws,

$$[I_n, I_m] = 0 = [I_n, H]. \tag{7.38}$$

Let us now consider a generic, local Hamiltonian with a bounded spectrum in a finite volume L, which we think of being asymptotically large (examples would be quantum spin models). Then the energy density varies in the finite range $e_{GS} \leq e \leq e_{max}$. States at the edges of the spectrum, i.e. with energies such that

$$|E - Le_{GS}| = \mathcal{O}(1), \quad \text{or } |E - Le_{max}| = \mathcal{O}(1), \tag{7.39}$$

are known to have unusual properties. For example, their entanglement entropies obey an area law [15]. States that are located away from the edges of the spectrum such that

$$|E - Le_{GS}| = \mathcal{O}(L), \quad \text{and } |E - Le_{max}| = \mathcal{O}(L) \tag{7.40}$$

are *locally thermal with probability one*. A state $|\Psi\rangle$ is locally thermal if for any local operator $\mathcal{O}$ one has

$$\frac{\langle\Psi|\mathcal{O}|\Psi\rangle}{\langle\Psi|\Psi\rangle} = \frac{1}{Z}\text{Tr}\left(e^{-\beta_{\text{eff}}H}\mathcal{O}\right), \quad Z = \text{Tr}\left(e^{-\beta_{\text{eff}}H}\right), \tag{7.41}$$

where the effective inverse temperature β_{eff} is fixed by requiring the energy density in the Gibbs ensemble (7.41) to coincide with the energy density of the eigenstate $|\Psi\rangle$ under consideration,

$$\frac{\langle\Psi|H_j|\Psi\rangle}{\langle\Psi|\Psi\rangle} = \frac{1}{Z}\text{Tr}\left(e^{-\beta_{\text{eff}}H}H_j\right). \tag{7.42}$$

The probabilistic aspect of the above statement means that if we randomly select a state in our energy window, it will be thermally local with a probability that scales to one as we approach the thermodynamic limit. Numerical studies show that in some models *all* energy eigenstates away from the edges of the spectrum are locally thermal, see e.g. [17–20]. However, models that have non-thermal energy eigenstates at finite energy densities can be constructed [21]. Equation (7.41) is closely related to the microcanonical ensemble in quantum statistical mechanics. The latter is defined as

$$\langle \mathcal{O} \rangle_{\text{MC}} = \sum_\alpha \frac{\langle \Psi_\alpha | \mathcal{O} | \Psi_\alpha \rangle}{\langle \Psi_\alpha | \Psi_\alpha \rangle}, \qquad (7.43)$$

where the sum is over an exponentially large number of energy eigenstates $H|\Psi_\alpha\rangle = E_\alpha|\Psi_\alpha\rangle$ in a small shell centred at energy E, i.e. $|E_\alpha - E| < \Delta E$. Equation (7.41) tells us that the flat average over an exponential number of states in (7.43) is in fact not required for local operators $\mathcal{O}$; it is sufficient to take the expectation value with respect to a single *typical* energy eigenstate (and picking an eigenstate in the energy window at random will provide a typical state with probability one in the thermodynamic limit). Up to finite-size corrections that vanish in the thermodynamic limit expectation values of local operators are the same in all typical states. This section is summarized as follows.

Finite energy density eigenstates in generic models

In generic models energy eigenstates at a given energy density are locally thermal with probability one, i.e. expectation values of local operators are indistinguishable from the corresponding thermal values at an effective temperature fixed by the energy density. Because of the lack of conservation laws other than energy, it is extremely difficult to access any non-thermal energy eigenstates away from the edges of the spectrum in a large volume.

7.2.1 Integrable models and conservation laws

A characteristic property of quantum integrable models in the infinite volume is the existence of infinitely many local conservation laws. These are commuting operators

$$[H^{(n)}, H^{(m)}] = 0, \qquad (7.44)$$

whose densities are local in the above sense (for continuum theories they would be of the form $H^{(n)} = \int dx\, \mathcal{H}^{(n)}(x)$). One of these conservation laws, say $H^{(1)}$, is the Hamiltonian. There is a standard method for constructing these conservation laws by means of the quantum inverse scattering method [16]. We will give a brief summary of this method for the example of the spin-1/2 Heisenberg chain shortly. I note in passing that in addition to these well-known local conservation laws there are others, which fulfil a weaker notion of locality [22]. As they are not needed for the following argument we will ignore them for now, and return to them later. The existence of local conservation laws leads to a dramatic change in the structure of energy eigenstates compared to the generic case. As we will show by explicit construction in Section 7.3, there are exponentially many (in system size) energy eigenstates $|\Psi_j\rangle$ such that

$$\lim_{\text{th}} \frac{1}{L} \langle \Psi_j | H | \Psi_j \rangle = \lim_{\text{th}} \frac{1}{L} \langle \Psi_k | H | \Psi_k \rangle,$$
$$\lim_{\text{th}} \frac{1}{L} \langle \Psi_j | H^{(n)} | \Psi_j \rangle \neq \lim_{\text{th}} \frac{1}{L} \langle \Psi_k | H^{(n)} | \Psi_k \rangle, \quad j \neq k, \ n > 1. \quad (7.45)$$

Using translational invariance we then have

$$\lim_{\text{th}} \frac{1}{L} \langle \Psi_j | H^{(n)} | \Psi_j \rangle = \lim_{\text{th}} \langle \Psi_j | H_\ell^{(n)} | \Psi_j \rangle, \quad (7.46)$$

where $H_\ell^{(n)}$ are the densities of the conservation laws. By construction these are local operators, and hence (7.45) establishes that not all energy eigenstates with the same energy density have identical local properties in the thermodynamic limit. This in turn implies that not all energy eigenstates away from the edges of the spectrum can be thermal. In fact, as we will see in the following, at a given energy density there are locally thermal as well as non-thermal energy eigenstates.

7.3 Macro states in the spin-1/2 Heisenberg chain

In order to understand the nature of energy eigenstates in integrable models we focus on the particular example of the spin-$\frac{1}{2}$ Heisenberg chain (which is the model Hans Bethe solved by means of his famous ansatz [23]).

$$H_{\text{Heis}} = \mathcal{J} \sum_{j=1}^{L} S_j^x S_{j+1}^x + S_j^y S_{j+1}^y + S_j^z S_{j+1}^z - \frac{1}{4} + h \sum_{j=1}^{L} S_j^z. \quad (7.47)$$

There are many excellent expositions of the Bethe ansatz solution of the Heisenberg model, see e.g. [13], [16], [24]. We will adopt the notations of [25]. The eigenstates of (7.47) can be expressed as

$$|\Psi_{\Lambda_1,\ldots,\Lambda_N}\rangle = \sum_{x_1 < x_2 < \ldots < x_N} \phi_{\Lambda_1,\ldots,\Lambda_N}(x_1,\ldots,x_N) S_{x_1}^- \ldots S_{x_N}^- |\uparrow \ldots \uparrow\rangle, \quad (7.48)$$

where the wave functions have Bethe ansatz form (*cf.* Jesper Jacobsen's lectures),

$$\phi_{\Lambda_1,\ldots,\Lambda_N}(x_1,\ldots,x_N) = \sum_{P \in S_N} \underbrace{\prod_{\substack{j<l \\ P_j > P_l}} \frac{\Lambda_{P_l} - \Lambda_{P_j} + 2i}{\Lambda_{P_l} - \Lambda_{P_j} - 2i} \prod_{j=1}^{N} \left(\frac{\Lambda_{P_j} + i}{\Lambda_{P_j} - i} \right)^{x_j}}_{A(P)} \quad (7.49)$$

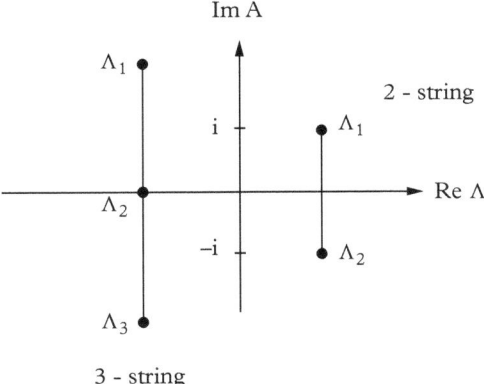

Figure 7.5 *String solutions of the BAE.*

and the corresponding energy is

$$E = \sum_{j=1}^{N} \left(-\frac{2\mathcal{J}}{\Lambda_j^2 + 1} + h \right) - h\frac{L}{2}. \tag{7.50}$$

The periodic boundary conditions impose quantization conditions known as *Bethe ansatz equations* (BAE),

$$\left(\frac{\Lambda_j + i}{\Lambda_j - i} \right)^L = \prod_{l \neq j}^{N} \frac{\Lambda_j - \Lambda_l + 2i}{\Lambda_j - \Lambda_l - 2i}, \quad j = 1, \ldots, N. \tag{7.51}$$

In order to proceed we need to know the structure of the solutions of these equations. To that end, let us take N fixed, L very large, and consider $\mathrm{Im}(\Lambda_j) > 0$. Then

$$\left| \frac{\Lambda_j + i}{\Lambda_j - i} \right|^L \gg 1 \Rightarrow \exists \Lambda_\ell \text{ such that } \Lambda_j - \Lambda_\ell \approx 2i. \tag{7.52}$$

If $N = 2$ this gives rise to so-called *two-string solutions* (Figure 7.5). This generalizes to n-string solutions of the form

$$\Lambda_\alpha^{n,j} = \Lambda_\alpha^n + i(n + 1 - 2j) + \delta_\alpha^{n,j}, \quad j = 1, \ldots, n. \tag{7.53}$$

Here $\delta_\alpha^{n,j}$ are deviations from 'ideal' strings. One can see by numerically solving (7.51) that most solutions are composed of strings. This gives rise to the string hypothesis:

String hypothesis

In the thermodynamic limit $L, N \to \infty$, N/L fixed, energy eigenstates at finite magnetization, and energy eigenstates at finite energy densities relative to the ground state, are believed to be described by solutions that are built from ideal strings.

It is known that the string hypothesis does not provide an exact description of the solutions to the BAE [23], [26]. However, thermodynamic properties can be determined using the string hypothesis [27].

Let us now consider a solution to (7.51) that contains M_n strings of length n with corresponding string centres Λ_α^n (this implies that $\sum_n M_n n = N$). Substituting (7.53) into (7.51) and neglecting the deviations we obtain a set of coupled equations for the set $\{\Lambda_\alpha^n\}$. Taking logarithms we arrive at

$$L\theta\left(\frac{\Lambda_\alpha^n}{n}\right) = 2\pi I_\alpha^n + \sum_{(m,\beta) \neq (n,\alpha)} \theta_{nm}(\Lambda_\alpha^n - \Lambda_\beta^m). \tag{7.54}$$

Here I_α^n are integer or half-odd integer numbers (arising from taking logarithms), $\theta(x) = 2\arctan(x)$, and

$$\theta_{nm}(x) = \begin{cases} \theta\left(\frac{x}{2n}\right) + 2\sum_{j=1}^{n-1} \theta\left(\frac{x}{2j}\right) & \text{for } m = n \\ \theta\left(\frac{x}{|n-m|}\right) + 2\theta\left(\frac{x}{|n-m|+2}\right) + \ldots + 2\theta\left(\frac{x}{n+m-2}\right) + \theta\left(\frac{x}{n+m}\right) & \text{for } m \neq n \end{cases}. \tag{7.55}$$

Equations (7.54) are called *Takahashi's equations*. They relate the solutions of the BAE to a set of integer or half-odd integer numbers, which therefore can be considered as quantum numbers of our problem. Within the framework of the string hypothesis we therefore have one-to-one correspondences

$$\{I_\alpha^n\} \Leftrightarrow \{\Lambda_\alpha^n\} \Leftrightarrow \{\Lambda_j\} \Leftrightarrow \phi_{\Lambda_1,\ldots,\Lambda_N}(x_1,\ldots,x_N). \tag{7.56}$$

The expression for the energy becomes

$$E = \sum_{n=1}^{\infty} \sum_{\alpha=1}^{M_n} \left(-\frac{2Jn}{n^2 + (\Lambda_\alpha^n)^2} + nh\right). \tag{7.57}$$

It is convenient to define *counting functions* for a given solution of Takahashi's equations,

$$z_n(\lambda) = \theta\left(\frac{\lambda}{n}\right) - \frac{1}{L}\sum_{(m,\beta)} \theta_{nm}(\lambda - \Lambda_\beta^m). \tag{7.58}$$

In terms of these Takahashi's equations (7.54) can be cast in the form

$$z_n(\Lambda_\alpha^n) = \frac{2\pi I_\alpha^n}{L}. \tag{7.59}$$

The branch cuts of the logarithms must be chosen in such a way that all counting functions are monotonically increasing functions of λ. The range of I_α^n can then be determined by considering the limits $\lambda \to \pm\infty$

$$\frac{L z_n(-\infty)}{2\pi} < I_\alpha^n < \frac{L z_n(\infty)}{2\pi}. \tag{7.60}$$

This gives

$$|I_\alpha^n| \le \frac{1}{2}\left[L - 1 - \sum_{m=1}^\infty \left(2\min(m,n) - \delta_{n,m}\right) M_m\right]. \tag{7.61}$$

The (half-odd) integers I_α^n are good 'quantum numbers' of our problem in the sense that every set $\{I_\alpha^n\}$ is in one-to-one correspondence with an energy eigenstate of the Hamiltonian. Let us consider a solution characterized by the set $\{I_\alpha^n\}$. The structure of the counting functions is sketched in Figure 7.6. We see that the (half-odd) integers $\{I_\alpha^n | \alpha = 1, \ldots, M_n\}$ define 'particles', while the 'missing' (half-odd) integers define 'holes'.

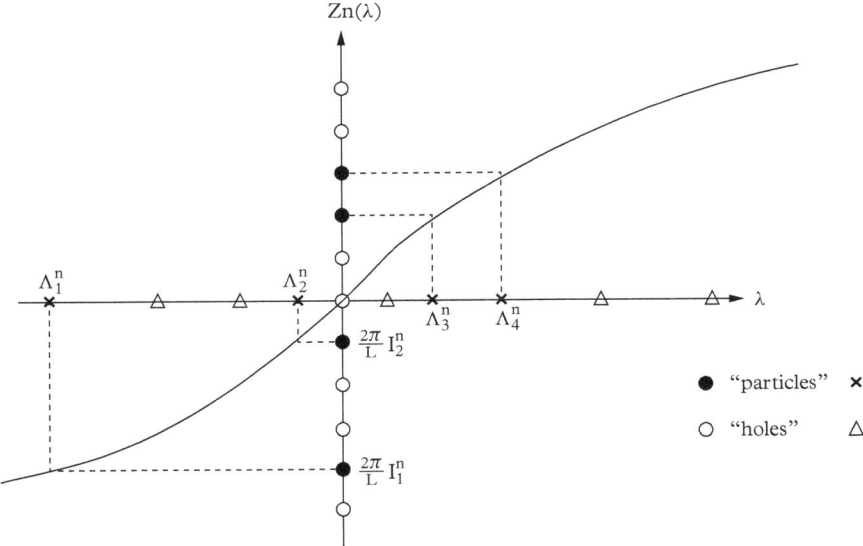

Figure 7.6 *Structure of the counting functions.*

7.3.1 Macro states

So far our discussion has focussed on 'micro states': in the framework of the string hypothesis each configuration of (half-odd) integers $\{I_\alpha^n | \alpha = 1, \ldots, M_n\}$ is in one-to-one correspondence with an energy eigenstate. We now want to construct 'macro states', which we define as classes of micro states that have the same local properties in the thermodynamic limit

$$L, N \to \infty, \quad \frac{N}{L} = n = \text{fixed}. \tag{7.62}$$

By local properties we mean expectation values of local operators, which we define as acting non-trivially only in a finite region of space in the thermodynamic limit. We now proceed in analogy to the description of macro states in ideal Bose and Fermi gases by means of distribution functions of particles and holes. To do so we focus on the particular subset of solutions to our quantization conditions (7.51) that have $M_n = \mathcal{O}(L)$ and

$$\Lambda_{\alpha+1}^n - \Lambda_\alpha^n = \mathcal{O}(L^{-1}). \tag{7.63}$$

In the thermodynamic limit we can describe these solutions in terms of the densities of particles $\rho_{n,p}(\lambda)$ and holes $\rho_{n,h}(\lambda)$. These are defined by

$$\rho_{n,p}(\lambda)\Delta\lambda = \text{number of particles in the interval } [\lambda, \lambda + \Delta\lambda],$$
$$\rho_{n,h}(\lambda)\Delta\lambda = \text{number of holes in the interval } [\lambda, \lambda + \Delta\lambda]. \tag{7.64}$$

Each such macro state can be described in terms of its counting functions, which are obtained by taking the thermodynamic limit of (7.58)

$$z_n(\lambda) = \theta\left(\frac{\lambda}{n}\right) - \sum_{m=1}^{\infty} \int_{-\infty}^{\infty} d\Lambda \, \theta_{nm}(\lambda - \Lambda) \, \rho_{m,p}(\Lambda). \tag{7.65}$$

By the 'counting property' of the $z_n(\lambda)$ we have

$$2\pi \left[\rho_{n,p}(\lambda) + \rho_{n,h}(\lambda)\right] \Delta\lambda = z_n(\lambda + \Delta\lambda) - z_n(\lambda) \simeq z_n'(\lambda)\Delta\lambda. \tag{7.66}$$

Taking the derivative of (7.65) we obtain the following.

Thermodynamic limit of the BAE

$$\rho_{n,p}(\lambda) + \rho_{n,h}(\lambda) = a_n(\lambda) - \sum_{m=1}^{\infty} \int_{-\infty}^{\infty} d\lambda' \, T_{nm}(\lambda - \lambda') \, \rho_{m,p}(\lambda'), \quad (7.67)$$

where

$$a_n(\lambda) = \frac{1}{2\pi} \frac{2n}{n^2 + \lambda^2}, \quad T_{nm}(\lambda) = (1 - \delta_{n,m})a_{|n-m|}(\lambda) + 2a_{|n-m|+2}(\lambda)$$
$$+ \cdots + 2a_{n+m-2}(\lambda) + a_{n+m}(\lambda). \quad (7.68)$$

Equations (7.67) express the hole densities in terms of the particle densities and are known as *thermodynamic limit of the BAE*. The energy per site in the thermodynamic limit is given by

$$e[\{\rho_{n,p}(\lambda)\}] = \sum_{n=1}^{\infty} \int_{-\infty}^{\infty} d\lambda \, \rho_{n,p}(\lambda) \underbrace{\left[\frac{-2n\mathcal{J}}{n^2 + \lambda^2} + hn \right]}_{\epsilon_n^{(0)}(\lambda)}. \quad (7.69)$$

The set $\{\rho_{n,p}(\lambda)\}$ identifies a *macro state* in the thermodynamic limit.

7.3.2 Thermal equilibrium

Having at hand a convenient formulation of macro states we are now in a position to construct the state of thermal equilibrium. This is defined as the most likely state for a given energy density, and is obtained by extremizing the Helmholtz free energy density (we are setting $k_B = 1$ in the following),

$$f = e - Ts, \quad (7.70)$$

where e is given by (7.69) and s is the entropy density,

$$e^{Ls} = \text{number of microstates}. \quad (7.71)$$

The number of micro states is readily calculated. The number of n-string states with rapidity in the interval $[\lambda, \lambda + \Delta\lambda]$ is given by the number of ways to distribute $L\rho_{n,p}(\lambda)\Delta\lambda$ particles among $L[\rho_{n,p}(\lambda) + \rho_{n,h}(\lambda)]\Delta\lambda$ vacancies,

$$dS_n = \ln \left[\frac{(L[\rho_{n,p}(\lambda) + \rho_{n,h}(\lambda)]\Delta\lambda)!}{(L\rho_{n,p}(\lambda)\Delta\lambda)! \, (L\rho_{n,h}(\lambda)\Delta\lambda)!} \right]$$
$$\simeq \left[(\rho_{n,p}(\lambda) + \rho_{n,h}(\lambda)) \ln (\rho_{n,p}(\lambda) + \rho_{n,h}(\lambda)) - \rho_{n,p}(\lambda) \ln (\rho_{n,p}(\lambda)) \right.$$
$$\left. - \rho_{n,h}(\lambda) \ln (\rho_{n,h}(\lambda)) \right] L\Delta\lambda, \quad (7.72)$$

where in the second step we have used Stirling's formula. This leads to the following expression for the entropy density:

$$s = \sum_{n=1}^{\infty} \int d\lambda \left[\left(\rho_{n,p}(\lambda) + \rho_{n,h}(\lambda) \right) \ln \left(\rho_{n,p}(\lambda) + \rho_{n,h}(\lambda) \right) - \rho_{n,p}(\lambda) \ln \left(\rho_{n,p}(\lambda) \right) \right.$$
$$\left. - \rho_{n,h}(\lambda) \ln \left(\rho_{n,h}(\lambda) \right) \right]. \tag{7.73}$$

In order to determine the root distributions of the state of thermal equilibrium we extremize

$$0 = \delta(e - Ts) = \sum_n \frac{\delta(e - Ts)}{\delta \rho_{n,h}(\lambda)} \delta \rho_{n,h}(\lambda) + \frac{\delta(e - Ts)}{\delta \rho_{n,p}(\lambda)} \delta \rho_{n,p}(\lambda), \tag{7.74}$$

where the variations of the hole densities are expressed in terms of the $\delta \rho_{n,p}$ by (7.67),

$$\delta \rho_{n,h}(\lambda') = -\delta \rho_{n,p}(\lambda') - \sum_{m=1}^{\infty} \int_{-\infty}^{\infty} d\lambda \, T_{nm}(\lambda' - \lambda) \, \delta \rho_{m,p}(\lambda). \tag{7.75}$$

Using that $\delta \rho_{n,p}$ are arbitrary we obtain an infinite system of coupled nonlinear integral equations as follows.

Thermodynamic Bethe ansatz equations

$$\ln \left(1 + e^{\epsilon_n(\lambda)/T} \right) = \frac{\epsilon_n^{(0)}(\lambda)}{T} + \sum_{m=1}^{\infty} \int_{-\infty}^{\infty} d\mu \underbrace{\left[\delta_{n,m} \delta(\lambda - \mu) + T_{nm}(\lambda - \mu) \right]}_{A_{nm}(\lambda - \mu)} \ln \left(1 + e^{-\epsilon_m(\mu)/T} \right), \tag{7.76}$$

where we have defined

$$e^{\epsilon_n(\lambda)/T} = \frac{\rho_{n,h}(\lambda)}{\rho_{n,p}(\lambda)}. \tag{7.77}$$

Equations (7.76) and (7.67) together determine the set $\{\rho_{n,p}^{(\text{th})}(\lambda)\}$ and hence the macro state corresponding to thermal equilibrium. Taking Fourier transforms [24] it is possible to cast the TBA equations in a 'tri-diagonal' form,

$$\epsilon_1(\lambda) = -2\pi \mathcal{J} s(\lambda) + T \int_{-\infty}^{\infty} d\mu \, s(\lambda - \mu) \ln\left[1 + e^{\epsilon_2(\mu)/T}\right],$$

$$\epsilon_n(\lambda) = T \int_{-\infty}^{\infty} d\mu \, s(\lambda - \mu) \left(\ln\left[1 + e^{\epsilon_{n-1}(\mu)/T}\right] + \ln\left[1 + e^{\epsilon_{n+1}(\mu)/T}\right]\right), \quad n \geq 2, \quad (7.78)$$

where

$$s(x) = \frac{1}{4\cosh\left(\frac{\pi x}{2}\right)}, \quad \lim_{n \to \infty} \frac{\epsilon_n(\lambda)}{n} = 2h. \quad (7.79)$$

The TBA equations can be solved in the limits $T \to 0$ and $T \to \infty$ [25].

7.3.2.1 Zero temperature limit

It follows from (7.78) that $\epsilon_n(\lambda) > 0$ for $n \geq 2$. This in turn implies that for $T = 0$ the particle densities $\rho_{n,p}(\lambda)$ for $n \geq 2$ are zero. The equilibrium state for $T \to 0$ is therefore described by the linear integral equations

$$\epsilon_1(\lambda) = \epsilon_1^{(0)}(\lambda) - \int_{-B}^{B} d\mu \, a_2(\lambda - \mu) \epsilon_1(\mu), \quad \epsilon_1(\pm B) = 0,$$

$$\rho_{1,p}(\lambda) = a_1(\lambda) - \int_{-B}^{B} d\mu \, a_2(\lambda - \mu) \rho_{1,p}(\mu), \quad |\lambda| \leq B. \quad (7.80)$$

Here we have used that $\text{sgn}(\epsilon_1(\lambda)) = \text{sgn}(|\lambda| - B)$ and concomitantly $\rho_{1,p}(\lambda) = 0$ for $|\lambda| > B$ (and similarly $\rho_{1,h}(\lambda) = 0$ for $|\lambda| < B$).

7.3.2.2 Infinite temperature limit

In this limit the TBA equations turn into a set of difference equations that can be solved exactly [24],

$$\epsilon_n(\lambda) = T \ln\left[\left(\frac{s_{n+1}}{s_1}\right)^2 - 1\right], \quad s_n = \sinh\left(\frac{nh}{T}\right). \quad (7.81)$$

The particle densities can then be determined from (7.67) [24],

$$\rho_{n,p}(\lambda) = \frac{s_1^2}{2 s_{n+1} \cosh(h/T)} \left[\frac{a_n(\lambda)}{s_n} - \frac{a_{n+2}(\lambda)}{s_{n+2}}\right]. \quad (7.82)$$

7.3.2.3 Negative 'temperatures'

By referring to the parameter T as temperature we have implicitly assumed that it is a positive quantity. However, what we have really done is to maximize the entropy of eigenstates at a given energy density e, where the parameter T^{-1} plays the role of a

Lagrange multiplier. Indeed, extremizing the cost function

$$\mathcal{C}\left(\beta, [\{\rho_{n,p}(\lambda)\}]\right) = \beta(e[\{\rho_{n,p}(\lambda)\}] - e_0) - s[\{\rho_{n,p}(\lambda)\}] \tag{7.83}$$

leads to the TBA equations (7.76) if we set $\beta = 1/T$. In this way of looking at things β can be either positive or negative. Negative values of T have a very simple meaning: they correspond to positive temperature $|T|$ for the Hamiltonian $-H$, i.e. the Heisenberg *ferromagnet*. Let us now consider eigenstates of our XXX Hamiltonian in a very large, but finite volume. The spectrum is bounded

$$E_{\min} \leq E \leq E_{\max}. \tag{7.84}$$

We then focus on an arbitrary slice away from the edges of the spectrum,

$$Le - \epsilon < E < Le + \epsilon, \tag{7.85}$$

where ϵ is small but fixed. We can always adjust our 'temperature' T that characterizes our equilibrium state via (7.76) and (7.67) such that $e = e(T)$. The variation of energy density with T determined in this way is shown schematically in Figure 7.7. The ground state has by definition the lowest energy and is obtained by taking the limit $T \to 0^+$.

Increasing T from zero covers the range in energy densities between the ground state and the value $\bar{e}$, which is the energy density of the infinite temperature state. The latter is finite for lattice models like the Heisenberg chain. In zero magnetic field we have

$$e_{\text{GS}} = -\mathcal{J}\left(\ln(2) - \frac{1}{4}\right), \quad e_{\max} = 0, \quad \bar{e} = \frac{-\mathcal{J}}{4}. \tag{7.86}$$

The top of the spectrum corresponds to the ferromagnetic state (more precisely to the degenerate SU(2) multiplet that contains the ferromagnetic state with all spins up). Negative temperatures in the above sense have been observed experimentally in a system of ultra-cold atoms in [28].

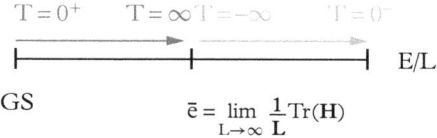

Figure 7.7 *Effective temperature as a function of the energy density. Note that T can be either positive or negative.*

7.3.2.4 Equilibrium thermodynamics

As mentioned in Section 7.1, thermodynamic measurements are standard experimental probes. In order to determine thermodynamic properties we need to compute the free energy density, which can be cast in the form

$$f = -\mathcal{J}\left(\ln(2) - \frac{1}{4}\right) - T\int_{-\infty}^{\infty} dx\, s(x) \ln\left(1 + e^{-\epsilon_1(x)/T}\right). \tag{7.87}$$

Once this has been determined, the magnetization per site, susceptibilities, and specific heat can be obtained by taking derivatives with respect to the magnetic field or temperature, cf. Section 7.1.4.1. In practice this can be done by truncating the infinite system of TBA equations at some finite string length and then solving them numerically by iteration. This is indeed how thermodynamics was analysed traditionally, until more efficient methods became available in the 1990s [29].

7.3.3 Atypical states with finite entropy densities

The state of thermal equilibrium is by construction the state with the largest entropy density at a given energy density $e(T)$, where the entropy density is determined by inserting the particle and hole densities obtained from the solution of the TBA equations into (7.73). Let us denote the corresponding value by s_{thermal}. It is clear from our discussion that there exist other macro states with the same energy density, but different physical properties. Indeed, all we have to do is to specify some positive functions $\rho_{n,p}(\lambda)$ such that we have

$$e(T) = \sum_{n=1}^{\infty} \int_{-\infty}^{\infty} d\lambda\, \rho_{n,p}(\lambda)\epsilon_n^{(0)}(\lambda). \tag{7.88}$$

The corresponding hole densities are then obtained by simply solving (7.67). All of these macro states will have finite entropy densities $s[\{\rho_{n,p}, \rho_{n,h}\}]$, which are however smaller than that of the state of thermal equilibrium,

$$s[\{\rho_{n,p}, \rho_{n,h}\}] < s_{\text{thermal}}. \tag{7.89}$$

Let us now imagine that we randomly select an energy eigenstate at an energy density $e(T)$. Then with overwhelming likelihood the state will be thermal. Let us consider non-thermal macro states characterized by root densities $\{\rho_{n,p}\}$ such that (7.88) holds. We know that their entropy densities are smaller than that of the thermal equilibrium macro state,

$$s[\{\rho_{n,p}\}] < s_{\text{thermal}}. \tag{7.90}$$

Hence the probability that our randomly selected eigenstate belongs to the macro state characterized by $\{\rho_{n,p}\}$ is exponentially smaller in the system size than the probability that it is thermal,

$$\frac{P(\{\rho_{n,p}\})}{P(\{\rho_{n,p}^{(\text{th})}\})} = \exp\left(Ls[\{\rho_{n,p}\}] - Ls_{\text{thermal}}\right). \tag{7.91}$$

As we will see in the Section 7.3.3.1, in spite of this atypical states are extremely important.

7.3.3.1 Local properties of atypical states

Crucially, the macro state $\{\rho_{n,p}\}$ has different local properties from the state of thermal equilibrium. The easiest way to see this is by considering the higher conservation laws, which are an inherent property of all integrable models. As we saw in Jesper Jacobsen's lectures, the Heisenberg XXZ Hamiltonian is intimately related to the 6-vertex model. He proved that the transfer matrix $t(u)$ of the 6-vertex model fulfils

$$[t(u), t(v)] = 0, \tag{7.92}$$

where u is the so-called spectral parameter. He also showed that the Heisenberg Hamiltonian is obtained by taking the logarithmic derivative of the transfer matrix,

$$H_{\text{XXZ}} = \frac{\mathcal{J} \sin \gamma}{2} \frac{\partial}{\partial u}\bigg|_{u=0} \ln\left[\frac{t(u)}{\sin(\gamma-u)^L}\right] = \mathcal{J} \sum_{j=1}^{L} S_j^x S_{j+1}^x + S_j^y S_{j+1}^y - \cos(\gamma)\left(S_j^z S_{j+1}^z - \frac{1}{4}\right). \tag{7.93}$$

By taking the limit $\gamma \to \pi$ we recover the isotropic Hamiltonian H_{Heis} we have been focussing on. It is useful to recall that $u=0$ corresponds to the so-called *shift point* of the transfer matrix, where $t(u)$ becomes proportional to the translation operator by one site

$$(t(0))^{\beta_1 \ldots \beta_L}_{\alpha_1 \ldots \alpha_L} = (\sin \gamma)^L \delta_{\alpha_1, \beta_2} \delta_{\alpha_2, \beta_3} \ldots \delta_{\alpha_L, \beta_1}. \tag{7.94}$$

It follows from (7.92) and (7.93) that H_{XXZ} and $t(u)$ have a basis of simultaneous eigenstates

$$t(u)|u_1, \ldots, u_N\rangle = \Lambda(u|\{u_j\})|u_1, \ldots, u_N\rangle. \tag{7.95}$$

The eigenvalue was also derived in Jesper's lectures,

$$\Lambda(u|\{u_j\}) = (\sin(\gamma - u))^L \prod_{j=1}^{N} \frac{\sin(\gamma - u_j + u)}{\sin(u_j - u)} + (\sin(u))^L \prod_{j=1}^{N} \frac{\sin(\gamma - u + u_j)}{\sin(u - u_j)}. \tag{7.96}$$

The point is that taking higher logarithmic derivatives produces higher conservation laws [16], [30],

$$H^{(n)}(\gamma) = i\left(\frac{-i\sin\gamma}{2}\right)^n \frac{\partial^n}{\partial u^n}\bigg|_{u=0} \ln\left[\frac{t(u)}{\sin(\gamma-u)^L}\right] = \sum_{j=1}^{L} f^{(n)}_{\alpha_0\ldots\alpha_n}(\gamma) S_j^{\alpha_0} S_{j+1}^{\alpha_1}\ldots S_{j+n}^{\alpha_n}. \quad (7.97)$$

Here the indices α_j are summed over and run from 0 to 3, where we have defined S_j^0 to be the identity operator on site j. Note that the density of the nth conservation law only contains spin interactions involving at most $n+1$ neighbouring sites. The coefficients $f^{(n)}_{\alpha_0\ldots\alpha_n}$ of the first few conservation laws have been worked out in [30]. As a result of (7.92) we have

$$[H^{(m)}(\gamma), H^{(n)}(\gamma)] = 0. \quad (7.98)$$

The conservation laws provide us with a simple means of determining certain local properties of eigenstates. To connect to our previous discussion we now take the limit $\gamma \to \pi$, i.e. focus on the case of the isotropic Heisenberg model H_{Heis}. Using translational invariance we have

$$\frac{1}{L}\langle\Psi_{\Lambda_1,\ldots,\Lambda_N}|H^{(n)}(\pi)|\Psi_{\Lambda_1,\ldots,\Lambda_N}\rangle = \langle\Psi_{\Lambda_1,\ldots,\Lambda_N}|f^{(n)}_{\alpha_0\ldots\alpha_n}(\pi) S_j^{\alpha_0} S_{j+1}^{\alpha_1}\ldots S_{j+n}^{\alpha_n}|\Psi_{\Lambda_1,\ldots,\Lambda_N}\rangle. \quad (7.99)$$

On the other hand, using that $|\Psi_{\Lambda_1,\ldots,\Lambda_N}\rangle$ is an eigenstate, we have

$$H^{(n)}(\pi)|\Psi_{\Lambda_1,\ldots,\Lambda_N}\rangle = \left(\sum_{j=1}^{N} \epsilon^{(n,0)}(\Lambda_j)\right)|\Psi_{\Lambda_1,\ldots,\Lambda_N}\rangle, \quad (7.100)$$

where the functions $\epsilon^{(n,0)}(\Lambda)$ can be worked out from the transfer matrix eigenvalue

$$\epsilon^{(n,0)}(\Lambda) = -\frac{\partial^{n-1}}{\partial\Lambda^{n-1}}\frac{2}{1+\Lambda^2}. \quad (7.101)$$

The string hypothesis can be implemented for the higher conservation laws in complete analogy to the energy. The eigenvalues of the nth conservation law for a m-string solution are

$$\epsilon_m^{(n,0)}(\Lambda) = \sum_{j=1}^{m} \epsilon_m^{(n,0)}(\Lambda + i(m+1-2j)). \quad (7.102)$$

Let us now consider a family of micro states $|\Psi_{\Lambda_1,\ldots,\Lambda_N}\rangle$ that in the thermodynamic limit are described by the macro state characterized by the set $\{\rho_{n,p}\}$ of root densities. Combining (7.99) with (7.100) for such a family of states we obtain

$$\lim_{\text{th}} \langle \Psi_{\Lambda_1,\ldots,\Lambda_N}| f^{(n)}_{\alpha_0\ldots\alpha_n}(0) S_j^{\alpha_0} S_{j+1}^{\alpha_1} \ldots S_{j+n}^{\alpha_n} |\Psi_{\Lambda_1,\ldots,\Lambda_N}\rangle$$

$$= \sum_{m=1}^{\infty} \int_{-\infty}^{\infty} d\Lambda \, \rho_{m,p}(\Lambda) \epsilon_m^{(n,0)}(\Lambda) \equiv h^{(n)}[\{\rho_{n,p}\}]. \tag{7.103}$$

The point is that for generic atypical macro states at a given energy density, we have

$$\{h^{(n)}[\{\rho_{n,p}\}]\} \neq \{h^{(n)}[\{\rho_{n,p}^{(\text{th})}\}]\}, \tag{7.104}$$

because the sets $\{\rho_{n,p}\}$ and $\{\rho_{n,p}^{(\text{th})}\}$ are different. This establishes that atypical states have different local properties from the state of thermal equilibrium at the same energy density. In the above the term 'generic' is important; we can construct particular atypical states that fulfil $\{h^{(n)}[\{\rho_{n,p}\}]\} = \{h^{(n)}[\{\rho_{n,p}^{(\text{th})}\}]\}$; see [31] for a related example. This implies that conservation laws $H^{(n)}$ are not sufficient to distinguish between any two different macro states. To do that one needs to include so-called *quasi-local* conservation laws as well [35]. At this point we have established the following.

Finite energy density eigenstates in integrable models

In integrable models, energy eigenstates at a given energy density can be either thermal or atypical. Thermal and atypical states have different local properties. In a large finite volume L there are exponentially (in L) more thermal than atypical states.

7.4 Quantum quenches

An obvious question is how to access atypical finite entropy states, given that they are rare compared to thermal ones. As we will now see, they naturally emerge as stationary states after *quantum quenches*. I will necessarily have to be very brief here, but there are a number of good reviews on the subject [32]. To define quantum quenches we consider a many-particle system with Hamiltonian $H(\Delta)$ that only contains short-ranged interactions, where Δ is some system parameter. An example is the integrable spin-$\frac{1}{2}$ Heisenberg XXZ chain,

$$H_{\text{XXZ}} = \mathcal{J} \sum_{j=1}^{L} S_j^x S_{j+1}^x + S_j^y S_{j+1}^y + \Delta S_j^z S_{j+1}^z. \tag{7.105}$$

A quantum quench is then defined as follows:

1. Prepare the system is the ground state $|GS, \Delta_0\rangle$ of $H(\Delta_0)$.
2. At time $t = 0$ suddenly quench the system parameter from Δ_0 to a different value Δ.
3. The time evolution of the system at $t > 0$ is given by the time-dependent Schrödinger equation

$$|\Psi(t)\rangle = e^{-iH(\Delta)t}|GS, \Delta_0\rangle. \tag{7.106}$$

It is crucial for our purposes that $|GS, \Delta_0\rangle$ is not an eigenstate of $H(\Delta)$, as otherwise the time evolution of the system is trivial; and that $|GS, \Delta_0\rangle$ has non-zero overlaps with exponentially many eigenstates of $H(\Delta)$ (for large, finite system size L).

4. The objective is to work out expectation values of local operators $\mathcal{O}$ in the thermodynamic limit

$$\langle \Psi(t)|\mathcal{O}|\Psi(t)\rangle. \tag{7.107}$$

By local we mean operators of the form $\prod_{j=1}^{k} S_{n_j}^{\alpha_j}$, where we keep k and n_j fixed and finite when taking the thermodynamic limit. These are local in the sense that they act non-trivially only in a finite sub-system.

To see why it is difficult to work out equal time expectation values like (7.107) it is useful to employ a spectral representation in terms of energy eigenstates $H(\Delta)|n\rangle = E_n|n\rangle$,

$$\langle \Psi(t)|\mathcal{O}|\Psi(t)\rangle = \sum_{n,m} \langle \Psi(0)|m\rangle \langle n|\Psi(0)\rangle \langle m|\mathcal{O}|n\rangle \, e^{i(E_n - E_m)t}. \tag{7.108}$$

We see that the time evolution is determined by complicated quantum mechanical many-particle interference encoded in the oscillating phase factors. Interestingly it appears that for all local operators $\mathcal{O}$ the following limit exists:

$$\lim_{t \to \infty} \lim_{L \to \infty} \langle \Psi(t)|\mathcal{O}|\Psi(t)\rangle = \langle \mathcal{O}\rangle_{\text{stat}}. \tag{7.109}$$

While there is no general proof for this assertion, there is ample analytical and numerical evidence in its favour, cf. [33] and references therein. The restriction to local operators in (7.109) is important, because one can always construct non-local operators that do not relax. A trivial example is operators of the form $\mathcal{A}_{nm} = |n\rangle\langle m| + |m\rangle\langle n|$, where $|m\rangle$ and $|n\rangle$ are energy eigenstates. Using the spectral representation (7.108) we have

$$\langle \Psi(t)|\mathcal{A}_{nm}|\Psi(t)\rangle = \langle \Psi(0)|m\rangle \langle n|\Psi(0)\rangle \, e^{i(E_n - E_m)t} + \text{c.c.} \tag{7.110}$$

This expectation value displays persistent oscillations at all times and does not relax. This is not a problem as $\mathcal{A}_{nm}$ are highly non-local operators.

Given that many-particle systems relax locally after quantum quenches, a natural question to ask is whether it is possible to describe their stationary states by statistical ensembles as we do in equilibrium statistical mechanics. An important observation is that as we are dealing with isolated systems, i.e. there is no environment, energy is always conserved,

$$\langle \Psi(t)|H(\Delta)|\Psi(t)\rangle = \langle \Psi(0)|H(\Delta)|\Psi(0)\rangle. \tag{7.111}$$

The energy density we have deposited into out system by means of the quantum quench is always larger than the ground state energy density,

$$e = \lim_{L\to\infty} \frac{\langle \Psi(0)|H(\Delta)|\Psi(0)\rangle}{L} > \lim_{L\to\infty} \frac{\langle \text{GS},\Delta|H(\Delta)|\text{GS},\Delta\rangle}{L}. \tag{7.112}$$

7.4.1 Thermalization

Let us first discuss the case of 'generic' systems, which have the defining property that energy is the only conserved quantity. According to the *Eigenstate Thermalization Hypothesis* [17], [34] the stationary state after quantum quenches is equal to a Gibbs ensemble, that is for local operators $\mathcal{O}$, we have

$$\langle \mathcal{O}\rangle_{\text{stat}} = \frac{1}{Z}\text{Tr}\left(e^{-\beta_{\text{eff}} H(\Delta)} \mathcal{O}\right), \quad Z = \text{Tr}\left(e^{-\beta_{\text{eff}} H(\Delta)}\right), \tag{7.113}$$

where the effective inverse temperature β_{eff} is fixed by requiring that the energy density in the Gibbs ensemble equals the energy density deposited into our system at time $t = 0$ by means of the quantum quench,

$$e = \lim_{\text{th}} \frac{1}{L}\frac{1}{Z}\text{Tr}\left(e^{-\beta_{\text{eff}} H(\Delta)} H(\Delta)\right). \tag{7.114}$$

7.4.2 Generalized thermalization

The situation in integrable models is fundamentally different. Integrable models cannot thermalize, because they have extensive sets of local (and quasi-local) conservation laws,

$$[I^{(\alpha)}, I^{(\beta)}] = 0 = [I^{(\alpha)}, H(\Delta)]. \tag{7.115}$$

The expectation values of all conservation laws are time independent,

$$\langle \Psi(t)|I^{(\alpha)}|\Psi(t)\rangle = \langle \Psi(0)|I^{(\alpha)}|\Psi(0)\rangle. \tag{7.116}$$

As the $I^{(\alpha)}$ are translationally invariant, (7.116) encodes information on certain local properties of the system, that are independent of time and set by the initial state. As a result the stationary state cannot be locally thermal (the only exception would be if we fine tune the initial state such that the initial values (7.116) happen to coincide with the respective thermal values at the relevant energy density). On the other hand, it follows from our previous discussion that there exist atypical finite entropy eigenstates such that the expectation values of higher conservation laws reproduce their initial values (7.116). It turns out that local properties in the stationary state after a quantum quench can be calculated in terms of a generalized micro-canonical ensemble built from these states; it was established in [35] and [36] that expectation values of local operators in the stationary state can be constructed as

$$\langle \mathcal{O} \rangle_{\text{stat}} = \langle \Phi | \mathcal{O} | \Phi \rangle, \quad (7.117)$$

where $|\Phi\rangle$ is a simultaneous eigenstate of $H(\Delta)$ and all local (and quasi-local) conservation laws such that

$$\lim_{\text{th}} \frac{1}{L} \langle \Phi | I^{(\alpha)} | \Phi \rangle = \lim_{\text{th}} \frac{1}{L} \langle \Psi(0) | I^{(\alpha)} | \Psi(0) \rangle. \quad (7.118)$$

Note that $|\Phi\rangle$ is precisely one of the atypical finite entropy density states we discussed earlier! The situation can be summarized as follows.

Stationary state after quantum quenches to integrable models

At late times after a quantum quench, integrable models relax locally to stationary states. The latter are locally indistinguishable from atypical finite entropy density eigenstates that are characterized by the initial values of higher conservation laws.

This shows that quantum quenches provide a natural framework for accessing atypical eigenstates. This is exciting, because it provides a mechanism for realizing exotic, novel states of matter, cf. [38] for a particularly nice example.

References

[1] A. C. Walters, T. G. Perring, J.-S. Caux, A. T. Savici, G. D. Gu, C.-C. Lee, W. Ku, and I. A. Zaliznyak, Nature Phys. 5, 867 (2009).
[2] A. H. MacDonald, S. M. Girvin, and D. Yoshioka, Phys. Rev. B 37, 16 (1988).
[3] I. Affleck, in *Fields, Strings and Critical Phenomena*, eds E. Brézin and J. Zinn-Justin, (Elsevier, Amsterdam, 1989); A. O. Gogolin, A. A. Nersesyan, and A. M. Tsvelik, *Bosonization and Strongly Correlated Systems* (Cambridge University Press, Cambridge, 1999); T. Giamarchi, *Quantum Physics in One Dimension* (Clarendon Press, Oxford, 2004).

[4] F. Essler and R. M. Konik in 'From Fields to Strings: Circumnavigating Theoretical Physics', ed. M. Shifman, A. Vainshtein, and J. Wheater, World Scientific, Singapore (2005) and references therein, cond-mat/0412421.
[5] E. H. Lieb and W. Liniger, Phys. Rev. **130**, 1605 (1963); E. H. Lieb, Phys. Rev. **130**, 1616 (1963).
[6] H. Moritz, T. Stöferle, M. Köhl, and T. Esslinger, Phys. Rev. Lett. **91**, 250402 (2003); T. Kinoshita, T. Wenger, and D. Weiss, Science **305**, 1125 (2004).
[7] T. Kinoshita, T. Wenger, and D. S. Weiss, Nature **440**, 900 (2006).
[8] S. Trotzky, Y.-A. Chen, A. Flesch, I. P. McCulloch, U. Schollwöck, J. Eisert, and I. Bloch, Nature Phys. **8**, 325 (2012).
[9] G. Pagano, M. Mancini, G. Cappellini, P. Lombardi, F. Schäfer, H. Hu, X.-J. Liu, J. Catani, C. Sias, M. Inguscio, and L. Fallani, Nature Phys. **10**, 198 (2014).
[10] M. Greiner, O. Mandel, T. Esslinger, T.W. Hänsch, and I. Bloch, Nature **415**, 39 (2002);
[11] I. Bloch, Nature Phys. **1**, 23 (2005).
[12] T. Esslinger, Ann. Rev. Cond. Matt. Phys. **1**, 129 (2010).
[13] L.D. Faddeev and L. Takhtadzhyan, J. Sov. Math. **24**, 241 (1984).
[14] See e.g. T. Kitagawa, A. Imambekov, J. Schmiedmayer, and E. Demler, New J. Phys. **13** 073018 (2011); T. Langen, T. Gasenzer, and J. Schmiedmayer, J. Stat. Mech. (2016) 064009.
[15] *Special Issue on Entanglement entropy in extended quantum systems*, eds P. Calabrese, J. Cardy, and B. Doyon, J. Phys. **A42** Number 50, (2009).
[16] V.E. Korepin, A.G. Izergin, and N.M. Bogoliubov, *Quantum Inverse Scattering Method, Correlation Functions and Algebraic Bethe Ansatz* (Cambridge University Press, 1993).
[17] L. D'Alessio, Y. Kafri, A. Polkovnikov, and M. Rigol, Adv. Phys. **65**, 239 (2016).
[18] M. Rigol, V. Dunjko, and M. Olshanii, Nature **452**, 854 (2008).
[19] W. Beugeling, R. Moessner, and M. Haque, Phys. Rev. E **89** 042112 (2014).
[20] H. Kim, T.N. Ikeda, and D.A. Huse, Phys. Rev. E **90**, 052105 (2014).
[21] N. Shiraishi and T. Mori, arXiv:1702.08227.
[22] E. Ilievski, M. Medenjak, T. Prosen, and L. Zadnik, arXiv:1603.00440 (2016).
[23] H. Bethe, Z. Physik **71**, 205 (1931).
[24] M. Takahashi, *Thermodynamics of one dimensional solvable models*, Cambridge University Press, Cambridge 1999.
[25] M. Takahashi, Prog. Theor. Phys. **46**, 401 (1971).
[26] A. A. Vladimirov, Phys. Lett. **A 105**, 418 (1984); F. H. L. Essler, V. E. Korepin, and K. Schoutens, J. Phys. **A25**, 4115 (1992); R. Hagemans and J.-S. Caux, J. Phys. **A40**, 14605 (2007).
[27] A.M. Tsvelik and P. Wiegmann, Adv. in Phys. **32**, 453 (1983), Appendix A to 6.2.1.
[28] S. Braun, J. P. Ronzheimer, M. Schreiber, S.S. Hodgman, T. Rom, I. Bloch, and U. Schneider, Science **339**, 52 (2013).
[29] A. Klümper, Z. Phys. **B91**, 507 (1993); A. Klümper, Eur. Phys. J. **B5**, 677 (1998).
[30] M.P. Grabowski, P. Mathieu, Ann. Phys. **243**, 299 (1995); N. Muramoto and M. Takahashi, J. Phys. Soc. Jpn **68**, 2098 (1999).
[31] M. Brockmann, B. Wouters, D. Fioretto, J. De Nardis, R. Vlijm, and J.-S. Caux, J. Stat. Mech. (2014) P12009.
[32] P. Calabrese, F.H.L. Essler, and G. Mussardo, *Special issue on Quantum Integrability in Out of Equilibrium Systems*, J. Stat. Mech. (2016) 064001.

[33] F.H.L. Essler and M. Fagotti, J. Stat. Mech. (2016) 064002.
[34] J. M. Deutsch, Phys. Rev. A **43**, 2046 (1991); M. Srednicki, Phys. Rev. E**50**, 888 (1994).
[35] E. Ilievski, J. De Nardis, B. Wouters, J.-S. Caux, F.H.L. Essler, and T. Prosen, Phys. Rev. Lett. **115**, 157201 (2015).
[36] J.-S. Caux and F.H.L. Essler, Phys. Rev. Lett. **110**, 257203 (2013).
[37] J.-S.Caux, J. Stat. Mech. (2016) 064006.
[38] L. Piroli, P. Calabrese, and F.H.L. Essler, SciPost Phys. **1**, 001 (2016).

8
Introduction to integrability and one-point functions in $\mathcal{N}=4$ supersymmetric Yang–Mills theory and its defect cousin

Marius de Leeuw, Asger C. Ipsen,
Charlotte Kristjansen, and Matthias Wilhelm

Niels Bohr Institute, Copenhagen University, Blegdamsvej 17,
2100 Copenhagen, Denmark

Chapter Contents

8 Introduction to integrability and one-point
   functions in $\mathcal{N}=4$ supersymmetric Yang–Mills
   theory and its defect cousin   352
   Marius de LEEUW, Asger C. IPSEN,
   Charlotte KRISTJANSEN, and Matthias WILHELM

   8.1 Introduction   354
   8.2 $\mathcal{N}=4$ supersymmetric Yang–Mills theory and the
        spectral problem   355
   8.3 The integrable Heisenberg spin chain in $\mathcal{N}=4$
        supersymmetric Yang–Mills theory   372
   8.4 $\mathcal{N}=4$ supersymmetric Yang–Mills theory with a defect,
        one-point functions, and integrability   380
   8.5 Outlook   394
        Acknowledgements   395

        References   395

8.1 Introduction

$\mathcal{N} = 4$ super Yang–Mills (SYM) theory is a distinguished quantum field theory carrying the maximal amount of supersymmetry for a non-gravitational theory in four dimensions and being conformal even at the quantum level. It plays the role of the CFT in the AdS/CFT correspondence and it exhibits integrability at the planar level. As a matter of fact, since its formulation almost forty years ago [1, 2] the theory has continuously revealed novel interesting features and keeps on doing so.

At the present time, the fundamentals of $\mathcal{N} = 4$ SYM theory are treated in several textbooks such as [3, 4], and there already exist a number of reviews discussing the integrability of the theory's planar spectral problem [5–8]. In this chapter, we will be brief about the basics of $\mathcal{N} = 4$ SYM theory and choose a slightly different perspective on its integrability properties, showing how the tools of integrability inherited from the planar spectral problem can be used to study one-point functions in a certain defect version of $\mathcal{N} = 4$ SYM theory. For a discussion of the role of $\mathcal{N} = 4$ SYM theory in the AdS/CFT correspondence, we refer to the lectures by G. Semenoff.

We start in Section 8.2.1 by presenting the action of $\mathcal{N} = 4$ SYM theory and briefly reviewing its symmetries. This allows us to introduce the key concept of the dilatation operator, the diagonalization of which constitutes the above-mentioned spectral problem. Furthermore, having at hand the explicit expressions for the symmetry generators will facilitate the discussion of symmetry breaking for the defect version of the theory.

Next, we move on to discussing in Section 8.2.2 the two-point functions of $\mathcal{N} = 4$ SYM theory. By extracting the logarithmically divergent pieces of the two-point functions, one can read off the dilatation generator of the theory. For the analysis of (quantum) one-point functions, however, one needs not only the logarithmically divergent pieces but also the finite parts of the two-point functions. Hence, we have chosen to present in quite some detail the perturbative calculation of two-point functions in the scalar sector of the theory, whereby, in addition, we fill some gaps in the earlier reviews.

The dilatation operator of $\mathcal{N} = 4$ SYM theory can be identified with the Hamiltonian of an integrable spin chain, and at one-loop order specializing to the simplest possible sector of the theory this spin chain reduces to the Heisenberg spin chain [9]. The Hamiltonian of the Heisenberg spin chain can be diagonalized either by coordinate-space or algebraic Bethe ansatz techniques. These techniques were explained in detail in Chapter 1 in this volume. Here, we will only highlight in Section 8.3 some features of the solution which will be of importance for the study of one-point functions, namely the parity properties of the eigenstates and their so-called Gaudin norm [10]. In addition, we will discuss, on a general level, how the spin-chain picture of $\mathcal{N} = 4$ SYM theory generalizes to higher loop orders. Finally, we will summarize various observations concerning the non-planar spectral problem.

When defects or boundaries are introduced in a conformal field theory such as $\mathcal{N} = 4$ SYM theory, novel features emerge. Hence, the theory can have non-trivial one-point functions, and two-point functions between operators with unequal conformal dimensions need no longer vanish. There exists a certain defect version of $\mathcal{N}=4$ SYM theory in which some of the scalar fields acquire a vacuum expectation value characterized by a certain representation label k and where one-point functions are non-trivial already at tree level. This defect conformal field theory (dCFT), moreover, has a holographic dual. The holographic dual consists of a so-called D5-D3 probe–brane system where a single probe D5 brane with geometry $AdS_4 \times S^2$ is embedded in the usual $AdS_5 \times S^5$ background of AdS/CFT and carries k units of background gauge field flux on the S^2. For details, we refer to Chapter 3 in this volume.

The remaining part of the chapter will be devoted to the study of this defect version of $\mathcal{N}=4$ SYM theory. Firstly, in Section 8.4.1 we will analyse its symmetry properties making explicit the surviving part of the $\mathcal{N}=4$ SYM symmetry algebra. Subsequently, we will demonstrate how the tools of integrability can be applied to the calculation of the one-point functions of the dCFT, firstly at tree level in Section 8.4.2 and subsequently in Section 8.4.3 at one-loop order. In particular, we will derive a closed expression for the one-point functions in the simplest so-called $SU(2)$ sector of $\mathcal{N}=4$ SYM theory valid for any operator and for any value of the representation label k. For the tree-level calculation, the transfer matrix of the Heisenberg spin chain will be shown to play a crucial role, and for the one-loop calculation we will make use of our explicit quantum-field-theoretical computations in Section 8.2.2. We will also briefly mention a proposal for an all loop so-called asymptotic one-point function formula. This formula correctly encodes all available one-loop data but whether or not the formula remains exhaustive at higher loop orders constitutes an open question.

We conclude with a discussion in Section 8.5 of this as well as other open questions related to the defect version of $\mathcal{N}=4$ SYM theory and in addition briefly list other recent applications of integrability in the context of $\mathcal{N}=4$ SYM theory.

Throughout the chapter, exercises are provided in order to help the interested student acquiring some hands-on knowledge of the different concepts.

8.2 $\mathcal{N}=4$ supersymmetric Yang–Mills theory and the spectral problem

8.2.1 Action and symmetries

8.2.1.1 Action

The maximally supersymmetric $\mathcal{N}=4$ SYM theory in four dimensions can be constructed from $\mathcal{N}=1$ SYM theory in ten dimensions via dimensional reduction. In this reduction, the ten-dimensional gauge fields split into the four-dimensional gauge field A_μ, $\mu = 0, 1, 2, 3$, and the six real scalars ϕ_i, $i = 1, 2, 3, 4, 5, 6$. Similarly, the ten-dimensional Majorana–Weyl fermion ψ splits into four four-dimensional Majorana fermions.

356 *Introduction to integrability and one-point functions*

The action of $\mathcal{N}=4$ SYM theory reads

$$S_{\mathcal{N}=4} = \frac{2}{g_{YM}^2} \int d^4x \, \text{tr}\left[-\frac{1}{4}F_{\mu\nu}F^{\mu\nu} - \frac{1}{2}D_\mu\phi_i D^\mu\phi_i \right.$$
$$\left. + \frac{i}{2}\bar{\psi}\Gamma^\mu D_\mu\psi + \frac{1}{2}\bar{\psi}\Gamma^i[\phi_i,\psi] + \frac{1}{4}[\phi_i,\phi_j][\phi_i,\phi_j]\right], \quad (8.1)$$

where the field strength $F_{\mu\nu}$ and the covariant derivatives D_μ are defined via

$$F_{\mu\nu} = \partial_\mu A_\nu - \partial_\nu A_\mu - i[A_\mu, A_\nu],$$
$$D_\mu\phi_i = \partial_\mu\phi_i - i[A_\mu, \phi_i], \quad D_\mu\psi = \partial_\mu\psi - i[A_\mu, \psi]. \quad (8.2)$$

Here, Γ denotes the ten-dimensional gamma matrices which govern the coupling of the ten-dimensional fermion ψ to the bosons. Exact expressions for Γ and the reductions of Γ and ψ to four dimensions can be found in [11] in our conventions; for the present discussion, they are however not required.

We consider $\mathcal{N}=4$ SYM theory with gauge group $U(N)$. All fields transform in the adjoint representation of the gauge group. We denote the colour components of, say, the scalars ϕ^i as $[\phi^i]_{ab}$, where $a, b = 1, \ldots, N$ are fundamental indices. We can build gauge-invariant local composite operators by taking traces of products of fields that transform covariantly under the gauge group.[1] Moreover, we can take products of such single-trace operators to obtain multi-trace operators.

Mostly, we are restricting ourselves to the 't Hooft limit, where $g_{YM} \to 0, N \to \infty$ while $\lambda = g_{YM}^2 N$ is kept fixed [12]. In this limit, only planar Feynman diagrams contribute to correlation functions, which is why it is also called the planar limit. Moreover, interactions that lead to splitting and joining of traces are suppressed, such that it is sufficient to look at operators with a fixed number of traces, typically single-trace operators. It is possible to go beyond the planar limit and do a double expansion in λ and $\frac{1}{N}$. We refer to Chapter 3 for a more detailed discussion of the 't Hooft limit and the large-N expansion.

From the action (8.1), we can derive the propagators. For example, the scalar propagator reads

$$\langle [\phi_i]_{ab}(x)[\phi_j]_{b'a'}(y)\rangle = \delta_{ij}\delta_{aa'}\delta_{bb'}\frac{g_{YM}^2}{8\pi^2}\frac{1}{(x-y)^2}. \quad (8.3)$$

8.2.1.2 Symmetries

$\mathcal{N}=4$ SYM theory exhibits an exceptional amount of symmetry. In its presentation, we follow the notation in [5]. The simplest of its symmetries is given by Poincaré symmetry, consisting of the six Lorentz transformations $M_{\mu\nu}$ and the four translations P_μ. When

---

[1] For the gauge fields, the covariant combinations are the field strength and covariant derivatives that can act on all fields.

treating fermions and dealing with supersymmetry, it is advantageous to exploit the decomposition of the Lorentz group $SO(1,3) \simeq SU(2)_L \times SU(2)_R$. The generators of the Lorentz group then are $L^\alpha{}_\beta$ and $\dot{L}^{\dot\alpha}{}_{\dot\beta}$. Moreover, the momentum generator can be written in terms of spinor indices as $P_{\alpha\dot\alpha} = P_\mu \sigma^\mu_{\alpha\dot\alpha}$, where $\sigma^\mu = (1, \sigma^1, \sigma^2, \sigma^3)$ with σ^i being the Pauli matrices. The commutation relations of the Lorentz generators among themselves and with the momentum generator $P_{\alpha\dot\alpha}$ follow some general rules. For any generator $\mathcal{J}$,

$$[L^\alpha{}_\beta, \mathcal{J}_\gamma] = \delta^\alpha_\gamma \mathcal{J}_\beta - \tfrac{1}{2}\delta^\alpha_\beta \mathcal{J}_\gamma, \qquad [L^\alpha{}_\beta, \mathcal{J}^\gamma] = -\delta^\gamma_\beta \mathcal{J}^\alpha + \tfrac{1}{2}\delta^\alpha_\beta \mathcal{J}^\gamma,$$
$$[\dot{L}^{\dot\alpha}{}_{\dot\beta}, \mathcal{J}_{\dot\gamma}] = \delta^{\dot\alpha}_{\dot\gamma} \mathcal{J}_{\dot\beta} - \tfrac{1}{2}\delta^{\dot\alpha}_{\dot\beta} \mathcal{J}_{\dot\gamma}, \qquad [\dot{L}^{\dot\alpha}{}_{\dot\beta}, \mathcal{J}^{\dot\gamma}] = -\delta^{\dot\gamma}_{\dot\beta} \mathcal{J}^{\dot\alpha} + \tfrac{1}{2}\delta^{\dot\alpha}_{\dot\beta} \mathcal{J}^{\dot\gamma}, \tag{8.4}$$

which is understood to be applied for each index.

Using these rules, write down the explicit commutation relations $[L^\alpha{}_\beta, L^\gamma{}_\delta]$.

Two translations commute,

$$[P_{\alpha\dot\alpha}, P_{\beta\dot\beta}] = 0. \tag{8.5}$$

In addition, $\mathcal{N} = 4$ SYM theory is conformally invariant. At the classical level, this follows from the absence of masses and dimensionful couplings in the action (8.1). The fact that this symmetry is preserved at the quantum level is however non-trivial [13–15]. Conformal symmetry in particular implies the invariance under scale transformations, generated by the dilatation operator D, and so-called special conformal transformations, generated by $K^{\alpha\dot\alpha} = K_\mu (\sigma^\mu)^{\alpha\dot\alpha}$. They satisfy the commutation relations

$$[D, P_{\alpha\dot\alpha}] = P_{\alpha\dot\alpha}, \qquad [D, L^\alpha{}_\beta] = [D, \dot{L}^{\dot\alpha}{}_{\dot\beta}] = 0, \qquad [D, K^{\alpha\dot\alpha}] = -K^{\alpha\dot\alpha},$$
$$[K^{\alpha\dot\alpha}, P_{\beta\dot\beta}] = \delta^{\dot\alpha}_{\dot\beta} L^\alpha{}_\beta + \delta^\alpha_\beta \dot{L}^{\dot\alpha}{}_{\dot\beta} + \delta^\alpha_\beta \delta^{\dot\alpha}_{\dot\beta} D. \tag{8.6}$$

The dilatations and special conformal transformations combine with the Poincaré transformations to form the conformal group $SO(2,4) \simeq SU(2,2)$.

Furthermore, Poincaré symmetry can be enhanced by supersymmetry, of which $\mathcal{N} = 4$ SYM theory has the maximal amount permitted in a theory without gravity. The supercharges Q^A_α and $\dot{Q}_{\dot\alpha A}$ have the following non-vanishing anticommutation relations among themselves:

$$\{\dot{Q}_{\dot\alpha A}, Q^B_\alpha\} = \delta^B_A P_{\alpha\dot\alpha}, \tag{8.7}$$

while their behaviour under Lorentz transformations is determined via (8.4). Maximal supersymmetry implies a bosonic R-symmetry with symmetry group $SU(4) \simeq SO(6)$. The behaviour of a general generator $\mathcal{J}$ under R-symmetry transformations is determined by the following rule in analogy to (8.4):

$$\left[R^A{}_B, \mathcal{J}_C\right] = \delta^A_C \mathcal{J}_B - \tfrac{1}{4}\delta^A_B \mathcal{J}_C, \qquad \left[R^A{}_B, \mathcal{J}^C\right] = -\delta^C_B \mathcal{J}^A + \tfrac{1}{4}\delta^A_B \mathcal{J}^C. \qquad (8.8)$$

Finally, supersymmetry and conformal symmetry combine to superconformal symmetry with the superconformal charges S^α_A and $\dot{S}^{\dot\alpha A}$. The additional non-vanishing (anti)commutation relations are

$$\left[D, Q^A_\alpha\right] = \tfrac{1}{2}Q^A_\alpha, \quad \left[D, \dot{Q}_{\dot\alpha A}\right] = +\tfrac{1}{2}\dot{Q}_{\dot\alpha A}, \quad \left[D, S^\alpha_A\right] = -\tfrac{1}{2}S^\alpha_A, \quad \left[D, \dot{S}^{\dot\alpha A}\right] = -\tfrac{1}{2}\dot{S}^{\dot\alpha A},$$

$$\left\{\dot{S}^{\dot\alpha A}, S^\alpha_B\right\} = \delta^A_B K^{\alpha\dot\alpha}, \quad \left[K^{\alpha\dot\alpha}, Q^A_\beta\right] = \delta^\alpha_\beta \dot{S}^{\dot\alpha A}, \quad \left[K^{\alpha\dot\alpha}, \dot{Q}_{\dot\beta A}\right] = \delta^{\dot\alpha}_{\dot\beta} S^\alpha_A,$$

$$\left\{S^\alpha_A, Q^B_\beta\right\} = \delta^B_A L^\alpha{}_\beta + \delta^\alpha_\beta R^B{}_A + \tfrac{1}{2}\delta^\alpha_\beta \delta^B_A D, \quad \left[S^\alpha_A, P_{\beta\dot\beta}\right] = \delta^\alpha_\beta \dot{Q}_{\dot\beta A},$$

$$\left\{\dot{S}^{\dot\alpha A}, \dot{Q}_{\dot\beta B}\right\} = \delta^A_B \dot{L}^{\dot\alpha}{}_{\dot\beta} - \delta^{\dot\alpha}_{\dot\beta} R^A{}_B + \tfrac{1}{2}\delta^{\dot\alpha}_{\dot\beta} \delta^A_B D, \quad \left[\dot{S}^{\dot\alpha A}, P_{\beta\dot\beta}\right] = \delta^{\dot\alpha}_{\dot\beta} Q^A_\beta. \qquad (8.9)$$

Together, they generate the superconformal group $PSU(2,2|4)$. For the action of $PSU(2,2|4)$ on composite operators, see e.g. [5].

Composite operators $\mathcal{O}_i$ that are primary states of $PSU(2,2|4)$ can be characterized via the charges $[\Delta, j_L, j_R, r_1, r_2, r_3]$. Primary means that the operators are annihilated by all lowering operators $\{K^{\alpha\dot\alpha}, S^\alpha_A, \dot{S}^{\dot\alpha A}, L^\alpha{}_\beta(\alpha < \beta), \dot{L}^{\dot\alpha}{}_{\dot\beta}(\dot\alpha < \dot\beta), R^A{}_B(A < B)\}$. All other operators, called descendents, can then be obtained by acting on the primaries with the raising operators $\{P_{\alpha\dot\alpha}, Q^A_\alpha, \dot{Q}_{\dot\alpha A}, L^\alpha{}_\beta(\alpha > \beta), \dot{L}^{\dot\alpha}{}_{\dot\beta}(\dot\alpha > \dot\beta), R^A{}_B(A > B)\}$. The conformal dimension of the operator, Δ, is measured by the dilatation operator D and defines the behaviour of the operator under a scale transformation,

$$x \to x' = \lambda x, \qquad \mathcal{O}_i(x) \to \mathcal{O}'_i(x) = \lambda^{\Delta_i} \mathcal{O}_i(\lambda x). \qquad (8.10)$$

It will play a particular role in the following, as it can receive quantum corrections. The other charges are the left and right spin j_L and j_R as well as the three charges r_1, r_2, and r_3 characterizing the $SU(4)$ representation in which the operator transforms.

A particular class of primary operators are also annihilated by some of the supercharges Q^A_α and $\dot{Q}_{\dot\alpha A}$, which are raising operators. Such primary operators are called BPS operators. From the anticommutation relations (8.9), it follows that their scaling dimensions are related to their spin and R-charge and hence protected from quantum corrections.

8.2.1.3 Correlation functions

In conformal field theories, conformal symmetry greatly restricts the form correlation functions can take. For instance, one-point functions of composite operators $\mathcal{O}_i$ have to be constant by conformal symmetry and are normally taken to vanish. More generally, all correlation functions are fixed in terms of the so-called conformal data (Δ, λ). The Δs are the conformal dimensions of the operators and the λs are called structure constants and describe three-point functions.

More precisely, the space-time dependence of two-point functions is completely fixed by the scaling dimensions of the operators

$$\langle \mathcal{O}_i(x)\bar{\mathcal{O}}_j(y)\rangle = \frac{M_{ij}}{|x-y|^{\Delta_i+\Delta_j}}, \qquad (8.11)$$

where $M_{ij} = 0$ for $\Delta_i \neq \Delta_j$. Moreover, conformal symmetry also fixes the three-point function up to the structure constant λ_{ijk}, which appears in the operator product expansion (OPE):

$$\mathcal{O}_i(x)\mathcal{O}_j(y) = \frac{M_{ij}}{|x-y|^{\Delta_i+\Delta_j}} + \sum_k \frac{\lambda_{ij}{}^k}{|x-y|^{\Delta_i+\Delta_j-\Delta_k}} C(x-y,\partial_y)\mathcal{O}_k(y), \qquad (8.12)$$

where the sum over k runs over conformal primary operators and the differential operator C in (8.12) accounts for the presence of conformal descendants. The indices on λ can be raised and lowered with the matrix M. The normalization of C is such that $C(x-y, \partial_y) = 1 + O(x-y)$. The scaling dimensions Δ_i and the structure constants λ_{ijk}, completely determine all four- and higher-point functions via repeated use of the OPE (8.12). Note that starting from four-point functions, a non-trivial dependence on conformal cross-ratios can occur.

8.2.2 Two-point functions and the spectral problem

Let us now calculate the two-point functions. For simplicity, we will in the following restrict ourselves to the leading large-N, planar limit. To keep the computation manageable, we further restrict to operators made only of the six scalar fields, the so-called $SO(6)$ sector.[2] In more detail, let $I = \{i_1, i_2, \ldots, i_L\}$, with $i_n = 1, \ldots, 6$. We then define our un-renormalized operators by

$$\mathcal{O}_I^{\text{bare}} = \text{tr}[\phi_{i_1}\phi_{i_2}\cdots\phi_{i_L}]. \qquad (8.13)$$

It is clear that $\Delta^{(0)} = L$. As correlation functions between single- and multi-trace operators are suppressed by powers of $\frac{1}{N}$, it is consistent to only consider single-trace operators. Since our operators do not contain derivatives, they are conformal primaries in the sense that $[K_\mu, \mathcal{O}_I^{\text{bare}}(0)] = 0$. It is obvious from the propagator (8.3) that the two-point functions take the predicted form (8.11) at tree level. More precisely, we can write

---

[2] Note that this sector is not closed beyond one-loop order.

$$\langle \mathcal{O}_I^{\text{bare}}(x) \bar{\mathcal{O}}_J^{\text{bare}}(y) \rangle_{\text{tree}} \propto \frac{1}{|x-y|^{2\Delta^{(0)}}} \delta_{IJ}, \tag{8.14}$$

where $\Delta^{(0)}$ is the common classical scaling dimension of $\mathcal{O}_I^{\text{bare}}$ and $\mathcal{O}_J^{\text{bare}}$, obtainable by standard power counting. Because of the cyclic invariance of the trace, we identify indices that are cyclic permutations of each other. The bar denotes Hermitian conjugation, which in the present case of real scalars only inverts the order of the fields in the trace.

At the quantum level, one observes the phenomenon of operator mixing, meaning that the two-point function between single-trace operators is no longer proportional to a delta-function. Furthermore, wave-function renormalization is needed in order to render the correlation functions finite and a regularization method has to be chosen. In the following, we will make use of dimensional regularisation, i.e.

$$S = \frac{2}{g_{\text{YM}}^2} \int d^4x\, \mathcal{L} \to S_\varepsilon = \frac{2}{(g_{\text{YM}}\mu^\varepsilon)^2} \int d^{4-2\varepsilon}x\, \mathcal{L}, \tag{8.15}$$

where μ is a parameter with the dimension of mass. With this choice of regulator, dimensional analysis shows that the full two-point function takes the form

$$\langle \mathcal{O}_I^{\text{bare}}(x) \bar{\mathcal{O}}_J^{\text{bare}}(y) \rangle_\varepsilon = \sum_{n=0}^{\infty} (g\mu^\varepsilon)^{2(\Delta^{(0)}+n)} \frac{\tilde{M}_{IJ}^{(n)}(\varepsilon)}{|x-y|^{2\Delta^{(0)}-2\varepsilon(\Delta^{(0)}+n)}}, \tag{8.16}$$

where we have defined the effective planar loop coupling

$$g^2 = \frac{g_{\text{YM}}^2 N}{16\pi^2}. \tag{8.17}$$

In general, the $\tilde{M}_{IJ}^{(n)}(\varepsilon)$ will have poles at $\varepsilon = 0$, so one cannot simply take the $\varepsilon \to 0$ limit of (8.16). Usually, such divergencies are dealt with by adding counterterms to the action. For $\mathcal{N} = 4$ SYM theory, it is not necessary to introduce such terms; instead one can render the correlation functions finite by a 'rescaling' of the operators alone.[3] We thus introduce renormalized operators by

$$\mathcal{O}_I^{\text{ren}} = \mathcal{Z}_{IJ}(g_{\text{YM}}, \varepsilon) \mathcal{O}_J^{\text{bare}}, \tag{8.18}$$

where $\mathcal{Z}_{IJ}(g_{\text{YM}}, \varepsilon)$ is some numerical matrix with poles at $\varepsilon = 0$ such that the correlation functions of $\mathcal{O}_I^{\text{ren}}$ are finite.

---

[3] Usually, this fact is expressed in the abbreviated form '$\mathcal{N} = 4$ SYM theory is finite'.

In perturbation theory, if there are no conformal anomalies, the two-point function of the renormalized operators takes the form (8.11), but with Δ_I a power series in g:

$$\Delta_I = \sum_{n=0}^{\infty} g^{2n} \Delta_I^{(n)}. \tag{8.19}$$

For historical reasons, the correction $\Delta - \Delta^{(0)}$ is called the anomalous dimension, even though there is nothing anomalous about it. Looking at (8.16), what must happen is that the $\log(x-y)^2$ terms resulting from the expansion of the summand in ε must exponentiate to form $[(x-y)^2]^\Delta$. To see the exact mechanism, let us consider the simplified situation with only one operator. We first rewrite the bare two-point function as

$$\langle \mathcal{O}^{\text{bare}}(x) \bar{\mathcal{O}}^{\text{bare}}(y) \rangle_\varepsilon = \frac{(g\mu^\varepsilon)^{2\Delta^{(0)}} \tilde{M}^{(0)}}{[(x-y)^2]^{(1-\varepsilon)\Delta^{(0)}}} \left(1 + \sum_{n=1}^{\infty} (g^2[\mu^2(x-y)^2]^\varepsilon)^n \frac{\tilde{M}^{(n)}}{\tilde{M}^{(0)}} \right)$$

$$= \frac{(g\mu^\varepsilon)^{2\Delta^{(0)}} \tilde{M}^{(0)}}{[(x-y)^2]^{(1-\varepsilon)\Delta^{(0)}}} \exp \left(\sum_{n=1}^{\infty} (g^2[\mu^2(x-y)^2]^\varepsilon)^n \mathcal{M}^{(n)} \right). \tag{8.20}$$

At the second equality, we take the formal logarithm of the series. It is easy to see that the $\mathcal{M}^{(n)}$ are expressible as polynomials in $\tilde{M}^{(n)}/\tilde{M}^{(0)}$:

$$\mathcal{M}^{(1)} = \frac{\tilde{M}^{(1)}}{\tilde{M}^{(0)}}, \qquad \mathcal{M}^{(2)} = \frac{\tilde{M}^{(2)}}{\tilde{M}^{(0)}} - \frac{1}{2}\left(\frac{\tilde{M}^{(1)}}{\tilde{M}^{(0)}}\right)^2, \qquad \text{etc.} \tag{8.21}$$

If we also write $\mathcal{Z}$ as an exponential,

$$\mathcal{Z} = \exp\left(\sum_{n=1}^{\infty} g^{2n} \mathcal{Z}^{(n)} \right), \tag{8.22}$$

the renormalized two-point function is[4]

$$\langle \mathcal{O}^{\text{ren}}(x) \bar{\mathcal{O}}^{\text{ren}}(y) \rangle_\varepsilon = \mathcal{Z} \langle \mathcal{O}^{\text{bare}}(x) \bar{\mathcal{O}}^{\text{bare}}(y) \rangle_\varepsilon \mathcal{Z}^\dagger$$

$$\propto \exp\left(\sum_{n=1}^{\infty} g^{2n} \left([\mu^2(x-y)^2]^{\varepsilon n} \mathcal{M}^{(n)} + 2\mathcal{Z}^{(n)} \right) \right). \tag{8.23}$$

For this to be finite, we must be able to cancel all divergences with some appropriate choice of $\mathcal{Z}$. If $\mathcal{M}^{(n)}$ had poles of higher degree than one, we would have divergent terms

---

[4] In this simple example with only one operator, we can clearly take $\mathcal{Z}^{(n)}$ to be real without loss of generality.

with a dependence on $(x-y)^2$, which could clearly not be cancelled. We conclude that we can expand $\mathcal{M}^{(n)}$ as

$$\mathcal{M}^{(n)} = -\frac{\Delta^{(n)}}{n\varepsilon} + \mathcal{M}^{(n),\text{fin}} + O(\varepsilon). \tag{8.24}$$

The choice of the name of the $1/\varepsilon$ coefficient will become clear in a moment.

From (8.23) and (8.24), it is obvious that a consistent choice of $\mathcal{Z}$ is

$$\mathcal{Z}^{(n)} = \frac{\Delta^{(n)}}{2n\varepsilon}, \tag{8.25}$$

which only cancels the pole (minimal subtraction). With this, we can now take the limit $\varepsilon \to 0$ and find

$$\langle \mathcal{O}^{\text{ren}}(x)\bar{\mathcal{O}}^{\text{ren}}(y)\rangle_{\varepsilon=0}$$
$$= \frac{g^{2\Delta^{(0)}}\tilde{M}^{(0)}(\varepsilon=0)}{[(x-y)^2]^{\Delta^{(0)}}} \exp\left(\sum_{n=1}^{\infty} g^{2n}\left(\mathcal{M}^{(n),\text{fin}} - \Delta^{(n)}\log\mu^2(x-y)^2\right)\right)$$
$$= \frac{(g\mu)^{2\Delta^{(0)}}\tilde{M}^{(0)}(\varepsilon=0)}{[\mu^2(x-y)^2]^{\Delta}} \exp\left(\sum_{n=1}^{\infty} g^{2n}\mathcal{M}^{(n),\text{fin}}\right), \tag{8.26}$$

with Δ as defined in (8.19). We see that the divergences of the bare correlation functions have transformed into corrections to the scaling dimensions of the renormalized operators. Note that the mass scale μ, which one could fear would spoil the conformality of the theory, ends up as a harmless overall normalization constant. For this reason, in the literature it is common to simply set $\mu = 1$. With more than one operator, the above considerations still apply, with the added technical complication that $\mathcal{M}^{(n)}$ and $\mathcal{Z}^{(n)}$ are matrices that do not necessarily commute.

Repeat, up to order g^4, the above analysis in the general case with several operators. For simplicity, assume that $\tilde{M}^{(0)}$ is proportional to the identity, and that the bare two-point function looks like

$$\langle \mathcal{O}^{\text{bare}}(x)\bar{\mathcal{O}}^{\text{bare}}(y)\rangle_{\varepsilon} \propto \exp\left(g^2[\mu^2(x-y)^2]^{\varepsilon}\left(-\frac{D^{(1)}}{\varepsilon} + \mathcal{M}^{(1),\text{fin}} + O(\varepsilon)\right)\right.$$
$$\left. -g^4[\mu^2(x-y)^2]^{2\varepsilon}\left(\frac{D^{(2)}}{2\varepsilon} + O(\varepsilon^0)\right) + O(g^6)\right), \tag{8.27}$$

where $D^{(1)}$ (but not $D^{(2)}$) is diagonal. Partial answer: Let R be a Hermitian matrix such that $D^{(2)} + i[R, D^{(1)}]$ is diagonal, and set

$$\mathcal{Z} = \exp\left(g^2 \left[\frac{1}{2\varepsilon}D^{(1)} + iR - \frac{1}{2}\mathcal{M}^{(1),\text{fin}}\right] + g^4 \left[\frac{1}{4\varepsilon}(D^{(2)} + i[R, D^{(1)}]) + O(\varepsilon)\right]\right). \tag{8.28}$$

Then, the renormalized two-point function is finite, and the anomalous dimensions are the entries of the diagonal matrix

$$g^2 D^{(1)} + g^4 (D^{(2)} + i[R, D^{(1)}]). \tag{8.29}$$

The higher loop calculations are most conveniently carried out in momentum space, where the contraction rules read

$$\langle [\phi_i]_{ab}[\phi_j]_{b'a'}\rangle = \frac{(g_{\text{YM}}\mu^\varepsilon)^2}{2}\frac{\delta_{ij}\delta_{aa'}\delta_{bb'}}{p^2}, \quad \langle [A_\mu]_{ab}[A_\nu]_{b'a'}\rangle = \frac{(g_{\text{YM}}\mu^\varepsilon)^2}{2}\frac{\delta_{\mu\nu}\delta_{aa'}\delta_{bb'}}{p^2}. \tag{8.30}$$

The transition from momentum space to configuration space is encoded in the formula

$$\int \frac{d^{4-2\varepsilon}p}{(2\pi)^{4-2\varepsilon}}\frac{e^{ip\cdot x}}{[p^2]^s} = \frac{\Gamma(2-\varepsilon-s)}{4^s \pi^{2-\varepsilon}\Gamma(s)}\frac{1}{[x^2]^{2-\varepsilon-s}}. \tag{8.31}$$

We denote by

$$K(x,y) = \frac{(g_{\text{YM}}\mu^\varepsilon)^2}{2}\int \frac{d^{4-2\varepsilon}p}{(2\pi)^{4-2\varepsilon}}\frac{e^{ip\cdot(x-y)}}{p^2} = \frac{(g_{\text{YM}}\mu^\varepsilon)^2}{2}\frac{\Gamma(1-\varepsilon)}{4\pi^{2-\varepsilon}}\frac{1}{[(x-y)^2]^{1-\varepsilon}} \tag{8.32}$$

the scalar propagator in $(4-2\varepsilon)$-dimensional position space.

This results in the following more specific form of the two-point function of bare operators:

$$\langle \mathcal{O}_I^{\text{bare}}(x)\bar{\mathcal{O}}_J^{\text{bare}}(y)\rangle_\varepsilon = \sqrt{c_I c_J} N^{\Delta^{(0)}} K(x,y)^{\Delta^{(0)}}\left(\delta_{IJ} + \sum_{n=1}^{\infty} g^{2n}\tilde{M}_{IJ}^{(n)}(\varepsilon)[\mu^2|x-y|^2]^{n\varepsilon}\right). \tag{8.33}$$

For convenience, we have pulled out the tree-level two-point function. The normalization constants c_I are easily seen to be the number of cyclic permutations that leave $I = \{i_1, i_2, \ldots, i_L\}$ invariant. For example,

$$c_{\{5,5,5,5\}} = 4, \quad c_{\{1,5,5,5\}} = 1, \quad c_{\{1,5,5,1,5,5\}} = 2. \tag{8.34}$$

364   Introduction to integrability and one-point functions

As we will explicitly show later, the divergence at one loop is a simple $1/\varepsilon$ pole. We can thus write

$$\tilde{M}^{(1)} = -\frac{1}{\varepsilon}D^{(1)} + \tilde{M}^{(1),\text{fin}} + O(\varepsilon), \tag{8.35}$$

with $D^{(1)}$ and $\tilde{M}^{(1),\text{fin}}$ independent of ε. We now choose our renormalization scheme as follows:

$$\mathcal{O}_I^{\text{ren}} = \mathcal{Z}_{IJ}(g_{\text{YM}},\varepsilon)\frac{\mathcal{O}_J^{\text{bare}}}{\sqrt{c_J}}, \tag{8.36}$$

with

$$\mathcal{Z} = 1 + \frac{g^2}{2\varepsilon}D^{(1)} - \frac{g^2}{2}\tilde{M}^{(1),\text{fin}}. \tag{8.37}$$

The motivation for this particular choice for the finite part is two-fold. Firstly, as will become evident, it leaves us with only one matrix to diagonalize and secondly, as will become clear in Section 8.4.3 it implies a convenient normalization of the renormalized operators. The choice of $\mathcal{Z}$ in (8.37) results in the following expression for the renormalized two-point function:

$$\langle \mathcal{O}_I^{\text{ren}}(x)\bar{\mathcal{O}}_J^{\text{ren}}(y)\rangle_{\varepsilon=0} = N^{\Delta^{(0)}}K(x,y)^{\Delta^{(0)}}\left(\delta_{IJ} - g^2 D_{IJ}^{(1)}\log(\mu^2|x-y|^2) + O(g^4)\right). \tag{8.38}$$

It is clear from (8.38) that, in order to determine the good conformal operators at the one-loop level and their corresponding conformal dimensions, one has to diagonalize the matrix $D_{IJ}^{(1)}$ which is denoted as diagonalizing the dilatation operator. More precisely, the eigenvectors of $D^{(1)}$ are the good conformal operators and the corresponding eigenvalues are the associated conformal dimensions at one-loop order.

Said more formally, we should really set

$$\mathcal{Z} = U\left(1 + \frac{g^2}{2\varepsilon}D^{(1)} - \frac{g^2}{2}\tilde{M}^{(1),\text{fin}}\right), \tag{8.39}$$

where U is a unitary matrix such that $UD^{(1)}U^\dagger$ is diagonal. Then, the renormalized two-point function takes the proper form (8.11), with

$$\Delta_I^{(1)} = [UD^{(1)}U^\dagger]_{II}. \tag{8.40}$$

Let us remark that it is necessary to know the two-point function at order g^2 to determine the g^0 piece of $\mathcal{Z}$ (i.e. U). This happens because we are really doing degenerate perturbation theory in the sense that several operators have the same value of $\Delta^{(0)}$. As one can explicitly see from (8.28), this pattern continues at higher loop order (i.e. $\mathcal{Z}$ at one loop depends on the two-loop correlation function). The remainder of this section will be dedicated to explicitly determining $D^{(1)}$ (and $\tilde{M}^{(1),\text{fin}}$) by a Feynman diagram calculation.

In the planar limit, the one-loop corrections to the two-point functions consist of three types of diagrams, see Figure 8.1. The colour structure is completely fixed by the planar limit, so the interesting part is the flavour structure. By $SO(6)$ symmetry, the self-energy diagram (Figure 8.1(a)) must be proportional to $\delta_{ii'}$. Similarly, since the gauge field is not charged under $SO(6)$, the gauge exchange diagram (Figure 8.1(b)) must have the structure $\delta_{ii'}\delta_{jj'}$. However, the four-point diagram (Figure 8.1(c)) allows for non-trivial flavour tensors. This is what leads to operator mixing at the one-loop level.

It happens that the most interesting diagram is also the easiest to compute, so let us begin by considering the diagram arising from the four-point interaction $\text{tr}\big([\phi_i,\phi_j][\phi_i,\phi_j]\big)$. Writing only the two fields of each operator that participate, we find the contribution

$$\langle [\phi_i\phi_j]_{ab}(x)[\phi_{j'}\phi_{i'}]_{b'a'}(y)\rangle_{(c)} = \frac{2N^2\delta_{ac'}\delta_{bb'}}{(g_{\text{YM}}\mu^\varepsilon)^2}(2\delta_{ij'}\delta_{ji'} - \delta_{ii'}\delta_{jj'} - \delta_{ij}\delta_{i'j'})$$
$$\times \int d^{4-2\varepsilon}z\, K(x,z)^2 K(z,y)^2. \tag{8.41}$$

We now insert the explicit form of the propagator using (8.31) and (8.32) to get

$$\cdots = \frac{2N^2\delta_{aa'}\delta_{bb'}}{(g_{\text{YM}}\mu^\varepsilon)^2}(2\delta_{ij'}\delta_{ji'} - \delta_{ii'}\delta_{jj'} - \delta_{ij}\delta_{i'j'})$$
$$\times \left[\frac{(g_{\text{YM}}\mu^\varepsilon)^2}{2}\frac{\Gamma(1-\varepsilon)}{4\pi^{2-\varepsilon}}\right]^4 \int d^{4-2\varepsilon}z\, \frac{1}{[(x-z)^2]^{2-2\varepsilon}[(z-y)^2]^{2-2\varepsilon}}. \tag{8.42}$$

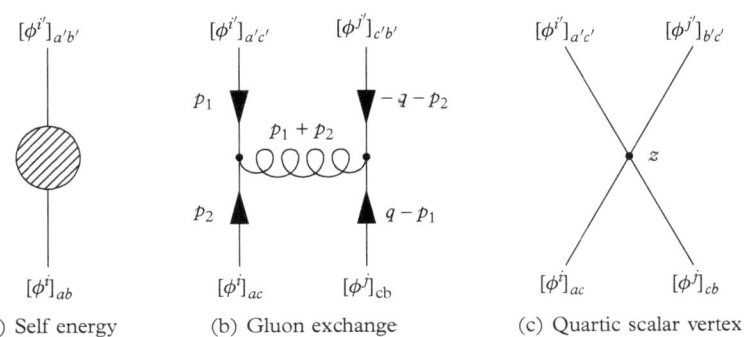

(a) Self energy   (b) Gluon exchange   (c) Quartic scalar vertex

Figure 8.1 *Interaction part of the Feynman diagrams contributing to the two-point function at one-loop order.*

This is a standard one-loop integral which can be evaluated using the formula (see e.g. [16])

$$\int d^{4-2\varepsilon} p \frac{1}{[p^2]^\alpha [(p-q)^2]^\beta} = \pi^{2-\varepsilon} G(\alpha,\beta) \frac{1}{[q^2]^{\alpha+\beta+\varepsilon-2}}. \tag{8.43}$$

Here, $G(\alpha,\beta)$ denotes the following combination of gamma functions,

$$G(\alpha,\beta) = \frac{\Gamma(\alpha+\beta+\varepsilon-2)\Gamma(2-\varepsilon-\alpha)\Gamma(2-\varepsilon-\beta)}{\Gamma(\alpha)\Gamma(\beta)\Gamma(4-\alpha-\beta-2\varepsilon)}. \tag{8.44}$$

We thus find

$$\begin{aligned}\langle [\phi_i \phi_j]_{ab}(x) [\phi_{j'} \phi_{i'}]_{b'a'}(y) \rangle_{(c)} &= \frac{(g_{YM}\mu^\varepsilon)^6 N^2 \Gamma(1-\varepsilon)^4 \delta_{aa'}\delta_{bb'}}{2^{11}\pi^{6-3\varepsilon}} \\ &\quad \times (2\delta_{ij'}\delta_{ji'} - \delta_{ii'}\delta_{jj'} - \delta_{ij}\delta_{i'j'}) \frac{G(2-2\varepsilon, 2-2\varepsilon)}{[(x-y)^2]^{2-3\varepsilon}} \\ &= g^2 N K(x,y)^2 \delta_{aa'}\delta_{bb'} (2\delta_{ij'}\delta_{ji'} - \delta_{ii'}\delta_{jj'} - \delta_{ij}\delta_{i'j'}) \\ &\quad \times \left(\frac{1}{\varepsilon} + 1 + \gamma_E + \log(\pi |x-y|^2) + O(\varepsilon)\right),\end{aligned} \tag{8.45}$$

where γ_E denotes the Euler–Mascheroni constant.

The second type of diagram which involves a pair of legs is the gauge boson exchange. The diagram is formed using two copies of the vertex $i\mathrm{tr}[(\partial_\mu \phi_i)[A_\mu,\phi_i]]$. Explicitly, we find the contribution

$$\begin{aligned}\langle [\phi_i \phi_j]_{ab}(x) [\phi_{j'} \phi_{i'}]_{b'a'}(y) \rangle_{(b)} &= -\frac{4\delta_{ii'}\delta_{jj'}\delta_{aa'}\delta_{bb'} N^2}{(g_{YM}\mu^\varepsilon)^4} \\ &\quad \times \int d^{4-2\varepsilon} z\, d^{4-2\varepsilon} w \left(K(x,z)\partial_{z_\mu} K(z,y) - [\partial_{z_\mu} K(x,z)] K(z,y)\right) K(z,w) \\ &\quad \times \left([\partial_{w_\mu} K(x,w)] K(w,y) - K(x,w) \partial_{w_\mu} K(w,y)\right).\end{aligned} \tag{8.46}$$

In this case, it is simpler to work in momentum space. We thus insert (8.32) and integrate over z,w and the momenta fixed by the resulting delta functions to find

$$\cdots = -\frac{(g_{YM}\mu^\varepsilon)^6 \delta_{ii'}\delta_{jj'}\delta_{aa'}\delta_{bb'} N^2}{2^3} \int \frac{d^{4-2\varepsilon}q}{(2\pi)^{4-2\varepsilon}} e^{iq\cdot(x-y)} H(q), \tag{8.47}$$

with

$$H(q) = \iint \frac{(p_1 - p_2)\cdot(2q - p_1 + p_2)}{V}, \qquad V = p_1^2 p_2^2 (p_1 - q)^2 (p_2 + q)^2 (p_1 + p_2)^2. \tag{8.48}$$

To save space, we abbreviate

$$\iint = \int \frac{d^{4-2\varepsilon} p_1 \, d^{4-2\varepsilon} p_2}{(2\pi)^{2(4-2\varepsilon)}}, \tag{8.49}$$

here and in the following.

We now have to perform the integrals over p_1 and p_2. First, we rewrite the numerator as a linear combination of the factors in the denominator,

$$(p_1 - p_2) \cdot (2q - p_1 + p_2) = -[p_1^2 + p_2^2 + (p_1 - q)^2 + (p_2 + q)^2] + (p_1 + p_2)^2 + 2q^2. \tag{8.50}$$

By the evident symmetries of the diagram, we then find

$$H = -4H_1 + H_2 + 2q^2 H_3, \tag{8.51}$$

with

$$H_1 = \iint \frac{p_1^2}{V}, \quad H_2 = \iint \frac{(p_1 + p_2)^2}{V}, \quad H_3 = \iint \frac{1}{V}. \tag{8.52}$$

The numerator of H_1 cancels one of the propagators, allowing us to perform the p_1 integral using (8.43). The remaining integral over p_2 then also follows from (8.43), and we obtain

$$H_1 = \frac{1}{(4\pi)^{4-2\varepsilon}} \frac{G(1,1)G(1,1+\varepsilon)}{[q^2]^{2\varepsilon}}. \tag{8.53}$$

The p_1 and p_2 integrals decouple for H_2, and we immediately find

$$H_2 = \frac{1}{(4\pi)^{4-2\varepsilon}} \frac{G(1,1)^2}{[q^2]^{2\varepsilon}}. \tag{8.54}$$

To get a closed expression for the final integral H_3 requires an extra trick. Here, we use integration by parts, following [16, 17]. We first observe that

$$0 = \iint (\partial_1 \cdot p_1 + p_2 \cdot \partial_1) \frac{1}{V} = -2\varepsilon H_3 + \iint \frac{p_2^2 - (p_1 + p_2)^2}{p_1^2 V} + \iint \frac{(p_2 + q)^2 - (p_1 + p_2)^2}{(p_1 - q)^2 V}. \tag{8.55}$$

We can now isolate H_3 and evaluate the remaining integrals by successive use of (8.43), with the result

$$H_3 = \frac{1}{\varepsilon} \iint \frac{p_2^2 - (p_1 + p_2)^2}{p_1^2 V} = \frac{1}{\varepsilon} \frac{1}{(4\pi)^{4-2\varepsilon}} \frac{G(1,1)[G(2,1+\varepsilon) - G(2,1)]}{[q^2]^{1+2\varepsilon}}. \tag{8.56}$$

After the transformation back to real space, the final result for the diagram is

$$\langle [\phi_i\phi_j]_{ab}(x)[\phi_{j'}\phi_{i'}]_{b'a'}(y)\rangle_{(b)} =$$
$$g^2 N K(x,y)^2 \delta_{ii'}\delta_{jj'}\delta_{aa'}\delta_{bb'}\left(\frac{1}{\varepsilon}+3+\gamma_E+\log(\pi|x-y|^2)+O(\varepsilon)\right). \quad (8.57)$$

Let us remark that H_3 is finite in four dimensions, in fact

$$H_3 = \frac{3\zeta(3)}{2^7\pi^4} + O(\varepsilon). \quad (8.58)$$

The Fourier transform yields an additional factor of ε, meaning that the contribution of H_3 to the two-point function is $O(\varepsilon)$. In a one-loop calculation, H_3 can thus be dropped.

Finally, we have the one-loop self-energy correction to the scalar propagator. The calculation again reduces to an application of (8.43). We omit the details, but see [18]. The one-loop corrected propagator is

$$\langle [\phi_i(x)]_{ab}[\phi_j(y)]_{b'a'}\rangle$$
$$= \delta_{ij}\delta_{aa'}\delta_{bb'}\left[K(x,y)-g_{YM}^4 N\int\frac{d^{4-2\varepsilon}q}{(2\pi)^{4-2\varepsilon}}e^{iq\cdot(x-y)}\frac{G(1,1)}{(4\pi)^{2-\varepsilon}}\frac{1}{[q^2]^{1+\varepsilon}}+O(g_{YM}^6)\right]$$
$$= \delta_{ij}\delta_{aa'}\delta_{bb'}K(x,y)\left[1-2g^2\left(\frac{1}{\varepsilon}+2+\gamma_E+\log(\pi|x-y|^2)+O(\varepsilon)\right)+O(g^4)\right]. \quad (8.59)$$

The planar two-point function of two single-trace operators now follows by inserting the corrections (8.45), (8.57), and (8.59) in the tree-level diagram. All told, we find

$$\langle O_I^{\text{bare}}(x)\bar{O}_J^{\text{bare}}(y)\rangle_\varepsilon = \sqrt{c_I c_J} N^{\Delta^{(0)}} K(x,y)^{\Delta^{(0)}}$$
$$\times\left[\delta_{IJ}-g^2\left(\frac{1}{\varepsilon}+1+\gamma_E+\log(\pi|x-y|^2)\right)D_{IJ}^{(1)}+g^2 O(\varepsilon)+O(g^4)\right], \quad (8.60)$$

with

$$D_{IJ}^{(1)} = \frac{1}{\sqrt{c_I c_J}}\sum_{n=1}^{L}\left(2-2\mathbb{P}_{n,n+1}+\mathbb{K}_{n,n+1}\right)\left(\delta_{i_1 j_1}\delta_{i_2 j_2}\cdots\delta_{i_L j_L}+\text{cyclic perm.}\right). \quad (8.61)$$

This expression requires some explanation. First of all, we identify $L+1=1$. In the last factor, one should add cyclic permutation of the j_n indices relative to the i_n indices, e.g.

$$\delta_{i_1 j_1}\delta_{i_2 j_2}\delta_{i_3 j_3} + \text{cyclic permutations}$$
$$= \delta_{i_1 j_1}\delta_{i_2 j_2}\delta_{i_3 j_3} + \delta_{i_1 j_2}\delta_{i_2 j_3}\delta_{i_3 j_1} + \delta_{i_1 j_3}\delta_{i_2 j_1}\delta_{i_3 j_2}. \quad (8.62)$$

The sum in (8.61) is understood to be an operator acting on the Kronecker deltas. Specifically, $\mathbb{P}_{n,n+1}$ acts on the factors involving i_n and i_{n+1} as

$$\mathbb{P}_{n,n+1}\left(\cdots \delta_{i_n,j_m}\delta_{i_{n+1},j_{m+1}}\cdots\right) = \cdots \delta_{i_{n+1},j_m}\delta_{i_n,j_{m+1}}\cdots, \qquad (8.63)$$

leaving all other factors invariant. Similarly, the action of $\mathbb{K}_{n,n+1}$ is

$$\mathbb{K}_{n,n+1}\left(\cdots \delta_{i_n,j_m}\delta_{i_{n+1},j_{m+1}}\cdots\right) = \cdots \delta_{i_n,i_{n+1}}\delta_{j_m,j_{m+1}}\cdots. \qquad (8.64)$$

In this way, one generates the two non-trivial flavour structures we found in (8.45).

There is an important subsector of the $SO(6)$ sector called the $SU(2)$ sector. Here, we are only allowed to build operators using the two complex scalar fields X and Y, defined by

$$X = \phi_1 + i\phi_4, \qquad Y = \phi_2 + i\phi_5. \qquad (8.65)$$

The propagators look like

$$\langle [X]_{ab}(x)[\bar{X}]_{b'a'}(y)\rangle = 2\delta_{aa'}\delta_{bb'}K(x,y), \quad \langle [X]_{ab}[X]_{b'a'}\rangle = \langle [\bar{X}]_{ab}[\bar{X}]_{b'a'}\rangle = 0, \qquad (8.66)$$

and similarly for Y. Note the extra factor of two. The $SU(2)$ sector is closed to all loop orders, in contrast to the $SO(6)$ sector. It is easy to deduce the $SU(2)$ dilatation operator from the above computation. For $S = \{s_1, s_2, \ldots, s_L\}$ with $s_n = \uparrow, \downarrow$, let us define the bare operator

$$\mathcal{O}_S^{\text{bare}} = \text{tr}[\phi_{s_1}\cdots\phi_{s_L}], \qquad \phi_\uparrow = X, \quad \phi_\downarrow = Y. \qquad (8.67)$$

We then find

$$\langle \mathcal{O}^{\text{bare}}{}_S(x)\bar{\mathcal{O}}^{\text{bare}}_{S'}(y)\rangle_\varepsilon = \sqrt{c_S c_{S'}}(2N)^{\Delta^{(0)}} K(x,y)^{\Delta^{(0)}}$$
$$\times \left[\delta_{SS'} - g^2\left(\frac{1}{\varepsilon} + 1 + \gamma_E + \log(\pi|x-y|^2)\right)D_{SS'}^{(1)} + g^2 O(\varepsilon) + O(g^4)\right], \qquad (8.68)$$

with

$$D_{SS'}^{(1)} = \frac{2}{\sqrt{c_S c_{S'}}}\sum_{n=1}^{L}(1 - \mathbb{P}_{n,n+1})\left(\delta_{s_1,s'_1}\delta_{s_2,s'_2}\cdots\delta_{s_L,s'_L} + \text{cyclic perm.}\right). \qquad (8.69)$$

Here, $\mathbb{P}_{n,n+1}$ acts as in (8.63), but with s's instead of i's. Terms originating from the $SO(6)$ trace operator $\mathbb{K}$ are seen to be proportional to $1 + i^2 = 0$.

Check the details of the reduction to the $SU(2)$ sector.

Since $D_{SS'}^{(1)}$ only involves the permutation operator $\mathbb{P}$, it is clear that the two-point function is only non-zero between operators with the same number of X and Y fields. Using the $SO(6)$ symmetry, it can be shown that this holds to all orders in g. When diagonalizing the dilatation matrix, one can thus restrict to operators with a fixed number of X's and Y's.

Using two X's and two Y's, one can form two bare single-trace operators. Construct the corresponding (one-loop) renormalized operators, and show that the anomalous dimensions are $\Delta^{(1)} = 0$ and $\Delta^{(1)} = 12$.

8.2.2.1 Spin chains

Now that we have computed the one-loop dilatation operator $D^{(1)}$ in the planar limit, the natural next problem is to find its eigenvectors and eigenvalues. This is called the spectral problem. Let us consider operators of length L, so that their classical conformal dimension is $\Delta^{(0)} = L$. These operators form a vector space on which the dilatation operator acts. A general operator in this space is of the form

$$\mathcal{O} = \Psi^S \mathcal{O}_S^{\text{bare}}, \tag{8.70}$$

where the coefficient Ψ^S is invariant under cyclic permutations,

$$\Psi^{\{s_1, s_2, \ldots, s_L\}} = \Psi^{\{s_L, s_1, s_2, \ldots, s_{L-1}\}}. \tag{8.71}$$

Now Ψ can be seen as a vector in $\mathbb{C}^{2L}$ (more precisely the cyclically invariant subspace of $\mathbb{C}^{2L}$) since each index $s_i = \uparrow, \downarrow$ takes two values. In this language, $D_{SS'}^{(1)}$ actually defines an operator $H : \mathbb{C}^{2L} \to \mathbb{C}^{2L}$. To see this precisely, we note that the two-point function of two operators of the form (8.70) is

$$\langle \mathcal{O}_1(x) \bar{\mathcal{O}}_2(y) \rangle_\varepsilon = L(2N)^{\Delta^{(0)}} K(x,y)^{\Delta^{(0)}}$$
$$\times \langle \Psi_2 | 1 - g^2 \left(\frac{1}{\varepsilon} + 1 + \gamma_E + \log(\pi |x-y|^2) \right) H + g^2 O(\varepsilon) + O(g^4) | \Psi_1 \rangle. \tag{8.72}$$

Here, the inner product is the usual

$$\langle \Psi_2 | \Psi_1 \rangle = (\Psi_2^S)^* \Psi_1^S, \tag{8.73}$$

and H is given by

$$H = 2\sum_{i=1}^{L}(1 - \mathbb{P}_{i,i+1}), \tag{8.74}$$

where $\mathbb{P}_{i,i+1}$ now denotes the operator which permutes two neighbouring spins,

$$(\mathbb{P}_{i,i+1}\Psi)^{\{s_1,\ldots,s_L\}} = \Psi^{\{s_1,\ldots,s_{i-1},s_{i+1},s_i,s_{i+2},\ldots,s_L\}}. \tag{8.75}$$

As a slight abuse of notation, we will also denote $D^{(1)} \equiv H$.

> Derive (8.72) from (8.68). To get the combinatorial factors right, it is important to use that Ψ_1 and Ψ_2 are invariant under cyclic permutations.

The fact that $\mathcal{O}$ is an eigenstate of the dilatation operator then simply translates to Ψ being an eigenstate of $D^{(1)}$. The anomalous dimension of $\mathcal{O}$ is then the eigenvalue of Ψ. It is obvious that the wave function $\Psi^{\{s_1,\ldots,s_L\}} = \delta_{s_1\uparrow}\ldots\delta_{s_L\uparrow}$ corresponding to all the spins in the spin chain state pointing upwards is an eigenstate of $D^{(1)}$ with eigenvalue zero and correspondingly the operator $\mathcal{O}$ built entirely from X-fields is an eigenstate of the dilatation operator and has anomalous dimension equal to zero.[5] This operator constitutes an example of a BPS operator and its anomalous dimension vanishes to all loop orders.

The operator H is a sum of terms for which only neighbouring sites interact. For this reason, it is called a nearest-neighbour operator. We can rewrite (8.74) in a more familiar way by using the Pauli matrices

$$H = L - 4\sum_{i=1}^{L}\vec{S}_i \cdot \vec{S}_{i+1}, \qquad \vec{S} = \frac{1}{2}(\sigma_1,\sigma_2,\sigma_3), \tag{8.76}$$

where $\vec{S}_i$ is the spin operator acting on the i'th vector space. The operator H is the Hamiltonian of the well-known Heisenberg spin chain. This Hamiltonian is integrable, which means that its eigenvalues and eigenvectors can be found by the Bethe ansatz, as we will discuss in Section 8.3.

8.2.2.2 Other sectors

Here, we have derived the one-loop dilatation generator for operators belonging to the $SU(2)$ sector of $\mathcal{N} = 4$ SYM theory, but the one-loop dilatation generator can be determined for any kind of composite operator [19]. When diagonalizing the dilatation

---

[5] Obviously, the argument can be repeated with $\uparrow$ replaced by $\downarrow$ and X replaced by Y.

generator, however, one typically considers only a sub-sector which then must be closed under renormalization. The closed sectors were classified in [19]. Most studied are the closed sectors of rank one of which there exist three. The simplest of these is the $SU(2)$ sector which we have already treated. With a step up in complexity one finds the $SL(2)$ sector, which is built from a complex scalar X with arbitrary many covariant derivatives $D_{1\dot{1}}$ polarized in a single light-like direction. In contrast with the $SU(2)$ sector, the $SL(2)$ sector is non-compact. The third sector of rank one is the so-called $SU(1|1)$ sector, which contains a bosonic as well as a fermionic field. An important phenomenon which occurs starting from two-loop order is that of the dilatation generator introducing length changing of the operators on which it acts. The smallest sector that exhibits length changing is $SU(2|3)$. For a discussion of this sector, we refer to [20]. In contrast, the largest sector in which length is preserved is $PSU(1,1|2)$. This sector is treated in [21, 22].

8.3 The integrable Heisenberg spin chain in $\mathcal{N}=4$ supersymmetric Yang–Mills theory

8.3.1 One loop

A simple way of expressing the fact that the Heisenberg spin chain constitutes an integrable system is by stating that there exist L local conserved charges $Q_1, Q_2, \ldots, Q_L$ which fulfil

$$[Q_i, Q_j] = 0, \qquad i, j = 1, \ldots, L. \tag{8.77}$$

The charges can be organized so that Q_l involves interactions between l neighbouring spins. The first charge Q_1 can be taken as the total momentum of the spin chain, the conservation of which obviously follows from the translational invariance of the chain. The second charge Q_2 can be taken as the Hamiltonian. The third charge Q_3, which will play a distinguished role in the following, can be chosen as

$$Q_3 = \sum_{i=1}^{L} [H_{i,i+1}, H_{i+1,i+2}], \tag{8.78}$$

and there exists a certain boosting procedure which makes it possible to construct the remaining higher conserved charges [23, 24].

Another way of expressing the integrability of the Heisenberg spin chain is by stating that it can be solved by the algebraic Bethe ansatz approach, explained in Chapter 1 of this volume. In this approach one starts from a reference state

$$|0\rangle_L \equiv |\underbrace{\uparrow \ldots \uparrow}_{L}\rangle, \tag{8.79}$$

with all spins pointing upwards, say, and obtains the other highest-weight eigenstates by acting on the reference state with a number of creation operators $\hat{B}(u)$, each of which generates an appropriate linear combination of states where one spin-up has been replaced by a spin-down, thus lowering the total spin by one unit, i.e.

$$|\mathbf{u}\rangle = \hat{B}(u_1)\ldots\hat{B}(u_M)|0\rangle_L. \tag{8.80}$$

The creation operators depend on rapidity variables u_i. In order for $|\mathbf{u}\rangle$ to be an eigenstate of the Heisenberg spin chain, the rapidities have to fulfil the Bethe equations

$$\left(\frac{u_k + \frac{i}{2}}{u_k - \frac{i}{2}}\right)^L = \prod_{\substack{j=1 \\ j\neq k}}^{M} \frac{u_k - u_j + i}{u_k - u_j - i}. \tag{8.81}$$

The corresponding energy eigenvalue is

$$E = 2\sum_{i=1}^{M} \frac{1}{u_i^2 + \frac{1}{4}}. \tag{8.82}$$

The algebraic Bethe ansatz simultaneously diagonalizes all the conserved charges of the spin chain. In particular, the total momentum of an eigenstate is

$$P = \sum_{i=1}^{M} p_i, \quad \text{where} \quad u_i = \frac{1}{2}\cot\frac{p_i}{2}. \tag{8.83}$$

As likewise explained in Chapter 1 of this volume, the spin-chain eigenstates can also be found by the coordinate-space Bethe approach, which leads to the eigenstates being expressible as a sum over plane waves. More precisely, the highest-weight eigenstates with M spins flipped compared to the reference state in this approach take the following form:

$$|\vec{p}\rangle := |p_1,\ldots,p_M\rangle = \sum_{\sigma \in S_M} \sum_{1 \leq n_1 < \ldots < n_M \leq L} e^{i\sum_m (p_{\sigma m} n_m + \frac{1}{2}\sum_{j<m}\theta_{\sigma j \sigma m})} S_{n_1}^- \ldots S_{n_M}^- |0\rangle, \tag{8.84}$$

where the sum runs over all permutations and where the variables $p_1,\ldots,p_M$ now clearly have the interpretation of the lattice momenta of the M excitations (flipped spins). Moreover, up to an overall phase, the wave function (8.84) only depends on the two-body S-matrix of the system,

$$S_{ij} := e^{\theta_{ij} - \theta_{ji}} = -\frac{1 + e^{ip_i + ip_j} - 2e^{ip_i}}{1 + e^{ip_i + ip_j} - 2e^{ip_j}}. \tag{8.85}$$

We are actually free to multiply (8.84) with an arbitrary phase since this neither affects the spectrum nor the orthonormality of the Bethe vectors. We will fix the phase when we discuss one-point functions in Section 8.4.

The momentum variables are related to the rapidity variables as in (8.83), but notice that the states $|\vec{p}\rangle$ and $|\mathbf{u}\rangle$ are not identical but only proportional to each other. The exact factor of proportionality was worked out in [25]. Whereas translational invariance tells us that the total momentum constitutes a good quantum number, the cyclicity of the single-trace operators tells us that this quantum number has to be a multiple of 2π for a Bethe state to qualify as a gauge-theory operator, i.e.

$$P = \sum_{k=1}^{M} p_k = 2\pi n, \quad \text{or} \quad \prod_{k=1}^{M} \left(\frac{u_k + \frac{i}{2}}{u_k - \frac{i}{2}}\right) = 1. \tag{8.86}$$

In what follows, we will need the norm of a Bethe state (8.84). There is an elegant closed expression of determinant type due to Gaudin [10], see also [26]. Let us rewrite the Bethe equations (8.81) and introduce the function Φ as their logarithm

$$1 = \left(\frac{u_k - \frac{i}{2}}{u_k + \frac{i}{2}}\right)^L \prod_{\substack{j=1 \\ j \neq k}}^{M} \frac{u_k - u_j + i}{u_k - u_j - i} \equiv \exp[i\Phi_k]. \tag{8.87}$$

Then, the norm is given in terms of the Jacobian matrix $G_{ij} = \partial_{u_i} \Phi_j$:

$$\langle \mathbf{u} | \mathbf{u} \rangle = \left[\prod_{i=1}^{M} \left(u_i^2 + \frac{1}{4}\right)\right] \det G. \tag{8.88}$$

As mentioned previously, the norm formula depends on the type of Bethe ansatz used and it will look different for the algebraic Bethe ansatz.

Now that we have described the eigenstates of the Heisenberg spin-chain Hamiltonian, let us spell out the identification between spin-chain states and field-theory operators explicitly. Let $|\mathbf{u}\rangle$ be a Bethe state, then we can write

$$\mathcal{O} \equiv \left(\frac{1}{2g}\right)^L \frac{\mathcal{Z}}{\sqrt{L}} \frac{\operatorname{tr} \prod_{l=1}^{L} (\langle \uparrow_l | \otimes X + \langle \downarrow_l | \otimes Y) |\mathbf{u}\rangle}{\sqrt{\langle \mathbf{u} | \mathbf{u} \rangle}}, \tag{8.89}$$

where we have normalized the operators such that $M_{ij} = \delta_{ij}$. The explicit normalization factor is most easily derived from (8.72) (since $|\mathbf{u}\rangle$ is an eigenstate, we effectively have $H \to \Delta^{(1)}$). Requiring that the only one-loop correction to the two-point function is $\propto \log(x-y)^2$ (we work in units where $\mu = 1$) fixes the renormalization constant to

$$\mathcal{Z} = 1 + g^2 \frac{\Delta^{(1)}}{2} \left(\frac{1}{\varepsilon} + 1 + \gamma_E + \log \pi\right) + O(g^4). \tag{8.90}$$

As an example, let us work out (8.89) for the Bethe states of length 2. The numerator of (8.89) becomes

$$\langle\uparrow\uparrow|\mathbf{u}\rangle\operatorname{tr}[XX]+\langle\uparrow\downarrow|\mathbf{u}\rangle\operatorname{tr}[XY]+\langle\downarrow\uparrow|\mathbf{u}\rangle\operatorname{tr}[YX]+\langle\downarrow\downarrow|\mathbf{u}\rangle\operatorname{tr}[YY]. \tag{8.91}$$

Then, for example, for $|\mathbf{u}\rangle = |0\rangle$ the corresponding operator becomes

$$\mathcal{O}_{|0\rangle} \equiv \frac{1}{4\sqrt{2}g^2}\operatorname{tr}[X^2], \tag{8.92}$$

which has a unit-normalized two-point function as can be seen from (8.68).

Let us define a parity operation $\mathcal{P}$ which acts on single-trace operators by inverting the orders of the fields inside the trace, i.e.

$$\mathcal{P}\cdot\operatorname{tr}[\phi_{i_1}\phi_{i_2}\cdots\phi_{i_L}] = \operatorname{tr}[\phi_{i_L}\phi_{i_{L-1}}\cdots\phi_{i_1}]. \tag{8.93}$$

Obviously, the Heisenberg Hamiltonian commutes with the parity operation. This means that the spin-chain eigenstates can be chosen to be parity eigenstates as well. However, the spin-chain eigenstates generated by the algebraic Bethe ansatz are not parity eigenstates as parity anti-commutes with all the odd charges, in particular Q_3, i.e.

$$[H, \mathcal{P}] = 0, \quad \{Q_3, \mathcal{P}\} = 0. \tag{8.94}$$

As usual, the parity operation changes the sign of all momenta and hence the sign of of all rapidities and it squares to the identity. Let us denote by $|-\mathbf{u}\rangle$ the state given by the right-hand side of (8.80) with each u_i being replaced by $-u_i$. Then, $\mathcal{P}|\mathbf{u}\rangle$ can differ from $|-\mathbf{u}\rangle$ by at most a phase factor. Since the Bethe equations (8.81), the cyclicity constraint (8.86), and the expression for the energy (8.82) are all invariant under $u_i \to -u_i$, the state $|-\mathbf{u}\rangle$ is again a cyclically invariant eigenstate of $\mathcal{H}$ with the same eigenvalue as $|\mathbf{u}\rangle$. It thus follows that the eigenstates of $\mathcal{H}$ can be separated into unpaired states for which $|\mathbf{u}\rangle = |-\mathbf{u}\rangle$ and paired states $(|\mathbf{u}\rangle, |-\mathbf{u}\rangle)$ for which $|\mathbf{u}\rangle \neq |-\mathbf{u}\rangle$. The unpaired states will play a distinguished role in Section 8.4. These states are parity eigenstates and can be shown to fulfil

$$\mathcal{P}|\mathbf{u}\rangle_{\text{unpaired}} = (-1)^{M(L+1)}|\mathbf{u}\rangle_{\text{unpaired}}, \quad Q_3|\mathbf{u}\rangle_{\text{unpaired}} = 0. \tag{8.95}$$

The degenerate states in a parity pair are not states of a definite parity but can be combined into parity eigenstates, so-called parity pairs $(|+\rangle, |-\rangle)$, in the following way:

$$|+\rangle = |\mathbf{u}\rangle + |-\mathbf{u}\rangle, \quad |-\rangle = |\mathbf{u}\rangle - |-\mathbf{u}\rangle, \tag{8.96}$$

where

$$\mathcal{P}|+\rangle = (-1)^{M(L+1)}|+\rangle, \quad \mathcal{P}|-\rangle = -(-1)^{M(L+1)}|-\rangle, \tag{8.97}$$

and

$$Q_3|+\rangle \propto |-\rangle, \quad Q_3|-\rangle \propto |+\rangle. \tag{8.98}$$

8.3.2 Higher loop orders

By doing explicit higher-loop computations following the same strategy as in Section 8.2.2, one can likewise derive a perturbative expression for the dilatation operator, i.e.

$$D = \sum_{n=0}^{\infty} g^{2n} D^{(n)}. \tag{8.99}$$

For the two-loop contribution, one finds [27]

$$D^{(2)} = -2\sum_{i=1}^{L} (\mathbb{P}_{i,i+2} - 1) + 8\sum_{i=1}^{L} (\mathbb{P}_{i,i+1} - 1). \tag{8.100}$$

We notice, in particular, that the anomalous dimension of the BPS operator $\text{tr} X^L$ stays zero at two-loop order as expected. When one diagonalized the dilatation operator including this correction term initially (by brute force), one observed that to order g^2 the spectrum still contained the same number of pairs of degenerate eigenstates with opposite parity. This fact was viewed as a smoking gun of higher-loop integrability as it hinted at the continued existence of a conserved third charge [27]. Indeed, it is possible to perturbatively modify the third and the higher order charges by terms of order g^2, i.e.

$$Q_i = Q_i^{(0)} + g^2 Q_i^{(1)}, \tag{8.101}$$

in such a way that the quantum corrected charges commute up to terms of order g^4:

$$[Q_i, Q_j] = O(g^4). \tag{8.102}$$

The same idea can be pursued at general loop order ℓ where one would have

$$Q_i = Q_i^{(0)} + g^2 Q_i^{(1)} + g^4 Q_i^{(2)} + \ldots + g^{2\ell} Q_i^{(\ell)}, \quad [Q_i, Q_j] = O(g^{(2\ell+2)}), \tag{8.103}$$

and one would denote the system as being perturbatively integrable. Here, the correction $Q_i^{(\ell)}$ is an operator which involves $i+\ell$ neighbouring spins. Concretely, the idea has been implemented up to four-loop order [28]. The algebraic Bethe ansatz approach does not apply to the quantum corrected system, where the interaction is no longer of nearest-neighbour type. Nevertheless, a modified, so-called asymptotic Bethe ansatz exists. It has been argued for in a long series of papers where the focus was shifted from the Hamiltonian and the conserved charges to the two-body scattering matrix of the theory [29], and the calculational effort was shifted from brute-force field-theoretical computations to symmetry considerations. The asymptotic Bethe equations read [30]

$$\left(\frac{x(u_k+\frac{i}{2})}{x(u_k-\frac{i}{2})}\right)^L = \prod_{\substack{j=1\\j\neq k}}^{M} \frac{u_k - u_j + i}{u_k - u_j - i} \exp(2i\theta(u_k, u_j)), \qquad (8.104)$$

where $\theta(u_k, u_j)$ is denoted as the dressing phase and is explicitly known [31]. The Zhukovski variable $x(u)$ is defined via

$$u = x + \frac{g^2}{x}. \qquad (8.105)$$

Furthermore, the cyclicity condition now reads

$$\prod_{k=1}^{M}\left(\frac{x(u_k+\frac{i}{2})}{x(u_k-\frac{i}{2})}\right) = 1, \qquad (8.106)$$

and the expression for the energy eigenvalues is modified to

$$E = \sum_{j=1}^{M} i\left(\frac{1}{x(u_j+\frac{i}{2})} - \frac{1}{x(u_j-\frac{i}{2})}\right). \qquad (8.107)$$

The dressing phase only plays a role at four-loop order and beyond. The Bethe equations (8.104) are asymptotic in the sense that they are only valid when expanded perturbatively to a given order n in g^2 for operators whose length is smaller than or equal to n. If this criterion is not fulfilled, one has to take into account wrapping corrections which, as indicated by their name, are corrections which occur when the spin-chain interaction wraps once or more around the operator [22], see Figure 8.2.

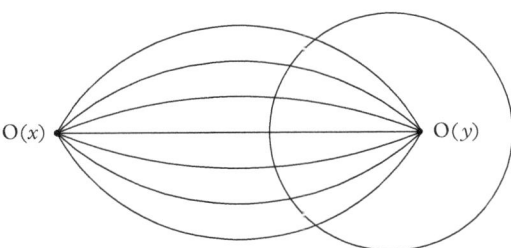

Figure 8.2 *Example of an interaction wrapping once around the operator.*

The asymptotic Bethe equations give access to the spectrum at higher loops but not to the corresponding wave functions. Wave functions at higher loop orders can be generated by a technique known as Θ-morphism [33].

8.3.3 Beyond the planar limit

The derivation in Section 8.2.2 can be generalized to give the full one-loop dilatation operator and not only its large-N limit. Including all terms, the action of the one-loop dilatation operator in the $SU(2)$ sector can be expressed in terms of an effective vertex, acting on an operator [34]:

$$V = \sum_{n=0}^{\infty} g^{2n} V^{(n)}. \tag{8.108}$$

The one-loop contribution reads

$$V^{(1)} = -\frac{2}{N} : \text{tr}[X, Y][\check{X}, \check{Y}] :, \quad (\check{X})_{ab} = \frac{\delta}{\delta X_{ab}}, \tag{8.109}$$

where the normal ordering symbol signifies that the derivatives are not allowed to act on fields belonging to the effective vertex itself. Going beyond the large-N limit, one cannot restrict oneself to single-trace operators but has to consider also multi-trace since the action of the dilatation operator now leads to splitting and joining of traces, as illustrated by the following example. Notice that we only show one out of four terms contributing to the dilatation generator and only one possible way of applying the derivatives:

$$\text{tr}(XY\check{X}\check{Y}) \cdot \text{tr}(YXYYX)\,\text{tr}(YX) = \text{tr}(XY\check{X}XYYX)\,\text{tr}(YX)$$

$$= N\text{tr}(XYYYX)\,\text{tr}(YX) + \text{tr}(XY)\,\text{tr}(XYY)\,\text{tr}(YX) + \text{tr}(XYXXXYYX).$$

From this example, it should be clear that we can decompose the vertex representing the full one-loop dilatation operator for finite N in the following way:

$$V^{(1)} = D^{(1)} + \frac{1}{N} D_{+}^{(1)} + \frac{1}{N} D_{-}^{(1)}, \tag{8.110}$$

where $D^{(1)}$, which was given in (8.74), conserves the number of traces, $D_{+}^{(1)}$ increases the trace number by one, and $D_{-}^{(1)}$ reduces the trace number by one. In the language of spin chains, $D_{+}^{(1)}$ splits a chain into two parts while $D_{-}^{(1)}$ joins two chains into one. In a similar manner, the full non-planar two-loop contribution to the dilatation operator can be expressed in terms of an effective vertex as [27]

$$V^{(2)} = D^{(2)} + \frac{1}{N}D^{(2)}_+ + \frac{1}{N}D^{(2)}_- + \frac{1}{N^2}D^{(2)}_{++} + \frac{1}{N^2}D^{(2)}_{--}, \tag{8.111}$$

where $D^{(2)}$ was given in (8.100) and where $D^{(2)}_{++}$ increases the trace number by two and $D^{(2)}_{--}$ reduces the trace number by two.

> By applying the effective vertex (8.109) to an operator from the $SU(2)$ sector, show that it reduces to the Hamiltonian of the Heisenberg spin chain in the limit $N \to \infty$.

The splitting and joining of traces or spin chains constitute highly non-local interactions for which the traditional tools of integrability are not applicable. One naive thing that one can do is to consider a finite, closed set of multi-trace operators of a given length L and with a given number of excitations, M, and diagonalize the dilatation operator including its non-planar terms by brute force in this subspace. Alternatively, one can at a slightly more advanced level start by diagonalizing the planar part of the Hamiltonian, still in a finite, closed set of multi-trace operators, treat the $\frac{1}{N}$ terms in the dilatation operator as a perturbation, and do quantum mechanical perturbation theory in $\frac{1}{N}$. From these types of simple analyses, there are few things that one can learn [27]. Firstly, one observes that the degeneracy between the planar parity pairs (the $(|+\rangle, |-\rangle)$ states) gets lifted when $\frac{1}{N}$ corrections are taken into account. Thus, the smoking gun of integrability is no longer present. Furthermore, for states which are degenerate at the planar level, such as the planar parity pairs, the leading non-planar correction to the energy behaves as $\frac{1}{N}$, whereas in the generic case the first non-planar correction behaves as $\frac{1}{N^2}$. This is a simple consequence of quantum mechanical perturbation theory. Finally, starting at the two-loop level, one finds that there are states which do not have a well-defined double expansion in g^2 and $\frac{1}{N}$.

Aiming at going beyond the planar level in a more systematic approach, a convenient basis of $SU(2)$ operators might be the so-called restricted Schur polynomials, which constitute a basis of multi-trace operators that are orthogonal for finite N. Studying the action of the one- and two-loop dilatation operator in this basis of operators, it is possible by imposing various limits on top of the large-N limit to find an integrable subsystem which, however, looks like a set of decoupled harmonic oscillators [35].

Another direction of investigation, which can be viewed as the first step in going beyond the planar level, is the study of three-point functions, as these can be seen as building blocks for non-planar correlation functions. Whereas in the study of the spectral problem of $\mathcal{N} = 4$ SYM theory one only needs the eigenvalues of the spin-chain Hamiltonian, in the study of the theory's three-point functions the explicit form of the spin-chain eigenfunctions plays a crucial role. The study of three-point functions has recently been boosted by the development of the so-called hexagon-program, which is covered in Chapter 10 in this volume.

8.4 $\mathcal{N}=4$ supersymmetric Yang–Mills theory with a defect, one-point functions, and integrability

Rather than computing three-point functions in $\mathcal{N}=4$ SYM theory, we will focus on a related problem which also requires the knowledge of the explicit form of the Bethe wave functions. We will compute the one-point functions in a certain defect version of $\mathcal{N}=4$ SYM theory.

8.4.1 $\mathcal{N}=4$ supersymmetric Yang–Mills theory with a defect

There exists a certain defect version of $\mathcal{N}=4$ SYM theory for which half of the supersymmetries are preserved and for which a holographic dual exists. In this theory, a codimension-one defect is positioned at $x_3 = 0$ and divides space into two regions, $x_3 > 0$ and $x_3 < 0$. In the bulk, one still has $\mathcal{N}=4$ SYM theory but with different gauge groups on the two sides of the defect. The gauge group for $x_3 < 0$ is $U(N-k)$, while the gauge group for $x_3 > 0$ is $U(N)$, see Figure 8.3. The $U(N)$ symmetry for $x_3 > 0$, however, is broken by some of the scalar fields acquiring a non-trivial vacuum expectation value (vev) so that the gauge symmetry there is also effectively $U(N-k)$. Because of the non-vanishing vevs, one-point functions can be non-trivial on one side of the defect already at tree level. In addition to the usual action of $\mathcal{N}=4$ SYM theory, the system has a three-dimensional action involving fields that are confined to the defect. These defect fields have self-interactions as well as interactions with the bulk fields of $\mathcal{N}=4$ SYM theory [36, 37]. In the remainder of the chapter, we will only work at tree level and at one-loop order where the defect field theory does not come into play.

8.4.1.1 Symmetries

Introducing the codimension-one defect at $x_3 = 0$ breaks several of the original symmetries of $\mathcal{N}=4$ SYM theory discussed in Section 8.2.1.2. To start with, let us analyse the minimal possible consequences of introducing the defect. The condition $x_3 = 0$ is preserved by the translations $P_{\hat{\mu}}$, $\hat{\mu} = 0, 1, 2$, but not by P_3. Similarly, the Lorentz transformations $M_{\hat{\mu}\hat{\nu}}$ preserve $x_3 = 0$, but $M_{\hat{\mu}3} = -M_{3\hat{\mu}}$ does not. The four-dimensional Poincaré symmetry is thus reduced to three-dimensional Poincaré symmetry. A scale

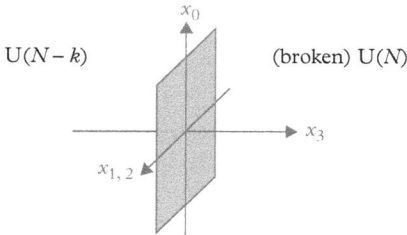

Figure 8.3 *The defect theory.*

transformation D preserves $x_3 = 0$ and so do the special conformal transformations $K_{\hat{\mu}}$ but not K_3. The four-dimensional conformal group $SO(4,2) \simeq SU(2,2)$ is thus reduced to the three-dimensional conformal group $SO(3,2) \simeq Sp(4)$.

While this analysis is straightforward in vector indices, let us now redo it in spinor indices as a preparation for understanding the influence on supersymmetry. For convenience, we will assume here that the defect is at $x_2 = 0$ instead of $x_3 = 0$.

Work out the similarity transformation that relates these cases.

Via the Pauli matrices $\sigma^{\mu}_{\alpha\dot{\alpha}}$, the Lorentz vector P_μ is translated to a 2×2 matrix. The condition that the component P_2 vanishes then translates to the matrix being symmetric. We thus have to determine the Lorentz transformations that yield symmetric matrices when applied to symmetric matrices. They are given by $\hat{L}^{\alpha}{}_{\beta} = L^{\alpha}{}_{\beta} + \dot{L}^{\dot{\alpha}}{}_{\dot{\beta}}$.[6] This explicitly shows how the four-dimensional Lorentz group $SO(1,3) \simeq SU(2)_L \times SU(2)_R$ is reduced to the three-dimensional Lorentz group $SO(1,2) \simeq SU(2)$.

Recalling that supercharges anticommute to translations (8.7), some of the supersymmetry is necessarily broken as well. As the supercharges have spinor indices, we now benefit from the previous analysis in spinor indices. Firstly, we observe that the preserved supercharges have to be spinors of $\hat{L}$. Secondly, they have to anticommute to a symmetric matrix in the spinor indices. This leads to half of the supercharges being preserved, namely $\hat{Q}^A_{\alpha} = Q^A_{\alpha} + \dot{Q}^A_{\dot{\alpha}}$. From this choice, we see that the anti-commutator of the $\hat{Q}$s is proportional to a symmetrized version of the momentum P, which does not contain P_2. In the same way, only half of the superconformal charges preserve $x_2 = 0$, as they anticommute to special conformal transformations. The preserved supercharges are manifestly real. Thus, the R-symmetry group $SU(4) \simeq SO(6)$ that acts on them is reduced to $SO(4) \simeq SO(3) \times SO(3)$. In total, the superconformal group $PSU(2,2|4)$ of $\mathcal{N} = 4$ SYM theory is thus reduced to $OSP(4|4)$.

So far, we have only considered the minimal effect of introducing a codimension-one defect into $\mathcal{N} = 4$ SYM theory. Depending on which fields occur on the defect and how they interact among themselves and with the fields of $\mathcal{N} = 4$ SYM theory, also more symmetry could be broken. However, there does indeed exist a defect action such that the (quantum) theory preserves $OSP(4|4)$, at least for $k = 0$ [36].

8.4.1.2 Correlation functions

The correlation functions in a CFT with a boundary or a codimension-one defect are less restricted than for a usual CFT. This is due to the fact that a defect breaks part of the conformal symmetry, as just discussed. Already one-point functions of composite operators $\mathcal{O}_i$ can be non-vanishing. The remaining conformal symmetry and the scaling dimension Δ_i of the operator fix the one-point functions up to a constant a_i [38]:

---

[6] Here, we identify dotted and undotted indices, so for example $\hat{L}^1{}_2 = L^1{}_2 + \dot{L}^{\dot{1}}{}_{\dot{2}}$.

$$\langle \mathcal{O}_i(x) \rangle = \frac{a_i}{x_3^{\Delta_i}}. \tag{8.112}$$

We see that one-point functions in a dCFT exhibit a complexity similar to three-point functions in a CFT. Two-point functions in a dCFT can be non-vanishing also for operators of unequal scaling dimensions and are fixed to be of the form

$$\langle \mathcal{O}_i(x)\mathcal{O}_j(y) \rangle = \frac{f(\xi)}{x_3^{\Delta_i} y_3^{\Delta_j}}, \tag{8.113}$$

where $f(\xi)$ is a function of the conformal ratio $\xi = \frac{|x-y|^2}{4x_3 y_3}$.

Finally, all correlation functions in a dCFT should reduce to the corresponding correlation functions in the absence of the defect if the distance to the defect is large compared to the distance between the insertion points.

8.4.1.3 Vacuum expectation values

For our specific model, the vacuum expectation values that the scalar fields pick up are described by $SU(2)$ representations [39]. More precisely, for $x_3 > 0$,

$$\phi_i^{\text{cl}} = -\frac{1}{x_3} \begin{pmatrix} (t_i)_{k \times k} & 0_{k \times (N-k)} \\ 0_{(N-k) \times k} & 0_{(N-k) \times (N-k)} \end{pmatrix}, \qquad i = 1, 2, 3, \tag{8.114}$$

$$\phi_i^{\text{cl}} = 0, \qquad i = 4, 5, 6, \tag{8.115}$$

where the three $k \times k$ matrices $t_i, i = 1, 2, 3$ constitute a k-dimensional unitary, irreducible representation of $SU(2)$; in particular,

$$[t_i, t_j] = i\varepsilon_{ijk} t_k. \tag{8.116}$$

For $x_3 < 0$, all classical fields are vanishing.

8.4.1.4 Representation of the algebra of $SU(2)$

To be explicit, let us spell out the k-dimensional irreducible representation. Introduce the standard $k \times k$ matrix unities $E^i{}_j$ that are zero everywhere except for a 1 at position (i,j). These matrices satisfy the relation $E^i{}_j E^k{}_l = \delta^k{}_j E^i{}_l$.

Next, we consider the following constants:

$$c_i = \sqrt{i(k-i)}, \qquad d_i = \frac{1}{2}(k - 2i + 1), \tag{8.117}$$

together with the following matrices:

$$t_+ = \sum_{i=1}^{k-1} c_i E^i{}_{i+1}, \qquad t_- = \sum_{i=1}^{k-1} c_i E^{i+1}{}_i. \qquad (8.118)$$

The usual k-dimensional $SU(2)$ representation is then given by

$$t_1 = \frac{t_+ + t_-}{2}, \qquad t_2 = \frac{t_+ - t_-}{2i}, \qquad t_3 = \sum_{i=1}^{k} d_i E^i{}_i. \qquad (8.119)$$

It is easy to check that these matrices satisfy the commutation relations (8.116). For the important special case $k=2$, the representation matrices are multiples of the Pauli matrices: $t_i|_{k=2} = \frac{1}{2}\sigma_i$.

8.4.2 Tree-level one-point functions in the $SU(2)$ sector

At tree level, the one-point function is obtained by inserting the classical solution (8.114) in an operator, see Figure 8.4. Clearly, only operators consisting solely of scalar operators can have a non-zero one-point function. In what follows, we will work in the planar limit so that we can apply the integrability techniques that were previously discussed. This means that we restrict to single-trace operators of the form

$$\mathcal{O} = \Psi^{i_1 \ldots i_L} \mathrm{tr}(\phi_{i_1} \ldots \phi_{i_L}). \qquad (8.120)$$

Inserting (8.114) into such an operator $\mathcal{O}$ then gives us at tree level

$$\langle \mathcal{O} \rangle^{cl} = (-1)^L \Psi^{i_1 \ldots i_L} \frac{\mathrm{tr}(t_{i_1} \ldots t_{i_L})}{x_3^L}. \qquad (8.121)$$

For any given operator, the above expression can straightforwardly be evaluated. However, this is hardly a constructive approach. For instance, to compute the one-point function of a scalar operator corresponding to a Bethe state, we would have to write

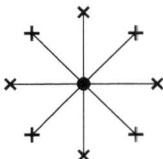

Figure 8.4 *The tree-level one-point function is obtained by inserting the classical solution into the operator. The operator is depicted by a dot and the insertion of the classical solution by a line with a cross at the end.*

384 Introduction to integrability and one-point functions

out its explicit wave function. Instead, we will now derive by integrability techniques a closed formula for $\langle \mathcal{O} \rangle^{cl}$ which is expressed entirely in terms of the length, the number of excitations, and the specific Bethe roots characterizing the operator.

8.4.2.1 The matrix product state

From now on, let us restrict to the $SU(2)$ sector. The first step to a more systematic approach is the realization that (8.121) can be written as an inner product between the Bethe state $|\mathbf{u}\rangle$ corresponding to our operator via (8.89) and a so-called matrix product state (MPS):

$$|\text{MPS}\rangle_L = \operatorname{tr} \prod_{n=1}^{L} [t_1 \otimes |\uparrow\rangle_n + t_2 \otimes |\downarrow\rangle_n]. \tag{8.122}$$

The subscript n stands for the usual embedding in the L-fold tensor product, while the trace is as usual in colour space. The MPS depends on the length L of the spin chain that we are considering, but in order to avoid cumbersome notation, we will from now on omit the subscript L.

Compute the MPS for $L = 2, 4$.

Using the explicit relation between the Bethe states and the field-theory operators (8.89), the problem of computing a one-point function then reduces to computing the following quantity:

$$\langle \mathcal{O} \rangle^{cl} = (-1)^L \left(\frac{1}{2g} \right)^L \frac{\mathcal{Z}}{\sqrt{L}} \frac{C_k}{x_3^L}, \qquad C_k = \frac{\langle \text{MPS} | \mathbf{u} \rangle}{\sqrt{\langle \mathbf{u} | \mathbf{u} \rangle}}, \tag{8.123}$$

where the various proportionality factors ensure that the operator is properly normalized.

There is the important subtlety that C_k is only defined up to a phase. In our identification of the field-theory operator and Bethe state, we can always insert an additional phase factor. This obviously leaves the two-point function invariant, but it will affect the overlap with the MPS. In order to fix this ambiguity, we will always choose the overall phase such that C_k is real and positive.

8.4.2.2 Generalities

It is easy to show that C_k is only non-vanishing if both L and M are even, where M is the number of excitations or equivalently the number of Bethe roots. Namely, the Lie algebra of $SU(2)$ admits an isomorphism where two of the t's are mapped to $-t$. This isomorphism is realized by a similarity transformation which leaves the MPS invariant due to cyclicity of the trace. For example, consider the case when $(t_1, t_2, t_3) \to (-t_1, -t_2, t_3)$. This immediately implies that

$$\langle \mathrm{MPS}|\mathbf{u}\rangle = (-1)^L \langle \mathrm{MPS}|\mathbf{u}\rangle, \tag{8.124}$$

which means that L has to be even. Similarly, it follows that M has to be even. For details, we refer to [40].

Apart from these restrictions on the quantum numbers, for a non-zero overlap with the MPS we also need some restrictions on the Bethe roots. It is easy to see that the MPS is parity even, so only parity even Bethe states can have a non-trivial overlap with the MPS. Finally, it can be shown that Q_3 and all the higher odd charges annihilate the MPS [40, 82]. From (8.98), we then see that the only possible states that have non-vanishing one-point functions are states that satisfy $|\mathbf{u}\rangle = |-\mathbf{u}\rangle$.

8.4.2.3 Vacuum

The first state to consider is the ferromagnetic vacuum (8.79), which corresponds to the operator $\mathrm{tr} X^L$. Its one-point function is given by

$$C_k = \frac{\langle \mathrm{MPS}|0\rangle}{\sqrt{\langle 0|0\rangle}} = \mathrm{tr}(t_1^L) = \sum_{i=1}^{k} d_i^L = -2\frac{B_{L+1}(\frac{1-k}{2})}{L+1}, \tag{8.125}$$

where d_i are the coefficients defining the $SU(2)$ representation (8.117) and B_{L+1} is the Bernoulli polynomial with index $L+1$. We see that the one-point function is a polynomial in k of degree $L+1$.

8.4.2.4 One-point functions for k=2

The simplest case that we can consider for $M > 0$ is the case $k = 2$. This actually turns out to be a fundamental building block for the general k case. For $k = 2$, the t-matrices are simple multiples of the Pauli matrices: $t_i = \frac{1}{2}\sigma_i$. They satisfy the following relations:

$$t_i^2 = \frac{1}{4}, \qquad t_i t_j = -t_j t_i \quad \text{for} \quad i \neq j. \tag{8.126}$$

This means that the inner product of the MPS with a Bethe state (8.84) dramatically simplifies. In particular, any trace factor can be easily evaluated:

$$\mathrm{tr}(t_1^{n_1-1} t_2 t_1^{n_2-n_1-1} t_2 \ldots) = (-1)^{n_1+n_2+\cdots} \mathrm{tr}(t_1^{L-M} t_2^M) = 2^{1-L}(-1)^{n_1+n_2+\cdots}. \tag{8.127}$$

Thus, the inner product of a Bethe state with the MPS takes the following form:

$$\langle \mathrm{MPS}|\mathbf{u}\rangle = 2^{1-L} \sum_{\sigma \in S_M} \sum_{n_i} e^{i\sum_m (p_{\sigma m} n_m + \frac{1}{2}\sum_{j<m} \theta_{\sigma j \sigma m})} (-1)^{n_1+\ldots+n_M}. \tag{8.128}$$

For two particles, compute the overlap $\langle \text{MPS}|u,-u\rangle$ and show that upon using the Bethe equations (8.81) it becomes

$$\langle \text{MPS}|u,-u\rangle = \frac{L}{2^{L-1}}\sqrt{\frac{u^2+\frac{1}{4}}{u^2}}. \tag{8.129}$$

Also show that $\langle u,-u|u,-u\rangle = L(L-1)$ such that $C_2 = \frac{1}{2^{L-1}}\sqrt{\frac{u^2+\frac{1}{4}}{u^2}}\sqrt{\frac{L}{L-1}}$.

To describe the overlap for a general number of excitations M, we introduce the following function:

$$K_{ij} := \frac{1}{2}\left[\frac{1+4u_i^2}{1+(u_i+u_j)^2} + \frac{1+4u_i^2}{1+(u_i-u_j)^2}\right], \tag{8.130}$$

and the following $\frac{M}{2} \times \frac{M}{2}$ matrix:

$$F_{ij} := \left(L - \sum_{n=1}^{M/2} K_{in}\right)\delta_{ij} + K_{ij}. \tag{8.131}$$

The overlap is then given by

$$\langle \text{MPS}|\mathbf{u}\rangle_{k=2} = 2^{1-L}(\det F)\sqrt{\prod_{i=1}^{M/2}\frac{u_i^2+\frac{1}{4}}{u_i^2}}. \tag{8.132}$$

In order to finally obtain the one-point function C_2, we need to divide by the norm of the Bethe state (8.88). For states with paired rapidities $|\mathbf{u}\rangle = |-\mathbf{u}\rangle$, the norm formula factorizes. Let us order the roots as $\{u_1,\ldots,u_{\frac{M}{2}},-u_1,\ldots,-u_{\frac{M}{2}}\}$ and introduce the following $\frac{M}{2} \times \frac{M}{2}$ dimensional matrices $G_\pm$:

$$G_\pm = \partial_{u_m}\Phi_n \pm \partial_{u_{m+\frac{M}{2}}}\Phi_n, \tag{8.133}$$

then $\det G = \det G_+ \det G_-$. In terms of these matrices, the one-point function for $k=2$ can finally be written as

$$C_2 = 2^{1-L}\sqrt{\frac{Q(\frac{i}{2})}{Q(0)}}\sqrt{\frac{\det G_+}{\det G_-}}, \tag{8.134}$$

where $Q(u) = \prod_{i=1}^{M}(u - u_i)$ is the Baxter polynomial. This means in particular that $\langle \text{MPS}|\mathbf{u}\rangle_{k=2} \sim \det G_+$. It is an interesting open question whether there is a state $|A\rangle$ such that $\langle A|\mathbf{u}\rangle_{k=2} \sim \det G_-$.

A formula of the same type as (8.132) has been obtained for the eigenstates of the $SU(3)$ spin chain [41], and this result has found application in the study of quantum quenches [42]. Furthermore, the expression (134) can be generalised to tree-level one-point functions of the $SU(3)$ sector of $N = 4$ SYM theory [41]. We note, however, that the $SU(3)$ sector is a closed sector only to one-loop order. Recently, both formulas have been generalized to the $SO(6)$ case as well [82].

8.4.2.5 Néel state

There is actually an interesting relation of one-point functions C_2 to the condensed-matter literature. It turns out that the MPS is cohomologically equivalent to the so-called Néel state,

$$|\text{Néel}\rangle = |\uparrow\downarrow\uparrow\downarrow\ldots\rangle + |\downarrow\uparrow\downarrow\uparrow\ldots\rangle. \tag{8.135}$$

The Néel state is a state at half-filling, i.e. it has $M = L/2$. It can be shown [40] that

$$2^L \left(\frac{i}{2}\right)^M |\text{MPS}\rangle\bigg|_{M=L/2} = |\text{Néel}\rangle + S^-|\ldots\rangle. \tag{8.136}$$

One of the remarkable properties of the Bethe ansatz is that the Bethe states are highest-weight states. This means that $S^+|\mathbf{u}\rangle = 0$ and thus for any Bethe state with $M = L/2$ the overlap of the MPS is the same as the overlap of the Bethe state with the Néel state, i.e. $2^{L-M}i^M\langle\mathbf{u}|\text{MPS}\rangle = \langle\mathbf{u}|\text{Néel}\rangle$. This is an overlap that has been studied in the condensed-matter literature [43].

This interesting relationship can be extended to general excitation numbers. Let $M = L/2 - 2m$, then

$$2^L \left(\frac{i}{2}\right)^M (2m)! |\text{MPS}\rangle\bigg|_{M=L/2-2m} = (S^+)^{2m}|\text{Néel}\rangle + S^-|\ldots\rangle. \tag{8.137}$$

The state $(S^+)^{2m}|\text{Néel}\rangle$ is called the $(2m)$-raised Néel state [44]. This means that the sought-after one-point functions can be rewritten in terms of a condensed-matter problem and the results from the condensed-matter literature then provide proof of the formulas that we have just presented.[7]

---

[7] See also [45] for an alternative proof.

8.4.2.6 General k

The one-point function for general k can be derived from the case $k = 2$ in a recursive way. This is due to the fact that there is a recursive relation between matrix product states with different values of k:

$$|\text{MPS}\rangle_{k+2} = T_1(\tfrac{ik}{2})|\text{MPS}\rangle_k - \left(\frac{k+1}{k-1}\right)^L |\text{MPS}\rangle_{k-2}, \qquad (8.138)$$

where $k \geq 2$ and $|\text{MPS}\rangle_0 = |\text{MPS}\rangle_1 = 0$.

Here, $T_1(v)$ is the transfer matrix of the XXX$_{1/2}$ Heisenberg spin chain, see Chapter 1 in this volume:[8]

$$T_1(v) := \text{tr}_a(R_{aL} \ldots R_{a1}), \qquad (8.139)$$

with the R-matrix

$$R(v) = 1 + \frac{\mathbb{P}}{v - \frac{i}{2}}, \qquad (8.140)$$

which is expressed in terms of the permutation operator $\mathbb{P}$. As usual, the label a refers to an auxiliary two-dimensional space, $\mathbb{C}^2$, which is traced over in the definition of $T_1(v)$.

The idea behind the proof of formula (8.138) is to consider the local action of the R-operator. The matrix product state is formed out of the local building blocks

$$\left(\langle \uparrow | \otimes t_1^{(k)} + \langle \downarrow | \otimes t_2^{(k)}\right) \in \mathbb{C}^2 \otimes \text{GL}(\mathbb{C}^k). \qquad (8.141)$$

Now, we add an additional auxiliary $\mathbb{C}^2$ space and consider the action of R on the physical space, which gives

$$R_{ia}(\tfrac{ik}{2})\left[\langle \uparrow_i | \otimes t_1^{(k)} + \langle \downarrow_i | \otimes t_2^{(k)}\right] =: \left(\langle \uparrow_i | \otimes \tau_1^{(k)} + \langle \downarrow_i | \otimes \tau_2^{(k)}\right) \in \mathbb{C}^2 \otimes \text{GL}(\mathbb{C}^{2k}),$$

where the matrices $\tau_{1,2}^{(k)}$ are given by

$$\tau_1^{(k)} = \begin{pmatrix} \frac{k+1}{k-1} t_1^{(k)} & 0 \\ \frac{2}{k-1} t_2^{(k)} & t_1^{(k)} \end{pmatrix}, \qquad \tau_2^{(k)} = \begin{pmatrix} t_2^{(k)} & \frac{2}{k-1} t_1^{(k)} \\ 0 & \frac{k+1}{k-1} t_2^{(k)} \end{pmatrix}. \qquad (8.142)$$

---

[8] This R-matrix is related to the one in Chapter 1 by a rescaling and by taking the appropriate $\cos \gamma \to 1$ limit.

The important observation is now that there exists a similarity transformation A such that

$$A\tau_i^{(k)} A^{-1} = \begin{pmatrix} t_i^{(k+2)} & 0 \\ \star_i & \frac{k+1}{k-1} t_i^{(k-2)} \end{pmatrix},\qquad (8.143)$$

where $\star_i$ stands for some irrelevant non-trivial entries [46]. This relation immediately proves the recursion relation (8.138).

Check that the transformation $U = \begin{pmatrix} 1 & 0 & \sqrt{3} & 0 \\ 0 & \sqrt{3} & 0 & 1 \\ i & 0 & -i\sqrt{3} & 0 \\ 0 & i\sqrt{3} & 0 & -i \end{pmatrix}$ identifies $\tau_{1,2}^{(4)} \sim t_{1,2}^{(4)}$.

As discussed in Chapter 1 in this volume, the Bethe states $|u\rangle$ are eigenvectors of the transfer matrix with eigenvalues

$$\Lambda(v|\mathbf{u}) = \left(\frac{v+\frac{i}{2}}{v-\frac{i}{2}}\right)^L \prod_{i=1}^{M} \frac{v-u_i-i}{v-u_i} + \prod_{i=1}^{M} \frac{v-u_i+i}{v-u_i}. \qquad (8.144)$$

The recursion relation (8.138) then fixes all overlap functions C_k for even k in terms of C_2 and $C_0 \equiv 0$ by the following recursion relation:

$$C_{k+2} = \Lambda\left(\frac{ik}{2}\Big|\mathbf{u}\right) C_k - \left(\frac{k+1}{k-1}\right)^L C_{k-2}. \qquad (8.145)$$

This then implies the following explicit form for the one-point function for $k > 2$:

$$C_k = i^L T_{k-1}(0) \sqrt{\frac{Q(\frac{i}{2})Q(0)}{Q^2(\frac{ik}{2})}} \sqrt{\frac{\det G_+}{\det G_-}}, \qquad (8.146)$$

where

$$T_n(u) = \sum_{a=-\frac{n}{2}}^{\frac{n}{2}} (u+ia)^L \frac{Q(u+\frac{n+1}{2}i)Q(u-\frac{n+1}{2}i)}{Q(u+(a-\frac{1}{2})i)Q(u+(a+\frac{1}{2})i)}. \qquad (8.147)$$

The function $T_n(u)$ can be identified as the transfer matrix of the Heisenberg spin chain where the auxiliary space is the $(n+1)$-dimensional representation.

Since the recursion relation (8.138) goes in steps of two, the $k=2$ result extends to all even k. Of course, equation (8.146) is well defined for any k and from numerical examples it is easily seen that it also works for odd k. Of course, by using (8.138), we see that for a proof of (8.146) for odd k we only need a proof for $k = 3$. This is still an open question. However, there seems to be a remarkable relation between C_2 and C_3. From (8.146), we find

$$C_3 = 2^L \frac{Q(0)}{Q(\frac{i}{2})} C_2. \qquad (8.148)$$

This suggest that C_3 and C_2 are related by Q-operators [47] rather than a transfer matrix, which has been checked for states with length up to 8 [46].

8.4.3 One-loop one-point functions in the $SU(2)$ sector

In order to calculate quantum corrections in the defect CFT, the action (8.1) has to be expanded around the classical solution (8.114):

$$\phi_i = \phi_i^{\text{cl}} + \tilde{\phi}_i. \qquad (8.149)$$

As the vacuum expectation values differ among the different flavours and are given by non-diagonal matrices in colour space, this leads to a mass matrix that mixes the different flavour and colour components of the fields. This mixing problem was solved in [11, 48]. Moreover, the vacuum expectation values are proportional to the inverse distance to the defect, $1/x_3$, such that the mass eigenvalues depend on $1/x_3$ as well. Via a Weyl transformation, this x_3-dependence can be absorbed to obtain standard propagators in an effective (auxiliary) AdS_4 space [11, 48, 49].

At one-loop order, two different diagrams have to be considered for the one-loop correction to a one-point function of a single-trace operator built from scalars, see Figure 8.5. The first of these diagrams, called the lollipop diagram, arises when expanding the composite operator to linear order in the quantum fields. It is given by

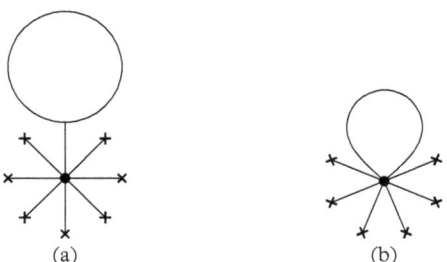

Figure 8.5 *Two diagrams have to be considered for the one-loop correction to a one-point function: the lollipop diagram (a) and the tadpole diagram (b).*

$$\langle\mathcal{O}\rangle_{\text{1-loop, tad}}(x) = \Psi^{i_1 i_2 \ldots i_L} \sum_{j=1} \text{tr}(\phi^{\text{cl}}_{i_1} \ldots \tilde{\phi}_{i_j} \ldots \phi^{\text{cl}}_{i_L})(x) \int d^4 y \sum_{\Phi_1, \Phi_2, \Phi_3} V_3(\Phi_1, \Phi_2, \Phi_3)(y). \tag{8.150}$$

The sum in this expression is over all cubic vertices of the defect CFT, i.e. the original cubic vertices of $\mathcal{N}=4$ SYM theory and the additional cubic vertices that arise from inserting one scalar vacuum expectation value into the quartic vertices of $\mathcal{N}=4$ SYM theory. As can be seen from Figure 8.5(a), the lollipop diagram is one-particle reducible and stems from the one-loop correction to the classical solution (8.114):

$$\langle\mathcal{O}\rangle_{\text{1-loop,lol}}(x) = \Psi^{i_1 i_2 \ldots i_L} \sum_{j=1}^{L} \text{tr}(\phi^{\text{cl}}_{i_1} \ldots \langle\phi_{i_j}\rangle_{\text{1-loop}} \ldots \phi^{\text{cl}}_{i_L})(x), \tag{8.151}$$

where

$$\langle\phi_i\rangle_{\text{1-loop}}(x) = \tilde{\phi}_i(x) \int d^4 y \sum_{\Phi_1, \Phi_2, \Phi_3} V_3(\Phi_1, \Phi_2, \Phi_3)(y). \tag{8.152}$$

In [11, 48], this correction was calculated and shown to vanish provided that the employed renormalization scheme preserves supersymmetry:

$$\langle\phi_i\rangle_{\text{1-loop}}(x) = 0. \tag{8.153}$$

This leaves us with the contribution of the diagram in Figure 8.5(b), called the tadpole diagram. The tadpole diagram arises from expanding the composite operator to quadratic order in the quantum fields and contracting these two quantum fields with a propagator:

$$\langle\mathcal{O}\rangle_{\text{1-loop, tad}}(x) = \sum_{j_1, j_2} \Psi^{i_1 \ldots i_{j_1} \ldots i_{j_2} \ldots i_L} \text{tr}(\phi^{\text{cl}}_{i_1} \ldots \tilde{\phi}_{i_{j_1}} \ldots \tilde{\phi}_{i_{j_2}} \ldots \phi^{\text{cl}}_{i_L})(x) \tag{8.154}$$

In addition to the above Feynman diagrams, the one-loop one-point function receives a contribution from the one-loop correction to the Bethe eigenstate, i.e. the two-loop eigenstate. As the $SO(6)$ sector is not closed under renormalization beyond one-loop order, we thus have to restrict ourselves to the $SU(2)$ sector, which is closed at all loop orders. We decompose the complex scalars in the $SU(2)$ sector as

$$X = [X]_{n,n'} E^n{}_{n'} + [X]_{n,a} E^n{}_a + [X]_{a,n} E^a{}_n + [X]_{a,a'} E^a{}_{a'}, \tag{8.155}$$

and similarly for Y. The indices n, n' take values $1, \ldots, k$ and the indices a, a' run from $k+1$ to N. In other words, $[X]_{n,n'}$ simply corresponds to the $k \times k$ block of the $N \times N$ $U(N)$ matrix.

The number of components in the $k \times k$ block does not scale with N and the components in the $(N-k) \times (N-k)$ block drop out when multiplied from the left or the right with a classical field, which is non-vanishing only in the $k \times k$ block. Hence, the only contribution in the large-N limit stems from the components in the $k \times (N-k)$ and $(N-k) \times k$ blocks. As these drop out unless they are neighbouring, we are back at the statement that only interactions among neighbouring fields contribute in the planar limit. Using dimensional regularization in the $3 - 2\varepsilon$ directions parallel to the defect, the required propagators read [11]

$$\langle [\tilde{X}]_{n,a}(x) [\tilde{X}]_{a',n'}(x) \rangle = \langle [\tilde{Y}]_{n,a}(x) [\tilde{Y}]_{a',n'}(x) \rangle = \delta_{a,a'} \delta_{n,n'} \frac{g^2}{N} \frac{1}{(x_3)^2} \tag{8.156}$$

and

$$\langle [X]_{n,a}(x) [Y]_{a',n'}(x) \rangle = -\langle [Y]_{n,a}(x) [X]_{a',n'}(x) \rangle$$
$$= \delta_{a,a'} \left[\left[X^{\text{cl}}, Y^{\text{cl}} \right] \right]_{n,n'}(x) \frac{2g^2}{N} \left(-\frac{1}{2\varepsilon} - \frac{1}{2} \log(4\pi) + \frac{1}{2} \gamma_E - \log(x_3) + \Psi(\tfrac{k+1}{2}) \right). \tag{8.157}$$

In these expressions, Ψ denotes Euler's digamma function. Inserting these propagators into (8.154) yields

$$\langle \mathcal{O} \rangle_{\text{1-loop,tad}}(x) = g^2 \frac{1}{(x_3)^2} \sum_j \delta_{s_j = s_{j+1}} \Psi^{s_1 \ldots s_j s_{j+1} \ldots s_L} \text{tr}(\phi^{\text{cl}}_{s_1} \ldots \phi^{\text{cl}}_{s_{j-1}} \phi^{\text{cl}}_{s_{j+2}} \ldots \phi^{\text{cl}}_{s_L})(x)$$
$$+ 2g^2 \left(-\frac{1}{2\varepsilon} - \frac{1}{2} \log(4\pi) + \frac{1}{2} \gamma_E - \log(x_3) + \Psi(\tfrac{k+1}{2}) \right) \tag{8.158}$$
$$\times \sum_j \Psi^{s_1 \ldots s_j s_{j+1} \ldots s_L} \text{tr}(\phi^{\text{cl}}_{s_1} \ldots \phi^{\text{cl}}_{s_{j-1}} [\phi^{\text{cl}}_{s_j}, \phi^{\text{cl}}_{s_{j+1}}] \phi^{\text{cl}}_{s_{j+2}} \ldots \phi^{\text{cl}}_{s_L})(x).$$

We observe that the first term, stemming from (8.156), is finite. The second term, stemming from (8.157), however, is (ultraviolet) divergent. The divergence has to be cancelled by the renormalization constant $\mathcal{Z}_{\mathcal{O}}$, providing us with a second way to derive the one-loop dilatation operator (8.74)![9] Using the renormalization constant (8.90) in

---

[9] The ultraviolet divergence occurring at an operator depends only on the operator but not on the quantity it occurs in. This way, the renormalization constant and thus the dilatation operator can be determined from

the renormalization scheme that leaves the one-loop two-point function normalized, we find

$$\langle \mathcal{Z}\mathcal{O}\rangle_{\text{1-loop,tad}}(x) = g^2 \frac{1}{(x_3)^2} \sum_j \delta_{s_j = s_{j+1}} \Psi^{s_1 \ldots s_j \, s_{j+1} \ldots s_L} \text{tr}(\phi^{\text{cl}}_{s_1} \ldots \phi^{\text{cl}}_{s_{j-1}} \phi^{\text{cl}}_{s_{j+2}} \ldots \phi^{\text{cl}}_{s_L})(x)$$

$$+ g^2 \left(\frac{1}{2} - \log 2 + \gamma_E - \log(x_3) + \Psi\left(\frac{k+1}{2}\right) \right) \Delta^{(1)} \langle \mathcal{O}\rangle_{\text{tree}}(x) \quad (8.159)$$

for a one-loop eigenstate with one-loop anomalous dimension $\Delta^{(1)}$. The term proportional to $\log(x_3)$ accounts for the correction to the scaling dimension expected from (8.112), whereas the other terms contribute to the correction to the coefficient a respectively C. While the second term in (8.159) is simply proportional to the tree-level one-point function, i.e. the overlap between the Bethe eigenstate and the MPS, the first term in (8.159) can be written as the overlap of the Bethe eigenstate with an amputated matrix product state (AMPS).

On the integrability side, the one-loop one-point function thus requires to calculate the overlap of the Bethe eigenstate with the AMPS and the overlap of the loop-corrected Bethe state [33] with the MPS. This calculation was done in [50] and shown to agree with the following conjecture for an all-loop asymptotic one-point function formula proposed there as well:

$$C_k = i^L \tilde{T}_{k-1}(0) \sqrt{\frac{Q(\frac{i}{2}) Q(0)}{Q^2(\frac{ik}{2})}} \sqrt{\frac{\det \tilde{G}_+}{\det \tilde{G}_-}} \mathbb{F}_k, \quad (8.160)$$

where the Bethe roots are assumed to satisfy the all-loop asymptotic Bethe equations (8.104), which are also used to define $\tilde{G}_\pm$ in analogy with (8.133), and

$$\tilde{T}_n(u) = g^L \sum_{a=-\frac{n}{2}}^{\frac{n}{2}} x(u+ia)^L \frac{Q(u + \frac{n+1}{2}i) Q(u - \frac{n+1}{2}i)}{Q(u + (a - \frac{1}{2})i) Q(u + (a + \frac{1}{2})i)} \quad (8.161)$$

is the quantum transfer matrix. Moreover, the introduction of a flux factor $\mathbb{F}_k$ was needed in (8.160), and it was found to be of the form

$$\mathbb{F}_k = 1 + g^2 \left[\Psi\left(\frac{k+1}{2}\right) + \gamma_E - \log 2 \right] \Delta^{(1)} + O(g^4). \quad (8.162)$$

The generalization of $\mathbb{F}_k$ to higher loops constitutes an open problem.

---

one-point functions, two-point functions, three- and higher-point functions, from form factors etc., or simply from calculating its vertex renormalization as one would for computing a beta function.

8.5 Outlook

Integrability continues to reveal itself in connection with yet more observables of $\mathcal{N}=4$ SYM theory. At the time of the major review [7], the integrability of the spectral problem was well understood and traces of integrability had been spotted in the form of a Yangian symmetry for tree-level and one-loop scattering amplitudes [51, 52]. Since then, the integrability properties of scattering amplitudes have been further elaborated via the introduction of a spectral parameter [53–56]. Furthermore, integrability techniques have been applied to the study of polygonal Wilson loops [57, 58], dual to planar scattering amplitudes, as well as to smooth Maldacena–Wilson loops [59]. Form factors of $\mathcal{N}=4$ SYM theory have been studied within the integrability language as well, both at weak coupling [60] and at strong coupling [61, 62]. In addition, it has been demonstrated that the Hagedorn temperature of $\mathcal{N}=4$ SYM [45] can be obtained within the integrability framework. Finally, the tools of integrability inherited from the planar spectral problem have been exploited in the calculation of higher-point correlation functions, more precisely of three-point functions [25, 33, 64] and of four-point functions [65–67]. These efforts have recently been boosted by the development of the so-called hexagon techniques [48, 69, 70] covered in Chapter 10 of this volume.

In this chapter, we have chosen to focus on the calculation of one-point functions in a certain defect version of $\mathcal{N}=4$ SYM theory, which constitutes yet another novel arena for the application of integrability methods. Combining the tools of integrability with field-theoretical computations, we have arrived at a closed expression for the tree-level and one-loop one-point functions of the defect version of $\mathcal{N}=4$ SYM theory dual to the D5-D3 probe-brane system with flux.

It would be interesting to carry out an integrability analysis of the one-point functions of the probe-brane system as well. For instance, one could imagine that an analysis of the classical equations of motion of open strings attached at one end to the boundary of AdS and at the other end to the probe brane would reveal some known or unknown system of integrable differential equations. So far, only a single classical solution of this type has been found, namely a solution corresponding to a point-like string [46, 71]. It should be possible to find solutions corresponding to various types of spinning strings as well (for a review see e.g. [72]), solutions which would correspond to non-protected operators in the field-theory language.

Obviously, a pressing question on the field-theory side is whether the closed one-point function formulae obtained at the two leading orders in perturbation theory can be extended to higher loop orders as in the case of the spectral problem. In [50], we took the first step in generalizing the results to higher loop orders by proposing an asymptotic all-loop formula (8.160), where in analogy with the situation for the spectral problem the Zhukovski transformation (8.105) plays a key role. This formula agrees with a string-theory prediction [71] in a double-scaling parameter up to wrapping order, but much more work is needed to confirm or possibly adjust the formula.

Another obvious question is whether integrability extends to other observables in the defect setup. So far only a few examples of other observables have been studied, namely some Maldacena–Wilson loops [49, 73, 74] and two-point functions involving either only BPS operators [75] or one operator of length two [76]. As mentioned earlier, the one- and two-point functions of the defect theory might have the virtue of providing input to the boundary conformal bootstrap programme [77–79].

There exists another defect version of $\mathcal{N}=4$ SYM theory which like the present one is dual to a probe-brane system with flux, more precisely a non-supersymmetric D7-D3 probe-brane setup [80]. For this setup, only tree-level one-point functions have been considered and so far signs of integrability have not been observed [81], see however [82]. It would be interesting to understand the apparent difference in the integrability properties of the two defect versions of $\mathcal{N}=4$ SYM theory at a more fundamental level.

$\mathcal{N}=4$ SYM theory has a three-dimensional somewhat close cousin, namely ABJM theory, which is an $\mathcal{N}=6$ supersymmetric Chern–Simons matter theory. For this theory, the planar spectral problem is likewise integrable and the theory has a holographic dual in the form of type IIA string theory on $AdS_4 \times CP^3$, see e.g. [83] for a review. In analogy with the AdS_5/CFT_4 situation, the dual string in the ABJM case allows for certain probe-brane systems with fluxes [84–86]. It would be interesting to study to which extent defect conformal field theories result from these brane constructions and if so carry through an analysis of their one-point functions and possibly reveal novel integrability structures.

Acknowledgements

C. Kristjansen thanks the organizers of the Les Houches Summer School for the invitation to lecture there and for creating an inspiring environment. All authors thank the organizers of the school for the invitation to contribute to the proceedings. We furthermore thank I. Buhl-Mortensen, G. Linardopoulos, S. Mori, G. Semenoff, K.E. Vardinghus, and K. Zarembo for useful discussions. The authors were supported by DFF-FNU through grant number DFF-4002-00037.

References

[1] F. Gliozzi, J. Scherk, and D. I. Olive, 'Supersymmetry, Supergravity Theories and the Dual Spinor Model,' *Nucl. Phys.* **B122** (1977) 253–290.
[2] L. Brink, J. H. Schwarz, and J. Scherk, 'Supersymmetric Yang-Mills Theories,' *Nucl. Phys.* **B121** (1977) 77–92.
[3] M. Ammon and J. Erdmenger, *Gauge/gravity duality*. Cambridge Univ. Pr., Cambridge, UK, 2015.
[4] J. Wess and J. Bagger, *Supersymmetry and supergravity*. Princeton Univ. Pr., Princeton, USA, 1992.
[5] N. Beisert, 'The Dilatation operator of $\mathcal{N}=4$ super Yang-Mills theory and integrability,' *Phys. Rept.* **405** (2004) 1–202, arXiv:hep-th/0407277 [hep-th].
[6] J. A. Minahan, 'A brief introduction to the Bethe ansatz in $\mathcal{N}=4$ super-Yang-Mills,' *J. Phys.* **A39** (2006) 12657–12677.

[7] N. Beisert et al., 'Review of AdS/CFT Integrability: An Overview,' *Lett. Math. Phys.* **99** (2012) 3–32, arXiv:1012.3982 [hep-th].
[8] D. Serban, 'Integrability and the AdS/CFT correspondence,' *J. Phys.* **A44** (2011) 124001, arXiv:1003.4214 [hep-th].
[9] J. A. Minahan and K. Zarembo, 'The Bethe ansatz for $\mathcal{N}=4$ superYang-Mills,' *JHEP* **03** (2003) 013, arXiv:hep-th/0212208 [hep-th].
[10] M. Gaudin, 'Diagonalization of a Class of Spin Hamiltonians,' *J. Phys. France* **37** (1976) 1086–1098.
[11] I. Buhl-Mortensen, M. de Leeuw, A. C. Ipsen, C. Kristjansen, and M. Wilhelm, 'A Quantum Check of AdS/dCFT,' *JHEP* **01** (2017) 098, arXiv:1611.04603 [hep-th].
[12] G. 't Hooft, 'A Planar Diagram Theory for Strong Interactions,' *Nucl. Phys.* **B72** (1974) 461.
[13] M. F. Sohnius and P. C. West, 'Conformal Invariance in $\mathcal{N}=4$ Supersymmetric Yang-Mills Theory,' *Phys. Lett.* **B100** (1981) 245.
[14] P. S. Howe, K. S. Stelle, and P. K. Townsend, 'Miraculous Ultraviolet Cancellations in Supersymmetry Made Manifest,' *Nucl. Phys.* **B236** (1984) 125–166.
[15] L. Brink, O. Lindgren, and B. E. W. Nilsson, 'The Ultraviolet Finiteness of the $\mathcal{N}=4$ Yang-Mills Theory,' *Phys. Lett.* **B123** (1983) 323–328.
[16] V. A. Smirnov, *Feynman integral calculus.* Springer, Berlin, Germany, 2006.
[17] D. I. Kazakov, 'Calculation of Feynman diagrams by the 'uniqueness' method,' *Theor. Math. Phys.* **58** (1984) 223–230. [Teor. Mat. Fiz.58,343(1984)].
[18] J. K. Erickson, G. W. Semenoff, and K. Zarembo, 'Wilson loops in $\mathcal{N}=4$ supersymmetric Yang-Mills theory,' *Nucl. Phys.* **B582** (2000) 155–175, arXiv:hep-th/0003055 [hep-th].
[19] N. Beisert, 'The complete one loop dilatation operator of $\mathcal{N}=4$ superYang-Mills theory,' *Nucl. Phys.* **B676** (2004) 3–42, arXiv:hep-th/0307015 [hep-th].
[20] N. Beisert, 'The su(2|3) dynamic spin chain,' *Nucl. Phys.* **B682** (2004) 487–520, arXiv:hep-th/0310252 [hep-th].
[21] N. Beisert and B. I. Zwiebel, 'On Symmetry Enhancement in the psu(1,1|2) Sector of $\mathcal{N}=4$ SYM,' *JHEP* **10** (2007) 031, arXiv:0707.1031 [hep-th].
[22] B. I. Zwiebel, 'Iterative Structure of the $\mathcal{N}=4$ SYM Spin Chain,' *JHEP* **07** (2008) 114, arXiv:0806.1786 [hep-th].
[23] M. Tetelman,' Lorentz group for two-dimensional integrable lattice systems.,' *Sov. Phys. JETP* **55(2)** (1982) 306–310.
[24] K. Sogo and M. Wadati, 'Boost Operator and Its Application to Quantum Gelfand-Levitan Equation for Heisenberg-Ising Chain with Spin One-Half,' *Prog. Theor. Phys.* **69(2)** (1983) 431–450.
[25] J. Escobedo, N. Gromov, A. Sever, and P. Vieira, 'Tailoring Three-Point Functions and Integrability,' *JHEP* **09** (2011) 028, arXiv:1012.2475 [hep-th].
[26] V. E. Korepin, 'Calculation of Norms of Bethe Wave Functions,' *Commun. Math. Phys.* **86** (1982) 391–418.
[27] N. Beisert, C. Kristjansen, and M. Staudacher, 'The Dilatation operator of conformal $\mathcal{N}=4$ superYang-Mills theory,' *Nucl. Phys.* **B664** (2003) 131–184, arXiv:hep-th/0303060 [hep-th].
[28] N. Beisert, T. McLoughlin, and R. Roiban, 'The Four-loop dressing phase of $\mathcal{N}=4$ SYM,' *Phys. Rev.* **D76** (2007) 046002, arXiv:0705.0321 [hep-th].
[29] M. Staudacher, 'The Factorized S-matrix of CFT/AdS,' *JHEP* **05** (2005) 054, arXiv:hep-th/0412188 [hep-th].
[30] N. Beisert and M. Staudacher, 'Long-range psu(2,2|4) Bethe Ansätze for gauge theory and strings,' *Nucl. Phys.* **B727** (2005) 1–62, arXiv:hep-th/0504190 [hep-th].

[31] P. Vieira and D. Volin, 'Review of AdS/CFT Integrability, Chapter III.3: The Dressing factor,' *Lett. Math. Phys.* **99** (2012) 231–253, arXiv:1012.3992 [hep-th].

[32] J. Ambjørn, R. A. Janik, and C. Kristjansen, 'Wrapping interactions and a new source of corrections to the spin-chain/string duality,' *Nucl. Phys.* **B736** (2006) 288–301, arXiv:hep-th/0510171 [hep-th].

[33] N. Gromov and P. Vieira, 'Tailoring Three-Point Functions and Integrability IV. Theta-morphism,' *JHEP* **04** (2014) 068, arXiv:1205.5288 [hep-th].

[34] N. Beisert, C. Kristjansen, J. Plefka, G. W. Semenoff, and M. Staudacher, 'BMN correlators and operator mixing in $\mathcal{N}=4$ superYang-Mills theory,' *Nucl. Phys.* **B650** (2003) 125–161, arXiv:hep-th/0208178 [hep-th].

[35] W. Carlson, R. de Mello Koch, and H. Lin, 'Nonplanar Integrability,' *JHEP* **03** (2011) 105, arXiv:1101.5404 [hep-th].

[36] O. DeWolfe, D. Z. Freedman, and H. Ooguri, 'Holography and defect conformal field theories,' *Phys. Rev.* **D66** (2002) 025009, arXiv:hep-th/0111135 [hep-th].

[37] J. Erdmenger, Z. Guralnik, and I. Kirsch, 'Four-dimensional superconformal theories with interacting boundaries or defects,' *Phys. Rev.* **D66** (2002) 025020, arXiv:hep-th/0203020 [hep-th].

[38] J. L. Cardy, 'Conformal Invariance and Surface Critical Behavior,' *Nucl. Phys.* **B240** (1984) 514–532.

[39] N. R. Constable, R. C. Myers, and O. Tafjord, 'The Noncommutative bion core,' *Phys. Rev.* **D61** (2000) 106009, arXiv:hep-th/9911136 [hep-th].

[40] M. de Leeuw, C. Kristjansen, and K. Zarembo, 'One-point Functions in Defect CFT and Integrability,' *JHEP* **08** (2015) 098, arXiv:1506.06958 [hep-th].

[41] M. de Leeuw, C. Kristjansen, and S. Mori, 'AdS/dCFT one-point functions of the $SU(3)$ sector,' *Phys. Lett.* **B763** (2016) 197, arXiv:1607.03123 [hep-th]

[42] M. Mestyan, B. Bertini, L. Piroli, and P. Calabrese, 'Exact solution for the quench dynamics of a nested integrable system,' *J. Stat. Mech.* **1708** (2017) no.8, 083103, arXiv:1705.00851 [cond-mat.stat-mech]

[43] B. Pozsgay, 'Overlaps between eigenstates of the XXZ spin-1/2 chain and a class of simple product states,' *Journal of Statistical Mechanics: Theory and Experiment* **2014** no. 6, (2014) P06011.

[44] M. Brockmann, 'Overlaps of q-raised Néel states with XXZ Bethe states and their relation to the Lieb-Liniger Bose gas,' *Journal of Statistical Mechanics: Theory and Experiment* **2014** no. 5, (2014) P05006.

[45] O. Foda and K. Zarembo, 'Overlaps of partial Néel states and Bethe states,' *J. Stat. Mech.* **1602** no. 2, (2016) 023107, arXiv:1512.02533 [hep-th].

[46] I. Buhl-Mortensen, M. de Leeuw, C. Kristjansen, and K. Zarembo, 'One-point Functions in AdS/dCFT from Matrix Product States,' *JHEP* **02** (2016) 052, arXiv:1512.02532 [hep-th].

[47] V. V. Bazhanov, T. Łukowski, C. Meneghelli, and M. Staudacher, 'A Shortcut to the Q-Operator,' *J. Stat. Mech.* **1011** (2010) P11002, arXiv:1005.3261 [hep-th].

[48] I. Buhl-Mortensen, M. de Leeuw, A. C. Ipsen, C. Kristjansen, and M. Wilhelm, 'One-loop one-point functions in gauge-gravity dualities with defects,' *Phys. Rev. Lett.* **117** no. 23, (2016) 231603, arXiv:1606.01886 [hep-th].

[49] K. Nagasaki, H. Tanida, and S. Yamaguchi, 'Holographic Interface-Particle Potential,' *JHEP* **01** (2012) 139, arXiv:1109.1927 [hep-th].

[50] I. Buhl-Mortensen, M. de Leeuw, A. C. Ipsen, C. Kristjansen, and M. Wilhelm, 'Asymptotic one-point functions in AdS/dCFT,' *Phys. Rev. Lett.* **119** (2017) no.26, 261604, arXiv:1704.07386 [hep-th].

[51] J. M. Drummond, J. M. Henn, and J. Plefka, 'Yangian symmetry of scattering amplitudes in $\mathcal{N}=4$ super Yang-Mills theory,' *JHEP* **05** (2009) 046, arXiv:0902.2987 [hep-th].

[52] N. Beisert, J. Henn, T. McLoughlin, and J. Plefka, 'One-Loop Superconformal and Yangian Symmetries of Scattering Amplitudes in $\mathcal{N}=4$ Super Yang-Mills,' *JHEP* **04** (2010) 085, arXiv:1002.1733 [hep-th].

[53] L. Ferro, T. Łukowski, C. Meneghelli, J. Plefka, and M. Staudacher, 'Harmonic R-matrices for Scattering Amplitudes and Spectral Regularization,' *Phys. Rev. Lett.* **110** no. 12, (2013) 121602, arXiv:1212.0850 [hep-th].

[54] L. Ferro, T. Łukowski, C. Meneghelli, J. Plefka, and M. Staudacher, 'Spectral Parameters for Scattering Amplitudes in $\mathcal{N}=4$ Super Yang-Mills Theory,' *JHEP* **01** (2014) 094, arXiv:1308.3494 [hep-th].

[55] D. Chicherin, S. Derkachov, and R. Kirschner, 'Yang-Baxter operators and scattering amplitudes in $\mathcal{N}=4$ super-Yang-Mills theory,' *Nucl. Phys.* **B881** (2014) 467–501, arXiv:1309.5748 [hep-th].

[56] J. Broedel, M. de Leeuw, and M. Rosso, 'A dictionary between R-operators, on-shell graphs and Yangian algebras,' *JHEP* **06** (2014) 170, arXiv:1403.3670 [hep-th].

[57] B. Basso, A. Sever, and P. Vieira, 'Spacetime and Flux Tube S-Matrices at Finite Coupling for $\mathcal{N}=4$ Supersymmetric Yang-Mills Theory,' *Phys. Rev. Lett.* **111** no. 9, (2013) 091602, arXiv:1303.1396 [hep-th].

[58] B. Basso, A. Sever, and P. Vieira, 'Space-time S-matrix and Flux tube S-matrix II. Extracting and Matching Data,' *JHEP* **01** (2014) 008, arXiv:1306.2058 [hep-th].

[59] D. Müller, H. Münkler, J. Plefka, J. Pollok, and K. Zarembo, 'Yangian Symmetry of smooth Wilson Loops in $\mathcal{N}=4$ super Yang-Mills Theory,' *JHEP* **11** (2013) 081, arXiv:1309.1676 [hep-th].

[60] R. Frassek, D. Meidinger, D. Nandan, and M. Wilhelm, 'On-shell diagrams, Graßmannians and integrability for form factors,' *JHEP* **01** (2016) 182, arXiv:1506.08192 [hep-th].

[61] J. Maldacena and A. Zhiboedov, 'Form factors at strong coupling via a Y-system,' *JHEP* **11** (2010) 104, arXiv:1009.1139 [hep-th].

[62] Z. Gao and G. Yang, 'Y-system for form factors at strong coupling in AdS_5 and with multi-operator insertions in AdS_3,' *JHEP* **06** (2013) 105, arXiv:1303.2668 [hep-th].

[63] T. Harmark and M. Wilhelm, 'The Hagedorn temperature of AdS5/CFT4 via integrability,' *Phys. Rev. Lett.* **120** (2018) no.7, 071605, arXiv:1706.03074 [hep-th].

[64] J. Escobedo, N. Gromov, A. Sever, and P. Vieira, 'Tailoring Three-Point Functions and Integrability II. Weak/strong coupling match,' *JHEP* **09** (2011) 029, arXiv:1104.5501 [hep-th].

[65] B. Eden and A. Sfondrini, 'Tessellating cushions: four-point functions in $\mathcal{N}=4$ SYM,' *JHEP* **1710** (2017) 098, arXiv:1611.05436 [hep-th].

[66] B. Basso, F. Coronado, S. Komatsu, H. T. Lam, P. Vieira, and D.-l. Zhong, 'Asymptotic Four Point Functions,' arXiv:1701.04462 [hep-th].

[67] T. Bargheer, 'Four-Point Functions with a Twist,' *J. Phys.* **A51** (2018) no.3, 035401, arXiv:1701.04424 [hep-th].

[68] B. Basso, S. Komatsu, and P. Vieira, 'Structure Constants and Integrable Bootstrap in Planar $\mathcal{N}=4$ SYM Theory,' arXiv:1505.06745 [hep-th].

[69] B. Basso, V. Goncalves, S. Komatsu, and P. Vieira, 'Gluing Hexagons at Three Loops,' *Nucl. Phys.* **B907** (2016) 695–716, arXiv:1510.01683 [hep-th].

[70] B. Basso, V. Goncalves, and S. Komatsu, 'Structure constants at wrapping order,' *JHEP* **05** (2017) 124, arXiv:1702.02154 [hep-th].

[71] K. Nagasaki and S. Yamaguchi, 'Expectation values of chiral primary operators in holographic interface CFT,' *Phys. Rev.* **D86** (2012) 086004, arXiv:1205.1674 [hep-th].

[72] A. A. Tseytlin, 'Review of AdS/CFT Integrability, Chapter II.1: Classical $AdS_5 \times S^5$ string solutions,' *Lett. Math. Phys.* **99** (2012) 103–125, arXiv:1012.3986 [hep-th].

[73] M. de Leeuw, A. C. Ipsen, C. Kristjansen, and M. Wilhelm, 'One-loop Wilson loops and the particle-interface potential in AdS/dCFT,' *Phys. Lett.* **B768** (2017) 192–197, arXiv:1608.04754 [hep-th].

[74] J. Aguilera-Damia, D. H. Correa, and V. I. Giraldo-Rivera, 'Circular Wilson loops in defect Conformal Field Theory,' *JHEP* **03** (2017) 023, arXiv:1612.07991 [hep-th].

[75] M. de Leeuw, A. C. Ipsen, C. Kristjansen, K. E. Vardinghus, and M. Wilhelm, 'Two-point functions in AdS/dCFT and the boundary conformal bootstrap equations,' *JHEP* **1708** (2017) 020, arXiv:1705.03898 [hep-th].

[76] E. Widen, 'Two-point functions of SU(2)-subsector and length-two operators in dCFT,' *Phys. Lett.* B **773** (2017) 435, arXiv:1705.08679 [hep-th].

[77] P. Liendo, L. Rastelli, and B. C. van Rees, 'The Bootstrap Program for Boundary CFT_d,' *JHEP* **07** (2013) 113, arXiv:1210.4258 [hep-th].

[78] F. Gliozzi, P. Liendo, M. Meineri, and A. Rago, 'Boundary and Interface CFTs from the Conformal Bootstrap,' *JHEP* **05** (2015) 036, arXiv:1502.07217 [hep-th].

[79] M. Billó, V. Gonçalves, E. Lauria, and M. Meineri, 'Defects in conformal field theory,' *JHEP* **04** (2016) 091, arXiv:1601.02883 [hep-th].

[80] R. C. Myers and M. C. Wapler, 'Transport Properties of Holographic Defects,' *JHEP* **12** (2008) 115, arXiv:0811.0480 [hep-th].

[81] M. de Leeuw, C. Kristjansen, and G. Linardopoulos, 'One-point functions of non-protected operators in the SO(5) symmetric D3-D7 dCFT,' *J. Phys.* **A50** no. 25, (2017) 254001, arXiv:1612.06236 [hep-th].

[82] M. De Leeuw, C. Kristjansen and G. Linardopoulos, 'Scalar one-point functions and matrix product states of AdS/dCFT,' *Phys. Lett.* **B781** (2018) 238, arXiv:1802.01598 [hep-th].

[83] T. Klose, 'Review of AdS/CFT Integrability, Chapter IV.3: N=6 Chern-Simons and Strings on $AdS_4 \times CP^3$,' *Lett. Math. Phys.* **99** (2012) 401–423, arXiv:1012.3999 [hep-th].

[84] S. Hohenegger and I. Kirsch, 'A Note on the holography of Chern-Simons matter theories with flavour,' *JHEP* **04** (2009) 129, arXiv:0903.1730 [hep-th].

[85] D. Gaiotto and D. L. Jafferis, 'Notes on adding D6 branes wrapping RP^3 in $AdS_4 \times CP^3$,' *JHEP* **11** (2012) 015, arXiv:0903.2175 [hep-th].

[86] M. Ammon, J. Erdmenger, R. Meyer, A. O'Bannon, and T. Wrase, 'Adding Flavor to AdS_4/CFT_3,' *JHEP* **11** (2009) 125, arXiv:0909.3845 [hep-th].

9
Spectrum of $\mathcal{N}=4$ supersymmetric Yang–Mills theory and the quantum spectral curve

Nikolay GROMOV

Mathematics Department, King's College London, The Strand, London WC2R 2LS, UK. & St.Petersburg INP, Gatchina, 188 300, St.Petersburg, Russia

Chapter Contents

9 Spectrum of $\mathcal{N}=4$ supersymmetric Yang–Mills theory and the quantum spectral curve   400
    Nikolay GROMOV

9.1	Introduction	402
9.2	From harmonic oscillator to QQ-relations	402
9.3	Classical string and strong coupling limit of QSC	418
9.4	QSC formulation	424
9.5	QSC—analytic examples	430
9.6	Solving QSC at finite coupling numerically	437
9.7	Applications, further reading, and open questions	443
	Acknowledgements	445
	References	445

9.1 Introduction

The importance of AdS/CFT correspondence in modern theoretical physics and the role of $\mathcal{N}=4$ SYM in it is hard to over-appreciate. In this chapter we try to give a pedagogical introduction to the quantum spectral curve (QSC) of $\mathcal{N}=4$ SYM, a beautiful mathematical structure which describes the non-perturbative spectrum of strings/anomalous dimensions of all single-trace operators. The historical development leading to the discovery of the QSC [7, 11] is a very long and interesting story by itself, and there are several reviews trying to cover the main steps on this route [6, 18]. For the purposes of this chapter we have taken another approach and try to motivate the construction by emphasizing numerous analogies between the QSC construction and basic quantum integrable systems such as the harmonic oscillator, Heisenberg spin chains, and classical sigma-models. In this way the QSC comes out naturally, bypassing extremely complicated and technical stages such as derivation of the S-matrix [19], dressing phase [20, 21], mirror theory [22], Y-system [2], thermodynamic Bethe ansatz [3–5, 23, 25], NLIE [7, 26], and finally derivation of the QSC [7, 11].

We also give examples of analytic solutions of the QSC and in the last section describe step-by-step the numerical algorithm allowing us to get the non-perturbative spectrum with almost unlimited precision [13]. We also briefly discuss the analytic continuation of the anomalous dimension to the Regge (BFKL) limit relevant for more realistic QCD.

The structure is the following: in Section 9.1 we re-introduce the harmonic oscillator and the Heisenberg spin chains in a way suitable for generalization to the QSC. Section 9.2 describes classical integrability of strings in a curved background, which give some important hints about the construction of the QSC. In Section 9.3 we give a clear formulation of the QSC. In Section 9.4 we consider some analytic examples. And in the last Section 9.5 we present the numerical method.

9.2 From harmonic oscillator to *QQ*-relations

9.2.1 Inspiration from the harmonic oscillator

To motivate the construction of the QSC we first consider the 1D harmonic oscillator and concentrate on the features which, as we will see later, have similarities with the construction for the spectrum of $\mathcal{N}=4$ SYM.

The harmonic oscillator is the simplest integrable model which at the same time exhibits non-trivial features surprisingly similar to $\mathcal{N}=4$ SYM. Our starting point is the Schrödinger equation

$$-\frac{\hbar^2}{2m}\psi''(x) + V(x)\psi(x) = E\psi(x), \qquad (9.1)$$

where $V(x) = \frac{m\omega^2 x^2}{2}$. Alternatively, it can be written in terms of the quasi-momentum

$$p(x) = \frac{\hbar}{i} \frac{\psi'(x)}{\psi(x)} \tag{9.2}$$

as

$$p^2 - i\hbar p' = 2m(E - V). \tag{9.3}$$

This non-linear equation is completely equivalent to (9.1). Instead of solving this equation directly let us make a simple ansatz for $p(x)$. We see that for large x the r.h.s. behaves as $-m^2\omega^2 x^2$ implying that at infinity $p \simeq im\omega x$. Furthermore, $p(x)$ should have simple poles at the position of zeros of the wave function which we denote x_i. All the residues at these points should be equal to $\hbar/i$ as one can see from (9.2). We can accommodate all these basic analytical properties with the following ansatz:

$$p(x) = im\omega x + \frac{\hbar}{i} \sum_{i=1}^{N} \frac{1}{x - x_i}. \tag{9.4}$$

We note that at large x the r.h.s. of (9.4) behaves as $im\omega x + \frac{\hbar}{i}\frac{N}{x} + O(1/x^2)$. Plugging this large x approximation of $p(x)$ into the exact equation (9.3) we get

$$\left(im\omega x + \frac{\hbar}{i} \frac{N}{x}\right)^2 + \hbar(m\omega) = 2m(E - m^2\omega^2 x^2/2) + O(1/x). \tag{9.5}$$

Comparing the coefficients in front of x^2 and x^0 we get $E = \hbar\omega(N + 1/2)$ which is the famous formula for the spectrum of the harmonic oscillator. In order to reconstruct the wave function we expand (9.3) near the pole $x = x_i$. Namely, we require

$$\operatorname{res}_{x=x_k}\left[\left(im\omega x + \frac{\hbar}{i}\sum_{i=1}^{N}\frac{1}{x-x_i}\right)^2 + i\hbar\frac{\hbar}{i}\sum_{i=1}^{N}\frac{1}{(x-x_i)^2}\right] = 0, \tag{9.6}$$

obtaining (from the first bracket)

$$x_k = \frac{\hbar}{\omega m} \sum_{\substack{j \neq i}}^{N} \frac{1}{x_i - x_k}, \quad k = 1, \ldots, N. \tag{9.7}$$

This set of equations determines all x_k in a unique way.

Exercise 9.1 Verify for 1 and 2 roots that there is a unique up to a permutation solution of equation (9.7), find the solution.

Finally, we can integrate (9.2) to obtain

$$\psi(x) = e^{-\frac{m\omega x^2}{2\hbar}} Q(x), \; Q(x) \equiv \prod_{i=1}^{N}(x - x_i). \quad (9.8)$$

It is here for the first time we see the Q-function, which is the analogue of the main building block of the QSC! We will refer to equation (9.7) for zeros of the Q-functions as the Bethe ansatz equation. We will call $\{x_i\}$ the Bethe roots.

Let us outline the main features which will be important for what follows:

- The asymptotic of $Q(x) \sim x^N$ contains quantum numbers of the state.
- Zeros of the $Q(x)$ function can be determined from the condition of cancellation of poles (9.6) (analogue of Baxter equation) which can be explicitly written as (9.7) (analogue of Bethe equations).
- The wave function can be completely determined from the Bethe roots or from $Q(x)$ (by adding a simple universal for all states factor).
- The Schrödinger equation has a second (non-normalizable) solution which behaves as $\psi_2 \simeq x^{-N-1} e^{+\frac{m\omega}{2\hbar}x^2}$. Together with the normalizable solution ψ_1 they form a Wronskian

$$W = \begin{vmatrix} \psi_1(x) & \psi_1'(x) \\ \psi_2(x) & \psi_2'(x) \end{vmatrix} \quad (9.9)$$

which is a constant.

Exercise 9.2 Prove that the Wronskian W is a constant for a general Schrödinger equation.

9.2.2 *SU(2)*-Heisenberg spin chain

In this section we discuss how the construction from Section 9.2.1. generalizes to integrable spin chains—a system with a large number of degrees of freedom. The simplest spin chain is the Heisenberg $SU(2)$ magnetic which is discussed in great detail in numerous reviews and lectures. We highly recommend Faddeev's 1982 Les Houches lectures [27] for that. We describe the results most essential for us below.

In short, the Heisenberg spin chain is a chain of L spin-1/2 particles with a nearest neighbour interaction. The Hamiltonian of the system can be written as

$$\hat{H} = 2g^2 \sum_{i=1}^{L}(1 - P_{i,i+1}), \quad (9.10)$$

Figure 9.1 *Two equivalent representations of the same state. In the first case we treat spin downs as excitations (magnons) moving with some momenta p_i and call spin ups correspond to the reference (vacuum) state. In the second case we treat spin ups as excitations moving with some momenta q_i.*

where $P_{i,i+1}$ is an operator which permutes the particles at the position i and $i+1$ and g is a constant. We introduce twisted boundary conditions by defining

$$P_{L,L+1}|\uparrow,\ldots,\uparrow\rangle = |\uparrow,\ldots,\uparrow\rangle, P_{L,L+1}|\uparrow,\ldots,\downarrow\rangle = e^{+2i\phi}|\downarrow,\ldots,\uparrow\rangle, \quad (9.11)$$

$$P_{L,L+1}|\downarrow,\ldots,\downarrow\rangle = |\downarrow,\ldots,\downarrow\rangle, P_{L,L+1}|\downarrow,\ldots,\uparrow\rangle = e^{-2i\phi}|\uparrow,\ldots,\downarrow\rangle. \quad (9.12)$$

The states can, again, be described by the Baxter function $Q_1(u) = e^{\phi u}\prod_{i=1}^{N_1}(u-u_i)$. The Bethe roots u_i have a physical meaning—they represent the momenta p_i of spin down 'excitations' moving in a sea of spin ups via $u_i = \frac{1}{2}\cot\frac{p_i}{2}$ (see Figure 9.1). We find the roots u_j from the equation similar to (9.7)[1]

$$\left(\frac{u_k+i/2}{u_k-i/2}\right)^L = e^{-2i\phi}\prod_{j\neq k}^{N_1}\frac{u_k-u_j+i}{u_k-u_j-i}, \quad k=1,\ldots,N_1 \quad (9.13)$$

Exercise 9.3 Take log and expand for large u_k. You should get exactly the same as (9.7) up to a rescaling and shift of u_j.

from where one gets a discrete set of solutions for $\{u_i\}$. The energy is then given by

$$E = \sum_j^{N_1}\frac{2g^2}{u_j^2+1/4}. \quad (9.14)$$

Exercise 9.4 Take $L=2$ and compute the energy spectrum in two different ways: (1) by directly diagonalizing the Hamiltonian (9.10), which becomes a 4×4 matrix of the form

$$2g^2\begin{pmatrix} 0 & 0 & 0 & 0 \\ 0 & 2 & -1-e^{-2i\phi} & 0 \\ 0 & -1-e^{2i\phi} & 2 & 0 \\ 0 & 0 & 0 & 0 \end{pmatrix}$$

Next solve the Bethe equation (9.13) for $N_1 = 0,1,2$ and compute the energy from the formula (9.14).

---

[1] One should assume all u_j to be different as in the case of the harmonic oscillator.

One could ask what the analogue of the Schrödinger equation is in this case. The answer is given by the Baxter equation of the form

$$T(u)Q_1(u) = (u+i/2)^L Q_1(u-i) + (u-i/2)^L Q_1(u+i), \tag{9.15}$$

where $T(u)$ is a polynomial which plays the role of the potential, but it is not fixed completely and has to be determined from the self-consistency of (9.15).

Exercise 9.5 Show that the leading large u coefficients of $T(u)$ are $T(u) \simeq 2\cos\phi u^L + u^{L-1}(N_2 - N_1)\sin\phi$, where $N_2 = L - N_1$.

In practice we do not even need to know $T(u)$ as it is sufficient to require polynomiality from $T(u)$ to get (9.13) as a condition of cancellation of the poles.

Exercise 9.6 For generic polynomial $Q(u)$ we see that $T(u)$ is a rational function with poles at $u = u_k$, where $Q(u_k) = 0$. Show that these poles cancel if the Bethe ansatz equation (9.13) is satisfied.

Notice that given some polynomial $T(u)$ there is another polynomial (up to a $e^{-u\phi}$ multiplier to 'twist' ϕ) solution to the Baxter equation, just like we had before for the Schrödinger equation. Its asymptotics are $Q_2 \simeq e^{-u\phi} u^{N_2}$ where $N_2 = L - N_1$. The roots of Q_2 also has a physical interpretation—they describe the $L - N_1$ spin up particles moving in the sea of the spin downs (i.e. opposite to Q_1 which described the reflected picture where the spin ups played the role of the observers and the spin downs were considered as particles). The second solution together with the initial one should satisfy the Wronskian relation (in the same way as for the Schrödinger equation)[2]

$$\begin{vmatrix} Q_1(u-i/2) & Q_1(u+i/2) \\ Q_2(u-i/2) & Q_2(u+i/2) \end{vmatrix} \propto Q_{12}(u) \tag{9.16}$$

where $Q_{12}(u)$ satisfies

$$\frac{Q_{12}(u+i/2)}{Q_{12}(u-i/2)} = \frac{(u+i/2)^L}{(u-i/2)^L}, \tag{9.17}$$

so we conclude that $Q_{12}(u) = -2i\sin\phi u^L$.

Exercise 9.7 Show that if Q_1 and Q_2 are two linearly independent solutions of (9.15), then (9.17) holds.

---

[2] The $\propto$ sign is used to indicate that the equality holds up to a numerical multiplier (which can be easily recovered from the large u limit).

We see that there are strict similarities with the harmonic oscillator. Furthermore, it is possible to invert the above logic and prove the following statement: equation (9.16) plus the polynomiality assumption (up to an exponential prefactor) by itself implies the Bethe equation, from which we departed. This logic is very close to the philosophy of the QSC.

Exercise 9.8 Show that the Baxter equation is the following 'trivial' statement:

$$\begin{vmatrix} Q(u-i) & Q(u) & Q(u+i) \\ Q_1(u-i) & Q_1(u) & Q_1(u+i) \\ Q_2(u-i) & Q_2(u) & Q_2(u+i) \end{vmatrix} = 0, \text{ for } Q = Q_1 \text{ or } Q = Q_2. \tag{9.18}$$

From that determine $T(u)$ in terms of Q_1 and Q_2.

9.2.3 Nested Bethe ansatz and QQ-relations

The symmetry of the Heisenberg spin chain from Section 9.2.2 is $SU(2)$. In order to get closer to $PSU(2,2|4)$ (the symmetry of $\mathcal{N}=4$ SYM) we now consider a generalization of the Heisenberg spin chain for the $SU(3)$ symmetry group. For that we just have to assume that there are three possible states per chain site instead of two, otherwise the construction of the Hamiltonian is very similar.

The spectrum of the $SU(3)$ spin chain can be found from the 'nested' Bethe ansatz equations [28], which now involve two different unknown (twisted) polynomials Q_A and Q_B. They can be written as[3]:

$$1 = -\frac{Q_A^{++} Q_B^-}{Q_A^{--} Q_B^+}, \quad u = u_{A,i} \tag{9.19}$$

$$\frac{Q_\theta^+}{Q_\theta^-} = -\frac{Q_A^- Q_B^{++}}{Q_A^+ Q_B^{--}}, \quad u = u_{B,i}$$

and the energy is given by

$$E = i\partial_u \log \frac{Q_B^+}{Q_B^-}\bigg|_{u=0}. \tag{9.20}$$

We denote $Q_\theta = u^L$. We have also introduced some very convenient notation,

$$f^\pm = f(u \pm i/2), \quad f^{\pm\pm} = f(u \pm i), \quad f^{[\pm a]} = f(u \pm ai/2). \tag{9.21}$$

Exercise 9.9 Show that the $SU(3)$ Bethe equations reduce to the $SU(2)$ equations (9.13) and (9.14) when $Q_A = 1$.

---

[3] By twisted polynomials we mean the functions of the form $e^{\psi u} \prod_i (u - u_i)$, for some number ψ.

9.2.3.1 Bosonic duality

From the $SU(2)$ Heisenberg spin chain we learned that the Baxter polynomial $Q_1(u)$ contains as many roots as arrow-downs we have in our state. In particular the trivial polynomial $Q_1(u) = e^{-u\phi}$ corresponds to the state $|\uparrow\uparrow\ldots\uparrow\rangle$. One can also check that there is only one solution of the Bethe equations where $Q_1(u)$ is a twisted polynomial of degree L and it satisfies

$$e^{i\phi/2}Q_1^- - e^{-i\phi/2}Q_1^+ = 2i\sin\phi \, u^L e^{-u\phi}. \tag{9.22}$$

Exercise 9.10 Solve this equation for $L = 1$ and $L = 2$ and check that Q_1 also solves the Bethe equations of the $SU(2)$ spin chain. Compute the corresponding energy.

As this equation produces a polynomial of degree L it must correspond to the maximally 'excited' state $|\downarrow\downarrow\ldots\downarrow\rangle$. It is clear that even though physically these states are very similar, our current description in terms of the Bethe ansatz singles out one of them. We will see that there is a 'dual' description where the Q-function corresponding to the state $|\downarrow\downarrow\ldots\downarrow\rangle$ is trivial. In the case of the $SU(3)$ spin chain where we have three different states per node of the spin chain, which we can denote $1,2,3$, there are three equivalent vacuum states $|11\ldots 1\rangle$, $|22\ldots 2\rangle$, and $|33\ldots 3\rangle$, but only one of them corresponds to the trivial solution of the $SU(3)$ nested Bethe ansatz. Below we concentrate on the $SU(3)$ case and demonstrate that there are several equivalent sets of Bethe ansatz equations (9.19).

To build a dual set of Bethe equations we first have to pick a Q-function which we are going to dualize. For example we can build a new set of Bethe equations by replacing Q_A, a twisted polynomial of degree N_A, with another twisted polynomial $Q_{\tilde{A}}$ of degree $N_{\tilde{A}} = N_B - N_A$, where N_B is the degree of the polynomial Q_B. For that we find a dual Q-function $Q_{\tilde{A}}$ from

$$\begin{vmatrix} Q_A^- & Q_A^+ \\ Q_{\tilde{A}}^- & Q_{\tilde{A}}^+ \end{vmatrix} \propto Q_B(u). \tag{9.23}$$

Let's see that $Q_{\tilde{A}}$ satisfies the same Bethe equation. By evaluating (9.23) at $u = u_{\tilde{A},i} + i/2$ and dividing by the same relation evaluated at $u = u_{\tilde{A},i} - i/2$ we get

$$\frac{Q_A Q_{\tilde{A}}^{++} - 0}{0 - Q_A Q_{\tilde{A}}^{--}} = \frac{Q_B^+}{Q_B^-}, \quad u = u_{\tilde{A},i}, \tag{9.24}$$

which is exactly the first equation (9.19) with A replaced by $\tilde{A}$! To accomplish our goal we should also exclude Q_A from the second equation. For that we notice that at $u = u_{B,i}$ the relation gives

$$\frac{Q_A^-}{Q_A^+} = \frac{Q_{\tilde{A}}^-}{Q_{\tilde{A}}^+}, \quad u = u_{B,i}, \tag{9.25}$$

which allows us to rewrite the whole set of equations (9.19) in terms of $Q_{\tilde{A}}$. We call this transformation a bosonic duality. Similarly one can apply the dualization procedure to Q_B. We determine $Q_{\tilde{B}}$ from

$$\begin{vmatrix} Q_B^- & Q_B^+ \\ Q_{\tilde{B}}^- & Q_{\tilde{B}}^+ \end{vmatrix} \propto Q_A(u) Q_\theta(u). \tag{9.26}$$

By doing this we will be able to replace B by $\tilde{B}$ in (9.19). Let us also show that we can use $Q_{\tilde{B}}$ instead of Q_B in the expression for the energy (9.14). We recall that $Q_\theta(u) \propto u^L$, so evaluating (9.26) at $u = 0$ we get

$$Q_B(-\tfrac{i}{2}) Q_{\tilde{B}}(+\tfrac{i}{2}) = Q_B(+\tfrac{i}{2}) Q_{\tilde{B}}(-\tfrac{i}{2}). \tag{9.27}$$

We can also differentiate (9.26) in u once and then set $u = 0$, so that

$$Q'_B(-\tfrac{i}{2}) Q_{\tilde{B}}(+\tfrac{i}{2}) + Q_B(-\tfrac{i}{2}) Q'_{\tilde{B}}(+\tfrac{i}{2}) = Q'_B(+\tfrac{i}{2}) Q_{\tilde{B}}(-\tfrac{i}{2}) + Q_B(+\tfrac{i}{2}) Q'_{\tilde{B}}(-\tfrac{i}{2}). \tag{9.28}$$

Dividing (9.28) by (9.27) we get

$$\frac{Q'_B(-\tfrac{i}{2})}{Q_B(-\tfrac{i}{2})} + \frac{Q'_{\tilde{B}}(+\tfrac{i}{2})}{Q_{\tilde{B}}(+\tfrac{i}{2})} = \frac{Q'_B(+\tfrac{i}{2})}{Q_B(+\tfrac{i}{2})} + \frac{Q'_{\tilde{B}}(-\tfrac{i}{2})}{Q_{\tilde{B}}(-\tfrac{i}{2})}, \tag{9.29}$$

which indeed gives

$$E = i\partial_u \log \frac{Q_B^+}{Q_B^-}\bigg|_{u=0} = i\partial_u \log \frac{Q_{\tilde{B}}^+}{Q_{\tilde{B}}^-}\bigg|_{u=0}. \tag{9.30}$$

Better notation for Q-functions. One can combine the above duality transformations and say dualize Q_A after dualizing Q_B and so on. In order to keep track of all possible transformations one should introduce some notation, as otherwise we can end up with multiple tildas. Another question we will try to answer in this part is how many equivalent BAs we will generate by applying the duality many times to various nodes.

In order to keep track of the dualities we place numbers $1, 2, 3$ in between the nodes of the Dynkin diagram. We place the Q-functions on the nodes of the diagram as in Figure 9.2. Then we interpret the duality as an exchange of the corresponding labels sitting on the links of the diagram, so if before the dualization of Q_A we had $1, 2, 3$, after the duality we have to exchange the indices 1 and 2 obtaining $2, 1, 3$. If instead we first dualized Q_B we would obtain $1, 3, 2$. Each duality produces a permutation of the numbers. We also use these numbers to label the Q-functions. Namely we assign the indexes to the Q function in accordance with the numbers appearing above the given node. So, in particular, in the new notation

$$Q_A = Q_1, \ Q_B = Q_{12}. \tag{9.31}$$

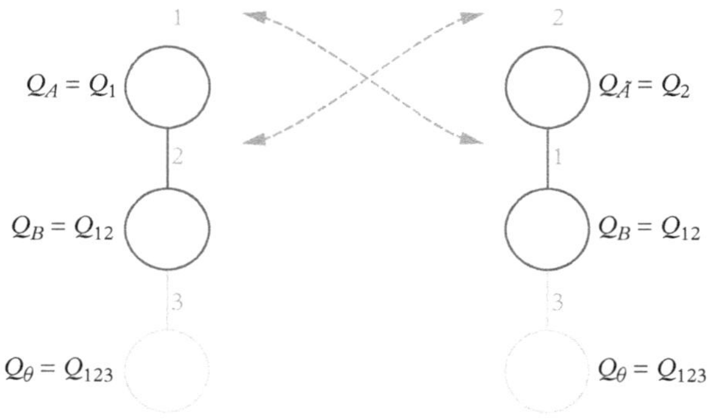

410 *Spectrum of $\mathcal{N}=4$ supersymmetric Yang–Mills theory and the quantum spectral curve*

Figure 9.2 *Bosonic duality applied to the first node of the BA.*

Each order of the indices naturally corresponds to a particular set of Bethe equations. For instance, the initial set of Bethe equations on Q_A, Q_B correspond to the order 1, 2, 3 and the Bethe ansatz (BA) for $Q_{\tilde{A}}, Q_B$ correspond to 2, 1, 3, and so on. Now we can answer the question of how many dual BA systems we could have; this is given by the number of permutations of 1, 2, 3 i.e. for the case of $SU(3)$ we get six equivalent systems of BA equations.

Following our prescription we also denote

$$Q_{\tilde{A}} = Q_2, \ Q_{\tilde{B}} = Q_{13}. \tag{9.32}$$

We note that we should not distinguish Qs which only differ by the order of indices. So, for instance, Q_{21} and Q_{12} is the same Q_B. We can count the total number of various Q-functions we could possibly generate with the dualities: $2^3 - 2 = 6$ different Q-functions, which are

$$Q_i, \ Q_{[ij]}, \ i,j = 1,\ldots,3, \tag{9.33}$$

For completeness we also add $Q_\emptyset \equiv 1$ and $Q_{123} = Q_{[ijk]} \equiv Q_\theta = u^L$ so that in total we have 2^3. For general $SU(N)$ we will find 2^N different Q-functions. We see that the number of the Q-functions grows rapidly with the rank of the symmetry group. For $PSU(2,2|4)$ we get 256 functions, and we should study the relations among them in more detail.

QQ-relations. Let us rewrite the bosonic duality in the new notation. The relation (9.23) becomes

$$\begin{vmatrix} Q_i^- & Q_i^+ \\ Q_j^- & Q_j^+ \end{vmatrix} \propto Q_{ij} Q_\emptyset, \tag{9.34}$$

where we added $Q_\emptyset = 1$ to the r.h.s. to make both l.h.s and r.h.s bilinear in Q. Very similarly (9.26) gives

$$\begin{vmatrix} Q_{12}^- & Q_{12}^+ \\ Q_{13}^- & Q_{13}^+ \end{vmatrix} \propto Q_1(u) Q_{123}(u). \tag{9.35}$$

We see that both identities can be written in one go as

$$\begin{vmatrix} Q_{Ii}^- & Q_{Ii}^+ \\ Q_{Ij}^- & Q_{Ij}^+ \end{vmatrix} \propto Q_I(u) Q_{Iij}(u), \tag{9.36}$$

where for general $SU(N)$ we would have $i = 1, \ldots, N$ and $j = 1, \ldots, N$ and I represents a set of indices such that in (9.35) it is an empty set $I = \emptyset$ and for the second identity (9.36) I contains only one element 1. Note that no indices inside I are involved with the relations and in the r.h.s. we get indices i and j glued together in the new function. We see that proceeding in this way we can build any Q-function starting from the basic Q_i with one index only. For that we can first take $I = \emptyset$ and build Q_{ij}, then take $I = i$ and build Q_{ijk}, and so on. It is possible to combine these steps together to get explicitly

$$Q_{ijk} Q_\emptyset^+ Q_\emptyset^- \propto \begin{vmatrix} Q_i^{--} & Q_i & Q_i^{++} \\ Q_j^{--} & Q_j & Q_j^{++} \\ Q_k^{--} & Q_k & Q_k^{++} \end{vmatrix}. \tag{9.37}$$

Whereas the first identity (9.34) is obvious from the definition, the second (9.37) is a simple exercise to prove from (9.36).

Exercise 9.11 Prove (9.37) using the following Mathematica code

```
(*define Q to be absolutely antisymmetric*)
Q[a___] := Signature[{a}] Q @@ Sort[{a}] /; ! OrderedQ[{a}]
(*program bosonic duality*)
Bosonic[J___, i_, j_] := Q[J, i, j][u_] -> (
Q[J, i][u + I/2] Q[J, j][u - I/2] -
Q[J, i][u - I/2] Q[J, j][u + I/2])/Q[J][u];
(*checking the identity*)
Q[1, 2, 3][u] Q[][u + I/2] Q[][u - I/2] /.
Bosonic[1, 2, 3] /. Bosonic[1, 2] /. Bosonic[1, 3] /.
Bosonic[2, 3] // Factor
```

Also derive a similar identity for Q_{ijkl} using the same code.

From the previous exercise it should be clear that we can generate any $Q_{ij\ldots k}$ as a determinant of the basic N Q-functions Q_i. In particular the 'full-set' Q-function $Q_{12\ldots N}$, which is also $Q_\theta = u^L$, can be written as a determinant of N basic polynomials Q_i.

Interestingly this identity by itself is constraining enough to give rise to the full spectrum of the $SU(N)$ spin chain! Indeed, $Q_{12...N}$ is a polynomial of degree L and thus we get L non-trivial relations on the coefficients of the (twisted) polynomials Q_i, which together contain exactly L Bethe roots. This means that this relation alone is equivalent to the whole set of nested Bethe ansatz equations. So we can put aside a non-unique BA approach, dependent on the choice of the vacuum, and replace it completely by a simple determinant like (9.37). In other words the QQ-relations and the condition of polynomiality is all we need to quantize this quantum integrable model. We will argue that for $N=4$ SYM we only have to replace the polynomiality with another slightly more complicated analyticity condition but otherwise keep the same QQ-relations. We will have to, however, understand what the QQ-relations look like for the case of super symmetries like $SU(N|M)$, which is described in Section 9.2.3.2.

9.2.3.2 Fermionic duality in $SU(N|M)$

We will see how the discussion in the previous section generalizes to the super-group case. Our starting point will be again the set of nested Bethe ansatz equations, which follow the pattern of the Cartan matrix. Let us discuss the construction of the Bethe ansatz. Below we wrote the Dynkin diagram, Cartan matrix and the Bethe ansatz equations for the $SU(3|3)$ super spin chain

$$
\begin{array}{l}
Q_A \ \bigcirc \\
Q_B \ \bigcirc \\
Q_C \ \otimes \\
Q_D \ \bigcirc \\
Q_E \ \bigcirc
\end{array}
\quad
\begin{array}{|ccccc|}
\hline
2 & -1 & 0 & 0 & 0 \\
-1 & 2 & -1 & 0 & 0 \\
0 & -1 & 0 & +1 & 0 \\
0 & 0 & +1 & -2 & +1 \\
0 & 0 & 0 & +1 & -2 \\
\hline
\end{array}
\quad
\begin{array}{l}
-1 = (Q_A^{++} Q_B^-)/(Q_A^{--} Q_B^+), u = u_{A,i} \\
-1 = (Q_A^- Q_B^{++} Q_C^-)/(Q_A^+ Q_B^{--} Q_C^+), u = u_{B,i} \\
+1 = (Q_B^- Q_D^+)/(Q_B^+ Q_D^-), u = u_{C,i} \\
-1 = (Q_C^+ Q_D^{--} Q_E^+)/(Q_C^- Q_D^{++} Q_E^-), u = u_{D,i} \\
-1 = (Q_D^+ Q_E^{--})/(Q_D^- Q_E^{++}), u = u_{E,i}
\end{array}
$$

(9.38)

The Q-functions still correspond to the nodes of the Dynkin diagrams and the shift of the argument of the Q-functions entering the numerators of the Bethe equations simply follow the pattern of the Cartan matrix (with the inverse shifts in numerators). Since the structure of the equations for the bosonic nodes is the same as before, one can still apply the bosonic duality transformation for instance on Q_B and replace it with $Q_{\tilde{B}}$. However for the fermionic type nodes (normally denoted by a crossed circle), such as Q_C, we get a new type of duality transformation,

$$Q_C Q_{\tilde{C}} \propto \begin{vmatrix} Q_B^- & Q_B^+ \\ Q_D^- & Q_D^+ \end{vmatrix}, \tag{9.39}$$

which look similar to the bosonic one with the difference that we can extract explicitly the dual Baxter polynomial $Q_{\tilde{C}}$.[4] Let us show that the middle Bethe equation can be

---

[4] Whereas for the bosonic duality (9.23) the dual Baxter polynomial occured in a complicated way and one had to solve a first-order finite difference equation in order to extract it.

obtained from the duality relation (9.39). Indeed we see again that for both $u = u_{C,i}$ and $u = u_{\hat{C},i}$ we get the middle equation,

$$1 = \frac{Q_B^+ Q_D^-}{Q_B^- Q_D^+}, \quad u = u_{\hat{C},i} \text{ or } u = u_{C,i}. \tag{9.40}$$

Next we should be able to exclude Q_C in the other two equations. For that we set $u = u_{B,i} + i/2$ and $u = u_{B,i} + i/2$ to get

$$Q_C^+ Q_{\hat{C}}^+ = c(0 - Q_D Q_B^{++}), \; Q_C^- Q_{\hat{C}}^- = +c(Q_D Q_B^{--} - 0) u = u_{B,i}. \tag{9.41}$$

Dividing one by the other

$$-1 = \frac{Q_C^+ Q_{\hat{C}}^+}{Q_C^- Q_{\hat{C}}^-} \frac{Q_B^{--}}{Q_B^{++}}, \quad u = u_{B,i} \tag{9.42}$$

which allows up to exclude Q_C from the second equation of (9.44). This then becomes

$$-1 = \frac{Q_A^- Q_B^{++} Q_C^-}{Q_A^+ Q_B^{--} Q_C^+} \leftrightarrow +1 = \frac{Q_A^- Q_{\hat{C}}^+}{Q_A^+ Q_{\hat{C}}^-}, \quad u = u_{B,i}. \tag{9.43}$$

As we see this changes the type of the equation from bosonic to fermionic. Thus we also change the type of the Dynkin diagram. This is expected since for super algebras the Dynkin diagram is not unique. Similarly the fourth equation also changes in a similar way. To summarize, after duality we get

$$
\begin{array}{l}
Q_A \; \bigcirc \\
Q_B \; \otimes \\
Q_{\hat{C}} \; \otimes \\
Q_D \; \otimes \\
Q_E \; \bigcirc
\end{array}
\quad
\begin{array}{|c|c|c|c|c|}
\hline
2 & -1 & 0 & 0 & 0 \\
\hline
-1 & 0 & +1 & 0 & 0 \\
\hline
0 & +1 & 0 & -1 & 0 \\
\hline
0 & 0 & -1 & 0 & +1 \\
\hline
0 & 0 & 0 & +1 & -2 \\
\hline
\end{array}
\quad
\begin{array}{l}
-1 = (Q_A^{++} Q_B^-)/(Q_A^{--} Q_B^+), \; u = u_{A,i} \\
-1 = (Q_A^- Q_{\hat{C}}^+)/(Q_A^+ Q_{\hat{C}}^-), \; u = u_{B,i} \\
-1 = (Q_B^+ Q_D^-)/(Q_B^- Q_D^+), \; u = u_{\hat{C},i} \\
-1 = (Q_{\hat{C}}^- Q_E^+)/(Q_{\hat{C}}^+ Q_E^-), \; u = u_{D,i} \\
-1 = (Q_D^+ Q_E^{--})/(Q_D^- Q_E^{++}), \; u = u_{E,i}
\end{array}
$$
(9.44)

Index notation. Again, in order to keep track of all possible combinations of dualities we have to introduce index notation. In the super case we label the links in the Dynkin diagram by two types of indices (with hat and without). The type of the index changes each time we cross a fermionic node. For instance our initial set of Bethe equations corresponds to the indices $123\hat{1}\hat{2}\hat{3}$. The fermionic duality transformation again simply exchanges the labels on the links of the Dynkin diagram (see Figure 9.3). So after duality we get $12\hat{1}3\hat{2}\hat{3}$, which is consistent with the $\bigcirc - \otimes - \otimes - \otimes - \bigcirc$ grading of the resulting Bethe ansatz equations. Finally, we label the Q-functions by two antisymmetric

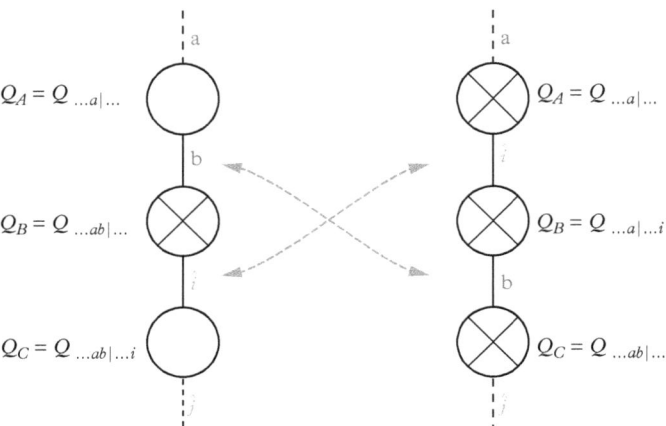

Figure 9.3 *Fermionic duality.*

groups of indices—with hat and without, again simply listing all indices appearing above the given node of the Dynkin diagram. In particular we get

$$Q_A = Q_1, \ Q_B = Q_{12}, \ Q_C = Q_{123}, \qquad (9.45)$$
$$Q_{\hat{C}} = Q_{12\hat{1}}, \ Q_D = Q_{123\hat{1}}, \ Q_E = Q_{123\hat{1}\hat{2}}.$$

An alternative notation is to omit hats and separate the two sets of indices by a vertical line:

$$Q_A = Q_{1|\emptyset}, \ Q_B = Q_{12|\emptyset}, \ Q_C = Q_{123|\emptyset}, \qquad (9.46)$$
$$Q_{\hat{C}} = Q_{12|1}, \ Q_D = Q_{123|1}, \ Q_E = Q_{123|12}.$$

Exercise 9.12 The fermionic duality transformation changes the type of the Dynkin diagram. The simplest way to understand which diagram one gets after the duality is to follow the indices attached to the links. Each time the type of the index changes (from hatted to non-hatted) you should draw a cross. List all possible Dynkin diagrams corresponding to $SU(3|3)$ Lie algebra.

Fermionic QQ-relations. In index notation (9.39) becomes

$$\boxed{Q_{Ib}Q_{I\hat{i}} \propto Q_I^- Q_{Ib\hat{i}}^+ - Q_I^+ Q_{Ib\hat{i}}^-}. \qquad (9.47)$$

For completeness let us write here the bosonic duality relations:

$$\boxed{Q_I Q_{Iab} \propto Q_{Ia}^+ Q_{Ib}^- - Q_{Ib}^+ Q_{Ia}^-, \ Q_I Q_{I\hat{i}\hat{j}} \propto Q_{I\hat{i}}^+ Q_{I\hat{j}}^- - Q_{I\hat{j}}^+ Q_{I\hat{i}}^-}. \qquad (9.48)$$

In the case of $SU(N|M)$ one could derive all Q functions in terms of $N+M$ functions Q_a and $Q_{\hat{\imath}}$. We will demonstrate this in Section 9.24 in the example of $SU(4|4)$.

9.2.4 QQ-relations for $PSU(2,2|4)$ spin chain

The global symmetry of $\mathcal{N}=4$ SYM is $PSU(2,2|4)$. The QQ-relations from Section 9.2.3.2 associated with this symmetry constitute an important part of the QSC construction. The symmetry (up to a real form and a projection) is the same as $SU(4|4)$. In this section we specialize the QQ-relations from the previous part to this case and derive all the most important relations among Q-functions. In particular, we show that all 256 various Q-functions can be derived from just $4+4$ Q-functions with one index,

$$Q_{a|\emptyset}, \quad Q_{\emptyset|i}. \tag{9.49}$$

which are traditionally denoted in the literature as

$$\mathbf{P}_a, \quad \mathbf{Q}_i. \tag{9.50}$$

These are the elementary Q-functions.

For us, another important object is $Q_{a|i}$. According to the general consideration above, it can be obtained from the fermionic duality relation (9.47) with $I = \emptyset$,

$$Q_{a|j}^+ - Q_{a|j}^- = \mathbf{P}_a \mathbf{Q}_j. \tag{9.51}$$

This is the first-order equation on $Q_{a|i}$ which one should solve; and the formal solution to this equation is[5]

$$Q_{a|j}(u) = -\sum_{n=0}^{\infty} \mathbf{P}_a(u + i\tfrac{2n+1}{2})\mathbf{Q}_j(u + i\tfrac{2n+1}{2}). \tag{9.52}$$

Exercise 9.13 Find a solution to equation (9.51) for $\mathbf{P}_a\mathbf{Q}_j = e^{\phi u}$ and also for $\mathbf{P}_a\mathbf{Q}_j = 1/u^2$.

Once we know $Q_{a|i}$ we can build any Q-function explicitly in terms of $Q_{a|i}, \mathbf{Q}_i$, and $\mathbf{P}_a$. For example, using the bosonic duality we can get

$$Q_{\mathbf{ab}|i} = \frac{Q_{\mathbf{a}|i}^+ Q_{\mathbf{b}|i}^- - Q_{\mathbf{a}|i}^- Q_{\mathbf{b}|i}^+}{\mathbf{Q}_i}. \tag{9.53}$$

---

[5] Note that there is a freedom to add a constant to $Q_{a|i}$. This freedom is fixed in the twisted case as we should require that $Q_{a|j}$ has a 'pure' asymptotics at large u i.e. $e^{\phi_{ai} u} u^\alpha (1 + A_1/u + A_2/u^2 + \cdots)$.

In this way we can build all Q-functions explicitly in terms of $Q_{a|i}, \mathbf{Q}_i$, and $\mathbf{P}_a$. There is a nice simplification taking place for Q-functions with equal numbers of indices:

$$Q_{ab|ij} = \begin{vmatrix} Q_{a|i} & Q_{a|j} \\ Q_{b|i} & Q_{b|j} \end{vmatrix}. \qquad (9.54)$$

Exercise 9.14 Prove (9.54) using the following Mathematica code:

```
(*define Q to be absolutely antisymmetric*)
Q[a___][b___][u_] := Signature[{a}] Signature[{b}]
Q[Sequence @@ Sort[{a}]][Sequence @@ Sort[{b}]][u]
/; ! (OrderedQ[{a}] && OrderedQ[{b}])
(*program bosonic and fermionic dualities*)
B1[J___, a_, b_][K___] := Q[J, a, b][K][u_] :>
(Q[J, a][K][u + I/2] Q[J, b][K][u - I/2] -
Q[J, a][K][u - I/2] Q[J, b][K][u + I/2])/Q[J][K][u];
B2[K___][J___, i_, j_] :=
Q[K][J, i, j][u_] :> (Q[K][J, i][u + I/2] Q[K][J, j][u - I/2] -
Q[K][J, i][u - I/2] Q[K][J, j][u + I/2])/Q[K][J][u];
F1[K___, a_][J___, i_][u_] := Q[K, a][J, i][u] :>
(Q[K, a][J, i][u - I] Q[K][J][u - I] +
Q[K, a][J][u - I/2] Q[K][J, i][u - I/2])/Q[K][J][u]
F2[K___, a_][J___, i_][u_] := Q[K, a][J, i][u] :>
(Q[K, a][J, i][u + I] Q[K][J][u + I] -
Q[K, a][J][u + I/2] Q[K][J, i][u + I/2])/Q[K][J][u]
(*deriving the identity*)
Q[a, b][i, j][u] /. B1[a, b][i, j] /. B2[a][i, j] /. B2[b][i, j] /.
Flatten[Table[F1[c][k][u + I], {c, {a, b}}, {k, {i, j}}]] /.
Flatten[Table[F2[c][k][u - I], {c, {a, b}}, {k, {i, j}}]] /.
B2[][i, j] // Simplify
```

Also derive a similar identity for $Q_{abc|ijk}$ using the same code. The general strategy is to use the bosonic duality to decompose Q's into Q-functions with fewer indices. Then use (9.51) to bring all $Q_{a|k}(u+in)$ to the same argument $Q_{a|k}(u)$. After that the expression should simplify enormously. Also show the following identities to hold:

$$Q_{abc|ijkl} = \mathbf{Q}_i Q^+_{abc|jkl} - \mathbf{Q}_j Q^+_{abc|kli} + \mathbf{Q}_k Q^+_{abc|lij} - \mathbf{Q}_l Q^+_{abc|ijk}, \qquad (9.55)$$

$$Q_{abcd|ijk} = \mathbf{P}_a Q^+_{bcd|ijk} - \mathbf{P}_b Q^+_{cda|ijk} + \mathbf{P}_c Q^+_{dab|ijk} - \mathbf{P}_d Q^+_{abc|ijk}. \qquad (9.56)$$

Also check (9.57) and (9.58) that follow.

In particular for the Q-function with all indices $Q_{1234|1234}$ (remember that the Q-function with all indices played an important role in the XXX spin chain giving the external 'potential' $Q_\theta = u^L$), we get

$$Q_{1234|1234} = \begin{vmatrix} Q_{1|1} & Q_{1|2} & Q_{1|3} & Q_{1|4} \\ Q_{2|1} & Q_{2|2} & Q_{2|3} & Q_{2|4} \\ Q_{3|1} & Q_{3|2} & Q_{3|3} & Q_{3|4} \\ Q_{4|1} & Q_{4|2} & Q_{4|3} & Q_{4|4} \end{vmatrix}. \tag{9.57}$$

Finally, one can show that

$$Q^+_{1234|1234} - Q^-_{1234|1234} = \sum_{a,\check{i}} \mathbf{Q}_i \mathbf{P}_a Q^-_{1234\check{a},1234\check{i}}, \tag{9.58}$$

where the check (inverse hat) denotes an 'index annihilator' i.e. for example $Q_{1234\check{4}|...} = Q_{123|...}$ and $Q_{123\check{4}\check{3}|...} = -Q_{123\check{3}\check{4}|...} = -Q_{124|...}$ and so on.

Hodge duality. The $SU(4|4)$ Dynkin diagram has an obvious symmetry—we can flip it upside down. At the same time the labelling of the Q-functions essentially breaks this symmetry as we agreed to list all indexes from above a given node and not below. To fix this we can introduce a Hodge dual set of Q-functions by defining

$$Q^{a_1...a_n|i_1...i_m} \equiv (-1)^{nm} \epsilon^{a_1...a_n b_1...b_{4-n}} \epsilon^{i_1...i_m j_1...j_{4-m}} Q_{b_1...b_{4-n}|j_1...j_{4-m}}, \tag{9.59}$$

with $b_1 < \cdots < b_{4-n}$ and $j_1 < \cdots < j_{4-n}$ so that there is only one term in the r.h.s. One can check that these Q-functions with upper indices satisfy the same QQ-relations as the initial Q-functions[6].

Finally, we have already set $Q_{\emptyset|\emptyset} = 1$ and considering the symmetry of the system we should also set $Q_{1234|1234} = Q^{\emptyset|\emptyset} = 1$. In fact that is indeed the case for $\mathcal{N} = 4$ SYM, whereas for the spin chains we have $Q_\theta = u^L$ attached to one of the ends of the Dynkin diagram, which breaks the symmetry.

Assuming $Q_{1234|1234} = 1$ we get some interesting consequences. In particular the l.h.s. of (9.58) vanishes and we get

$$\mathbf{Q}_i \mathbf{P}_a Q^{a|i} = 0. \tag{9.60}$$

Also we can rewrite (9.55) and (9.56) in our new notation,

$$\mathbf{P}^a \equiv Q^{a|\emptyset} = Q^{a|i}(u+i/2)\mathbf{Q}_i, \tag{9.61}$$

$$\mathbf{Q}^i \equiv Q^{\emptyset|i} = Q^{a|i}(u+i/2)\mathbf{P}_a. \tag{9.62}$$

Combining that with (9.60) we get

$$\mathbf{P}_a \mathbf{P}^a = \mathbf{Q}_i \mathbf{Q}^i = 0. \tag{9.63}$$

---

[6] In particular (9.59) implies $Q^{\emptyset|1} = +Q_{1234|234}$ and $Q^{\emptyset|2} = -Q_{1234|134}$ and so on.

Finally, we can expand the determinant of the 4×4 matrix in (9.57) in the first row to get

$$1 = Q_{1|1}Q_{234|234} - Q_{1|2}Q_{234|134} + Q_{1|3}Q_{234|124} - Q_{1|4}Q_{234|123}, \qquad (9.64)$$

which is equivalent to $-1 = Q_{1|a}Q^{1|a}$. Also, we can replace the first row in (9.57) by $Q_{2|i}$ instead of $Q_{1|i}$ to get the zero determinant. At the same time, expanding this determinant in the first row will result in $0 = Q_{2|a}Q^{1|a}$. At the end we will get the following general expression

$$Q_{i|a}Q^{j|a} = -\delta_i^j, \qquad (9.65)$$

which implies that $Q^{i|a}$ is inverse to $Q_{i|a}$.

With these relations we have completed the task of building the QQ-relations for $SU(4|4)$ symmetry (with an additional condition that $Q_{1234|1234} = 1$, which can be associated with 'P' in $PSU(2,2|4)$). The next step is to understand the analytical properties of the Q-functions. For the case of the spin chain all Q-functions are simply polynomials and it was sufficient to produce the spectrum from the QQ-relations. However, in that construction there is no room for a continuous parameter—the 't Hooft coupling $g = \frac{\sqrt{\lambda}}{4\pi}$, and thus for $N = 4$ SYM the analytical properties should be more complicated and we will motivate the analyticity in Section 9.3. The analytical properties are the missing ingredients in the construction and to deduce them we will have to revise the strong coupling limit.

9.3 Classical string and strong coupling limit of QSC

In this section we briefly describe the action of the super-string in $AdS^5 \times S_5$, following closely [1]. We also advise study of Chapter 5 in this volume.

9.3.1 Classical string action

The classical action is similar to the action of the principal chiral field (PCF), so let us briefly review this. The fields $g(\sigma, \tau)$ in PCF belong to the $SU(N)$ group. One builds 'currents' out of them by

$$\mathcal{J}_\mu \equiv -g^{-1}\partial_\mu g \qquad (9.66)$$

and then the classical action is simply

$$S = \frac{\sqrt{\lambda}}{4\pi} \int \mathrm{tr}(\mathcal{J} \wedge \mathcal{J}). \qquad (9.67)$$

The global symmetry of this action is $SU_L(N) \times SU_R(N)$ since we can change $g(\sigma, \tau) \to h_L g(\sigma, \tau) h_R$ for arbitrary $h_L, h_R \in SU(N)$ without changing the action.

The construction for the Green–Schwartz superstring action is very similar. We take $g \in SU(2,2|4)$ and then the current $\mathcal{J}$ (taking values in the $su(2,2|4)$ algebra) is built in the same way as in (9.66). The only new ingredient is that we have to decompose the current into four components in order to ensure an extra local $sp(2,2) \times sp(4)$ symmetry in the way we will now describe.

The superalgebra $su(2,2|4)$ can be represented by 8×8 supertraceless supermatrices,

$$M = \left(\begin{array}{c|c} A & B \\ \hline C & D \end{array} \right), \tag{9.68}$$

where $A \in u(2,2)$ and $B \in u(4)$ and the fermionic components are related by

$$C = B^\dagger \left(\begin{array}{cc} 1_{2\times 2} & 0 \\ 0 & -1_{2\times 2} \end{array} \right). \tag{9.69}$$

An important property of the $su(2,2|4)$ superalgebra is that there is a Z_4 automorphism (meaning that one should act four times to get a trivial transformation). This Z_4 automorphism has its counterpart in the QSC construction, as we discuss later. Its action on an element of the algebra is defined in the following way:

$$\phi[M] \equiv \left(\begin{array}{c|c} EA^T E & -EC^T E \\ \hline EB^T E & ED^T E \end{array} \right), \quad E = \left(\begin{array}{cccc} 0 & -1 & 0 & 0 \\ 1 & 0 & 0 & 0 \\ 0 & 0 & 0 & -1 \\ 0 & 0 & 1 & 0 \end{array} \right). \tag{9.70}$$

It is easy to see that $\phi^4 = 1$. The consequence of this is that any element of the algebra can be decomposed into the sum $M = M^{(0)} + M^{(2)} + M^{(3)} + M^{(4)}$, such that $\phi[M^{(n)}] = i^n M^{(n)}$.

Exercise 9.15 Find $M^{(n)}$ for $n = 0, 1, 2, 3$ explicitly in terms of A, B, C, D, E.

The invariant part $M^{(0)}$ is exactly $sp(2,2) \times sp(4)$. In particular we can decompose the current $\mathcal{J} = \mathcal{J}^{(0)} + \mathcal{J}^{(1)} + \mathcal{J}^{(2)} + \mathcal{J}^{(3)}$ and define the action as

$$S = \frac{\sqrt{\lambda}}{4\pi} \int \mathrm{str}\left(\mathcal{J}^{(2)} \wedge *\mathcal{J}^{(2)} - \mathcal{J}^{(1)} \wedge \mathcal{J}^{(3)} \right). \tag{9.71}$$

Exercise 9.16 Show that $M^{(0)} \in sp(2,2) \times sp(4)$.

Exercise 9.17 The fact that the action does not contain $\mathcal{J}^{(0)}$ guarantees the local invariance of the action w.r.t. $sp(2,2) \times sp(4)$. Explain why.

The equations of motion which one can derive from the action (9.71) are

$$\partial_\mu k_\mu = 0, \quad k_\mu = g K_\mu g^{-1}, \quad K = \mathcal{J}^{(2)} + \frac{1}{2} * \mathcal{J}^{(1)} - \frac{1}{2} * \mathcal{J}^{(3)}. \tag{9.72}$$

One can also interpret k_μ as a Noether charge w.r.t to the global $PSU(2,2|4)$ symmetry $g \to hg$.

Exercise 9.18 Derive k_μ from Noether's theorem.

9.3.2 Classical integrability

The equations of motion (9.72) and the flatness condition,

$$d\mathcal{J} - \mathcal{J} \wedge \mathcal{J} = 0, \tag{9.73}$$

can be packed into the flatness condition of the 1-form,

$$A(u) = \mathcal{J}^{(0)} + \frac{u}{\sqrt{u^2 - 4g^2}} \mathcal{J}^{(2)} - \frac{2g}{\sqrt{u^2 - 4g^2}} * \mathcal{J}^{(2)}, \quad u \in \mathbb{C}, \tag{9.74}$$

where we use that classically we can set $\mathcal{J}^{(1)} = \mathcal{J}^{(3)} = 0$, as these fermionic parts only become relevant at 1-loop level.

Exercise 9.19 By expanding in Taylor series in u show that each term in the expansion is zero as a consequence of (9.72) and (9.73), i.e.

$$dA(u) - A(u) \wedge A(u) = 0, \quad \forall u. \tag{9.75}$$

Hint: First verify (9.75) for $u = 0$. For that you will have to firstly project the equation (9.73) into Z_4 components. For example,

$$d\mathcal{J}^{(0)} - \mathcal{J}^{(0)} \wedge \mathcal{J}^{(0)} - \mathcal{J}^{(2)} \wedge \mathcal{J}^{(2)} = 0. \tag{9.76}$$

The existence of the flat connection $A(u)$, depending on a spectral parameter u, implies integrability of the model at least at the classical level. Note that (9.75)[7] implies that $A(u)$ is a *pure gauge* i.e. there exists a matrix valued function $G(\sigma, \tau, u)$ such that

$$A_\mu(u) = -G^{-1} \partial_\mu G. \tag{9.77}$$

---

[7] Which in more familiar notations becomes $F_{\mu\nu} = \partial_\mu A_\nu - \partial_\nu A_\mu + [A_\mu, A_\nu] = 0$.

Classical string and strong coupling limit of QSC 421

A way to build G is to compute the Wilson line from some fixed point to (σ,τ),

$$G(\sigma,\tau,u) = \text{Pexp} \int^{(\sigma,\tau)} A(u). \tag{9.78}$$

Using G we can build the monodromy matrix (which is a supermatrix $(4+4)\times(4+4)$):

$$\Omega(u,\tau) = G^{-1}(0,\tau,u)G(2\pi,\tau,u) = \text{Pexp}\oint_\gamma A(u), \tag{9.79}$$

where γ is a closed path starting and ending at some point on the world-sheet and wrapping around once. The flatness condition allows us to deform the contour freely provided the endpoints are fixed. Shifting the whole path in time will produce a similarity transformation of $\Omega(u,\tau)$.

Exercise 9.20 Show that the eigenvalues of $\Omega(u,\tau)$ do not depend on τ if A is flat.

We denote the eigenvalues of $\Omega(u,\tau)$ as

$$\{e^{ip_1}, e^{ip_2}, e^{ip_3}, e^{ip_4} | e^{i\hat{p}_1}, e^{i\hat{p}_2}, e^{i\hat{p}_3}, e^{i\hat{p}_4}\}. \tag{9.80}$$

These functions of the spectral parameter u are called quasimomenta. Since they do not depend on time they represent a generating function for conserved quantities. One can, for instance, expand $p_i(u)$ in the Taylor series at large u to obtain inifinitely many integrals of motion, which leads to integrability of string theory. We will now study the analytic properties of the quasimomenta.

'Zhukovsky' square roots. All the quasimomenta have a square root singularity with the branch points at $\pm 2g$ (inherited from the definition of A (9.74)). Note that the analytic continuation under the cut changes the sign of the terms with $\mathcal{J}^{(2)}$ in (9.74), which is in fact equivalent to applying the Z_4 automorphism.

At the same time one can show that

$$C^{-1}\Omega(u)C = \tilde{\Omega}^{-ST}(u), \quad C = \left(\begin{array}{c|c} E & 0 \\ \hline 0 & E \end{array}\right), \tag{9.81}$$

where $\tilde{\Omega}(u)$ denotes the analytic continuation of $\Omega(u)$ under the cut $[-2g,2g]$.

Exercise 9.21 Show that
$$C^{-1}MC = -\phi[M]^{ST}, \tag{9.82}$$
where ST denotes super transpose, is defined as $\left(\begin{array}{c|c} A & B \\ \hline C & D \end{array}\right)^{ST} \equiv \left(\begin{array}{c|c} A^T & C^T \\ \hline -B^T & D^T \end{array}\right)$.
Use this to show that $C^{-1}AC = -\tilde{A}^{ST}$ (where the tilde denotes analytic continuation under the branch cut $[-2g,2g]$). Then prove (9.81).

Equation (9.81) implies that the eigenvalues of $\Omega(u)$ are related to the eigenvalues of $\tilde{\Omega}(u)$ by inversion and possible permutation. This statement in terms of the quasimomenta (9.80) tells us that the analytic continuation of the quasimomenta i.e. $\tilde{p}_a(u)$ and $\tilde{p}_{\hat{\imath}}(u)$ results in the change of sign and possible reshuffling. The exact way they reshuffle can be determined by considering some particular classical solutions and building the quasimomenta explicitly. Some examples can be found in [1]. Since all the classical solutions are related to each other continuously one finds that[8]

$$\tilde{p}_{\hat{1}}(u) = -p_{\hat{2}}(u), \quad \tilde{p}_{\hat{2}}(u) = -p_{\hat{1}}(u), \quad \tilde{p}_{\hat{3}}(u) = -p_{\hat{4}}(u), \quad \tilde{p}_{\hat{4}}(u) = -p_{\hat{3}}(u). \quad (9.83)$$

This property will play a crucial role in the QSC construction as we discuss in the next section. One can consider (9.83) as a manifestation of Z_4 symmetry of the action.

Large u asymptotics and quantum numbers. Another important property of the quasimomenta is that the quantum numbers of the state can be read off from their values at infinity. To see this, notice the following property:

$$A = -g^{-1}\left(d + *k\frac{2g}{u}\right)g, \quad (9.84)$$

where k_μ is the Noether current defined in (9.72). This implies that

$$\Omega = -g^{-1}\left(1 + \frac{2g}{u}\int_0^{2\pi} d\sigma\, k_\tau\right)g \quad (9.85)$$

using that the charge $Q_{\text{Noether}} = 2g\int_0^{2\pi} k_\tau\, d\sigma$, we immediately get

$$\begin{pmatrix} p_{\hat{1}} \\ p_{\hat{2}} \\ p_{\hat{3}} \\ p_{\hat{4}} \\ p_1 \\ p_2 \\ p_3 \\ p_4 \end{pmatrix} \simeq \frac{1}{2u} \begin{pmatrix} +\Delta - S_1 + S_2 \\ +\Delta + S_1 - S_2 \\ -\Delta - S_1 - S_2 \\ -\Delta + S_1 + S_2 \\ +\mathcal{J}_1 + \mathcal{J}_2 - \mathcal{J}_3 \\ +\mathcal{J}_1 - \mathcal{J}_2 + \mathcal{J}_3 \\ -\mathcal{J}_1 + \mathcal{J}_2 + \mathcal{J}_3 \\ -\mathcal{J}_1 - \mathcal{J}_2 - \mathcal{J}_3 \end{pmatrix}, \quad (9.86)$$

where the r.h.s. comes from the diagonalization of Q_{Noether}/u (in the fundamental representation). Here $\mathcal{J}_i$ are integer R-charges (which map to the scalar fields in gauge theory), S_1, S_2 are integer Lorentz charges (corresponding to the covariant derivatives), and Δ is the dimension of the state, i.e. its energy. Again we will see the quantum counterpart of this formula when we discuss QSC construction in Section 9.3.3.

---

[8] It is also possible to shift the quasimomenta by $2\pi m$ where m is an integer. This is indeed the case for p_i for the classical solutions which wind in S^5 and m gives their winding number. The AdS^5 quasimomenta still satisfy (9.83).

Action variables and WKB quantization. Another reason the quasimomenta were introduced is because they allow us to define the action variables very easily. For non-trivial solutions the quasimomenta have additional quadratic branch cuts, which come from the diagonalization procedure. The integrals around these cuts give the action variables [30][9]

$$I_C = \frac{1}{2\pi i} \oint_C p_A(u) du, \tag{9.87}$$

where C is some branch cut of $p_A(u)$. Here A can take any of eight values. In the Bohr–Sommerfeld quasi-classical quantization procedure one simply imposes $I_C \in \mathbb{Z}$ to get the first quantum correction. For example in [1] this property was used to obtain the 1-loop quantum spectrum of the string.

9.3.3 Quasimomenta and the strong coupling limit of QSC

To understand how the quansimomenta we introduced previously are related to the Q-functions from Section 9.3.2 we are going to first get an insight from the harmonic oscillator. Reconstructing the ψ from p, but inverting the relation (9.2), we get

$$\psi(x) = e^{-\frac{m\omega x^2}{2\hbar}} Q(x) = e^{\frac{i}{\hbar}\int^x p(x)dx}. \tag{9.88}$$

Similarly to what we found in (9.87) we also had

$$N = \frac{1}{2\pi i}\frac{i}{\hbar} \oint_C p(x)dx, \tag{9.89}$$

which allows us to identify $\frac{ix}{\hbar} \to u$ so that (9.89) and (9.87) become really identical. Under this identification we can deduce from (9.88),

$$Q_A \simeq \exp\left(\int^u p_A(v)dv\right). \tag{9.90}$$

This naive argument indeed produces the right identification for the strong coupling limit (i.e. $g \to \infty$) of $\mathbf{P}_a$ and $\mathbf{Q}_i$ functions introduced earlier. More precisely we get

$$\mathbf{P}_a \sim \exp\left(-\int^u p_a(v)dv\right), \quad \mathbf{P}^a \sim \exp\left(+\int^u p_a(v)dv\right) \tag{9.91}$$

$$\mathbf{Q}_i \sim \exp\left(-\int^u p_{\hat{i}}(v)dv\right), \quad \mathbf{Q}^i \sim \exp\left(+\int^u p_{\hat{i}}(v)dv\right). \tag{9.92}$$

---
[9] This property fixes the choice of the spectral parameter u, which otherwise can be replaced by any $f(u)$.

Note that at the leading classical level we do not control the pre-exponential factors and they may contain some order 1 powers of u. From that we can immediately draw a number of important consequences:

- We can deduce that large u asymptotics of $\mathbf{P}_a$ and $\mathbf{Q}_i$ from (9.86) are of the form $u^{Q_{\text{Noether}}/2}$.
- We can no longer expect that $\mathbf{P}_a$ or $\mathbf{Q}_i$ are polynomials, as the expressions (9.91) have Zhukovsky branch cuts $[-2g, 2g]$.
- From (9.83) we can deduce the following analytic continuation under the branch cut:

$$\tilde{\mathbf{Q}}_i \sim \exp\left(+\int^u \tilde{p}_i(v) dv\right) = \exp\left(-\int^u p_{\hat{\phi}_i}(v) dv\right) \sim \mathbf{Q}^{\phi_i}, \qquad (9.93)$$

where ϕ_i is determined by (9.83) to be $\phi_1 = 2, \phi_2 = 1, \phi_3 = 4, \phi_4 = 3$. So more explicitly we should have the following monodromies:

$$\tilde{\mathbf{Q}}_1 = \mathbf{Q}^2, \ \tilde{\mathbf{Q}}_2 = \mathbf{Q}^1, \ \tilde{\mathbf{Q}}_3 = \mathbf{Q}^4, \ \tilde{\mathbf{Q}}_4 = \mathbf{Q}^3. \qquad (9.94)$$

These relations remain almost intact at the quantum level. The only improvement one should make is to complex conjugate the r.h.s., as at the quantum level the Q_i are not real,

$$\tilde{\mathbf{Q}}_1 = \bar{\mathbf{Q}}^2, \ \tilde{\mathbf{Q}}_2 = \bar{\mathbf{Q}}^1, \ \tilde{\mathbf{Q}}_3 = \bar{\mathbf{Q}}^4, \ \tilde{\mathbf{Q}}_4 = \bar{\mathbf{Q}}^3. \qquad (9.95)$$

The reason for the complex conjugation will become clear in Section 9.4.

To conclude this section we notice that we have managed to get all the crucial additional information we have to add to the QQ-relations from just the classical limit. Namely, the existence of the Zhukovsky cut and the 'gluing' conditions (9.127). In Section 9.4 we combine all the information together and give the complete description of the spectrum of $N = 4$ SYM by means of the QSC.

9.4 QSC formulation

The goal of this section is to summarize the insights we got from the classical limit and from the spin chains and to motivate further the analytic properties of the basic $\mathbf{P}_a, \mathbf{P}^a, \mathbf{Q}_i, \mathbf{Q}^i$ Q-functions.

9.4.1 Main QQ-relations

The Q-functions of the $\mathcal{N} = 4$ SYM satisfy exactly the same QQ-relations as those of the $SU(4|4)$ spin chain. So we simply summarize the most important relations from Section 9.2.4 here to make this section self-contained:

$$Q_{a|i}^+ - Q_{a|i}^- = \mathbf{P}_a \mathbf{Q}_i, \tag{9.96}$$

$$\mathbf{P}_a \mathbf{P}^a = \mathbf{Q}_i \mathbf{Q}^i = 0, \tag{9.97}$$

$$\mathbf{Q}_i = -\mathbf{P}^a Q_{a|i}^+, \tag{9.98}$$

$$\mathbf{Q}^i = +\mathbf{P}_a Q^{a|i-}, \tag{9.99}$$

$$Q^{a|i} = -(Q_{a|i})^{-t}. \tag{9.100}$$

We also note that the first identity (9.96) can be combined with (9.98) into

$$Q_{a|i}^+ - Q_{a|i}^- = -\mathbf{P}_a \mathbf{P}^b Q_{b|i}^+. \tag{9.101}$$

This relation tells us that we can use eight functions $\mathbf{P}_a$ and $\mathbf{P}^a$ as the basis to reconstruct all other Q-functions i.e. we can in principle solve (9.101) in terms of $\mathbf{P}$'s (we will have an example in the following sections). Then we can use $Q_{a|i}$ to find $Q^{a|i}$ as its inverse (9.100). Then one can reconstruct $\mathbf{Q}_i$ and $\mathbf{Q}^i$ using (9.98) and (9.99).

The advantage of this choice of basis is, as we explain below, due to the fact that the analytic properties of $\mathbf{P}_a$ and $\mathbf{P}^a$ are the simplest among all Q-functions and they can be very efficiently parameterized.

9.4.2 Large u asymptotic and the quantum numbers of the state

The large u asymptotics of $\mathbf{P}$'s and $\mathbf{Q}$'s can be deduced from their classical limit (9.91) and (9.86). The main complication here is that in the non-twisted theory there are some additional powers of u coming from the pre-exponent of (9.91), which modify the asymptotic by ± 1. To fix the asymptotic completely one can make a comparison with the asymptotic Bethe ansatz of Beisert–Staudacher, which can be derived as a limit of QSC. We don't discuss this calculation here but this was done in detail in the original paper [11]. Here we just quote the result,

$$\mathbf{P}_a \simeq A_a u^{-\tilde{M}_a}, \quad \mathbf{Q}_i \simeq B_i u^{\hat{M}_i - 1}, \quad \mathbf{P}^a \simeq A^a u^{\tilde{M}_a - 1}, \quad \mathbf{Q}^i \simeq B^i u^{-\hat{M}_i}, \tag{9.102}$$

where

$$\tilde{M}_a = \left\{ \frac{\mathcal{J}_1 + \mathcal{J}_2 - \mathcal{J}_3 + 2}{2}, \frac{\mathcal{J}_1 - \mathcal{J}_2 + \mathcal{J}_3}{2}, \frac{-\mathcal{J}_1 + \mathcal{J}_2 + \mathcal{J}_3 + 2}{2}, \frac{-\mathcal{J}_1 - \mathcal{J}_2 - \mathcal{J}_3}{2} \right\} \tag{9.103}$$

$$\hat{M}_i = \left\{ \frac{\Delta - S_1 - S_2 + 2}{2}, \frac{\Delta + S_1 + S_2}{2}, \frac{-\Delta - S_1 + S_2 + 2}{2}, \frac{-\Delta + S_1 - S_2}{2} \right\}; \tag{9.104}$$

we see that indeed the asymptotics are consistent with what we found in the classical limit. Another way to understand the shift by ± 1 in the asymptotic is to consider a more general twisted theory. The twists (like the parameter ϕ we introduced in the spin chain section) remove many degeneracies.[10] For example without the twist the leading asymptotic in the l.h.s. of (9.96) cancels and one needs to know the subleading term to deduce the asymptotic of the r.h.s. This does not happen in the twisted case when $Q_{a|i} \sim e^{\phi_{a,i} u} u^{M_{a,i}}$ and the asymptotic behaves more predictably. As a result, in the twisted theory there are no ± 1 shifts w.r.t. the classical limit asymptotic and one can alternatively derive (9.102) by considering first the twisted $\mathcal{N} = 4$ SYM and then removing them.

Finding normalization of P and Q. We will see in the Section 9.4.3 that in the near BPS limit **P** and **Q** become small which will allow us to solve the QSC exactly at finite coupling. In order to see this we derive a more general result for the coefficients A_a, A^a and B_i, B^i from (9.102).

Exercise 9.22 Use (9.96) and (9.102) to show that

$$Q_{a|j} \simeq -i A_a B_j \frac{u^{-\tilde{M}_a + \hat{M}_j}}{-\tilde{M}_a + \hat{M}_j}. \tag{9.105}$$

From (9.105) we can fix the constants A^a and B^i in terms of $\tilde{M}_a$ and $\hat{M}_j$. Substituting the asymptotic (9.105) into (9.98) we get

$$-A^a u^{\tilde{M}_a - 1} \left(-i A_a B_j \frac{u^{-\tilde{M}_a + \hat{M}_j}}{-\tilde{M}_a + \hat{M}_j} \right) = B_j u^{\hat{M}_j - 1}, \tag{9.106}$$

which simplifies to

$$-1 = i \sum_{a=1}^{4} \frac{A^a A_a}{\tilde{M}_a - \hat{M}_j}, \tag{9.107}$$

which allows us to find the combinations $A^1 A_1$, $A^2 A_2$ and so on. Solving this linear system, we find

$$A^{a_0} A_{a_0} = i \frac{\prod_j (\tilde{M}_{a_0} - \hat{M}_j)}{\prod_{b \neq a_0} (\tilde{M}_{a_0} - \tilde{M}_b)}, \quad B^{j_0} B_{j_0} = i \frac{\prod_a (\hat{M}_{j_0} - \tilde{M}_a)}{\prod_{k \neq j_0} (\hat{M}_{j_0} - \hat{M}_k)}, \quad a_0, j_0 = 1 \ldots 4 \tag{9.108}$$

(with no summation over a_0 or j_0 in the l.h.s.).

Exercise 9.23 Derive the relations (9.108).

---

[10] See [31] for more details about the twisted version of QSC.

QSC formulation

Interestingly, the condition that all $A^{a_0}A_{a_0} = 0$ for all a_0 singles out the BPS states with protected dimension (which works for physical and even non-physical operators like in the BFKL regime).

Exercise 9.24 Find all solutions of $A^{a_0}A_{a_0} = 0, \forall a_0$ in terms of $\mathcal{J}_i, S_i$, and Δ.

Next we investigate the cut structure of $\mathbf{P}_a$ and $\mathbf{Q}_i$.

9.4.3 Analytic structure of Q-functions

In this section we deduce the analytic properties of $\mathbf{P}_a$ and $\mathbf{Q}_i$ functions following a maximal simplicity principle, i.e. we assume simplest possible analytical properties which do not contradict the classical limit and the structure of the QQ-system. In Section 9.3, from the strong coupling analysis we deduced that $\mathbf{P}_a$ and $\mathbf{Q}_i$ should have cuts with branch points at $\pm 2g$ (to recall $g = \frac{\sqrt{\lambda}}{4\pi}$ where λ is the 't Hooft coupling). We can assume that $\mathbf{P}_a$ should have just one single cut $[-2g, 2g]$. Note that since $\mathbf{P}^a$ is related to $\mathbf{P}_a$ by the symmetry of flipping the Dynkin diagram upside down it should also have the same analytic properties.

Note that $\mathbf{Q}_i$ (and $\mathbf{Q}^i$) cannot have the same analytic properties as $\mathbf{P}$'s. Indeed, in general, Δ in the asymptotic of $\mathbf{Q}_i$ is not-integer and thus we must have a non-trivial monodromy around infinity.[11] The simplest way to gain such a monodromy is to choose the branch cut to close through infinity i.e. we can assume that $\mathbf{Q}_i$ and $\mathbf{Q}^i$ have a "long" branch-cut $(-\infty, -2g] \cup [+2g, +\infty)$. This simple argument leads us to the simple analyticity picture Figure 9.4, which historically was derived using the TBA approach in [11].

Note that $\mathbf{P}$ and $\mathbf{Q}$ are additionally constrained to be a part of the same Q-system. This makes it very inconvenient to have different conventions for the choice of the branch cuts so we may need to look under the long cut of $\mathbf{Q}$. A simple way to explore the space under the cut of $\mathbf{Q}$ is to use the QQ-relation (9.101) written in the form

$$Q_{a|i}^+ + \mathbf{P}_a \mathbf{P}^b Q_{b|i}^+ = Q_{a|i}^-, \qquad (9.109)$$

Figure 9.4 *Cut structure of* $\mathbf{P}_a, \mathbf{P}^a$ *and* $\mathbf{Q}_i, \mathbf{Q}^i$.

---

[11] Depending on the values of $\mathcal{J}_i$ there could be a similar issue with $\mathbf{P}_a$ as the asymptotic could contain half-integer numbers. Strictly speaking $\mathbf{P}_a$ could have an extra cut going to infinity which would disappear in any bi-linear combinations of $\mathbf{P}$s.

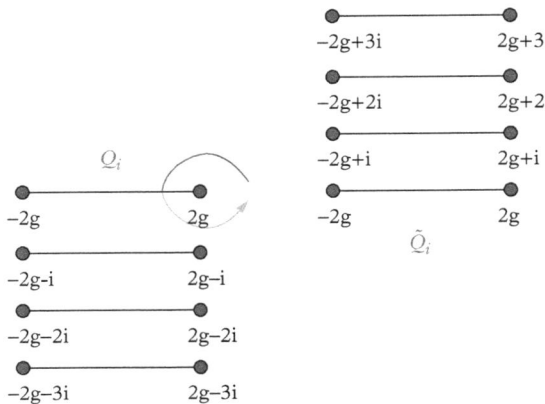

Figure 9.5 *Analytic structure of Q_i under its long cut.*

which implies that

$$Q_i = -\mathbf{P}^a Q_{a|i}^+ = -\mathbf{P}^a(\delta_a^b + \mathbf{P}_a^{[+2]}\mathbf{P}^{b[+2]})Q_{b|i}^{[3]}$$
$$= -\mathbf{P}^a(\delta_a^b + \mathbf{P}_a^{[+2]}\mathbf{P}^{b[+2]})(\delta_b^c + \mathbf{P}_b^{[+4]}\mathbf{P}^{c[+4]})Q_{c|i}^{[5]} = \ldots. \qquad (9.110)$$

Firstly we note that from the formal solution (9.52) we can always assume that $Q_{a|i}$ is regular in the upper half plane. From that we see that (9.110) implies that $\mathbf{Q}_i$ has an infinite ladder of cuts. The first term in the last line of (9.110) has a cut at $[-2g, 2g]$, the second has the cut at $[-2g - i, 2g - i]$ and so on. See Fig. 9.5. The puzzle is how to make this structure of cuts compatible with the initial guess that $\mathbf{Q}_i$ has only one cut going to infinity. In fact there is no contradiction so far as in order to see the infinite ladder of the cuts we should go to the right from the branch point at $2g$ i.e. under the long cut. At the same time if we want to go to the lower half plane avoiding the long cut we should go under the first short cut. What is expected to be seen under the first short cut is no branch point singularities below the real axis. Thus if we denote by $\tilde{\mathbf{Q}}_i$ the analytic continuation of $\mathbf{Q}_i$ under the first cut it will have no branch cut singularities below the real axis (see Figure 9.5).

From Figure 9.5 we notice an obvious asymmetry of the upper half plane and the lower half plane. Indeed the function $\mathbf{Q}_i$, which is a part of the Q-system, is analytic in the upper half plane, whereas $\tilde{\mathbf{Q}}_i$, which does not necessarily satisfy any QQ-relation is analytic in the lower half plane. That is, building the Q-system we decided to keep all Q-functions analytic above the real axis and now we have potentially lost the symmetry under complex conjugation which can be linked to unitarity of the theory. To reconstruct the symmetry we have to impose the 'gluing conditions'.

9.4.4 Gluing conditions

In this section we address an imperfection of our construction where the upper half plane plays a more important role from a QQ-relations point of view. To exchange the upper and lower halves we can complex conjugate Q-functions. This procedure does not affect $\mathbf{P}_a$ and $\mathbf{P}^a$ much, depending on the normalization constant they can at most change their signs. For $\mathbf{Q}_i$ the complex conjugation seems to be more dramatic as the ladder of branch cuts going down will now go up. Simply multiplying $\mathbf{Q}_i$ by a constant would not undo the complex conjugation, however, if we also analytically continue $\bar{\mathbf{Q}}_i$ under the first branch cut, we actually get a very similar analytic structure to the initial $\mathbf{Q}_i$! That is, the complex conjugation and the analytic continuation should give us back either some $\mathbf{Q}_i$ or $\mathbf{Q}^i$. To determine which of $\mathbf{Q}$ could do the job we recall that in the classical limit we obtained (9.127) and in accordance with that we impose the following gluing conditions:[12]

$$\boxed{\tilde{\mathbf{Q}}_1 \propto \bar{\mathbf{Q}}^2, \tilde{\mathbf{Q}}_2 \propto \bar{\mathbf{Q}}^1, \tilde{\mathbf{Q}}_3 \propto \bar{\mathbf{Q}}^4, \tilde{\mathbf{Q}}_4 \propto \bar{\mathbf{Q}}^3.} \qquad (9.111)$$

Together with (9.96)–(9.100), (9.134) constitutes the closed system of QSC equations. It is rather non-trivial that these equations only have a discrete set of solutions (and so far there is no mathematically rigorous proof of this). To demonstrate this we consider some simple examples in Section 9.4.5 and also implement an algorithm which allows us to find solutions numerically.

9.4.5 Left–right symmetric subsector

In many situations it is sufficient to restrict ourselves to a subset of all states which has an additional symmetry. The left–right (LR) symmetric subsector, which includes $su(2)$ and $sl(2)$ subsectors, contains the states preserving the upside down symmetry of the Dynkin diagram, i.e. the states which should have $\mathcal{J}_3 = 0, S_2 = 0$. To understand what we should expect in this case consider the bosonic subgroup $SO(4,2) \times SO(6)$. The $SO(6)$ Dynkin diagram has three nodes and imposing the LR symmetry would imply that the nodes 1 and 3 are indistinguishable, which reduces the symmetry to $SO(5)$ (see Figure 9.6). In order to break $SO(6)$ to $SO(5)$ it is sufficient to select some preferable direction in the vector 6D representation. Our $\mathbf{P}_a$ and $\mathbf{P}^a$ are in 4D fundamental and anti-fundamental representations of $SO(6)$. The vector representation can be realized as antisymmetric tensors with two fundamental indices A_{ab}, and so we can pick a direction to break $SO(6)$

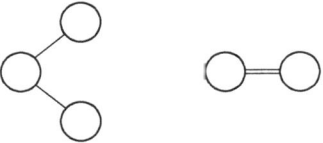

Figure 9.6 *Under identification of the upper and lower nodes, the $SO(6)$ Dynkin diagram (on the left) becomes the $SO(5)$ Dynkin diagram (on the right).*

---

[12] For physical operators $\mathbf{Q}_2$ and $\mathbf{Q}_4$ can mix with $\mathbf{Q}_1$ and $\mathbf{Q}_3$ as they are growing faster and have the same non-integer part in the asymptotic (similarly $\mathbf{Q}^3$ and $\mathbf{Q}^1$ are defined modulo $\mathbf{Q}^2$ and $\mathbf{Q}^4$).

to $SO(5)$ by picking a particular antisymmetric tensor χ_{ab} which can be used to relate fundamental and anti-fundamental representations, i.e. can be used to lower the indices. In this subsector we will get

$$\mathbf{P}_a = \chi_{ab}\mathbf{P}^b. \qquad (9.112)$$

Since we have already selected the order of **P**'s by assigning their asymptotic we can see that the only non-zero components of χ, consistent with the asymptotic of **P**, are 14, 23, 32, 41. Finally we still have freedom to rescale $\mathbf{P}_a$ to bring χ_{ij} to the conventional form,

$$\chi_{ab} = \begin{pmatrix} 0 & 0 & 0 & 1 \\ 0 & 0 & -1 & 0 \\ 0 & 1 & 0 & 0 \\ -1 & 0 & 0 & 0 \end{pmatrix}. \qquad (9.113)$$

By the same argument we should impose

$$\mathbf{Q}_i = \chi_{ij}\mathbf{Q}^j \qquad (9.114)$$

for the same tensor χ_{ij}.

As a result they are ambiguously defined. Fortunately, we don't have to impose all four of the gluing conditions and it is sufficient to use another pair of equations.

9.5 QSC—analytic examples

In this section we consider an example where the QSC can be solved analytically at finite coupling. It is unfortunate that the analytical solutions for physical operators are rather complicated. It is possible to get the solution perturbatively at weak coupling, but this already involves computer algebra. Here instead we consider a non-local operator which can be understood as an analytic continuation of twist-$\mathcal{J}$ states. The twist operators are the states with $\mathcal{J}_1 = \mathcal{J}, \mathcal{J}_2 = \mathcal{J}_3 = 0$, and $S_1 = S, S_2 = 0$. They belong to the LR symmetric subsector described in Section 9.4 and we will give a description of the $sl(2)$ sector to which these states also belong in Section 9.5.1.

9.5.1 $sl(2)$ sector

We will discuss the simplifications which arise in the $sl(2)$ sector. As the $sl(2)$ sector is inside the LR subsector we can restrict ourselves to the Q-functions with lower indices due to (9.114) and (9.112). The asymptotics of $\mathbf{P}_a$ (9.102) become

$$\mathbf{P}_1 = A_1 u^{-L/2-1},\ \mathbf{P}_2 = A_2 u^{-L/2},\ \mathbf{P}_3 = A_3 u^{+L/2-1},\ \mathbf{P}_4 = A_4 u^{+L/2}. \qquad (9.115)$$

Similarly for $\mathbf{Q}_i$,

$$\mathbf{Q}_1 = B_1 u^{+(\Delta-S)/2}, \quad \mathbf{Q}_2 = B_2 u^{+(\Delta+S)/2-1},$$
$$\mathbf{Q}_3 = B_3 u^{-(\Delta+S)/2}, \quad \mathbf{Q}_4 = B_4 u^{-(\Delta-S)/2-1}. \quad (9.116)$$

Also we write (9.108) explicitly for this case,

$$A_1 A_4 = -\frac{i(-\Delta+L-S+2)(-\Delta+L+S)(\Delta+L-S+2)(\Delta+L+S)}{16L(L+1)}$$
$$A_2 A_3 = -\frac{i(-\Delta+L-S)(-\Delta+L+S-2)(\Delta+L-S)(\Delta+L+S-2)}{16(L-1)L} \quad (9.117)$$

and

$$B_1 B_4 = -\frac{i(-\Delta+L+S-2)(-\Delta+L+S)(\Delta+L-S)(\Delta+L-S+2)}{16\Delta(S-1)(-\Delta+S-1)}$$
$$B_2 B_3 = +\frac{i(\Delta-L+S-2)(\Delta-L+S)(\Delta+L+S-2)(\Delta+L+S)}{16\Delta(S-1)(\Delta+S-1)}. \quad (9.118)$$

We can see that both $A_a A^a$ and $B_a B^a$ vanish for $\Delta \to L$, $S \to 0$. The reason for this is that $S=0$ is the BPS protected state and the vanishing of the coefficients indicates the shortening of the multiplet. At the same time when $\mathbf{P}$ and $\mathbf{Q}$ are small we get an enormous simplification, as we show in Section 9.5.2, where we consider a near BPS limit where S is small.

9.5.2 Analytic continuation in S

In this section we introduce an analytic continuation in the Lorentz spin $S_1 = S$, which for local operators must be integer. The analytic continuation in the spin plays an important role as it links BFKL and DGLAP regimes or high energy scattering in QCD.[13] We leave aside questions related to the physics of light energy scattering and describe in detail the analytic continuation in S from the QSC point of view.

The simplest way to describe the analytic continuation is by considering the gluing conditions (9.134), which for a LR-symmetric sector reduce to just two,

$$\tilde{\mathbf{Q}}_1 \propto \bar{\mathbf{Q}}_3, \quad \tilde{\mathbf{Q}}_2 \propto \bar{\mathbf{Q}}_4, \quad (9.119)$$

since the two other gluing conditions follow by taking complex conjugate and analytically continue the previous two conditions.

Also, as we will see that from the numerical analysis in Section 9.5.3, these two conditions are not independent and only one of them is sufficient to build the spectrum.

---
[13] For the applications of QSC in this regime, see [14, 40].

At the same time imposing both conditions (9.127) leads to the quantization of the charge S_1 whereas keeping only the first condition $\tilde{\mathbf{Q}}_1 \propto \bar{\mathbf{Q}}_3$ allows us to have S_1 non-integer![14] However, this will modify the second gluing condition. To constrain the possible form of the modified gluing conditions we denote

$$\tilde{\mathbf{Q}}_i(u) = M_i{}^j(u)\bar{\mathbf{Q}}_j(u), \; M_i{}^j(u) = \begin{pmatrix} 0 & 0 & M_1{}^3 & 0 \\ M_2{}^1 & M_2{}^2 & M_2{}^3 & M_2{}^4 \\ M_3{}^1 & 0 & 0 & 0 \\ M_4{}^1 & M_4{}^2 & M_4{}^3 & M_4{}^4 \end{pmatrix}_{ij}. \quad (9.120)$$

Since the gluing condition tells us that $\tilde{\mathbf{Q}}_i$ is essentially the same as $\mathbf{Q}^i$ up to a possible symmetry of Q-system transformation, we should assume that $M_i{}^j(u)$ is an i-periodic function of u: $M_i{}^j(u+i) = M_i{}^j(u)$. Furthermore, since $M_i{}^j(u)$ relates two functions which are both analytic in the lower half plane it should be analytic.

Exercise 9.25 Use a periodicity of $M_i{}^j$ and equation (9.120) to find M_k^j explicitly in terms of $\tilde{\mathbf{Q}}_k(u), \tilde{\mathbf{Q}}_k(u+i), \tilde{\mathbf{Q}}_k(u+2i), \tilde{\mathbf{Q}}_k(u+3i)$, and $\bar{\mathbf{Q}}_j(u), \bar{\mathbf{Q}}_j(u+i), \bar{\mathbf{Q}}_j(u+2i), \bar{\mathbf{Q}}_j(u+3i)$. From that relation you can see that $M_k{}^j$ does not have any branch-cuts, but could possibly have poles. However, existence of poles would contradict the power-like asymptotic of $\bar{\mathbf{Q}}_j(u)$ and analyticity of $\tilde{\mathbf{Q}}_i(u)$ as we will have to conclude that $\bar{\mathbf{Q}}_j$ has infinitely many zeros in the lower half plane, which is impossible with a power-like asymptotic.

Armed with the new knowledge of regularity of M we can analytically continue both sides of (9.120) and complex conjugate them to find the following condition on the matrix M:

$$\bar{M}(u) = M^{-1}(u). \quad (9.121)$$

Another constraint comes from the LR-symmetry of the state, which tells us that $\mathbf{Q}_i = \chi_{ij}\mathbf{Q}^j$, where $\mathbf{Q}^j$ is a tri-linear combination of $\mathbf{Q}_i$ as in (9.37). So using what we get from (9.120),

$$\tilde{\mathbf{Q}}^l(u) = (\chi^{-1})^{li} M_i{}^j \chi_{jk} \bar{\mathbf{Q}}^k(u), \quad (9.122)$$

at the same time we can use (9.37) and (9.59) to rewrite the r.h.s. as a combination of 3 $\bar{\mathbf{Q}}_i$ and then apply the initial (9.120); this results in the following equation:

$$\tilde{\mathbf{Q}}^i(u) = -\det(M)(M^{-1})_j{}^i \bar{\mathbf{Q}}^j(u). \quad (9.123)$$

---

[14] One can show that the second condition necessary leads to the quantization of S [53]. It could be simpler to check this numerically with the code we explain in Section 9.6.

Comparing (9.122) and (9.123) we get

$$(\chi^{-1})^{li} M_i{}^j \chi_{jk} M_n{}^k = -\det(M)\delta_n^l, \qquad (9.124)$$

Exercise 9.26 Derive (9.124) by combining (9.122) and (9.123).

or in matrix form,

$$M \chi M^T = -\chi \det(M), \qquad (9.125)$$

which implies in particular that $\det(M) = \pm 1$. Imposing (9.124) and (9.121) we obtain that M should reduce to the following form:

$$M_i{}^j(u) = \begin{pmatrix} 0 & 0 & \alpha & 0 \\ \beta & 0 & \gamma & -\bar{\alpha} \\ \frac{1}{\bar{\alpha}} & 0 & 0 & 0 \\ \frac{\gamma}{\alpha\bar{\alpha}} & -\frac{1}{\alpha} & \bar{\beta} & 0 \end{pmatrix}_{ij}, \qquad (9.126)$$

with real γ, which results in the following two independent gluing conditions:

$$\begin{aligned} \tilde{\mathbf{Q}}_1 &= \alpha \bar{\mathbf{Q}}_3, \\ \tilde{\mathbf{Q}}_2 &= \beta \bar{\mathbf{Q}}_1 + \gamma \bar{\mathbf{Q}}_3 - \bar{\alpha} \bar{\mathbf{Q}}_4. \end{aligned} \qquad (9.127)$$

Since α appears in both the numerator and denominator it cannot be a non-trivial function of u as it would create poles. At the same time β and γ can be non-trivial periodic functions of u. For the case of the twist two operators $\mathrm{tr} Z D_-^S Z$[15] with non-integer Lorentz spin S we will verify numerically that γ is a constant and $\beta = \beta_1 + \beta_2 \cosh(2\pi u) + \beta_3 \sinh(2\pi u)$. For integer S both γ and β vanish. Using this gluing matrix one can compute the BFKL pomeron/odderon eigenvalue by analytically continuing to $S \sim -1$.

9.5.3 Slope function

The possibility of having the Lorentz spin S non-integer allows us to study the near-BPS regime $S \to 0$ analytically. In this section we will compute the first term linear in S called the slope function [29] analytically to all orders in g. This calculation was presented originally in [9] in a slightly different form, where also the next term in the small S expansion was derived. Here we adopt a more widely accepted notation from [11], which is different from [9].

---

[15] Z is a complex scalar of the theory, D_- is a light-cone covariant derivative.

The main simplification in this limit is due to the scaling of $\mathbf{P}_a$ and $\mathbf{Q}_i$ with $S \to 0$ and $\Delta = L + eS$ where $e \sim 1$, which can be deduced from the scaling of A_a and B_i (9.117) and (9.118):

$$A_1 A_4 \simeq -B_1 B_4 \simeq -\frac{i}{2}(1-e)S, \quad A_2 A_3 \simeq -B_2 B_3 \simeq -\frac{i}{2}(1+e)S. \tag{9.128}$$

From that we can deduce that $\mathbf{P}_a$ and $\mathbf{Q}_i$ both scale as $\sqrt{S}$. This assumption in the main simplification—the equation for $Q_{a|i}$ (9.96) becomes simply

$$Q_{a|i}^+ - Q_{a|i}^- \simeq 0, \tag{9.129}$$

i.e. $Q_{a|i}$ is a constant! To find which constants they are we can simply use the general formula (9.106), which in our limit gives

$$Q_{a|j} = \begin{pmatrix} -\frac{2iA_1 B_1}{(e-1)S} & 0 & 0 & 0 \\ 0 & -\frac{2iA_2 B_2}{(e+1)S} & 0 & 0 \\ 0 & 0 & \frac{2iA_3 B_3}{(e+1)S} & 0 \\ 0 & 0 & 0 & \frac{2iA_4 B_4}{(e-1)S} \end{pmatrix}. \tag{9.130}$$

Using the rescaling symmetry[16] we can set $B_1 = iA_4, B_2 = iA_3, B_3 = iA_2, B_4 = iA_1$ giving

$$Q_{a|j} = \begin{pmatrix} i & 0 & 0 & 0 \\ 0 & -i & 0 & 0 \\ 0 & 0 & i & 0 \\ 0 & 0 & 0 & -i \end{pmatrix}. \tag{9.131}$$

This implies that $\mathbf{Q}_i$ and $\mathbf{P}_a$ are essentially equal in this limit due to (9.99):

$$\mathbf{Q}_1 = i\mathbf{P}_4, \quad \mathbf{Q}_2 = i\mathbf{P}_3, \quad \mathbf{Q}_3 = i\mathbf{P}_2, \quad \mathbf{Q}_4 = i\mathbf{P}_1. \tag{9.132}$$

This makes our calculations much easier as we can write the gluing condition (9.127) directly on $\mathbf{P}$,

$$\tilde{\mathbf{P}}_4 = -\alpha \bar{\mathbf{P}}_2 \tag{9.133}$$

$$\tilde{\mathbf{P}}_3 - \bar{\alpha}\bar{\mathbf{P}}_1 = -[\beta_1 + \beta_2 \cosh(2\pi u) - \beta_3 \sinh(2\pi u)]\bar{\mathbf{P}}_4 - \gamma \bar{\mathbf{P}}_2. \tag{9.134}$$

---

[16] We can rescale $\mathbf{P}_1 \to f\mathbf{P}_1$ and $\mathbf{P}_2 \to g\mathbf{P}_2$, rescaling simultaneously $\mathbf{P}_3 \to 1/f\mathbf{P}_1$ and $\mathbf{P}_4 \to 1/g\mathbf{P}_4$ and similar for $\mathbf{Q}_i$. In addition for $\mathbf{P}$'s only we have the freedom freedom $\mathbf{P}_3 \to \mathbf{P}_3 + \gamma_2 \mathbf{P}_2 - \gamma_1 \mathbf{P}_1$ and $\mathbf{P}_4 \to \mathbf{P}_4 + \gamma_3 \mathbf{P}_1 + \gamma_1 \mathbf{P}_2$ for some constants γ_n, this ambiguity is resolved in the twisted theory. These transformations are the most general which preserve χ_{ij} tensor and do not modify the asymptotic of $\mathbf{P}$'s.

To solve these equations we have to impose the asymptotic on $\mathbf{P}_a$. For simplicity we consider the $L=2$ case only, leaving general L as an exercise. For $L=2$ (9.115) gives

$$\mathbf{P}_1 \simeq A_1 \frac{1}{u^2},\ \mathbf{P}_2 \simeq A_2 \frac{1}{u},\ \mathbf{P}_3 \simeq A_3,\ \mathbf{P}_4 \simeq A_4 u. \tag{9.135}$$

Since $\mathbf{P}_a$ is a function with only one branch-cut which can be resolved with the help of the Zhukovsky variable $x(u) = \frac{u+\sqrt{u^2-4g^2}}{2g}$, we can use the following general ansatz[17]:

$$\mathbf{P}_1 = \sum_{n=2}^{\infty} \frac{c_{1,n}}{x^n},\ \mathbf{P}_2 = \sum_{n=1}^{\infty} \frac{c_{2,n}}{x^n},\ \mathbf{P}_3 = \sum_{n=0}^{\infty} \frac{c_{3,n}}{x^n},\ \mathbf{P}_4 = \sum_{n=-1}^{\infty} \frac{c_{4,n}}{x^n}. \tag{9.136}$$

Note that under analytic continuation $\tilde{x} = 1/x$. Now we can use the condition (9.133) to deduce $\mathbf{P}_4$ and $\mathbf{P}_1$. Plugging the ansatz (9.136) into (9.133) we get

$$\frac{c_{4,-1}}{x} + c_{4,0} + c_{4,1} x + c_{4,2} x^2 + \cdots = -\alpha \left(\frac{\bar{c}_{2,1}}{x} + \frac{\bar{c}_{2,2}}{x^2} + \frac{\bar{c}_{2,3}}{x^3} + \cdots \right). \tag{9.137}$$

We see that the l.h.s. contains infinitely many positive powers of x whereas in the r.h.s. there are only negative powers, which implies that $c_{4,n\geq 0} = 0$ and $c_{2,n\geq 2} = 0$ and thus

$$\mathbf{P}_2 = \frac{c_{2,1}}{x},\ \mathbf{P}_4 = -\alpha \bar{c}_{2,1} x. \tag{9.138}$$

In order to deal with the second equation is a similar way we should use the identities

$$\cosh(2\pi u) = \sum_{n=-\infty}^{\infty} I_{2n}\left(\sqrt{\lambda}\right) x^{2n}(u),\ \sinh(2\pi u) = \sum_{n=-\infty}^{\infty} I_{2n+1}\left(\sqrt{\lambda}\right) x^{2n+1}(u), \tag{9.139}$$

where $I_n(z)$ is the Bessel function of the second kind defined as

$$I_n(y) = \oint \frac{e^{y/2(z+1/z)}}{z^{1-n}} \frac{dz}{2\pi i}. \tag{9.140}$$

Exercise 9.27 Prove identities (9.139), you will have to use that $u = g(x+1/x)$ and that $g = \frac{\sqrt{\lambda}}{4\pi}$.

---

[17] Which can be interpreted as a Laurent series expansion in the x plane, where the functions $\mathbf{P}$ are analytic in the exterior of the unit circle and the first singularity lies inside the unit circle, ensuring good convergence of the series expansion.

After that we can express both sides of (9.134) as a power series in x and match the coefficients. In particular comparing the coefficients of x^0 and x^{-2} we get

$$c_{3,0} = -\alpha\beta_3 c_{2,1} I_1(\sqrt{\lambda}), \quad c_{1,2} = \frac{\bar{\alpha}\bar{\beta}_3 \bar{c}_{2,1}}{\alpha} I_3(\sqrt{\lambda}). \tag{9.141}$$

Finally, forming the combinations

$$A_1 A_4 = g c_{4,-1} c_{1,2} = -g\bar{\alpha}\bar{\beta}_3 \bar{c}_{2,1}^2 I_3(\sqrt{\lambda}) \tag{9.142}$$

$$A_2 A_3 = g c_{2,1} c_{3,0} = -g\alpha\beta_3 c_{2,1}^2 I_1(\sqrt{\lambda}) \tag{9.143}$$

which are also given in (9.128) in terms of a real quantity $e = (\Delta - L)/S$ and S, we conclude that

$$\Delta - L = S \frac{I_1(\sqrt{\lambda}) + I_3(\sqrt{\lambda})}{I_1(\sqrt{\lambda}) - I_3(\sqrt{\lambda})}. \tag{9.144}$$

reproducing the result from [29]!

Exercise 9.28 Repeat the above calculation for arbitrary L. You have to obtain

$$\Delta - L = S \frac{\sqrt{\lambda} I_{L+1}(\sqrt{\lambda})}{L I_L(\sqrt{\lambda})}. \tag{9.145}$$

In the derivation you can assume that $\gamma = \beta_1 = \beta_2 = 0$. We explain below why that is the case for $L = 2$.

In order to fix the solution for $\mathbf{P}_a$ we notice that we also get

$$c_{2,1}^2 = \frac{iS}{g\alpha\beta_3 (I_1(\sqrt{\lambda}) - I_3(\sqrt{\lambda}))}. \tag{9.146}$$

Even though the constants γ, β_1, and β_2 did not enter into the calculation, leading to the dimension Δ, they will still appear in the solution for $\mathbf{P}_a$. Here we will fix them further from the reality conditions. Let us also show that $\gamma = \beta_2 = 0$. First the coefficients x and $1/x$ from (9.134) give the following combination:

$$\beta_1 = -\beta_2 I_0(\sqrt{\lambda}), \quad \gamma = -\alpha\beta_2 \frac{c_{2,1}}{\bar{c}_{2,1}} I_2(\sqrt{\lambda}). \tag{9.147}$$

Since $c_{2,1}$ is already fixed (9.146) we obtain

$$\frac{\gamma^2 |\beta_3|^2}{|\alpha|^2 I_2^2(\sqrt{\lambda})} = -\beta_2^2 \bar{\beta}_3^2, \tag{9.148}$$

where the l.h.s. is real and positive. At the same time if we compare the coefficients of x^2 and x^3 in (9.134) we get

$$\frac{c_{3,3}}{c_{3,2}}\frac{I_1(\sqrt{\lambda})}{I_2(\sqrt{\lambda})} = \frac{\beta_2}{\beta_3} = \frac{\bar{\beta}_2}{\bar{\beta}_3}, \qquad (9.149)$$

where again the l.h.s. should be real due to the complex conjugation property of $\mathbf{P}_a$ which allows us to complex conjugate the r.h.s.[18] Combining (9.149) with (9.148) we obtain

$$\frac{\gamma^2|\beta_3|^2}{|\alpha|^2 I_2^2(\sqrt{\lambda})} = -\beta_2^2 \bar{\beta}_2^2, \qquad (9.150)$$

which is only possible if $\beta_2 = \gamma = 0$.

9.6 Solving QSC at finite coupling numerically

9.6.1 Description of the method

In this part we describe the numerical algorithm and analyse some of the numerical results. We illustrate the general method initially proposed in [13] by considering the same states as in Section 9.5.3, $\mathrm{tr} Z D_-^S Z$ i.e. twist-2 operators. First we consider the $S = 2$ case—the Konishi operator. Additionally, from the beginning we impose the parity symmetry which these states have i.e. symmetry under $u \to -u$ which reflects in the parity of $\mathbf{P}_a$ functions. The *Mathematica* code that we used can be found as an ancillary file for the arXiv submission 1504.06640.

Firstly, we describe the main steps and ideas for the numerical procedure.

- Parameterize the system in terms of the truncated series in x of $\mathbf{P}_a$ as follows:

$$\mathbf{P}_a = (xg)^{-\tilde{M}_a} \mathbf{p}_a, \quad \mathbf{p}_a = \left(A_a + \sum_{n=1}^{\infty} \frac{c_{a,n}}{x^{2n}} \right), \qquad (9.151)$$

where in the code we cut the sum at some finite value `Pcut`. We will see that to get six digits of precision with need `Pcut` as small as 3 (for relatively small $g = 1/5$). This series converges very well even for $|x| = 1$. Note that under the analytic continuation to the next sheet we simply replace $x \to 1/x$ so that

$$\tilde{\mathbf{P}}_a = (x/g)^{\tilde{M}_a} \left(A_a + \sum_{n=1}^{\infty} c_{a,n} x^{2n} \right). \qquad (9.152)$$

---

[18] Under the complex conjugation $\mathbf{P}_a \to e^{i\phi_a} \mathbf{P}_a$, for some real ϕ_a. This implies that the ratios $c_{a,n}/c_{a,m}$ are real.

- Given $\mathbf{P}_a$ in terms of $c_{a,n}$ find $Q_{a|i}(u)$ as a series expansion in large u:

$$Q_{a|i}(u) = u^{-\tilde{M}_a + \hat{M}_j} \sum_{n=0} \frac{B_{a,i,n}}{u^{2n}}. \qquad (9.153)$$

We find the coefficients $B_{a,i,n}$ by plugging the expansion (9.153) into the finite difference equation (9.101). The term $B_{a,i,0}$ we found before in (9.105). Expanding it at large u we get a linear system on the coefficients $B_{a,i,n}$. The series (9.153) is asymptotic and works well as long as u is large enough. In our numerical implementation we keep around twelve terms.

- Starting from the expansion (9.153) at large Imu we can move down to the real axis using (9.101) in the form

$$Q_{a|i}(u - \tfrac{i}{2}) = \left(\delta_a^b - \mathbf{P}_a(u)\mathbf{P}_c(u)\chi^{cb}\right) Q_{b|i}(u + \tfrac{i}{2}). \qquad (9.154)$$

Applying (9.154) recursively we can decrease the imaginary part of u from the asymptotic area to reach finite values of u (see Figure 9.7). We will mostly need values of $Q_{a|i}$ at Im$u = 1/2$ with $-2g < $ Re$u < 2g$.

- With $Q_{a|i}$ computed we reconstruct $\mathbf{Q}_i$ and $\tilde{\mathbf{Q}}_i$ from

$$\mathbf{Q}_i(u) = Q_{a|i}(u + i/2)\chi^{ab}\mathbf{P}_b(u), \quad \tilde{\mathbf{Q}}_i(u) = Q_{a|i}(u + i/2)\chi^{ab}\tilde{\mathbf{P}}_b(u), \qquad (9.155)$$

where $\mathbf{P}_a$ and $\tilde{\mathbf{P}}_a$ are given in terms of $c_{a,n}$ in (9.151) and (9.152).

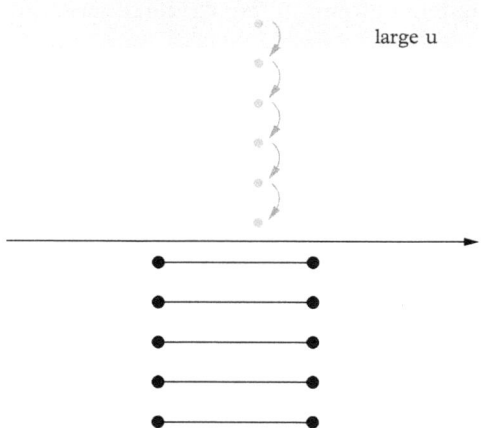

Figure 9.7 To reconstruct $Q_{a|i}$ at the values of $u \sim 1$ we perform several jumps by i using (9.154) into the region on the upper half plane where the asymptotic expansion (9.153) is applicable.

- Finally, we constrain $c_{a,n}$ from the gluing conditions (9.127). We will see that it is sufficient to impose only half of them. In our numerical implementation we build a function

$$F(\Delta, c_{a,n}, u) = \tilde{Q}_3 - \alpha \bar{Q}_1 \tag{9.156}$$

and then adjust Δ and $c_{a,n}$ to minimize $F(\Delta, c_{a,n}, u)$ at some set of probe points $u_k \in (2g, 2g)$. For this we use standard numerical optimization methods.

In Section 9.6.2 we give more details about the *Mathematica* implementation of our method.

9.6.2 Implementation in *Mathematica*

The *Mathematica* notebook we describe below, with slight improvements, can be downloaded from arXiv [13].

First we make basic definitions. We define $x(u)$ in such a way as to ensure that it has only one cut $[-2g, 2g]$,

```
X[u_] = (u + g*Sqrt[u/g - 2]*Sqrt[u/g + 2])/(2*g);
chi = {{0, 0, 0, -1}, {0, 0, 1, 0}, {0, -1, 0, 0}, {1, 0, 0, 0}};
```

Exercise 9.29 What is the branch-cut structure of the naive definition
`X[u_]=(u+Sqrt[u^2-4g^2])/(2g)`? Consider also the case of complex g.

Next we define $\hat{M}$ and $\tilde{M}$ as in (9.103). We also specialize to the Konishi operator in the $sl(2)$ sector with $\mathcal{J}_1 = S = 2$. The variable **d** denotes the full dimension Δ:

```
J1 = 2; J2 = 0; J3 = 0; S1 = 2; S2 = 0;
Mt = {(J1+J2-J3+2)/2,(J1-J2+J3)/2,(-J1+J2+J3+2)/2,(-J1-J2-J3)/2}
Mh = {(d-S1-S2+2)/2,(d+S1+S2)/2,(-d-S1+S2+2)/2,(-d+S1-S2)/2}
powp = -Mt;
powq = Mh - 1;
(*setting the value for the coupling*)
g = 1/5;
```

The variables **powp** and **powq** give the powers of $\mathbf{P}_a$ and $\mathbf{Q}_i$. We also set the coupling to a particular value $g = 1/5$.

Parameters. There are several parameters which are responsible for the precision of the result.

```
cutP = 3;(* number of terms we keep in the expansion of P_a *)
cutQai = 12;(* number of powers in expansion of Qai at large *)
shiftQai = 20;  (* Number of jumps from asymptotic region *)
WP = 50;(* Working precision *)
PO = 12;(* Number of the sampling points on the cut to use *)
```

Ansatz for $\mathbf{P}_a$ and parameters of the problem. The set of $\mathbf{p}_a$ from (9.151) we define as follows:

```
ps = {A[1] + I*Sum[c[1, n]/x^(2*n), {n, cutP}],
      A[2] + I*Sum[c[2, n]/x^(2*n), {n, cutP}],
      A[3] +   Sum[c[3, n]/x^(2*n), {n, cutP}],
      A[4] +   Sum[c[4, n - 1]/x^(2*n), {n, 2, cutP + 1}]};
```

Note that we set the first sub-leading coefficient in $\mathbf{P}_4$ to zero, this is always possible to do due to the residual symmetry (see footnote 16). Whereas the coefficients $c_{a,n}$ will serve as parameters in the optimization problem, the leading coefficients A_a are fixed in terms of the quantum numbers of the state via (9.117),

```
A[1]=-I Product[(Mt[[1]]-Mh[[j]])/If[j==1,1,Mt[[1]]-Mt[[j]]],{j,4}]
A[2]=+I Product[(Mt[[2]]-Mh[[j]])/If[j==2,1,Mt[[2]]-Mt[[j]]],{j,4}]
A[3] = 1; A[4] = 1;
```

Similarly, we code the leading coefficients of $\mathbf{Q}_i$ and $Q_{a|i}$,

```
B[1]=-I Product[(Mh[[1]]-Mt[[j]])/If[j==1,1,Mh[[1]]-Mh[[j]]],{j,4}]
B[2]=+I Product[(Mh[[2]]-Mt[[j]])/If[j==2,1,Mh[[2]]-Mh[[j]]],{j,4}]
B[3] = 1; B[4] = 1;
(* leading order coefficients in Q_ai *)
Do[B[a,i,0] = -I(A[a]B[i])/(powq[[i]]+powp[[a]]+1),{a,4},{i,4}]
```

The whole Q-system, which we are going to partially reconstruct, is thus parameterized by a set of $c_{i,n}$ and d. The substitute **sb** will replace these variables by their values stored in the list **params**,

```
prm := {d}~Join~Flatten[Table[c[i, n], {i, 4}, {n, cutP}]];
sb := Rule @@@ (Transpose[{prm, SetPrecision[params, WP]}])
```

We will update the list **params** at each iteration with its better approximation. As we are going to solve it with a Newton-like method, which is very sensitive to the starting points, one should roughly know where to look for the solution. A perturbative solution, available in some cases, could be good to start with, but sometimes even a very rough estimate of d and a few first coefficients will lead to a convergent procedure. For **cutP**= 3 we need in total $1 + 4*3 = 13$ parameters.

Finding $Q_{a|i}$ at large u. Having finished with defining the basics we can finally accomplish the first step in the algorithm—find the large u expansion of $Q_{a|i}$ in the form (9.153). First we re-expand $\mathbf{P}_a$ at large u:

```
psu = Series[(g x/u)^powp ps/.x->X[u],{u,Infinity,cutQai+2}];
```

Next we define separately the non-integer power $u^{-\tilde{M}_a+\hat{M}_j}$ and the series in inverse negative powers (9.153),

```
qaipow = Table[u^powq[[i]]*u^powp[[a]]*u, {a, 4}, {i, 4}];
Bpart = Table[Sum[B[a,i,n]/u^(2*n),{n,0,cutQai/2}],{a,4},{i,4}];
```

For optimization purposes we pre-expand these parts of the expansion separately with shifts $u \to u \pm i/2$,

```
powP=Series[(qaipow/.u->u+I/2)/qaipow/.(u+a_)^(b_):>u^b*(1+a/u)^b
    ,{u, Infinity, cutQai + 2}];
powM=Series[(qaipow/.u->u-I/2)/qaipow/.(u+a_)^(b_):>u^b*(1+a/u)^b
    ,{u, Infinity, cutQai + 2}];
BpartP = Series[Bpart /. u -> u + I/2, {u, Infinity, cutQai + 2}];
BpartM = Series[Bpart /. u -> u - I/2, {u, Infinity, cutQai + 2}];
```

Finally we code the function which computes the coefficients $B_{a,i,n}$,

```
FindQlarge := Block[{},
PP=Series[KroneckerProduct[psu,chi.psu],{u,Infinity,cutQai+2}]/.sb;
eqs=ExpandAll[Series[Normal[
(BpartP/.sb)*powP-(BpartM/.sb)*powM+(1/u)*PP.(BpartP*powP) /. sb]
,{u,Infinity,cutQai+2}]];
slB = Last[Solve[LogicalExpand[eqs == 0]]];
Qailarge = qaipow*Bpart/.slB/.sb]
```

The function computes the expansion and stores it in the variable `Qailarge`.

Finding $Q_{a|i}, Q_i$, and $\tilde{Q}_i$ on the real axis. In order to impose the gluing conditions on the Zhukovsky branch cut (9.127) we will use a set of sampling points (`points`), chosen so that their density increases near the ends of the interval $[-2g, 2g]$ to guarantee maximal efficiency (we use Chebyshev nodes):

```
points = N[Table[-2*g*Cos[Pi*((n - 1/2)/PO)], {n, PO}], WP];
```

Now for each of the sampling points we have to climb up to the asymptotic region using (9.154):

```
SolveQPP[n0_] := Block[{}, Clear[Qai, PP, PS];
PS[uu_] := PS[uu] = SetPrecision[Expand[(x*g)^powp*ps/.sb]
    /.x^(a_.)->X[uu]^a /. sb, WP];
PP[(uu_)?NumericQ] := PP[uu] = IdentityMatrix[4]+
    KroneckerProduct[PS[uu], chi.PS[uu]];
Qai[n0][uu_] = Qailarge /. u -> uu + I*n0 - I/2;
Qai[n_][u_] := Qai[n][u] = SetPrecision[PP[u+I*n].Qai[n+1][u],WP];
Qaiplist = Table[Qai[1][p], {p, points}]];
```

This function creates `Qaiplist` which contains values of $Q_{a|i}$ at the sampling points. This allows us to compute Q_i using simple matrix multiplication via (9.155):

```
DoQlist := Block[{},
Qilist = Transpose[Table[((x*g)^powp*ps/.x->X[u]/.sb
    /.u->points[[i]]).chi.Qaiplist[[i]], {i, PO}]];
Qitlist = Transpose[Table[((x*g)^powp*ps/.x->1/X[u]/.sb
    /.u->points[[i]]).chi.Qaiplist[[i]], {i, PO}]];];
```

Now, when we have the values of Q_i and also $\tilde{Q}_i$ we can define the function F, which depends on the parameters $d, c_{a,n}$ and computes the mismatch of the gluing condition at the sampling points,

```
F[Plist_List] := (F[Plist] = Block[{}, Print[Plist];
params = Plist;
FindQlarge;
SolveQPP[shiftQai];
DoQlist;
C1list = Qilist[[1]]/Conjugate[Qilist[[3]]];
C2list = Qitlist[[1]]/Conjugate[Qitlist[[3]]];
c = Mean[Join[C1list, C2list]];
Flatten[{Re[{C1list-c, C2list-c}/c], Im[{C1list-c, C2list-c}/c]}]
)/; NumericQ[Total[Plist]];
```

Finally, we have to tune the values of parameters so that the square of the function F is minimized.

```
(*setting the starting configuration*)
params0 = SetPrecision[{4.5,0,0,0,-1,0,0,1,0,0,0,0,0}, WP];
(*finding optimal parameters*)
FindMinimum[(1/2)*F[prm].F[prm],
  Transpose[{prm, params0}],
  Method -> {"LevenbergMarquardt", "Residual" -> F[prm]},
  WorkingPrecision -> 30,
  AccuracyGoal -> 7]
```

The in-built function **FindMinimum** is rather slow and takes around ten minutes to run. It is much better to use the implementation from the notebook attached to the arXiv submission [13] which uses parallel computing and gives the result in about one minute. It is possible to further improve the above basic code performance by a factor of between ten and one hundred, but that will also make it more cumbersome.

Exercise 9.30 Use the above code to get the dimension of the Konishi operator at $g = 1/5$. Compare your result with the high precision evaluation $\Delta = 4.41885988080235096225036287$ taken from [13].

Exercise 9.31 Use the result for $g = 1/5$ as a starting point to compute $g = 3/10$. You should get $\Delta = 4.826949$. Note that the convergence radius of the perturbation theory is $g^* = 1/4$,[19] so this value is already outside the range accessible with perturbation theory.

---

[19] The finite convergence radius of the perturbation theory is due to the branch-cut singularity of the spectrum at $g_* = \pm i/4$. This is the value of the coupling when branch points of the Zhukovsky cuts $2g + in$ and $-2g + in \pm i$ become equal.

Exercise 9.32 Check that the same code will work perfectly for non-integer values of the Lorentz spin S. Analytic continuation in the spin is very important for between ten and one hundred, BFKL applications [14, 40]. Try to change $S = S_1$ gradually until it reaches $S = 3/2$ for $\Delta = 2/10$. You should get $\Delta = 3.85815$. Verify (9.127) numerically and show that $\gamma \simeq 0.0030371$ is indeed a real constant and $\beta = \beta_1 + \beta_2 \cosh(2\pi u) + \beta_3 \sinh(2\pi u)$ for some constants β_k.

9.7 Applications, further reading, and open questions

In this section we attempt to cover most of the recent applications of the QSC methods and offer some open questions.

QSC for ABJ(M) theory. The QSC was also developed for ABJ(M) theory (which is a 3D $\mathcal{N} = 6$ Chern–Simons theory) in [10, 17]. A non-trivial specific feature of this theory is that the positions of the branch points are related to the 't Hooft coupling in a very non-trivial way and is called the interpolation function $h(\lambda)$. By comparing the results of localization with the analytic calculation of the slope function (similar to what we did in Section.9.5.3), it was possible to obtain an expression for the interpolation function for ABJM theory [32] and for a more general ABJ theory [33]. The detailed proof of these expressions is still an open question and would likely require the QSC formulation for the cusped Wilson-line in these theories, which is not yet known.

QSC for Wilson line with a cusp. The anomalous dimension for the Maldacena–Wilson line with a cusp was shown to be integrable in [34, 35]. In [15] the QSC construction for this observable was formulated, which allowed for precise numerical analysis and non-perturbative analytic results. In [16], by taking an appropriate limit of the cusp anomalous dimension, the potential between heavy quark–anti-quarks was studied in detail with the help of QSC.

QSC for high order perturbative expansion. The QSC method allows for very efficient analytic perturbative expansion. A very nice and powerful method for the $sl(2)$ sector was developed in [37] allowing the computation of 10-loop analytical coefficients on a standard laptop in just three hours. An alternative method, which can be applied in the general situation was developed in [14]. In [53] the project of creating a database perturbative expansion of low lying anomalous dimensions was initiated.

QSC for QCD pomeron. As we discuss in Section 9.5.2 the QSC enables a very simple analytic continuation in the quantum numbers such as Lorentz spin S. As was explained in [12, 38], one can approach the regime where $\mathcal{N} = 4$ SYM becomes similar to QCD. This regime can also be studied with the QSC [40]. In particular, the most complicated highest transcendentality parts of the planar QCD result at three loops was obtained for the first time in [14], by using the QSC. It was later confirmed by an independent calculation in [39].

QSC for deformations of $\mathcal{N} = 4$ **SYM.** The $\mathcal{N} = 4$ admits numerous deformations. Some of them are analogous to the twists we discussed in Section 9.2.2 and can easily be introduced into the QSC formalism simply by modifying the asymptotic of the Q-functions. For some examples see [15, 31]. Another deformation is called η-deformation [36], which can most likely be described by the QSC as well, by replacing a simple cut in the $\mathbf{P}_a$ function with a periodized set of cuts.

QSC for fishnet graphs. In the limit when one of the twist parameters becomes large and the 't Hooft coupling simultaneously scales to zero, one gets a significant simplification in the perturbation theory, which is dominated by the 'fishnet' scalar graphs. At first this limit was considered for the cusp anomalous dimension in [34, 41] and it was possible to reproduce the result analytically from the QSC. A more systematic study of the 'fishnet' limit of $\mathcal{N} = 4$ was then initiated by [42], where it was demonstrated that many more observables can be studied by considering a special type of diagram. In [43] it was shown how the QSC methods can be used to evaluate this type of Feynman graph.

Open questions. Even though a number of longstanding problems have been resolved with the help of the QSC there are still a number of open questions which could potentially be solved using the QSC. Some of them are likely to be solved soon, others may never be solved. Below we give an incomplete list of such problems, focussing on those more likely to be solved before the next Ice Age.

It would be very useful to be able to extract the strong coupling expansion of the spectrum analytically from the QSC. Some first steps were done in [44].

The structure of the QSC is very constraining and at the moment we only know two QSCs for SYM and ABJ(M). It would be useful to make a complete classification of the QSCs starting from the symmetry group. This way one should find the QSC for AdS_3/CFT_2 and also possibly for a mysterious $6D$ theory—a mother theory of $6D$ integrable fishnet graphs. Similarly, different asymptotic and gluing conditions represent different observables in $\mathcal{N} = 4$ SYM, it would be useful to have a complete classification of such asymptotics and gluing conditions. For even more mathematically oriented readers there is the question of proving existence/countability of the solution of the QSC.

A big open conceptual question is how to derive the QSC from the gauge theory perspective, without a reference to AdS/CFT correspondence, as this would allow us to prove to some extent AdS/CFT by taking the classical limit of the QSC, deriving the Green–Schwartz classical spectral curve.

Some of the problems which are within immediate reach include: studying oderon dimension in a similar way to the BFKL pomeron [14] (see for settings [52]); constructing the QSC for the recently proposed integrability framework for the Hagedorn phase transition [45] which would enable analytic weak coupling expansion and numerical analysis for this observable; integrable boundary problems with non-diagonal twist, as recently considered in [46] could most likely be treated in a way similar to [16], this problem seems also to be related to the problem of finding the spectrum of tachyons [47], which is another problem where QSC reformulation could help to provide further advances.

A more complicated but very important problem is to extend the QSC formalism to the problem of computing the n-point correlation function. The existing beautiful integrability-based hexagon formalism [48] should give important hints on re-summing wrapping corrections. This problem seems to be linked to the problem of finding separated variables in the AdS/CFT; for some first steps at weak coupling see [49]. The one-point function [51] could be the perfect framework for developing a new QSC-based formalism for the correlators. See also [50] for more exotic observables which could also be potentially governed by integrability.

Finally, the other main open questions are whether we could also use integrability to get non-planar corrections and also get closer to real world QCD.

If you have questions, please feel free to email to nikgromov@gmail.com. You are also welcome to email any answers to the above questions to nickgromov@mail.ru!

Acknowledgements

In memory of Ludvig Dmitrievich Faddeev.

I am very grateful to M. Almov, A. Cavaglia, S. Leurent, F. Levkovich-Malyuk, G. Sizov, D. Volin, and especially to V. Kazakov and P. Vieira for numerous discussions on closely related topics. I am thankful to D.Grabner, D. Lee and J.[20] for carefully reading the manuscript. The work was supported by the European Research Council (Programme \Ideas" ERC-2012-AdG 320769 AdS-CFT-solvable). We are grateful to Humboldt University (Berlin) for the hospitality and nancial support of this work in the framework of the \Kosmos" programe. We wish to thank STFC for support from Consolidated grant number ST/J002798/1. This work has received funding from the People Programme (Marie Curie Actions) of the European Union's Seventh Framework Programme FP7/2007-2019/ under REA Grant Agreement No 317089 (GATIS).

References

[1] N. Gromov and P. Vieira, 'The AdS(5) x S**5 superstring quantum spectrum from the algebraic curve,' Nucl. Phys. B **789** (2008) 175 doi:10.1016/j.nuclphysb.2007.07.032 [hep-th/0703191 [HEP-TH]].

[2] N. Gromov, V. Kazakov, and P. Vieira, 'Exact Spectrum of Anomalous Dimensions of Planar N=4 Supersymmetric Yang-Mills Theory,' Phys. Rev. Lett. **103** (2009) 131601 doi:10.1103/PhysRevLett.103.131601 [arXiv:0901.3753 [hep-th]].

[3] N. Gromov, V. Kazakov, A. Kozak, and P. Vieira, 'Exact Spectrum of Anomalous Dimensions of Planar N = 4 Supersymmetric Yang-Mills Theory: TBA and excited states,' Lett. Math. Phys. **91** (2010) 265 doi:10.1007/s11005-010-0374-8 [arXiv:0902.4458 [hep-th]].

[4] N. Gromov, V. Kazakov, and P. Vieira, 'Exact Spectrum of Planar $\mathcal{N} = 4$ Supersymmetric Yang-Mills Theory: Konishi Dimension at Any Coupling,' Phys. Rev. Lett. **104** (2010) 211601 doi:10.1103/PhysRevLett.104.211601 [arXiv:0906.4240 [hep-th]].

---

[20] i.e. Julius, who only has a first name

[5] N. Gromov, 'Y-system and Quasi-Classical Strings,' JHEP **1001** (2010) 112 doi:10.1007/JHEP01(2010)112 [arXiv:0910.3608 [hep-th]].
[6] N. Beisert et al., 'Review of AdS/CFT Integrability: An Overview,' Lett. Math. Phys. **99**, 3 (2012) doi:10.1007/s11005-011-0529-2 [arXiv:1012.3982 [hep-th]].
[7] N. Gromov, V. Kazakov, S. Leurent, and D. Volin, 'Solving the AdS/CFT Y-system,' JHEP **1207**, 023 (2012) doi:10.1007/JHEP07(2012)023 [arXiv:1110.0562 [hep-th]].
[8] N. Gromov, V. Kazakov, S. Leurent, and D. Volin, 'Quantum Spectral Curve for Planar $\mathcal{N} =$ Super-Yang-Mills Theory,' Phys. Rev. Lett. **112**, no. 1, 011602 (2014) doi:10.1103/PhysRevLett.112.011602 [arXiv:1305.1939 [hep-th]].
[9] N. Gromov, F. Levkovich-Maslyuk, G. Sizov, and S. Valatka, 'Quantum spectral curve at work: from small spin to strong coupling in $\mathcal{N} = 4$ SYM,' JHEP **1407**, 156 (2014) doi:10.1007/JHEP07(2014)156 [arXiv:1402.0871 [hep-th]].
[10] A. Cavaglià, D. Fioravanti, N. Gromov, and R. Tateo, 'Quantum Spectral Curve of the $\mathcal{N} = 6$ Supersymmetric Chern-Simons Theory,' Phys. Rev. Lett. **113**, no. 2, 021601 (2014) doi:10.1103/PhysRevLett.113.021601 [arXiv:1403.1859 [hep-th]].
[11] N. Gromov, V. Kazakov, S. Leurent, and D. Volin, 'Quantum spectral curve for arbitrary state/operator in AdS_5/CFT_4,' JHEP **1509**, 187 (2015) doi:10.1007/JHEP09(2015)187 [arXiv:1405.4857 [hep-th]].
[12] M. Alfimov, N. Gromov, and V. Kazakov, 'QCD Pomeron from AdS/CFT Quantum Spectral Curve,' JHEP **1507**, 164 (2015) doi:10.1007/JHEP07(2015)164 [arXiv:1408.2530 [hep-th]].
[13] N. Gromov, F. Levkovich-Maslyuk, and G. Sizov, 'Quantum Spectral Curve and the Numerical Solution of the Spectral Problem in AdS5/CFT4,' JHEP **1606**, 036 (2016) doi:10.1007/JHEP06(2016)036 [arXiv:1504.06640 [hep-th]].
[14] N. Gromov, F. Levkovich-Maslyuk, and G. Sizov, 'Pomeron Eigenvalue at Three Loops in $\mathcal{N} = 4$ Supersymmetric Yang-Mills Theory,' Phys. Rev. Lett. **115**, no. 25, 251601 (2015) doi:10.1103/PhysRevLett.115.251601 [arXiv:1507.04010 [hep-th]].
[15] N. Gromov and F. Levkovich-Maslyuk, 'Quantum Spectral Curve for a cusped Wilson line in $\mathcal{N} = 4$ SYM,' JHEP **1604**, 134 (2016) doi:10.1007/JHEP04(2016)134 [arXiv:1510.02098 [hep-th]].
[16] N. Gromov and F. Levkovich-Maslyuk, 'Quark-anti-quark potential in $\mathcal{N} = 4$ SYM,' JHEP **1612**, 122 (2016) doi:10.1007/JHEP12(2016)122 [arXiv:1601.05679 [hep-th]].
[17] D. Bombardelli, A. Cavaglià, D. Fioravanti, N. Gromov, and R. Tateo, 'The full Quantum Spectral Curve for AdS_4/CFT_3,' arXiv:1701.00473 [hep-th].
[18] D. Bombardelli et al., 'An integrability primer for the gauge-gravity correspondence: An introduction,' J. Phys. A **49** (2016) no.32, 320301 doi:10.1088/1751-8113/49/32/320301 [arXiv:1606.02945 [hep-th]].
[19] N. Beisert, 'The SU(2|2) dynamic S-matrix,' Adv. Theor. Math. Phys. **12** (2008) 945 doi:10.4310/ATMP.2008.v12.n5.a1 [hep-th/0511082].
[20] R. A. Janik, 'The AdS(5) x S**5 superstring worldsheet S-matrix and crossing symmetry,' Phys. Rev. D **73**, 086006 (2006) doi:10.1103/PhysRevD.73.086006 [hep-th/0603038].
[21] N. Beisert, B. Eden, and M. Staudacher, 'Transcendentality and Crossing,' J. Stat. Mech. **0701**, P01021 (2007) doi:10.1088/1742-5468/2007/01/P01021 [hep-th/0610251].
[22] J. Ambjorn, R. A. Janik, and C. Kristjansen, 'Wrapping interactions and a new source of corrections to the spin-chain/string duality,' Nucl. Phys. B **736** (2006) 288 doi:10.1016/j.nuclphysb.2005.12.007 [hep-th/0510171].
[23] A. Cavaglia, D. Fioravanti, and R. Tateo, 'Extended Y-system for the AdS_5/CFT_4 correspondence,' Nucl. Phys. B **843** (2011) 302 doi:10.1016/j.nuclphysb.2010.09.015 [arXiv:1005.3016 [hep-th]].

[24] D. Bombardelli, D. Fioravanti, and R. Tateo, 'Thermodynamic Bethe Ansatz for planar AdS/CFT: A Proposal,' J. Phys. A **42** (2009) 375401 doi:10.1088/1751-8113/42/37/375401 [arXiv:0902.3930 [hep-th]].

[25] G. Arutyunov and S. Frolov, 'Thermodynamic Bethe Ansatz for the AdS(5) x S(5) Mirror Model,' JHEP **0905**, 068 (2009) doi:10.1088/1126-6708/2009/05/068 [arXiv:0903.0141 [hep-th]].

[26] J. Balog and A. Hegedus, 'Hybrid-NLIE for the AdS/CFT spectral problem,' JHEP **1208**, 022 (2012) doi:10.1007/JHEP08(2012)022 [arXiv:1202.3244 [hep-th]].

[27] L. D. Faddeev, 'How algebraic Bethe ansatz works for integrable model,' hep-th/9605187.

[28] P. P. Kulish and N. Y. Reshetikhin, 'Diagonalization Of Gl(n) Invariant Transfer Matrices And Quantum N Wave System (lee Model),' J. Phys. A **16** (1983) L591. doi:10.1088/0305-4470/16/16/001

[29] B. Basso, 'An exact slope for AdS/CFT,' arXiv:1109.3154 [hep-th].

[30] N. Dorey and B. Vicedo, 'On the dynamics of finite-gap solutions in classical string theory,' JHEP **0607**, 014 (2006) doi:10.1088/1126-6708/2006/07/014 [hep-th/0601194].

[31] V. Kazakov, S. Leurent, and D. Volin, 'T-system on T-hook: Grassmannian Solution and Twisted Quantum Spectral Curve,' JHEP **1612**, 044 (2016) doi:10.1007/JHEP12(2016)044 [arXiv:1510.02100 [hep-th]].

[32] A. Cavaglià, N. Gromov, and F. Levkovich-Maslyuk, 'On the Exact Interpolating Function in ABJ Theory,' JHEP **1612**, 086 (2016) doi:10.1007/JHEP12(2016)086 [arXiv:1605.04888 [hep-th]].

[33] N. Gromov and G. Sizov, 'Exact Slope and Interpolating Functions in N=6 Supersymmetric Chern-Simons Theory,' Phys. Rev. Lett. **113** (2014) no.12, 121601 doi:10.1103/PhysRevLett.113.121601 [arXiv:1403.1894 [hep-th]].

[34] D. Correa, J. Maldacena, and A. Sever, 'The quark anti-quark potential and the cusp anomalous dimension from a TBA equation,' JHEP **1208**, 134 (2012) doi:10.1007/JHEP08(2012)134 [arXiv:1203.1913 [hep-th]].

[35] N. Drukker, 'Integrable Wilson loops,' JHEP **1310** (2013) 135 doi:10.1007/JHEP10(2013)135 [arXiv:1203.1617 [hep-th]].

[36] G. Arutyunov, R. Borsato, and S. Frolov, 'S-matrix for strings on η-deformed AdS5 x S5,' JHEP **1404** (2014) 002 doi:10.1007/JHEP04(2014)002 [arXiv:1312.3542 [hep-th]].

[37] C. Marboe and D. Volin, 'Quantum spectral curve as a tool for a perturbative quantum field theory,' Nucl. Phys. B **899** (2015) 810 doi:10.1016/j.nuclphysb.2015.08.021 [arXiv:1411.4758 [hep-th]].

[38] A. V. Kotikov, L. N. Lipatov, A. Rej, M. Staudacher, and V. N. Velizhanin, 'Dressing and wrapping,' J. Stat. Mech. **0710**, P10003 (2007) doi:10.1088/1742-5468/2007/10/P10003 [arXiv:0704.3586 [hep-th]].

[39] S. Caron-Huot and M. Herranen, 'High-energy evolution to three loops,' arXiv:1604.07417 [hep-ph].

[40] M. Alfimov, N. Gromov, and V. Kazakov, 'QCD Pomeron from AdS/CFT Quantum Spectral Curve,' JHEP **1507**, 164 (2015) doi:10.1007/JHEP07(2015)164 [arXiv:1408.2530 [hep-th]].

[41] J. K. Erickson, G. W. Semenoff, R. J. Szabo, and K. Zarembo, 'Static potential in N=4 supersymmetric Yang-Mills theory,' Phys. Rev. D **61** (2000) 105006 doi:10.1103/PhysRevD.61.105006 [hep-th/9911088].

[42] Ö. Gürdoğan and V. Kazakov, 'New Integrable 4D Quantum Field Theories from Strongly Deformed Planar $\mathcal{N} = 4$ Supersymmetric Yang-Mills Theory,' Phys. Rev. Lett. **117** (2016) no.20, 201602 Addendum: [Phys. Rev. Lett. **117** (2016) no.25, 259903] doi:10.1103/PhysRevLett.117.201602, 10.1103/PhysRevLett.117.259903 [arXiv:1512.06704 [hep-th]].

[43] N. Gromov, V. Kazakov, G. Korchemsky, S. Negro, and G. Sizov, 'Integrability of Conformal Fishnet Theory,' arXiv:1706.04167 [hep-th].
[44] Á. Hegedűs and J. Konczer, 'Strong coupling results in the AdS_5 /CF T_4 correspondence from the numerical solution of the quantum spectral curve,' JHEP **1608** (2016) 061 doi:10.1007/JHEP08(2016)061 [arXiv:1604.02346 [hep-th]].
[45] T. Harmark and M. Wilhelm, 'The Hagedorn temperature of AdS5/CFT4 via integrability,' arXiv:1706.03074 [hep-th].
[46] M. Guica, F. Levkovich-Maslyuk, and K. Zarembo, 'Integrability in dipole-deformed N=4 super Yang-Mills,' arXiv:1706.07957 [hep-th].
[47] Z. Bajnok, N. Drukker, Á. Hegedüs, R. I. Nepomechie, L. Palla, C. Sieg, and R. Suzuki, 'The spectrum of tachyons in AdS/CFT,' JHEP **1403**, 055 (2014) doi:10.1007/JHEP03(2014)055 [arXiv:1312.3900 [hep-th]].
[48] B. Basso, S. Komatsu, and P. Vieira, arXiv:1505.06745 [hep-th].
[49] N. Gromov, F. Levkovich-Maslyuk, and G. Sizov, arXiv:1610.08032 [hep-th].
[50] I. Buhl-Mortensen, M. de Leeuw, A. C. Ipsen, C. Kristjansen, and M. Wilhelm, 'Asymptotic one-point functions in AdS/dCFT,' arXiv:1704.07386 [hep-th].
[51] I. Buhl-Mortensen, M. de Leeuw, C. Kristjansen, and K. Zarembo, 'One-point Functions in AdS/dCFT from Matrix Product States,' JHEP **1602** (2016) 052 doi:10.1007/JHEP02(2016)052 [arXiv:1512.02532 [hep-th]].
[52] R. C. Brower, M. S. Costa, M. Djurić, T. Raben, and C. I. Tan, 'Strong Coupling Expansion for the Conformal Pomeron/Odderon Trajectories,' JHEP **1502** (2015) 104 doi:10.1007/JHEP02(2015)104 [arXiv:1409.2730 [hep-th]].
[53] M. Alfimov, N. Gromov, G. Sizov to appear.
[54] C. Marboe and D. Volin, 'The full spectrum of AdS5/CFT4 I: Representation theory and one-loop Q-system,' arXiv:1701.03704 [hep-th].

10
Three-point functions in $\mathcal{N}=4$ supersymmetric Yang–Mills theory

Shota KOMATSU

School of Natural Sciences, Institute for Advanced Study, Princeton, New Jersey USA

Chapter Contents

10	**Three-point functions in $\mathcal{N}=4$ supersymmetric Yang–Mills theory** Shota KOMATSU	**449**
	Preface	451
10.1	Three-point functions at weak coupling	451
10.2	Hexagon approach	472
10.3	Conclusion and prospects	498
	Acknowledgements	499
	References	499

Preface

In the past ten years, there has been significant development in our understanding of $\mathcal{N} = 4$ supersymmetric Yang–Mills theory ($\mathcal{N} = 4$ SYM). This has been achieved largely by the use of integrability, a powerful technique for studying a certain class of two-dimensional quantum field theories. Such study was initiated in the seminal paper by Minahan and Zarembo [1], in which they discovered a link between computation of the anomalous dimension and the diagonalization of the Hamiltonians of certain integrable spin chains. Rapid progress in subsequent years has culminated in the powerful and elegant method called the quantum spectral curve [2].

In a parallel line of development, the integrability method has been applied also to other observables. One notable success in this regard is in computation of the expectation value of the null polygonal Wilson loop [3], which, through the T-duality, is related to the scattering amplitude [4]. The basic idea of the computation is to break it down into building blocks called pentagons and characterize them as form factors in a two-dimensional integrable theory.

Quite recently, a similar approach was proposed also for the three-point function [5]. It provides a powerful framework which allows us to compute the structure constant at the non-perturbative level. The main goal of this chapter is to give a pedagogical introduction to and explain necessary backgrounds for that approach. In Section 10.2 we describe the computation of the three-point function at tree level and rewrite them as a sum over partitions. In Section 10.3 we explain how the sum-over-partition expression can be generalized to finite coupling and then introduce a non-perturbative approach, called the *hexagon approach*. We also explain how the symmetry and the integrability can be used to determine the hexagon form factor, which is the fundamental object in that approach.

10.1 Three-point functions at weak coupling

10.1.1 Motivations

Before plunging into the computation, let us give several motivations for why we wish to study three-point functions of $\mathcal{N} = 4$ SYM using integrability.

Firstly they are simply interesting to study. By studying them, one can appreciate the beautiful interplay of physics in two and four dimensions. For instance, it tells us how two-dimensional integrable models emerge from four-dimensional gauge theories and how four-dimensional physics is reflected in those two-dimensional models. As we will see in Section 10.3, the two-dimensional integrable model arising from $\mathcal{N} = 4$ SYM is rather special and is highly constrained by the fact that it comes out of a *gauge* theory.

Secondly, the integrability might be helpful in understanding the AdS/CFT correspondence. $\mathcal{N} = 4$ SYM is the most typical example of AdS/CFT correspondence and its three-point function describes the interaction of strings in AdS space-time. The study of such objects will undoubtedly give us deeper insight into the holography.

Thirdly, solving $\mathcal{N}=4$ SYM can be an important stepping stone towards understanding four-dimensional interacting conformal field theories. As is well known two-dimensional conformal field theory is highly constrained by infinite-dimensional Virasoro algebra. On the other hand, not much is known about theories in higher dimensions partly because the conformal group is finite-dimensional. The hope is that $\mathcal{N}=4$ SYM would give us important insights, with which we could start exploring more general CFTs in higher dimensions.

Let us make an additional remark on this point. In recent years, the conformal bootstrap is applied to higher-dimensional conformal field theories and has been yielding impressive results[6]. The conformal bootstrap and the integrability-based approach share a common feature that both of them are non-perturbative approaches, but they are, of course, different in numerous respects. One obvious difference is that, while the integrability-based method works only for special theories in the large N limit, the conformal bootstrap can work, in principle, for any theory. However, a more important difference which we wish to emphasize is the role of gauge symmetry. In the conformal bootstrap, one always deals with gauge-invariant quantities and the gauge symmetry is completely invisible (and unnecessary). In the integrability-based approach, on the other hand, gauge symmetry plays an essential role, as we will see in what follows.

10.1.2 Brief review of $\mathcal{N}=4$ supersymmetric Yang–Mills theory

Let us start with a lightning review of the basic properties of $\mathcal{N}=4$ SYM.

10.1.2.1 Symmetries and field contents

$\mathcal{N}=4$ SYM has the following global symmetries:

- As the name suggests, it has four sets of supercharges,

$$\{Q_\alpha^A, \bar{Q}_{\dot{\alpha}A}\} \qquad (\alpha, \dot{\alpha} = 1, 2, \quad A = 1, \ldots, 4). \tag{10.1}$$

  Here, α and $\dot{\alpha}$ are the spinor indices which take 1 or 2, and A is the index distinguishing different sets of supercharges and runs from 1 to 4. In total it has $4 \times 4 = 16$ supercharges.
- The theory also has global $SU(4) \simeq SO(6)$ symmetry which rotates these four sets of supercharges.
- In addition, the theory is known to have conformal symmetry $SO(4,2) \simeq SU(2,2)$, whose generators are given by the dilatation D, special conformal transformations K_μ, translations P^μ, and Lorentz transformations $L_{\mu\nu}$.
- The commutators of the supersymmetry generators and the special conformal generators yield an extra sixteen fermionic generators $[K, Q] \sim S$, which are superconformal generators:

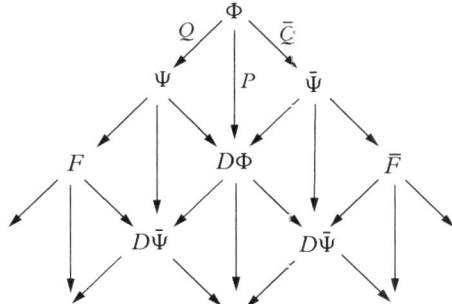

Figure 10.1 *The structure of the field mutiplet in $\mathcal{N}=4$ SYM. $\mathcal{D}$ denotes the covariant derivative and $\mathcal{F}$ and $\bar{\mathcal{F}}$ are the field strength.*

$$\{S_A^\alpha, \bar{S}^{\dot\alpha A}\} \qquad (\alpha,\dot\alpha = 1,2, \quad A=1,\ldots,4). \tag{10.2}$$

These symmetries combine into the $\mathcal{N}=4$ superconformal group,[1] which is isomorphic to the supergroup PSU(2,2|4).

Since the $\mathcal{N}=4$ superconformal group is sufficiently large, the fields appearing in the Lagrangian all fit in a single superconformal multiplet. It consists of six scalars, four Weyl fermions, and one gauge field (see Figure 10.1 for the structure of the multiplet):

$$\begin{array}{ll} \Phi_I & (I=1,\ldots,6) \\ \Psi_\alpha^A, \bar{\Psi}_{\dot\alpha A} & (\alpha=1,2, \quad A=1,\ldots,4) \\ A_\mu & \end{array} \tag{10.3}$$

As a consequence, all the fields belong to the adjoint representation of the gauge group $SU(N_c)$.

10.1.2.2 Two- and three-point functions

In conformal field theories, the basic objects to study are the correlation functions of local operators. The most fundamental among them are two- and three-point functions, since any higher-point correlation functions can be decomposed into two- and three-point functions using the operator product expansion.

The space-time dependence of the two- and three-point functions are constrained by the conformal symmetry. For instance, the correlators of the scalar operators[2] take the following form:

---

[1] For those who want to know more about superconformal symmetry, we recommend the paper [7].
[2] Here we also assume that the operators are conformal primary.

$$\langle \mathcal{O}_i(x_1)\mathcal{O}_j(x_2)\rangle = \frac{\delta_{ij}\mathcal{N}_i}{|x_1-x_2|^{2\Delta_i}},$$

$$\langle \mathcal{O}_i(x_1)\mathcal{O}_j(x_2)\mathcal{O}_k(x_3)\rangle = \frac{C_{ijk}\sqrt{\mathcal{N}_i\mathcal{N}_j\mathcal{N}_k}}{|x_1-x_2|^{\Delta_i+\Delta_j-\Delta_k}|x_2-x_3|^{\Delta_j+\Delta_k-\Delta_i}|x_3-x_1|^{\Delta_k+\Delta_i-\Delta_j}}. \quad (10.4)$$

Here Δ_i is called the conformal dimension and is the eigenvalue of the dilatation operator,

$$D \cdot \mathcal{O}_i = \Delta_i \mathcal{O}_i, \quad (10.5)$$

whereas C_{ijk} is the structure constant of the operator product expansion,

$$\mathcal{O}_i(x_1)\mathcal{O}_j(x_2) \sim \frac{C_{ijk}}{|x_1-x_2|^{\Delta_i+\Delta_j-\Delta_k}}\mathcal{O}_k(x_2) + \cdots. \quad (10.6)$$

The constants $\mathcal{N}_i$'s denote the normalization factors of the operators. They can always be set to 1 by re-normalizing the operators as $\mathcal{O}_i \to \mathcal{O}_i/\sqrt{\mathcal{N}_i}$. For actual computation, however, it is sometimes useful to keep such factors.

10.1.2.3 Single-trace operators and the spin chain

As mentioned previously, all the fields in $\mathcal{N}=4$ SYM belong to the adjoint representation of the gauge group $SU(N_c)$. Therefore, the gauge invariant quantities can be built easily by multiplying the fields and taking traces.

In the large N_c limit, the most important set of local operators are *single-trace operators*, the operators which consist of a single trace:

$$\text{tr}\left(\Phi\Psi\Phi F_{\mu\nu}\cdots\right) \quad \text{etc.} \quad (10.7)$$

In addition to the single-trace operators, there exist other operators that consist of multiple traces. However, in the large N_c limit, the conformal dimension of a multi-trace operator is simply given by a sum of conformal dimensions of the constituent single-trace operators:

$$\mathcal{O} = \underbrace{\text{tr}(\cdots)}_{\mathcal{O}_1}\underbrace{\text{tr}(\cdots)}_{\mathcal{O}_2} \quad \longrightarrow \quad \Delta_\mathcal{O} = \Delta_{\mathcal{O}_1} + \Delta_{\mathcal{O}_2} + O(1/N_c). \quad (10.8)$$

Because of this property, the single-trace operators are the most basic quantity in the large N_c theories.

As found by Minahan and Zarembo [1], there is a non-trivial link between the single-trace operators in the large N_c limit and the two-dimensional integrable system. We will not review it in detail here but the upshot is that one can identify the action of the dilatation operator on the single-trace operators with the action of the Hamiltonian in a certain $1+1$-dimensional integrable system. Under this identification, the conformal dimension is mapped to the energy of the integrable system. We can then use various

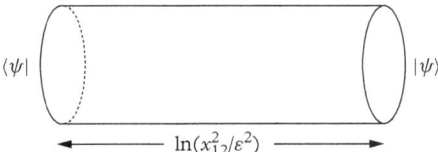

Figure 10.2 *The cylinder partition function with length* $\ln\left(x_{12}^2/\epsilon^2\right)$ *corresponds to the two-point function in* $\mathcal{N} = 4$ *SYM.*

techniques of integrability, such as the Bethe ansatz and the thermodynamic Bethe ansatz, to determine the conformal dimension:

$$D \leftrightarrow H_{2d}, \quad \Delta \leftrightarrow E_{2d}. \tag{10.9}$$

It is also possible to relate the two-point function itself, not just the conformal dimension, to a 2d integrable system. The idea is to consider a (Euclidean) two-dimensional cylinder with length $\ln(x_{12}^2/\epsilon^2)$, as depicted in Figure 10.2. The partition function of this cylinder can be computed as[3]

$$\langle \psi_{2d} | e^{-H_{2d} \ln(x_{12}^2/\epsilon^2)} | \psi_{2d} \rangle = \left(\frac{\epsilon}{x_{12}}\right)^{2\Delta}. \tag{10.10}$$

Obviously this reproduces the correct two-point function, including the space-time dependence. This map, however, may seem too ad hoc since it gives an impression that we intentionally chose the length of the cylinder so that the final result comes out correctly. However, it actually has a natural physical interpretation. To further clarify this point, it is useful to work out the following small exercise.

Exercise 1 Consider two operators inserted at $x_1^\mu = (0,0,0,0)$ and $x_2^\mu = (a,0,0,0)$, respectively, and cut out small spheres with radius ϵ around these points (see Figure 10.3). Our claim is that $\ln(x_{12}^2/\epsilon^2)$ is related to how much 'dilatation transformation' we need in order to bring one sphere to the other. In order to make this statement precise, we perform a special conformal transformation and bring x_2 to infinity. This is because the dilatation transformation does not map points near x_1 to points near x_2. Instead it maps points near the origin to points near infinity.

After the special conformal transformation, we get a small sphere around the origin and a large sphere near infinity. What this exercise requires is for you to figure out the radius of these spheres after the transformation and read off how much dilatation you need in order to map them to each other. You will then find that the length $\ln(x_{12}^2/\epsilon^2)$ appears naturally.

---

[3] Here $\epsilon (\ll 1)$ is the cut-off we introduced.

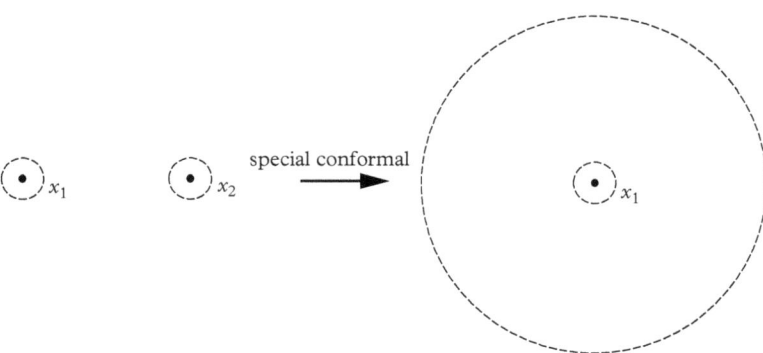

Figure 10.3 *The action of the special conformal transformation.*

10.1.3 BPS correlators at tree level

Having reviewed the minimal backgrounds, we now start studying the correlation functions at weak coupling.

Among the single-trace operators, the simplest are 1/2 BPS operators. Those are operators of the following form:

$$\mathrm{tr}\left((P^I \Phi_I)^L\right) =: \mathrm{tr}\left((P \cdot \Phi)^L\right), \tag{10.11}$$

where P^I is a complex six-dimensional null vector, called the *polarization vector*, satisfying[4] $P^2 = 0$. For instance, if we choose P^I to be $(1, i, 0, 0, 0, 0)$, we get

$$\mathrm{tr}\left(Z^L\right), \quad Z := \Phi_1 + i\Phi_2, \tag{10.12}$$

which is the operator commonly used in the literature. The 1/2 BPS operators have several important properties:

- They are annihilated by all sixteen superconformal charges S and eight supercharges Q. The combination of the supercharges which annihilates the operator depends on the choice of P^I.
- They belong to the short multiplet of the superconformal group and its conformal dimension is fixed to be $\Delta = L$. Namely, they do not receive quantum corrections.

10.1.3.1 Two-point function

Let us now compute the two-point function of the BPS operators:

$$\langle \mathrm{tr}\left((P_1 \cdot \Phi)^L\right)(x_1) \ \mathrm{tr}\left((P_2 \cdot \Phi)^L\right)(x_2) \rangle. \tag{10.13}$$

---

[4] This condition is necessary to make the operator BPS.

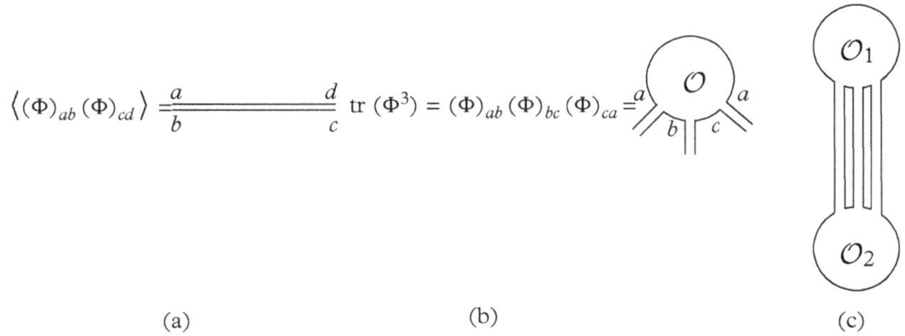

Figure 10.4 *The double-line notations. (a) Double-line notation for the propagator of the scalar fields. Each line signifies the Kronecker delta, δ_{ad} etc. (b) Double-line representation of the single-trace operator. (c) Planar Wick contraction for the two-point function.*

At tree level, we just need to perform the Wick contraction. For two scalar fields, the Wick contraction reads

$$\langle \Phi_I(x_1)\Phi_J(x_2) \rangle = \frac{\delta_{IJ}}{|x_1-x_2|^2}$$
$$\Rightarrow \langle (P_1\cdot\Phi)(x_1)(P_2\cdot\Phi)(x_2) \rangle = \frac{(P_1\cdot P_2)}{|x_1-x_2|^2}. \tag{10.14}$$

In (10.14) we suppressed the indices for the gauge group SU(N_c), which is given by[5]

$$\langle (\Phi)_{ab}(\Phi)_{cd} \rangle \propto \delta_{ad}\delta_{bc}. \tag{10.18}$$

As is well known, the best way to take into account the index structure of SU(N_c) is to use the double-line notation, as depicted in Figure 10.4(a).

Now, using the double-line notation, each single trace operator can be denoted as in Figure 10.4(b). Then the two-point function can be computed by drawing all the possible

---

[5] To derive (10.18), we first rewrite the kinetic term $\text{tr}(\partial_\mu\Phi\partial^\mu\Phi)$ in terms of the orthogonal basis as $\Phi = \sum_A \Phi_A T^A$ with $\text{tr}(T^A T^B) = \delta^{AB}$. Then, the kinetic term and the Wick contraction read

$$\sum_A \partial_\mu\Phi_A\partial^\mu\Phi_A, \quad \langle\Phi_A\Phi_B\rangle \propto \delta_{AB}. \tag{10.15}$$

To go back to the matrix representation, we simply multiply T^A and T^B and use the relation

$$\sum_A (T^A)_{ab}(T^B)_{cd} \propto \delta_{ad}\delta_{bc} - \frac{1}{N_c}\delta_{ab}\delta_{cd}. \tag{10.16}$$

In the large N_c limit, the second term can be neglected and we get

$$\langle(\Phi)_{ab}(\Phi)_{cd}\rangle = \sum_{A,B}(T^A)_{ab}(T^B)_{cd}\langle\Phi_A\Phi_B\rangle \propto \delta_{ad}\delta_{bc}. \tag{10.17}$$

planar diagrams connecting two operators. For the length L operator, there are L different ways of planar contractions which are related to each other by the cyclic permutation of one of the operators. As a result, we get

$$\langle \mathrm{tr}\left((P_1\cdot\Phi)^L\right)(x_1)\ \mathrm{tr}\left((P_2\cdot\Phi)^L\right)(x_2)\rangle \sim L N_c^L \frac{(P_1\cdot P_2)^L}{|x_1-x_2|^{2L}}. \tag{10.19}$$

The dependence on N_c can be easily deduced from Figure 10.4(c). In order to normalize the two-point function, we should redefine the operator as

$$\mathcal{O}_i \to \frac{1}{N_c^{L_i/2}\sqrt{L_i}}\mathcal{O}_i. \tag{10.20}$$

10.1.3.2 Three-point function

We now move onto computation of the three-point function,

$$\langle \mathrm{tr}\left((P_1\cdot\Phi)^{L_1}\right)(x_1)\ \mathrm{tr}\left((P_2\cdot\Phi)^{L_2}\right)(x_2)\ \mathrm{tr}\left((P_3\cdot\Phi)^{L_3}\right)(x_3)\rangle. \tag{10.21}$$

An example of the planar diagram connecting all three operators is given in Figure 10.5. In the case of three-point functions, there are $L_1 L_2 L_3$ inequivalent ways of planar contractions which are related to each other by the cyclic permutations of the individual operators. Therefore the result reads

$$\begin{aligned}&\langle \mathrm{tr}\left((P_1\cdot\Phi)^{L_1}\right)(x_1)\ \mathrm{tr}\left((P_2\cdot\Phi)^{L_2}\right)(x_2)\ \mathrm{tr}\left((P_3\cdot\Phi)^{L_3}\right)(x_3)\rangle \\ &= N_c^{\frac{L_1+L_2+L_3}{2}-1} L_1 L_2 L_3 \frac{(P_1\cdot P_2)^{\ell_{12}}(P_2\cdot P_3)^{\ell_{23}}(P_3\cdot P_1)^{\ell_{31}}}{|x_1-x_2|^{2\ell_{12}}|x_2-x_3|^{2\ell_{23}}|x_3-x_1|^{2\ell_{31}}}. \end{aligned} \tag{10.22}$$

Here, $\ell_{ij}=(L_i+L_j-L_k)/2$ denotes the number of Wick contractions between the operators $\mathcal{O}_i$ and $\mathcal{O}_j$. In what follows, we call it the *bridge length*.

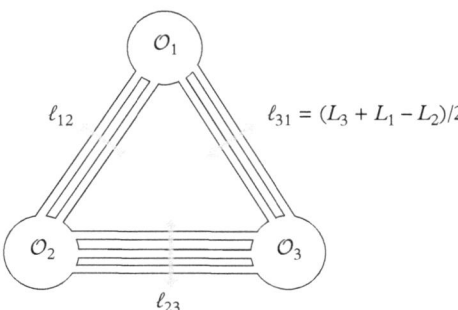

Figure 10.5 Three-point functions at tree level. L_i denotes the length of the operator $\mathcal{O}_i$ and ℓ_{ij} denotes the number of Wick contractions between operators $\mathcal{O}_i$ and $\mathcal{O}_j$.

After the normalization (10.20), the three-point function is given by

$$\frac{(10.22)}{N_c^{L_1/2}\sqrt{L_1}N_c^{L_2/2}\sqrt{L_2}N_c^{L_3/2}\sqrt{L_3}} \propto \frac{\sqrt{L_1 L_2 L_3}}{N_c}. \qquad (10.23)$$

As is manifest from (10.23), the three-point function is $O(1/N_c)$. In this sense, studying the three-point function already requires us to go slightly beyond the strict large N_c limit. This is easier to understand on the dual string-theory side. As mentioned in Section , the three-point function corresponds to the splitting and joining processes of the strings in AdS and therefore it must come with the string coupling constant $g_s \sim 1/N_c$.

Since the factor (10.23) is common to all three-point functions, it is convenient to strip it off from the structure constant as

$$C_{123} = \frac{\sqrt{L_1 L_2 L_3}}{N_c} c_{123}. \qquad (10.24)$$

In Section 10.2.4, we compute c_{123} using perturbation theory.

Exercise 2 Confirm the N_c dependence of the three-point function (10.22).

Before finishing this section, let us make a short remark on the BPS three-point function. As is the case with two-point functions, the BPS three-point functions turns out to be a protected quantity. This can be shown[6] by taking a derivative of the three-point function with respect to the coupling constant,

$$\frac{\partial}{\partial g_{YM}^2} \langle \mathcal{O}_1 \mathcal{O}_2 \mathcal{O}_3 \rangle = \int d^4x \langle \mathcal{O}_1 \mathcal{O}_2 \mathcal{O}_3 \mathcal{L} \rangle. \qquad (10.25)$$

Here $\mathcal{L}$ is the Lagrangian. In $\mathcal{N} = 4$ SYM, the Lagrangian is a superconformal descendant of the length 2 BPS operator $\hat{\mathcal{O}}$,

$$\mathcal{L} = (Q)^4 \hat{\mathcal{O}}, \qquad (10.26)$$

where $(Q)^4$ denotes an appropriate linear combination of products of four supercharges. Using this property, we can express the Lagrangian as

$$\mathcal{L} = \tilde{Q}\tilde{\mathcal{C}}, \qquad (10.27)$$

---

[6] For details, see [8]. It provides a simple and concise argument on this point.

with $\tilde{Q}$ being a linear combination of the supercharges[7] which annihilates all three operators

$$\tilde{Q}\mathcal{O}_i = 0. \tag{10.28}$$

Now, using the Ward identity, we can prove

$$\langle \mathcal{O}_1 \mathcal{O}_2 \mathcal{O}_3 \mathcal{L} \rangle = \langle \mathcal{O}_1 \mathcal{O}_2 \mathcal{O}_3 \tilde{Q}\tilde{\mathcal{O}} \rangle = -\langle \tilde{Q}(\mathcal{O}_1 \mathcal{O}_2 \mathcal{O}_3) \tilde{\mathcal{O}} \rangle = 0. \tag{10.29}$$

This shows that the three-point function of a BPS operator does not depend on the coupling constant.

10.1.4 Non-BPS three-point functions at tree level

We now set out the computation of the non-BPS three-point functions.

10.1.4.1 Set-up

To illustrate the basic physical picture, avoiding unnecessary complication, we focus on the following configurations, where one operator is a non-BPS operator in the so-called SU(2) sector and the rest are BPS (see also Figure 10.6):

$$\mathcal{O}_1 = (ZYZZY\cdots) + \cdots, \quad \mathcal{O}_2 = \operatorname{tr}\left(\bar{Z}^{L_2}\right), \quad \mathcal{O}_3 = \operatorname{tr}\left((Z+\bar{Z}+Y-\bar{Y})^{L_3}\right). \tag{10.30}$$

Here Z and Y are given by $Z = \Phi_1 + i\Phi_2$ and $Y = \Phi_3 + i\Phi_4$, respectively. In our notation, it amounts to choosing $P_2 = (1, -i, 0, 0, 0, 0)$ and $P_3 = (1, 0, 0, i, 0, 0)$. A simplifying feature of this configuration is that all Y fields contained in $\mathcal{O}_1$ contract only with $\mathcal{O}_3$

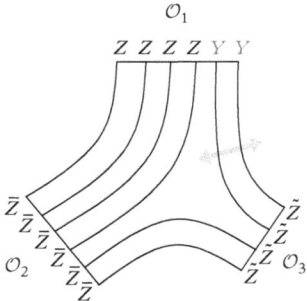

Figure 10.6 *The non-BPS three-point function we discuss in the main text. Here $\tilde{Z}$ denotes a linear combination of fields $\tilde{Z} = Z + \bar{Z} + Y - \bar{Y}$. A simplifying feature of this configuration is that all the magnon excitations in $\mathcal{O}_1$ are contracted with the operator $\mathcal{O}_3$; namely the magnons can only live in the region indicated by the arrow.*

---

[7] An example of such supercharges will be given explicitly in Section 10.3.

since Y cannot contract with $\bar{Z}$. This substantially simplifies the combinatorics of the Wick contraction.

To perform the contraction, we need to know the precise form of $\mathcal{O}_1$. It is determined by the condition that $\mathcal{O}_1$ is the eigenvector of the dilatation operator:

$$D \cdot \mathcal{O}_1 = \Delta_1 \mathcal{O}_1. \tag{10.31}$$

By expanding both sides with respect to the coupling constant, we obtain

$$(D^{(0)} + \lambda D^{(1)} + \cdots)(\mathcal{O}_1^{(0)} + \lambda \mathcal{O}_1^{(1)} + \cdots) = (\Delta_1^{(0)} + \lambda \Delta_1^{(0)} + \cdots)(\mathcal{O}_1^{(0)} + \lambda \mathcal{O}_1^{(1)} + \cdots).$$

The leading term in the expansion yields the relation $D^{(0)} \mathcal{O}_1^{(0)} = \Delta_1^{(0)} \mathcal{O}_1^{(0)}$. This relation, however, is not constraining enough to fix the form of $\mathcal{O}_1^{(0)}$. The tree-level dilatation operator $D^{(0)}$ only counts the number of fields and therefore it is proportional to the identity operator if we restrict ourselves to the space of operators with a fixed number of fields. Obviously, there is no way to determine the eigenvector of the identity operator uniquely. The situation is quite analogous to the degenerate perturbation theory in quantum mechanics. As in that case, what we need to do is to consider the next term in the expansion to lift the degeneracy.[8] In the present case, this amounts to considering the eigenvector of the *one-loop* dilatation operator:

$$D^{(1)} \cdot \mathcal{O}_1^{(0)} = \Delta_1^{(1)} \mathcal{O}_1^{(0)}. \tag{10.33}$$

As mentioned in Section 10.1.2, the one-loop dilatation operator can be identified with the integrable spin-chain Hamiltonian. In the present case, the map between the operator and the spin is

$$Z \mapsto \uparrow, \qquad Y \mapsto \downarrow, \tag{10.34}$$

and the Hamiltonian is given by

$$H_{\text{Heisenberg}} \propto \sum_j (I_{j,\,j+1} - P_{j,\,j+1}), \tag{10.35}$$

---

[8] To understand this point, it might be helpful to consider the eigenvectors of the following matrix:

$$M = \begin{pmatrix} 1 & 0 \\ 0 & 1 \end{pmatrix} + \lambda \begin{pmatrix} a & b \\ c & d \end{pmatrix}. \tag{10.32}$$

Clearly, the eigenvectors of the matrix M are determined by the second term on the right hand side, not the first term. In addition, although the matrix M depends linearly on λ, the eigenvectors are independent of λ. In other words, the 'one-loop part' of M determines the 'tree-level part' of the eigenvector.

where $I_{j,j+1}$ and $P_{j,j+1}$ are the identity and the permutation operators acting on the j-th and $(j+1)$-th spins. Once we map the problem to the spin chain, we can systematically construct the eigenvector using the coordinate Bethe ansatz, as we will see shortly.[9]

10.1.4.2 Single-magnon state

Let us first consider the operator with a single Y (called "magnon" a *magnon* in spin chain terminology) in the sea of Z's. In this case, the eigenvector is simply given by the plane wave,

$$|\psi\rangle = \sum_{n=1}^{L_1} e^{ipn} |\cdots \underset{n}{\downarrow} \cdots\rangle. \quad (10.36)$$

Since Y excitations can only contract with $\mathcal{O}_3$ and all different contractions yield the same space-time dependence, the structure constant c_{123} is simply given by the sum of the wave function on the segment with length ℓ_{31}:

$$c_{123} \propto \sum_{1 \leq n \leq \ell_{31}} e^{ipn}. \quad (10.37)$$

This is just a geometric series and one can easily perform the summation as

$$c_{123} \propto \sum_{1 \leq n \leq \ell_{31}} e^{ipn} = g(p)(1 - e^{ip\ell_{31}}), \quad (10.38)$$

with $g(p)$ defined by

$$g(p) := \frac{1}{e^{-ip} - 1}. \quad (10.39)$$

To understand the physical meaning of the geometric sum, it is convenient to introduce a pictorial notation and denote (10.38) as

$$\sum_{1 \leq n \leq \ell_{31}} \underset{\bullet}{\overset{n}{\text{———}}} = g(p) \left(\bullet\text{———} - \text{———}\bullet \right). \quad (10.40)$$

In this notation, the magnon is denoted by the black dot. The second term on the right-hand side, in which the magnon lives at the right edge of the segment, corresponds to the factor $e^{ip\ell_{31}}$. This is the phase shift acquired by the magnon when it propagates from the left edge to the right edge. In the first term on the other hand, the magnon lives at the left edge of the segment. Therefore, it corresponds to the factor $1(= e^{ip0})$.

---

[9] A concise summary of the coordinate Bethe ansatz of the Heisenberg spin chain is given in Section 2 of [9]. For a more pedagogical introduction, see Section 3 of [10].

10.1.4.3 Two-magnon state

We next discuss the two-magnon state. The wave function for the two-magnon state is composed of two terms,

$$|\psi(p_1,p_2)\rangle = \sum_{1\leq n<m\leq L_1} \underbrace{e^{ip_1 n + ip_2 m}}_{\psi_{12}} |\cdots \downarrow_n \cdots \downarrow_m \cdots\rangle + \underbrace{S(p_1,p_2) e^{ip_2 n + ip_1 m}}_{\psi_{21}} |\cdots \downarrow_n \cdots \downarrow_m \cdots\rangle, \quad (10.41)$$

where $S(p_1,p_2)$ is the S-matrix of the Heisenberg spin chain. It takes a simple form in terms of the so-called rapidity variables u_1 and u_2:

$$S(p_1,p_2) := \frac{u_1 - u_2 - i}{u_1 - u_2 + i}, \qquad \left(\frac{u_1 + i/2}{u_1 - i/2} = e^{ip_1}, \quad \frac{u_2 + i/2}{u_2 - i/2} = e^{ip_2}\right). \quad (10.42)$$

In what follows, we denote the first term in (10.41) by ψ_{12} and the second term by ψ_{21}. The contribution from ψ_{12} can be computed as

$$\psi_{12}: \sum_{1\leq n<m\leq L_1} e^{ip_1 n + ip_2 m} \underset{\text{sum over } n}{=} g(p_1) \sum_{1\leq m\leq L_1} \left(1 - e^{ip_1(m-1)}\right) e^{ip_2 m}$$

$$\underset{\text{sum over } m}{=} g(p_1)\left[g(p_2)(1 - e^{ip_2 \ell_{31}}) - g(p_1+p_2)e^{-ip_1}(1 - e^{i(p_1+p_2)\ell_{31}})\right]. \quad (10.43)$$

To make clear the physical picture, let us rewrite this summation using the pictorial notation introduced previously:

$$\sum_{1\leq n<m\leq \ell_{31}} \underset{n\quad m}{\bullet\!-\!\!-\!\bullet} = g(p_1) \sum_m \left(\underset{m}{\bullet\!-\!\!-\!\bullet} - \underset{m}{\bullet\!-\!\!-\!\longrightarrow}\right)$$

$$= g(p_1)\Big(g(p_2)\,\bullet\!-\!\!-\!\bullet\; -\; g(p_2)\,\bullet\!-\!\!-\!\bullet$$

$$-\; g(p_1+p_2)e^{-ip_1}\,\bullet\!-\!\!-\!\bullet\; +\; g(p_1+p_2)e^{-ip_1}\,\bullet\!-\!\!-\!\bullet\Big). \quad (10.44)$$

As is clear from this expression, the result of the geometric sum can be written as the sum over different ways of distributing magnons between the left and the right edges of the segment. This structure persists also for the multi-magnon state and it is the origin of the sum-over-partition expression we derive at the end of this section.

The contribution from ψ_{21} can be computed similarly. Aside from the S-matrix factor, the result can be obtained by exchanging p_1 and p_2 in (10.43). Now, by summing two contributions, we obtain the following expression:

$$-(u_1+i/2)(u_2+i/2)\left[h(u_1,u_2) - e^{ip_2\ell_{31}} - e^{ip_1\ell_{31}} + h(u_1,u_2)e^{i(p_1+p_2)\ell_{31}}\right]. \quad (10.45)$$

Here $h(u,v)$ is a function defined by

$$h(u,v) = \frac{u-v}{u-v+i}. \tag{10.46}$$

Exercise 3 Check (10.45).

The function $h(u,v)$ appearing in the formula satisfies an important relation,

$$S(u,v) = \frac{h(u,v)}{h(v,u)}, \tag{10.47}$$

whose physical meaning will become clear in Section 10.3.

Before moving onto the multi-magnon generalization, let us explain a simple trick to efficiently compute this geometric series. In (10.43), we first summed over n and then summed over m. In this way of summation, we obtain two different terms which do not contain a length-dependent phase shift, $g(p_1)g(p_2)$ and $-g(p_1)g(p_1+p_2)e^{-ip_1}$. However, if we instead sum over m first and then over n, there will be only a single term without the phase shift and we can immediately get $g(p_2)g(p_1+p_2)$:

$$\sum_{n,m} \underset{nm}{\underline{\bullet\bullet}} \overset{\text{sum over } m}{=} g(p_2) \sum_m \underset{n}{\underline{\bullet}} + \cdots \tag{10.48}$$

$$\overset{\text{sum over } n}{=} g(p_2)g(p_1+p_2) \,\bullet\underline{} + \cdots$$

In general, for each different phase shift, there exists an appropriate order of doing the summation and, if we choose that order, we can immediately get the term of interest. This trick is quite useful when we discuss the multi-magnon generalization.

10.1.4.4 Generalization to multi-magnon state

We are now in a position to generalize the previous results to muti-magnon states.[10] The coordinate Bethe wave function for the multi-particle state is given by a superposition of the plane waves:

$$|\psi(p_1,\ldots,p_M)\rangle = \sum_{n_1<\cdots<n_M} \psi(n_1,\ldots,n_M) |\cdots \underset{n_1}{\downarrow} \cdots \underset{n_M}{\downarrow} \cdots\rangle, \tag{10.49}$$

with

$$\psi(n_1,\ldots,n_M) = \sum_{\sigma\in S_M} \prod_{\substack{j<k \\ \sigma_k<\sigma_j}} S(p_{\sigma_k},p_{\sigma_j}) \prod_{j=1}^{M} e^{ip_{\sigma_j} n_j}. \tag{10.50}$$

---

[10] The derivation to be explained was first worked out independently by Naoki Kiryu and Ho Tat Lam.

At first sight, the expression for the S-matrix factor may seem cryptic. However, there is a simple and physically intuitive way to figure out this factor. As an example, let us take a look at the three-particle wave function:

$$\psi(n_1, n_2, n_3) = e^{i(p_1 n_1 + p_2 n_2 + p_3 n_3)} + S(p_1, p_2) e^{i(p_2 n_1 + p_1 n_2 + p_3 n_3)} \\ + \cdots + S(p_1, p_3) S(p_2, p_3) e^{i(p_3 n_1 + p_1 n_2 + p_2 n_3)} + \cdots \quad (10.51)$$

The latter term comes from the permutation $\{1, 2, 3\} \to \{3, 1, 2\}$. The S-matrix factor associated with this permutation can be determined by drawing a diagram in Figure 10.7. For each intersection in the diagram, we associate an S-matrix and the S-matrix factor in (10.50) can be obtained simply by multiplying such S-matrices.

To proceed, it is useful to slightly rewrite the wave function. The idea is to rewrite the S-matrix factor using (10.47) as follows:

$$\prod_{\substack{j<k \\ \sigma_k<\sigma_j}} S(p_{\sigma_k}, p_{\sigma_j}) = \left(\prod_{\substack{j<k \\ \sigma_k<\sigma_j}} h(u_{\sigma_k}, u_{\sigma_j}) \right) \left(\prod_{\substack{j<k \\ \sigma_k<\sigma_j}} \frac{1}{h(u_{\sigma_j}, u_{\sigma_k})} \right)$$

$$= \left(\prod_{\sigma_k<\sigma_j} h(u_{\sigma_k}, u_{\sigma_j}) \right) \left(\prod_{\substack{j>k \\ \sigma_k<\sigma_j}} \frac{1}{h(u_{\sigma_k}, u_{\sigma_j})} \right) \left(\prod_{\substack{j<k \\ \sigma_k<\sigma_j}} \frac{1}{h(u_{\sigma_j}, u_{\sigma_k})} \right) \quad (10.52)$$

$$= \left(\prod_{j<k} h(u_j, u_k) \right) \left(\prod_{\substack{j<k \\ \sigma_j<\sigma_k}} \frac{1}{h(u_{\sigma_j}, u_{\sigma_k})} \right) \left(\prod_{\substack{j<k \\ \sigma_k<\sigma_j}} \frac{1}{h(u_{\sigma_j}, u_{\sigma_k})} \right)$$

$$= \prod_{j<k} h(u_j, u_k) \prod_{j<k} \frac{1}{h(u_{\sigma_j}, u_{\sigma_k})}$$

Figure 10.7 *The structure of the Bethe wave function. The S-matrix factor associated with the permutation* $\{1, 2, 3\} \to \{3, 1, 2\}$ *can be obtained by multiplying the S-matrices at the black dots in the figure.*

In the second line, we decomposed the first term in the right-hand side into a product of two factors. In the third line, we relabelled the indices of the first and the second term. To go to the last line, we combined the second and the third terms in the line above. After doing so, the wave function takes a concise form:

$$\psi(n_1,\ldots,n_M) = \prod_{j<k} h(u_j,u_k) \sum_{\sigma \in S_M} \prod_{j<k} \frac{1}{h(u_{\sigma_j},u_{\sigma_k})} e^{i\sum_j p_{\sigma_j} n_j}. \tag{10.53}$$

Now using this expression, we can compute the structure constant as

$$c_{123} \propto \sum_{1 \le n_1 < \cdots < n_M \le \ell_{31}} \psi(n_1,\ldots,n_M)$$

$$= \prod_{j<k} h(u_j,u_k) \sum_{\sigma \in S_M} \left(\prod_{j<k} \frac{1}{h(u_{\sigma_j},u_{\sigma_k})} \right) M(p_{\sigma_1},\ldots,p_{\sigma_M}), \tag{10.54}$$

with

$$M(p_1,\ldots,p_M) := \sum_{1 \le n_1 < \cdots < n_M \le \ell_{31}} e^{i\sum_j p_j n_j}. \tag{10.55}$$

The function M has the following expression in terms of a sum over (ordered) partitions:

Lemma 1

$$M(p_1,\ldots,p_M) :=$$

$$\sum_{\substack{\alpha=\{1,\ldots,m\} \\ \bar{\alpha}=\{m+1,\ldots,M\}}} (-1)^{|\bar{\alpha}|} \left(\prod_{j\in\bar{\alpha}} e^{ip_j \ell_{31}} \right) \left(\prod_{j\in\alpha} \frac{1}{e^{-i\sum_{k=j}^m p_k} - 1} \right) \left(\prod_{j\in\bar{\alpha}} \frac{e^{ip_j}}{1 - e^{i\sum_{k=j+1}^M p_k}} \right). \tag{10.56}$$

Here $|\bar{\alpha}|$ is the number of elements in $\bar{\alpha}$.

Proof. Here we only sketch the proof and leave it for the reader to fill in the details. In terms of the pictorial notation we introduced, the set α ($\bar{\alpha}$) denotes a set of magnons living on the left (right) edge of the segment.

The basic idea is to use the trick explained previously: to compute the contribution corresponding with $\alpha = \{1,\ldots,m\}$, we should first sum over the position of the m-th magnon. This produces two terms, one of which corresponds to $n_m = n_{m-1}$ and the other corresponds to $n_m = n_{m+1}$. However, the second term cannot contribute to the partition of interest, where the m-th magnon and the $(m+1)$-th are separate. Thus, we just need to keep the first term:

$$\sum_{n_m} \frac{\overset{n_{m-1}\ n_m\ n_{m+1}}{\bullet\ \bullet\ \bullet}}{} = g(p_m) \left(\underbrace{\rule{2cm}{0.4pt}}_{\text{Only this term contributes}} - \rule{2cm}{0.4pt} \right). \tag{10.57}$$

We then sum over the position of the $(m-1)$-th magnon. Also in this case, we only need to keep one of two terms that arise from the geometric sum. Performing the same analysis up until the first magnon, we obtain $\prod_{j\in\alpha}\cdots$ term in (10.56).

After doing so, we then sum over the positions of the rest of the magnons starting from the $(m+1)$-th magnon up to the M-th magnon. This reproduces the rest of the terms. □

The next task is to rewrite (10.54) and simplify the expression. From the structure of (10.54) and (10.56), it is not hard to see that the final expression is given in terms of the sum over partition $\alpha\cup\bar{\alpha} = \{1,\cdots,M\}$. To a given partition, several different permutations can contribute and we should sum over such permutations to get a simple formula. For instance, if we consider four particles, the partition $\alpha = \{14\}$, $\bar{\alpha} = \{23\}$ receives a contribution from four different permutations:

$$\{1,2,3,4\} \mapsto \quad \{1,4,2,3\}, \quad \{4,1,2,3\}, \quad \{1,4,3,2\}, \quad \{4,1,3,2\}. \tag{10.58}$$

As is clear from this example, these permutations are related to each other by the permutation inside α and the permutation inside $\bar{\alpha}$. This observation leads to the following formula:

$$\sum_{\sigma\in S_M} \left(\prod_{j<k} \frac{1}{h(u_{\sigma_j}, u_{\sigma_k})} \right) M(p_{\sigma_1},\ldots,p_{\sigma_M})$$

$$= \sum_{\alpha\cup\bar{\alpha}=\{1,\ldots,M\}} (-1)^{\bar{\alpha}} \left(\prod_{j\in\bar{\alpha}} e^{ip_j\ell_{31}} \right) \left(\prod_{\substack{i,j \\ i\in\alpha, j\in\bar{\alpha}}} \frac{1}{h(u_i, u_j)} \right) F(\alpha) \bar{F}(\bar{\alpha}), \tag{10.59}$$

with

$$F(1,2,\ldots,m) := \sum_{\sigma\in S_m} \left(\prod_{i<j} \frac{1}{h(u_{\sigma_i}, u_{\sigma_j})} \right) \left(\prod_{j=1}^{m} \frac{1}{e^{-i\sum_{k=j}^{m} p_{\sigma_k}} - 1} \right),$$

$$\bar{F}(1,2,\ldots,m) := \sum_{\sigma\in S_m} \left(\prod_{i<j} \frac{1}{h(u_{\sigma_i}, u_{\sigma_j})} \right) \left(\prod_{j=1}^{m} \frac{e^{ip_{\sigma_j}}}{1 - e^{i\sum_{k=j}^{m} p_{\sigma_k}}} \right). \tag{10.60}$$

Although the definitions of F and $\bar{F}$ involve a sum over permutations, it turns out that they evaluate to a simple product:

Lemma 2

$$F(1,\cdots,m) = \bar{F}(1,\cdots,m) = \prod_{k=1}^{m} i(u_k + i/2). \tag{10.61}$$

Proof. We focus on $F(1,\cdots,m)$ since the derivation for $\bar{F}(1,\cdots,m)$ is very similar. The basic strategy is to use mathematical induction. Firstly, it is trivial to check the relation for $m=1$. Next, suppose the relation holds for $F(1,\ldots,m-1)$. Then, we decompose $F(1,\cdots,m)$ with respect to σ_m as follows:

$$F(1,\cdots,m) = \frac{1}{e^{-i\sum_{k=1}^{m} p_k} - 1} \sum_{j=1}^{m} \left(\prod_{k \neq j} \frac{1}{h(u_k, u_j)} \right) \tag{10.62}$$
$$\times \sum_{\sigma' \in S_{\{1,\ldots,m\}/j}} \left(\prod_{1 \leq i < j \leq m-1} \frac{1}{h(u_{\sigma'_i}, u_{\sigma'_j})} \right) \left(\frac{1}{e^{-i\sum_{k=j}^{m} p_{\sigma'_k}} - 1} \right).$$

It turns out that the second line in (10.62) is nothing but $F(1,\ldots,\check{j},\ldots,m)$ ($F(1,\ldots,m)$ with j omitted). Thus using the assumption of the induction, we obtain

$$F(1,\cdots,m) = \frac{\prod_{k=1}^{m} i(u_k + i/2)}{e^{-i\sum_{k=1}^{m} p_k} - 1} \sum_{j=1}^{m} \frac{1}{i(u_j + i/2)} \prod_{k \neq j} \frac{1}{h(u_k, u_j)}. \tag{10.63}$$

To proceed, we use the following identity:

$$\left(e^{-i\sum_{k=1}^{m} p_k} - 1 = \right) \prod_{k=1}^{m} \frac{u_k - i/2}{u_k + i/2} - 1 = \sum_{j=1}^{m} \frac{1}{i(u_j + i/2)} \prod_{k \neq j} \frac{1}{h(u_k, u_j)}. \tag{10.64}$$

This identity can be shown by first expressing the left-hand side using the contour integral as

$$\prod_{k=1}^{m} \frac{u_k - i/2}{u_k + i/2} - 1 = \oint_{z=0} \frac{dz}{2\pi i} \frac{1}{z} \left(\prod_{k=1}^{m} \frac{u_k - z - i/2}{u_k - z + i/2} - 1 \right), \tag{10.65}$$

then deforming the contour and picking up other poles in the integrand. From this identity, the statement of the lemma follows immediately. $\square$

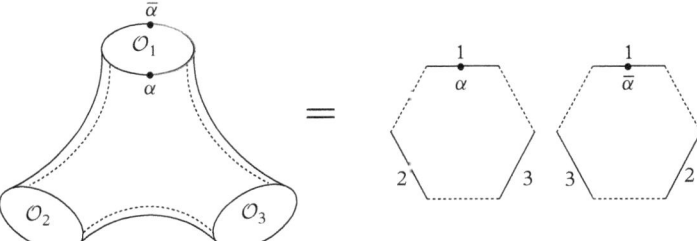

Figure 10.8 *The planar diagram for the three-point function has a topology of a pair of pants. The darker curve on the left-hand side denotes the segment on which we summed the geometric series. To separate two ends of the segment, we cut the pair of pants into two hexagons as depicted on the right-hand side.*

Now using Lemma 2, we arrive at the following expression for the strcture constant:

$$c_{123} \propto \prod_k i(u_k + i/2) \prod_{j<k} h(u_j, u_k) \sum_{\alpha \cup \bar{\alpha} = \{1,\ldots,M\}} (-1)^{\bar{\alpha}} \left(\prod_{j \in \bar{\alpha}} e^{ip_j \ell_{31}} \right) \left(\prod_{\substack{i,j \\ i \in \alpha, j \in \bar{\alpha}}} \frac{1}{h(u_i, u_j)} \right).$$
(10.66)

To compute the structure constant precisely, we need to normalize the operators. The plane-wave state we have been using has a non-trivial norm $\langle \psi | \psi \rangle$, and we need to divide by its square root in order to get a correct structure constant. Fortunately, for integrable spin chains, there is a compact expression for the norm of the state, which is called the *Gaudin formula*.[11] Using the Gaudin formula, we find the following expression for the structure constant:

$$c_{123} = \frac{\mathcal{A}}{\sqrt{\prod_{i<j} S(u_i, u_j) \det \partial_{u_j} \phi_k}},$$
(10.67)

$$\mathcal{A} = \prod_{j<k} h(u_j, u_k) \sum_{\alpha \cup \bar{\alpha} = \{1,\ldots,M\}} (-1)^{|\bar{\alpha}|} \left(\prod_{j \in \bar{\alpha}} e^{ip_j \ell_{31}} \right) \left(\prod_{\substack{i,j \\ i \in \alpha, j \in \bar{\alpha}}} \frac{1}{h(u_i, u_j)} \right).$$

Here ϕ_j is defined by the Bethe equation as

$$e^{i\phi_j} := e^{ip_j L} \prod_{k \neq j} S(u_j, u_k).$$
(10.68)

---

[11] We will not discuss the Gaudin formula. For details, see for instance [11].

10.1.5 Physical interpretation

Let us now interpret the result (10.67) physically. For this purpose, we rewrite $\mathcal{A}$ as

$$\mathcal{A} = \sum_{\alpha \cup \bar{\alpha} = \{1,\ldots,M\}} (-1)^{|\bar{\alpha}|} \prod_{j \in \bar{\alpha}} e^{ip_j \ell_{31}} \prod_{\substack{j<k \\ j \in \bar{\alpha}, k \in \alpha}} S(u_i, u_j) \mathcal{H}(\alpha) \mathcal{H}(\bar{\alpha}), \qquad (10.69)$$

with

$$\mathcal{H}(\alpha) = \prod_{\substack{i<j \\ i,j \in \alpha}} h(u_i, u_j), \qquad \mathcal{H}(\bar{\alpha}) = \prod_{\substack{i<j \\ i,j \in \bar{\alpha}}} h(u_i, u_j). \qquad (10.70)$$

Exercise 4 Show the equivalence of (10.67) and (10.69) using the relation $S(u,v) = h(u,v)/h(v,u)$.

We now examine this expression more closely. Firstly, it contains a sum over bi-partite partitions. As explained the bi-partite partition comes from the geometric series and α ($\bar{\alpha}$) denotes a set of magnons living at the left (right) edge of the segment. If we draw the segment in the full planar diagram, which has a topology of a pair of pants, the two ends of the segment sit at the front and the back of the pair of pants, respectively. This naturally leads to the interpretation that $\mathcal{H}(\alpha)$ ($\mathcal{H}(\bar{\alpha})$) corresponds to the contribution coming from the front (back) side of the pair of pants. In order to separate the front and the back sides, it is natural to cut the pair of pants along the dashed lines, as denoted in Figure 10.8. After doing so, we obtain two hexagons; the one from the front the other from the back. The expression (10.69) suggests that these hexagons (which we also call *hexagon form factors*) are the building blocks for three-point functions.

The remaining factors admit natural physical interpretation as well. The factor $e^{ip_j \ell_{31}}$ is the phase shift, required for moving the magnons from the first (front) hexagon to the second (back) hexagon. When we move magnons, we sometimes need to change their orders. When this happens, the phase shift also receives the contribution from the S-matrix. The S-matrix factor present in (10.69) precisely accounts for this effect (see Figure 10.9 for an illustrative example).

The only remaining factor is $(-1)^{|\bar{\alpha}|}$. It is clear why such a factor arises from the geometric series. However, unfortunately, we have not yet figured out their geometric origin. It would be desirable to understand this factor more physically.

This concludes our analysis at weak coupling. Before moving on, let us briefly mention what happens if we go to one loop. At one loop, there are two sources of corrections:[12]

---

[12] For detail on one-loop computation, see [12, 13].

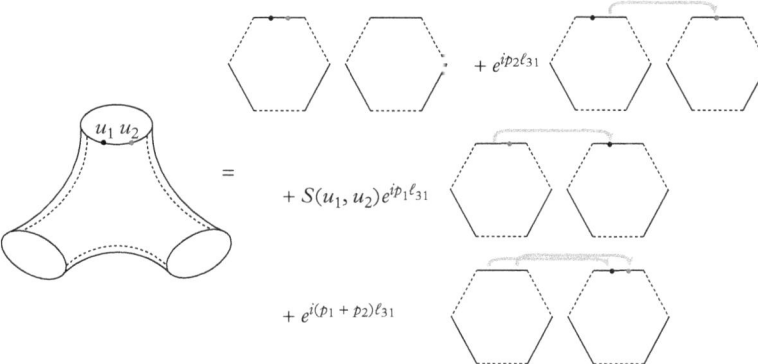

Figure 10.9 *A pictorial explanation of the sum over partitions. Each partition comes with the weight factor given by a product of the propagation factor ($e^{ip\ell_{31}}$) and the S-matrices.*

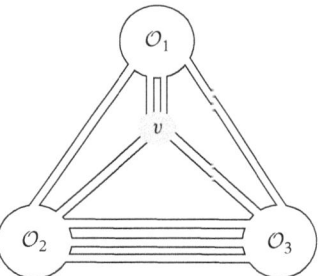

Figure 10.10 *One-loop correction to the three-point function. It amounts to inserting an extra spin chain operator $\mathcal{V}$ at the positions where the operator splits into two. There are six such splitting points (two for each operator) but here we depict only one.*

Firstly the wave function should now be the eigenstate of the two-loop dilatation operator. Secondly there are corrections coming from the one-loop planar diagrams. This results in adding an operator insertion at the point where the operator splits into two (see Figure 10.10). Each of these effects leads to a complicated correction. However, what is truly remarkable is that, if we combine these two effects, the resulting expression simplifies substantially and can be recast into a sum over partition, now with $h(u,v)$ modified slightly [14, 15]. This suggests that the sum over partitions and the physical picture associated with it has deeper meaning. In Section 10.3, we will indeed see that this physical picture can be transformed into a non-perturbative framework which allows us to compute the structure constant at finite coupling.

10.2 Hexagon approach

Based on the observations we made in Section 10.2, we now introduce a non-perturbative framework to study the three-point function, called the hexagon approach.

10.2.1 Asymptotic three-point function

Let us first state our conjecture for the finite-coupling generalization of (10.69) for correlators of one non-BPS SU(2) operator and two BPS operators.

Claim (asymptotic three-point function): When the bridge lengths satisfy $\ell_{ij} \gg 1$, the sum-over-partition expression (10.69) holds also at finite coupling, under the following replacements:

$$e^{ip(u)} = \frac{x^+(u)}{x^-(u)}, \qquad S_{\text{su}(2)}(u,v) = \frac{u-v-i}{u-v+i}\frac{1}{\sigma^2(u,v)},$$

$$H(\alpha) = \prod_{i \in \alpha}\sqrt{\mu(u_i)} \prod_{\substack{i<j \\ i,j \in \alpha}} h_{\text{su}(2)}(u_i, u_j), \qquad \mu(u) = \frac{\left(1 - 1/x^+(u)x^-(u)\right)^2}{\left(1 - 1/(x^-(u))^2\right)\left(1 - 1/(x^+(u))^2\right)}$$

$$h_{\text{su}(2)}(u_i, u_j) = \frac{u-v}{u-v+i}\frac{1 - 1/x^+(u)x^+(v)}{1 - 1/x^+(u)x^-(v)}\frac{1 - 1/x^-(u)x^-(v)}{1 - 1/x^-(u)x^+(v)}\frac{1}{\sigma(u,v)}. \qquad (10.71)$$

The normalization factor (the factor inside the square root in (10.67)) will also be replaced accordingly.

Here and below, the notation $f^\pm(u)$ denotes the shift of the argument $f^\pm(u) = f(u \pm i/2)$ and the function $x(u)$ is called the Zhukowski variable, defined by

$$u = g\left(x + \frac{1}{x}\right), \qquad (10.72)$$

where g is related to the 't Hooft coupling constant λ as

$$g := \frac{\sqrt{\lambda}}{4\pi}. \qquad (10.73)$$

The function $\sigma(u,v)$ is called the dressing phase. We will not write down their explicit forms since they are complicated and not necessary for the purpose of this chapter. For those who want to know the definitions, see [16] for instance. Note also that $H(\alpha)$ now includes the factor $\prod \sqrt{\mu}$, which we call the *measure* factor.

Exercise 5 The energy of a magnon is defined by

$$E(u) := \frac{1}{2}\frac{1+1/x^+x^-}{1-1/x^+x^-}. \tag{10.74}$$

Show that, at weak coupling, it gives

$$E(u) = \frac{1}{2} + \frac{g^2}{u^2+1/4} + O(g^4). \tag{10.75}$$

Exercise 6 Show the following useful identity:

$$\frac{x^+(u)-x^+(v)}{1-1/x^-(u)x^-(v)} = \frac{x^-(u)-x^-(v)}{1-1/x^+(u)x^+(v)}. \tag{10.76}$$

Regarding the conjecture above, there are several questions that one would come up with immediately:

- How do we determine varous coupling-dependent quantities such as $h(u,v)$?
- What about other sectors (beyond the SU(2) sector)?
- The function $h_{\text{su}(2)}(u,v)$ still satisfies[13] the relation $S_{\text{su}(2)}(u,v) = h_{\text{su}(2)}(u,v)/h_{\text{su}(2)}(v,u)$. What is the meaning of this relation?

The rest of the section will be devoted to answering these questions. We will see that the symmetries (namely the superconformal symmetry and the gauge symmetry) and the integrability are key players in this story.

10.2.2 Symmetries of two-point functions

Understanding the symmetry is always an important starting point for solving the problem. A textbook example is the classification of particles using the Poincaré group. The key steps in that case are (see also Figure 10.11):

1. Fix the centre of mass motion by going to the *rest frame*, in which the 4-momentum takes $P^\mu = (M,0,0,0)$.
2. Consider the *little group*, which leaves the momentum invariant.
3. Classify the 'internal motion' (namely spin) using the representation theory of the little group.

It turns out that a similar idea also works for our problem. The analogue of step 1 is to consider correlation functions of BPS operators. The correlators of BPS operators often preserve a certain amount of symmetry, which plays the role of the little group in step 2. The non-BPS operators can be obtained by adding magnons to BPS correlators,

---

[13] One can verify this by using $\sigma(u,v)\sigma(v,u) = 1$.

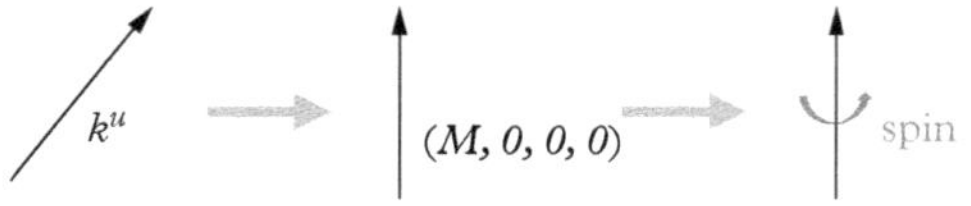

Figure 10.11 *Classification of particles using the Poincaré symmetry. To classify particles in flat space using the Poincaré symmetry, we first bring the system to a 'canonical configuration' (namely, the rest frame), and then classify internal motion such as spin by using the little group, which leaves the canonical configuration invariant.*

which correspond to exciting the 'internal motion'. As in step 3, they are constained by the little group. In what follows, we will see in greater detail how it works for two-point functions.

10.2.2.1 Global symmetry

The 'rest frame' for two-point functions is the following correlator of two BPS operators:

$$\langle \underbrace{\mathrm{tr}\left(Z^L\right)(0)}_{\mathcal{O}_1} \ \underbrace{\mathrm{tr}\left(\bar{Z}^L\right)(\infty)}_{\mathcal{O}_2} \rangle. \tag{10.77}$$

As alluded to previously, there is a non-trivial 'little group', which leaves this two-point function invariant. Its bosonic part is particularly easy to understand: since the operators are inserted at the origin and at infinity, the SO(4) rotation around the origin clearly leaves the correlator invariant. In addition to this, there is an SO(4) R-symmetry, which rotates the scalars $\phi_2 \ldots \phi_6$.

To identify its fermionic part, we need to know properties of the operators $\mathcal{O}_1$ and $\mathcal{O}_2$. Since the first operator $\mathcal{O}_1$ is a 1/2-BPS operator, it is annihilated by eight supercharges (Q's). In addition, it is a *superconformal primary operator* and is annihilated by *all* sixteen superconformal charges (S's). On the other hand, the second operator $\mathcal{O}_2$ is annihilated by sixteen S's and eight Q's, since the operator is at infinity and the roles of supercharges and superconformal charges are swapped. As a consequence, we have eight supercharges and eight superconformal charges which annihilate both $\mathcal{O}_1$ and $\mathcal{O}_2$. Together with the bosonic symmetries we just described, they form a super group $\mathfrak{psu}(2|2)^2$.

There is also an extra symmetry generator $D - \mathcal{J}$, where D is the dilatation and $\mathcal{J}$ is the U(1) rotation $Z \to e^{i\alpha} Z$. This generator commutes with the $\mathfrak{psu}(2|2)^2$ generators and therefore is a central charge of this group.

Having identified the little group, the last step is to classify the 'internal motions' using the symmetry group. In this case, the internal motions correspond to adding magnons,

$$\cdots ZZ \cdots \to \cdots Z\Psi ZDZXZ \cdots \quad \text{etc.} \tag{10.78}$$

As it turns out, magnons belong to a bi-fundamental representation of the $\mathfrak{psu}(2|2)^2$ symmetry. To make clear the structure of the group and the representation, it is customary to represent the magnons $\mathcal{X}$ as follows:

$$\mathcal{X} = \chi^A \dot\chi^{\dot A}. \qquad (10.79)$$

Here χ and $\dot\chi$ are the fundamental representations of the left and the right $\mathfrak{psu}(2|2)$ respectively and are given by

$$\chi^A = (\varphi^1, \varphi^2, \psi^1, \psi^2), \qquad \dot\chi_{\dot A} = (\dot\varphi^1, \dot\varphi^2, \dot\psi^1, \dot\psi^2). \qquad (10.80)$$

They are related to the fields in $\mathcal{N} = 4$ SYM as

$$\begin{aligned}\varphi^1 \dot\varphi^1 &= X, \quad \varphi^1 \dot\varphi^2 = Y, \quad \varphi^2 \dot\varphi^1 = \bar Y, \quad \varphi^2 \dot\varphi^2 = -\bar X, \\ \psi^\alpha \dot\psi^{\dot\alpha} &= D^{\alpha\dot\alpha} Z, \qquad\quad \psi^\alpha \dot\varphi^{\dot a}, \varphi^a \dot\psi^{\dot\alpha} = \text{fermion}.\end{aligned} \qquad (10.81)$$

Here D is a covariant derivative and $X, \bar X, Y,$ and $\bar Y$ are complex scalars.

From the above analysis, one can derive certain constraints on the dynamics of magnons. They are however not strong enough. The main problem is that they know nothing about the dependence on the coupling constant g.

The key idea in overcoming this difficulty is to consider an infinitely long operator and cut open the trace (Figure 10.12):

$$\text{tr}(\cdots) \to \cdots Z \cdots Z \cdots \qquad (10.82)$$

There are two advantages of doing this. Firstly, since the spin chain now has an infinite length, we can consider the S-matrix *on the spin chain*. Secondly, since the trace is cut open, there exist large gauge transformations, namely the transformations which do not die off at infinity. As is often the case, these transformations should be regarded as part of the global symmetries.

10.2.2.2 'Large gauge' symmetry

The gauge transformation always comes with the coupling constant as $\delta\Phi \sim g[\Phi, \Lambda]$. This seemingly simple fact leads to a powerful non-perturbative method when combined with other symmetries.

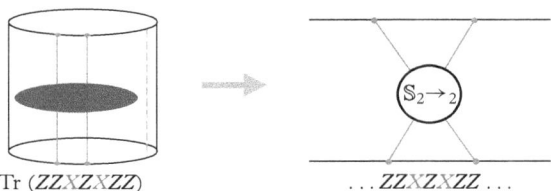

Figure 10.12 *The decompactification of the spin chain. At higher loops, the spin chain Hamiltonian can take quite a complicated form. By considering an infinitely long spin chain and focussing on the S-matrix, we can avoid dealing with the Hamiltonian directly and constrain the dynamics by symmetry.*

To see this, let us recall the transformation laws of the supersymmetry in $\mathcal{N} = 4$ SYM:[14]

$$\delta\Phi = i\bar{\epsilon}\Gamma_i\Psi, \qquad \delta\Psi = \cdots + \frac{i}{2}g[\Phi_i,\Phi_j]\Gamma^{ij}\epsilon + \cdots. \tag{10.83}$$

It is straightforward to see that, acting the supersymmetry transformations twice to a fermion, we get a term of the form

$$\delta^2\Psi \sim g[\Phi,\Psi]. \tag{10.84}$$

This is a (field-dependent) gauge transformation and therefore is proportional to the coupling constant. In particular, for the supercharges belonging to the $\mathfrak{psu}(2|2)^2$, we have

$$Q^2\mathcal{X} \sim g[Z,\mathcal{X}] \quad (:= P \cdot \mathcal{X}). \tag{10.85}$$

In what follows, we denote this field-dependent transformation as P.

Since P is a gauge transformation, it acts trivially on the gauge-invariant quantities:

$$P \cdot \mathrm{tr}(\cdots) \sim \mathrm{tr}([Z,\cdots]) = 0. \tag{10.86}$$

It however can act non-trivially once we cut open the trace and consider an infinitely long operator, since it becomes the 'large gauge transformation'. For instance, the action on the plane-wave state (of infinite length) reads[15]

$$P \cdot \left(\sum_n e^{ipn} |\cdots \underbrace{\mathcal{X}}_{n\text{-th site}} \cdots \rangle \right) \sim g \sum_n e^{ipn} \left(|\cdots Z \underbrace{\mathcal{X}}_{n+1\text{-th}} \cdots \rangle - |\cdots \underbrace{\mathcal{X}}_{n\text{-th}} Z \cdots \rangle \right)$$

$$= g(e^{-ip} - 1) \sum_n e^{ipn} |\cdots \underbrace{\mathcal{X}}_{n\text{-th}} \cdots \rangle. \tag{10.87}$$

This shows that the transformation P is the shift (difference) operator of the spin chain.

The symmetry algebra $\mathfrak{psu}(2|2)^2$ contains the superconformal generator S's in addition to supercharges. The study of two-loop dilatation operators [17] suggests that, at finite coupling, $(S)^2$ also produces a field-dependent gauge transformation,

---

[14] Here we only write down the terms necessary for our purposes. The full transformation laws can be found for instance in [17].
[15] In the second line, we have neglected the extra Z field since the operator is infinitely long.

$$(S)^2 \mathcal{X} \sim g[Z^{-1}, \mathcal{X}] \quad (:= K \cdot \mathcal{X}). \tag{10.88}$$

Unlike the transformation P, the transformation K defined above is purely quantum. Namely, it is not visible at the level of the classical transformation properties of the fields. Their existence is however an inevitable consequence of the anti-commutation relation $\{Q, S\} \sim D$: Since D receives the quantum correction (which yields anomalous dimensions), the generators Q and S must also contain the coupling-dependent part. Once one determines the coupling-dependent part of S, one can then confirm that $(S)^2$ yields a field-dependent gauge transformation (10.88). An important property of these extra generators P and K is that they commute with all the other generators in $\mathfrak{psu}(2|2)^2$. Namely, they are central charges of the algebra.

10.2.2.3 Centrally extended $\mathfrak{psu}(2|2)$

Based on these observations, Beisert proposed the centrally extended $\mathfrak{psu}(2|2)^2$ as the algebra governing the scattering on the spin chain for $\mathcal{N} = 4$ SYM [18, 19]. It consists of generators[16] of $\mathfrak{psu}(2|2)^2$,

$$\begin{aligned} &\text{Supersymmetry: } Q^\alpha{}_a, \dot{Q}^{\dot\alpha}{}_{\dot a}, \quad &\text{Superconformal: } S^a{}_\alpha, \dot{S}^{\dot a}{}_{\dot\alpha}, \\ &\text{R-symmetry: } R^a{}_b, \dot{R}^{\dot a}{}_{\dot b}, \quad &\text{Lorentz: } L^\alpha{}_\beta, \dot{L}^{\dot\alpha}{}_{\dot\beta}, \end{aligned} \tag{10.89}$$

and three central charges,

$$P, \quad K, \quad C(:= (D - \mathcal{J})/2). \tag{10.90}$$

As mentioned previously, these central charges appear in the anti-commutators of the fermionic generators as

$$\begin{aligned} \{Q^\alpha{}_a, Q^\beta{}_b\} &= \epsilon^{\alpha\beta} \epsilon_{ab} P, \\ \{S^a{}_\alpha, S^b{}_\beta\} &= \epsilon_{\alpha\beta} \epsilon^{ab} K, \\ \{Q^\alpha{}_a, Q^\beta{}_b\} &= \delta^\alpha_\beta R^a{}_b + \delta^b_a L^\alpha{}_\beta + \delta^b_a \delta^\alpha_\beta C. \end{aligned} \tag{10.91}$$

Here and below we only write down the formulae for the left $\mathfrak{psu}(2|2)$ since the ones for the right $\mathfrak{psu}(2|2)$ are similar.[17]

These generators act on magnons (10.79) as[18]

---

[16] Here undotted generators belong to the left $\mathfrak{psu}(2|2)$ and dotted generators belong to the right $\mathfrak{psu}(2|2)$.
[17] Namely, we just need to put dots to the $\mathfrak{psu}(2|2)$ generators in (10.91).
[18] The actions of the bosonic generators of $\mathfrak{psu}(2|2)$ are standard: they merely acts on indices.

$$Q^\alpha{}_a|\varphi^b(u)\rangle = \mathbf{a}\delta^b_a|\psi^\alpha(u)\rangle, \qquad Q^\alpha{}_b|\psi^\beta(u)\rangle = \mathbf{b}\epsilon^{\alpha\beta}\epsilon_{ab}|Z\varphi^b(u)\rangle,$$
$$S^a{}_\alpha|\varphi^b(u)\rangle = \mathbf{c}\epsilon^{ab}\epsilon^{\alpha\beta}|Z^{-1}\psi^\beta(u)\rangle, \qquad S^a{}_\alpha|\psi^\beta(u)\rangle = \mathbf{d}\delta^\beta_\alpha|\varphi^a(u)\rangle,$$
$$P|\mathcal{X}^{A\dot{A}}(u)\rangle = g\left(1 - \frac{x^+(u)}{x^-(u)}\right)|Z\mathcal{X}^{A\dot{A}}\rangle, \qquad K|\mathcal{X}^{A\dot{A}}(u)\rangle = g\left(1 - \frac{x^-(u)}{x^+(u)}\right)|Z^{-1}\mathcal{X}^{A\dot{A}}\rangle,$$
(10.92)

with

$$\mathbf{a} = \sqrt{g}\gamma, \quad \mathbf{b} = \frac{\sqrt{g}}{\gamma}\left(1 - \frac{x^+}{x^-}\right), \quad \mathbf{c} = \frac{i\sqrt{g}\gamma}{x^+}, \quad \mathbf{d} = \frac{\sqrt{g}x^+}{i\gamma}\left(1 - \frac{x^+}{x^-}\right), \quad (10.93)$$
$$\gamma := \sqrt{i(x^- - x^+)}.$$

Here u is the rapidity of the magnon and $Z^\pm$ accounts for the change of length caused by a field-dependent gauge transformation. Note that for P and K we have used the full expression of a magnon, $\mathcal{X}^{A\dot{A}}$, instead of its left/right part, since the central charges are shared between two $\mathfrak{psu}(2|2)$ factors and they act on both the left and the right parts. The positions of $Z^\pm$ inside the state can be changed according to the rule,

$$|\mathcal{X}^{A\dot{A}}Z^\pm\rangle = \left(\frac{x^+}{x^-}\right)^\pm |Z^\pm \mathcal{X}^{A\dot{A}}\rangle. \tag{10.94}$$

As shown in (10.93), **b** and **d** contain a factor $(1 - e^{ip(u)})$ since they correspond to field-dependent gauge tranformations as described previously.[19]

Exercise 7 Derive the actions of P and K from actions of fermionic generators and the anti-commutation relations (10.91).

Exercise 8 Check that $\mathbf{a}, \mathbf{c} \sim O(1)$ and $\mathbf{b}, \mathbf{d} \sim O(g)$ at weak coupling. (See also (10.72).) This reflects the fact that the transformations associated with **b** and **d** are gauge transformations whereas others are not.

As shown previously, the centrally extended $\mathfrak{psu}(2|2)$ 'knows' the coupling constant g. Therefore, any quantities that are determined by this symmetry group entail full coupling-constant dependence. For instance, the $2 \to 2$ S-matrix of magnons are fixed up to overall scalar factor by imposing the symmetry condition

---
[19] This can also be understood from the fact that these transformations involve $Z^\pm$.

$$[h, \mathbb{S}_{2\to 2}(u,v)] = 0 \quad (h \in \mathfrak{psu}(2|2)) \quad \Rightarrow \quad \mathbb{S}_{2\to 2} = S_0 \left(\hat{S}(u,v) \otimes \hat{S}(u,v) \right). \tag{10.95}$$

Here $\hat{S}$ is the matrix structure determined by the symmetry[20] and S_0 is the overall scalar factor. As shown in [20, 21], S_0 can be determined by imposing the crossing symmetry of the S-matrix.

A notable feature of the two-body S-matrix $\mathbb{S}_{2\to 2}$ is that it satisfies the Yang–Baxter equation. Note that this is just a necessary condition for the existence of the integrability and the factorization of the multi-particle S-matrices. One can nevertheless proceed by *assuming* the existence of integrability and express the multi-particle S-matrices in terms of $\mathbb{S}_{2\to 2}$'s. After doing so, one can then follow a standard recipe for solving integrable systems to determine the spectrum: namely, writing down the Bethe equation to determine the asymptotic spectrum and using the thermodynamic Bethe ansatz to compute the finite-size spectrum. The results turn out to match direct perturbative computation up to surprisingly high orders, providing strong evidence for the integrability of this model.

10.2.3 Symmetries of three-point functions

Let us now move on to the discussion on the three-point function.

10.2.3.1 *Twisted translations and diagonal* $\mathfrak{psu}(2|2)$

Again the starting point is to consider the 'rest frame' and identify the 'little group'. As in the two-point function, we first consider the correlator of three BPS operators,

$$\langle \underbrace{\operatorname{tr}(P_1 \cdot \Phi)^{L_1}(x_1)}_{\mathcal{O}_1} \underbrace{\operatorname{tr}(P_2 \cdot \Phi)^{L_2}(x_2)}_{\mathcal{O}_2} \underbrace{\operatorname{tr}(P_3 \cdot \Phi)^{L_3}(x_3)}_{\mathcal{O}_3} \rangle, \tag{10.96}$$

To discuss the symmetry, it is convenient to perform the conformal + R-symmetry transformations and bring them to the so-called 'twisted translation frame',

$$\mathcal{O}_1 = \operatorname{tr}(\mathfrak{Z}^{L_1}(a_1)), \quad \mathcal{O}_2 = \operatorname{tr}(\mathfrak{Z}^{L_2}(a_2)), \quad \mathcal{O}_3 = \operatorname{tr}(\mathfrak{Z}^{L_3}(a_3)), \tag{10.97}$$

where $\mathfrak{Z}(a)$ is a twisted-translated scalar defined by

$$\mathfrak{Z}(a) = e^{\mathcal{T}^a} Z(0) = \left(Z + a^2 \bar{Z} + a(Y - \bar{Y}) \right)(0, a, 0, 0), \tag{10.98}$$

---

[20] For explicit expressions for $\hat{S}$, see [18, 19].

with

$$\mathcal{T} := -i\epsilon_{\alpha\dot{\alpha}} P^{\dot{\alpha}\alpha} + \epsilon_{\dot{a}a} R^{a\dot{a}}. \tag{10.99}$$

This configuration is invariant under the subgroup h of $\mathfrak{psu}(2|2)^2$ that commutes with the twisted-translation generator $\mathcal{T}$, namely $[h, \mathcal{T}] = 0$. It turns that that this 'little group' is a diagonal $\mathfrak{psu}(2|2)$ whose generators are given by

$$\begin{aligned}
\mathcal{L}^\alpha{}_\beta &:= L^\alpha{}_\beta + \dot{L}^\alpha{}_\beta, & \mathcal{R}^a{}_b &:= R^a{}_b + \dot{R}^a{}_b, \\
\mathcal{Q}^\alpha{}_a &:= Q^\alpha{}_a + i\epsilon^{\alpha\dot{\beta}} \epsilon_{ab} \dot{S}^b{}_{\dot{\beta}}, & \mathcal{S}^a{}_\alpha &:= S^a{}_\alpha + i\epsilon^{a\dot{b}} \epsilon_{\alpha\dot{\beta}} \dot{Q}^{\dot{\beta}}{}_{\dot{b}}.
\end{aligned} \tag{10.100}$$

In addition to these generators, the commutator between $\mathcal{Q}$ and $\mathcal{S}$ yields one central charge $\mathcal{P} := P - K$:

$$\begin{aligned}
\{\mathcal{Q}^\alpha{}_a, \mathcal{Q}^\beta{}_b\} &= \epsilon^{\alpha\beta} \epsilon_{ab} \mathcal{P}, & \{\mathcal{S}^a{}_\alpha, \mathcal{Q}^b{}_\beta\} &= -\epsilon_{\alpha\beta} \epsilon^{ab} \mathcal{P}, \\
\{\mathcal{Q}^\alpha{}_a, \mathcal{S}^b{}_\beta\} &= \delta^\alpha_\beta \mathcal{R}^b{}_a + \delta^b_a \mathcal{L}^\alpha{}_\beta.
\end{aligned} \tag{10.101}$$

10.2.3.2 Adding 'internal motions'

Having understood the symmetry for BPS three-point functions, the next step is to add the 'internal motion' to describe more general correlators, which involve non-BPS operators. The best way to do this while keeping the underlying symmetries manifest is to first add magnons to the 'vacuum state' and then perform the twisted translation:

$$\text{tr}\left(Z^L\right)(0) \xrightarrow{\text{add magnons}} \underbrace{\text{tr}(\cdots ZXZ\Psi Z \cdots)(0)}_{\mathcal{O}(0)} \xrightarrow{\text{twisted translate}} \mathcal{O}(a) := e^{\mathcal{T}a} \cdot \mathcal{O}(0). \tag{10.102}$$

When the operator $\mathcal{O}(0)$ is a conformal primary and is highest weight in R-symmetry, the space-time dependence of the three-point function is fixed to be as follows:

$$\langle \mathcal{O}_1(a_1) \mathcal{O}_2(a_2) \mathcal{O}_3(a_3) \rangle = \frac{\mathbb{C}_{123}}{(a_1 - a_2)^{\delta_{12|3}} (a_2 - a_3)^{\delta_{23|1}} (a_3 - a_1)^{\delta_{31|2}}}. \tag{10.103}$$

Here $\delta_{ij|k}$ is given by

$$\delta_{ij|k} = 2(C_i + C_j - C_k), \tag{10.104}$$

where C_i is the central charge,

$$C_i = (\Delta_i - \mathcal{J}_i)/2. \tag{10.105}$$

What the hexagon approach computes is the proportionality factor[21] $\mathbb{C}_{123}$.

10.2.4 Constraining the hexagon form factors from symmetries

We are now in a position to constrain the hexagon form factor

$$\mathcal{H}(\Psi) = \langle \mathcal{H} | \Psi \rangle \tag{10.106}$$

using the diagonal $\mathfrak{su}(2|2)$ (which we will denote as $\mathfrak{psu}_D(2|2)$). In (10.106), $|\Psi\rangle$ is the spin chain state for which we compute the form factor. In this representation, the symmetry constraints read

$$\langle \mathcal{H} | h = 0, \qquad h \in \mathfrak{su}_D(2|2). \tag{10.107}$$

10.2.4.1 One-particle form factor

Let us first consider the case with a single magnon.

As stated previously, each magnon belongs to a bi-fundamental representation of $\mathfrak{psu}(2|2)^2$ and thus is denoted by $\chi^A \chi^{\dot{A}}$. By imposing the bosonic part of $\mathfrak{su}_D(2|2)$, we can show

$$\begin{aligned} \langle \mathcal{H} | \psi^\alpha \psi^{\dot{\beta}} \rangle &= \sqrt{\mu(u)} \epsilon^{\alpha \dot{\beta}}, \\ \langle \mathcal{H} | \varphi^a \psi^{\dot{\alpha}} \rangle &= \langle \mathcal{H} | \psi^\alpha \varphi^{\dot{a}} \rangle = 0, \\ \langle \mathcal{H} | \varphi^a \varphi^{\dot{b}} \rangle &= \sqrt{v(u)} \epsilon^{a \dot{b}}, \end{aligned} \tag{10.108}$$

where μ and ν are functions of the rapidity u, which are not fixed by the symmetry. (We are defining them with square roots for later convenience.) The derivation is pretty much straightforward. For instance by imposing $\langle \mathcal{H} | \mathcal{R}^1{}_2 = 0$, we get

$$\begin{aligned} 0 &= \langle \mathcal{H} | \mathcal{R}^1{}_2 | \varphi^1 \varphi^2 \rangle = \langle \mathcal{H} | \left(R^1{}_2 + \dot{R}^1{}_2 \right) | \varphi^1 \varphi^2 \rangle = \langle \mathcal{H} | \varphi^1 \varphi^2 \rangle + \langle \mathcal{H} | \varphi^2 \varphi^1 \rangle, \\ &\Leftrightarrow \quad \langle \mathcal{H} | \varphi^1 \varphi^2 \rangle = -\langle \mathcal{H} | \varphi^2 \varphi^1 \rangle. \end{aligned} \tag{10.109}$$

We next consider the constraint from the central charge $\mathcal{P}$. The Ward identity for $\mathcal{P}$ reads

$$0 = \langle \mathcal{H} | \mathcal{P} | \chi^A \chi^{\dot{A}} \rangle = g \left(1 - e^{ip} \right) \langle \mathcal{H} | Z \chi^A \chi^{\dot{A}} \rangle - g \left(1 - e^{-ip} \right) \langle \mathcal{H} | Z^- \chi^A \chi^{\dot{A}} \rangle. \tag{10.110}$$

---

[21] In general, the factor $\mathbb{C}_{123}$ contains a tensor structure when the operators have spin. What is important however is that it does not depend on the positions a_i.

The 'square root' of this relation leads to

$$\langle \mathcal{H}|Z^{\pm}\chi^A\chi^{\dot{A}}\rangle = \left(iU^{-1}\right)^{\pm}\langle \mathcal{H}|\chi^A\chi^{\dot{A}}\rangle \qquad (10.111)$$

with $U = e^{ip/2}$.

Finally we have contraints coming from fermionic charges. The action of $\mathcal{Q}^{\beta}{}_b$ on $|\varphi^a\psi^{\dot{\alpha}}\rangle$ is given by

$$\mathcal{Q}^{\beta}{}_b|\varphi^a\psi^{\dot{\alpha}}\rangle = \mathbf{a}\delta^a_c|\psi^{\beta}\psi^{\dot{\alpha}}\rangle + i\mathbf{d}\epsilon^{\beta\dot{\alpha}}\epsilon_{b\dot{c}}|\varphi^a\varphi^{\dot{c}}\rangle. \qquad (10.112)$$

The Ward identity associated with this action gives

$$\sqrt{\nu(u)} = \frac{\mathbf{a}}{i\mathbf{d}}\sqrt{\mu(u)} = -i\sqrt{\mu(u)}. \qquad (10.113)$$

Exercise 9 Derive (10.113) from $\langle \mathcal{H}|\mathcal{Q}^{\beta}{}_b|\varphi^a\psi^{\dot{\alpha}}\rangle = 0$.

We can also consider the Ward identity associated with $\mathcal{S}^b{}_{\beta}$, but it turns out that it also yields the same relation.

Exercise 10 Check that the Ward identity for $\mathcal{S}^b{}_{\beta}$ also gives (10.113).

In summary, the one-particle form factor is constrained by the symmetries as

$$\langle \mathcal{H}|\psi^{\alpha}\psi^{\dot{\beta}}\rangle = \sqrt{\mu(u)}\epsilon^{\alpha\dot{\beta}},$$
$$\langle \mathcal{H}|\varphi^a\psi^{\dot{\alpha}}\rangle = \langle \mathcal{H}|\psi^{\alpha}\varphi^{\dot{a}}\rangle = 0, \qquad (10.114)$$
$$\langle \mathcal{H}|\varphi^a\varphi^{\dot{b}}\rangle = -i\sqrt{\mu(u)}\epsilon^{a\dot{b}}.$$

The overall factor $\mu(u)$ is often called the *measure factor*.

It is convenient to represent this structure of the one-particle form factor as in Figure 10.13. In this pictorial representation, a line connecting two indices denotes a product of the *measure factor* $\sqrt{\mu}$ and the invariant tensor of $\mathfrak{psu}(2|2)$ ($\epsilon^{\alpha\dot{\beta}}$ and $\epsilon^{a\dot{b}}$).

10.2.4.2 Two-particle form factor

We can also determine the matrix structure of two-particle form factors using the symmetry. Since the analysis is straightforward but more complicated than one-particle cases, here we simply state the final result.

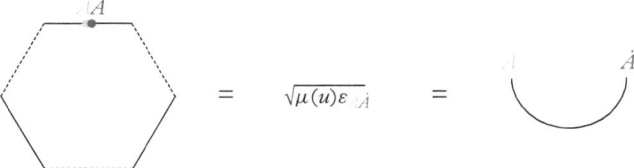

Figure 10.13 *A pictorial representation of the one-particle hexagon form factor. The line connecting two indices denotes a product of the measure factor and the invariant tensor (denoted by $\epsilon_{A\dot{A}}$ in the figure).*

Two-particle form factor

The two-particle form factor consists of the *matrix part* $\mathcal{H}_{\text{mat}}$, which is constrained by the symmetry, and the *dynamical factor h*, which is an overall scalar factor and not fixed by the symmetry:

$$\langle \mathcal{H} | \left(\chi^{A_1} \chi^{\dot{A}_1} \right)(u_1) \left(\chi^{A_2} \chi^{\dot{A}_2} \right)(u_2) \rangle = h(u_1, u_2) \mathcal{H}_{\text{mat}}(u_1, u_2). \tag{10.115}$$

The matrix part $\mathcal{H}_{\text{mat}}$ can be computed as follows:

1. First rearrange the indices $(A_1 \dot{A}_1)(A_2 \dot{A}_2)$ to $(A_1 A_2)(\dot{A}_1 \dot{A}_2)$. This rearrangement produces a sign factor $(-1)^{|\dot{A}_1||A_2|}$, where $|A| = 0$ for φ's and $|A| = 1$ for ψ's.

2. After the rearrangement, we obtain the state

$$|\chi^{A_1}(u_1) \chi^{A_2}(u_2)\rangle \otimes |\chi^{\dot{A}_1}(u_1) \chi^{\dot{A}_2}(u_2)\rangle. \tag{10.116}$$

Then apply the S-matrix $\hat{S}$ to the left psu(2|2) sector, and get the state of the form

$$\hat{S}_{12} |\chi^{A_1}(u_1) \chi^{A_2}(u_2)\rangle \otimes |\chi^{\dot{A}_1}(u_1) \chi^{\dot{A}_2}(u_2)\rangle. \tag{10.117}$$

3. After doing so, contract the resulting state with the *gluing vertex* $\langle \mathcal{G}|$, which produces a product of one-point functions. (The precise definition will be given shortly).

In summary, it can be expressed as follows:

$$\langle \mathcal{H} | \left(\chi^{A_1} \chi^{\dot{A}_1} \right) \left(\chi^{A_1} \chi^{\dot{A}_1} \right) \rangle$$
$$= (-1)^{|\dot{A}_1||A_2|} h(u_1, u_2) \langle \mathcal{G} | \left(\hat{S}_{12} | \chi^{A_1}(u_1) \chi^{A_2}(u_2) \rangle \otimes | \chi^{\dot{A}_1}(u_1) \chi^{\dot{A}_2}(u_2) \rangle \right). \tag{10.118}$$

484   *Three-point functions in $\mathcal{N} = 4$ supersymmetric Yang–Mills theory*

The contraction between the gluing vertex and a multi-magnon state is defined as follows.

Definition of the gluing vertex

$$\langle \mathcal{G} | \left(| \chi^{A_M} \cdots \chi^{A_2} \chi^{A_1} \rangle \otimes | \chi^{\dot{A}_1} \chi^{\dot{A}_2} \cdots \chi^{\dot{A}_M} \rangle \right) := \prod_{j=1}^{M} \langle \mathcal{H} | \chi^{A_1} \chi^{\dot{A}_1} \rangle, \qquad (10.119)$$

$$\langle \mathcal{G} | \left(| Z \chi^{A_M} \cdots \chi^{A_2} \chi^{A_1} \rangle \otimes | \chi^{\dot{A}_1} \chi^{\dot{A}_2} \cdots \chi^{\dot{A}_M} \rangle \right)$$

$$:= \left(\prod_{j=1}^{M} \frac{i}{U_i} \right) \langle \mathcal{G} | \left(| \chi^{A_M} \cdots \chi^{A_2} \chi^{A_1} \rangle \otimes | \chi^{\dot{A}_1} \chi^{\dot{A}_2} \cdots \chi^{\dot{A}_M} \rangle \right), \qquad (10.120)$$

$$\langle \mathcal{G} | \left(| \chi^{A_M} \cdots \chi^{A_2} \chi^{A_1} \rangle \otimes | Z \chi^{\dot{A}_1} \chi^{\dot{A}_2} \cdots \chi^{\dot{A}_M} \rangle \right)$$

$$:= \left(\prod_{j=1}^{M} \frac{i}{U_i} \right) \langle \mathcal{G} | \left(| \chi^{A_M} \cdots \chi^{A_2} \chi^{A_1} \rangle \otimes | \chi^{\dot{A}_1} \chi^{\dot{A}_2} \cdots \chi^{\dot{A}_M} \rangle \right),$$

where U_i is defined by $U_i = e^{ip_i/2}$.

To discuss the symmetry property of the two-particle hexagons, it is convenient to first study the symmetry of the gluing vertex. Firstly the definition (10.120) guarantees that $\langle \mathcal{G} |$ is annihilated by the central charge $\mathcal{P}$. Secondly, it is annihilated by all the generators of $\mathfrak{psu}(2|2)_D$: It is relatively easy to see that $\langle \mathcal{G} |$ is annihilated by $\mathcal{R}$'s and $\mathcal{L}$'s. Since these transformations act locally in the spin chain (namely they do not produce any Z's), the factorized form of the gluing vertex (10.119) guarantees the invariance under these symmetry transformations.

Therefore, we just need to worry about $\mathcal{Q}$'s and $\mathcal{S}$'s. As it turns out, owing to the factorized form of our definition (10.119), the only non-trivial cases we need to check are

$$\langle \mathcal{G} | \mathcal{Q} \left(| \chi^{A_1} \cdots \psi^{A_N} \cdots \rangle \otimes | \cdots \varphi^{\dot{A}_N} \cdots \chi^{\dot{A}_1} \rangle \right) = 0, \qquad (10.121)$$

$$\langle \mathcal{G} | \mathcal{S} \left(| \chi^{A_1} \cdots \varphi^{A_N} \cdots \rangle \otimes | \cdots \psi^{\dot{A}_N} \cdots \chi^{\dot{A}_1} \rangle \right) = 0. \qquad (10.122)$$

In the first case (10.121), the action of $\mathcal{Q}$ produces a state of the form

$$\mathbf{c}(u_N) | \cdots Z \varphi^{B_N} \cdots \rangle \otimes | \cdots \varphi^{\dot{A}_N} \cdots \rangle + \mathbf{d}(u_N) | \cdots \psi^{B_N} \cdots \rangle \otimes | \cdots Z^{-1} \psi^{\dot{A}_N} \cdots \rangle, \qquad (10.123)$$

with u_N being the rapidity of the N-th magnon. We can then use the rules to move and remove Z's ((10.94) and (10.119)) to show that they vanish when contracted with $\langle \mathcal{G} |$:

$$\mathbf{c}(u_N)\langle \mathcal{G}|\left(|\cdots Z\varphi^{B_N}\cdots\rangle \otimes |\cdots \varphi^{\dot{A}_N}\cdots\rangle\right)$$
$$= \mathbf{c}(u_N)\frac{i\prod_{k=1}^{N-1}U_i}{\prod_{i=N}^{M}U_i}\langle \mathcal{G}|\left(|\cdots \varphi^{B_N}\cdots\rangle \otimes |\cdots \varphi^{\dot{A}_N}\cdots\rangle\right).$$
$$\mathbf{d}(u_N)\langle \mathcal{G}|\left(|\cdots \psi^{B_N}\cdots\rangle \otimes |\cdots Z^{-1}\psi^{\dot{A}_N}\cdots\rangle\right)$$
$$= \mathbf{d}(u_N)\frac{\prod_{k=1}^{N}U_i}{i\prod_{i=N+1}^{M}U_i}\langle \mathcal{G}|\left(|\cdots \psi^{B_N}\cdots\rangle \otimes |\cdots \psi^{\dot{A}_N}\cdots\rangle\right).$$
(10.124)

Then, the sum can be expressed as

$$\frac{\prod_{i=1}^{N-1}U_i}{\prod_{i=N+1}^{M}U_i}\left[\frac{i\mathbf{c}(u_N)}{U_N}\langle \mathcal{G}|\left(|\cdots \varphi^{B_N}\cdots\rangle \otimes |\cdots \varphi^{\dot{A}_N}\cdots\rangle\right)\right.$$
$$\left.+\frac{\mathbf{d}(u_N)U_N}{i}\langle \mathcal{G}|\left(|\cdots \psi^{B_N}\cdots\rangle \otimes |\cdots \psi^{\dot{A}_N}\cdots\rangle\right)\right].$$
(10.125)

Importantly, all the non-local effects (dependence on the momenta of other excitations) can be factorized out and the prefactors inside the square brackets only depend on quantum numbers of the N-th excitation. We can therefore recycle the proof for the one-particle case and easily confirm that the gluing vertex is annihilated by $\mathcal{Q}$.

Exercise 11 Show the second identity (10.122) in a similar manner.

Once the symmetry of the gluing vertex is understood, it is straightforward to check that the two-particle hexagon (in particular the matrix part $\mathcal{H}_{\text{mat}}$) satisfies the symmetry constraints. Since the two-particle hexagon is defined by the two-particle S-matrix $\hat{S}$, which commutes with $\mathfrak{psu}(2|2)$, and the gluing vertex, which is annihilated by $\mathfrak{psu}(2|2)_D$, it is obviously annihilated by $\mathfrak{psu}(2|2)_D$.

This argument also provides a quick proof for the uniqueness of the matrix part $\mathcal{H}_{\text{mat}}$. Without loss of generality, a general two-particle form factor can be expressed as

$$\langle \mathcal{H}|\left(\chi^{A_1}\chi^{\dot{A}_1}\right)\left(\chi^{A_1}\chi^{\dot{A}_1}\right)\rangle$$
$$= (-1)^{|\dot{A}_1||A_2|}\langle \mathcal{G}|\left(\hat{F}_{12}|\chi^{A_1}(u_1)\chi^{A_2}(u_2)\rangle \otimes |\chi^{\dot{A}_1}(u_1)\chi^{\dot{A}_2}(u_2)\rangle\right),$$
(10.126)

where $\hat{F}$ is some operator acting on the left $\mathfrak{psu}(2|2)$ state. Owing to the symmetry property of the gluing vertex, the operator $\hat{F}$ has to satisfy

$$[\hat{F},h]=0, \qquad h \in \mathfrak{psu}(2|2) + \text{central charges}. \qquad (10.127)$$

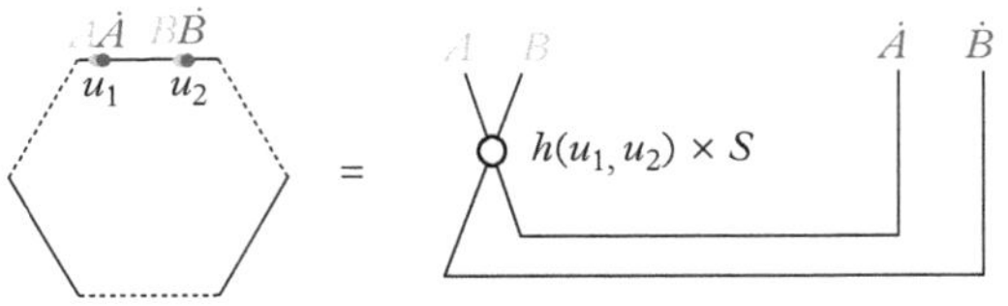

Figure 10.14 *A pictorial representation of the two-particle hexagon form factor. As in Figure 10.13, the line connecting indices denotes a product of the measure factor and the invariant tensor. The circle dot in the middle represents the action of the* $\mathfrak{psu}(2|2)$ *S-matrix and the multiplication of the dynamical factor h.*

(This follows from the condition $\langle \mathcal{H}|h_D = 0$ with $h_D \in \mathfrak{psu}(2|2)_D$.) As discussed in [18, 19], the only two-particle tensor which satisfies the property (10.127) (and swaps the order of magnons) is the S-matrix $\hat{S}_{12}$. This completes the proof for the uniqueness.

The two-particle hexagon also admits a pictorial representation, as shown in Figure 10.14. In the figure, the intersection of two lines denotes a product of the S-matrix $\hat{S}_{12}$ and a dynamical factor h, and the contraction of two indices corresponds to a product of the measure factor $\sqrt{\mu}$ and the invariant tensor. As alluded to previously, the dynamical factor is not constrained by the symmetry analysis we performed here. To determine its structure, one has to use the integrability, as we will see shortly.

10.2.4.3 Multi-particle form factor

Unlike the two-particle form factor, the structures of the multi-particle form factors are not determined by the symmetries. In principle, they can have very complicated tensor structures and their dependence on rapidities can be highly non-trivial. This is in fact the case for general form factors in integrable models and is what hinders the application of integrability beyond the spectral problem.

However, in the case at hand, we know that the multi-particle form factor is no more complicated than two-particle form factors at least in the weak coupling regime. Namely, they are given by a product of two-particle form factors as we saw in (10.70):

$$\mathcal{H}(\{u_j\}) \overset{\text{weak coupling}}{=} \prod_{i<j} h(u_i, u_j) \qquad (10.128)$$

Motivated by this simple structure, we conjecture that the multi-particle form factor at finite coupling is also given by a 'product' of two-particle form factors.

Multi-particle form factors
The multi-particle form factor is given by a product of the *matrix part* $\mathcal{H}_{\text{mat}}$ and the *dynamical part* $\mathcal{H}_{\text{dyn}}$ as

$$\langle \mathcal{H}| \left(\chi^{A_1}\chi^{\dot{A}_1}\right)(u_1) \cdots \left(\chi^{A_M}\chi^{\dot{A}_M}\right)(u_M)\rangle = \mathcal{H}_{\text{dyn}}(\{u_j\})\mathcal{H}_{\text{mat}}(\{u_j\}). \qquad (10.129)$$

The dynamical part is a simple product of the two-particle dynamical factors,

$$\mathcal{H}_{\text{dyn}}(\{u_j\}) = \prod_{i<j} h(u_i, u_j), \tag{10.130}$$

whereas the matrix part is a natural generalization of the matrix part for the two-particle form factor.

1. First rearrange the indices $(A_1 \dot{A}_1) \cdots (A_M \dot{A}_M)$ to $(A_1 A_2 \cdots A_M)(\dot{A}_1 \dot{A}_2 \cdots \dot{A}_M)$. This rearrangement produces a sign factor $(-1)^{\sum_{i<j} |\dot{A}_i||A_j|}$.

2. After the rearrangement, we obtain the state

$$|\chi^{A_1}(u_1) \cdots \chi^{A_M}(u_M)\rangle \otimes |\chi^{\dot{A}_1}(u_1) \cdots \chi^{\dot{A}_M}(u_M)\rangle. \tag{10.131}$$

Then apply the S-matrix $\hat{S}$ to the left $\mathfrak{psu}(2|2)$ sector to reverse the order of the particles,

$$\prod_{j<k} \hat{S}_{jk} |\chi^{A_1}(u_1) \cdots \chi^{A_M}(u_M)\rangle \otimes |\chi^{\dot{A}_1}(u_1) \cdots \chi^{\dot{A}_M}(u_M)\rangle \tag{10.132}$$

$$\sim |\chi^{B_M}(u_M) \cdots \chi^{B_1}(u_1)\rangle \otimes |\chi^{\dot{A}_1}(u_1) \cdots \chi^{\dot{A}_M}(u_M)\rangle.$$

3. After doing so, contract the resulting state with the *gluing vertex* $\langle \mathcal{G}|$, which produces a product of one-point functions.

The structure of the multi-particle form factor is summarized neatly in Figure 10.15. We should emphasize again that this structure is just a conjecture and is not dictated by the symmetry. However, as we will see later, it nicely realizes the constraints from integrability and also reproduces various perturbative data.

10.2.5 Constraints from integrability

In Section 10.3.4.3, we determined the matrix part of the hexagon form factor by imposing the symmetry. However, as alluded to previously, the overall scalar factors

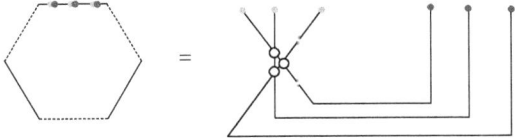

Figure 10.15 *A pictorial representation of the multi-particle hexagon form factor. Notations are the same as in previous figures.*

$\sqrt{\mu}$ and $h(u,v)$ cannot be fixed just from the symmetry analysis. To determine these, one has to take into account constraints from integrability, or more precisely, one has to impose the *form factor axioms* of the integrable quantum field theories.

10.2.5.1 Watson equation

The first constraint is what is called the Watson equation. It essentially tells us that, if we swap two particles, the form factor is multiplied by the S-matrix. More precisely it can be expressed as the following equality:

$$\langle \mathcal{H} | \mathbb{S}_{i,i+1} | \mathcal{X}^{A_1 \dot{A}_1}(u_1) \cdots \mathcal{X}^{A_M \dot{A}_M}(u_M) \rangle = \langle \mathcal{H} | \mathcal{X}^{A_1 \dot{A}_1}(u_1) \cdots \mathcal{X}^{A_M \dot{A}_M}(u_M) \rangle. \quad (10.133)$$

Here $\mathbb{S}_{i,i+1}$ is the S-matrix which swaps the order of $\mathcal{X}^{A_i \dot{A}_i}$ and $\mathcal{X}^{A_{i+1} \dot{A}_{i+1}}$. It can also be expressed pictorially as shown in Figure 10.16. The equation (10.133) is in principle a highly non-trivial matrix equality since the S-matrix $\mathbb{S}$ has quite a complicated structure. However, thanks to our conjecture on the matrix structure of the multi-particle form factor (10.129), it boils down to a single scalar equality for the dynamical factor $h(u,v)$.

To see this in the simplest example, let us consider the two-particle form factor. A crucial observation is that the gluing vertex defined in (10.119) and (10.120) has a nice property in that it 'commutes' with the S-matrix:

$$\langle \mathcal{G} | \left(\hat{S}_{12} | \chi_1^A \chi_2^B \rangle \otimes | \chi_1^{\dot{A}} \chi_2^{\dot{B}} \rangle \right) = \langle \mathcal{G} | \left(| \chi_1^A \chi_2^B \rangle \otimes \hat{S}_{12} | \chi_1^{\dot{A}} \chi_2^{\dot{B}} \rangle \right). \quad (10.134)$$

This relation can be verified by computing both sides explicitly.

Now, using the definition of the hexagon form factor (10.118), we can express the left-hand side of (10.133) (for two particles) as

$$\langle \mathcal{H} | \mathbb{S}_{12} | \mathcal{X}^{A_1 \dot{A}_1}(u) \mathcal{X}^{A_2 \dot{A}_2}(v) \rangle$$
$$= (-1)^{|\dot{A}_1||A_2|} h(v,u) S_0(u,v) \langle \mathcal{G} | \left(\hat{S}_{21} \hat{S}_{12} | \chi^{A_1} \chi^{A_2} \rangle \otimes \hat{S}_{12} | \chi^{\dot{A}_1} \chi^{\dot{A}_2} \rangle \right). \quad (10.135)$$

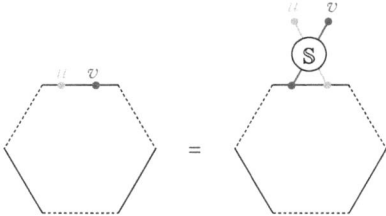

Figure 10.16 Watson equation. The Watson equation states that the form factor does not change if one first changes the order of the particle by acting the S-matrix and then computes the form factor.

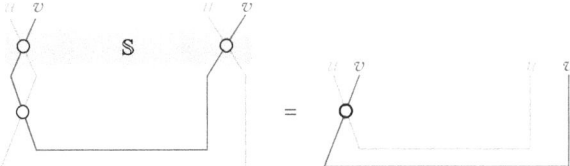

Figure 10.17 *A pictorial explanation for how the Watson equation is realized in the hexagon form factor. The left-hand side denotes the form factor with the action of the S-matrix. By using the unitarity of the S-matrix, we can eliminate two circle dots and obtain the right figure, which is the form factor without the action of the S-matrix.*

Here we have denoted the factors associated with the action of $\mathbb{S}_{12}$ in red. We can then use the unitarity relation $\hat{S}_{21}\hat{S}_{12} = 1$ and the relation (10.134) to rewrite it further as

$$\langle \mathcal{H}|\mathbb{S}_{12}|\mathcal{X}^{A_1\dot{A}_1}(u)\mathcal{X}^{A_2\dot{A}_2}(v)\rangle = (-1)^{|\dot{A}_1||A_2|}h(v,u)S_0(u,v)\langle\mathcal{G}|\left(\hat{S}_{12}|\chi^{A_1}\chi^{A_2}\rangle \otimes |\chi^{\dot{A}_1}\chi^{\dot{A}_2}\rangle\right)$$

$$= \frac{h(v,u)}{h(u,v)}S_0(u,v)\langle\mathcal{H}|\mathcal{X}^{A_1\dot{A}_1}(u)\mathcal{X}^{A_2\dot{A}_2}(v)\rangle. \qquad (10.136)$$

Therefore, the Watson equation boils down to a single equality,

$$\frac{h(u,v)}{h(v,u)} = S_0(u,v). \qquad (10.137)$$

There is also a pictorial way to understand how the Watson equation works; see Figure 10.17.

10.2.5.2 Crossing/mirror transformations

Before considering the other constraint, it is useful to introduce the notion of crossing and mirror transformations. In terms of Zhukowski variables, introduced in (10.72), the energy and momentum of a magnon at finite coupling read as

$$e^{ip} = \frac{x^+(u)}{x^-(u)}, \qquad E = \frac{1}{2}\frac{1+\frac{1}{x^+x^-}}{1-\frac{1}{x^+x^-}}. \qquad (10.138)$$

Since the Zhukowski variables have cuts as a function of the rapidity u,

$$x(u) = \frac{u + \sqrt{u^2 - 4g^2}}{2g}, \qquad (10.139)$$

the functions (10.138) live on a complex plane with two branch cuts (see Figure 10.18). Since we are equipped with a complex plane with cuts, a natural question to ask is what happens if we pass through a cut and go to a different sheet. It turns out

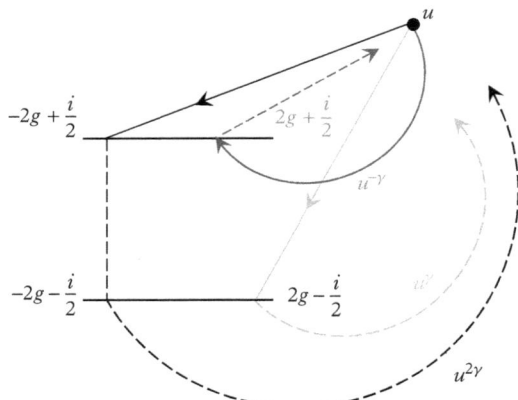

Figure 10.18 *Analytic structure of the magnon energy/momentum and the crossing and mirror transformations. The energy and the momentum of the magnon has two cuts $[-2g \pm i/2, 2g \pm i/2]$. The upper and the lower cuts correspond to the cuts of x^- and x^+ respectively. The crossing and the mirror transformations are defined as analytic continuations along the paths depicted in the figure.*

that the effect of this analytic continuation on the Zhukowski variable is quite simple:

$$x(u) \stackrel{\text{analytic continuation}}{\rightarrow} \frac{1}{x(u)}. \tag{10.140}$$

From this transformation property, we can immediately conclude that going through two cuts in (10.138) (which we denote by 2γ) leads to

$$e^{ip} \stackrel{u \to u^{2\gamma}}{\rightarrow} e^{-ip}, \quad E \stackrel{u \to u^{2\gamma}}{\rightarrow} -E. \tag{10.141}$$

Exercise 12 Check (10.141).

As can be seen from (10.141), the transformation 2γ inverts the signs of the energy and the momentum. Physically this should be interpreted as a crossing transformation which takes a particle in the past and converts it into an anti-particle in the future. See Figure 10.19 for a pictorial explanation. We therefore call it the *crossing transformation*.

We can also consider a 'half-crossing' transformation, by going through only one of two cuts:

$$e^{ip} \stackrel{u \to u^{\gamma}}{\rightarrow} \frac{1}{x^+ x^-}, \quad E \stackrel{u \to u^{\gamma}}{\rightarrow} \frac{1}{2} \frac{1 + \frac{x^+}{x^-}}{1 - \frac{x^+}{x^-}},$$

$$e^{ip} \stackrel{u \to u^{-\gamma}}{\rightarrow} x^+ x^-, \quad E \stackrel{u \to u^{-\gamma}}{\rightarrow} \frac{1}{2} \frac{1 + \frac{x^-}{x^+}}{1 - \frac{x^-}{x^+}}, \tag{10.142}$$

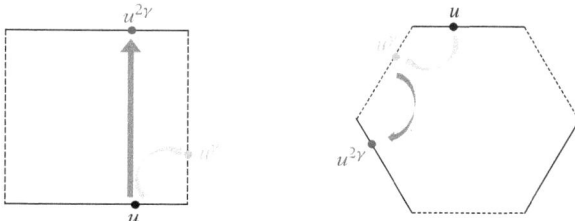

Figure 10.19 *Physical interpretation of the crossing and the mirror transformations. For two-point functions, the crossing transformation sends an anti-particle in the past to a particle in the future, while the mirror transformation rotates the position of the particle by ninety degrees and brings it to a theory with space and time swapped. The same transformation can be performed in three-point functions. In that case, we can move a particle inside a hexagon by successive applications of mirror transformations*

Here γ denotes the transformation which inverts x^+ while the transformation $-\gamma$ inverts x^-. To understand its physical meaning, let us recall how the crossing transformation is realized in ordinary relativistic 2d QFT's. In relativistic 2d QFT's, the momenta of particles can be labelled by the rapidity θ given by

$$E = m\cosh\theta, \quad p = m\sinh\theta. \tag{10.143}$$

Then, the crossing transformation corresponds to adding $i\pi$ to θ:

$$E \overset{\theta \to \theta+i\pi}{\to} -E, \quad p \overset{\theta \to \theta+i\pi}{\to} -p. \tag{10.144}$$

On the other hand, the half-crossing transformation, namely adding $i\pi/2$ to θ, leads to

$$E \overset{\theta \to \theta+i\pi/2}{\to} ip, \quad p \overset{\theta \to \theta+i\pi/2}{\to} iE. \tag{10.145}$$

This means that addition by $i\pi/2$ exchanges the space and the time directions. Such a space-time exchanged theory is often called a *mirror* theory.

Let us now go back to our problem. Also in our case, it is known[22] that the half-crossing transformation (now to be called the mirror transformation) is interpreted as a map from the original theory to a theory with space and time swapped:

$$e^{ip} \overset{u \to u^\gamma}{\to} e^{-\tilde{E}} = \frac{1}{x^+ x^-}, \quad E \overset{u \to u^\gamma}{\to} i\tilde{p} = \frac{1}{2}\frac{1+\frac{x^+}{x^-}}{1-\frac{x^+}{x^-}},$$
$$e^{ip} \overset{u \to u^{-\gamma}}{\to} e^{\tilde{E}} = x^+ x^-, \quad E \overset{u \to u^{-\gamma}}{\to} -i\tilde{p} = \frac{1}{2}\frac{1+\frac{x^-}{x^+}}{1-\frac{x^-}{x^+}}. \tag{10.146}$$

---

[22] A more detailed explanation from the point of view of string theory in AdS$_5$ is given in [22].

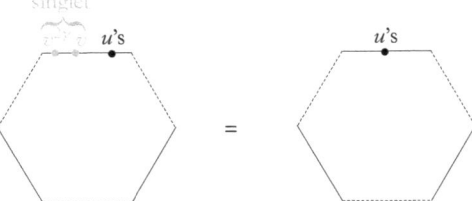

Figure 10.20 Decoupling condition. The addition of a particle–antiparticle pair should not change the form factor.

Here $\tilde{E}$ and $\tilde{p}$ are the energy and the momentum in the mirror theory. Unlike the relativistic QFT's, the dispersion relation changes under the mirror transformation.

Exercise 13 Compute the mirror energy and momentum at weak coupling and show

$$\tilde{E} = \frac{g^2}{u^2 + 1/4} + O(g^4), \qquad \tilde{p} = u + O(g^2). \qquad (10.147)$$

For the case of two-point functions, the mirror transformation moves the excitations from one edge to a neighbouring edge, see Figure 10.19. This is also true for three-point functions; by applying the mirror transformation, one can move magnons from one edge to another[23] inside a hexagon, see Figure 10.19.

10.2.5.3 Decoupling

The other constraint from integrability is called the decoupling relation.[24] Physically it means that the addition of a particle–antiparticle pair, which can be generated by vacuum fluctuation, does not affect the hexagon form factor; see Figure 10.20. It can be expressed in formulae as

$$-i\underset{u=v}{\mathrm{Res}} \left[\langle \mathcal{H} | \underbrace{\mathcal{X}(u^{2\gamma}) \mathcal{X}(v)}_{\text{singlet}} \mathcal{X}(u_1) \cdots \mathcal{X}(u_M) \rangle \right] = \langle \mathcal{H} | \mathcal{X}(u_1) \cdots \mathcal{X}(u_M) \rangle. \qquad (10.148)$$

Here we have suppressed the writing of indices for simplicity.

---

[23] There is one subtle point that we have not discussed in this section. When we perform the crossing transformation inside a hexagon, the excitations sometime change their flavour. For instance, a derivative along one direction gets transformed into a derivative along another direction. The detail of this transformation rule can be found in Appendix D of [5]. Its symmetry origin was recently discussed in Appendix A of [23].

[24] In the context of form factor axioms in integrable quantum field theories, this relation is sometimes called the *kinematical pole axiom*.

Let us first consider the simplest case, where there are no extra particles. In this case, (10.148) boils down to

$$-i\operatorname*{Res}_{u=v}\left[\underbrace{\langle\mathcal{H}|\mathcal{X}(u^{2\gamma})\mathcal{X}(v)\rangle}_{\text{singlet}}\right]=1. \tag{10.149}$$

The simplest choice of the particle–antiparticle pair is given by a pair of derivatives,

$$|D_{1\dot{2}}(u^{2\gamma})D_{2\dot{1}}(v)\rangle. \tag{10.150}$$

Inserting this state into equation (10.149), one can compute the left-hand side as

$$\begin{aligned}-i\operatorname*{Res}_{u=v}\left[\langle\mathcal{H}|D_{1\dot{2}}(u^{2\gamma})D_{2\dot{1}}(v)\rangle\right]&=-i\sqrt{\mu(v)\mu(u^{2\gamma})}\operatorname*{Res}_{u=v}\left[h(u^{2\gamma},v)\mathcal{H}_{\text{mat}}(u^{2\gamma},v)\right]\\ &=-i\mu(u)\operatorname*{Res}_{u=v}\left[h(u^{2\gamma},v)\mathcal{H}_{\text{mat}}(u^{2\gamma},v)\right].\end{aligned} \tag{10.151}$$

Here $\mathcal{H}_{\text{mat}}$ is the matrix part for the state given by (10.150). In the second line of (10.151), we have used the fact that the measure is invariant under the crossing transformation. (This is because two hexagons, one with u and one with $u^{2\gamma}$, are related by the rotation by 120 degrees.) The matrix part $\mathcal{H}_{\text{mat}}$ can be computed by the definition of the two-particle form factor and the structure of the S-matrix given in [19].

Exercise 14 Using the S-matrix given in [19], derive the following expression for the matrix part for (10.150):

$$\mathcal{H}_{\text{mat}}(u^{2\gamma},v)=\frac{v-u+i}{u-v}\frac{(1-1/x_u^-x_v^-)(1-1/x_u^+x_v^+)}{(1-1/x_u^-x_v^+)(1-1/x_u^+x_v^-)}. \tag{10.152}$$

Here $x_u=x(u)$ and $x_v=x(v)$.

Equation (10.152) shows that there is a simple pole at $u=v$ in the matrix part. Since this is the only singularity we expect, we can conclude that the dynamical factor $h(u^{2\gamma},v)$ is not singular in the limit $u\to v$. Therefore, one can rewrite the decoupling condition (10.149) as

$$\mu(u)=\frac{1}{h(u^{2\gamma},u)}\frac{(1-1/x^+x^-)^2}{(1-1/(x^-)^2)(1-1/(x^+)^2)}. \tag{10.153}$$

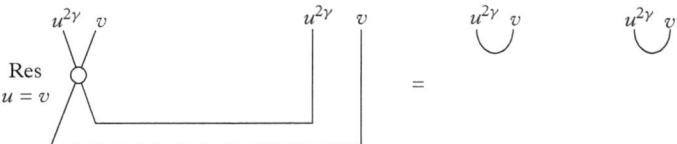

Figure 10.21 *A pictorial explanation of the decoupling condition with no extra particles. When the rapidities u and v are set to be equal, the matrix part simplifies and we obtain the two disconnected lines. We are thus left with the condition on the measure factor given in (10.153).*

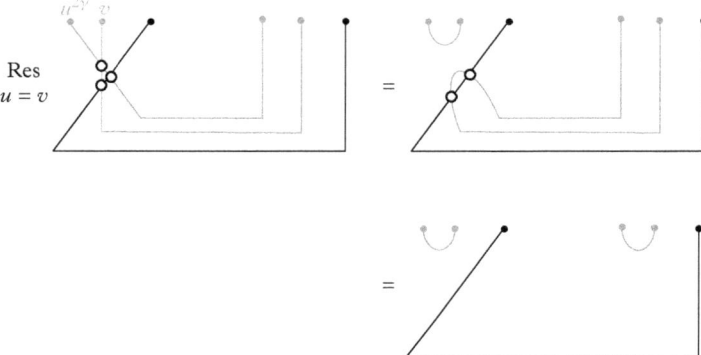

Figure 10.22 *A pictorial explanation of the decoupling condition with one extra particle. By setting the rapidities u and v to be equal, we can resolve the crossing of the lines and transform it into the second figure above. Then, the decoupling condition reduces to the second equality in the figure.*

In terms of the pictorial representation of the matrix part, the decoupling condition we have just discussed can be interpreted as a resolution of intersecting lines. See Figure 10.21.

Let us now discuss the next complicated case, that with a single spectator excitation,

$$-i\operatorname*{Res}_{u=v}\left[\langle \mathcal{H}| \underbrace{\mathcal{X}(u^{2\gamma})\mathcal{X}(v)}_{\text{singlet}} \mathcal{X}(w) \rangle\right] = \langle \mathcal{H}|\mathcal{X}(w)\rangle. \qquad (10.154)$$

Since working it out in full detail would take up too much space, we will resort to the pictorial representation at the cost of accuracy. The three-particle hexagon form factor can be depicted as the first element in Figure 10.22. Since the three-particle form factor is essentially a product of two-particle form factors, taking the residue at $u = v$ corresponds to locally resolving the scattering of the particles u and v, as shown in the second element in Figure 10.22. Then, the decoupling condition boils down to the equality relating the second and the third figures in Figure 10.22. This relation is reminiscent of a version of the crossing equation for the S-matrix that was reformulated by Beisert [19]. It physically means that the scattering of the singlet against a test particle is trivial. The only difference from the usual S-matrix story is that we are now dealing with only one copy of $\mathfrak{psu}(2|2)$ S-matrix. This case, however, has also already been discussed in the paper [19], and, as was shown there, it produces the following constraint on the dynamical factors:

$$h(u^{2\gamma},w)h(u,w) = \frac{x_u^- - x_w^-}{x_u^- - x_w^+} \frac{1 - 1/x_u^+ x_w^-}{1 - 1/x_u^+ x_w^+}. \tag{10.155}$$

In principle, we can go on to more complicated cases where there are several spectator excitations. However, it turns out that they do not yield any new constraints. Thus, the constraints coming from decoupling are exhausted by (10.153) and (10.155).

Exercise 15 Using the pictorial representation of the hexagon form factor and studying the cases with more than one spectator excitation, convince yourself that there are no other constraints.

10.2.5.4 *A solution to the constraints*

Let us now present a solution to the constraints coming from integrability. We will not discuss its uniqueness, but instead we show *a posteriori* that the solution satisfies all the constraints and also matches with the weak coupling expressions.

Definitions of the dynamical part and the measure
The dynamical part $h(u,v)$ and the measure $\mu(u)$ are given by

$$h(u,v) = \frac{x_u^- - x_v^-}{x_u^- - x_v^+} \frac{1 - 1/x_u^- x_v^+}{1 - 1/x_u^+ x_v^+} \frac{1}{\sigma(u,v)},$$

$$\mu(u) = \frac{(1 - 1/x_u^+ x_u^-)^2}{(1 - 1/(x^+)^2)(1 - 1/(x^-)^2)}. \tag{10.156}$$

Exercise 16 Check that they satisfy all the constraints (10.137), (10.153), and (10.155) using the properties of the dressing phase:

$$\sigma(u,u) = 1, \qquad \sigma(u^{2\gamma},v)\sigma(u,v) = \frac{(1 - 1/x_u^+ x_v^+)(1 - x_u^-/x_v^+)}{(1 - 1/x_u^+ x_v^-)(1 - x_u^-/x_v^-)}. \tag{10.157}$$

Exercise 17 Show that the hexagon form factor in the SU(2) sector is given in (10.71). (One has to use the Beisert S-matrix [19].)

10.2.6 Finite-size correction

The conjecture given in Section 10.2.1 is valid only when all the bridge lengths are large. In general, it also receives extra contributions coming from the fact that the bridge lengths are finite. In what follows, we discuss very briefly how to compute such finite-size corrections.[25]

Let us first explain why there can be finite-size corrections from the point of view of gauge theory. As mentioned below Figure 10.10, one has to insert an operator at the splitting points in order to compute loop corrections to the three-point function. Owing to the general properties of loop expansions in the planar limit, we expect that the operator at n loops acts on n consecutive spin sites. Therefore, if the bridge length is smaller than $2n$, two operators (the operator inserted at the front and at the back) start to talk to each other. This indicates that one has to take into account some new effect when the bridge length is small. In addition, Feynman diagrams corresponding to such processes generally yield complicated numbers, including zeta functions, and it is rather easy to see that such transcendental numbers cannot arise from our conjecture on the asymptotic three-point function.

In spectral problems, similar effects, called wrapping effects, have already been discussed. From the integrability point of view, those wrapping effects were understood as the corrections arising from mirror particles (magnons living in the mirror theory). Also in the present case, we propose that the mirror particles are responsible for these finite-size corrections. More precisely, we claim that when we cut the pair of pants into hexagons according to our proposal, we need to insert a complete set of states along the 'seams' (dashed line in Figure 10.23), and hexagons with such non-trivial states on the seams yield the finite-size corrections.

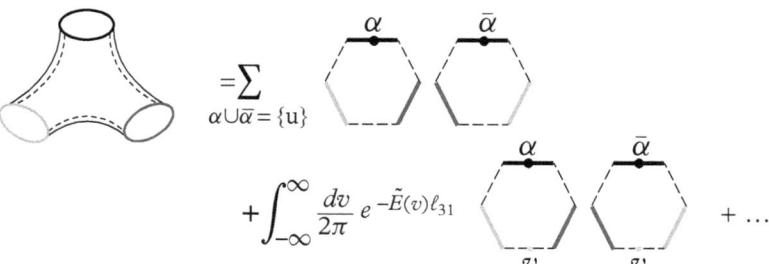

Figure 10.23 *The mirror particle corrections to the hexagon approach. When we cut the pair of pants into two, we need to insert a complete basis of states on the 'seams' (dashed lines in the figure). This leads to a series expansion in terms of the number of mirror particles in each edge. Because of the propagation factor $e^{-\tilde{E}\ell}$, these corrections are suppressed when the operators are long.*

---

[25] This material was not covered at all during the summer school, and the explanation given below is not self-contained by any means. For details, see the original article [5].

The states appearing on the 'seams' are the states in the mirror theory. Therefore, they are also labelled by the number and rapidities of magnons as well as their flavour indices. Thus, the series we obtain from this procedure looks like (see also Figure 10.23)

$$C_{123} \sim \sum_{\alpha \cup \bar{\alpha} = \{u\}} w_{\alpha, \bar{\alpha}} \left(\mathcal{H}(\alpha) \mathcal{H}(\bar{\alpha}) + \int \frac{dv}{2\pi} e^{-\tilde{E}(v) \ell_{31}} \sum_{\text{flavor}} \mathcal{H}(\alpha; v) \mathcal{H}(\bar{\alpha}; v) + \cdots \right). \quad (10.158)$$

Here $w_{\alpha, \bar{\alpha}}$ is the partition-dependent weight factor ($e^{ip\ell}$ and a product of S-matrices). We have only written down the leading correction from one mirror particle living on the bottom edge (see Figure 10.23). An important point in this series expression is that each mirror particle comes with a factor $e^{-\tilde{E}\ell}$. Since $e^{-\tilde{E}}$ is of order g^2, the mirror particle correction is suppressed when the bridge lengths are large, as expected. $\mathcal{H}(\alpha; v)$ and $\mathcal{H}(\bar{\alpha}; v)$ are the hexagon form factors with one mirror particle at the bottom edge, and they depend on the flavours of the mirror particle (although the dependence was not written explicitly). In principle, they can be computed by utilizing the crossing/mirror transformations and collecting all the particles at the top edge, but depending on the flavour of the mirror magnon, the expression can be quite complicated in practice.

However, something nice happens when we sum over the flavour indices. As shown in Figure 10.24, if we sum over the flavour indices of the mirror magnon, the line corresponding to the mirror particle forms a loop in the pictorial representation of the hexagon form factor. This means that the matrix part can be simply expressed as a transfer matrix. Since the transfer matrix is a well-studied object, this feature greatly simplifies the actual computation. In fact, utilizing such observations, the hexagon formalism was tested at two [5], three [24, 25], and four loops [26].

An extremely interesting, but challenging question to ask, is whether we can resume such finite-size corrections. In the case of the spectrum, such resummation led to the thermodynamic Bethe ansatz, which was later transformed into an elegant formalism called the quantum spectral curve. The ultimate goal in the study of three-point functions in $\mathcal{N} = 4$ SYM would be to find an analogue of the quantum spectral curve, although it is not clear at the moment whether such a notion exists or not.

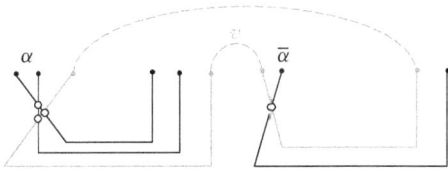

Figure 10.24 *The mirror particle corrections and the transfer matrix. The dashed line in the figure denotes the contribution from the mirror particle. If we sum over the flavour indices of the mirror particle, it effectively adds dashed curves, shown above, and transform the mirror particle constrinution into a transfer matrix.*

10.3 Conclusion and prospects

In this chapter, we have discussed how to compute the three-point function using integrability. Let us summarize the basic strategy we took to solve the problem.

1. Perform the weak coupling computation.
2. Identify the basic objects.
3. Constrain the 1- and 2-particle contributions using symmetries.
4. Guess the multi-particle structure (using the weak coupling results).
5. Formulate and solve the crossing equation by imposing the decoupling of a pair (a particle and an anti-particle).
6. Include the finite-size correction using the mirror transformation.

As it turns out, the same strategy works also for two-point functions. In this sense, the procedure outlined above may be viewed as a universal recipe for solving $\mathcal{N} = 4$ SYM using integrability.

One point we wished to emphasize in this chapter is the importance of gauge symmetry. In a sense, gauge symmetry is a mixed blessing: It is, on the one hand, a source of complication since it inevitably introduces length-changing processes to the spin chain description. On the other hand, a seemingly simple fact that the gauge transformation is always accompanied by the coupling constant proves to be very powerful when combined with educated guesses, and allows us to solve the problem.

To conclude the chapter, let us indicate several future directions. Firstly it would be interesting to study other observables following the strategy as summarized. For instance, a similar idea may be applicable to the one-point function in the presence of the domain wall, which was discussed recently in [27]. Another interesting direction is to study the non-planar dilatation operator. In the so-called BMN limit, it was shown that there exists a relation between the matrix element of the non-planar dilatation operator and the string vertex [28, 29]. Given that the string vertex was studied recently using integrability [30], it would be worthwhile revisiting this problem now using the full machinery of integrability.

Secondly, as previously mentioned, the building blocks of our construction can be interpreted as the form factor of the twist operator. A similar idea of using the twist operator for studying string interaction showed up previously in the IR limit of matrix string theory [31, 32]. Since matrix string theory describes type IIA string theory, whereas the theory discussed here corresponds to type IIB string theory, it is not that we expect direct physical connection between the two. Nevertheless the intriguing similarity certainly calls for further investigation.

Thirdly some of the ideas discussed in this chapter, such as the idea of constraining the quantities using a combination of global symmetry and 'large gauge' symmetry, do not rely strongly on the existence of integrability. It would thus be interesting to see how many such ideas could be applied to a broader class of theories including non-integrable ones.

Acknowledgements

First of all, I would like to thank the organizers of Les Houches Summer School, *Integrability: From statistical systems to gauge theory*, for the invitation to give lectures about the three-point functions. I also thank the participants of the school for attending the lectures and giving feedback. I am indebted to Benjamin Basso and Pedro Vieira for the collaboration which led to the main result in Section 10.3. My gratitude goes also to Naoki Kiryu, Ho Tat Lam, and Takuya Nishimura; discussion with them helped a lot in shaping Section 10.2. Lastly, I thank the mountains in Les Houches and Chamonix for enjoyable hikes during the summer school. This research was supported in part by Perimeter Institute for Theoretical Physics. Research at Perimeter Institute is supported by the Government of Canada through the Department of Innovation, Science and Economic Development Canada and by the Province of Ontario through the Ministry of Research, Innovation and Science.

References

[1] J. A. Minahan and K. Zarembo, 'The Bethe ansatz for N=4 superYang-Mills,' JHEP **0303**, 013 (2003) doi:10.1088/1126-6708/2003/03/013 hep-th/0212208.
[2] N. Gromov, V. Kazakov, S. Leurent, and D. Volin, 'Quantum Spectral Curve for Planar $\mathcal{N} =$ Super-Yang-Mills Theory,' Phys. Rev. Lett. **112**, no. 1, 011602 (2014) doi:10.1103/PhysRevLett.112.011602 arXiv:1305.1939.
[3] B. Basso, A. Sever, and P. Vieira, 'Spacetime and Flux Tube S-Matrices at Finite Coupling for N=4 Supersymmetric Yang-Mills Theory,' Phys. Rev. Lett. **111**, no. 9, 091602 (2013) doi:10.1103/PhysRevLett.111.091602 arXiv:1303.1396.
[4] L. F. Alday and J. M. Maldacena, 'Gluon scattering amplitudes at strong coupling,' JHEP **0706**, 064 (2007) doi:10.1088/1126-6708/2007/06/064 arXiv:0705.0303.
[5] B. Basso, S. Komatsu, and P. Vieira, 'Structure Constants and Integrable Bootstrap in Planar N=4 SYM Theory,' arXiv:1505.06745.
[6] S. El-Showk, M. F. Paulos, D. Poland, S. Rychkov, D. Simmons-Duffin, and A. Vichi, 'Solving the 3D Ising Model with the Conformal Bootstrap,' Phys. Rev. D **86**, 025022 (2012) doi:10.1103/PhysRevD.86.025022 arXiv:1203.6064.
[7] S. Minwalla, 'Restrictions imposed by superconformal invariance on quantum field theories,' Adv. Theor. Math. Phys. **2**, 781 (1998) hep-th/9712074.
[8] M. Baggio, J. de Boer, and K. Papadodimas, 'A non-renormalization theorem for chiral primary 3-point functions,' JHEP **1207**, 137 (2012) doi:10.1007/JHEP07(2012)137 arXiv:1203.1036.
[9] J. Escobedo, N. Gromov, A. Sever, and P. Vieira, 'Tailoring Three-Point Functions and Integrability,' JHEP **1109**, 028 (2011) doi:10.1007/JHEP09(2011)028 arXiv:1012.2475.
[10] M. Takahashi, *Thermodynamics of one-dimensional solvable models*. Cambridge University Press, 2005.

[11] M. Gaudin, *The Bethe Wavefunction*. Cambridge University Press, 2014.
[12] K. Okuyama and L. S. Tseng, 'Three-point functions in N = 4 SYM theory at one-loop,' JHEP **0408**, 055 (2004) doi:10.1088/1126-6708/2004/08/055 hep-th/0404190.
[13] L. F. Alday, J. R. David, E. Gava, and K. S. Narain, 'Structure constants of planar N = 4 Yang Mills at one loop,' JHEP **0509**, 070 (2005) doi:10.1088/1126-6708/2005/09/070 hep-th/0502186.
[14] N. Gromov and P. Vieira, 'Quantum Integrability for Three-Point Functions of Maximally Supersymmetric Yang-Mills Theory,' Phys. Rev. Lett. **111**, no. 21, 211601 (2013) doi:10.1103/PhysRevLett.111.211601 arXiv:1202.4103.
[15] N. Gromov and P. Vieira, 'Tailoring Three-Point Functions and Integrability IV. Theta-morphism,' JHEP **1404**, 068 (2014) doi:10.1007/JHEP04(2014)068 arXiv:1205.5288.
[16] P. Vieira and D. Volin, 'Review of AdS/CFT Integrability, Chapter III.3: The Dressing factor,' Lett. Math. Phys. **99**, 231 (2012) doi:10.1007/s11005-011-0482-0 1012.3992.
[17] N. Beisert, 'The Dilatation operator of N=4 super Yang-Mills theory and integrability,' Phys. Rept. **405**, 1 (2004) doi:10.1016/j.physrep.2004.09.007 hep-th/0407277.
[18] N. Beisert, 'The SU(2|2) dynamic S-matrix,' Adv. Theor. Math. Phys. **12**, 948 (2008) doi:10.4310/ATMP.2008.v12.n5.a1 hep-th/0511082.
[19] N. Beisert, 'The Analytic Bethe Ansatz for a Chain with Centrally Extended su(2|2) Symmetry,' J. Stat. Mech. **0701**, P01017 (2007) doi:10.1088/1742-5468/2007/01/P01017 nlin/0610017.
[20] R. A. Janik, 'The AdS(5) x S**5 superstring worldsheet S-matrix and crossing symmetry,' Phys. Rev. D **73**, 086006 (2006) doi:10.1103/PhysRevD.73.086006 hep-th/0603038.
[21] N. Beisert, B. Eden, and M. Staudacher, 'Transcendentality and Crossing,' J. Stat. Mech. **0701**, P01021 (2007) doi:10.1088/1742-5468/2007/01/P01021 hep-th/0610251.
[22] G. Arutyunov and S. Frolov, 'On String S-matrix, Bound States and TBA,' JHEP **0712**, 024 (2007) doi:10.1088/1126-6708/2007/12/024 arXiv:0710.1568.
[23] T. Fleury and S. Komatsu, 'Hexagonalization of Correlation Functions,' arXiv:1611.05577.
[24] B. Eden and A. Sfondrini, 'Three-point functions in $\mathcal{N} = 4$ SYM: the hexagon proposal at three loops,' JHEP **1602**, 165 (2016) doi:10.1007/JHEP02(2016)165 arXiv:1510.01242.
[25] B. Basso, V. Goncalves, S. Komatsu, and P. Vieira, 'Gluing Hexagons at Three Loops,' Nucl. Phys. B **907**, 695 (2016) doi:10.1016/j.nuclphysb.2016.04.020 arXiv:1510.01683.
[26] B. Basso, V. Goncalves, and S. Komatsu, 'Structure constants at wrapping order,' JHEP **1705**, 124 (2017) doi:10.1007/JHEP05(2017)124 arXiv:1702.02154.
[27] M. de Leeuw, C. Kristjansen, and K. Zarembo, 'One-point Functions in Defect CFT and Integrability,' JHEP **1508**, 098 (2015) doi:10.1007/JHEP08(2015)098 arXiv:1506.06958.
[28] D. J. Gross, A. Mikhailov and R. Roiban, 'A Calculation of the plane wave string Hamiltonian from N=4 superYang-Mills theory,' JHEP **0305**, 025 (2003) doi:10.1088/1126-6708/2003/05/025 hep-th/0208231.
[29] J. Gomis, S. Moriyama and J. w. Park, 'SYM description of SFT Hamiltonian in a PP wave background,' Nucl. Phys. B **659**, 179 (2003) doi:10.1016/S0550-3213(03)00220-7 hep-th/0210153.
[30] Z. Bajnok and R. A. Janik, 'String field theory vertex from integrability,' JHEP **1504**, 042 (2015) doi:10.1007/JHEP04(2015)042 arXiv:1501.04533.
[31] L. Motl, 'Proposals on nonperturbative superstring interactions,' hep-th/9701025.
[32] R. Dijkgraaf, E. P. Verlinde, and H. L. Verlinde, 'Matrix string theory,' Nucl. Phys. B **500**, 43 (1997) doi:10.1016/S0550-3213(97)00326-X hep-th/9703030.

11
Localization and $\mathcal{N}=2$ supersymmetric field theory

Vasily PESTUN

Institut des Hautes Etudes Scientifiques, Bures-sur-Yvette, France

Chapter Contents

11 Localization and $\mathcal{N}=2$ supersymmetric field theory   501
    Vasily PESTUN

    11.1 Gauge theory   503
    11.2 Kähler, special Kähler, and hyperKähler geometry   507
    11.3 Supersymmetry   512
    11.4 Supersymmetric gauge theory   517
    11.5 Seiberg–Witten integrable system   531

        References   539

11.1 Gauge theory

Let G be a Lie group and let X be a real smooth oriented space-time manifold of dimension d. A *G-gauge theory* on X is a field theory in which the space of fields is a space of principal smooth G-bundles P on X with connections. In a given trivialization (choice of gauge) of a principal G-bundle, a connection $\nabla_A = d + A$ is represented by a $\mathfrak{g}$-valued one-form on X, the section $A \in \Gamma(X, T_X^* \otimes \mathfrak{g})$, where $\mathfrak{g} = Lie(G)$.

A matter field ϕ in a G-gauge theory is defined by a choice of a vector bundle $\mathcal{E}_S$ over X associated to the spin and a vector bundle $\mathcal{E}_R$ over X associated to the principal G-bundle by a representation $\rho: G \to R$. The matter field ϕ is a section $\phi \in \Gamma(X, \mathcal{E}_S \otimes_X \mathcal{E}_R)$. Most often in physical constructions the bundle $\mathcal{E}_S$ with a fixed connection comes from a (super)gravity background in which we consider gauge theory. The matter field ϕ could be chosen to have an even (bosonic, commuting) or odd (fermionic, anticommuting) statistics. For example, the matter bundle $\mathcal{E}_S$ could be a trivial bundle $X \times W$, where W is a fixed vector space if ϕ is a scalar field valued in W, a bundle of p-forms on X, a *Spin* or *Spin^c* bundle over X whenever X is a *Spin* or *Spin^c* manifold, etc.

11.1.1 Yang–Mills theory

The Yang–Mills theory is a gauge theory on X equipped with Yang–Mills functional S_{YM} on the space of gauge fields defined by a choice of the Hodge star operator on degree 2 forms $\star: \Omega^2(X) \to \Omega^{d-2}(X)$ (where $\Omega^p(X)$ denotes the space of p-forms on X), a G-invariant symmetric bilinear form $\langle,\rangle$ on $\mathfrak{g} = Lie(G)$, and a parameter called the Yang–Mills coupling constant g_{YM},

$$S_{YM} = \frac{1}{g_{YM}^2} \int_X \langle F_A \wedge \star F_A \rangle, \qquad (11.1)$$

Remark. This action functional assumes that G is a compact simple Lie group. If G is a product of simple factors then there is an independent coupling constant g_{YM} for each simple factor in G.

Remark. A choice of a non-degenerate metric $g \in \Gamma(X, Sym^2(T_X^*))$ on X is sufficient to construct S_{YM} functional since g defines the Hodge star operator $\star_g$, but it is not necessary. For example, in dimension 2 the Hodge star $\star$ on 2-forms is defined by a choice of volume form on X (therefore 2d Yang–Mills is invariant under area-preserving diffeomorphisms of X), while in dimension 4 the Hodge star on 2-forms depends only on the conformal class of metric and is invariant under local Weyl transformation $g \to e^{2\Omega}g$, hence 4d Yang–Mills is classically a conformal theory. The quantization breaks the conformal invariance of the pure YM theory (without matter fields) in 4d, but not the area-preserving diffeomorphism invariance in 2d.

Remark. When X is a space-time of Minkowski signature $(d-1, 1)$, the assumptions of unitarity and energy positivity of quantum theory require G to be a compact Lie group, the invariant form $\langle,\rangle$ on $\mathfrak{g}$ to be a positive definite metric on $\mathfrak{g}$, and g_{YM} to be a real number, so that after Wick rotation (time $t \mapsto \sqrt{-1}t$) the S_{YM} is a positive definite functional for a theory on a Euclidean manifold X.

Remark. Yang–Mills functional is often complemented by topological characteristic classes of the G-bundle. Let G be a simple compact Lie group. The most familiar is the second Chern class for a 4d gauge theory,

$$S_{YM} = \frac{1}{g_{YM}^2} \int_X \langle F_A \wedge \star F_A \rangle + \frac{i\theta}{8\pi^2} \int_X \langle F_A \wedge F_A \rangle. \tag{11.2}$$

If $\langle \rangle$ on $\mathfrak{g}$ is normalized such that the long coroot length squared is 2, then

$$k = -\frac{1}{8\pi^2} \int \langle F_A \wedge F_A \rangle \tag{11.3}$$

is integral and is called the instanton charge. The instanton charge is positive for self-dual fields $F_A = -\star F_A$, e.g. $F_A^+ = 0$.

11.1.2 Electro-magnetic duality

Electro-magnetic duality that exchanges the electric and magnetic fields as well as electric and magnetic charges is the basic symmetry of abelian Maxwell equations. We consider the electro-magnetic duality for $U(1)^r$ theory in the presence of θ-term couplings $trF \wedge F$. Let F^i for $i = 1,\ldots,r$ label the 2-forms of $U(1)^r$ field strengh that take value in the Lie algebra $\mathbb{R}^r$ of $U(1)^r$ and let $\star$ be the Hodge dual operator on the Euclidean space-time X. The abelian Yang–Mills action is now (we induce the metric on the Lie algebra by $-tr$ in the fundamental representation)

$$S_{YM} = \frac{\iota}{4\pi} \int_X F^i \wedge (\tau'_{ij} F^j - \iota \tau''_{ij} \star F^j), \tag{11.4}$$

where τ'_{ij} and positive definite τ''_{ij} are real symmetric matrices that combine into the complex symmetric matrix

$$\tau_{ij} = \tau'_{ij} + \iota \tau''_{ij}. \tag{11.5}$$

For $r = 1$ and

$$\tau = \frac{4\pi \iota}{g_{YM}^2} + \frac{\theta}{2\pi} \tag{11.6}$$

the above action reduces to (11.2). The field configuration in the presence of electric charge $n \in \mathbb{Z}^r$ that moves along the contour $\gamma \subset X$ is defined by extremizing the action (11.4) in the presence of the source term generated by electric charge,

$$S_e = -\imath \int_\gamma n_i A^i. \tag{11.7}$$

For abelian theory $F^i = dA^i$ and extremization of $S = S_{YM} + S_e$ is elementary,

$$\int_X \frac{\imath}{2\pi} \delta A^i \wedge d(\tau'_{ij} F^j - \imath \tau''_{ij} \star F^j) = \imath n_i \int \delta A^i \tag{11.8}$$

implies

$$\frac{1}{2\pi} d(\tau'_{ij} F^j - \imath \tau''_{ij} \star F^j) = n_i \tilde{\gamma}, \tag{11.9}$$

where $\tilde{\gamma}$ is the delta-function three-form that is Hodge dual to γ, that is $\int_\gamma A = \int_X \gamma^\vee$. Let S^2 be a two-sphere canonically linked with ν. Then the Gauss flux of electric field associated to electric charge $n = \{n_i\}$ is found to be

$$\frac{1}{2\pi} \int_{S^2} (\tau'_{ij} F^j - \imath \tau''_{ij} \star F^j) = n_i. \tag{11.10}$$

A magnetic charge on a contour γ creates the magnetic flux of F^i, by definition

$$\frac{1}{2\pi} \int_{S^2} F^i = m^i. \tag{11.11}$$

Electric-magnetic duality (EM duality), by definition, is a linear integral transformation for the lattice $(n, m) \in \mathbb{Z}^{2r}$ of electro-magnetic charges that preserves the canonical symplectic structure ω on the space of charges,

$$\omega((n,m);(\tilde{n},\tilde{m})) = r_i \tilde{m}^i - \tilde{n}_i m^i. \tag{11.12}$$

All electro-magnetic duality transformations form an infinite discrete group $\Gamma = Sp(2r, \mathbb{Z})$. Let an element $g \in \Gamma$ be denoted as

$$g = \begin{pmatrix} A & B \\ C & D \end{pmatrix}, \tag{11.13}$$

where A, B, C, D are $r \times r$ matrices. Let

$$T_{ij} = \tau'_{ij} - \imath \tau''_{ij} \star \tag{11.14}$$

be a matrix of linear operator that acts on $\Gamma(X, \Lambda^2 T_X^* \otimes \mathbb{R}^r)$. Then electric and magnetic flux (11.10) equations are

$$\frac{1}{2\pi} \int_X T_{ij} F^j = n_i$$
$$\frac{1}{2\pi} \int_X 2\pi F^i = m^i. \quad (11.15)$$

After we apply electro-magnetic duality $g \in \Gamma$ to the charge lattice we find new charges $(\tilde{n}, \tilde{m})$ to be given by

$$\begin{pmatrix} \tilde{n} \\ \tilde{m} \end{pmatrix} = \begin{pmatrix} A & B \\ C & D \end{pmatrix} \begin{pmatrix} n \\ m \end{pmatrix}. \quad (11.16)$$

And we want to satisfy the equations

$$\frac{1}{2\pi} \int_X \tilde{T}_{ij} \tilde{F}^j = \tilde{n}_i$$
$$\frac{1}{2\pi} \int_X \tilde{F}^i = \tilde{m}^i \quad (11.17)$$

on the linearly transformed field strength ($\tilde{F}^i$). Then we find

$$\tilde{T} = (AT + B)(CT + D)^{-1} \quad (11.18)$$

and

$$\tilde{F} = (CT + D)F. \quad (11.19)$$

Since algebraically the operator $-\iota \star$ in the definition of $T = \tau'_{ij} - \iota \tau''_{ij} \star$ satisfies $(-\iota \star)^2 = -1$, it is equivalent to $\sqrt{-1}$ in the formula (11.14). That is, define $\tilde{T} = \tilde{\tau}'_{ij} - \iota \tilde{\tau}''_{ij} \star$ and $r \times r$ complex matrix $\tilde{\tau} = \tilde{\tau}' + \iota \tilde{\tau}''$, then

$$\tilde{\tau} = (A\tau + B)(C\tau + D)^{-1}, \quad (11.20)$$

exactly like the integral symplectic transformation for change of special coordinates in the special Kähler manifold (11.48), whose geometry is reviewed in Section 11.2.

The main message learned as a consequence of electric-magnetic duality is that the Yang-Mill coupling constant is not just a number but a principally polarized abelian variety.

See more in [1–3].

11.2 Kähler, special Kähler, and hyperKähler geometry

The Kähler geometry on the target space is the common feature of the supersymmetric theory with four supercharges such as 4d $\mathcal{N}=1$ or related by dimensional reduction 2d (2,2) supersymmetric models.

In the theories with eight supercharges such as 6d $\mathcal{N}=1$ or its dimensional reduction to 4d $\mathcal{N}=2$ the geometry of the target space is more constrained. The geometry of the 4d vector multiplet is *integral special Kähler* and the geometry of the 4d hypermultiplet is *hyperKähler*.

11.2.1 Complex manifold

A complex structure on $V = \mathbb{R}^{2n}$ is $I \in End(V)$ such that $I^2 = -1$. An eigen subspace $V^{(1,0)} \subset V \otimes \mathbb{C}$ with eigenvalue $+i$ of I is called holomorphic, and with eigenvalue $-i$ is called antiholomorphic. An almost complex structure I on a smooth $2n$-dimensional manifold M is a complex structure $I \in End(T_M)$. An almost complex structure I is integrable if the Lie bracket for any two $(1,0)$ vector fields is a $(1,0)$ vector field.

11.2.2 Kähler manifold

A metric g (symmetric positive definite bilinear form on T_M) is compatible with complex structure I if the bilinear form

$$\omega_I(u,v) = g(Iu,v) \tag{11.21}$$

is a symplectic structure, i.e. ω_I is antisymmetric, non-degenerate, and closed. This implies that ω_I is a $(1,1)$ two-form for complex structure I called a *real Kähler symplectic form*. For any Kähler metric there is always a local function, called a *Kähler potential* $K(a,\bar{a})$ such that in local holomorphic coordinates $\{a^i\}$

$$g_{i\bar{j}} = \partial_i \partial_{\bar{j}} K(a,\bar{a}), \qquad g = g_{i\bar{j}}(da^i \otimes da^{\bar{j}} + da^{\bar{j}} \otimes da^i) \tag{11.22}$$

and

$$\omega = \imath \partial \bar{\partial} K(a,\bar{a}), \qquad \omega_{i\bar{j}} = \imath \partial_i \partial_{\bar{j}} K(a,\bar{a}), \qquad \omega = \omega_{i\bar{j}} da^i \wedge da^{\bar{j}}. \tag{11.23}$$

For example $K = \frac{1}{2}|a|^2$ gives the standard metric on $\mathbb{C}$. The holonomy group of the Kähler manifold is $U(n)$.

If compact group G acts on Kähler space M in a Hamiltonian way with respect to symplectic structure ω_I and the $\mathfrak{g}^*$ valued function μ_I called the *moment map* (that means for any $\xi \in \mathfrak{g}$ it holds that $(\xi, d\mu_I) = i_\xi \omega$ where i_ξ is a contraction with the vector field ξ on M generated by G action), and $\zeta \in \mathfrak{g}^*$ is a coadjoint-invariant element, the Kähler quotient $M//G$ is

$$M//G = \{\mu^{-1}(\zeta)\}/G. \tag{11.24}$$

The space $M//G$ is also Kähler space of real dimension $\dim_\mathbb{R} M - 2\dim_\mathbb{R} G$. As a complex manifold, $M//G$ is isomorphic to the complex quotient $M^s/G^\mathbb{C}$ of an open set $M^s \subset M$ of stable points on M by complexified group $G^\mathbb{C}$. A point is stable if its $G^\mathbb{C}$ orbit intersects $\mu^{-1}(\zeta)$.

For example, $\mathbb{CP}^1$ is the Kähler quotient of $\mathbb{C}^2$ with moment map $\mu = \frac{1}{2}(|z_1|^2 + |z_2|^2)$ and $\zeta = 1$ for $U(1)$ action $z_i \to t z_i$, where t is the defining character of $U(1)$, and equivalently $\mathbb{CP}^1$ is the complex quotient of $\mathbb{C}^2 \setminus (0,0)$ by $\mathbb{C}^\times$ action $z_i \to \lambda z_i$ for non-zero complex numbers $\lambda \in \mathbb{C}^\times$. The point $\{0,0\} \in \mathbb{C}^2$ is not stable because its orbit under $\mathbb{C}^\times$ action does not intersect the locus $\mu^{-1}(1) = \{(z_1, z_2) : \frac{1}{2}(|z_1|^2 + |z_2|^2) = 1\}$, so this point is discarded in the complex quotient definition. For Kähler quotient by an abelian group $U(1)^r$ the parameters $\zeta \in \mathbb{R}^r \in \mathrm{Lie}(U(1)^r)$ are called FI parameters in physics literature.

For example, the moduli space of G-flat connections $M^G_{X_2}$ on a smooth orientable 2d manifold X_2 (this moduli space is associated with the theory coming from the reduction of $d=4$ $\mathcal{N}=1$ to 2d as described in Section 11.1.2) is the symplectic quotient of the (infinite-dimensional) affine space $\mathcal{A}$ of connection 1-forms A on a G-principal bundle P on X_2 by the (infinite-dimensional) group $\mathcal{G}$ of automorphisms of the bundle (gauge group action). The symplectic structure ω on the space of connection $\mathcal{A}$ is defined by the formula

$$\omega(\delta A, \delta A') = \int_{X_2} \langle \delta A \wedge \delta A' \rangle, \tag{11.25}$$

(where $\langle, \rangle$ is an invariant bilinear form on $\mathfrak{g}$). Notice that the definition of ω requires only orientation of X_2, not a complex structure or a metric.

The space $\mathrm{Lie}(\mathcal{G})^*$, dual to the Lie algebra of $\mathcal{G}$, is the space of linear forms on sections of $\phi \in \Gamma(X_2, \mathrm{ad}P)$. We can identify it with the space $F \in \Gamma(X_2, \Lambda^2(T^*_{X_2}) \otimes \mathrm{ad}P)$ of adjoint valued two-forms. The value of F on ϕ is given by

$$\int_{X_2} \langle F, \phi \rangle. \tag{11.26}$$

The moment map function

$$\mu : \mathcal{A} \to \mathrm{Lie}(\mathcal{G})^* \tag{11.27}$$

for the symplectic structure (11.25) is simply the curvature two-form of the connection

$$\mu = F_A. \tag{11.28}$$

Indeed, it satisfies the definition of the moment map

$$\int_\Sigma \langle \delta F_A, \phi \rangle = \int_\Sigma \langle D_A \delta A, \phi \rangle = -\int_\Sigma \langle D_A \phi \wedge \delta A \rangle \qquad (11.29)$$

since the vector field on the space of connections $\mathcal{A}$ generated by the element ϕ of Lie algebra $\mathcal{G}$ is $\delta A = -D_A \phi$, see e.g. [4].

11.2.3 HyperKähler

A hyperKähler structure on a Riemannian manifold M of real dimension $4n$ is a triplet $(I, \mathcal{J}, K)$ of complex structures such that $I\mathcal{J} = K$ and the Riemannian metric g that is Kähler with respect to each of complex structures $I, \mathcal{J}, K$. HyperKähler structure implies the existence of S^2-worth of the complex structures: $I_\zeta = \zeta_I I + \zeta_\mathcal{J} \mathcal{J} + \zeta_K K$ for $\zeta \in S^2 \subset \mathbb{R}^3 | \zeta_I^2 + \zeta_\mathcal{J}^2 + \zeta_K^2 = 1$. The holonomy group of the hyperKähler manifold is $USp(2n) = U(2n) \cap Sp(2n, \mathbb{C})$.

In addition, to each complex structure $I, \mathcal{J}, K$ there is an associated $(2, 0)$ holomorphic two-form,

$$\Omega_I = \omega_\mathcal{J} + \imath \omega_K \qquad \Omega_\mathcal{J} = \omega_K + \imath \omega_I \qquad \Omega_K = \omega_I + \imath \omega_\mathcal{J}. \qquad (11.30)$$

If compact group G acts on hyperKähler space M in a Hamiltonian way with respect to symplectic structures $(\omega_I, \omega_\mathcal{J}, \omega_K)$ and the $\mathfrak{g}^*$ valued function $(\mu_I, \mu_\mathcal{J}, \mu_K)$ called the *moment map*, and $\xi \in \mathfrak{g}^* \otimes \mathbb{R}^3$ is a coadjoint-invariant element, the hyperKähler quotient $M//G$ is

$$M////G = \{\mu^{-1}(\xi)\}/G. \qquad (11.31)$$

The space $M////G$ is again a hyperKähler space of real dimension $\dim_\mathbb{R} M - 4 \dim_\mathbb{R} G$. For example, the moduli space of self-dual connection on hyperKähler manifold X_4 is the hyperKähler quotient of the space of all connections by the action of the (infinite-dimensional) gauge group. The hyperKähler moment map functional is F_A^+. (The moment map functional is valued in $\Gamma(X_4, \mathbb{R}^3 \otimes \text{ad}(\mathfrak{g}))$ where $\mathbb{R}^3 \simeq \Lambda^2(T^*_{X_2})$.)

Now break the hyperKähler symmetry by taking $\mathbb{R}^3 \simeq \mathbb{R} \oplus \mathbb{C}$ and define the *complex moment map* $\mu_\mathbb{C} : M \to \mathfrak{g}^* \times \mathbb{C}$ and real moment map $\mu_\mathbb{C} : M \to \mathfrak{g}^* \times \mathbb{R}$ by

$$\mu_\mathbb{C} = \mu_\mathcal{J} + \imath \mu_K, \qquad \mu_\mathbb{R} = \mu_I. \qquad (11.32)$$

The complex moment map satisfies

$$d\mu_I^\mathbb{C}(\chi) = i_\chi \Omega_I \qquad (11.33)$$

for $\chi \in \mathfrak{g}$ and where Ω_I is a holomorphic (2,0)-form. Consequently, $i_\chi \Omega_I$ is the (1,0) form, therefore the complex moment map $\mu_I^{\mathbb{C}}$ is an *I-holomorphic* function on M. Let $\xi_{\mathbb{R}} = \xi_I, \xi_{\mathbb{C}} = \xi_J + \imath \xi_K$. Then the hyperKähler quotient is

$$M////G = \{\mu_{\mathbb{R}}^{-1}(\xi_{\mathbb{R}}) \cap \mu_{\mathbb{C}}^{-1}(\xi_{\mathbb{C}})\}/G, \tag{11.34}$$

which is a symplectic quotient of the space $\mu_{\mathbb{C}}^{-1}(\xi_{\mathbb{C}})$, and consequently, as a complex manifold it is isomorphic to

$$M////G = (\mu_{\mathbb{C}}^{-1}(\xi_{\mathbb{C}}))^s/G^{\mathbb{C}}, \tag{11.35}$$

which is precisely the complexified definition of the symplectic quotient; instead of real symplectic structure we are using (2,0) holomorphic symplectic structure and the action by the compact group is replaced by the holomorphic action of the complex group $G^{\mathbb{C}}$.

See more in [5] for integration over hyperKähler quotients.

11.2.4 Special Kähler

A coordinate-free definition of special Kähler manifold M can be given [6]. Namely, a special Kähler manifold M is a Kähler manifold M equipped with a real flat torsion-free symplectic connection ∇ on T_M such that

$$d_\nabla I = 0. \tag{11.36}$$

In the above equation, $d_\nabla : \Gamma(M, T_M^* \otimes T_M) \to \Gamma(M, \Lambda^2 T_M^* \otimes T_M)$ is the exterior differential defined by the connection ∇. It acts on complex structure I viewed as a 1-form on M valued in vectors.

Equivalently, a special Kähler manifold (of real dimension $2n$) can be defined as a Kähler manifold in which there exists locally a set of holomorphic coordinates $\{a^i\}_{i=1...n}$ and a locally holomorphic function $F(a)$, called *prepotential*, such that the Kähler potential is

$$K = \frac{1}{2}\Im(\bar{a}^i \partial_i F). \tag{11.37}$$

Consequently, the Kähler metric is

$$g_{i\bar{j}} = \frac{1}{2}\Im(\partial_i \partial_j F) \tag{11.38}$$

and the real symplectic form is

$$\omega = \frac{\imath}{2}\partial\bar{\partial}\Im(\bar{a}^i \partial_i F), \qquad \omega_{i\bar{j}} = \frac{1}{2}\imath\Im(\partial_i \partial_j F). \tag{11.39}$$

The second derivative $\partial_i \partial_j F$ is called the *period matrix* τ_{ij},

$$\tau_{ij} = \partial_i \partial_j F \tag{11.40}$$

for reasons explained shortly. The τ_{ij} is a symmetric $n \times n$ matrix whose imaginary part needs to be positive definite so that the Kähler metric $g_{i\bar{j}} = \Im \tau_{ij}$ is a Riemannian metric. The simplest example is $M = \mathbb{C}$ with

$$F = \frac{1}{2} \tau a^2, \tag{11.41}$$

which gives

$$\omega = \frac{\imath}{2} \Im \tau \, da \wedge d\bar{a} = \Im \tau (dx \wedge dy) \tag{11.42}$$

for $a = x + \imath y$.

Given a function $F(a)$ of special holomorphic coordinates $\{a^i\}$, define another set of special holomorphic coordinates $\{b_i\}$ called *dual* coordinates,

$$b_i = \partial_i F \equiv \frac{\partial F}{\partial a^i}. \tag{11.43}$$

The symplectic form ω is actually flat in terms of the real coordinates $\Re a^i, -\Re b_i$. Indeed, first notice that because of (11.43) it holds that

$$da^i \wedge db_i = 0, \qquad d\bar{a}^{\bar{i}} \wedge db_{\bar{i}} = 0 \tag{11.44}$$

and from (11.39) we find

$$\omega = -\frac{1}{4}(d\bar{a}^{\bar{i}} \wedge db_i + da^i \wedge db_{\bar{i}}) = -\frac{1}{4}((da^i + d\bar{a}^{\bar{i}}) \wedge (db_i + db_{\bar{i}})) = d(\Re a^i) \wedge d(-\Re b^i). \tag{11.45}$$

Therefore, besides the holomorphic coordinates $\{a^i\}$, the special Kähler manifold is equipped with Darboux flat coordinates $\{\Re a^i, -\Re b_i\}$ for the symplectic structure ω.

How ambiguous is the choice of special coordinates that we are using to describe the special Kähler geometry? Instead of the coordinate system $\{a^i\}$, consider the double set $\{a^i\} \cup \{b_i\}$ and notice that there is a pointwise linear relation between the differentials,

$$db_i = d(\partial_i F) = \partial_{ij} F da^j = \tau_{ij} da^j. \tag{11.45}$$

The flat symplectic structure $\omega = -\Re da^i \wedge \Re db_i$ is invariant under symplectic transformation in $Sp(2n, \mathbb{R})$. Let the constant matrix from $Sp(2n, \mathbb{R})$ that acts on $(da^1, \ldots da^n, db_1, \ldots, db_n)$ be denoted as

$$\begin{pmatrix} A & B \\ C & D \end{pmatrix} \in Sp(2n, \mathbb{R}), \tag{11.47}$$

where A, B, C, D are $n \times n$ real matrices. The tilded holomorphic double set $\{\tilde{a}^i\} \cup \{\tilde{b}_i\}$ is

$$\begin{pmatrix} d\tilde{b}^i \\ d\tilde{a}_i \end{pmatrix} = \begin{pmatrix} A & B \\ C & D \end{pmatrix} \begin{pmatrix} db^i \\ da_i \end{pmatrix}. \tag{11.48}$$

It is clear that

$$d\tilde{b}_i = \tilde{\tau}_{ij} d\tilde{a}^j, \qquad \tilde{\tau} = (A\tau + B)(C\tau + D)^{-1} \tag{11.49}$$

In tilded coordinate system $\{\tilde{a}^i\} \cup \{\tilde{b}_i\}$ the tilded prepotential $\tilde{F}$ is defined by integrating the matrix of its second derivatives τ_{ij} (which is of course symmetric, from (11.48)). The constant $Sp(2n, \mathbb{R})$ transformation to $\{\tilde{a}^i\} \cup \{\tilde{b}_i\}$ keeps the symplectic structure ω and the complex structure, and hence the metric invariant.

11.2.5 Integral special Kähler

An integral special Kähler manifold of real dimension $2n$ is a special Kähler manifold with additional data: a bundle of rank $2n$ integral lattice $\Lambda \subset T_M$ with respect to which the symplectic form ω is integral. All constructions in Section 11.2.4 hold literally, except that now the ambiguity in the special coordinate system is restricted to a subgroup of $\Gamma \subset Sp(2n, \mathbb{Z}) \subset Sp(2n, \mathbb{R})$ that preserves the lattice Λ. If the integral symplectic form is principal (e.g. there is a basis α^i, β_j on lattice such that $\omega(\alpha^i, \alpha^j) = 0, \omega(\alpha^i, \beta_j) = \delta^i_j, \omega(\beta_i, \beta_j) = 0$), then $\Gamma = Sp(2n, \mathbb{Z})$.

A full lattice $\Lambda \subset T_M$ induces full dual lattice $\Lambda^* \subset T_M^*$.

Now since τ_{ij} is a symmetric matrix with a positive definite imaginary part, we can think about it as a period matrix of rank n abelian variety attached to each point of a special Kähler manifold M. The defining data of integral special Kähler manifold M is equivalent to the fibration $A \to P \to M$, where at each point $u \in M$ the fibre $A_u = (T_M^*)_u / \Lambda$ is a polarized abelian variety with period matrix τ_{ij}. The lattice Λ is the lattice of 1-cycles $H_1(A, \mathbb{Z})$, and the ambiguity $\Gamma \subset Sp(2n, \mathbb{Z})$ precisely corresponds to the ambiguity in the choice of basis of Λ.

See more in [7], [3].

11.3 Supersymmetry

Suppose that X is a d-dimensional affine space $X \simeq \mathsf{V} = \mathbb{R}^d$ with a constant flat metric g, the $SO_g(\mathsf{V})$ is the Lie group of special orthogonal transformations of V relative to the metric g, and the *Poincaré group* $ISO_g(\mathsf{V}) = SO_g(\mathsf{V}) \ltimes \mathsf{V}$ is the extension of $SO_g(\mathsf{V})$ by the Lie translation (abelian) group V of the affine space V (we are using the same letter V for the affine space and its vector space, hopefully, without confusion),

$$V \to ISO_g(V) \to SO_g(V). \tag{11.50}$$

In particular, the translation group $V \subset ISO_g(V)$ is a normal subgroup of the Poincaré group and the orthogonal group $SO_g(V) = ISO_g(V)/V$ is the quotient. Obviously, the YM functional S_{YM} on $X = (V,g)$ is invariant under $ISO_g(V)$.

11.3.1 SuperPoincaré symmetry

The notion of supersymmetry refers to the Lie supergroups that are symmetries of $\mathbb{Z}_2$-graded spaces also known as supermanifolds.

The Lie supegroups are easier to analyse in terms of their Lie superalgebras over the ground field $\mathbf{k} = \mathbb{C}$ to then specialize the global and the real structure.

11.3.2 The spinor modules of $Spin(V)$

Let $V \simeq \mathbb{C}^d$ a complex vector space with a symmetric bilinear form g. Let $Spin(V)$ be the $\mathbb{Z}_2$ extension of $SO(V)$,

$$\mathbb{Z}_2 \to Spin(V) \to SO(V), \tag{11.51}$$

and let S be the complex Dirac spinor module of $Spin(V)$ of $\dim_\mathbb{C} S = 2^{\lfloor \frac{d}{2} \rfloor}$. If d is odd then S is an irreducible $Spin(V)$ module, if d is even then S decomposes $S = S^+ \oplus S^-$ to the direct sum of irreducible $Spin(V)$-modules called Weyl spinors of positive and negative chirality with $\dim_\mathbb{C} S^+ = \dim_\mathbb{C} S^- = \frac{1}{2} \dim_\mathbb{C} S = 2^{\frac{d-2}{2}}$.

In terms of Dynkin weights, if $d = 2k+1$ then $\mathfrak{so}(V) = B_k$ and S is the irreducible highest weight B_k-module with the highest weight ω_k associated to the last node (the short root) in the B_k Dynkin diagram,

- - -o--o--o==>o

If $d = 2k$ then $\mathfrak{so}(V) = D_k$ and $S^\pm$ are the irreducible highest weight D_k modules with highest weights ω_{k-1}-th and the ω_k corresponding to the last two nodes of the D_k Dynkin graph where it splits,

```
- - -o--o--o--o
           |
           o
```

11.3.3 Lie superalgebras

A Lie superalgebra is a $\mathbb{Z}_2$-graded Lie algebra $\mathfrak{p} = \mathfrak{p}_0 \oplus \mathfrak{p}_1$, the $\mathfrak{p}_0$ is called *the bosonic subalgebra* of the superalgebra $\mathfrak{p}$, and $\mathfrak{p}_1$ is called *the fermionic* extension. A superalgebra ($\mathbb{Z}_2$ graded algebra) means that there is an extra sign $(-1)^{|x||y|}$ in all relations every time the position of the two elements x and y is exchanged, where $|x|$ denotes the $\mathbb{Z}_2$ grade of an element x. In particular, the Lie algebra bracket $[,] : \mathfrak{p} \otimes \mathfrak{p} \to \mathfrak{p}$ decomposes as

$$[,] : \mathfrak{p}_0 \otimes \mathfrak{p}_0 \to \mathfrak{p}_0, \quad [,] ; \mathfrak{p}_0 \otimes \mathfrak{p}_1 \to \mathfrak{p}_1, \quad [,] : \mathfrak{p}_1 \otimes \mathfrak{p}_1 \to \mathfrak{p}_0. \tag{11.52}$$

The super antisymmetry of the Lie superbracket $[,]$ implies that the fermionic–fermionic bracket $[,] : \mathfrak{p}_1 \otimes \mathfrak{p}_1 \to \mathfrak{p}_0$ is symmetric. The Jacobi identity on $\mathfrak{p}$ implies that $\mathfrak{p}_0$ is a Lie algebra itself, the $\mathfrak{p}_1$ is a $\mathfrak{p}_0$-module, and that the symmetric fermionic–fermionic bracket $[,] : \mathfrak{p}_1 \otimes \mathfrak{p}_1 \to \mathfrak{p}_0$ is $\mathfrak{p}_0$-equivariant.

11.3.4 Poincaré superalgebras

A Poincaré Lie superalgebra $\mathfrak{p}$ is a fermionic(odd) extension of Poincaré Lie algebra $\mathfrak{p}_0 = \mathfrak{iso}(V)$ by a *Spin*(V) module $\mathfrak{p}_1 = S$. If S is a *Spin*(V) module then it is automatically an *ISO*(V) module on which translation subgroup $V \subset ISO(V)$ acts trivially. To complement definition of the Lie superalgebra we need to specify a *Spin*(V)-equivariant symmetric map

$$[,] : S \otimes S \to V. \tag{11.53}$$

Such a symmetric map induces a super anti-symmetric Lie bracket, and the resulting Lie superalgebra $\mathfrak{siso}(V|S) := \mathfrak{iso}(V) \oplus_s S$ is called a superPoincaré algebra of the superspace (V|S).

Remark. Caution on notations. The *Spin*(V) module S in a superPoincaré algebra $\mathfrak{siso}(V|S)$ is not always just a single copy of a Dirac or Weyl spinor modules S or $S^\pm$, but may contain several copies of S, $S^\pm$ in various combinations under the condition that symmetric map (11.53) exists.

Remark. Dualizing the map (11.53) and using the metric g on V to identify $V \to V^\vee$, we can define the action by vectors from V on elements from S as $\cdot$ (γ-map)

$$\gamma, \cdot : V \otimes S \to S^\vee. \tag{11.54}$$

This is not exactly a Clifford action, because V sends S to its dual space $S^\vee$ rather than to itself. However, since *Spin*(V) acts on S, we have map

$$\Lambda^{\text{even}} V \otimes S \to S. \tag{11.55}$$

11.3.5 Clifford algebra

For a vector space with bilinear form (V, g), the Clifford algebra $Cl(\mathsf{V})$ is a free tensor algebra over V modulo relation $v \cdot v = g(v,v) \cdot 1$. If γ_μ for $\mu = 1 \dots d$, where $d = \dim \mathsf{V}$, denotes a basis in V and $g_{\mu\nu}$ is the matrix of g, then γ_μ generate $Cl(\mathsf{V})$ modulo relation

$$\gamma_\mu \gamma_\nu + \gamma_\nu \gamma_\mu = 2 g_{\mu\nu}. \tag{11.56}$$

If the vector space V is a complex vector space of $\dim_\mathbb{C} \mathsf{V} = d$, then $Cl(\mathsf{V})$ is represented by $2^{\lfloor \frac{d}{2} \rfloor} \times 2^{\lfloor \frac{d}{2} \rfloor}$ matrices, and Dirac spinor module S is a module for $Cl(\mathsf{V})$. The action by $\mathsf{V} \subset Cl(\mathsf{V})$ on S is called a Clifford action,

$$(v \in \mathsf{V}, \psi \in S) \mapsto v \cdot \psi, \tag{11.57}$$

and in terms of γ-matrices is written as

$$(v^\mu, \psi) \mapsto v^\mu \gamma_\mu \psi. \tag{11.58}$$

This action of V on S naturally extends to the action of the r-th external power $\Lambda^r(\mathsf{V})$ on S. Let $\gamma_{\underline{r}}$ be the basis in $\Lambda^r(\mathsf{V})$ where $\underline{r}$ is a multi-index notation for an r-tuple of distinct indices from $(1, 2, \ldots, d)$, and $\gamma_{\mu_1 \ldots \mu_r} := \gamma_{\{\mu_1} \gamma_{\mu_2} \cdots \gamma_{\mu_r\}}$, where $\{,\}$ denotes complete antisymmetrization. Then $v^{\underline{r}} \in \Lambda^r \mathsf{V}$ acts on S as

$$(v^{\underline{r}}, \psi) \mapsto v^{\underline{r}} \gamma_{\underline{r}} \psi. \tag{11.59}$$

Notice that in even dimension $\gamma_{\underline{r}} : S^\pm \to S^\mp$ if r is odd and $\gamma_{\underline{r}} : S^\pm \to S^\pm$ if r is even.

11.3.6 Invariant bilinear form on S

Lemma. (An exercise is to prove it using an explicit recursive construction of γ-matrices). Let $\mathsf{V} \simeq \mathbb{C}^d$. The Clifford algebra module S (Dirac spinor module) can be always equipped with an invariant bilinear form,

$$(,)_{\mathrm{II}} : S \otimes S \to \mathbb{C}, \tag{11.60}$$

which is symmetric if $d \in \{0, 1, 6, 7\} \mod 8$ and antisymmetric otherwise, e.g. for $d \in \{2, 3, 4, 5\}$, and if d is even, with another bilinear form,

$$(,)_{\mathrm{I}} : S \otimes S \to \mathbb{C}, \tag{11.61}$$

which is symmetric if $d \in \{0, 2\} \mod 8$ and antisymmetric for $d \in 4, 6 \mod 8$.

If we compose the bilinear form $(,)$ with the Clifford action $\Lambda^r(V)$ on S we obtain maps

$$(,)_{I,II}^{(r)} : S \otimes S \mapsto \Lambda^r(V^*) \qquad (11.62)$$

which are bilinear in spinors and valued in exterior forms on V. In the basis $\gamma_{\underline{r}}$ in $\Lambda^r(V)$ the above map is written as

$$(\psi, \chi) \mapsto (\psi \gamma_{\underline{r}} \chi). \qquad (11.63)$$

Verify that $(,)_{II}^{(r)}$ is antisymmetric in its arguments for $r \in \{2,3\} + \lfloor \frac{d}{2} \rfloor$ mod 4 and symmetric for $r \in \{0,1\} + \lfloor \frac{d-2}{2} \rfloor$ mod 4, and that $(,)_{I}^{(r)}$ is antisymmetric for $r \in \{1,2\} + \frac{d-2}{2}$ mod 4 and symmetric for $r \in \{0,3\} + \frac{d-2}{2}$ mod 4.

In addition, verify that for even d the map (11.62) projects non-trivially on

$$S^{\pm} \otimes S^{\pm} \to \Lambda^r(V^*) \qquad (11.64)$$

for $\frac{d}{2} + r$ even and on

$$S^{\pm} \otimes S^{\mp} \to \Lambda^r(V^*) \qquad (11.65)$$

for $\frac{d}{2} + r$ odd.

Using the metric on V we can identify $V \to V^*$ and hence at $r = 1$ the definition (11.62) provides a map $S \otimes S \to V$ that can be used in the construction of superPoincaré Lie algebra *if this map is symmetric*.

11.3.7 Minimal supersymmetry

Now we examine the various cases of d mod 8 and define minimal superPoincaré algebra.

1. For $d \in \{0,1,2,3\}$ mod 8 the map $(,)_{II}^{(1)}$ is symmetric and therefore we can define $[,] : S \otimes S \to V$ using $(,)_{II}^{(1)}$.

2. For $d \in \{2,4\}$ mod 8 the map $(,)_{II}^{(1)}$ is symmetric and we can define $[,] : S \otimes S \to V$ using $(,)_{I}^{(1)}$.

3. For $d \in \{5,6,7\}$ mod 8 both maps $(,)_{I,II}^{(1)}$ are antisymmetric and cannot be used to define an odd–odd bracket $[,] : S \otimes S \to V$. This can be fixed by taking $\mathsf{S} = S \otimes W$, where W is an even-dimensional fixed vector space equipped with a non-degenerate symplectic form and extending $(,)$ on S by tensoring $(,)$ on S with the symplectic form on W. The elements of $\mathsf{S} = S \otimes W$ are called symplectic spinors, and the symmetry $Sp(W)$ is called *symplectic R-symmetry*.

What is the minimal irreducible S in each of these cases? We need to examine additionally only the cases of even dimension $d \in \{0,2,4,6\}$ mod 8, where we can

Table 11.1 Minimal modules for defining superPoincare algebra over a complex field.

d mod 8	0	1	2	3	4	5	6	7	8	9	10	11	12
S	S	S	S^+	S	S	$S\otimes\mathbb{C}^2$	$S^+\otimes\mathbb{C}^2$	$S\otimes\mathbb{C}^2$	S	S	S^+	S	S
$\mathcal{N}$	1	1	(0,1)	1	1	1	(0,1)	1	1	1	(0,1)	1	1
dim S	1	1	1	2	4	8	8	16	16	16	16	32	64
$S_{V=\mathbb{R}^{d-1,1}}$				$2\mathbb{R}$	$2\mathbb{C}$		$2\mathbb{H}$				$2\mathbb{O}$		

possibly project S to the chiral submodule S^+. The (11.64) implies that such projection is possible if $d=\{2,6\}$ mod 8. Then $\mathsf{S}=S^+$ for $d=2$ and $\mathsf{S}=S^+\otimes W$ for $d=6$. In these cases the superPoincaré algebra is called *chiral*.

We conclude with a table (Table 11.1) of the minimal modules S for defining the v superPoincaré algebra over a complex field $\mathbb{C}$.

The table is redundantly extended (the first two lines are mod 8) to the dimension up to $d=12$ to capture all the important dimensions of superstring theory, M-theory and F-theory. The notation $\mathcal{N}$ reflects the number of copies of the minimal extension module S used to construct the superPoincaré Lie algebra. In the dimensions $d=2,6,10,\ldots$ the minimal superPoincaré Lie algebra is chiral, hence in these dimensions the notation $(\mathcal{N}_-,\mathcal{N}_+)$ is used to denote the number of copies of the corresponding minimal extension modules $\mathsf{S}^\pm$. (The notation also applies to the minimal superPoincaré algebra in dimension 6, where $\mathsf{S}=S^+\otimes\mathbb{C}^2$ involves two copies of Weyl spinors S^+ and is called $\mathcal{N}=(0,1)$). The dim S, i.e. the fermionic dimension of $\mathsf{siso}(\mathsf{V}|\mathsf{S})$, is called the *number of supercharges* of superPoincaré algebra.

11.4 Supersymmetric gauge theory

11.4.1 Minimal super Yang–Mills: $\mathcal{N}=1$ SYM

For a given dim $\mathsf{S}=2^p$ notice the maximal dimension d in which such an S is possible. For $p=1,2,3,4$ the list of dimensions goes as $d=3,4,6,10$ and up to $p=4$ satisfies

$$d-2 = 2^{p-1}. \tag{11.66}$$

The superPoincaré algebra in dimensions $d=3,4,6,10$ has the following list of special properties, related to each other.

1. In $d=3,4,6,10$ the Yang–Mills functional has minimal supersymmetric extension by the fermionic fields $\psi \in \Gamma(X,\mathsf{S})$ [8]. This supersymmetric gauge theory is called $\mathcal{N}=1$ SYM. The equivalent condition on superPoincaré algebra bracket (11.53)(11.54) is the following three-cycle rule. The map

$$\mathsf{S}\otimes\mathsf{S}\otimes\mathsf{S}\to\mathsf{S}^\vee$$
$$(\psi_1,\psi_2,\psi_3) \mapsto [\psi_1,\psi_2]\cdot\psi_3, \tag{11.67}$$

where we use the map $[,]$ (11.53) to get a vector from two spinors and then map· (11.54) to act by the vector $[\psi_1, \psi_2]$ on ψ_3, vanishes after 3-cyclic symmetrization,

$$\sum_{123 \to 231 \to 312} [\psi_1, \psi_2] \cdot \psi_3 = 0. \tag{11.68}$$

Remark. The above 3-cyclic identity is not the Jacobi identity of the superPoincaré Lie algebra. The map (11.54) is the bracket of superPoincaré Lie algebra. The bracket operation of the superPoincaré Lie algebra between translations in V and elements of S is zero.

2. Let $H^\bullet(\mathfrak{p}, \mathbf{k})$ be the Chevalley–Eilenberg cohomology of minimal superPoincaré Lie algebra $\mathfrak{p} = \mathsf{siso}(V|S)$, with values in the trivial representation $\mathbf{k}$ (ground field, e.g. $\mathbf{k} = \mathbb{C}$ or $\mathbf{k} = \mathbb{R}$). In dimensions $3, 4, 6, 10$ the $H^3(\mathfrak{p}, \mathbf{k}) = \mathbf{k}$ and $H^3(\mathfrak{p}, \mathbf{k}) = 0$ otherwise. In dimensions $d = 3, 4, 6, 10$ the non-trivial 3-cocycle in $H^3(\mathfrak{p}, \mathbf{k})$ can be used to extend the minimal superPoincaré Lie algebra $\mathfrak{p}$ to a 2-Lie superalgebra in a sense of L_∞-algebras. Namely, using a symmetric bracket $[,] : S \otimes S \to V$ and metric $\langle, \rangle : V \otimes V \to \mathbf{k}$ on V, define the map $\alpha : S \otimes S \otimes V \to \mathbf{k}$ by

$$(\psi, \chi, v) \mapsto \langle [\psi, \chi], v \rangle. \tag{11.69}$$

Then α is a non-trivial 3-cocycle in $d = 3, 4, 6, 10$ generating $H^3(\mathfrak{p}, \mathbf{k}) = \mathbf{k}$.

See more in [9].

3. In $d = 3, 4, 6, 10$ in Minkowski signature $V = \mathbb{R}^{d-1,1}$ the minimal superPoincaré algebra $\mathsf{siso}(V|S)$ over the ground field $\mathbf{k} = \mathbb{R}$ relates to the list of real normed division algebras $\mathbb{R}, \mathbb{C}, \mathbb{H}, \mathbb{O}$ (isomorphic to $\mathbb{R}, \mathbb{R}^2, \mathbb{R}^4, \mathbb{R}^8$ as vector spaces) and equivalently to the list of parallizable spheres S^0, S^1, S^3, S^7 (isomorphic to the elements of the real normed division algebras of norm 1).

See more in [10].

Let $\mathbb{K}$ be one of $\{\mathbb{R}, \mathbb{C}, \mathbb{H}, \mathbb{O}\}$ algebras. The module S is identified with the two copies of $\mathbb{K}$,

$$S_\psi = \mathbb{K}_{\psi_1} \oplus \mathbb{K}_{\psi_2} \tag{11.70}$$

and the vector space $V \simeq \mathbb{R}^{d-1,1}$ is decomposed into $\mathbb{R}^{d-1,1} \simeq \mathbb{R}_t^{0,1} \oplus \mathbb{R}_x^{1,0} \oplus \mathbb{R}_y^{d-2,0}$, where $\mathbb{R}_y^{d-2,0} \simeq \mathbb{K}_y$ and identified with Hermitian 2×2 matrices valued in $\mathbb{K}$,

$$V = \mathrm{Mat}_{2 \times 2}^H(\mathbb{K}), \tag{11.71}$$

by the following rule:

$$\begin{aligned} \hat{} &: V \to \mathrm{Mat}_{2 \times 2}^H(\mathbb{K}), \\ \hat{} &: v \mapsto \hat{v} \qquad (t, x, y) \mapsto \begin{pmatrix} t+x & y \\ y^* & t-x \end{pmatrix}. \end{aligned} \tag{11.72}$$

The Minkowski norm of $v \in \mathsf{V}$ corresponds to the minus determinant of $\hat{v}$,

$$-v^2 = t^2 - x^2 - y^2 = \det(\hat{v}). \tag{11.73}$$

The operation (11.54) that takes $\mathsf{V} \otimes \mathsf{S} \to \mathsf{S}^{\vee}$ is just matrix multiplication,

$$\cdot : \mathsf{V} \otimes \mathsf{S} \to \mathsf{S}^{\vee}, \qquad v, \psi \mapsto \hat{v} \begin{pmatrix} \psi_1 \\ \psi_2 \end{pmatrix}. \tag{11.74}$$

However, notice that the image lands in the dual space $\mathsf{S}^{\vee}$. Define the action of V on the dual space $\mathsf{S}^{\vee}$ by the rule

$$\cdot : \mathsf{V}^{\vee} \otimes \mathsf{S} \to \mathsf{S}, \qquad v, \boldsymbol{\psi} \mapsto \check{v} \begin{pmatrix} \psi_1 \\ \psi_2 \end{pmatrix}, \tag{11.75}$$

where $\check{v}$ is

$$\begin{aligned} \check{} &: \mathsf{V} \to \mathrm{Mat}^{\mathrm{H}}_{2 \times 2}(\mathbb{K}), \\ \check{} &: v \mapsto \check{v} \qquad (t,x,y) \mapsto \begin{pmatrix} -t+x & y \\ y^* & -t-x \end{pmatrix}. \end{aligned} \tag{11.76}$$

It is easy to check that

$$\check{v}\hat{v}\psi = -v^2 \psi \tag{11.77}$$

and hence $\mathsf{S} \oplus \mathsf{S}^{\vee}$ is the Clifford algebra module with the action of v represented by

$$\begin{pmatrix} 0 & \check{v} \\ \hat{v} & 0 \end{pmatrix}. \tag{11.78}$$

11.4.2 Minkowski flat space-time

Let $\mathfrak{siso}(\mathsf{V}|\mathsf{S})$ be the minimal superPoincaré Lie algebra in one of the special dimensions $d = 3, 4, 6, 10$ over the real ground field $\mathbb{R}$. Let $X = \mathsf{V} = \mathbb{R}^{d-1,1}$ be the flat Minkowski space-time. Let $\mathcal{E}_\mathsf{S}$ denote the product bundle $X \times \mathsf{S}$. Let ψ be a fermionic field $\psi \in \Gamma(X, \mathcal{E}_\mathsf{S} \otimes_X \mathcal{E}_\mathfrak{g})$, called the gluino superpartner of gauge connection A, and the other notations as in 11.1.1. Then

$$S_{SYM} = \frac{1}{g_{YM}^2} \int_X \langle F_A \wedge \star F_A \rangle + \langle \psi \slashed{D} \psi \rangle \mu_X, \tag{11.79}$$

where the Dirac operator $\slashed{D} : \Gamma(X, \mathcal{E}_\mathsf{S}) \to \Gamma(X, \mathcal{E}_{\mathsf{S}^{\vee}})$ is a composition of the covariant derivative $D_A \psi$ and the γ-map (11.54), and μ_X is the volume form. In components $\langle \psi \slashed{D} \psi \rangle = \langle \psi \gamma^\mu D_\mu \psi \rangle$ where $\langle \psi \gamma^\mu \psi \rangle$ denotes the symmetric map $\mathsf{S} \otimes \mathsf{S} \to \mathsf{V}$ evaluated on $\psi, \bar\psi$. Modulo equations of motion the SYM action functional is invariant under the following supersymmetry variation of the fields, called *vector multiplet representation*:

$$\delta_\epsilon A = \frac{1}{2}(\psi \gamma_\mu \epsilon)$$
$$\delta_\epsilon \lambda = -\frac{1}{4} F_{\mu\nu} \gamma^{\mu\nu} \epsilon. \tag{11.80}$$

Our conventions are that fermionic(odd) elements of the $\mathsf{S} \subset \mathfrak{siso}$ are presented as $\epsilon^\alpha Q_\alpha$ where fermionic(odd) Q_α elements form a basis in S and bosonic(even) spinors ϵ^α are coordinates of an element $\epsilon^\alpha Q_\alpha \in \mathsf{S}$.

11.4.3 Euclidean space-time and complexification

It is not possible to impose the real structure on the minimal complex superPoincaré superalgebra $\mathfrak{siso}(\mathsf{V}|\mathsf{S})$ in special dimensions $3, 4, 6, 10$ such that $\mathsf{V} \simeq \mathbb{R}^d$ would have a Euclidean metric. An easy way to see this is to notice that the vector v in the image of the symmetric odd–odd bracket $v = [\epsilon, \epsilon]$ is always null-like,

$$v = [\epsilon, \epsilon] \quad \Rightarrow \quad v^2 = 0. \tag{11.81}$$

This follows immediately from the algebraic property (11.68). In Euclidean space the only null-like vector is $v = 0$, and therefore a non-trivial bracket $[\epsilon, \epsilon]$ does not exist in real Euclidean superspace.

Therefore, the $\mathcal{N} = 1$ SYM gauge theory in Euclidean space-time, algebraically, has to be defined over the ground field $\mathbb{C}$, while the action functional and other structures (like representation of the superPoincaré algebra on the space of fields) are locally analytic (i.e. holomorphic away from singularities). The path integral of quantum theory over bosonic (even) fields is understood as a choice of an integration half-dimensional contour in the space of complex fields, over which a top holomorphic form is integrated. Since the action functional and the supersymmetry algebra are holomorphic, the result does not depend precisely on the integration contour, but only on its homotopy class. Integration over fermionic fields is an algebraic operation: evaluation of the top form. Therefore it is not necessary to impose a real structure on fermions in quantum theory.

The standard integration contour for bosonic fields in Euclidean gauge theory is chosen in such a way that the gauge group G of gauge theory is a compact form of the complexified gauge group $G_\mathbb{C}$, and the action functional on the space of bosonic fields is positive definite.

11.4.4 Off-shell closure of the supersymmetry algebra

By the definition of superPoincaré algebra the operator δ_ϵ^2 is the translation operator

$$\delta_\epsilon^2 = \mathcal{L}_v, \tag{11.82}$$

where $\mathcal{L}_v$ is the vector field

$$v = [\epsilon, \epsilon], \qquad v^\mu = (\epsilon \gamma^\mu \epsilon). \tag{11.83}$$

The transformation rules (11.80) of the full space $\mathcal{A}$ of fields (A, ψ) is not a representation in a regular sense of the superPoincaré algebra, because the algebraic relations such as $\delta_v^2 = \mathcal{L}_v$ hold in (11.80) only modulo equations of motion of the super Yang-Mills functional (in a flat space-time $X \simeq V$). In other words, the transformation rules (11.80) is a representation of the superPoincaré algebra in the critical (*on-shell*) locus $\mathcal{A}_{\text{crit}} \subset \mathcal{A}$ defined by $dS = 0$.

In quantum theory it is useful (for example for localization) to have a representation of the supersymmetry algebra on the full space of fields $\mathcal{A}$ over which the path integral is computed. Often this can be done by introducing the auxiliary fields which appear in the action as free quadratic terms. This procedure is called *off-shell* closure of the representation of the supersymmetry algebra.

The *off-shell* closure of the *vector multiplet* representation of $\mathcal{N} = 1$ superPoincaré algebra is easy to describe in the case of dimensions $3, 4$, and 6 related to $\mathbb{R}, \mathbb{C}$, and $\mathbb{H}$, but is much more complicated in dimension 10 related to $\mathbb{O}$ because of the non-associativity of $\mathbb{O}$. What is easy to describe is the *off-shell* closure of the $(1|1)$ supersymmetry subalgebra of the $\mathcal{N} = 1$ superPoincaré algebra generated by a fixed element δ_ϵ for $\epsilon \in \mathsf{S}$.

Let us fix an element $\epsilon \in \mathsf{S}$ such that $v = [\epsilon, \epsilon] \neq 0$. Then the off-shell closure is convenient to describe by the following exact sequence of the vector spaces:

$$0 \to \mathsf{K} \otimes \mathfrak{g} \to \mathsf{S} \otimes \mathfrak{g} \stackrel{[\epsilon, -]}{\to} \mathsf{V} \otimes \mathfrak{g} \stackrel{([\epsilon, \epsilon], -)}{\to} \mathfrak{g} \to 0. \tag{11.84}$$

The arrow from spinors to vectors S to V is an odd–odd bracket with ϵ. The last arrow from vectors to scalars is a convolution with the non-zero vector $v = [\epsilon, \epsilon]$. The vector space $\mathsf{K} \subset \mathsf{S}$ is the kernel of the map $[\epsilon, -]$ and $\mathsf{K} \to \mathsf{S}$ is the inclusion. The space K is called the space of auxiliary fields.

Aside from the dimension of $\mathfrak{g}$, the dimensions of the vector spaces in the above complex are as follows:

$$\begin{aligned}
d &= 3 & 0 &\to 0 \to 2 & &\to 3 \to 1 \to 0 \\
d &= 4 & 0 &\to 1 \to 4 & &\to 4 \to 1 \to 0 \\
d &= 6 & 0 &\to 3 \to 8 & &\to 6 \to 1 \to 0 \\
d &= 10 & 0 &\to 7 \to 16 & &\to 10 \to 1 \to 0
\end{aligned} \tag{11.85}$$

This means that the off-shell closure of the $\mathcal{N} = 1$ vector multiplet requires respectively $0, 1, 3, 7$ auxiary fields in $d = 3, 4, 6, 10$ (recall the list S^0, S^1, S^3, S^7 of parallizable spheres).

The space $\mathsf{K} = \ker_{[\epsilon, -]}$ is easy to describe explicitly in dimensions $3, 4, 6$.

In dimension 3 we have $\mathsf{K} = 0$.

In dimension 4 let $\gamma^* = -\gamma_1\gamma_2\gamma_3\gamma_4$ be the chirality operator, and $\mathsf{S} = S \simeq \mathbb{C}^4$ be the space of the 4d Dirac spinors equipped with invariant bilinear form (,) such that $(-,\gamma^\mu-)$ is symmetric map $\mathsf{S} \otimes \mathsf{S} \to \mathsf{V}$. It is easy to check that $(-,\gamma^*\gamma^\mu-)$ is an antisymmetric map. Hence $[\gamma^*\epsilon, \epsilon] = 0$, and therefore $\mathsf{K} = \mathbb{C}\gamma^*\epsilon$. The generator γ^* is actually a generator of the automorphism of the $d = 4$ minimal superPoincaré algebra called $U(1)$ R-symmetry. (In a complexified description an element $t \in GL(1)$ of R-symmetry acts on $\mathsf{S} = S^+ \oplus S^-$ by $(S^+, S^-) \to (tS^+, t^{-1}S^-)$).

In dimension 6, recall that $\mathsf{S} = S^+ \otimes W$ where $W \simeq \mathbb{C}^2 \simeq \mathbb{C} \oplus \mathbb{C}^\vee$ is acted by the symplectic group $Sp(W) \simeq Sp(2,\mathbb{C})$. Let $\sigma_I, I = 1, 2, 3$ be the generators of $\mathfrak{sp}(2,\mathbb{C})$ (the traceless 2×2 matrices). Then, again it is easy to check that $[\sigma_I\epsilon, \epsilon] = 0$, and therefore $\mathsf{K} = \oplus_{I=1}^3 \mathbb{C}\sigma_I\epsilon$. The generators σ_I for $I = 1\ldots 3$ are actually generators of the $Sp(W)$ R-symmetry automorphism group of the minimal $d = 6$ superPoincaré algebra.

In dimension 10 there is no R-symmetry. The shortest geometrical description of the space K is just $\mathsf{K} = \ker_{[\epsilon,-]}$.

In all of these cases $d = 3, 4, 6, 10$ the space $\mathsf{K} = \ker_{[\epsilon,-]}$ can also be computed as the quotient of the $\mathbb{C}$-cone over the orbit of ϵ by the stabilizer of the v subgroup of $Spin(\mathsf{V})$,

$$K = (\mathbb{C}\mathrm{Stab}_{Spin(\mathsf{V})}(v) \cdot \epsilon)/(\mathbb{C}\epsilon). \tag{11.86}$$

In terms of the normed division algebra presentation (11.72), using a $Spin((V))$ transformation assume that an element ϵ is of the form $(k,0) \in (\mathbb{K} \oplus \mathbb{K})$. Then the vector space of auxiliary fields K is spanned by $(jk, 0)$, where j is one of $(0,1,3,7)$ standard imaginary units in $(\mathbb{R}, \mathbb{C}, \mathbb{H}, \mathbb{O})$ respectively for $d = 3, 4, 6, 10$.

If $k \in \mathbb{K}$ is of unit norm, then k is a point in a sphere S^0, S^1, S^3, S^7. The line spanned by k is the normal bundle to the sphere. The vector space spanned by elements jk, where j runs over imaginary units of $\mathbb{K}$, is the tangent bundle to the sphere. This also illustrates how $\mathcal{N} = 1$ SYM in $d = 3, 4, 6, 10$ relates to the list of parallizable spheres.

11.4.5 The $\mathcal{N} = 1$ supersymmetry algebra in $d = 4$ in superspace

The $\mathcal{N} = 1$ supersymmetry has a particularly simple realization by fields on the superspace $(\mathsf{V}|\mathsf{S})$. Let index α label the basis in S, the index μ label the basis in V, and let $\gamma^\mu_{\alpha\beta}$ be the symmetric map $\mathsf{S} \otimes \mathsf{S} \to \mathsf{V}$ that defines the odd–odd bracket of the superPoincaré Lie algebra. Then we have anti-commutator relations,[1]

$$\{Q_\alpha, Q_\beta\} = 2\gamma^\mu_{\alpha\beta}\partial_\mu. \tag{11.87}$$

We can represent the superPoincaré generators Q_α by the differential operators on the superspace $(\mathsf{V}, \mathsf{S})_{\langle x,\theta\rangle}$,

---

[1] In this presentation we consider the superPoincaré algebra over $\mathbb{C}$ and consequently we are not keeping track of conventional ι-factors related to the choice of real structure.

$$Q_\alpha = \frac{\partial}{\partial \theta^\alpha} + \gamma^\mu_{\alpha\beta}\theta^\beta \partial_\mu. \tag{11.88}$$

A function $F(x,\theta)$ on a superspace (V,S) is called a superfield. The operator $(1+\epsilon^\alpha Q_\alpha)$ acts on a function $F(x,\theta)$ as a differential operator

$$(1+\epsilon^\alpha Q_\alpha)F(x,\theta) = F(x+\epsilon\gamma\theta, \theta+\epsilon). \tag{11.89}$$

Chiral fields. To define an interesting representation of the $\mathcal{N}=1$ superPoincaré algebra it is convenient to introduce conjugated operators,

$$D_\alpha = \frac{\partial}{\partial \theta^\alpha} - \gamma^\mu_{\alpha\beta}\theta^\beta \partial_\mu, \tag{11.90}$$

whose defining property is that D_α and Q_β anti-commute,

$$\{D_\alpha, Q_\beta\} = 0. \tag{11.91}$$

Recall that in $d=4$ the spin-module $\mathsf{S} = S = S^+ + S^-$ where S are Dirac spinors and $S^\pm$ are the $\pm$-chiral spinors, as well we have splitting $Q = Q^+ + Q^-$ and $D = D^+ + D^-$. Given operators D_α anti-commuting with Q_β we can define a subspace in the space of superfields by the condition

$$D_\alpha^+ \Phi^-(x,\theta) = 0 \qquad \text{:chiral constraint:} \tag{11.92}$$

Notice that

$$D^+(x+\theta^+\gamma\theta^-) = 0, \qquad D^+\theta^- = 0. \tag{11.93}$$

Define

$$y^- \equiv x+\theta^+\gamma\theta^- \tag{11.94}$$

and notice that

$$D^+ y^- = 0. \tag{11.95}$$

Then a general solution of the chirality constraint (11.92) can be represented as $\Phi(y,\theta^-)$:

$$D^+ \Phi^-(y^-,\theta^-) = 0. \tag{11.96}$$

The field $\Phi^-(y^-,\theta^-)$ is called a chiral field.

524   *Localization and $\mathcal{N}=2$ supersymmetric field theory*

D-term: Kähler kinetic term. From a generic superfield $F(x,\theta)$ a supersymmetric action can be constructed by taking

$$\int [dx] \int [d\theta] F(x,\theta) \qquad : D-term: \qquad (11.97)$$

where $\int [d\theta]$ is the Berezian integral; by definition $\int [d\theta] F(x,\theta)$ picks up the top form of $F(x,\theta)$. Indeed, supersymmetric variation vanishes,

$$\int [dx] \int [d\theta] Q_\alpha F(x,\theta) = 0, \qquad (11.98)$$

because the term with $\frac{\partial}{\partial \theta} F(x,\theta)$ has zero top form (by θ-degree counting), and the term with $\gamma^\mu_{\alpha\beta} \theta^\beta \partial_\mu$ is the total derivative in the x-space that vanishes after integration over x. The non-chiral actions as above are called *D-term* and are mostly used in sigma-models for construction of the kinetic term for the scalar fields using the Kähler potential $K(z,\bar{z})$ for the Kähler metric on the target space,

$$\int [dx] \int [d\theta] K(\Phi^-, \Phi^+) \qquad :Kähler\ D\text{-}term: \qquad (11.99)$$

where Φ^+ is an *anti-chiral* field defined in the opposite way (switching $+$ with $-$) to Φ^-.

F-term: holomorphic superpotential. As well, we can notice that the chiral action defined using chiral field (11.92),

$$\int [dx] \int [d\theta^-] \Phi(y,\theta^-), \qquad : F-term: \qquad (11.100)$$

is also supersymmetric invariant. Indeed,

$$Q^+ y^-_\mu = 2\theta^- \gamma_\mu, \qquad Q^- y^-_\mu = 0, \qquad (11.101)$$

hence

$$Q^+ \Phi(y^-,\theta^-) = 2\theta^- \gamma^\mu \partial_\mu \Phi(y^-,\theta^-), \qquad Q^- \Phi(y^-,\theta^-) = \partial_{\theta^-} \Phi(y^-,\theta^-), \qquad (11.102)$$

so the Q-variation of (11.100) is a total derivative in (x,θ) space and thus vanishes after integration over $[dx]$ and $[d\theta]$. The chiral action of the form (11.100) is called the *F-term* and is usually generated by an arbitrary holomorphic function, called the *superpotential*, $W(z)$,

$$\int [dx] \int [d\theta^-] W(\Phi(y^-,\theta^-)): \qquad :superpotential\ F\text{-}term: \qquad (11.103)$$

Gauge fields. We consider the $U(1)$ gauge field to simplify presentation. The supersymmetrization of the gauge connection is a superfield $V(x,\theta^+,\theta^-)$. The gauge transformation is

$$V \to V + \Phi^- - \Phi^+, \qquad (11.104)$$

where Φ^-, Φ^+ are chiral and antichiral fields. The supersymmetrization of the field strength is

$$\begin{aligned} W_\alpha^- &= (D^+)^2 D_\alpha^- V \\ W_\alpha^+ &= (D^-)^2 D_\alpha^+ V. \end{aligned} \qquad (11.105)$$

The superfield W_α^- is chiral because $(D^+)^3 = 0$ (indeed $\dim S^- = 2$), and the superfield W_α^+ is antichiral because $(D^-)^3 = 0$ (indeed $\dim S^+ = 2$). And the superfields $W^\pm$ are obviously invariant under supergauge transformation (11.104). The superfield $W^\pm$ is called the gauge superfield strength with the expansion

$$W^- = \psi^- + \theta^- K + F^- \theta^- + (\theta\theta) \slashed{D} \psi^+, \qquad (11.106)$$

where in all coefficient fields $(\psi^-, K^-, F^-, \psi^+)$ we have omitted arguments (y^-), the operator $\slashed{D} = \gamma^\mu D_\mu$ is a Dirac differential operator, the $\slashed{F} = F_{\mu\nu}\gamma^{\mu\nu}$ is the Clifford action by the 2-form $F_{\mu\nu}$, the F^- denotes the anti-selfdual part of the curvature, the *gluino field* ψ is the fermionic superpartner of the gauge connection whose curvature is F, and K is the *auxiliary* scalar field.[2] In the same way,

$$W^+ = \psi^+ + \theta^+ K + F^+ \theta^+ + (\theta\theta) \slashed{D} \psi^-. \qquad (11.107)$$

The gauge field action functionals $S_{SYM}^\pm$ are defined by complexified coupling constants τ^- and τ^+ as follows:

$$\begin{aligned} S_{SYM}^- &= \frac{\imath}{4\pi}\tau^- \int [dx] \int [d\theta^-](W^-)^2 = \frac{\imath}{4\pi}\tau^- \int [dx] (F^- \wedge F^- - \frac{1}{2}K^2 - \frac{1}{2}\psi \slashed{D}\psi) \\ S_{SYM}^+ &= \frac{\imath}{4\pi}\tau^+ \int [dx] \int [d\theta^+](W^+)^2 = \frac{\imath}{4\pi}\tau^+ \int [dx] (F^+ \wedge F^+ + \frac{1}{2}K^2 + \frac{1}{2}\psi \slashed{D}\psi). \end{aligned}$$
$$(11.108)$$

---

[2] Sometimes this auxiliary field K is denoted in supersymmetric literature as the D-field, but we are already overusing the symbol D.

If a real structure for fields is chosen in Euclidean signature, then the standard convention is to denote $\tau^- = \tau$ and $\tau^+ = \bar{\tau}$ where τ is a complexified coupling constant,

$$\tau = \frac{4\pi \imath}{g_{YM}^2} + \frac{\theta}{2\pi}, \qquad (11.109)$$

then

$$S_{SYM} = S_{SYM}^- + S_{SYM}^+ \qquad (11.110)$$

is the standard SYM action (11.79).

11.4.6 The 4d $\mathcal{N} = 2$ susy from 6d $\mathcal{N} = 1$ susy

Recall from the supersymmetry (Table 11.1, Section 11.3.7) that an $\mathcal{N} = 1$ super-Poincaré in $d = 6$ uses a spinor module $\mathsf{S}_{(6)} = S_{(6)}^+ \otimes \mathbb{C}^2$. Consider a decomposition of the 6d space-time into 4d and 2d spaces: $\mathsf{V}_6 = \mathsf{V}_4 \oplus \mathsf{V}_2$, and respectively the subgroup $Spin(4) \times Spin(2) \subset Spin(6)$. The chiral Weyl spinors S_6^+ (irreducible $Spin(6)$-module of dimension 4) transform as Dirac spinors $S_4 = S_4^+ + S_4^-$ (reducible $Spin(4)$-module of dimension 4) under $Spin(4) \subset Spin(6)$, so

$$S_6^+ \simeq S_4 \qquad \text{as } Spin(4) \subset Spin(6) \text{ modules.} \qquad (11.111)$$

The $\mathcal{N} = 1$ $d = 6$ uses spinors in $\mathsf{S}_6 = S_{(6)}^+ \otimes \mathbb{C}^2$, that is, the two copies of $Spin(6)$ Weyl spinors S_6^+. Under the reduction to $Spin(4) \subset Spin(6)$ we obtain two copies of $Spin(4)$ Dirac spinors: $S_4 \otimes \mathbb{C}^2$. The minimal supersymmetry $\mathcal{N} = 1$ in $d = 4$ uses one copy of Dirac spinors S_4 (see Table 11.1). Therefore, with respect to the decomposition $\mathsf{V}_6 = \mathsf{V}_4 \oplus \mathsf{V}_2$ the minimal 6d superPoincaré algebra $\mathfrak{iso}(\mathsf{V}_6|\mathsf{S}_6)$ reduces to the 4d extended superPoincaré algebra $\mathfrak{iso}(\mathsf{V}_4|\mathsf{S}_4 \otimes \mathbb{C}^2)$. Recall that the $\mathbb{C}^2$ factor in S_6 is $\mathbb{C}^2 = \mathbb{C} \oplus \mathbb{C}^\vee$ with canonical symplectic form and that we had to introduce this factor to have a symmetric odd–odd bracket in the superPoincaré Lie algebra. The automorphism group $Sp(\mathbb{C}^2) \simeq SL(2,\mathbb{C})$ is a symplectic group, called R-*symmetry*. Its compact subgroup $SU(2)$ remains after the appropriate choice of the real structure for the superPoincaré Lie algebra is made. This $SU(2) \subset SL(2,\mathbb{C})$ acting on the internal space $\mathbb{C}^2 = \mathbb{C} \oplus \mathbb{C}^\vee$ is called the $SU(2)_R$-*symmetry* group of the 4d $\mathcal{N} = 2$ supersymmetry; it acts naturally on the $\mathsf{S}_4 \otimes \mathbb{C}^2$ spinors of the 4d $\mathcal{N} = 2$ superPoincaré $\mathfrak{iso}(\mathsf{V}_4|\mathsf{S}_4 \otimes \mathbb{C}^2)$.

Another symmetry that commutes with superPoincaré $\mathfrak{iso}(\mathsf{V}_4|\mathsf{S}_4 \otimes \mathbb{C}^2)$ comes from the rotation group $Spin(\mathsf{V}_2) \simeq SO(2)$ of the 2-plane V_2 on which we reduce the 6d $\mathcal{N} = 1$ superPoincaré to get a 4d $\mathcal{N} = 2$ superPoincaré. This $SO(2) \simeq U(1)_R$ is called a $U(1)_R$ R-symmetry group.

Finally, the translations in V_2 also commute with $\mathfrak{iso}(\mathsf{V}_4|\mathsf{S}_4 \otimes \mathbb{C}^2)$; these translation generators in V_2 are scalars from the 4d perspective and are called central charges.

If real structure is chosen such that $V_2 \simeq \mathbb{R}^2$ as a Euclidean 2-plane, it is customary to denote $V_2 \simeq \mathbb{R}^2 \simeq \mathbb{C}$ and then the central charge $Z \in V_2$ is a complex-valued scalar operator commuting with the rest of the 4d $\mathcal{N} = 2$ superPoincaré (hence the name 'central charge').

Expicitly, all generators of $\mathcal{N} = 2$ super Lie algebra can be listed as follows:

$$\left| \begin{array}{c} V_4 \\ P_\mu \\ (\tfrac{1}{2},\tfrac{1}{2},0,0) \end{array} \right| \begin{array}{c} \Lambda^2 V_4 \\ M_{\mu\nu} \\ (1,0,0,0) \oplus (0,1,0,0) \end{array} \left| \begin{array}{c} S^+ \otimes \mathbb{C}^2 \\ Q^+_{\alpha i} \\ (\tfrac{1}{2},0,\tfrac{1}{2},-\tfrac{1}{2}) \end{array} \right| \begin{array}{c} S^- \otimes \mathbb{C}^2 \\ Q^-_{\dot\alpha i} \\ (0,\tfrac{1}{2},\tfrac{1}{2},\tfrac{1}{2}) \end{array} \left| \begin{array}{c} V_2 \\ Z \\ (0,0,0,1) \end{array} \right., \quad (11.112)$$

where the last line describes the representation with respect to $SU(2)_- \times SU(2)_+ \times SU(2)_R \times U(1)_R$, where $SU(2)_- \times SU(2)_+ \simeq Spin(4)$

11.4.7 The 4d $\mathcal{N} = 2$ vector mutliplet

The $U(1)^r$ abelian $\mathcal{N} = 2$ supersymmetric vectormultiplet is composed of $\mathcal{N} = 1$ supergauge field W_α and the $\mathcal{N} = 1$ chiral field Φ. The $\mathcal{N} = 2$ abelian Lagrangian contains only kinetic terms,

$$S_{SYM,\mathcal{N}=2} = \frac{\imath}{4\pi} \left(\int [dx] \int [d\theta^-] \mathcal{F}_{ij}(\Phi) W^{i-} W^{j-} \right.$$
$$\left. + \int [dx] \int [d\theta] (\Phi^{i+} e^V \mathcal{F}_i(\Phi^-)) + :\text{anti-chiral terms}: \right). \quad (11.113)$$

Here $i,j = 1 \ldots r$ labels $U(1)$ factors, and Φ^{i-} and W_α^{i-} denote respectively chiral scalar field and chiral gauge field strength, $\mathcal{F}_i = \partial_{a_i} \mathcal{F}, \mathcal{F}_{ij} = \partial_{a_i} \partial_{a_j} \mathcal{F}$. The first term is the chiral type F-term gauge field strength action, the second term is a non-chiral D-term Kähler potential type action for scalars. The Kähler potential is

$$K(\bar{a}^{\bar\imath}, a^i) = \frac{1}{2} \Im(\bar{a}^{\bar\imath} \partial_i \mathcal{F}(a)), \quad (11.114)$$

where $\mathcal{F}$ is the prepotential of the special Kähler geometry on r-complex dimensional target space $\mathcal{U}$ with local special coordinates a^i. The coordinates a^i are the scalar components of the chiral superfield Φ^i. The $r \times r$ matrix

$$\tau_{ij}(a) = \frac{\partial^2 \mathcal{F}}{\partial a^i \partial a^j} \quad (11.115)$$

is the dynamical (a-dependent) complexified coupling constant.

The $\mathcal{N} = 2$ supersymmetry mixes up fields from the $\mathcal{N} = 1$ vector W and $\mathcal{N} = 1$ chiral Φ. For example, the spinors in W and the spinors in Φ form together the $SU(2)_R$ doublet. The action (11.113) ensures $\mathcal{N} = 2$ supersymmetry; a single holomorphic function $\mathcal{F}(a)$ determines both the coupling constant $\tau_{ij}(a)$ for the gauge field strength and the Kähler metric with the Kähler potential (11.114) on the target space for the scalars, and the same kinetic term is proportional to $\tau_{ij}(a)$ for the spinors in W and Φ.

The special Kähler manifold $\mathcal{U}$ of complex dimension r is the target space of the non-linear $U(1)^r$ $d=4$ $\mathcal{N}=2$ abelian sigma model $\mathrm{Maps}(\mathbb{R}^4, \mathcal{U})$.

11.4.8 The 4d $\mathcal{N}=2$ hypermultiplet

The 4d $\mathcal{N}=2$ hypermultiplet is a pair of 4d $\mathcal{N}=1$ chiral multiplets. The target space for the scalars of the hypermultiplet must be hyperKähler manifold to ensure $\mathcal{N}=2$ supersymmetry. If the hypermultiplet transforms in representation R of gauge group G, the representation R has to be a quaternionic representation of G of real dimension $4n$. Starting from any complex G-representation ρ of real dimension $2n$ we can construct a quaternionic representation $R = \rho \oplus \rho^*$. The associated supermultiplet (with scalars in $\rho \oplus \rho^*$) is called a *full-hypermultiplet of ρ* or equivalently a *half-hypermultiplet* of $R = \rho \oplus \rho^*$. Then the minimal supermultiplet associated to an abstract quaternionic representation R (with $4n$ real scalars for $\dim_{\mathbb{R}} R = 4n$) is called a *half-hypermultiplet of R*:

$$\text{half-hyper}(\rho \oplus \rho^*) \simeq \text{full-hyper}(\rho), \qquad (11.116)$$

and then for any complex representation ρ of G we have conventions

$$\dim_{\mathbb{R}}(\text{hyper}(\rho)) = 2 \dim_{\mathbb{R}} \rho = 4 \dim_{\mathbb{C}} \rho, \qquad (11.117)$$

where $\dim_{\mathbb{R}}(\text{full-hyper}(\rho))$ denotes the real dimension of the hyperKähler target space for real scalars in the hypermultiplet. In most literature hyper $\equiv$ full-hyper. If ρ is an irreducible complex representation of G, the hypermultiplet associated to N_F copies of ρ, that is, to representation $\rho \otimes \mathbb{C}^{N_F}$, is said to have N_F *flavours*. The $U(N_F)$ action on $\mathbb{C}^{N_F}$ commutes with the R-action and is called the *flavour symmetry* group $U(N_F)$.

More generally, for any compact Lie group G (gauge group) and any compact Lie group F (flavour group), a quaternionic representation R of $G \times F$ gives rise to a half-hyper hypermultipet associating R with flavour symmetry group F. The hypermultiplet masses are the background complex scalars of the vector multiplet associated with the flavour symmetry group F.

Hence, a mass parameter m of a half-hypermultiplet in a quaternionic $G \times F$-module R is an element of the complexified Lie algebra $\mathfrak{f}_{\mathbb{C}}$ of F,

$$m \in \mathrm{Lie}(F_{\mathbb{C}}). \qquad (11.118)$$

The theory is invariant under symmetry F, hence by F-adjoint transformation the mass parameter can be taken to be in the complexified Cartan algebra of F,

$$m \in \mathrm{Lie}(T_{F_{\mathbb{C}}}). \qquad (11.119)$$

For example, for a theory with N_F flavours we have $F = U(N_F)$ and $\text{Lie}(T_{F_{\mathbb{C}}}) \simeq \mathbb{C}^{N_f}$ so that the mass parameter of such a hypermultiplet is an N_f-tuple of complex numbers.

11.4.9 Lagrangian $\mathcal{N} = 2$ gauge theories

Here we summarize the data that describes the Lagrangian of 4d $\mathcal{N} = 2$ gauge theory. The representation of theoretical data is a choice between:

1. Compact Lie group G: the gauge group.
2. Compact Lie group F: the flavour group.
3. Quaternionic $G \times F$ representation R,

with the condition that
the (G, F, R)-theory is UV complete.

The triplet (G, F, R) needs to satisfy the UV-complete condition of the non-positive β-function: absence of a Landau pole at large energies, so that the QFT is asymptotically free or conformal in the limit of UV = small distances = large momenta. In other words, under renormalization group flow the theory must be weakly coupled as small distances. Such a renormalizable theory is called UV-complete.

Let $\mathfrak{g} = Lie(G)$ be the reducible Lie algebra of compact Lie group G, and let

$$\mathfrak{g} = \oplus_{i \in I} \mathfrak{g}_i \tag{11.120}$$

be the decomposition of $\mathfrak{g}$ into irreducible Lie algebras $\mathfrak{g}_i$ and let $\langle \rangle_i$ be an adjoint invariant Killing form on $\mathfrak{g}_i$.

For any representation ρ_i of irreducible Lie algebra $\mathfrak{g}_i$ let $C_2(\rho_i)$ be the eigenvalue of the second Casimir operator[3] of $\mathfrak{g}_i$, and let $c_2(\rho_i)$ be the coefficient that relates bilinear form $tr_{\rho_i} xy$ with $\langle xy \rangle$,

$$tr_{\rho_i} xy = -c_2(\rho_i) \langle xy \rangle, \qquad x, y \in \mathfrak{g}_i. \tag{11.121}$$

The numbers C_{ρ_i} and $c_2(\rho_i)$ are related,

$$c_2(\rho_i) = \frac{\dim \rho_i}{\dim \mathfrak{g}} C_2(\rho_i). \tag{11.122}$$

For example, if $\mathfrak{g}_i$ is simple and the Killing form is chosen such that the length of the long root squared is 2, then $c_2(\mathfrak{g}) = 2h^\vee$, where $h^\vee$ is the dual Coxeter number of $\mathfrak{g}$. For $\mathfrak{g}_i = SU(N)$ and ρ_i fundamental representation the $c_2(\rho_i) = 1$.

Now let $\rho_i : \mathfrak{g}_i \to \text{End}(\mathbb{H}^n)$ be the quaternionic half-hypermultiplet representation $\rho_i \simeq R$ induced from the inclusion $\mathfrak{g}_i \subset \mathfrak{g} \oplus \mathfrak{f}$ and represented over quaternions $\mathbb{H}^n$ (for $\mathcal{N} = 2$ theory with $4n$ real scalars in hypermultiplets). For each i define

---

[3] For a Lie algebra $\mathfrak{g}$, a basis (t_α) in $\mathfrak{g}$ and Killing form $\langle \rangle$ on $\mathfrak{g}$, the second Casimir operator is $C_2 = -(\langle t_\alpha t_\beta \rangle)^{-1} t_\alpha t_\beta$.

530 *Localization and $\mathcal{N} = 2$ supersymmetric field theory*

$$\beta_i = c_2(\mathfrak{g}_i) - c_2(\rho_i|\mathbb{H}). \tag{11.123}$$

Our conventions are such that for $\rho_i = \tilde{\rho}_i \oplus \tilde{\rho}_i^*$ where $\tilde{\rho}_i$ is a complex representation of $G \times F$ the half-hyper of $\rho_i = \tilde{\rho}_i \oplus \tilde{\rho}_i^*$ is hyper of $\tilde{\rho}_i$, and $c_2(\rho_i|\mathbb{H}) = c_2(\tilde{\rho}_i|\mathbb{C})$. For example, the case $\tilde{\rho} = \mathbb{C}^N$ for $\mathfrak{g} = SU(N)$ is one fundamental hyper, that is the $SU(N)$ theory with $N_F = 1$, so for complex hypers defined by complex rep $\tilde{\rho}_i$ we have equivalently

$$\beta_i = c_2(\mathfrak{g}_i) - c_2(\tilde{\rho}_i|\mathbb{C}). \tag{11.124}$$

In the standard physical conventions the β-function has an opposite sign to the quantity we have called β, we want $\beta_i \geq 0$ for a good UV-complete theory.

Finally, the condition of UV-completeness is

4. For each irreducible factor $\mathfrak{g}_i$ for $\mathfrak{g} = \oplus_{i \in I} \mathfrak{g}_i$ it should hold that

$$c_2(\mathfrak{g}_i) - c_2(\rho_i|\mathbb{H}) \geq 0, \tag{11.125}$$

where $c_2(\mathfrak{g}_i)$ denotes c_2 in the adjoint representation. Since abelian factors $\mathfrak{g}_i$ have $c_2 = 0$ it follows that R should be a trivial representation for all abelian factors $\mathfrak{g}_i$, and consequently they decouple (except for possibly global topological effects).

For example, if $G = SU(N)$ and $F = U(N_F)$, we consider N_F fundamental flavours for the $SU(N)$ gauge group, we compute $\beta_i = 2N - N_F$, and obtain the UV-completeness bound $N_F \leq 2N$.

The continious parameters of (G, F, R) theory are

1. Complex coupling constant $\tau = (\tau_i)$ for each irreducible $\mathfrak{g}_i$ term in $\mathfrak{g} = \oplus_i \mathfrak{g}_i$ with $\Im(\tau_i) > 0$.
2. Mass parameter $m \in Lie(T_{F_\mathbb{C}})$.

Summary

The Lagrangian $\mathcal{N} = 2$ theory is defined by a compact Lie gauge group G, a compact Lie flavour group F, and a quaternionic representation R of $G \times F$ that satisfies inequalities (11.125). The complex parameters are the complex coupling constant $\tau = (\tau_i)$, where i labels irreducible terms in $\mathfrak{g} = \oplus_i \mathfrak{g}_i$, and complex mass $m \in Lie(T_{F_\mathbb{C}})$.

See [11] for complete classification of $\mathcal{N} = 2$ theories with Lagrangian.

11.4.10 Lagrangian $\mathcal{N} = 2$ quiver gauge theories

Let Γ be a directed graph called a quiver with the set of nodes I. To each node $i \in I$ we associate a vector space $N_i = \mathbb{C}^{N_i}$ (we use notation N_i to denote the vector space $\mathbb{C}^{N_i}$ as well as its dimension $\dim_\mathbb{C} N^i$, but the meaning should be clear from the context) and consider the $\mathcal{N} = 2$ gauge theory with gauge group $G = \times SU(N_i)$ and full hypermultiplet in representation

$$R = \oplus_{i \to j} Hom(N_i, N_j), \tag{11.126}$$

where the sum is over all arrows in quiver Γ. This is called minimal quiver gauge theory. We can extend the minimal quiver by matter in a fundamental representation: connect each node of the quiver with two extra nodes $(M_i, \tilde{M}_i)$ called respectively M_i fundamental and $\tilde{M}_i$ anti-fundamental matter multiplets for the i-th node. Then we take a full hyper defined by the complex rep of G as follows:

$$R = \oplus_{i\to j} Hom(N_i, N_j) \oplus_i Hom(N_i, M_i) \oplus_i Hom(\tilde{M}_i, N_i). \tag{11.127}$$

The flavour symmetry F is

$$F = \otimes_{i\to j} U(1) \otimes_i U(M_i) \otimes_i U(\tilde{M}_i). \tag{11.128}$$

Let C_{ij} be the Cartan matrix of the graph Γ where

$$C_{ij} = 2\delta_{ij} - (\#i \to j + \#j \to j) \tag{11.129}$$

and $\#i \to j$ denotes the number of arrows from the node i to the node j. The UV-completeness, i.e. the β-function inequality, leads to the bound

$$C_{ij} N_j \geq M_i + \tilde{M}_i \quad \forall i. \tag{11.130}$$

The solution of these bounds is equivalent to the classification of simply-laced finite and affine generalized Cartan matrices. The result is that the quiver Γ is the Dynkin diagram of ADE or affine ADE. Moreover, if Γ is affine ADE, then $M_i = \tilde{M}_i = 0$ and $N_i = a_i^\vee N$, where $a_i^\vee$ are dual Dynkin labels on Γ and N is a positive integer ($N > 1$ for $\Gamma = \hat{A}_r$ when all $a_i^\vee = 1$ in order for the theory with the gauge group $\times SU(N_i)$ to be non-empty).

See more in [12], [13].

11.5 Seiberg–Witten integrable system

See more in [1], [2], [14], [3], [7].

The deformations of a given vacua of $\mathcal{N} = 2$ theory correspond either to massless vectormultiplet or massless hypermultiplet excitations. The deformations associated with massless vector multiplets give rise to the space $\mathcal{U}$ called the *Coulomb branch of the moduli space of vacua*, for the reason that the low-energy theory of excitations around such vacua is $U(1)^r$ abelian theory in which charged particles interact by the Coulomb law (equivalent to (11.10)(11.11)). Algebraically, $\mathcal{U}$ is the spectrum of the 1/2-BPS chiral ring operators of the vector-multiplet type, like $\langle P(\phi) \rangle$ where ϕ is the complex G-adjoint valued scalar of a G-gauge vector-multiplet and P is an adjoint invariant polynomial on $\mathfrak{g}_\mathbb{C}$.

For example, in the simplest case of $G = SU(2)$ theory, the only independent 1/2-BPS chiral ring operator is $u = tr\phi^2$. The variable u is a coordinate on the one-dimensionl affine space $\mathcal{U}$: the moduli space of vacua of the 4d $\mathcal{N} = 2$ theory.

From the analysis of $\mathcal{N}=2$ supersymmetry for the abelian $U(1)^r$ theory we have found that such a theory is a non-linear $\mathcal{N}=2$ sigma-model of maps from the 4d space-time to the target space $\mathcal{U}$. The target space $\mathcal{U}$ must be equipped with a special Kähler structure to satisfy $\mathcal{N}=2$ supersymmetry, and moreover, this special Kähler structure must be integral so that electric–magnetic duality holds.

The special Kähler structure on $\mathcal{U}$ implies that there are local special coordinates a^i and holomorphic function $F(a)$, called the prepotential, which determines symmetric $r \times r$ matrix $\tau_{ij}(a) = \partial_{ij} F(a)$ of coupling constants for $U(1)^r$ theory, moreover, the $\Im(\tau_{ij})$ is positive definite; it determines the positive-definite action functional for the gauge fields and the positive definite Kähler metric on $\mathcal{U}$.

Therefore, at a given point $u \in \mathcal{U}$, the τ_{ij} can be interpreted as a period matrix of rank r of a principally polarized abelian variety. A rank r polarized abelian variety A can be viewed as an r-dimensional complex torus $\mathbb{C}^r/\Lambda$, where Λ is a full rank $2r$ lattice that has symmetric period matrix τ_{ij} with positive definite $\Im(\tau_{ij})$. The condition on τ_{ij} implies that one can easily define θ-functions on A which are quasi-periodic, that is, holomorphic sections of a holomorphic line bundle on A, and then use these θ-functions to map A into a projective space. Hence A is a projective algebraic variety and an abelian group (complex torus), thus A is an abelian variety. A principal polarization means a symplectic form on $H_1(A, \mathbb{Z})$ for which there exists a basis of 1-cycles α^i, β_i in canonical intersection $\alpha^i \cap \alpha^j = 0, \beta_i \cap \beta_j = 0, \alpha^i \cap \beta_j = \delta^i_j$.

The electric–magnetic duality implies that $\tau_{ij}(a)$ is defined only up to $Sp(2r, \mathbb{Z})$ which naturally acts on the lattice of electric–magnetic charges (n_i, m^i).

Hence, the defining data of $\mathcal{N}=2$ low-energy theory for the $U(1)^r$ vector multiplet can be recast in the form of the fibration $\mathcal{P}$ of rank r abelian varieties A over the moduli space of vacua $\mathcal{U}$:

$$A \to \mathcal{P} \to \mathcal{U}, \tag{11.131}$$

where $\mathcal{P}$ of complex dimension $2r$ denotes the total space of the fibration.

The lattice of electric–magnetic charges is identified with $H_1(A, \mathbb{Z}) \simeq \mathbb{Z}^{2r}$.

The central charge Z evaluated on a state with electric–magnetic charge (n, m) is

$$Z = n_i a^i + m^i b_i. \tag{11.132}$$

The central charge is a locally holomorphic funtion on $\mathcal{U}$. Since lattice of charges (n, m) is identified with $H_1(A, \mathbb{Z})$ we can represent the central charge function Z of a given (n, m) state by

$$Z(n, m) = \int_{n_i \alpha^i + m^i \beta_i} \lambda, \tag{11.133}$$

where λ is a meromorphic 1-form. Hence

$$a^i = \int_{\alpha_i} \lambda, \qquad b_i = \int_{\beta^i} \lambda. \tag{11.134}$$

Moreover, since on the base $\mathcal{U}$ it holds that

$$db_i = \tau_{ij}(a)da^j, \qquad (11.135)$$

we find that the α_i and β^i periods of $d\lambda$ from the equations

$$da^i = \int_{\alpha_i} d\lambda, \quad db_i = \int_{\beta^i} d\lambda \qquad (11.136)$$

should be related by the period matrix τ_{ij} of the abelian variety A at a given base point in $\mathcal{U}$.

Therefore, the 2-form

$$\Omega = d\lambda \qquad (11.137)$$

is a closed holomorphic non-degenerate $(2,0)$ form on the total space $\mathcal{P}$ that turns $\mathcal{P}$ into a holomorphic symplectic manifold. Moreover, the fibration $\mathcal{P} \to \mathcal{U}$ is Lagrangian, which means that Ω evaluates to zero when restricted to the fibre directions.

We see that the defining data of the $\mathcal{N}=2$ abelian $U(1)^r$ sigma-model with target $\mathcal{U}$ are exactly equivalent to the structure of an algebraic completely integrable system,

$$\mathcal{P} \to \mathcal{U}. \qquad (11.138)$$

The ring of Poisson commuting functions on $\mathcal{P}$ is the ring $\mathcal{O}(\mathcal{U})$ of holomorphic functions on the base $\mathcal{U}$; these functions are holomorphic Hamiltonians of a completely integrable system. In the $\mathcal{N}=2$ gauge theory language the ring $\mathcal{O}(\mathcal{U})$ is the chiral ring of 1/2-BPS operators of the vectormultiplet type.[4]

For example, for $SU(2)$ theory the base $\mathcal{U} \simeq \mathbb{C}$ is a one-dimensional complex space, and the only independent Hamiltonian function on the complex phase space $\mathcal{P} \to \mathcal{U}$ is u. Without proof, let us quote the result for an algebraic integrable system associated to $G = SU(2), F = 1, R = 0$ theory, that is the minimal $SU(2)$ 4d $\mathcal{N}=2$ theory with no hypers. Let $q = \exp(2\pi i \tau_{YM})$ be the exponentiated coupling constant.[5]

The base is $\mathcal{U} \simeq \mathbb{C}$ with, say, complex coordinate u. The fibres are rank 1-abelian varieties, i.e. elliptic curves. At a point $u \in \mathcal{U}$ the elliptic curve Σ_u is defined by an algebraic equation in complex space $\mathbb{C} \times \mathbb{C}^\times$ with the coordinates (x, y) by the equation

$$\Sigma_u: \quad y + \frac{q}{y} = T(x; u), \qquad (11.139)$$

---

[4] The 1/2-BPS operators of hypermultiplet type parameterize the Higgs branch of the moduli space of vacua that we do not consider here.

[5] Of course the theory is asymptotically free, therefore the coupling constant q is dimension-full parameter defined at a certain scale.

where
$$T(x;u) = x^2 - u. \tag{11.140}$$

The symplectic structure on the $\mathbb{C} \times \mathbb{C}^\times$ surface is $dx \wedge dy/y$ and the meromorphic Liouville–Seiberg–Witten 1-form λ on a curve Σ_u is

$$\lambda = x\frac{dy}{y}. \tag{11.141}$$

In this example, the total space $\mathcal{P}$ can be identified with a $\mathbb{C} \times \mathbb{C}^\times$ space which we have used to describe the spectral curve Σ_u by algebraic equation (11.139). Denote $y = -e^\phi$ and $x = p$. Then the symplectic form on $\mathcal{P}$ is $\Omega = dp \wedge d\phi$, while the equation of the spectral curve Σ_u gives Hamiltonian function u,

$$u = p^2 + e^\phi + \mathfrak{q}e^{-\phi}. \tag{11.142}$$

This is an example of the famous Toda integrable system, and is the simplest non-relativistic closed 2-particle Toda Hamiltonian reduced by the centre of mass motion.

Remark. For real Hamiltonian mechanics, on a phase space of real dimension 2 any real Hamiltonian is automatically integrable. However, for complex Hamiltonian mechanics, for a phase space of complex dimension 2, there is not any complex holomorphic Hamiltonian u that defines an algebraic integrable system. Indeed, compared to the real case, a non-trivial condition for algebraic integrability is that the complex curve that describes the energy level should be a one-dimensional abelian variety (elliptic curve in this case) rather than a generic algebraic curve.

11.5.1 $\mathcal{N} = 2$ gauge theory on $\mathbb{R}^3 \times \mathbb{R}$

The special coordinates a^i on the base $\mathcal{U}$ of the fibration $A \to \mathcal{P} \to \mathcal{U}$ are holomorphic action variables. Is there a gauge theory interpretation of the angle variables parametrizing the holomorphic Arnold–Liouville toric fibres A? The answer is yes, and to see the angle variables in gauge theory we need to put the $\mathcal{N} = 2$ theory on $\mathbb{R}^3 \times S^1$ periodic space-time. Each $U(1)$-abelian vector multiplet after compactification on S^1 produces two additional periodic scalars: one scalar comes from the holonomy of the gauge field around the compactification circle S^1, the other periodic sclar is the 3d dual photon. See more in [14].

Therefore, the holomorphic symplectic phase space $\mathcal{P}$ is the moduli space of vacua of the $\mathcal{N} = 2$ QFT on a $\mathbb{R}^3 \times S^1$ background. In fact, after Kaluze–Klein reduction such a theory can be viewed as a 3d theory (with an infinite tower of particles coming from Kaluze–Klein modes), and such a 3d theory has eight supercharges. The relevant massless multiplet of 3d theory with eight supercharges is a hypermultiplet. The original complex scalar of the 4d $\mathcal{N} = 2$ vector multiplet combines with two other real periodic scalars that came from the reduction on S^1 into a quaternionic scalar parameterizing the hyperKähler moduli space of vacua. Hence, the symplectic space $\mathcal{P}$ carries not only a holomorphic symplectic structure, but a full hyperKähler metric. For a finite radius of the compactification circle S^1 the metric is highly non-trivial, receiving non-perturbative contributions from the massive states wrapping S^1. Only in the limit of S^1 of infinite radius can we neglect the non-perturbative contributions and approximate the metric on $\mathcal{P}$ by the naive metric, flat in the fibre directions, defined by the canonical flat along fibres metric on the total space $T^*\mathcal{U}/\Lambda$. The space $\mathcal{P} = T^*\mathcal{U}/\Lambda$ defined on a special Kähler manifold carries a canonical hyperKähler metric, but this metric is not the exact metric of the moduli space of vacua $\mathcal{P}_R$ of the $\mathcal{N}=2$ theory reduced on S^1 of finite radius R.

Even the exact metric depends non-trivially on the radius R of the S^1 compactification, one of the hyperKähler complex structures, customarily denoted by I, is independent of the radius R. As an I-complex holomorphic symplectic manifold the space $\mathcal{P}_R$ does not depend on R. This can be shown by analysis of the supersymmetry: I-holomorphic quantities are protected from R-corrections.

11.5.2 Spectral curves

In practice, often a Seiberg–Witten integrable system is described not by a family of abelian varieties fibred over the base $\mathcal{U}$, but by a family of curves Σ fibred over the same base $\mathcal{U}$ with the idea that the abelian variety is the Jacobian of a spectral curve (or more generally Prym subspace of a Jacobian for some correspondence symmetry on the curve),

$$A_u = \mathcal{J}ac(\Sigma_u). \tag{11.143}$$

Let S be a holomorphic symplectic surface and (Σ, L) be a pair: a holomorphic curve $\Sigma_u \subset S$ of a fixed homology class and a line bundle on L. The moduli space of such pairs (Σ, L) is equivalent to the moduli space of rank 1 sheaves on S of a fixed cohomology class determined by the homology class of Σ. It is clear that the holomorphic symplectic structure on S induces the holomorphic symplectic structure on the moduli space of sheaves on S. Thus the moduli space $\mathcal{P} = \{(\Sigma, L)\}$ for $\Sigma \subset S$ for a holomorphic surface S carries holomorphic symplectic structure as well as the projection $\mathcal{P} \to \mathcal{U}$ to the moduli space $\mathcal{U}$ of holomorphic curves $\Sigma \subset S$ of a fixed homology class. The fibre of the projection map $\mathcal{P} \to \mathcal{U}$ is the moduli space of line bundles on Σ, that is $A_u = \mathcal{J}ac(\Sigma_u)$.

See more in [15], [16], [7].

11.5.3 Class S theories and Hitchin system

One well-known class of algebraic integrable system is associated with the symplectic surface $S = K_C$, which is the total space of the canonical bundle K (holomorphic cotangent bundle) to a holomorphic curve C, possibly with marked points. On C we consider a moduli space of Higgs bundles. Let $G^{\mathbb{C}}$ be a complex (algebraic) reductive group. A Higgs bundle is a pair,

(stable principal $G^{\mathbb{C}}$-bundle on C, holomorphic section $\phi \in \Gamma(K \otimes \mathrm{ad}_{G^{\mathbb{C}}}))$. (11.144)

The moduli space

$$\mathfrak{M}_{G^{\mathbb{C}}}^{Hit}(C) = \{G^{\mathbb{C}}\text{-Higgs bundles on C}\} \quad (11.145)$$

is called a Hitchin moduli space. Let $\mathrm{Bun}_{G^{\mathbb{C}}}(C)$ denote the moduli space of principal $G^{\mathbb{C}}$-holomorphic bundles on C. It is clear that in a generic point,

$$\mathfrak{M}_{G^{\mathbb{C}}}^{Hit}(C) \simeq T^* \mathrm{Bun}_{G^{\mathbb{C}}}(C) \quad (11.146)$$

and hence $\mathcal{P} = \mathfrak{M}_{G^{\mathbb{C}}}^{Hit}(C)$ is a holomorphic symplectic variety. Let the collection of $\{P_{d_k}\}$ be the polynomial generators (of degree d_k) of the ring of adjoint invariant polynomials on $\mathfrak{g}_{\mathbb{C}}$. Then the Higgs field ϕ induces a holomorphic section $P_{d_k}(\phi) \in \Gamma(C, K^{\otimes d_k})$. This defines projection $\mathcal{P} \to \mathcal{U}$:

$$\mathfrak{M}_{G^{\mathbb{C}}}^{Hit}(C) \to \oplus_k H^0(C, K^{\otimes d_k}) \quad (11.147)$$

for $\mathcal{U} = \oplus_k H^0(C, K^{\otimes d_k})$. Hitchin has shown that $\dim_{\mathbb{C}} \mathcal{U} = \frac{1}{2} \dim_{\mathbb{C}} \mathcal{P}$ and that the fibres of the projection are holomorphic symplectic Lagrangian varieties. Hence (11.147) is an algebraic integrable system. Let R be a representation of $G_{\mathbb{C}}$. The curve $\Sigma \subset T^*C$ defined by the spectral equation

$$\Sigma : \det_R(\phi - t) = 0, \quad (11.148)$$

where t is a coordinate in the cotangent fibre of T^*C is called the *spectral curve* for a Hitchin system. For $G_{\mathbb{C}} = GL(n)$ and fundamental representation R the fibration $\mathcal{P} \to \mathcal{U}$ is precisely of the type reviewed in section 11.5.2. The base $\mathcal{U}$ is the family of n-fold spectral covers $\Sigma \to C$ in the total space $S = T^*C$, and the fibre A_u at a given point $u \in \mathcal{U}$ is the moduli space of line bundles on Σ_u, that is $\mathcal{J}ac(\Sigma_u)$. For a general G and R one needs to deal with correspondence on Σ and define a Prym subvariety of $\mathcal{J}ac(\Sigma_u)$, see [16].

If $\mathfrak{g}$ is a simple Lie algebra of ADE type, and $\Gamma_{\mathfrak{g}} \subset SU(2)$ is the McKay dual to $\mathfrak{g}$ discrete subgroup, consider the 4d gauge theory on $\mathbb{R}^4$ obtained from the IIB string theory on $\mathbb{R}^4 \times C \times \mathbb{C}^2/\Gamma_{\mathfrak{g}}$. This 4d gauge theory is called the 4d $\mathcal{N} = 2$ gauge theory of class $S_{C,\mathfrak{g}}$. The phase space of the G-Hitchin integrable system on C is isomorphic to the Coulomb branch of the moduli space of vacua of the class $S_{C,\mathfrak{g}}$ theory on $\mathbb{R}^3 \times S^1$.

See more in [17], [18], [19], [20].

11.5.4 Quiver theories and multiplicative Hitchin system (periodic monopoles)

What happens if we try to compactify the $\mathbb{C}$-fibre of T^*C to a cylinder $\mathbb{C}^\times$? Namely, let C be a genus 0 or genus 1 curve with flat structure dz, which means that C is either of $\mathbb{C}, \mathbb{C}^\times, \mathcal{E}$ (where $\mathcal{E}$ is an elliptic curve) with a fixed holomorphic differential dx.

We consider the surface $S = C \times \mathbb{C}^\times$. The vertical curve $\mathbb{C}^\times$ is equipped with $\mathbb{C}^\times$ invariant holomorphic differential $\frac{dy}{y}$. The surface S is holomorphic symplectic with symplectic structure,

$$\Omega = dx \wedge \frac{dy}{y}. \tag{11.149}$$

The compactification of S in the limiting points 0 and ∞ of $\mathbb{C}^\times$ has singularities for Ω along the curves C_0 and C_∞ covering C. Define the holomorphic symplectic phase space $\mathcal{P}$,

$$\mathcal{P} = \{G_\mathbb{C} - \text{principal bundles on } S \text{ with fixed trivialization at } C_0 \text{ and } C_1\}, \tag{11.150}$$

and define the base

$$\mathcal{U} = \text{Maps}(C_0, \text{Bun}_{G_\mathbb{C}}(C^\times)). \tag{11.151}$$

It is clear we have projection

$$\mathcal{P} \to \mathcal{U}. \tag{11.152}$$

Moreover, it turns out that the fibres of this projection are holomorphic Lagrangian varieties. The space $\mathcal{P}$ can be described as a group-valued Higgs bundle,

$$\mathcal{P} = \{G_\mathbb{C} - \text{holomorphic bundle on } C, \text{holomorphic section } \phi \in \Gamma(C, Ad_{G_\mathbb{C}})\}, \tag{11.153}$$

with certain asymptotic conditions at the infinite points of the base curve C.

The base $\mathcal{U}$ can be characterized similarly to the base of a Hitchin system. Let χ_i be the generators of the ring of adjoint invariant functions on the group $G_\mathbb{C}$. For example, for a simple $G_\mathbb{C}$ we take χ_i to be the characters of the highest weight fundamental $G_\mathbb{C}$ modules with fundamental highest weight ω_i. Then $\chi_i(\phi)$ is a holomorphic function on C with certain asymptotic growth conditions at infinity points of C; let the space of such holomorphic functions be denoted by $\mathcal{O}_{d_i}(C)$. Later we will characterize the degrees d_i,

$$\mathcal{U} = \oplus_i \mathcal{O}_{d_i}(C). \tag{11.154}$$

The vertical curve $C^\times$ can be contracted to S^1. Therefore, the group-valued (multiplicative) Hitchin system is equivalent to the integrable system on the moduli space of monopoles on $C \times S^1$.

The group-valued Higgs field ϕ is the monodromy of the complexified connection $A + i\Phi$, where (A, Φ) are solving the BPS monopole equations $F_A = \star d_A \Phi$ on $C \times S^1$ for gauge connection $d + A$ and the Lie algebra-valued scalar field Φ.

The definition of a group Hitchin system can be extended for the ramified case when a group-valued Higgs field ϕ develops poles. In the language of monopoles one chooses a point in $C \times S^1$ and a coweight $\lambda^\vee : U(1) \to G$ and requests the monopole configuration toi develop the standard Dirac type $U(1)$ monopole singularity with $\frac{1}{2\pi} \int_{S^2} F = 1$ (where S^2 surrounds the point supporting a singular monopole) that embeds into the G-monopole configuration by the coweight $\lambda^\vee$.

For such a monopole system with singularities the group Higgs field $\phi(x)$ has poles at the points on C obtained by the projection of the points of monopole singularities on $C \times S^1$, and for a given $\lambda^\vee$-Dirac-type singularity the pole of $\phi(x)$ is characterized by $\lambda^\vee$.

A local characterization of the $\lambda^\vee$ singularity contribution to $\mathcal{P}$ is the $\lambda^\vee$ orbit in affine $G_\mathbb{C}$ Grassmanian which has complex dimension $2(\rho, \lambda^\vee)$, where ρ is the Weyl vector in the weight space of $\mathfrak{g}_\mathbb{C}$.

For rigid boundary conditions at infinity of C we find

$$\dim \mathcal{P} = \sum_i 2(\rho, \lambda_i^\vee), \qquad (11.155)$$

where the sum is over singular points i and $\lambda_i^\vee$ denotes the type of $U(1)$ Dirac singularity at a given point. The above formula is a well-known dimension [Pauly] of the moduli space of monopoles with singularities.

The integrable system of $G_\Gamma = ADE$ quiver gauge theories in 4d, 5d on S^1, or 6d on $\mathcal{E}^\vee$ is the $G_\Gamma^\mathbb{C}$-group Hitchin system on $C = \mathbb{C}, \mathbb{C}^\times, \mathcal{E}$ respectively. Each fundamental multiplet attached to the i-th node of the quiver Γ with mass m inserts Dirac singularity at point $m \in C$ and of type $\lambda_i^\vee$.

The affine Kac–Moody group-valued Hitchin system (double periodic instantons). The affine Kac–Moody group-valued Hitchin system is defined in a similar way, but now the vertical curve $\mathbb{C}^\times$ is replaced by a non-degenerate elliptic curve $\mathcal{E}_v$. The phase space $\mathcal{P}$ is the moduli space of $G_\Gamma^\mathbb{C}$-bundles on $C \times \mathcal{E}_v$, and the base $\mathcal{U}$ is the space of holomorphic maps from C to $\mathrm{Bun}_{G_\Gamma^\mathbb{C}}(\mathcal{E}_v)$.

For an affine Kac–Moody Hitchin group-valued system we do not allow singularities. The $\mathcal{N} = 2$ gauge theory is an affine ADE quiver theory that does not carry fundamental hypermultiplets.

The integrable system of $G_\Gamma = \hat{A}\hat{D}\hat{E}$ quiver gauge theories on $\mathbb{R}^4, \mathbb{R}^4 \times S^1, \mathbb{R}^4 \times E^\vee$ is the $G_\Gamma^\mathbb{C}$ affine Kac–Moody group Hitchin system (double periodic instantons) on $C \times \mathcal{E}_v$ for $C = \mathbb{C}, \mathbb{C}^\times, E$. The 'vertical' elliptic curve $\mathcal{E}_v$ has elliptic modulus defined by $\times_{i \in I_\Gamma} \mathfrak{q}_i^{a_i^\vee}$, where $\mathfrak{q}_i$ is an exponentiated coupling constant for the $SU(N_i)$ gauge group of $\mathcal{N} = 2$

QFT, and the horizontal elliptic curve E is dual to the curve $E^\vee$ that defines the geometry of the space-time $\mathbb{R}^4 \times E^\vee$ on which we consider the 6d $\mathcal{N}=1$ affine $\hat{A}\hat{D}\hat{E}$ quiver gauge theory.

See more details in [21], [22], [23], [13], [24].

References

[1] N. Seiberg and E. Witten, 'Monopoles, duality and chiral symmetry breaking in N=2 supersymmetric QCD,' *Nucl. Phys.* **B431** (1994) 484–550, arXiv:hep-th/9408099.

[2] N. Seiberg and E. Witten, 'Electric - magnetic duality, monopole condensation, and confinement in N=2 supersymmetric Yang-Mills theory,' *Nucl. Phys.* **B426** (1994) 19–52, arXiv:hep-th/9407087.

[3] R. Y. Donagi, 'Seiberg-Witten integrable systems,' arXiv:alg-geom/9705010.

[4] E. Witten, 'Two-dimensional gauge theories revisited,' *J. Geom. Phys.* **9** (1992) 303–368, arXiv:hep-th/9204083.

[5] G. W. Moore, N. Nekrasov, and S. Shatashvili, 'Integrating over Higgs branches,' *Commun. Math. Phys.* **209** (2000) 97–121, arXiv:hep-th/9712241.

[6] D. S. Freed, 'Special Kähler manifolds,' *Comm. Math. Phys.* **203** (1999), no. 1 31–52.

[7] R. Donagi and E. Witten, 'Supersymmetric Yang-Mills theory and integrable systems,' *Nucl. Phys.* **B460** (1996) 299–334, arXiv:hep-th/9510101.

[8] L. Brink, J. H. Schwarz, and J. Scherk, 'Supersymmetric Yang-Mills theories,' *Nucl. Phys.* **B121** (1977) 77.

[9] J. C. Baez and J. Huerta, 'Division Algebras and Supersymmetry I,' in *Superstrings, Geometry, Topology, and C*-algebras, eds. R. Doran, G. Friedman and J. Rosenberg, Proc. Symp. Pure Math. 81, AMS, Providence, 2010, 65-80*, pp. 65–80, 2009. arXiv:0909.0551.

[10] T. Kugo and P. K. Townsend, 'Supersymmetry and the division algebras,' *Nucl. Phys.* **B221** (1983) 357.

[11] L. Bhardwaj and Y. Tachikawa, 'Classification of 4d N=2 gauge theories,' *JHEP* **12** (2013) 100, arXiv:1309.5160.

[12] M. R. Douglas and G. W. Moore, 'D-branes, quivers, and ALE instantons,' arXiv:hep-th/9603167.

[13] N. Nekrasov and V. Pestun, 'Seiberg-Witten geometry of four dimensional N=2 quiver gauge theories,' arXiv:1211.2240. 197 pages.

[14] N. Seiberg and E. Witten, 'Gauge dynamics and compactification to three-dimensions,' arXiv:hep-th/9607163.

[15] R. Donagi and E. Markman, 'Spectral curves, algebraically completely integrable Hamiltonian systems, and moduli of bundles,' *arXiv* (July, 1995) 7017, arXiv:alg-geom/9507017.

[16] R. Donagi, 'Spectral Covers,' *arXiv* (May, 1995) 5009, arXiv:alg-geom/9505009.

[17] D. Gaiotto, 'N=2 dualities,' *JHEP* **1208** (2012) 034, arXiv:0904.2715.

[18] D. Gaiotto, G. W. Moore, and A. Neitzke, 'Four-dimensional wall-crossing via three-dimensional field theory,' *Commun. Math. Phys.* **299** (2010) 163–224, arXiv:0807.4723.

[19] D. Gaiotto, G. W. Moore, and A. Neitzke, 'Wall-crossing, Hitchin Systems, and the WKB Approximation,' arXiv:0907.3987.

[20] L. F. Alday, D. Gaiotto, and Y. Tachikawa, 'Liouville Correlation Functions from Four-dimensional Gauge Theories,' *Lett. Math. Phys.* **91** (2010) 167–197, arXiv:0906.3219.

[21] J. C. Hurtubise and E. Markman, 'Elliptic Sklyanin integrable systems for arbitrary reductive groups,' *Adv. Theor. Math. Phys.* **6** (2002), no. 5 873–978 (2003), arXiv:math/0203031.

[22] B. Charbonneau and J. Hurtubise, 'Singular Hermitian-Einstein monopoles on the product of a circle and a Riemann surface,' *ArXiv e-prints* (Dec., 2008) arXiv:0812.0221.

[23] S. A. Cherkis and A. Kapustin, 'Periodic monopoles with singularities and N=2 super QCD,' *Commun.Math.Phys.* **234** (2003) 1–35, arXiv:hep-th/0011081.

[24] N. Nekrasov, V. Pestun, and S. Shatashvili, 'Quantum geometry and quiver gauge theories,' arXiv:1312.6689.

Appendix A
Lectures given at the school but no written contribution

Conformal field theory in higher dimensions

Slava Rychkov

Laboratoire de Physique Théorique, École Normale Supérieure, 24 Rue Lhomond, 75005 Paris, France

542 *Appendix A: Lectures given at the school but no written contribution*

Introduction to 4d N = 2 SUSY theories

Leonardo Rastelli

C. N. Yang Institute for Theoretical Physics, Stony Brook University,
Stony Brook, 11794, NY, USA

Integrability and amplitudes in N = 4 SYM

Benjamin Basso

Laboratoire de Physique Théorique, École Normale Supérieure,
24 Rue Lhomond, 75005 Paris, France

Pedro Vieira

Perimeter Institute for Theoretical Physics, Waterloo, Ontario N2L 2Y5, Canada

Integrability from topology

Paul Fendley

The Rudolf Peierls Centre for Theoretical Physics, Oxford University, Oxford, OX1 3NP, United Kingdom

Random partitions in gauge and string theory

Andrei OKOUNKOV

Columbia University, New York, USA

Appendix B
Seminars given during the school

Quantum geometry

Vladimir Bazhanov

Department of Theoretical Physics, Research School of Physics and Engineering,
Australian National University, Canberra, ACT 2601, Australia

New integrable 3D and 4D QFTs from strongly twisted N = 4 SYM

Vladimir Kazakov

Laboratoire de Physique Theorique, Ecole Normale Superieure, 24 Rue Lhomond, 75005 Paris, France

Correlation functions of heavy states in N = 4 SYM

Ivan Kostov

Institut de Physique Theorique, Saclay, UMR3681 CNRS, bat 774 Saclay,
F-91191 Gif-sur-Yvette, France

Symmetric functions and quantum integrability

Paul ZINN-JUSTIN

School of Mathematics and Statistics, The University of Melbourne, Parkville, Victoria 3010, Australia

The manufacturer's authorised representative in the EU for product safety is
Oxford University Press España S.A. of el Parque Empresarial San Fernando de
Henares, Avenida de Castilla, 2 – 28830 Madrid (www.oup.es/en or product.
safety@oup.com). OUP España S.A. also acts as importer into Spain of products
made by the manufacturer.

www.ingramcontent.com/pod-product-compliance
Ingram Content Group UK Ltd.
Pitfield, Milton Keynes, MK11 3LW, UK
UKHW062306230426
12049UKWH00005B/119